D0945326

VOLUME FOUR HUNDRED AND THIRTY-SIX

METHODS IN ENZYMOLOGY

Globins and Other Nitric Oxide–Reactive Proteins, Part A

METHODS IN ENZYMOLOGY

VOLUME FOUR HUNDRED AND THIRTY-SIX

METHODS IN ENZYMOLOGY

Globins and Other Nitric Oxide–Reactive Proteins, Part A

EDITED BY

ROBERT K. POOLE
Department of Molecular Biology and Biotechnology
University of Sheffield
Sheffield, United Kingdom

AMSTERDAM • BOSTON • HEIDELBERG • LONDON
NEW YORK • OXFORD • PARIS • SAN DIEGO
SAN FRANCISCO • SINGAPORE • SYDNEY • TOKYO
Academic Press is an imprint of Elsevier

Academic Press is an imprint of Elsevier
525 B Street, Suite 1900, San Diego, California 92101-4495, USA
84 Theobald's Road, London WC1X 8RR, UK

This book is printed on acid-free paper. ♾

For information on all Elsevier Academic Press publications
visit our Web site at www.books.elsevier.com

ISBN-13: 978-0-12-374277-3

PRINTED IN THE UNITED STATES OF AMERICA
08 09 10 11 9 8 7 6 5 4 3 2 1

Contents

Contributors

Rubina G. Aga
Department of Chemistry, King's College London, London, United Kingdom

Michael Angelo
Department of Chemistry and Biochemistry, Montana State University, Bozeman, Montana

Pedro Aparicio-Tejo
Instituto de Agrobiotecnologia, Universidad Pública de Navarra-CSIC-Gobierno de Navarra, Pamplona, Nararre, Spain

Idoia Ariz
Instituto de Agrobiotecnologia, Universidad Pública de Navarra-CSIC-Gobierno de Navarra, Pamplona, Nararre, Spain

Raúl Arredondo-Peter
Laboratorio de Biofísica y Biología Molecular, Facultad de Ciencias, Universidad Autónoma del Estado de Morelos, Cuernavaca, Morelos, Mexico

Paolo Ascenzi
National Institute for Infectious Diseases IRCCS Lazzaro Spallanzani, Rome, Italy, and Department of Biology and Interdepartmental Laboratory for Electron Microscopy, University Roma Tre, Rome, Italy

Iñigo Auzmendi
Instituto de Agrobiotecnologia, Universidad Pública de Navarra-CSIC-Gobierno de Navarra, Pamplona, Nararre, Spain

Luca Bargelloni
Department of Public Health, Comparative Pathology, and Veterinary Hygiene, University of Padova, Legnaro, Italy

S. Van Berloo
DEMO, Delft University of Technology, Delft, The Netherlands

Alberto Boffi
Department of Biochemical Sciences, University of Rome La Sapienza, Rome, Italy

Christian J. T. Bollinger
Institute of Microbiology, ETH Zürich, Zürich, Switzerland

Martino Bolognesi
Department of Biomolecular Sciences and Biotechnology CNR-INFM, University of Milan, Milan, Italy

Alessandra Bonamore
Department of Biochemical Sciences, University of Rome La Sapienza, Rome, Italy

Lindsay J. Cole
Department of Biology, University of York, Heslington, York, United Kingdom

Guo-Qiang Chen
Multidisciplinary Research Center, Shantou University, Guangdong, China

Hazel A. Corker
Syntopix Group Plc, Institute of Pharmaceutical Innovation, University of Bradford, Bradford, United Kingdom

Ian R. Davies
World Precision Instruments Limited, Aston, United Kingdom

Heinz Decker
Institute for Molecular Biophysics, Johannes Gutenberg University, Mainz, Germany

Agnes Dettaï
UMR, Département Systématique et Evolution, Muséum National d'Histoire Naturelle, Paris, France and Institute of Protein Biochemistry, CNR, Naples, Italy

Sylvia Dewilde
Department of Biomedical Sciences, University of Antwerp, Antwerp, Belgium

Alexander D. Frey
Institute of Microbiology, ETH Zürich, Zürich, Switzerland

Paul R. Gardner
Department of Chemistry, University of Dayton, Dayton, Ohio

Verónica Garrocho-Villegas
Laboratorio de Biofísica y Biología Molecular, Facultad de Ciencias, Universidad Autónoma del Estado de Morelos, Cuernavaca, Morelos, Mexico

Antonio Di Giulio
Department of Science and Biomedical Technology, University of L'Aquila, L'Aquila, Italy

Sara Goldstein
Department of Physical Chemistry, The Hebrew University of Jerusalem, Jerusalem, Israel

Bridget Gollan
Centre for Molecular Microbiology and Infection, Department of Infectious Diseases, Imperial College of Science, London, United Kingdom

Vera L. Gonçalves
Instituto de Tecnologia Química e Biológica, Universidade Nova de Lisboa, Oeiras, Portugal

Sabarinathan Kuttalingam Gopalasubramaniam
Laboratorio de Biofísica y Biología Molecular, Facultad de Ciencias, Universidad Autónoma del Estado de Morelos, Cuernavaca, Morelos, Mexico

Puspita Halder
Department of Biochemistry, Biophysics, and Molecular Biology, Iowa State University, Ames, Iowa

Mark S. Hargrove
Department of Biochemistry, Biophysics, and Molecular Biology, Iowa State University, Ames, Iowa

Alfred Hausladen
Department of Chemistry and Biochemistry, Montana State University, Bozeman, Montana

Nadja Hellmann
Institute for Molecular Biophysics, Johannes Gutenberg University, Mainz, Germany

Robert D. Hill
Department of Plant Science, University of Manitoba, Winnipeg, Manitoba, Canada

Martin N. Hughes
Royal Free and University College Medical School, Centre for Hepatology, London, United Kingdom

Martin N. Hughes
Centre for Hepatology, Royal Free and University College Medical School, Centre for Hepatology, London, United Kingdom

Wilhelmina M. Huston
Department of Biology, University of York, Heslington, York, United Kingdom

Abir U. Igamberdiev
Department of Plant Science, University of Manitoba, Winnipeg, Manitoba, Canada

Andrea Ilari
CNR Institute of Molecular Biology and Pathology, University of Rome La Sapienza, Rome, Italy

Pauli T. Kallio
Institute of Microbiology, ETH Zürich, Zürich, Switzerland

J. H. Kattenberg
Section of Enzymology, Department of Biotechnology, Delft University of Technology, Delft, The Netherlands

Laurent Kiger
Le Kremlin-Bicetre, France

Taija Koskenkorva
Institute of Microbiology, ETH Zürich, Zürich, Switzerland

Olga V. Kosmachevskaya
A. N. Bach Institute of Biochemistry, Russian Academy of Sciences, Moscow, Russia

Jay R. Laver
Academic Unit of Infection and Immunity, School of Medicine and Biomedical Sciences, University of Sheffield, Sheffield, United Kingdom

Christophe Lechauve
Le Kremlin-Bicetre, France

Guillaume Lecointre
UMR, Département Systématique et Evolution, Muséum National d'Histoire Naturelle, Paris, France and Institute of Protein Biochemistry, CNR, Naples, Italy

Megan E. S. Lewis
Department of Molecular Biology and Biotechnology, University of Sheffield, Sheffield, United Kingdom

Beth Y. Lin
Biology Department, Brookhaven National Laboratory, Upton, New York

Michael C. Marden
Le Kremlin-Bicetre, France

Lelio Mazzarella
Institute of Biostructures and Bioimaging, CNR and Department of Chemistry and Consorzio Bioteknet, University of Naples Federico II, Naples, Italy

Kirsten Mees
Department of Biomedical Sciences, University of Antwerp, Antwerp, Belgium

Gabor Merényi
Department of Chemistry, Nuclear Chemistry, The Royal Institute of Technology, Stockholm, Sweden

Mario Milani
Department of Biomolecular Sciences and Biotechnology CNR-INFM, University of Milano, Milan, Italy

Luc Moens
Department of Biomedical Sciences, University of Antwerp, Antwerp, Belgium

James W. B. Moir
Department of Biology, University of York, Heslington, York, United Kingdom

Jose F. Moran
Instituto de Agrobiotecnologia, Universidad Pública de Navarra-CSIC-Gobierno de Navarra, Pamplona, Nararre, Spain

Marco Nardini
Department of Biomolecular Sciences and Biotechnology CNR-INFM, University of Milano, Milan, Italy

Enrico Negrisolo
Department of Public Health, Comparative Pathology, and Veterinary Hygiene, University of Padova, Legnaro, Italy

Lígia S. Nobre
Instituto de Tecnologia Química e Biológica, Universidade Nova de Lisboa, Oeiras, Portugal

Catherine Ozouf-Costaz
Department Systematics and Evolution, Paris, France

Elio Parisi
UMR, Départment Systématigue at Evolution, Merríum National d'histoire Naturelle, Paris, France, and Institute of Protein Biochemistry, CNR, Naples, Italy

Nina Pastor
Facultad de Ciencias, Universidad Autónoma del Estado de Morelos, Cuernavaca, Morelos, Mexico

Tomaso Patarnello
Department of Public Health, Comparative Pathology, and Veterinary Hygiene, University of Padova, Legnaro, Italy

Alessandra Pesce
Department of Physics CNR-INFM, and Center for Excellence in Biomedical Research, University of Genoa, Genoa, Italy

James L. Pickford
Department of Molecular Biology and Biotechnology, University of Sheffield, Sheffield, United Kingdom

Eva Pisano
Department of Biology, University of Genoa, Genoa, Italy

Robert K. Poole
Department of Molecular Biology and Biotechnology, University of Sheffield, Sheffield, United Kingdom

L. A. M. Pouvreau
Section of Enzymology, Department of Biotechnology, Delft University of Technology, Delft, The Netherlands

Guido di Prisco
Institute of Protein Biochemistry, CNR, Naples, Italy

Robert C. Read
Academic Unit of Infection and Immunity, School of Medicine and Biomedical Sciences, University of Sheffield, Sheffield, United Kingdom

Austen F. Riggs
Section of Neurobiology, University of Texas at Austin, Austin, Texas

Genoveva Bustos Rivera
Laboratorio de Biofísica y Biología Molecular, Facultad de Ciencias, Universidad Autónoma del Estado de Morelos, Cuernavaca, Morelos, Mexico

Selene Rol
Instituto de Agrobiotecnologia, Universidad Pública de Navarra-CSIC-Gobierno de Navarra, Pamplona, Nararre, Spain

Lígia M. Saraiva
Instituto de Tecnologia Química e Biológica, Universidade Nova de Lisboa, Oeiras, Portugal

Konstantin B. Shumaev
A. N. Bach Institute of Biochemistry, Russian Academy of Sciences, Moscow, Russia

Martha N. Simon
Biology Department, Brookhaven National Laboratory, Upton, New York

David J. Singel
Department of Chemistry and Biochemistry, Montana State University, Bozeman, Montana

Benoit J. Smagghe
Immune Disease Institute, Harvard Medical School, Boston, Massachusetts

Jonathan S. Stamler
Department of Medicine and Department of Biochemistry, Duke University Medical Center, Durham, North Carolina

Tânia M. Stevanin
Academic Unit of Infection and Immunity, School of Medicine and Biomedical Sciences, University of Sheffield, Sheffield, United Kingdom

M. J. F. Strampraad
Section of Enzymology, Department of Biotechnology, Delft University of Technology, Delft, The Netherlands

Alexander A. Timoshin
Russian Cardiology Scientific Research Complex, Moscow, Russia

Alexey F. Topunov
A. N. Bach Institute of Biochemistry, Russian Academy of Sciences, Moscow, Russia

Estibaliz Urarte
Instituto de Agrobiotecnologia, Universidad Pública de Navarra-CSIC-Gobierno de Navarra, Pamplona, Nararre, Spain

Thomas L. Vandergon
Natural Science Division, Pepperdine University, Malibu, California

Anatoly F. Vanin
N. N. Semenov Institute of Chemical Physics, Russian Academy of Sciences, Moscow, Russia

Cinzia Verde
Institute of Protein Biochemistry, CNR, Naples, Italy

Alessandro Vergara
Institute of Biostructures and Bioimaging, CNR and Department of Chemistry and Consorzio Bioteknet, University of Naples Federico II, Naples, Italy

Serge N. Vinogradov
Department of Biochemistry and Molecular Biology, School of Medicine, Wayne State University, Detroit, Michigan

Paolo Visca
National Institute for Infectious Diseases IRCCS Lazzaro Spallanzani, Rome, Italy, and Department of Biology and Interdepartmental Laboratory for Electron Microscopy, University Roma Tre, Rome, Italy

Luigi Vitagliano
Institute of Biostructures and Bioimaging, CNR, Naples, Italy

S. de Vries
Section of Enzymology, Department of Biotechnology, Delft University of Technology, Delft, The Netherlands

Laura Wainwright
Queen Alexandra Hospital, Cosham, Portsmouth, United Kingdom

Joseph S. Wall
Biology Department, Brookhaven National Laboratory, Upton, New York

Roy E. Weber
Zoophysiology, Institute of Biological Sciences, University of Aarhus, Denmark

Xiao-Xing Wei
Department of Biological Sciences and Biotechnology, Tsinghua University, Beijing, China

Guanghui Wu
Food and Environmental Safety, Veterinary Laboratories Agency-Weybridge, New Haw, Addlestone, Surrey, United Kingdom

Xueji Zhang
World Precision Instruments Inc., Sarasota, Florida

Preface

The genesis of ideas for these two volumes of *Methods in Enzymology* appears to be a talk (subtitled *Bloody Bacteria*) that I presented at the Agouron Institute meeting in Santa Fe, New Mexico, in April of 2006. The topic of the meeting was *Oxygen,* but my message was not how microbial hemoglobins manage oxygen but rather how the primary function of many such hemoglobins is nitric oxide detoxification. Despite my straying from my brief, John Abelson and Mel Simon generously invited me to consider editing a volume of *Methods in Enzymology* to cover these emerging aspects of such a well-studied protein family. Further discussion of the proposal at the XIVth International Conference on Dioxygen Binding and Sensing Proteins at Stazione Zoologica Anton Dohrn in beautiful Napoli later that year—warmly hosted by Cinzia Verde and Guido di Prisco—generated much interest and support. The result was a two-volume heterodimer: I hope cooperativity can be found in Volumes 436 and 437.

Just as the organizers of the Agouron Institute conference interpreted *Oxygen* with commendable flexibility, *Methods in Enzymology* has allowed some freedom in the definition of an enzyme. In 1994, when the topic *Hemoglobins (Part C)* was last covered explicitly in this series (Volume 232), some justification for labeling a hemoglobin as an enzyme might have been warranted. But as Maurizio Brunori pointed out in 1999 (*Trends in Biochemical Sciences*, 24, 158–161), the promotion of hemoglobin to the status of "honorary enzyme" had been conferred decades earlier by Monod, Wyman, and Changeux. In 2007, the idea that certain hemoglobins, even those not displaying allosteric heme–heme interactions, have enzymatic functions is well established; the most obvious examples being those hemoglobins that transform substrates into products, such as nitric oxide into nitrate.

Other topics covered in these volumes are not new to the *Methods in Enzymology* series either. The most recent coverage of overtly related topics was *Nitric Oxide (Part E)* in Volume 396 (2005) and *Oxygen Sensing* in Volume 381 (2004). I hope, however, that the particular juxtaposition of topics in these two volumes will draw attention to the intimate links between globins, their gaseous ligands (nitric oxide, oxygen, and carbon monoxide), and the sensing and detoxification of these biologically critical small molecules. There is a strong microbial flavor in these volumes, reflecting some of the most exciting developments in recent years. Volume 436 deals with some chemical and analytical aspects of nitric oxide and methods for bacterial and archaeal hemoglobins, as well as diverse (especially "newer") hemoglobins in plants and animals. Volume 437 covers various non-hemoglobin nitric oxide-detoxifying proteins, sensors for gaseous ligands, advanced spectroscopic tools, and aspects of the functions of these proteins in microbial and plant physiology. In each volume, some chapters serve not as methodological recipes but short reviews to place the methods in a proper framework.

These volumes would not have been possible without the tremendous enthusiasm of so many colleagues, contributors, and friends around the world. I thank them all, and also Tari Broderick and Cindy Minor (Elsevier, San Diego, California), for their help and encouragement in leading these volumes to a successful and timely outcome.

Robert K. Poole

Methods in Enzymology

VOLUME LIII. Biomembranes (Part D: Biological Oxidations)
Edited by SIDNEY FLEISCHER AND LESTER PACKER

VOLUME LIV. Biomembranes (Part E: Biological Oxidations)
Edited by SIDNEY FLEISCHER AND LESTER PACKER

VOLUME LV. Biomembranes (Part F: Bioenergetics)
Edited by SIDNEY FLEISCHER AND LESTER PACKER

VOLUME LVI. Biomembranes (Part G: Bioenergetics)
Edited by SIDNEY FLEISCHER AND LESTER PACKER

VOLUME LVII. Bioluminescence and Chemiluminescence
Edited by MARLENE A. DELUCA

VOLUME LVIII. Cell Culture
Edited by WILLIAM B. JAKOBY AND IRA PASTAN

VOLUME LIX. Nucleic Acids and Protein Synthesis (Part G)
Edited by KIVIE MOLDAVE AND LAWRENCE GROSSMAN

VOLUME LX. Nucleic Acids and Protein Synthesis (Part H)
Edited by KIVIE MOLDAVE AND LAWRENCE GROSSMAN

VOLUME 61. Enzyme Structure (Part H)
Edited by C. H. W. HIRS AND SERGE N. TIMASHEFF

VOLUME 62. Vitamins and Coenzymes (Part D)
Edited by DONALD B. MCCORMICK AND LEMUEL D. WRIGHT

VOLUME 63. Enzyme Kinetics and Mechanism (Part A: Initial Rate and Inhibitor Methods)
Edited by DANIEL L. PURICH

VOLUME 64. Enzyme Kinetics and Mechanism (Part B: Isotopic Probes and Complex Enzyme Systems)
Edited by DANIEL L. PURICH

VOLUME 65. Nucleic Acids (Part I)
Edited by LAWRENCE GROSSMAN AND KIVIE MOLDAVE

VOLUME 66. Vitamins and Coenzymes (Part E)
Edited by DONALD B. MCCORMICK AND LEMUEL D. WRIGHT

VOLUME 67. Vitamins and Coenzymes (Part F)
Edited by DONALD B. MCCORMICK AND LEMUEL D. WRIGHT

VOLUME 68. Recombinant DNA
Edited by RAY WU

VOLUME 69. Photosynthesis and Nitrogen Fixation (Part C)
Edited by ANTHONY SAN PIETRO

VOLUME 70. Immunochemical Techniques (Part A)
Edited by HELEN VAN VUNAKIS AND JOHN J. LANGONE

VOLUME 71. Lipids (Part C)
Edited by JOHN M. LOWENSTEIN

VOLUME 72. Lipids (Part D)
Edited by JOHN M. LOWENSTEIN

VOLUME 73. Immunochemical Techniques (Part B)
Edited by JOHN J. LANGONE AND HELEN VAN VUNAKIS

VOLUME 74. Immunochemical Techniques (Part C)
Edited by JOHN J. LANGONE AND HELEN VAN VUNAKIS

VOLUME 75. Cumulative Subject Index Volumes XXXI, XXXII, XXXIV–LX
Edited by EDWARD A. DENNIS AND MARTHA G. DENNIS

VOLUME 76. Hemoglobins
Edited by ERALDO ANTONINI, LUIGI ROSSI-BERNARDI, AND EMILIA CHIANCONE

VOLUME 77. Detoxication and Drug Metabolism
Edited by WILLIAM B. JAKOBY

VOLUME 78. Interferons (Part A)
Edited by SIDNEY PESTKA

VOLUME 79. Interferons (Part B)
Edited by SIDNEY PESTKA

VOLUME 80. Proteolytic Enzymes (Part C)
Edited by LASZLO LORAND

VOLUME 81. Biomembranes (Part H: Visual Pigments and Purple Membranes, I)
Edited by LESTER PACKER

VOLUME 82. Structural and Contractile Proteins (Part A: Extracellular Matrix)
Edited by LEON W. CUNNINGHAM AND DIXIE W. FREDERIKSEN

VOLUME 83. Complex Carbohydrates (Part D)
Edited by VICTOR GINSBURG

VOLUME 84. Immunochemical Techniques (Part D: Selected Immunoassays)
Edited by JOHN J. LANGONE AND HELEN VAN VUNAKIS

VOLUME 85. Structural and Contractile Proteins (Part B: The Contractile Apparatus and the Cytoskeleton)
Edited by DIXIE W. FREDERIKSEN AND LEON W. CUNNINGHAM

VOLUME 86. Prostaglandins and Arachidonate Metabolites
Edited by WILLIAM E. M. LANDS AND WILLIAM L. SMITH

VOLUME 87. Enzyme Kinetics and Mechanism (Part C: Intermediates, Stereo-chemistry, and Rate Studies)
Edited by DANIEL L. PURICH

VOLUME 88. Biomembranes (Part I: Visual Pigments and Purple Membranes, II)
Edited by LESTER PACKER

VOLUME 105. Oxygen Radicals in Biological Systems
Edited by LESTER PACKER

VOLUME 106. Posttranslational Modifications (Part A)
Edited by FINN WOLD AND KIVIE MOLDAVE

VOLUME 107. Posttranslational Modifications (Part B)
Edited by FINN WOLD AND KIVIE MOLDAVE

VOLUME 108. Immunochemical Techniques (Part G: Separation and Characterization of Lymphoid Cells)
Edited by GIOVANNI DI SABATO, JOHN J. LANGONE, AND HELEN VAN VUNAKIS

VOLUME 109. Hormone Action (Part I: Peptide Hormones)
Edited by LUTZ BIRNBAUMER AND BERT W. O'MALLEY

VOLUME 110. Steroids and Isoprenoids (Part A)
Edited by JOHN H. LAW AND HANS C. RILLING

VOLUME 111. Steroids and Isoprenoids (Part B)
Edited by JOHN H. LAW AND HANS C. RILLING

VOLUME 112. Drug and Enzyme Targeting (Part A)
Edited by KENNETH J. WIDDER AND RALPH GREEN

VOLUME 113. Glutamate, Glutamine, Glutathione, and Related Compounds
Edited by ALTON MEISTER

VOLUME 114. Diffraction Methods for Biological Macromolecules (Part A)
Edited by HAROLD W. WYCKOFF, C. H. W. HIRS, AND SERGE N. TIMASHEFF

VOLUME 115. Diffraction Methods for Biological Macromolecules (Part B)
Edited by HAROLD W. WYCKOFF, C. H. W. HIRS, AND SERGE N. TIMASHEFF

VOLUME 116. Immunochemical Techniques
(Part H: Effectors and Mediators of Lymphoid Cell Functions)
Edited by GIOVANNI DI SABATO, JOHN J. LANGONE, AND HELEN VAN VUNAKIS

VOLUME 117. Enzyme Structure (Part J)
Edited by C. H. W. HIRS AND SERGE N. TIMASHEFF

VOLUME 118. Plant Molecular Biology
Edited by ARTHUR WEISSBACH AND HERBERT WEISSBACH

VOLUME 119. Interferons (Part C)
Edited by SIDNEY PESTKA

VOLUME 120. Cumulative Subject Index Volumes 81–94, 96–101

VOLUME 121. Immunochemical Techniques (Part I: Hybridoma Technology and Monoclonal Antibodies)
Edited by JOHN J. LANGONE AND HELEN VAN VUNAKIS

VOLUME 122. Vitamins and Coenzymes (Part G)
Edited by FRANK CHYTIL AND DONALD B. MCCORMICK

Volume 123. Vitamins and Coenzymes (Part H)
Edited by Frank Chytil and Donald B. McCormick

Volume 124. Hormone Action (Part J: Neuroendocrine Peptides)
Edited by P. Michael Conn

Volume 125. Biomembranes (Part M: Transport in Bacteria, Mitochondria, and Chloroplasts: General Approaches and Transport Systems)
Edited by Sidney Fleischer and Becca Fleischer

Volume 126. Biomembranes (Part N: Transport in Bacteria, Mitochondria, and Chloroplasts: Protonmotive Force)
Edited by Sidney Fleischer and Becca Fleischer

Volume 127. Biomembranes (Part O: Protons and Water: Structure and Translocation)
Edited by Lester Packer

Volume 128. Plasma Lipoproteins (Part A: Preparation, Structure, and Molecular Biology)
Edited by Jere P. Segrest and John J. Albers

Volume 129. Plasma Lipoproteins (Part B: Characterization, Cell Biology, and Metabolism)
Edited by John J. Albers and Jere P. Segrest

Volume 130. Enzyme Structure (Part K)
Edited by C. H. W. Hirs and Serge N. Timasheff

Volume 131. Enzyme Structure (Part L)
Edited by C. H. W. Hirs and Serge N. Timasheff

Volume 132. Immunochemical Techniques (Part J: Phagocytosis and Cell-Mediated Cytotoxicity)
Edited by Giovanni Di Sabato and Johannes Everse

Volume 133. Bioluminescence and Chemiluminescence (Part B)
Edited by Marlene DeLuca and William D. McElroy

Volume 134. Structural and Contractile Proteins (Part C: The Contractile Apparatus and the Cytoskeleton)
Edited by Richard B. Vallee

Volume 135. Immobilized Enzymes and Cells (Part B)
Edited by Klaus Mosbach

Volume 136. Immobilized Enzymes and Cells (Part C)
Edited by Klaus Mosbach

Volume 137. Immobilized Enzymes and Cells (Part D)
Edited by Klaus Mosbach

Volume 138. Complex Carbohydrates (Part E)
Edited by Victor Ginsburg

Volume 139. Cellular Regulators (Part A: Calcium- and Calmodulin-Binding Proteins)
Edited by Anthony R. Means and P. Michael Conn

Volume 140. Cumulative Subject Index Volumes 102–119, 121–134

Volume 141. Cellular Regulators (Part B: Calcium and Lipids)
Edited by P. Michael Conn and Anthony R. Means

Volume 142. Metabolism of Aromatic Amino Acids and Amines
Edited by Seymour Kaufman

Volume 143. Sulfur and Sulfur Amino Acids
Edited by William B. Jakoby and Owen Griffith

Volume 144. Structural and Contractile Proteins (Part D: Extracellular Matrix)
Edited by Leon W. Cunningham

Volume 145. Structural and Contractile Proteins (Part E: Extracellular Matrix)
Edited by Leon W. Cunningham

Volume 146. Peptide Growth Factors (Part A)
Edited by David Barnes and David A. Sirbasku

Volume 147. Peptide Growth Factors (Part B)
Edited by David Barnes and David A. Sirbasku

Volume 148. Plant Cell Membranes
Edited by Lester Packer and Roland Douce

Volume 149. Drug and Enzyme Targeting (Part B)
Edited by Ralph Green and Kenneth J. Widder

Volume 150. Immunochemical Techniques (Part K: *In Vitro* Models of B and T Cell Functions and Lymphoid Cell Receptors)
Edited by Giovanni Di Sabato

Volume 151. Molecular Genetics of Mammalian Cells
Edited by Michael M. Gottesman

Volume 152. Guide to Molecular Cloning Techniques
Edited by Shelby L. Berger and Alan R. Kimmel

Volume 153. Recombinant DNA (Part D)
Edited by Ray Wu and Lawrence Grossman

Volume 154. Recombinant DNA (Part E)
Edited by Ray Wu and Lawrence Grossman

Volume 155. Recombinant DNA (Part F)
Edited by Ray Wu

Volume 156. Biomembranes (Part P: ATP-Driven Pumps and Related Transport: The Na, K-Pump)
Edited by Sidney Fleischer and Becca Fleischer

Volume 173. Biomembranes [Part T: Cellular and Subcellular Transport: Eukaryotic (Nonepithelial) Cells]
Edited by Sidney Fleischer and Becca Fleischer

Volume 174. Biomembranes [Part U: Cellular and Subcellular Transport: Eukaryotic (Nonepithelial) Cells]
Edited by Sidney Fleischer and Becca Fleischer

Volume 175. Cumulative Subject Index Volumes 135–139, 141–167

Volume 176. Nuclear Magnetic Resonance (Part A: Spectral Techniques and Dynamics)
Edited by Norman J. Oppenheimer and Thomas L. James

Volume 177. Nuclear Magnetic Resonance (Part B: Structure and Mechanism)
Edited by Norman J. Oppenheimer and Thomas L. James

Volume 178. Antibodies, Antigens, and Molecular Mimicry
Edited by John J. Langone

Volume 179. Complex Carbohydrates (Part F)
Edited by Victor Ginsburg

Volume 180. RNA Processing (Part A: General Methods)
Edited by James E. Dahlberg and John N. Abelson

Volume 181. RNA Processing (Part B: Specific Methods)
Edited by James E. Dahlberg and John N. Abelson

Volume 182. Guide to Protein Purification
Edited by Murray P. Deutscher

Volume 183. Molecular Evolution: Computer Analysis of Protein and Nucleic Acid Sequences
Edited by Russell F. Doolittle

Volume 184. Avidin-Biotin Technology
Edited by Meir Wilchek and Edward A. Bayer

Volume 185. Gene Expression Technology
Edited by David V. Goeddel

Volume 186. Oxygen Radicals in Biological Systems (Part B: Oxygen Radicals and Antioxidants)
Edited by Lester Packer and Alexander N. Glazer

Volume 187. Arachidonate Related Lipid Mediators
Edited by Robert C. Murphy and Frank A. Fitzpatrick

Volume 188. Hydrocarbons and Methylotrophy
Edited by Mary E. Lidstrom

Volume 189. Retinoids (Part A: Molecular and Metabolic Aspects)
Edited by Lester Packer

Volume 206. Cytochrome P450
Edited by Michael R. Waterman and Eric F. Johnson

Volume 207. Ion Channels
Edited by Bernardo Rudy and Linda E. Iverson

Volume 208. Protein–DNA Interactions
Edited by Robert T. Sauer

Volume 209. Phospholipid Biosynthesis
Edited by Edward A. Dennis and Dennis E. Vance

Volume 210. Numerical Computer Methods
Edited by Ludwig Brand and Michael L. Johnson

Volume 211. DNA Structures (Part A: Synthesis and Physical Analysis of DNA)
Edited by David M. J. Lilley and James E. Dahlberg

Volume 212. DNA Structures (Part B: Chemical and Electrophoretic Analysis of DNA)
Edited by David M. J. Lilley and James E. Dahlberg

Volume 213. Carotenoids (Part A: Chemistry, Separation, Quantitation, and Antioxidation)
Edited by Lester Packer

Volume 214. Carotenoids (Part B: Metabolism, Genetics, and Biosynthesis)
Edited by Lester Packer

Volume 215. Platelets: Receptors, Adhesion, Secretion (Part B)
Edited by Jacek J. Hawiger

Volume 216. Recombinant DNA (Part G)
Edited by Ray Wu

Volume 217. Recombinant DNA (Part H)
Edited by Ray Wu

Volume 218. Recombinant DNA (Part I)
Edited by Ray Wu

Volume 219. Reconstitution of Intracellular Transport
Edited by James E. Rothman

Volume 220. Membrane Fusion Techniques (Part A)
Edited by Nejat Düzgüneş

Volume 221. Membrane Fusion Techniques (Part B)
Edited by Nejat Düzgüneş

Volume 222. Proteolytic Enzymes in Coagulation, Fibrinolysis, and Complement Activation (Part A: Mammalian Blood Coagulation Factors and Inhibitors)
Edited by Laszlo Lorand and Kenneth G. Mann

Volume 223. Proteolytic Enzymes in Coagulation, Fibrinolysis, and Complement Activation (Part B: Complement Activation, Fibrinolysis, and Nonmammalian Blood Coagulation Factors)
Edited by Laszlo Lorand and Kenneth G. Mann

Volume 224. Molecular Evolution: Producing the Biochemical Data
Edited by Elizabeth Anne Zimmer, Thomas J. White, Rebecca L. Cann, and Allan C. Wilson

Volume 225. Guide to Techniques in Mouse Development
Edited by Paul M. Wassarman and Melvin L. DePamphilis

Volume 226. Metallobiochemistry (Part C: Spectroscopic and Physical Methods for Probing Metal Ion Environments in Metalloenzymes and Metalloproteins)
Edited by James F. Riordan and Bert L. Vallee

Volume 227. Metallobiochemistry (Part D: Physical and Spectroscopic Methods for Probing Metal Ion Environments in Metalloproteins)
Edited by James F. Riordan and Bert L. Vallee

Volume 228. Aqueous Two-Phase Systems
Edited by Harry Walter and Göte Johansson

Volume 229. Cumulative Subject Index Volumes 195–198, 200–227

Volume 230. Guide to Techniques in Glycobiology
Edited by William J. Lennarz and Gerald W. Hart

Volume 231. Hemoglobins (Part B: Biochemical and Analytical Methods)
Edited by Johannes Everse, Kim D. Vandegriff, and Robert M. Winslow

Volume 232. Hemoglobins (Part C: Biophysical Methods)
Edited by Johannes Everse, Kim D. Vandegriff, and Robert M. Winslow

Volume 233. Oxygen Radicals in Biological Systems (Part C)
Edited by Lester Packer

Volume 234. Oxygen Radicals in Biological Systems (Part D)
Edited by Lester Packer

Volume 235. Bacterial Pathogenesis (Part A: Identification and Regulation of Virulence Factors)
Edited by Virginia L. Clark and Patrik M. Bavoil

Volume 236. Bacterial Pathogenesis (Part B: Integration of Pathogenic Bacteria with Host Cells)
Edited by Virginia L. Clark and Patrik M. Bavoil

Volume 237. Heterotrimeric G Proteins
Edited by Ravi Iyengar

Volume 238. Heterotrimeric G-Protein Effectors
Edited by Ravi Iyengar

Volume 239. Nuclear Magnetic Resonance (Part C)
Edited by Thomas L. James and Norman J. Oppenheimer

Volume 240. Numerical Computer Methods (Part B)
Edited by Michael L. Johnson and Ludwig Brand

Volume 241. Retroviral Proteases
Edited by Lawrence C. Kuo and Jules A. Shafer

Volume 242. Neoglycoconjugates (Part A)
Edited by Y. C. Lee and Reiko T. Lee

Volume 243. Inorganic Microbial Sulfur Metabolism
Edited by Harry D. Peck, Jr., and Jean LeGall

Volume 244. Proteolytic Enzymes: Serine and Cysteine Peptidases
Edited by Alan J. Barrett

Volume 245. Extracellular Matrix Components
Edited by E. Ruoslahti and E. Engvall

Volume 246. Biochemical Spectroscopy
Edited by Kenneth Sauer

Volume 247. Neoglycoconjugates (Part B: Biomedical Applications)
Edited by Y. C. Lee and Reiko T. Lee

Volume 248. Proteolytic Enzymes: Aspartic and Metallo Peptidases
Edited by Alan J. Barrett

Volume 249. Enzyme Kinetics and Mechanism (Part D: Developments in Enzyme Dynamics)
Edited by Daniel L. Purich

Volume 250. Lipid Modifications of Proteins
Edited by Patrick J. Casey and Janice E. Buss

Volume 251. Biothiols (Part A: Monothiols and Dithiols, Protein Thiols, and Thiyl Radicals)
Edited by Lester Packer

Volume 252. Biothiols (Part B: Glutathione and Thioredoxin; Thiols in Signal Transduction and Gene Regulation)
Edited by Lester Packer

Volume 253. Adhesion of Microbial Pathogens
Edited by Ron J. Doyle and Itzhak Ofek

Volume 254. Oncogene Techniques
Edited by Peter K. Vogt and Inder M. Verma

Volume 255. Small GTPases and Their Regulators (Part A: Ras Family)
Edited by W. E. Balch, Channing J. Der, and Alan Hall

Volume 256. Small GTPases and Their Regulators (Part B: Rho Family)
Edited by W. E. Balch, Channing J. Der, and Alan Hall

Volume 275. Viral Polymerases and Related Proteins
Edited by Lawrence C. Kuo, David B. Olsen, and Steven S. Carroll

Volume 276. Macromolecular Crystallography (Part A)
Edited by Charles W. Carter, Jr., and Robert M. Sweet

Volume 277. Macromolecular Crystallography (Part B)
Edited by Charles W. Carter, Jr., and Robert M. Sweet

Volume 278. Fluorescence Spectroscopy
Edited by Ludwig Brand and Michael L. Johnson

Volume 279. Vitamins and Coenzymes (Part I)
Edited by Donald B. McCormick, John W. Suttie, and Conrad Wagner

Volume 280. Vitamins and Coenzymes (Part J)
Edited by Donald B. McCormick, John W. Suttie, and Conrad Wagner

Volume 281. Vitamins and Coenzymes (Part K)
Edited by Donald B. McCormick, John W. Suttie, and Conrad Wagner

Volume 282. Vitamins and Coenzymes (Part L)
Edited by Donald B. McCormick, John W. Suttie, and Conrad Wagner

Volume 283. Cell Cycle Control
Edited by William G. Dunphy

Volume 284. Lipases (Part A: Biotechnology)
Edited by Byron Rubin and Edward A. Dennis

Volume 285. Cumulative Subject Index Volumes 263, 264, 266–284, 286–289

Volume 286. Lipases (Part B: Enzyme Characterization and Utilization)
Edited by Byron Rubin and Edward A. Dennis

Volume 287. Chemokines
Edited by Richard Horuk

Volume 288. Chemokine Receptors
Edited by Richard Horuk

Volume 289. Solid Phase Peptide Synthesis
Edited by Gregg B. Fields

Volume 290. Molecular Chaperones
Edited by George H. Lorimer and Thomas Baldwin

Volume 291. Caged Compounds
Edited by Gerard Marriott

Volume 292. ABC Transporters: Biochemical, Cellular, and Molecular Aspects
Edited by Suresh V. Ambudkar and Michael M. Gottesman

Volume 293. Ion Channels (Part B)
Edited by P. Michael Conn

VOLUME 312. Sphingolipid Metabolism and Cell Signaling (Part B)
Edited by ALFRED H. MERRILL, JR., AND YUSUF A. HANNUN

VOLUME 313. Antisense Technology (Part A: General Methods, Methods of Delivery, and RNA Studies)
Edited by M. IAN PHILLIPS

VOLUME 314. Antisense Technology (Part B: Applications)
Edited by M. IAN PHILLIPS

VOLUME 315. Vertebrate Phototransduction and the Visual Cycle (Part A)
Edited by KRZYSZTOF PALCZEWSKI

VOLUME 316. Vertebrate Phototransduction and the Visual Cycle (Part B)
Edited by KRZYSZTOF PALCZEWSKI

VOLUME 317. RNA–Ligand Interactions (Part A: Structural Biology Methods)
Edited by DANIEL W. CELANDER AND JOHN N. ABELSON

VOLUME 318. RNA–Ligand Interactions (Part B: Molecular Biology Methods)
Edited by DANIEL W. CELANDER AND JOHN N. ABELSON

VOLUME 319. Singlet Oxygen, UV-A, and Ozone
Edited by LESTER PACKER AND HELMUT SIES

VOLUME 320. Cumulative Subject Index Volumes 290–319

VOLUME 321. Numerical Computer Methods (Part C)
Edited by MICHAEL L. JOHNSON AND LUDWIG BRAND

VOLUME 322. Apoptosis
Edited by JOHN C. REED

VOLUME 323. Energetics of Biological Macromolecules (Part C)
Edited by MICHAEL L. JOHNSON AND GARY K. ACKERS

VOLUME 324. Branched-Chain Amino Acids (Part B)
Edited by ROBERT A. HARRIS AND JOHN R. SOKATCH

VOLUME 325. Regulators and Effectors of Small GTPases (Part D: Rho Family)
Edited by W. E. BALCH, CHANNING J. DER, AND ALAN HALL

VOLUME 326. Applications of Chimeric Genes and Hybrid Proteins (Part A: Gene Expression and Protein Purification)
Edited by JEREMY THORNER, SCOTT D. EMR, AND JOHN N. ABELSON

VOLUME 327. Applications of Chimeric Genes and Hybrid Proteins (Part B: Cell Biology and Physiology)
Edited by JEREMY THORNER, SCOTT D. EMR, AND JOHN N. ABELSON

VOLUME 328. Applications of Chimeric Genes and Hybrid Proteins (Part C: Protein–Protein Interactions and Genomics)
Edited by JEREMY THORNER, SCOTT D. EMR, AND JOHN N. ABELSON

Volume 346. Gene Therapy Methods
Edited by M. Ian Phillips

Volume 347. Protein Sensors and Reactive Oxygen Species (Part A: Selenoproteins and Thioredoxin)
Edited by Helmut Sies and Lester Packer

Volume 348. Protein Sensors and Reactive Oxygen Species (Part B: Thiol Enzymes and Proteins)
Edited by Helmut Sies and Lester Packer

Volume 349. Superoxide Dismutase
Edited by Lester Packer

Volume 350. Guide to Yeast Genetics and Molecular and Cell Biology (Part B)
Edited by Christine Guthrie and Gerald R. Fink

Volume 351. Guide to Yeast Genetics and Molecular and Cell Biology (Part C)
Edited by Christine Guthrie and Gerald R. Fink

Volume 352. Redox Cell Biology and Genetics (Part A)
Edited by Chandan K. Sen and Lester Packer

Volume 353. Redox Cell Biology and Genetics (Part B)
Edited by Chandan K. Sen and Lester Packer

Volume 354. Enzyme Kinetics and Mechanisms (Part F: Detection and Characterization of Enzyme Reaction Intermediates)
Edited by Daniel L. Purich

Volume 355. Cumulative Subject Index Volumes 321–354

Volume 356. Laser Capture Microscopy and Microdissection
Edited by P. Michael Conn

Volume 357. Cytochrome P450, Part C
Edited by Eric F. Johnson and Michael R. Waterman

Volume 358. Bacterial Pathogenesis (Part C: Identification, Regulation, and Function of Virulence Factors)
Edited by Virginia L. Clark and Patrik M. Bavoil

Volume 359. Nitric Oxide (Part D)
Edited by Enrique Cadenas and Lester Packer

Volume 360. Biophotonics (Part A)
Edited by Gerard Marriott and Ian Parker

Volume 361. Biophotonics (Part B)
Edited by Gerard Marriott and Ian Parker

Volume 362. Recognition of Carbohydrates in Biological Systems (Part A)
Edited by Yuan C. Lee and Reiko T. Lee

Volume 363. Recognition of Carbohydrates in Biological Systems (Part B)
Edited by Yuan C. Lee and Reiko T. Lee

Volume 364. Nuclear Receptors
Edited by David W. Russell and David J. Mangelsdorf

Volume 365. Differentiation of Embryonic Stem Cells
Edited by Paul M. Wassauman and Gordon M. Keller

Volume 366. Protein Phosphatases
Edited by Susanne Klumpp and Josef Krieglstein

Volume 367. Liposomes (Part A)
Edited by Nejat Düzgüneş

Volume 368. Macromolecular Crystallography (Part C)
Edited by Charles W. Carter, Jr., and Robert M. Sweet

Volume 369. Combinational Chemistry (Part B)
Edited by Guillermo A. Morales and Barry A. Bunin

Volume 370. RNA Polymerases and Associated Factors (Part C)
Edited by Sankar L. Adhya and Susan Garges

Volume 371. RNA Polymerases and Associated Factors (Part D)
Edited by Sankar L. Adhya and Susan Garges

Volume 372. Liposomes (Part B)
Edited by Nejat Düzgüneş

Volume 373. Liposomes (Part C)
Edited by Nejat Düzgüneş

Volume 374. Macromolecular Crystallography (Part D)
Edited by Charles W. Carter, Jr., and Robert W. Sweet

Volume 375. Chromatin and Chromatin Remodeling Enzymes (Part A)
Edited by C. David Allis and Carl Wu

Volume 376. Chromatin and Chromatin Remodeling Enzymes (Part B)
Edited by C. David Allis and Carl Wu

Volume 377. Chromatin and Chromatin Remodeling Enzymes (Part C)
Edited by C. David Allis and Carl Wu

Volume 378. Quinones and Quinone Enzymes (Part A)
Edited by Helmut Sies and Lester Packer

Volume 379. Energetics of Biological Macromolecules (Part D)
Edited by Jo M. Holt, Michael L. Johnson, and Gary K. Ackers

Volume 380. Energetics of Biological Macromolecules (Part E)
Edited by Jo M. Holt, Michael L. Johnson, and Gary K. Ackers

Volume 381. Oxygen Sensing
Edited by Chandan K. Sen and Gregg L. Semenza

Volume 382. Quinones and Quinone Enzymes (Part B)
Edited by Helmut Sies and Lester Packer

Volume 383. Numerical Computer Methods (Part D)
Edited by Ludwig Brand and Michael L. Johnson

Volume 384. Numerical Computer Methods (Part E)
Edited by Ludwig Brand and Michael L. Johnson

Volume 385. Imaging in Biological Research (Part A)
Edited by P. Michael Conn

Volume 386. Imaging in Biological Research (Part B)
Edited by P. Michael Conn

Volume 387. Liposomes (Part D)
Edited by Nejat Düzgüneş

Volume 388. Protein Engineering
Edited by Dan E. Robertson and Joseph P. Noel

Volume 389. Regulators of G-Protein Signaling (Part A)
Edited by David P. Siderovski

Volume 390. Regulators of G-Protein Signaling (Part B)
Edited by David P. Siderovski

Volume 391. Liposomes (Part E)
Edited by Nejat Düzgüneş

Volume 392. RNA Interference
Edited by Engelke Rossi

Volume 393. Circadian Rhythms
Edited by Michael W. Young

Volume 394. Nuclear Magnetic Resonance of Biological Macromolecules (Part C)
Edited by Thomas L. James

Volume 395. Producing the Biochemical Data (Part B)
Edited by Elizabeth A. Zimmer and Eric H. Roalson

Volume 396. Nitric Oxide (Part E)
Edited by Lester Packer and Enrique Cadenas

Volume 397. Environmental Microbiology
Edited by Jared R. Leadbetter

Volume 398. Ubiquitin and Protein Degradation (Part A)
Edited by Raymond J. Deshaies

Volume 399. Ubiquitin and Protein Degradation (Part B)
Edited by Raymond J. Deshaies

Volume 400. Phase II Conjugation Enzymes and Transport Systems
Edited by Helmut Sies and Lester Packer

Volume 401. Glutathione Transferases and Gamma Glutamyl Transpeptidases
Edited by Helmut Sies and Lester Packer

Volume 402. Biological Mass Spectrometry
Edited by A. L. Burlingame

Volume 403. GTPases Regulating Membrane Targeting and Fusion
Edited by William E. Balch, Channing J. Der, and Alan Hall

Volume 404. GTPases Regulating Membrane Dynamics
Edited by William E. Balch, Channing J. Der, and Alan Hall

Volume 405. Mass Spectrometry: Modified Proteins and Glycoconjugates
Edited by A. L. Burlingame

Volume 406. Regulators and Effectors of Small GTPases: Rho Family
Edited by William E. Balch, Channing J. Der, and Alan Hall

Volume 407. Regulators and Effectors of Small GTPases: Ras Family
Edited by William E. Balch, Channing J. Der, and Alan Hall

Volume 408. DNA Repair (Part A)
Edited by Judith L. Campbell and Paul Modrich

Volume 409. DNA Repair (Part B)
Edited by Judith L. Campbell and Paul Modrich

Volume 410. DNA Microarrays (Part A: Array Platforms and Web-Bench Protocols)
Edited by Alan Kimmel and Brian Oliver

Volume 411. DNA Microarrays (Part B: Databases and Statistics)
Edited by Alan Kimmel and Brian Oliver

Volume 412. Amyloid, Prions, and Other Protein Aggregates (Part B)
Edited by Indu Kheterpal and Ronald Wetzel

Volume 413. Amyloid, Prions, and Other Protein Aggregates (Part C)
Edited by Indu Kheterpal and Ronald Wetzel

Volume 414. Measuring Biological Responses with Automated Microscopy
Edited by James Inglese

Volume 415. Glycobiology
Edited by Minoru Fukuda

Volume 416. Glycomics
Edited by Minoru Fukuda

Volume 417. Functional Glycomics
Edited by Minoru Fukuda

Volume 418. Embryonic Stem Cells
Edited by Irina Klimanskaya and Robert Lanza

Volume 419. Adult Stem Cells
Edited by Irina Klimanskaya and Robert Lanza

VOLUME 436. Globins and Other Nitric Oxide-Reactive Proteins, Part A
Edited by ROBERT K. POOLE

VOLUME 437. Globins and Other Nitric Oxide-Reactive Proteins, Part B (in preparation)
Edited by ROBERT K. POOLE

VOLUME 438. Small GTPases in Diseases, Part A (in preparation)
Edited by WILLIAM E. BALCH, CHANNING J. DER, AND ALAN HALL

SECTION ONE

NITRIC OXIDE: CHEMICAL AND ANALYTICAL METHODS

CHAPTER ONE

Chemistry of Nitric Oxide and Related Species

Martin N. Hughes

Contents

Abstract

Nitric oxide (NO) has essential roles in a remarkable number of diverse biological processes. The reactivity of NO depends upon its physical properties, such as its small size, high diffusion rate, and lipophilicity (resulting in its accumulation in hydrophobic regions), and also on its facile but selective chemical reactivity toward a variety of cellular targets. NO also undergoes reactions with oxygen, superoxide ions, and reducing agents to give products that themselves show distinctive reactivity toward particular targets, sometimes with the manifestation of toxic effects, such as nitrosative stress. These include nitroxyl (HNO), the oxides NO_2/N_2O_4, and N_2O_3, peroxynitrite, and S-nitrosothiols (RSNO). HNO is attracting considerable attention due to its pharmacological properties, which appear to be distinct from those of NO, and that may be significant in the treatment of heart failure.

Royal Free and University College Medical School, Centre for Hepatology, Royal Free/Hampstead Campus, Rowland Hill Street, London, United Kingdom

Methods in Enzymology, Volume 436
ISSN 0076-6879, DOI: 10.1016/S0076-6879(08)36001-7

1. Introduction

The gas nitric oxide (NO, nitrogen monoxide) is exceedingly well known for its major roles in many physiological systems, for example, as a messenger and as a defense molecule in the immune system (Ignarro, 2000). In addition, a breakdown in NO homeostasis is associated with a number of disorders, such as high blood pressure, diabetes, and neurodegenerative diseases.

NO is formed by the oxidation of the amino acid *L*-arginine to give NO and citruline, in a reaction that is catalyzed by the nitric oxide synthases. This involves the intermediate formation of *N*-hydroxy-*L*-arginine, and is a five-electron process (Stuehr, 1999). However, there is evidence to suggest that under some conditions the product is nitrous oxide, N_2O, rather than NO (Ishimura *et al.*, 2005). The formation of nitrous oxide suggests that nitroxyl HNO is the final product of the NOS-catalyzed reaction, as nitroxyl dimerizes to form N_2O. Formation of HNO from *L*-arginine involves a four-electron oxidation (Schmidt *et al.*, 1996). HNO may then be oxidized to NO by superoxide dismutase or other agents.

There are three NO synthases, namely endothelial (eNOS), neuronal (nNOS), and inducible NOS (iNOS) (Alderton *et al.*, 2001). The eNOS and nNOS enzymes are constitutively expressed in many tissues, regulated mainly by Ca^{2+}, and produce low concentrations of NO. iNOS is associated with the immune system, independent of Ca^{2+}, and catalyzes the production of NO in high concentrations in response to various stimuli such as cytokines. It should be noted that NO may also be formed by a variety of other routes. Many reactions of HNO give NO as a product, for example, while the long history of use of amyl nitrite and nitroglycerin in the alleviation of chest pains is a consequence of the formation of NO from these compounds.

Two other potentially toxic gases, CO and H_2S, also exert biological effects, such as signaling (Pryor *et al.*, 2006). NO differs significantly from these two molecules in the quite remarkable diversity of its biological roles. This reflects the complex reactivity of NO in cellular environments and the availability of many reactive targets (Gow and Ischiropoulos, 2001). The reactivity of NO is emphasized by the fact that its function as a messenger involves direct chemical reactions with target molecules rather than just physical interactions. Thus, NO may be reduced to nitroxyl HNO, be oxidized to higher oxides, or bind to Fe centers with a number of distinctive biological consequences in each of these cases. But NO is still relatively stable for a radical species, allowing time for it to demonstrate selectivity in its reactions. It is a fine balance.

2. Chemistry of NO

The reactivity of NO reflects the general chemistry of the element nitrogen itself. This is characterized by facile redox reactions, while the reactions of nitrogen compounds are often subject to kinetic rather than thermodynamic control. This may lead to complex reaction stoichiometries, with several minor products. Indeed, the chemistry of simple, apparently well-understood compounds of nitrogen has so often turned out to be very surprising.

2.1. Properties of the NO molecule

NO is a radical, with one unpaired electron in an anti-bonding π molecular orbital, and a bond order of 2.5. NO is a relatively stable radical although it reacts with other radicals, such as superoxide ion and dioxygen. However, it does not dimerize in the gas phase or in solution, although a dimer has been observed at very low temperatures. This lack of reactivity probably results from the delocalization of the unpaired electron in the π^* molecular orbital between the nitrogen and oxygen atoms, rather than being centered on the nitrogen atom. However, under high pressure (such as in a gas cylinder) over a time period of a few weeks, NO undergoes disproportionation to give nitrous oxide and nitrogen dioxide (Eq. (1.1)). This reaction probably involves the formation of a dimer $(NO)_2$ (Eq. (1.2)), which then reacts with a third molecule of NO in the rate-determining step (Eq. (1.3)), to give nitrous oxide and nitrogen dioxide. Both these gases, particularly NO_2, are toxic. The preparation of NO free from nitrogen dioxide is considered in Chapter 3:

$$3NO \rightarrow N_2O + NO_2 \tag{1.1}$$

$$2NO \Leftrightarrow N_2O_2 \tag{1.2}$$

$$N_2O_2 + NO \rightarrow N_2O + NO_2 \tag{1.3}$$

NO is only slightly soluble in water (1.57 mmol dm^{-3} at 35°). This low solubility is probably a consequence of its small dipole moment. NO does not react with water and so is not an acid anhydride. NO has a high diffusion coefficient of 3300 μm^2 s^{-1} (Malinski *et al.*, 1993). The diffusion distance of NO from an NO-releasing cell is 100–200 μm. Lancaster (1997) has discussed the significance of these properties and has pointed out that the "actions of NO are surprisingly long-range."

Radical reactions of NO in biology may be protective or oxidizing. NO may react with other radicals to give species of greater reactivity and toxicity.

Notable examples are the reaction with the superoxide radical to give peroxynitrite, and with oxygen to give nitrogen dioxide. An example of a protective reaction of NO is that with potentially toxic peroxyl radicals, with values of the second-order rate constant $k_2 = 1\text{–}3 \times 10^{-9}\ \text{dm}^3\ \text{mol}^{-1}\ \text{s}^{-1}$, close to the diffusion rate, and results in the inhibition of chain reactions associated with peroxidation of lipids.

NO undergoes a one-electron reduction to give nitroxyl, HNO, a molecule of considerable current interest from chemical and pharmacological viewpoints, which has recently been very comprehensively reviewed by Miranda (2005). Bartberger *et al.* (2002) have determined the potential for the reduction of NO to $^3NO^-$ to be about -0.8 V ($\pm$ 0.2), suggesting NO is inert to reduction. This is consistent with the formation of stable nitrosyl complexes of cytochrome oxidase, which reduces oxygen to water. It should be noted that the compound NaNO formed by passing NO into sodium in liquid ammonia is not a salt of nitroxyl, but is probably a dimeric species, such as sodium *cis*-hyponitrite.

The physiological functions of HNO and their pharmacological consequences (Paolocci *et al.*, 2003), in particular, are attracting attention, and will be discussed later. It is noteworthy, though, that NO and HNO exert different effects from each other, suggesting that they function through mutually exclusive pathways (Ma *et al.*, 1999; Wink *et al.*, 2003). A further consequence is that HNO must survive long enough to exert these effects, and that the lifetime of HNO may therefore require reinvestigation (Miranda *et al.*, 2003).

The relationship between the biological activity of NO and its chemistry is particularly complex, as NO will react in the cellular environment to give a range of products with different reactivities and biological effects. Thus, while the reaction of NO with oxygen in aqueous solution at concentrations typical of those that exist in cells is quite slow, with a half-life for NO of about an hour, the lifetime of NO in cells is a matter of seconds only. It is evident that the reactions of NO in the cell are more complex than just those expected in terms of a conventional reaction with oxygen in water.

The lipophilicity of NO is a significant factor in its role as a messenger, as a high lipophilicity together with its small size will allow NO to enter and accumulate in membranes. Lipophilicity has traditionally been estimated from the water–octanol partition coefficient, *P*, although the water–hexane partition coefficient may be a better estimate of this factor. The properties of NO relevant to its passive transport through biological membranes have been studied by Abraham *et al.* (2000), who have calculated P_{octanol} for NO to be 5.5, an unexpectedly low value, but quite similar to P_{octanol} for other gases, for example, nitrogen (4.68). This value suggests that NO is not especially lipophilic compared to other gaseous molecules. This is consistent with its low hydrogen bond basicity and its essentially nonpolar character.

However, Shaw and Vosper (1977) have measured the partition coefficient for NO between hexane and water to be 9.9, which is considerably higher and suggests that NO will be concentrated in hydrophobic regions in the cell.

Liu *et al.* (1998) have measured the rate of disappearance of NO in an aerobic buffered solution in the presence of various hydrophobic materials such as biological or phospholipid membranes and detergent micelles. They calculated that the rate of reaction between NO and oxygen is accelerated by a factor of about 300 in these hydrophobic environments, as a consequence of the high concentration of NO (and O_2) present. This reaction occurs about 300 times faster than in the aqueous medium, so that about 90% of this reaction occurs in this hydrophobic environment, despite the small percentage of the total cell volume utilized by hydrophobic regions. This will lead to the formation of high concentrations of reactive nitrogen species such as NO_2 and N_2O_3 in these hydrophobic environments, and eventually to high concentrations of the products of their reactions with amines and thiols, such as various N- and S-nitroso compounds. The high diffusion rate of NO in water and biological material reported by Lancaster (1997) will also enhance the reactivity of NO, as this will increase the number of collisions with potential reactants.

More recently, the partitioning of NO and oxygen into liposomes and low-density lipoprotein has been measured directly by Moller *et al.* (2005). If the reaction is diffusion controlled then the extent of reaction can be estimated from the value of the product of the partition coefficient and the diffusion rate, $Kp \times D$. Moller *et al.* showed that there was a 4.4-fold increase in the concentration of NO in liposomes and a 3.4 increase in low-density lipoprotein, with slightly smaller effects for oxygen. These data are similar to the partition coefficients calculated by Abraham *et al.* (2000). However, these conclusions all support the view that these lipophilic phases would be associated with enhanced production of both nitrosating and nitrating/oxidizing agents. Moller *et al.* (2005) also showed that the diffusion rates of O_2 and NO in buffer are similar (4.5×10^{-5} cm^2 s^{-1}) and that they are also similar in organic solvents. In contrast, in liposomes, the diffusion rates for the two gases are different, values for NO and O_2 being 3.1×10^{-6} and 6.4×10^{-6} cm^2 s^{-1}, respectively. This interesting difference must be a consequence of the structure of the liposome group.

2.2. Reaction of NO with oxygen

This is an exceptionally well-studied reaction, although there may still be some uncertainty about the detailed mechanism. The first product of the reaction of NO with oxygen is NO_2, which in aqueous solution reacts with NO to give N_2O_3, the anhydride of nitrous acid (Eq. (1.4)). The products of this reaction will therefore be exclusively nitrite. When the reaction occurs in air, the nitrogen dioxide dimerizes to give N_2O_4, which dissolves in water to give nitrite and nitrate ions:

$$2HNO_2 \Leftrightarrow N_2O_3 + H_2O \tag{1.4}$$

As noted earlier, the reaction of NO with oxygen will be accelerated in hydrophobic regions, with various consequences for the cell resulting from nitrosative stress. Toxic species formed from NO, such as dinitrogen trioxide and nitrogen dioxide, may also be concentrated in these hydrophobic areas of cells; it would be important to assess the effect that this will have on their reactivity and stability. Dinitrogen trioxide is an effective nitrosating agent, while elevated concentrations of NO_2 will also increase the possibility of toxic peroxonitrite being formed.

3. Compounds Related to NO

3.1. Nitroxyl HNO

The biological and medical properties of nitroxyl, the product of the one-electron reduction of NO, are attracting considerable attention at present, largely because HNO appears to be a promising treatment for cardiovascular diseases (Favaloro and Kemp-Harper, 2007; Feelisch, 2003; Miranda, 2005; Paolocci *et al.*, 2003). The pharmacological properties of NO and HNO (strictly HNO donors) in the cardiovascular system differ from each other, and are almost mutually exclusive (Miranda *et al.*, 2003). The simplest explanation of this phenomenon is that HNO and NO bind to different sites to trigger these responses. NO binds to Fe(II) in guanylyl cyclase and is well known to bind to other Fe(II) centers. In contrast, it is believed that HNO binds to Fe(III) or possibly Cu(II). HNO is also directly reactive with thiols (Bartberger *et al.*, 2001), whereas NO only reacts with thiols after oxidation to an NO^+ carrying species, that then acts as a nitrosating agent. The demonstration of discrete pharmacological behavior for NO and HNO also suggests that interconversion of these species is not occurring over these time scales.

An important development in understanding the chemistry of nitroxyl was the correction by Shafirovich and Lymar (2002) of a previous value for the pKa of nitroxyl of 4.70. Their value of 11.4 shows that nitroxyl will be fully present as HNO at physiological pH values, with significant consequences for its biological activity.

Nevertheless, nitroxyl, is an unstable molecule, decomposing to give nitrous oxide. Accordingly, various nitroxyl-releasing compounds have been used in studies of its reactions. By far the most widely used source of HNO is Angeli's salt ($Na_2N_2O_3$, sodium trioxodinitrate). This decomposes at pH values of 4 and upward via the monoprotonated trioxodinitrate ion, $HN_2O_3^-$, to give HNO and nitrite as products (Eq. (1.5)) (Hughes and Wimbledon, 1976; Miranda *et al.*, 2005):

$$HN_2O_3^- \Leftrightarrow HNO + NO_2^- \tag{1.5}$$

$$2HNO \rightarrow [HONNOH] \rightarrow N_2O + H_2O \tag{1.6}$$

The formation of nitroxyl in this reaction was subsequently confirmed by the observation that photolysis of solutions of Angeli's salt under aerobic conditions at high pH led to the formation of peroxynitrite (Donald *et al.*, 1986). Later studies confirmed the formation of peroxynitrite almost quantitatively on photolysis at high pH and also at neutral pH on UV flash photolysis (Shafirovich and Lymar, 2002). Hughes and Wimbledon (1977) studied the decomposition of Angeli's salt in the presence of added sodium nitrite and found that the decomposition was reversible. They suggested therefore that the rate-determining step must involve the cleavage of the N–N single bond in a tautomeric form of Angeli's salt. Such a reaction gives singlet state nitroxyl, which will react with the added nitrite.

The reactions of HNO are difficult to study due to its extremely rapid dimerization to give NO, with $k_2 = 8 \times 10^6$ dm^3 mol^{-1} s^{-1} (Shafirovich and Lymar, 2002). This reaction is usually postulated to occur through the formation of hyponitrous acid (HON=NOH) as a transient intermediate (Eq. (1.6)). However, the half-life of *trans*-hyponitrous acid at 35° and pH 7 is about 10 min (Hughes and Stedman, 1963; Hughes, 1968), so the intermediacy of a species such as the as yet unreported *cis*-hyponitrous acid is a more likely option.

Piloty's acid, (N-hydroxybenzenesulfonamide, $C_6H_5SO_2NHOH$) is another well-known source of nitroxyl. However, its decomposition is very slow at biological pH values. The kinetically active species is the deprotonated form, but the pKa of Piloty's acid is 9.29, so the rate constant for the reaction is a maximum at pH values of about 11. At pH 7.4 and 25°, the half-life for the decomposition of Piloty's acid is about 36 h. Thus, Piloty's acid is only useful as a nitroxyl releaser at higher pH values. Nevertheless, Piloty's acid is able to bring about vasodilation in aortic rings, indicative of release of nitroxyl. The synthesis of compounds related to Piloty's acid but with lower pKa values has been a target in recent times, although in some cases these have been only of very limited solubility in water.

Other classes of compounds synthesized as nitroxyl releasers have included N-substituted hydroxylamines, NH(X)OH, where HX is a good leaving group (King and Nagasawa, 1999) (e.g., methylsulfonylhydroxylamine, MSHA), and acyl nitroso species (Zeng *et al.*, 2004). It is important that new HNO releasers are synthesized, which reproducibly release HNO at various rates. These compounds will be important as experimental tools and as potential therapeutic agents.

HNO can be formed by several routes, which may occur intracellularly. These may include H atom abstraction by NO (Akhtar *et al.*, 1985), the reaction between S-nitrosothiols and thiols (Wong *et al.*, 1998), and reduction of NO by xanthine oxidase (Saleem and Ohshima, 2004).

Nitroxyl is formed during the autoxidation of hydroxylamine, NH_2OH, at high pH values. Under these conditions, nitroxyl will be present as the anion, NO^-, and will react with dioxygen to give the peroxynitrite ion, which is stable at high pH. The species NO^- may also be trapped by metal centers to form colored nitrosyl complexes, such as the conversion of the anion $[Ni(CN)_4]^{2-}$ to the purple $[Ni(CN)_3(NO)]^{2-}$ species in the autoxidation of hydroxylamine (Hughes and Nicklin, 1971). Cytochrome *d* is readily nitrosylated by nitroxyl released from trioxodinitrate, and can be detected by UV-vis spectrometry (Bonner *et al.*, 1991).

Concentrations of HNO have been determined quantitatively by trapping with nitrosobenzene to produce cupferron, which can be complexed to Fe(III), with a characteristic spectrum. This will allow nitroxyl to be detected down to 5×10^{-6} mol dm^{-3} (Shoeman and Nagasawa, 1998). HNO has also been trapped by reaction with *N*-acetyl-*L*-cysteine to give *N*-acetyl-*L*-cysteinesulfinamide (Shoeman *et al.*, 2000).

As noted briefly earlier, it appears that HNO rather than NO may be the product of the oxidation of *L*-arginine by the enzyme NO synthase in a four-electron reaction. Superoxide dismutase converts HNO to NO in accord with the observation that this enzyme enhances the formation of NO from *L*-arginine by NOS. It has previously been suggested that HNO formation from arginine occurs through the use of damaged NOS. Indeed, the oxidation of arginine catalyzed by tetrahydrobiopterin-free neuronal NOS is known to give HNO. HNO rapidly dimerizes to give N_2O, with a second-order rate constant of 8×10^6 dm^3 mol^{-1} s^{-1} (Shafirovich and Lymar, 2002). Ishimura *et al.* (2005) have shown by GCMS that significant amounts of NO are formed in the catalysis of oxidation of arginine by neuronal NO synthase. These results certainly strengthen the view that the nNOS-catalyzed oxidation of arginine gives HNO as a product. Ishimura *et al.* (2005) also have pointed out that if HNO is the final product, then the stoichiometry of the reaction will involve two moles each of oxygen and of NADPH, which is the standard stoichiometry observed for monooxygenases. This is in contrast to the case for NOS catalysis of the oxidation of *L*-arginine in a five-electron process to give NO, which would involve the unique stoichiometry of reaction of 2 moles of oxygen and 1.5 moles of NADH. This seems to be a significant argument in support of the view that nitroxyl is the main product of NOS (Schmidt *et al.*, 1996) and that HNO is then reduced to NO by various electron acceptors such as superoxide dismutase (Murphy and Sies, 1991).

3.2. Reactions of HNO

HNO reacts with thiols via electrophilic attack, leading to the formation of compounds with S–S bonds. This could be a significant toxic effect in light of the key roles sometimes played by thiol groups in the activity of enzymes (Arnelle and Stamler, 1995; Doyle *et al.*, 1988; Wong *et al.*, 1998). HNO can react with oxygen to give peroxynitrite, with many consequences, such as hydroxylation and nitration of ring systems, effects on proteins, and DNA-strand breakage.

An interesting reaction is the rapid trapping by Fe(II) deoxymyoglobin of HNO, released by decomposition of Angeli's salt or methylsulfonylhydroxylamine (MSHA) (Sulc *et al.*, 2004). This gives a stable MbHNO species in 60 to 80% yield. The formation of this HNO compound is established from the ^{1}H NMR signal for HNO at 14.8 ppm. At pH 10, the observed rate of formation of MbHNO is close to the rate of decomposition of MSHA. This facile binding of HNO to reduced myoglobin suggests that ferrous heme proteins could play an important role in the mediation of the biological activity of HNO.

4. Nitrous Acid, Nitrosation, and S-Nitrosothiols

The process of nitrosation formally involves the transfer of the NO^+ group from a nitrous-acid-derived compound to nucleophilic centers, particularly nitrogen (such as amines) or sulphur (thiols) to give N-nitrosamines and S-nitrosothiols, respectively (Williams, 1988, 1999).

Mechanisms of nitrosation in aqueous solution have been described by Williams (1988). NO itself cannot act as a nitrosating agent but may do so in the presence of oxygen as a consequence of the formation of NO_2 and N_2O_3. However, a direct reaction of NO with thiols will occur under anaerobic conditions to give the disulfide and nitrous oxide, N_2O (Eq. (1.7)) (DeMaster *et al.*, 1995).

$$2RSH + 2NO \rightarrow RSSR + N_2O + H_2O \tag{1.7}$$

Nitrosation can cause toxic effects particularly through S-nitrosation of key thiol groups in enzymes or the deamination of amine groups in DNA. Nitrosating agents well known to occur in aqueous solution are protonated nitrous acid $H_2NO_2^+$ (at lower pH values), nitrosyl species ONX (where X is a halide or an anion of an organic acid), and the oxides N_2O_3 and N_2O_4. The rate-determining step in nitrosation is usually the reaction of the nitrosating agent with the target, although, in the case of N_2O_3, formation of the nitrosating agent may be rate determining, with a characteristic second-order rate constant. Carboxylic acid anions will catalyze the formation of N_2O_3.

The pKa value for nitrous acid of 3.45 indicates that concentrations of nitrous acid will be very low at biological pH values, and that nitrosation could be expected to be very slow. However, there are other pathways for nitrosation. Nitrogen oxides will bring about nitrosation at neutral and higher pH values. N_2O_3, the anhydride of nitrous acid, formed by the reaction of NO with NO_2, is likely to be important under these conditions, while heme nitrosyl complexes, where the nitrosyl group is NO^+ (indicated by high values of the N–O stretching frequency in the infrared spectrum), could in principle bring about nitrosation.

Oxygenated solutions of NO are well known to nitrosate thiols and amines at physiological pH values. Goldstein and Czapski (1996) studied the nitrosation of thiols and morpholine under these conditions and found the rate law to be independent of the substrate, with $-d[NO]/dt = 4k_1[NO]^2[O_2]$, identical to the familiar rate law for the oxidation of NO to NO_2. This shows that the rate-determining step is the formation of the nitrosating agent, followed by fast reaction with thiol or amine. While the rate-limiting step in these reactions is the formation of NO_2, the nitrosating agent could still be N_2O_3, formed by fast reaction of NO_2 with NO. Goldstein and Czapski (1996) also pointed out that under physiological conditions, where $[NO] < 1\ \mu mol\ dm^{-3}$ and $[O_2] < 200\ \mu mol\ dm^{-3}$, the half-life for nitrosation of thiols is over 7 min, and therefore unlikely to provide a route for the rapid biosynthesis of S-nitrosothiols. Other investigators subsequently have also argued that the reaction between NO and oxygen is too slow to lead to significant formation of S-nitrosoalbumin. It is suggested that this observation casts doubt on the view (Stamler *et al.*, 1992) that NO is circulated in plasma mainly as an S-nitroso adduct of serum albumin, although it is clear that some S-nitrosothiols do have similar biological activity to NO, such as vasodilation and inhibition of platelet aggregation. More recently it has been shown (Goldman *et al.*, 1998; Marley *et al.*, 2000) that S-nitrosoalbumin is formed in plasma in the concentration range 30 to 120 nmol dm^{-3}, much lower than previously thought. Marley *et al.* (2001) have measured the concentrations of S-nitrosoalbumin formed on exposure of plasma to various concentrations of NO. At physiological levels of NO, S-nitrosoalbumin was formed at concentrations in the 400 to 1000 nmol dm^{-3} range. Levels were decreased in whole blood to about 80 nmol dm^{-3}, presumably by binding to hemoglobin, suggesting that physiologically relevant concentrations of NO can be formed in blood from the reaction of NO with oxygen (Marley *et al.*, 2001).

5. Reactions of Peroxynitrite, $ONOO^-$

Peroxynitrite may be formed by the very rapid reaction between NO and superoxide ion (Goldstein and Czapski, 1995; Huie and Padmaja, 1993) or by the reaction of nitroxyl with oxygen. Peroxynitrous acid has a pKa of

6.5 (Logager and Sehested, 1993), so HOONO will be present at physiological pH values. Peroxynitrite formally contains an NO^+ group which, through its inductive effect, causes a weakening of the O–O bond compared to that in hydrogen peroxide. Thus, bond strengths are 90 and 170 kJ mol^{-1}, respectively. This facilitates the homolysis of peroxynitrite to give the hydroxyl and nitrogen dioxide radicals.

Peroxynitrous acid is a strong oxidizing agent and readily oxidizes thiols, ascorbate, and lipids. It can bring about the nitration and hydroxylation of aromatic centers and cause oxidative damage to biological tissues, leading to various pathological conditions. The mechanism of decomposition of peroxynitrous acid is now accepted to involve homolysis of the peroxo bond to give NO_2 and $\cdot OH$, which recombine to give nitrate. At physiological pH values, peroxynitrous acid has a lifetime of about 1 s. At higher pH values, decomposition of peroxynitrite leads to the formation of oxygen and nitrite in a 1:2 ratio. At pH 9 about 40% oxygen is formed.

Methods of preparation of peroxynitrous acid are discussed in Chapter 4 of this volume. The concentration of peroxynitrite may be determined from the molar absorptivity at 302 nm of 1670 $dm^3\ mol^{-1}$ (Hughes and Nicklin, 1968).

The peroxynitrite ion reacts very rapidly with CO_2 to give the adduct $ONOOCO_2^-$ (Eq. (1.8)). Carbon dioxide concentrations in physiological environments are greater than 1 mmol dm^{-3}, so the formation of this adduct is an important factor. The nitrosoperoxycarbonate anion is unstable and reacts to give nitrogen dioxide and the carbonate radical anion $CO_3^{\cdot-}$ (Eq. (1.9)). The nitrosoperoxycarbonate ion also isomerizes to the nitrocarbonate ion, $O_2NOCO_2^-$, which decomposes to nitrate and bicarbonate ion (Eq. (1.10)). The decomposition of this bicarbonate to give CO_2 leads to the catalytic decomposition of peroxynitrite to give nitrogen dioxide and the carbonate radical anion:

$$ONOO^- + CO_2 \rightarrow ONOOCO_2^- \quad (1.8)$$

$$ONOOCO_2^- \rightarrow NO_2 + CO_3^{\cdot-} \quad (1.9)$$

$$O_2NOCO_2^- + H_2O \rightarrow NO_3^- + HCO_3^- + H^+ \quad (1.10)$$

Gow *et al.* (1996) have shown that the peroxynitrite-mediated nitration of protein tyrosine groups is enhanced by carbon dioxide.

6. Nitrogen Dioxide

This familiar brown gas, formed on exposure of NO to air, exists in equilibrium with its dimer N_2O_4 in the gas phase, although in water N_2O_4 reacts rapidly to give nitrite and nitrate. Biological routes for the formation of NO_2 include the oxidation of NO, the decomposition of peroxynitrous acid, formed from the reaction between NO and superoxide, the oxidation of nitrite by hemoglobin, and the reaction of NO with organic peroxyl radicals to give an organic peroxynitrite intermediate that decomposes to give NO_2 and alkoxyl radicals (Eq. (1.11)) (Huie, 1994) It is also, of course, a well-known component of polluted air. Exposure of human plasma to NO_2 results in rapid loss of thiol, urate, and ascorbate, which are antioxidants (Ford *et al.*, 2002). The reaction with thiol is shown in Eq. (1.12):

$$ROO + NO \rightarrow ROONO \rightarrow RO^{\cdot} + NO_2 \quad (1.11)$$

$$NO_2 + RSH \rightarrow NO_2^- + RS^{\cdot} + H^+ \quad (1.12)$$

The thyl radical $RS^{\cdot}$ undergoes a number of reactions, for example with thiol to give $RSSR^{\cdot}$ and with oxygen to give $RSOO^{\cdot}$.

7. Nitrosative Stress and Reactive Nitrogen Species

Reference has already been made to the toxicity of peroxynitrite and nitrogen dioxide, and to the toxic consequences of nitrosation, particularly in the context of the overproduction of NO by inducible NOS in macrophages, which have been implicated in various pathogenic conditions, such as neurodegenerative disease, inflammation, and cardiovascular disorders (Dedon and Tannenbaum, 2004). The oxidation of NO to nitrite, via N_2O_3 formation, will lead to high levels of NO_2 through further oxidation by myeloperoxidase, while peroxynitrite will be formed from superoxide plus NO, whose decomposition will be catalyzed by carbon dioxide. Alongside this there will also be oxidative stress.

NO is protective against infections but damage will be caused to tissue, proteins, and DNA. DNA will be damaged by N_2O_3 and by peroxynitrite, as noted earlier. There is evidence that chronic inflammation is linked to increased levels of cancer in a number of organs, including the gastrointestinal tract, liver, and lung.

8. NO Complexes with Metal Centers

The binding of NO to iron centers in biological molecules, particularly to ferrous heme proteins, in the process of nitrosylation, is very important and is linked to the signaling functions of NO. The best known example of this process is the binding of NO to the Fe(II) center in the heme prosthetic group of guanylate cyclase in the formation of messenger cGMP. NO and CO are unique ligands in having a strong *trans*-effect in octahedral metal centers, which results in the labilizing of the ligand in the axial position *trans* to NO. This may contribute to the catalytic activity of guanylate cyclase and other reactions involving NO (Traylor *et al.*, 1993). Ferric hemes will also bind NO, although less strongly than do ferrous hemes. Thus, metmyoglobin [Fe(III)] reversibly binds NO while myoglobin [Fe(II)] binds NO irreversibly. NO inhibits many oxidases, oxygenases, and dioxygenases. It is well known to bind to hemoglobin.

NO is not especially toxic to iron–sulfur proteins that catalyze electron transfer, as these exclude oxygen effectively from the Fe–S center and will also exclude NO. Enzymes such as aconitase that have a channel for substrates to access the Fe–S active site will be inhibited by NO, which will at high concentrations disrupt the cluster, with the formation of the well-known dinitrosyl dithiolate complex of iron (Cooper, 1999).

The nitrosyl group in metal nitrosyl complexes metal may be present as either NO^+ or NO^-. Complexes with an N–O stretching frequency higher than about 1880 cm^{-1} in the infrared spectrum will function as nitrosating agents and transfer the NO^+ group to an appropriate acceptor site. Appropriate nitrosyl complexes are often able to carry out nitrosation reactions, sometimes at neutral pH values. Sodium nitroprusside, $Na_2[Fe(CN)_5NO]$, is probably the best known example of such a species. Sodium nitroprusside acts as a vasodilator and was once used in the treatment of high blood pressure. It exerts this effect by nitrosating a thiol group to give an S-nitroso product that then decomposes to give NO. This process also causes release of cyanide from the nitroprusside and so is no longer used. Compounds such as Roussin's black salt $NH_4[Fe_4S_3(NO)_7]$ are exceedingly toxic to microorganisms, but the mechanism of their action is still not fully clarified. This may involve nitrosation but alternatively the toxicity may arise from inhibition of electron transfer (Cammack *et al.*, 1999).

9. Mobilization of Metal Ions from Biological Sites by NO

Pearce *et al.* (2000) have shown that NO interacts with metallothioneins, with displacement of bound zinc or cadmium, and so may enhance the toxicity of cadmium (Misra *et al.*, 1996). Binet *et al.* (2002) have demonstrated that NO causes the release of intracellular zinc from prokaryotic metallothionein in *Escherichia coli*, which might contribute to the toxic effects of NO on the function of the cell. These observations also have implications for zinc homeostasis in bacteria. NO indirectly brings about the mobilization of iron in cells by binding to Fe in a cellular pool and so preventing it being incorporated into ferritin (Watts and Richardson, 2002).

10. Scavenging NO with Metal Complexes

Certain disease states arise from a dysfunction in NO metabolism. Overproduction of NO through an upregulation of iNOS causes septic shock due to a dramatic drop in blood pressure. In addition, overproduction of NO plays a role in inflammatory diseases, such as rheumatoid arthritis and inflammatory bowel disease. One solution to this problem is to use inhibitors with appropriate selectivity for iNOS. An alternative strategy (Cameron *et al.*, 2003) has involved the use of metal complexes as scavengers for NO, that is, to remove NO by binding to metal complexes to give metal nitrosyls. Ruthenium(II) complexes are especially interesting in this context as they have high affinity for NO to form inert nitrosyl complexes. Futhermore, the ligands can be varied to provide different lipophilicities and charges. Ruthenium(III) polyaminocarboxylate complexes in which one coordination position is filled by a water molecule have been successfully evaluated for NO scavenging ability using murine macrophage cells.

References

Abraham, M. H., Gola, J. M. R., Cometto-Muniz, J. E. E., and Cain, W. S. (2000). The solvation properties of nitric oxide. *J. Chem. Soc., Perkin Trans.* **2,** 2067–2070.

Akhtar, M. J., Bonner, F. T., and Hughes, M. N. (1985). Reaction of nitric oxide with hyponitrous acid: A hydrogen atom abstraction reaction. *Inorg. Chem.* **24,** 1934–1935.

Alderton, W. K., Cooper, C. E., and Knowles, R. G. (2001). Nitric oxide synthases: Structure, function and inhibition. *Biochem. J.* **357,** 593–615.

Arnelle, D. R., and Stamler, J. S. (1995). NO^+, NO and NO^- donation by S-nitrosothiols: Implications for regulation of physiological functions by S-nitrosylation and acceleration of disulfide formation. *Arch. Biochem. Biophys.* **318,** 279–285.

Bartberger, M. D., Fukuto, J. M., and Houk, K. N. (2001). On the acidity and reactivity of HNO in aqueous solution and biological systems. *Proc. Natl. Acad. Sci. USA* **98,** 2194–2198.

Bartberger, M. D., Liu, W., Ford, E., Miranda, K. M., Switzer, C., Fukuto, J. M., Farmer, P. J., Wink, D. A., and Houk, K. N. (2002). The reduction potential of nitric oxide (NO) and its importance to NO biochemistry. *Proc. Natl. Acad. Sci. USA* **99,** 10958–10963.

Binet, M. R. B., Cruz-Ramos, H., Laver, J., Hughes, M. N., and Poole, R. K. (2002). Nitric oxide elicited intracellular zinc release from prokaryotic metallothionein in *Escherichia coli*. *FEMS Microbiol. Lett.* **213,** 121–126.

Bonner, F. T., Hughes, M. N., Poole, R. K., and Scott, R. I. (1991). Kinetics of the reaction of trioxodinitrate and nitrite ions with cytochrome *d* from *Escherichia coli*. *Biochim. Biophys. Acta* **1056,** 133–140.

Cameron, B. R., Darkes, M. C., Yee, H., Olsen, M., Fricker, S. P., Skerlj, R. T., Bridger, G. J., Davies, N. A., Wilson, M. T., Rose, D. J., and Zubieta, J. (2003). Ruthenium(III) polyaminocarboxylate complexes: Efficient and effective nitric oxide scavengers. *Inorg. Chem.* **42,** 1868.

Cammack, R., Jouannou, C., Cui, X.-Y., Maraj, S. M., Torres-Martinez, C., and Hughes, M. N. (1999). Nitrite and nitrosyl complexes in food preservation. *Biochim. Biophys. Acta* **1411,** 475–488.

Cooper, C. E. (1999). Nitric oxide and iron proteins. *Biochim. Biophys. Acta* **1411,** 290–309.

Dedon, P. C., and Tannenbaum, S. R. (2004). Reactive nitrogen species in the chemical biology of inflammation. *Arch. Biochem. Biophys* **423,** 12–22.

DeMaster, E. G., Quast, B. J., Redfern, B., and Nagasawa, H. T. (1995). Reaction of nitric oxide with the free sulphydryl group of human serum albumin yields a silfenic acid and nitrous oxide. *Biochemistry* **34,** 11494–11499.

Donald, C. E., Hughes, M. N., Thompson, J. M., and Bonner, F. T. (1986). Photolysis of the N=N bond in trioxodinitrate. Reaction between triplet NO^- and O_2 to form peroxynitrite. *Inorg. Chem.* **25,** 2676–2677.

Doyle, M. P., Mahapatro, S. N., Broene, R. D., and Guy, J. K. (1988). Oxidation and reduction of hemoproteins by trioxodinitrate (II). The role of nitrosyl hydride and nitrite. *J. Am. Chem. Soc.* **110,** 593–599.

Favaloro, J. L., and Kemp-Harper, B. K. (2007). Nitroxyl (HNO) is a potent dilator of rat coronary vasculature. *Cardiovasc. Res.* **73,** 587–596.

Feelisch, M. (2003). Nitroxyl gets to the heart of the matter. *Proc. Natl. Acad. Sci. USA* **100,** 4978–4980.

Ford, E., Hughes, M. N., and Wardman, P. (2002). Kinetics of the reactions of nitrogen dioxide with glutathione, cysteine and uric acid at physiological pH. *Free Radical Biol. Med.* **32,** 1314–1323.

Goldman, R. K., Vlessis, A. A., and Trunkey, D. D. (1998). Nitrosothiol quantification in human plasma. *Anal. Biochem.* **259,** 98–103.

Goldstein, S., and Czapski, G. (1995). The reaction of NO with O_2^-: A pulse radiolysis study. *Free Radical Biol. Med.* **19,** 505–510.

Goldstein, S., and Czapski, G. (1996). Mechanism of the nitrosation of thiols and amines by oxygenated NO solutions: The nature of the nitrosating intermediates. *J. Am. Chem. Soc.* **118,** 3419–3425.

Gow, A. J., Duran, D., Thom, S. R., and Ischiropoulos, H. (1996). Carbon dioxide enhancement of peroxynitrite-mediated protein tyrosine nitration. *Arch. Biochem. Biophys.* **333,** 42–48.

Gow, A. J., and Ischiropoulos, H. (2001). Nitric oxide chemistry and cellular signalling. *J. Cell. Physiol.* **187,** 277–282.

Hughes, M. N., and Nicklin, H. G. (1968). The chemistry of peroxynitrites. Part 1. The kinetics of decomposition of peroxynitrous acid. *J. Chem. Soc.A* **1968,** 450.

Hughes, M. N., and Nicklin, H. G. (1971). The autoxidation of hydroxylamine in alkaline solutions. *J. Chem. Soc.* **(A)**, 164–169.

Hughes, M. N., and Stedman, G. (1963). Kinetics and mechanism of the decomposition of hyponitrous acid. Part 1. *J. Chem. Soc.* **1963,** 1239–1243.

Hughes, M. N. (1968). Hyponitrites. *Quart. Rev. Chem. Soc.* **22,** 1–13.

Hughes, M. N., and Wimbledon, P. E. (1976). The chemistry of trioxodinitrates. Part 1. The decomposition of sodium trioxodinitrate in aqueous solution. *J. Chem. Soc., Dalton. Trans.* **1976,** 703–707.

Hughes, M. N., and Wimbledon, P. E. (1977). The chemistry of trioxodinitrates. Part 2. The effect of added nitrite on the stability of sodium trioxodinitrate in aqueous solution. *J. Chem. Soc., Dalton. Trans.* **1977,** 1650–1653.

Huie, R. E. (1994). The reaction kinetics of NO_2. *Toxicology* **89,** 193–216.

Huie, R. E., and Padmaja, S. (1993). The reaction of NO with superoxide. *Free Radical Res. Commun.* **18,** 195–199.

Ignarro, L. J. (2000). "Nitric Oxide: Biology and Pathobiology" Academic Press, San Diego.

Ishimura, Y., Gao, Y. T., Panda, S. P., Roman, L. J., Masters, B. S. S., and Weintraub, S. T. (2005). Detection of nitrous oxide in the neuronal nitric oxide synthase reaction by gas chromatography-mass spectrometry. *Biochem. Biophys. Res. Commun.* **338,** 543–549.

King, S. B., and Nagawasa, H. T. (1999). Chemical approaches towards generating nitroxyl. *Meth. Enzymol.* **301,** 211–219.

Lancaster, J. R., Jr. (1997). A tutorial on the diffusibility and reactivity of free nitric oxide. *Nitric Oxide* **1,** 18–30.

Liu, X., Miller, M. J. S., Joshi, M. S., Thomas, D. D., and Lancaster, J. R. (1998). Accelerated reaction of nitric oxygen with O_2 within the hydrophobic interior of biological membranes. *Proc. Natl. Acad. Sci. USA* **95,** 2175–2179.

Logager, T., and Sehested, K. (1993). Formation and decay of peroxynitrous acid: A pulse radiolysis study. *J. Phys. Chem.* **97,** 6664–6669.

Ma, X. L., Gao, F., Liu, G.-L., Lopez, B. L., Christopher, T. A., Fukuto, J. M., Wink, D. A., and Feelisch, M. (1999). Opposite effects of nitric oxide and nitroxyl on postischemic myocardial injury. *Proc. Natl. Acad. Sci. USA* **96,** 14617–14622.

Malinski, T., Taha, Z., Grunfeld, S., Patton, S., Kapturczac, M., and Tomboulian, P. (1993). Diffusion of nitric oxide in the aorta wall monitored *in situ* by porphyrinic microsensors. *Biochem. Biophys. Res. Commun.* **193,** 1076–1082.

Marley, R., Patel, R. P., Orie, N., Ceaser, E., Darley-Usmer, V., and Moore, K. (2001). Formation of nanomolar concentrations of S-nitrosoalbumin in human plasma by nitric oxide. *Free Radical Biol. Med.* **31,** 688–696.

Marley, R., Feelisch, M., Holt, S., and Moore, K. (2000). A chemiluminescence-based assay for S-nitrosoalbumin and other plasma S-nitrosoalbumins. *Free Radical Res.* **32,** 1–9.

Miranda, K. M. (2005). The chemistry of nitroxyl (HNO) and implications in biology. *Coord. Chem. Rev.* **249,** 433–455.

Miranda, K. M., Dutton, A. S., Ridnour, L. A., Foreman, C. A., Ford, E., Paolocci, N., Katori, T., Tocchetti, C. G., Mancardi, D., Thomas, D. D., Espey, M. G., Houk, K. N., *et al.* (2005). Mechanism of aerobic decomposition of Angeli's Salt (sodium trioxodinitrate) at physiological pH. *J. Am. Chem. Soc.* **127,** 722–731.

Miranda, K. M., Paolocci, N., Katori, T., Thomas, D. D., Ford, E., Bartberger, M. D., Espey, M. G., Kass, D. A., Feelisch, M., Fukuto, J. M., and Wink, D. A. (2003). A biochemical rationale for the discrete behaviour of nitroxyl and nitric oxide in the cardiovascular system. *Proc. Natl. Acad. Sci. USA* **100,** 9196–9201.

Misra, R. R., Hochadel, J. F., Smith, G. T., Cook, J. C., Waalkes, M. P., and Wink, D. A. (1996). Evidence that nitric oxide enhances cadmium toxicity by displacing the metal from metallothionein. *Chem. Res. Toxicol.* **9,** 326–332.

Moller, M., Botti, H., Batthyany, C., Rubbo, C. H., Radi, R., and Denicola, A. (2005). Direct measurement of nitric oxide and oxygen partitioning into liposomes and low density lipoprotein. *J. Biol. Chem.* **280,** 8850–8854.

Murphy, M. E., and Sies, H. (1991). Reversible conversion of nitroxyl anion to nitric oxide by superoxide dismutase. *Proc. Natl. Acad. Sci. USA* **88,** 10860–10864.

Paolocci, N., Katori, T., Champion, H. C., St. John, M. E., Miranda, K. M., Fukuto, J. M., Wink, D. A., and Kass, D. A. (2003). *Proc. Natl. Acad. Sci. USA* **100,** 5537–5542.

Pearce, L. L., Gandley, R. E., Han, W., Wasserloos, K., Stitt, M., McLaughlin, A. J., Pitt, B. R., and Levitan, E. S. (2000). Role of metallothionein in nitric oxide signalling as revealed by a green fluorescent fusion protein. *Proc. Natl. Acad. Sci. USA* **97,** 477–482.

Pryor, W. A., Houk, K. N., Foote, C. S., Fukuto, J. M., Ignarro, L. J., Squadrito, G. L., and Davies, K. J. A. (2006). Free radical biology and medicine: It's a gas, man! *Am. J. Physiol. Regul. Integr. Comp. Physiol.* **291,** R491–R511.

Saleem, M., and Ohshima, H. (2004). Xanthine oxidase converts nitric oxide to nitroxyl that inactivates the enzyme. *Biochem. Biophys. Res. Commun.* **315,** 455–462.

Schmidt, H. H., Hofmann, H., Schindler, U., Shutenko, Z. S., Cunningham, D. D., and Feelisch, M. (1996). No NO from NO synthase. *Proc. Natl. Acad. Sci. USA* **93,** 14492–14497.

Shafirovich, V., and Lymar, S. V. (2002). Nitroxyl and its anion in aqueous solutions: Spin states, protic equilibria, and reactivities towards oxygen and nitric oxide. *Proc. Natl. Acad. Sci. USA* **99,** 7340–7345.

Shaw, A. W., and Vosper, A. J. (1977). Solubility of nitric oxide in aqueous and non-aqueous solvents. *J. Chem. Soc., Faraday Trans.* **8,** 1239–1244.

Shoeman, D. W., and Nagasawa, H. T. (1998). The reaction of nitroxyl (HNO) with nitrosobenzene gives Cupferron (N-nitrosophenylhydroxylamine). *Nitric Oxide* **2,** 66–72.

Shoeman, D. W., Shirota, F. N., DeMaster, E. G., and Nagasawa, H. T. (2000). Reaction of nitroxyl, an aldehyde dehydrogenase inhibitor with *N*-acetyl-L-cysteine. *Alcohol* **20,** 55–59.

Stamler, J. S., Jaraki, O., Osborne, J., Simon, D. L., Keaney, J., Vita, J., Singel, D., Valeri, C. R., and Loscalzo, J. (1992). Nitric oxide circulates in mammalian plasma primarily as an S-nitroso adduct of serum albumin. *Proc. Natl. Acad. Sci. USA* **89,** 7674–7677.

Stuehr, D. J. (1999). Mammalian nitric oxide synthases. *Biochim. Biophys. Acta* **1411,** 217–230.

Sulc, F., Imoos, C. E., Pervitsky, D., and Farmer, P. J. (2004). Efficient trapping of HNO by deoxymyoglobin. *J. Am. Chem. Soc.* **126,** 1096–1101.

Traylor, T. G., Duprat, A. F., and Sharma, V. S. (1993). Nitric oxide-triggered heme-mediated hydrolysis: A possible model for biological reactions of NO. *J. Am. Chem. Soc.* **115,** 810–811.

Watts, R. N., and Richardson, D. R. (2002). The mechanism of nitrogen monoxide (NO)-mediated iron mobilization from cells. *Eur. J. Biochem.* **269,** 3383–3392.

Williams, D. L. H. (1988). "Nitrosation," Cambridge University Press, Cambridge, U.K.

Williams, D. L. H. (1999). The chemistry of S-nitrosothiols. *Acc. Chem. Res.* **32,** 869–876.

Wink, D. A., Miranda, K. M., Katori, T., Mancardi, D., Thomas, D. D., Ridnour, L., Espey, M. G., Feelisch, M., Colton, C. A., Fukuto, J. M., Pagliaro, P., Kass, D. A., *et al.* (2003). Orthogonal properties of the redox siblings nitroxyl and nitric oxide in the cardiovascular system: A novel redox paradigm. *Am. J. Physiol. Heart Circ. Physiol.* **285,** H2264–H2276.

Wong, S.-Y., Hyun, J., Fukuto, J. M., Shiroto, F. N., DeMaster, E. G., Shoeman, D. W., and Nagasawa, H. T. (1998). Reaction between S-nitrosothiols and thiols: Generation of nitroxyl (HNO) and subsequent chemistry. *Biochemistry* **37,** 5362–5371.

Zeng, B.-B., Huang, J., Wright, M. W., and King, S. B. (2004). Nitroxyl (HNO) release from new functionalised N-hydroxyurea-derived acyl nitroso-9,10-dimethylanthracene cycloadducts. *Bioorg. Med. Chem. Lett.* **14,** 5565–5568.

CHAPTER TWO

Delivery of Nitric Oxide for Analysis of the Function of Cytochrome *c*′

Lindsay J. Cole, Wilhelmina M. Huston, *and* James W. B. Moir

Contents

Abstract

On delivery of nitric oxide (NO) to protein samples (e.g., cytochrome *c*′), for spectroscopic experiments it is important to avoid exposure to oxygen and to remove contaminants from the NO gas. We describe a number of techniques for steady-state UV/Vis spectrophotometry and pre-steady-state stopped-flow spectrophotometry analysis of cytochrome *c*′.

1. Introduction

The cytochromes *c*′ are a family of proteins that have been characterized from a range of bacteria, including photosynthetic bacteria, denitrifiers, sulfur oxidizers, and methanogens. They are 4-helix bundle, C-type cytochromes. The heme iron lacks a sixth coordination residue, and has a mixed intermediate and high-spin state, which gives these proteins interesting magnetic and spectroscopic properties. The cytochromes *c*′ are distinct from other related high-spin heme proteins, such as myoglobin, because the hydrophobic cage around the sixth coordinate position allows only small, uncharged ligands to access and bind the heme.

Department of Biology, University of York, Heslington, York, United Kingdom

Methods in Enzymology, Volume 436
ISSN 0076-6879, DOI: 10.1016/S0076-6879(08)36002-9

The interaction of this family of proteins with nitric oxide (NO) is of particular interest for two primary reasons. The spectral features of many of the cytochromes *c′* closely resemble those of the eukaryotic protein guanylate cyclase (Andrew *et al.*, 2002), an important protein that detects NO as a diffusible signaling molecule in eukaryotic organisms. Cytochrome *c′* has been used as a comparative model for the NO heme chemistry of guanylate cyclase (Ballou *et al.*, 2002; Berks *et al.*, 1995; Zhao *et al.*, 1999). More recently, the cytochrome *c′* in *Rhodobacter capsulatus* (Cross *et al.*, 2000, 2001) and *Neisseria meningitidis* (Anjum *et al.*, 2002) has been shown to protect the bacteria against NO stress, although the exact mechanism is unclear. Two mechanisms have been proposed: either the protein acts as a simple NO carrier protein binding free cellular NO and transporting it to the membrane, where it is turned over by NO reductase, or the protein directly participates in catalytic turnover with another redox partner in the cell.

The importance of the formation of nitrosyl complexes between the heme of cytochrome *c′* and NO has been extensively studied. The major issue in studies of interactions of NO with heme proteins is the delivery of controlled amounts of NO free of higher oxides and other contaminants. In this paper, we describe a number of techniques for handling protein samples and achieving anaerobiosis, as well as for reliable delivery of NO to proteins in solution.

2. Steady-State Titration and Addition of NO into Anaerobic Cytochrome *c′*

Many species of cytochrome *c′* can form heme nitrosyl complexes in both the ferric and the ferrous forms. The ferric form generally appears as a 6-coordinate species, while the ferrous form is either a 5 coordinate species or a mixture of 5 and 6 coordinate.

Cytochrome *c′* from *R. capsulatus* (RCCP) and from *N. meningitidis* (NMCP) has been the main target of research within our lab, primarily because of the evidence that the presence of cytochrome *c′* protects these organisms from NO damage. Both proteins have been successfully heterologously over-expressed in *E. coli* with coexpression of the pST2 plasmid (Turner *et al.*, 2003). This expression system enables us to express and purify significant quantities of both proteins and to carry out site-directed mutagenesis. The experiments described herein use example experiments with samples of these proteins.

Commercially supplied cylinders containing NO gas are frequently contaminated with nitrous oxide (N_2O) and nitrogen dioxide (NO_2), arising from a disproportionation reaction. NO_2 dimerizes to give N_2O_4, which is a mixed anyhdride and, when dissolved in water, gives nitrous and nitric acids equally. Consequently, a rapid acidification of any aqueous solution occurs when nitrogen dioxide is bubbled through it. NO gas has to be delivered without exposure to oxygen, and any contaminating higher oxides have to be removed from the gas stream by passage through solid sodium hydroxide or a highly alkaline solution. Further experimental details are provided in the chapter by Aga and Hughes (Chapter 3).

When there is insufficient care in preparation of dissolved NO solutions, significant acidification of the solutions can occur outside of the buffering range of the buffer used. If the acidification of the solution goes unnoticed during characterization of a heme nitrosyl complex, erroneous conclusions due to the changes in pH and the concentration of contaminating NO_2^- and NO_3^- species can result. An example is the oxidation of ferrous iron to the ferric form by contaminating and protonated nitrous acid at low pH (<5) (Doyle *et al.*, 1981). Thus, features of the acidified ferric-NO form present in the sample may erroneously be attributed to the neutral pH ferrous-NO form.

Figure 2.1 shows the example of a sample of reduced NMCP protein in a sealed quartz cuvette at pH 7.4 (10 m*M* phosphate, 137 m*M* NaCl, 2.7 m*M* KCl). This solution was made anaerobic by sparging the headspace of the cuvette with N_2 and then reduced by an excess of dithionite. NO from a lecture bottle (Sigma) was sparged into the headspace of the cuvette without sufficient removal of contaminating NO_2. The final spectrum had the features of a ferric-NO bound form. The spectral features were characteristic of the 6-coordinate ferric-NO bound form, and clearly the heme had undergone reoxidation to the ferric form before NO binding.

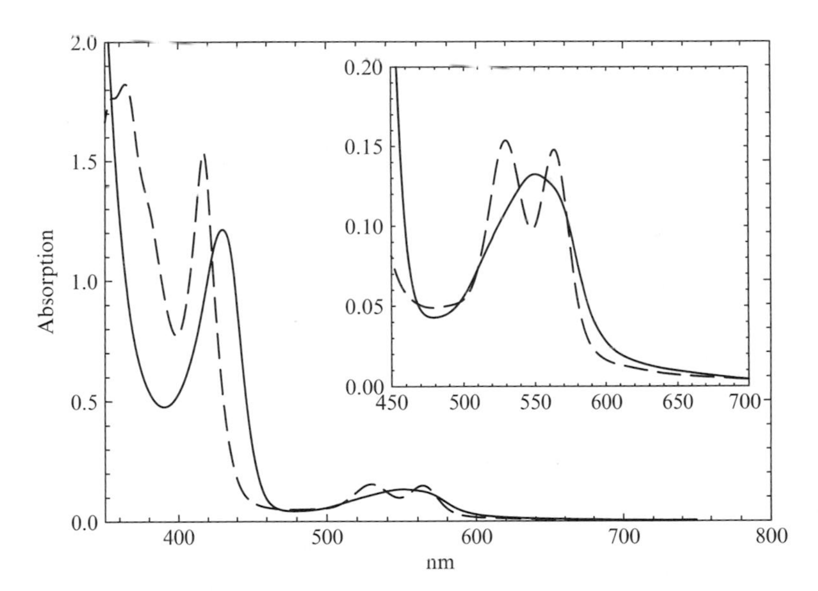

Figure 2.1 UV/Vis absorption spectrum of reduced NMCP before sparging (solid line) and after sparging (dashed line) of the cuvette headspace with an NO2-contaminated NO stream. The spectrum after sparging is characteristic of the ferric-NO bound form.

Solutions of saturated NO can be produced by the following method. First, the system is assembled as illustrated in Fig. 2.2. sixty ml of 4M H_2SO_4 in a pressure-equalizing dropping funnel, 200 ml of 2M $NaNO_2$ solution in a 500-ml two-necked RB flask, and the rest of the system as illustrated is flushed with oxygen-free N_2 for 20 to 30 min. The N_2 stream is shut off and the H_2SO_4 solution added to the $NaNO_2$ solution over the course of 30 min with stirring. The evolution of NO is steady and not violent at room temperature. The evolved NO gas stream is sparged through 300 ml of 4 *M* NaOH in a gas wash bottle to remove any contaminating NO_2, and then the NO gas stream is sparged for a period of 15 min either through a buffer solution in septum-sealed vials or into the headspace of a septum-sealed sample of protein. All contact surfaces must be borosilicate glass, quartz glass, Viton tubing, Teflon, or stainless steel to minimize corrosion of components by the NO stream. All components are contained in a fume cupboard to vent away NO and NO_2. Opening of sealed NO-sparged vessels should always be carried out in the fume cupboard to vent excess NO gas.

The concentration of a saturated solution of NO at 25° and 1 atm in pure water is 1.96 m*M*. This is the limit of the concentration for dissolved NO without increasing the pressure of the system significantly above 1 atm. Application of significantly higher pressures over 1 atm can greatly increase the solubility of NO, but this significantly increases the complexity of the experimental setup. Also, the use of standard glassware and quartz cuvettes

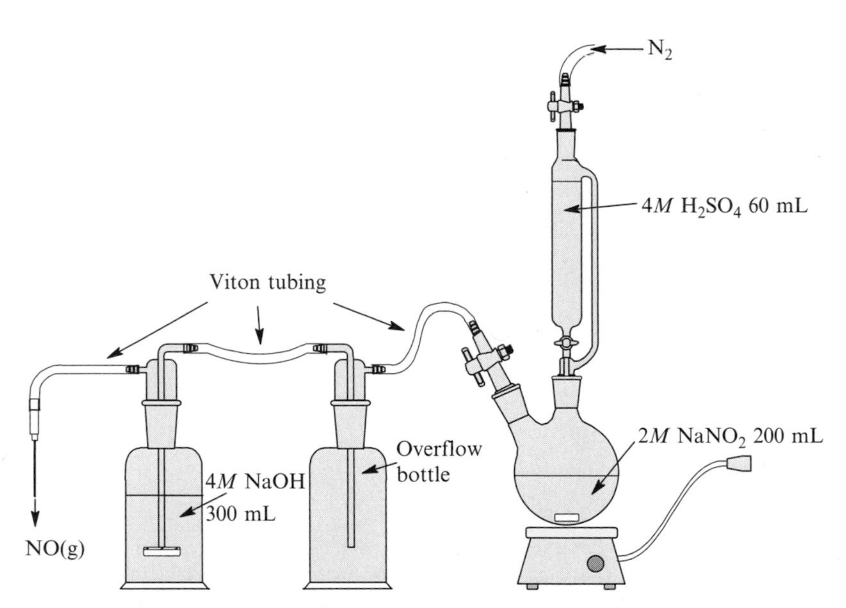

Figure 2.2 NO gas formation and NO gas scrubbing system. All components are borosilicate glass, Teflon, stainless steel for needles, or Viton tubing.

limits the range of pressure that can be used safely. On an experimental validity note, *in vivo* NO concentrations are never likely to reach the level of 100% saturation. For all practical considerations, concentrations of NO above 100 μM are unlikely to be biologically relevant.

The concentration of the NO standard solution can be confirmed by measuring a small aliquot of the NO solution by one of the many standard techniques, such as the common chemiluminescent or amperometric techniques. In our lab, we used a calibrated iso-NOP electrode (World Precision Instruments, Stevenage, UK) using a dilution (1/500) of the NO stock solution into 5 ml of N_2-sparged buffer; the setup and use of this electrode is described in a later chapter (5).

A solution of oxidized RCCP in a septum-sealed quartz cuvette (1.0 ml), in the buffer of interest, was made anaerobic by sparging the headspace of the cuvette with an oxygen-free nitrogen stream. The nitrogen had been sparged through 300 ml of distilled water to saturate the solution with water vapor, which prevents evaporation of the buffer solution and a subsequent decrease in volume and increase in solute concentration. The sparging was carried out for 15 to 20 min, by either stirring the protein solution in the cuvette with a magnetic flea or regularly shaking the cuvette. At this point, a UV/V is spectrum was taken (Fig. 2.3).

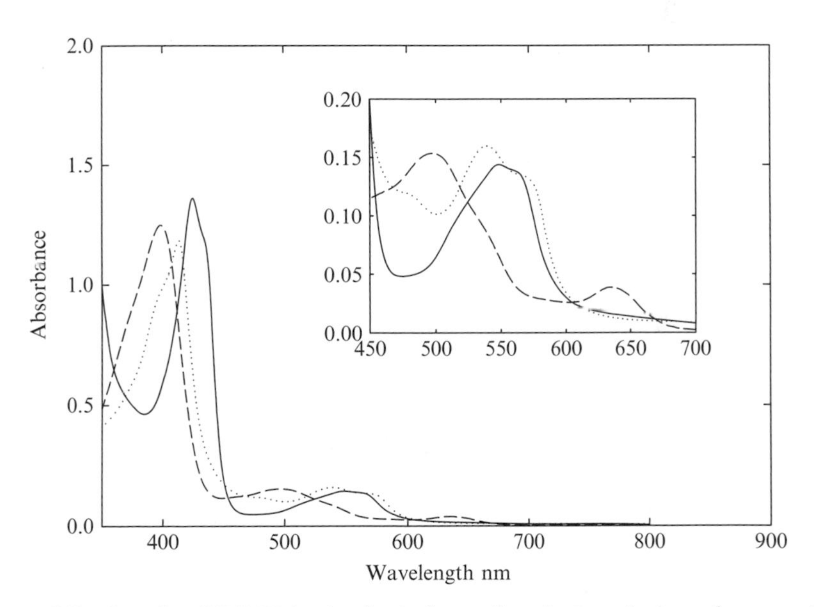

Figure 2.3 Sample of RCCP in the ferric form after the protein is made anaerobic (dashed line), in the ferrous form after reduction by dithionite (solid line), and in the ferrous-NO bound forms after sparging the headspace of the protein solution in the cuvette with saturated NO (dotted line). Protein is in pH 7.0 buffer (100 mM potassium phosphate), and the temperature is 25°.

To reduce the protein heme, a small excess of dithionite was added via a gas-tight syringe (Hamilton) through the septum and into the protein. The dithionite stock solution was made by sparging ≈50 ml of buffer in a Mininert valve (Valco, Supelco) sealed bottle that had been sparged for 10 min with oxygen-free nitrogen. The septum bottle was quickly opened and solid dithionite was added to a final concentration of between 1 and 2 m*M*; then the bottle was resealed and nitrogen sparging continued for another 5 min. The dithionite solution was extracted using a gas-tight syringe (Hamilton) through the Mininert valve septum seal. After the addition of dithionite to the RCCP solution, a UV/Vis spectrum of the protein solution in the cuvette was taken (Fig. 2.4).

Another consideration when sparging into the headspace of a protein sample is that the dissolution rate of NO gas can be slow. NO gas streams over a protein solution will take considerable time to equilibrate to full saturation. The rate of equilibration can be increased by increasing the size of the gas-liquid boundary and by applying significant agitation while avoiding the formation of bubbles in the protein solution. Alternatively, titration in buffer saturated with NO, using gas-tight syringes, into an anaerobic sample under a septum seal is another effective way to administer NO.

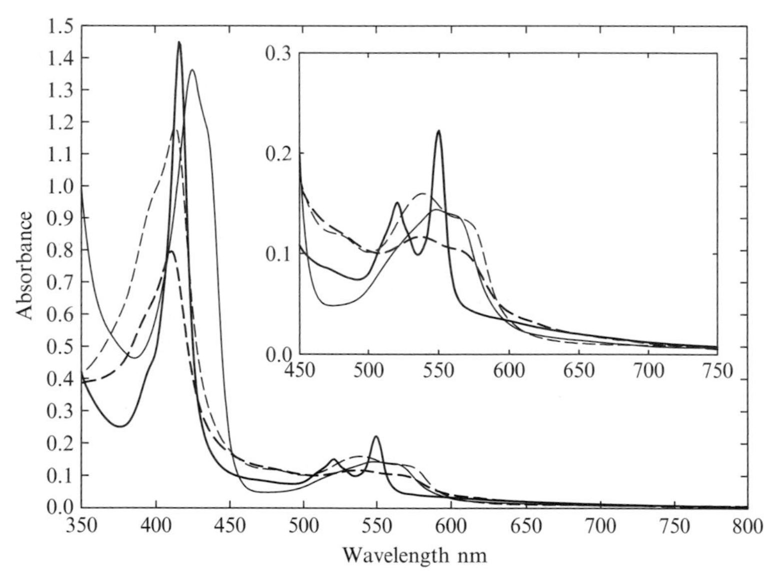

Figure 2.4 UV/Vis spectrum of ferrous RCCP wild-type protein (thin solid line), ferrous-NO bound RCCP wild-type protein (thin dashed line), ferrous F14A mutant RCCP protein (thick solid line), and ferrous-NO bound F14A mutant RCCP protein (thick dashed line) displaying the marked effect on the UV/Vis spectral properties of the F14A mutation. The inset is a magnification of the region between 450 and 700 nm. All are in a 100-m*M* pH 7.0 potassium phosphate buffer at 25°.

Protein samples can also be prepared in an anaerobic hood (Coy Labs, Grass Lake, MI), reduced with an excess of dithionite to reduce the protein, and the excess dithionite removed by desalting through a PD-10 desalting column (Amersham, Little Chalfont, UK) before sealing in septum-sealed cuvettes. When working with anaerobic ferric forms of the protein in an anaerobic hood, the reducing environment of the hood (5% H_2 typically) can reduce the heme of some cytochrome *c′* partially over a period of a day or more.

Although UV/Vis spectroscopy is able to rapidly and conveniently follow heme coordination state and redox state, the species involved can only be inferred from similarities to other similar proteins and experiments. More direct information about the heme coordination state and redox state of each species involved allows for confidence in assigning spectral features to each species. EPR is the most common method for directly confirming the redox state and environment around the heme iron. Only systems with noninteger spins are detectable by EPR; thus, ferric cytochrome *c′* (unliganded) is EPR detectable, but ferrous cytochrome *c′* (unliganded) is EPR silent. Similarly, ferrous-NO complexes are EPR detectable, but ferric NO complexes are EPR silent.

NO donors, such as NONOates and GSNO, are effective ways to introduce NO into protein solutions without many of the problems involved in dealing with NO gas. These nitrosyl compounds decompose at predictable rates to form NO when dissolved in aqueous solutions at physiological pH or via catalytic metal ions. NO donors have had a great impact on NO in the fields of physiology, cell biology, and microbiology because their slow-release characteristics make it possible to administer an amount of the NO donor where the decomposition time can, to some degree, be tuned to maintain a low-level, physiologically relevant concentration of NO. In *in vitro* studies of NO protein complexes, where the studied events have time scales of milliseconds to seconds, the slow decomposition time and the effect on the decomposition rate of temperature, pH, and contaminating catalysts make it difficult to be quantitative with NO concentration. However, for preliminary work and to obtain qualitative information, NO donors' ease of use makes them attractive.

Site-directed mutagenesis around the heme binding site has many long- and short-range influences on the heme environment. For instance, there has been much speculation on the extent of the effect the protein structure of cytochrome *c′* has on the heme environment, especially the small space available for axial ligands to bind to the heme, and the effect this has on the specificity for ligands and the spectral features of cytochrome *c′*. It is predicted that removing phenylalanine-14 in RCCP, with its close proximity to the axial side of the heme, by mutation to alanine opens up the axial side of the heme. Both protein samples (WT-RCCP and RCCP F14A) were prepared using the method of sparging a sealed cuvette containing

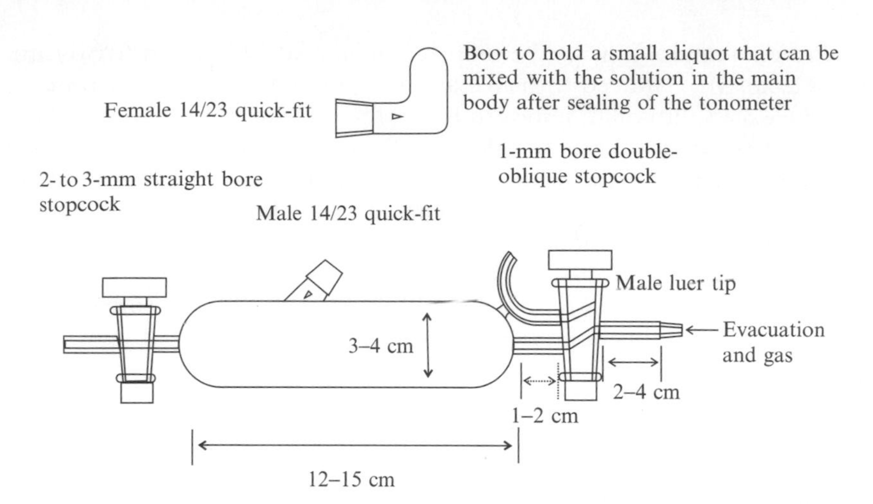

Figure 2.5 Diagram of a stopped-flow tonometer, a piece of glassware that allows solutions under controlled atmospheres to be easily loaded into the drive syringes of a stopped-flow instrument.

the protein sample with oxygen-free N_2 gas, reducing the protein by titration of an excess of dithionite, and then sparging the headspace of the protein sample with pure NO gas. Notably, the UV/Vis spectral features are markedly different between the F14A mutant and WT RCCP, as Fig. 2.4 shows.

3. Stopped-Flow Analysis of NO Ferrous Cytochrome *c′* Formation

Pre-steady-state kinetics can provide a great deal of mechanistic insight. For instance, stopped-flow analysis of the kinetics of NO binding to the ferrous cytochrome *c′* from *Alcaligenes xylosoxidans* (AXCP), which has been shown by crystallography to bind NO in 5-coordinate proximal form (Lawson *et al.*, 2003), showed that NO binding occurs via a 6-coordinate intermediate before the final 5-coordinate form (Andrew *et al.*, 2005). The proposed mechanism is that the 5-coordinate proximal histidine form of cytochrome *c′* binds NO transiently to form the 6-coordinate NO histidine nitrosyl complex. Subsequently, another NO molecule displaces the histidine ligand to form the di-nitrosyl complex and releases the axial NO to form the 5-coordinate proximal NO bound form.

It is highly instructive to compare the kinetics of NO by other ferrous cytochromes *c′*. Stopped-flow spectrophotometry enables us to observe absorbance changes as the heme changes coordination state or redox state.

The main experimental problems to overcome in this type of study are reducing the cytochrome c' to the ferrous form; removing or minimizing excess reducing agent that may react with NO when the two components are mixed, and thus significantly affect the NO concentration; and protecting the protein from reoxidation by oxygen in the time frame of the experiment (a few hours to half a day). Also, NO solutions have to be prepared with known concentrations and loaded on the instrument, without significant reaction of the NO with oxygen. The techniques used are widely applicable to a number of different systems or experimental setups.

Samples of purified ferrous NMCP were prepared with absorbance at 405 nm of between 1.0 and 1.5. A volume of about 1.5 to 2 ml was taken into an anaerobic hood and left to equilibrate with the atmosphere of the hood by stirring for 2 h.

After equilibration, the heme was reduced by addition of an excess of dithionite by the addition of 100 μl of a 50-mM dithionite solution. The protein was then passaged down a PD-10 desalting column equilibrated with the experimental buffer (e.g., 100 mM Tris-Cl pH 8.0) that had also been equilibrated in the anaerobic hood. This both removes the excess dithionite, leaving the reduced protein free of excess dithionite, and equilibrates the protein into the correct buffer.

The NMCP solution was then made up to a final volume of $\approx$4 ml with the experimental buffer and placed into a glass anaerobic tonometer equilibrated with the atmosphere of the anaerobic hood and sealed.

Stopped-flow tonometers are specialized pieces of glassware made to maintain a solution under a controlled atmosphere and then load it into a drive syringe of a stopped flow without exposure to the outside atmosphere; Fig. 2.5 shows the general design. The tonometer is connected to the stopped-flow system through a male ground-glass luer joint through a 1-mm capillary to a double-oblique two-way stopcock selecting between the main body of the tonometer and a waste outlet. The main body is usually long and cylindrical, which allows for two to be placed side by side on most stopped-flow instruments (possibly with adapters available from the manufacturer). The double-oblique stopcock allows for the solution already in the drive syringe of the instrument to be pushed out of a waste outlet and the solution in the main body to be drawn down into the drive syringe and easily exchanged with the solution in the drive syringe by repeated flushing. Another useful addition to the design is a side-arm quick-fit joint on the side of the main body, to which side arms, cuvettes, or other accessories can be added as needed for various applications. The curved side arm shown in Fig. 2.5 allows for small volumes held in the side arm to be added to the main body, after the tonometer has been sealed, by mixing into a solution from the main body.

The tonometer can be equilibrated with any atmosphere up to $\approx$2 psi pressure by sealing the filled tonometer inside a glove box equilibrated with

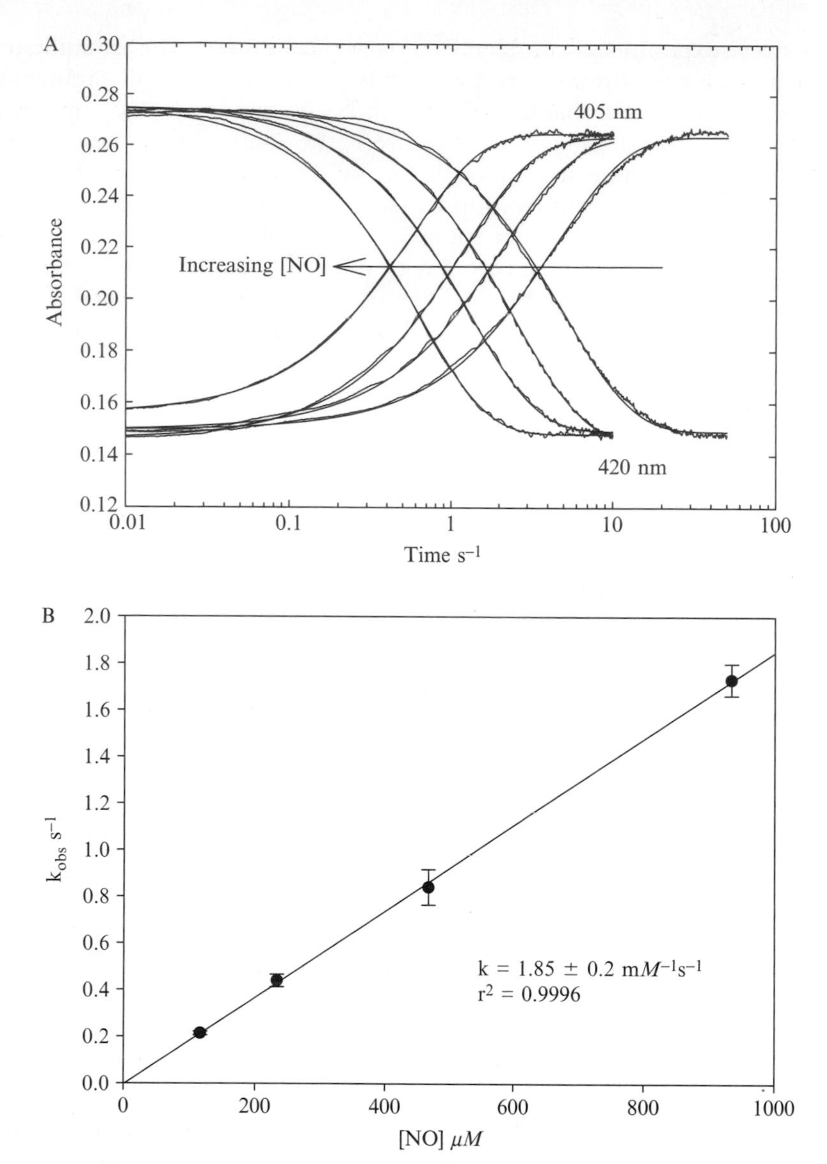

Figure 2.6 Ferrous NMCP at pH 8.0 (100 m*M* Tris-Cl) and 25° binding NO. (A) Traces of NMCP binding NO. Binding displays a second-order dependence on NO concentration. Overlaid are single exponential fits to the data. Note the logarithmic time scale, which helps highlight multiple phases in a reaction when they are present. (B) Linear fit of the observed time constants, with a second-order dependence on NO of 1.85 $mM^{-1}s^{-1}$.

the appropriate atmosphere (e.g., NO gas, O_2-free N_2). Alternatively, the tonometer can be connected to a two-way valve, selecting between the gas stream that the tonometer is to be equilibrated with and vacuum, through the double-oblique stopcock and luer fitting. The solution is equilibrated with the gas stream by repeated cycles of vacuum and gas (15–20 cycles), with periods of between 30 s and 1 min in between cycles to let the solution in the tonometer equilibrate with the new atmosphere. When doing this type of experiment, it is important to maintain only a volume of solution in the main body of less than 1/3 of the total volume to prevent accidental loss of sample.

A sample of ≈20 ml of experimental buffer was saturated with NO by the method described in Section 1, in 50-ml Mininert valve and septum-sealed bottles. Dilutions of the saturated solution (1.96 m*M*) were made into ground-glass syringes (5 or 10 ml) with N_2-sparged buffer, and the syringes were capped with either plastic luer caps or with luer valves. Similarly, 10 ml of anaerobic buffer was drawn into a ground-glass syringe and capped.

Because the polymer-based syringe seals and valves inside the flow path of a stopped-flow instrument often retain significant amounts of oxygen, a scrubbing solution is needed to remove this excess oxygen. A common scrubbing solution is the glucose (10 m*M*) glucose oxidase (0.1 mg/ml) and catalase (0.01 mg/ml) in anaerobic phosphate buffer (pH 7.0; 10 to 100 m*M*). For stopped-flow use, the solution of 10 to 20 ml is placed into an anaerobic tonometer and sealed. The tonometer is placed on the instrument and the scrubbing solution flushed through the instrument and left to sit for at least 2 h, and often overnight.

The system was flushed with a sample of an anaerobic buffer and a spectral baseline obtained. Subsequently, a sample of the reduced protein in a tonometer was loaded onto the system, and anaerobic solution was flushed through the outlet of the tonometer and then displaced by flushing through the system. Several shots of reduced protein and the anaerobic buffer without collecting data were performed to mix and prime the system, after which a single spectrum of the protein was taken. This confirmed the final concentration of the protein and demonstrated that the protein had been transferred into the stopped-flow system without any exposure to oxygen or subsequent reoxidation.

To load solutions of NO, the syringe that contains them was uncapped, the nonprotein reservoir syringe quickly swapped, and the NO solution drawn into the drive syringe and flushed through the system. Care should be taken to minimize the exposure to oxygen and to prevent air bubbles from being introduced into the drive syringe.

Both diode array and single-wavelength studies are useful tools for studying absorbing species, and each has advantages and disadvantages. Diode array allows the complete visible spectral changes to be collected in a single shot, but at the expense of a much sparser time data set and increased noise than single-wavelength measurement with a photomultiplier detector. Thus, spectra of intermediates are easier to identify using diode array data, but the accurate analysis of the kinetics to determine a mechanism often requires the better signal to noise and time data set of single-wavelength measurements.

For single-wavelength measurements, a selection of wavelengths at each NO concentration should be chosen to perform shots at and to compare across NO concentrations. From steady-state spectra, it is often obvious which absorbance wavelengths provide the largest changes from the unbound to the bound NO spectra, but several other wavelengths should be chosen, especially at isosbestic points between the initial and the final spectra. This makes identification of transient intermediates easier, because they will not be swamped by the large absorbance changes from the bound to the unbound form at these wavelengths

Analysis of selected wavelengths of the reaction of ferrous NMCP with various concentrations of NO shows a second-order dependence on concentration (Fig. 2.6). The protein binds NO in a primarily 5-coordinate ferrous NO spectrum. It is important to note that even though the data appears to suggest a simple single-step binding of NO with little 6-coordinate intermediate, the absence of a 6-coordinate nitrosyl intermediate does not mean that it is not present transiently before a rapid binding of a second NO and rearrangement to the 5-coordinate NO form. Thus, much care has to be taken in interpreting the meaning of pre-steady-state kinetic data.

REFERENCES

Andrew, C. R., George, S. J., Lawson, D. M., and Eady, R. R. (2002). Six- to five-coordinate heme-nitrosyl conversion in cytochrome c′ and its relevance to guanylate cyclase. *Biochemistry* **41,** 2353–2360.

Andrew, C. R., Kemper, L. J., Busche, T. L., Tiwari, A. M., Kecskes, M. C., Stafford, J. M., Croft, L. C., Lu, S., Moenne-Loccoz, P., Huston, W., Moir, J. W., and Eady, R. R. (2005). Accessibility of the distal heme face, rather than Fe-His bond strength, determines the heme-nitrosyl coordination number of cytochromes c′: Evidence from spectroscopic studies. *Biochemistry* **44,** 8664–8672.

Anjum, M. F., Stevanin, T. M., Read, R. C., and Moir, J. W. (2002). Nitric oxide metabolism in *Neisseria meningitidis*. *J. Bacteriol.* **184,** 2987–2993.

Ballou, D. P., Zhao, Y., Brandish, P. E., and Marletta, M. A. (2002). Revisiting the kinetics of nitric oxide (NO) binding to soluble guanylate cyclase: The simple NO-binding model is incorrect. *Proc. Natl. Acad. Sci. USA* **99,** 12097–12101.

Berks, B. C., Ferguson, S. J., Moir, J. W., and Richardson, D. J. (1995). Enzymes and associated electron transport systems that catalyse the respiratory reduction of nitrogen oxides and oxyanions. *Biochim. Biophys. Acta* **1232,** 97–173.

Cross, R., Aish, J., Paston, S. J., Poole, R. K., and Moir, J. W. (2000). Cytochrome c′ from *Rhodobacter capsulatus* confers increased resistance to nitric oxide. *J. Bacteriol.* **182,** 1442–1447.

Cross, R., Lloyd, D., Poole, R. K., and Moir, J. W. (2001). Enzymatic removal of nitric oxide catalyzed by cytochrome c′ in *Rhodobacter capsulatus*. *J. Bacteriol.* **183,** 3050–3054.

Doyle, M. P., Pickering, R. A., DeWeert, T. M., Hoekstra, J. W., and Pater, D. (1981). Kinetics and mechanism of the oxidation of human deoxyhemoglobin by nitrites. *J. Biol. Chem.* **256,** 12393–12398.

Lawson, D. M., Stevenson, C. E., Andrew, C. R., George, S. J., and Eady, R. R. (2003). A two-faced molecule offers NO explanation: The proximal binding of nitric oxide to haem. *Biochem. Soc. Trans.* **31,** 553–557.

Turner, S., Reid, E., Smith, H., and Cole, J. (2003). A novel cytochrome c peroxidase from *Neisseria gonorrhoeae*: A lipoprotein from a Gram-negative bacterium. *Biochem. J.* **373,** 865–873.

Zhao, Y., Brandish, P. E., Ballou, D. P., and Marletta, M. A. (1999). A molecular basis for nitric oxide sensing by soluble guanylate cyclase. *Proc. Natl. Acad. Sci. USA* **96,** 14753–14758.

CHAPTER THREE

The Preparation and Purification of NO Gas and the Use of NO Releasers: The Application of NO Donors and Other Agents of Nitrosative Stress in Biological Systems

Rubina G. Aga* *and* Martin N. Hughes†

Contents

Abstract

Cylinders and lecture bottles are often the source of nitric oxide (NO) in studies of the biological chemistry of this remarkable molecule. The NO from both sources will probably contain NO_2 (and N_2O) formed by disproportionation of NO. The NO_2 must be removed by passing the NO through a thoroughly deoxygenated sequence of traps containing sodium hydroxide solution and then water. The presence of NO_2 in aqueous solutions of NO may be determined readily using 2,2′ azino(3-ethylbenzothiazoline-6-sulfonic acid). NO_2 oxidizes

* Department of Chemistry, King's College London, London, United Kingdom
† Centre for Hepatology, Royal Free and University College Medical School, Centre for Hepatology, London, United Kingdom

Methods in Enzymology, Volume 436
ISSN 0076-6879, DOI: 10.1016/S0076-6879(08)36003-0

this compound to a long-lived radical anion, the concentration of which may be determined spectrophotometrically. The formation of NO by the decomposition of nitrous acid (via its disproportionation to nitrate and NO) and by the use of commercially available NO-releasing compounds with defined half-lives are also discussed. Other reactions that lead indirectly to the formation of NO are noted. In all cases, care must be taken to exclude oxygen to minimize as much as possible the formation of NO_2 (and, consequently, the nitrosating agent N_2O_3). The uses of these methods for generating NO and the reactivity of related compounds are illustrated with examples of studies of nitrosative stress.

1. Introduction

The controlled delivery of NO to biological and biochemical targets is not a trivial exercise. In particular, considerable care is necessary to ensure that oxygen is completely excluded from the reaction system. Failure to achieve this can result in various complications and potentially misleading results. NO is oxidized by oxygen to give the brown gas nitrogen dioxide (NO_2) (Eq. 3.1), a well-known toxic species. If this reaction occurs in aqueous solution, then there is a further reaction between NO and NO_2 to give dinitrogen trioxide (Eq. 3.2), the anhydride of nitrous acid, and an effective nitrosating agent. If NO is exposed to oxygen in the headspace of a reaction vessel, for example, then the NO_2 so formed dimerizes to give dinitrogen tetroxide, N_2O_4 (Eq. 3.3). This mixed anhydride reacts with water to give nitrous and nitric acids (Eq. 3.4). Nitric acid is a strong acid, so its formation in sufficient concentration leads to the acidification of the reaction solution, despite the presence of buffer. This has a variety of consequences and in extreme cases leads to the denaturation of proteins. Smaller effects on pH and/or the formation of oxidation products of NO may give rise to more subtle consequences, such as nitrosative stress. This term collectively describes the toxic effects resulting from the formation of nitrosating agents, notably N_2O_3, and the resulting transfer of the NO^+ group to sulfur and nitrogen centers. Oxidative stress is primarily a consequence of the action of superoxide and peroxide, but the formation of NO_2 and peroxynitrite will also contribute to oxidative stress.

$$2NO + O_2 \rightarrow 2NO_2 \quad (3.1)$$

$$NO_2 + NO \rightarrow N_2O_3 \quad (3.2)$$

$$2NO_2 \rightarrow N_2O_4 \quad (3.3)$$

$$N_2O_4 + H_2O \rightarrow HNO_2 + H^+ + NO_3^- \quad (3.4)$$

2. Supply of NO

NO can be (a) obtained from a cylinder, (b) generated through the decomposition of nitrous acid, (c) released at known and widely varying rates from a variety of commercially available series of related compounds, such as the NONOates, and (d) released (sometimes indirectly) from a variety of compounds, such as sodium nitroprusside, which may have an extended tradition of use for this purpose.

2.1. NO from cylinders and lecture bottles

The use of NO cylinders has complications due to the disproportionation of NO under high pressure in the cylinder to give nitrous oxide and nitrogen dioxide (Eq. 3.5). Lecture bottles of nitric oxide involve substantially lower pressures of NO, and it is therefore sometimes suggested that disproportionation of NO will not occur under these conditions. In our experience (and that of others), this is not the case. NO from a lecture bottle may contain substantial amounts of NO_2. The presence of the one-electron oxidizing agent NO_2 (Stanbury, 1989) in solutions containing NO may be determined by use of 2,2′ azino(3-ethylbenzothiazoline-6-sulfonic acid), $ABTS^{2-}$ (Ford, 2001). A known volume of the NO solution, obtained using an air-tight syringe, is added to a nitrogen-saturated solution of $ABTS^{2-}$ (concentration 1 mmol dm^{-3}), which is shaken and left to stand for an hour. The presence of NO_2 is shown by the formation of a green color that results from the oxidized species $ABTS^{\cdot -}$. The concentration of this species (and thus of NO_2) may be obtained from its absorption at 417 nm, with a molar absorptivity of 34700 dm^3 mol^{-1} cm^{-1} (Scott *et al.*, 1993). If another sample of the NO solution is added to an air-saturated solution of $ABTS^{2-}$, then all the NO will be oxidized to NO_2 and the concentration of $ABTS^{\cdot -}$ in this case will be equal to the sum of NO and NO_2 in the original solution. Thus, the concentration of NO in the solution may also be determined.

$$3NO \rightarrow N_2O + NO_2 \tag{3.5}$$

NO_2 may be removed from the NO gas supplied from cylinders and lecture bottles by passing the NO through a sequence of connected Dreschler bottles containing 1 mol dm^{-3} sodium hydroxide solution and finally through bottles containing distilled water or buffer to ensure that droplets of the alkaline solution are not carried forward. The Dreschler bottles are usually connected by rubber or synthetic tubing. This tubing is permeable to a limited extent and should be changed regularly. This procedure is described more fully for the case of the generation of NO by decomposition of nitrous acid.

2.2. Generation of NO in the laboratory

NO may be prepared by the self-decomposition through disproportionation of nitrous acid in aqueous solution at high concentration (Eq. 3.6). The experimental details of this procedure vary with laboratory, but in all cases the need to exclude oxygen as far as possible and then to remove any NO_2 that has been formed are paramount.

$$3HNO_2 \rightarrow 2NO + HNO_3 + H_2O \qquad (3.6)$$

In our laboratory, this procedure involved the careful, dropwise addition of a concentrated solution of sodium nitrite (120 g in 200 ml of water) from a pressure-equalizing dropping funnel into a 500-ml round-bottomed flask containing approximately 200 ml of about 6 mol dm^{-3} sulfuric acid. The dropping funnel and the reaction flask are connected by an adaptor with a side arm. These connections should be secured with plastic clips. Clear plastic tubing is attached to the side arm and leads the NO into a sequence of six Dreschler bottles that serve as traps. In each case, the entry tube into the bottle terminates in a fine glass sinter to generate small bubbles of NO gas. The sinter heads attached to the Dreschler bottles are connected by clear plastic tubing. Alternatively, all-glass connections via ball-and-socket joints could be used. All glass-to-plastic tubing connections were wrapped around with plastic film.

The first Dreschler bottle is empty and is intended to contain droplets carried through on the nitrogen stream and, if necessary, to collect any suck-back from the second and third Dreschler bottles. These two bottles contain a solution of sodium hydroxide (1 mol dm^{-3}) to trap any NO_2 and to neutralize acid carried over from the reaction flask. The next two bottles contain distilled water to trap droplets of sodium hydroxide solution, and the final Dreschler bottle contains a molecular sieve to dry the outgoing gas. The sinter heads of the Dreschler bottles may rise if there is increased pressure in the system and so air may enter the system. The sinter heads should therefore be fixed to the bottle, possibly by clips but ideally by springs attached to a neck brace.

Before generation of NO is commenced, oxygen-free nitrogen must be passed through the entire experimental system for at least two hours to deoxygenate the whole apparatus and the solutions. This is achieved by passing the nitrogen into the sodium nitrite solution in the funnel from where it passes through the rest of the apparatus via the pressure-equalizing component of the dropping funnel. Nevertheless, when the sodium nitrite solution is first run into the sulfuric acid, it is likely that brown fumes of NO_2 will be seen, as it is difficult to fully deoxygenate the sulfuric acid solution. This NO_2 will be removed in the traps. The NO is passed through the train of Dreschler bottles into thoroughly deoxygenated distilled water or buffer contained in a tube closed with a Suba seal cap with two needles, one used for connection to the reaction train and one to release gas to the atmosphere so as to avoid a pressure buildup of NO in the collection vessel. NO gas was usually collected over a period of 30 min, and the collecting tube, containing water or buffer solution, was kept in ice. At this temperature, the concentration of NO is 3.27 mmol dm^{-3}. These solutions of NO were allowed to warm up to the required experimental temperature before use. Samples of NO may be withdrawn from the tube using a syringe, but this operation should ideally be carried out carefully under nitrogen (using a plastic bag, through which nitrogen had been passed for about 20 min, and with the flow continuing during sampling). This is necessary because air may enter the tube on removing the needle from the septum. Table 3.1 shows solubilities of NO in water at different temperatures.

Sharpe and Cooper (1998) and Ishimura *et al.* (2005) have generated NO using a Kipps apparatus (once famous for the generation of H_2S). Sharpe and Cooper (1998) prepared NO in the Kipps apparatus by the addition of 2 mol dm^{-3} sulfuric acid to solid sodium nitrite. The NO gas was passed through four NaOH traps (20%) to remove the NO_2 and then through a solid CO_2 trap. The gas was collected in a buffer solution that had previously been thoroughly deoxygenated by four vacuum/nitrogen cycles. The concentration of the NO solution was in the range of 1.2 to 2 mmol dm^{-3}, with a nitrite concentration of about 0.3 mmol dm^{-3}.

Strategies for adding NO to a reaction system will vary, depending upon whether a bolus addition is required to start a reaction or whether the concentration of NO is to be maintained at a fixed value, as far as possible, or if the NO is added to the headspace. Appropriate volumes of the NO solutions prepared may be used to initiate reactions, using a syringe for transfer, as noted previously. The NO-generation system can also be used to pass NO directly into an appropriately sealed suitable container, from which NO gas may subsequently be removed as required using an air-tight gas syringe and transferred directly into a reaction system.

2.3. Determination of NO concentration

The concentration of NO in aqueous solution may most conveniently be determined by the use of an NO electrode. Alternative methods include the use of 2,2′ azino(3-ethylbenzothiazoline-6-sulfonic acid), as already noted,

Table 3.1 Solubilities of NO in water at various temperatures

T °C	0	10	20	25	37
Conc. (mmol dm^{-3})	3.27	2.52	2.08	1.88	1.58

while samples of NO in water may be oxidized stoichiometrically to nitrite by exposure to air. The nitrite may then be analyzed colorimetrically by standard methods involving diazotization and coupling to give an azo dye.

The Saltzman method (1960) for the determination of NO and NO_2 in atmospheric air has been modified to enable the simultaneous determination of NO, NO_2, nitrite, and nitrate in aqueous solutions of NO (Ohkawa *et al.*, 2001).

3. NO Releasers

As a consequence of the difficulties involved in the preparation and use of aqueous solutions of NO, there has been great interest in the development and use of compounds that generate NO in solution in a controlled manner. A variety of compounds have been used for this purpose. Some of these have a long history of use in a medical context (Wang *et al.*, 2002). Selection of an NO releasing compound should be made carefully (Feelisch, 1998). Some compounds decompose chemically, while others need enzyme catalysis. Some react with thiols to release NO. In this review, NO releasers that are useful tools in laboratory work will be considered, in particular those that release NO directly on decomposition without requiring the addition of reducing or other agents.

3.1. S-nitrosothiols

Various S-nitrosothiols (RSNO) decompose in solution to generate NO and RS radicals, the latter dimerizing to give RSSR. Recent studies have suggested that the RS-NO homolytic bond dissociation energies are about 30 kcal mol^{-1}, which suggests that homolysis is not likely to be a major pathway for decomposition of S-nitrosothiols under physiological conditions (Bartberger *et al.*, 2001; Lu *et al.*, 2001). The rate of decomposition of S-nitrosothiols depends on the nature of the R group, but notably also depends on the presence of trace, catalytic amounts of Cu(I), which may be readily generated from Cu(II) by reaction with thiols (Dicks *et al.*, 1996).

S-nitrosothiols commonly used to generate NO in solution include S-nitrosoglutathione (GSNO), S-nitrosoacetylpenicillamine (SNAP), and S-nitrosocysteine (SNOC). These compounds may also act as trans-nitrosating agents rather than NO releasers, provided that strongly nucleophilic species are present to act as receptors for the NO^+ group. It is possible, therefore, that *in vivo* production of NO by decomposition of a typical S-nitrosothiol could be much faster than that anticipated if a fast trans-nitrosation step to another thiol group occurs to give an unstable S-nitrosothiol.

3.2. Metal nitrosyl complexes

Sodium nitroprusside ($Na_2[Fe(CN)_5(NO)]$) is well known to function as an NO donor, a property that was once exploited in its use as an antihypertension drug. However, NO is released indirectly from the nitroprusside group, probably via the S-nitrosation of thiolate groups by the nitroprusside, followed by the homolytic decomposition of the resulting S-nitroso species to give NO.

The efficiency of sodium nitroprusside as a nitrosating agent reflects the fact that the nitrosyl ligand in the nitroprusside group is an NO^+ species, bound to Fe(II), with an N–O stretching frequency in the infrared spectrum of 1938 cm^{-1}, typical of the NO^+ group and at the top end of the normal range for (NO^+) nitrosyl complexes. There is a reasonable correlation between the N–O stretching frequency and the efficiency of nitrosyl complexes as nitrosating agents. The description of the nitroprusside group as an Fe(II) (NO^+) species is further supported by structural studies that show the FeNO entity to be essentially linear, with an angle of 175.7° (Bottomley and White, 1979). Complications arise from the reduction of SNP, which leads to the release of cyanide. SNP is also very photosensitive. Aqueous solutions of sodium nitroprusside need to be prepared immediately before use and stored in a container covered with foil.

Roussin's black salt ($Na[Fe_4S_3(NO)_7]$) releases six moles of NO on photolysis in aqueous aerobic solution (Kudo *et al.*, 1997) and so offers a convenient method of rapidly releasing NO to initiate a reaction (Bourassa *et al.*, 1997). This procedure has been used in the treatment of several medical conditions. Roussin's black salt is exceedingly toxic to a range of microorganisms, although the mechanisms of the action are not fully understood.

Dinitrosyl iron thiol complexes (DNICs) are thought to act in the storage and transport of NO *in vivo*. They can be prepared in aqueous solution under oxygen-free conditions by the reaction between Fe(II) sulphate, cysteine, or glutathione and NO (Boese *et al.*, 1995; Feelisch *et al.*, 1994). These solutions may be stored at −80° for long periods, but they decompose readily to give NO at ambient temperatures.

3.3. NOR, NOC, and NONOate compounds

The structures of these three types of NO releasers, reviewed by Wang *et al.* (2002), are shown in Fig. 3.1. Details of half-lives for the release of NO from these compounds are listed in Table 3.2. In each case, a number of related compounds are available, with defined half-lives (from minutes to hours) for release of NO.

The NOR compounds are alkyl and aryl oximes and release NO under oxidative conditions (Hino *et al.*, 1989). NOR-3 decomposes with

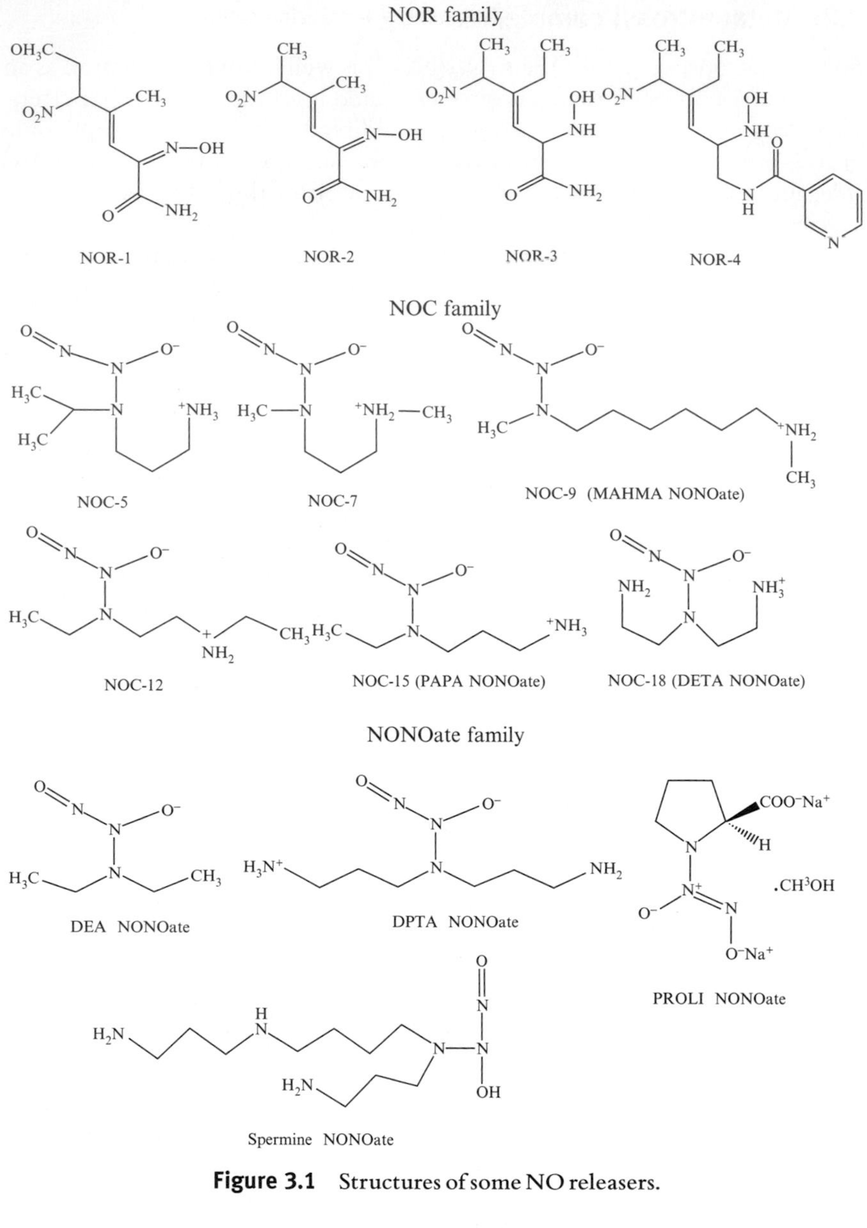

NOR family

NOR-1 NOR-2 NOR-3 NOR-4

NOC family

NOC-5 NOC-7 NOC-9 (MAHMA NONOate)

NOC-12 NOC-15 (PAPA NONOate) NOC-18 (DETA NONOate)

NONOate family

DEA NONOate DPTA NONOate PROLI NONOate

Spermine NONOate

Figure 3.1 Structures of some NO releasers.

first-order kinetics to give a ketone and NO, with a half-life of about 30 min at 37°C. Structural modification of NOR-3, as in methoxy substitution at carbon atom 1 to give NOR-1 (see Fig. 3.1), gives a compound with much faster release of NO, with a half-life at 37° of 1 to 2 min (see Table 3.2).

Table 3.2 Half-lives and solubilities of some NO-releasing compounds

NO releaser	$t_{\{1/2\}}$ (min) 37°	$t_{\{1/2\}}$ (min) 22°	Solvent
PROLI NONOate	–	1.8 s	H_2O and MeOH
MAHMA NONOate (NOC-9)	1 – 2	3	H_2O
NOR-1	1.8	–	0.1 M PBS* (pH 7.4)
NOC-7	5	–	0.1M PBS (pH 7.4)
PAPA NONOate (NOC-15)	15	77	H_2O
DEA NONOate	–	16	H_2O, PBS
NOC-5	25	93	0.1M PBS (pH 7.4)
NOR-2	28	–	0.1M PBS (pH 7.4)
NOR-3	30	–	0.1M PBS (pH 7.4)
Spermine NONOate	39	230	H_2O
DPTA NONOate	39	300	–
SIN-1A	40	–	H_2O, isotonic saline
NOR-4	60	–	PBS
NOC-12	100	327	0.1M PBS (pH 7.4)
DETA NONOate (NOC-18)	1260	3400	0.1M PBS (pH 7.4)

The NOC releasers are a series of stabilized NO-amine complexes (Hrabie *et al.*, 1993). Amines can form a complex that contains two NO molecules and is usually unstable, decomposing immediately with release of NO (Drago and Panlik, 1960). This decomposition is triggered by protonation of the oxygen of the NO group. The NOC releasers are therefore considered Drago-type complexes and are intramolecular zwitterions stabilized by intramolecular hydrogen bonding, as dispersion of the negative charge prevents protonation. The rate of release of NO follows first-order kinetics and depends on the weakness of the hydrogen bond. Decomposition of NOC compounds is faster in acidic solution, and therefore a stock solution should be prepared in 0.1 mol dm^{-3} NaOH solution. These stock solutions are stable over a day. However, the alkaline stock solution decomposes by about 5% overnight at −20°, but the solid is stable for about a year at this temperature.

The NONOates (diazeniumdiolates) are synthesized by reacting NO with various nucleophiles, as Drago (1962) first reported. These compounds are stable as solids but decompose in solution to give up to 2 mols of NO per mol of reactant. The decomposition of these compounds follows first-order kinetics. A range of NO releasers of this type are available, with half-lives for the decomposition in the range from seconds to many hours at 37°

(see Table 3.2). Rate constants are pH dependent, increasing with decrease in pH (Maragos *et al.*, 1991).

It is important to carry out control experiments on the biological activity of the products of decomposition of these compounds. As noted previously, it may be helpful to monitor the concentration of the NO released from these donors under the experimental conditions in use to ensure that the results fit in with the expected kinetics of decomposition of the NO releaser. The release of NO from the NONOates is similar to the rate of decomposition of these compounds, and the NO concentrations in experiments using these releasers can readily be calculated. This may not necessarily be true for all NO releasers.

Sometimes it may be useful to work with mixtures of several compounds of one class to maintain approximately constant and appropriate levels of NO over a time period rather than with a bolus addition of a solution of NO gas. An example of the use of NOC releasers in this way is the use by Cruz-Ramos *et al.* (2002) of a combination of NOC-5 (half-life of 25 min at pH 7.0 and 37°) and NOC-7 (half-life of 7 min at pH 7 and 37°) to provide and maintain NO in the range of concentrations sufficient to inactivate the O_2-responsive regulator FNR and thus the up-regulation of synthesis of the flavohemoglobin Hmp of *Escherichia coli*, a protein with key roles in resistance to NO. The calculated total NO released by the two NOC compounds matched well the values determined by use of an NO electrode in an experiment with an *E. coli* strain, until the levels of NO decreased due to take-up of NO by the cells.

4. Applications to Nitrosative Stress

The release of NO in aerobic environments may lead to nitrosative stress via the formation of NO_2 and then N_2O_3, which is an efficient nitrosating agent. The concentration of NO_2 and NO in cells is critical to the development of nitrosative stress.

4.1. Cellular fate of NO_2 and the level of nitrosative stress

NO itself does not participate directly in cellular nitrosative stress. As noted previously, NO is not an especially reactive radical. However, the rate of the reaction between NO and oxygen to give NO_2 is accelerated in lipid or hydrophobic phases in cells (Lui *et al.*, 1998). Higher concentrations of NO_2 favor reaction with NO and so may lead to higher levels of N_2O_3. This enhances the possibility of nitrosation occurring, in which the NO^+ group from a carrier such as N_2O_3 is transferred to sulfur and nitrogen centers, with toxic effects (Williams, 1988). Thus, the concentration and lifetime of NO_2 is likely to be a major factor in determining the extent of cellular nitrosative stress. It is important, therefore, to assess the pathways for formation and decomposition of this molecule.

In addition to the route involving the oxidation of NO by oxygen, NO_2 may also be formed from the reaction between NO and superoxide via the intermediacy of peroxynitrite, which reacts with carbon dioxide to give NO_2 and carbonate radicals (Radi *et al.*, 2001). However, NO_2 can also be scavenged by cellular antioxidants such as glutathione, cysteine, or uric acid, which diminishes the possibility of nitrosation occurring. These issues have been studied by Ford *et al.* (2002), who used pulse radiolysis to generate NO_2 and study its reactions with these antioxidants. In summary, the implications of this work are (a) that the lifetime of NO_2 in cytosol is less than 10 μs, (b) thiols are the major sink for NO_2 in cells and tissue, but urate is also a major scavenger in plasma, (c) the diffusion distance of NO_2 is about 0.2 μm in the cytoplasm and less than 0.8 μm in plasma, (d) urate protects GSH against depletion on oxidative challenge from NO_2, and (e) the reactions between NO_2 and thiols and urate will severely diminish the possibility that NO_2 will react with NO to form N_2O_3 in the cytoplasm, and thus reduce the importance of nitrosative stress.

4.2. Microbes and nitrosative stress

An area of particular interest is the toxicity of NO and related reactive nitrogen species to bacteria. These may react with a number of important biomolecules, including nitrosylation of metal centers in metalloproteins and the nitrosation of thiol groups in proteins and low-molecular-weight thiols such as glutathione and homocysteine (Poole and Hughes, 2000). Nitrosation of metallothioneins can bring about release of coordinated zinc or cadmium (Pearce *et al.*, 2000), while Binet *et al.* (2002) have shown that NO (supplied using NOC-5 and NOC-7) can bring about release of zinc from *E. coli*. Bacteria are able to develop resistance to the stresses brought about by NO and nitrosation (Poole and Hughes, 2000). Reference has already been made to the flavohemoglobin Hmp of *E. coli* in this context. Flavohemoglobin is a key factor in terms of resistance to NO and nitrosative stress (Membrillo–Hernandez *et al.*, 1999). Thus, the structural gene *hmp* is up-regulated by NO and nitrosating agents, while *hmp* mutants are hypersensitive to these factors. Under aerobic conditions, NO is oxidized to nitrate; under anaerobic conditions, it is reduced to N_2O, presumably via the intermediacy of nitroxyl, HNO. It has been shown that FNR, the O_2-responsive regulator, also acts as a sensor for NO. NO, supplied in this case by proline NONOate, a rapid releaser of NO with a half-life of 13 seconds, reacts anaerobically with the $(4Fe\text{-}4S)^{2+}$ cluster of purified FNR. The UV/Vis spectra are consistent with the formation of a dinitrosyl-iron-cysteine complex. NO inactivation of FNR will then lead to up-regulation of Hmp synthesis and resistance to NO (Cruz-Ramos *et al.*, 2002).

The yeast *Saccharomyces cerevisiae* also contains a flavohemoglobin that protects against nitrosative stress from NO and *S*-nitrosothiols (Liu *et al.*, 2000). Deletion of the flavohemoglobin gene resulted in the loss of uptake of NO by yeast cells, while nitrosation of proteins after incubation with NO donors was more than 10 times higher in the *yhb1* mutant yeast than in isogenic wild-type cells. The resistance of mutant cells to oxidative stress was unimpaired. The flavohemoglobin protected yeast cells against the effects of NO and S-nitrosothiols under anaerobic and aerobic conditions. This suggests it has a primary role in NO detoxification.

The use of NO releasers and compounds such as sodium nitroprusside and Roussin's black salt in the study of nitrosative stress is very widespread, for example, on food spoilage organisms (Joannou *et al.*, 1998) and *Giardia intestinalis* (Lloyd *et al.*, 2003). Joannou *et al.* (1998) have shown a good correlation between the N–O stretching frequency of nitroprusside and the pentacyanonitrosyl complexes of other metals with their toxicity toward *Clostridium sporogenes*. This shows that the toxic effect is due to the nitrosative ability of these complexes. In general, Roussin's black salt is an especially toxic species, for example to *G. intestinalis* with major toxic effects at micromolar concentrations. However, the mechanism of its toxic action is still unclear. Roussin's black salt does not release NO under normal conditions, and its toxic effects may involve disruption of electron transfer rather than function as a nitrosating agent.

REFERENCES

Bartberger, M. D., Mannion, J. D., Powell, S. C., Stamler, J. S., Houk, K. N., and Toone, E. J. (2001). S-N dissociation energies of S-nitrosothiols: On the origins of nitrosothiol decomposition rates. *J. Am. Chem. Soc.* **123,** 8868–8869.

Binet, M. R. B., Cruz-Ramos, H., Laver, J., Hughes, M. N., and Poole, R. K. (2002). Nitric oxide releases intracellular zinc from prokaryotic metallothionein in *Escherichia coli*. *FEMS Microbiology Letters* **213,** 121–126.

Boese, M., Mordvintcev, P. I., Vanin, A. F., Busse, R., and Mulsch, A. (1995). S-nitrosation of serum albumin by dinitrosyl iron complex. *J. Biol. Chem.* **270,** 29244–29249.

Bottomley, F., and White, P. S. (1979). Redetermination of the structure of disodium pentacyanonitrosylferrate (sodium nitroprusside). *Acta Cryst.* **B35,** 2193–2195.

Bourassa, J., DeGraff, W., Kudo, S., Wink, D. A., Mitchell, J. B., and Ford, P. C. (1997). Photochemistry of Roussin's red salt, $Na_2[Fe_2S_2(NO)_4]$, and of Roussin's black salt, $NH_4[Fe_4S_3(NO)_7]$. (1997) *In situ* nitric oxide generation to sensitize gamma-radiation induced cell death. *J. Amer. Chem. Soc.* **119,** 2853–2860.

Cruz-Ramos, H., Crack, J., Wu, G., Hughes, M. N., Scott, C., Thomson, A. J., Green, J., and Poole, R. K. (2002). NO sensing by FNR: Regulation of the *Escherichia coli* NO-detoxifying flavohaemoglobin, Hmp. *EMBO J.* **21,** 3235–3244.

Dicks, A. P., Swift, H. R., Williams, D. L. H., Butler, A. R., Al-Sadoni, H. H., and Co, B. G. (1996). Identification of Cu^{2+} as the effective reagent in nitric oxide formation from S-nitrosothiols (RSNO). *J. Chem. Soc. Perkin Trans.* **248,** 1–2487.

Drago, R. S. (1962). Reactions of nitrogen(II) oxide: Free radicals in inorganic chemistry. Advances in Chemistry Series, Number 36. American Chemical Society, Washington, DC, 143–149.

Drago, R. S., and Paulik, F. E. (1960). The reaction of nitrogen (II) oxide with diethylamine. *J. Am. Chem. Soc.* **82,** 96–99.

Feelisch, M. (1998). The use of nitric oxide donors in pharmacological studies. *Naunyn-Schmiedeberg's Arch. Pharmacol.* **358,** 113–122.

Feelisch, M., Te Poel, M., Zamora, R., Deussen, A., and Moncada, S. (1994). Understanding the controversy over EDRF. *Nature* **368,** 62–65.

Ford, E. (2001). Ph.D. thesis. Mechanisms of reaction of nitrogen oxides with biological molecules and the chemical basis for cellular oxidative stress. University of London, United Kingdom, pp. 47–96.

Ford, E., Hughes, M. N., and Wardman, P. (2002). Kinetics of the reaction of nitrogen dioxide with glutathione, cysteine and uric acid at physiological pH. *Free Radic. Biol. Med.* **32,** 1314–1323.

Hino, N., Iwami, M., Okamato, M., Yoshida, K., Haruta, H., Okhura, M., Hosoda, J., Kofsak, M., Aoki, H., and Imanaka, H. (1989). *J. Antibiot.* **42,** 1578–1584.

Hrabie, J. A., Klose, J. R, Wink, D. A., and Keefer, L. K. (1993). New nitric oxide-releasing zwitterions derived from polyamines. *J. Org. Chem.* **58,** 1472–1476.

Ishimura, Y., Gao, Y. T., Panda, S. P., Roman, L. J., Masters, B. S., and Weintraub, S. T. (2005). Detection of nitrous oxide in the neuronal nitric oxide synthase reaction by gas chromatography-mass spectrometry. *Biochem. Biophys. Res. Commun.* **338,** 543–549.

Joannou, C. L., Cui, X. I., Rogers, N., Violette, N., Torres-Martinez, C. L., Vugman, N. V., Hughes, M. N., and Cammack, R. (1998). Characterisation of the bacteriocidal effects of sodium nitroprusside and other pentacyanonitrosyl complexes on the food spoilage organism *Clostridium sporogenes*. *Appl. Environ. Microbiology* **64,** 30195–30201.

Kudo, S., Bourassa, J. L., Boggs, S. E., Sato, Y., and Ford, P. C. (1997). *In situ* nitric oxide (NO) measurement by modified electrodes: NO labilised by photolysis of metal nitrosyl complexes. *Anal. Biochem.* **247,** 193–202.

Lloyd, D., Harris, J. C., Maroulis, S., Mitchell, A., Hughes, M. N., Wadley, R. B., and Edwards, M. R. (2003). Nitrosative stress induced cytotoxicity in *Giardia intestinalis*. *J. Appl. Microbiology* **95,** 576–583.

Liu, L., Zeng, M., Hausladen, A., Heitman, J., and Stamler, J. S. (2000). Protection from nitrosative stress by yeast flavohemoglobin. *Proc. Natl. Acad. Sci. USA* **97,** 4672–4676.

Lu, J.-M., Wittbrodt, J. M., Wang, K., Wen, Z., Schlegel, H. B., Wang, P. G., and Cheng, J.-P. (2001). NO affinities of S-nitrosothiols: A direct experimental and computational investigation of RS NO bond dissociation energies. *J. Am. Chem. Soc.* **123,** 2903–2905.

Lui, X., Miller, M. J. S., Joshi, M. S., Thomas, D. D., and Lancaster, J. R., Jr. (1998). Accelerated reaction of nitric oxide with O_2 within the hydrophobic interior of biological membranes. *Proc. Natl. Acad. Sci. USA* **97,** 2175–2179.

Maragos, C. M., Morley, D., Wink, D. A., Dunams, T. M., Saavedra, J. E., Hoffman, A., Bove, A. A., Isaac, L., Hrabie, J. A., and Keefer, L. K. (1991). Complexes of NO with nucleophiles as agents for the controlled biological release of nitric oxide. *J. Med. Chem.* **34,** 3242–3247.

Membrillo-Hernandez, J., Coopamah, M. D., Anjum, M. F., Stevanin, T. M., Kelly, A., Hughes, M. N., and Poole, R. K. (1999). The flavohemoglobin of *Escherichia coli* confers resistance to a nitrosating agent, a "nitric oxide releaser" and paraquat and is essential for transcriptional responses to oxidative stress. *J. Biol. Chem.* **274,** 748–754.

Ohkawa, T., Hiramoto, K., and Kikugawa, K. (2001). Standardisation of nitric oxide aqueous solutions by modified Saltzman method. *Nitric oxide: Biol. Chem.* **5,** 515–524.

Pearce, L. L., Gandley, R. E., Han, W., Wasserloos, K., Stitt, M., Kanai, A. J., McLaughlin, M. K., Pitt, B. R., and Levitan, E. S. (2000). Role of metallothionein in nitric oxide signalling as revealed by a green fluorescent fusion protein. *Proc. Natl. Acad. Sci. USA* **97,** 477–482.

Poole, R. K., and Hughes, M. N. (2000). New functions for the ancient globin family: Bacterial responses to nitric oxide and nitrosative stress. *Molecular Microbiology* **36,** 775–783.

Radi, R., Peluffo, G., Alvarez, M. N., Naviliat, M., and Cayota, A. (2001). Unravelling peroxynitrite formation in biological systems. *Free Radic. Biol. Med.* **30,** 463–488.

Saltzman, B. E. (1960). Colorometric microdetermination of nitrogen dioxide in the atmosphere. *Anal. Chem.* **32,** 135–136.

Scott, S. L., Chen, W.-J., Bakac, A., and Espenson, J. H. (1993). Spectroscopic parameters, electrode potential, acid ionisation constants and electron exchange rates of the 2,2′-azinobis(3-ethylbenzothiaoline-6-sulfonate) radicals and ions. *J. Phys. Chem.* **97,** 6710–6714.

Sharpe, M. A., and Cooper, C. E. (1998). Reactions of nitric oxide with mitochondrial cytochrome *c*′: A novel mechanism for the formation of nitroxyl anion and peroxynitrite. *Biochem. J.* **332,** 9–19.

Stanbury, D. M. (1989). Reduction potentials involving inorganic free radicals in aqueous solution. *Adv. Inorg. Chem.* **33,** 69–138.

Wang, G. W., Xian, M., Tang, X., Wu, X., Wen, Z., Cai, T, and Janczuk, A. J. (2002). Nitric oxide donors: Chemical activities and biological applications. *Chem. Rev.* **102,** 1091–1134.

Williams, D. L. H. (1988). Nitrosation Cambridge, UK: Cambridge University Press, Cambridge, UK.

CHAPTER FOUR

The Chemistry of Peroxynitrite: Implications for Biological Activity

Sara Goldstein* *and* Gabor Merényi†

Contents

Abstract

In biological systems, nitric oxide (NO) combines rapidly with superoxide (O_2^-) to form peroxynitrite ion ($ONOO^-$), a substance that has been implicated as a culprit in many diseases. Peroxynitrite ion is essentially stable, but its protonated form (ONOOH, $pK_a = 6.5$ to 6.8) decomposes rapidly via homolysis of the O–O bond to form about 28% free NO_2 and OH radicals. At physiological pH and in the presence of large amounts of bicarbonate, $ONOO^-$ reacts with CO_2 to produce about 33% NO_2 and carbonate ion radicals (CO_3^-) in the bulk of the solution. The quantitative role of OH/CO_3^- and NO_2 radicals during the decomposition of peroxynitrite (ONOOH/$ONOO^-$) under physiological conditions is described in detail. Specifically, the effect of the peroxynitrite dosage rate on the yield and distribution of the final products is demonstrated. By way of an example, the detailed mechanism of nitration of tyrosine, a vital aromatic amino acid, is delineated, showing the difference in the nitration yield between the addition of authentic peroxynitrite and its continuous generation by NO and O_2^- radicals.

* Department of Physical Chemistry, The Hebrew University of Jerusalem, Jerusalem, Israel
† Department of Chemistry, Nuclear Chemistry, The Royal Institute of Technology, Stockholm, Sweden

Methods in Enzymology, Volume 436
ISSN 0076-6879, DOI: 10.1016/S0076-6879(08)36004-2

1. Introduction

Superoxide (HO_2/O_2^-) is formed in the mitochondria during normal cellular respiration, and its steady-state concentration is maintained low by endogenous superoxide dismutases (SODs). These enzymes, which are present in all aerobic organisms, provide a defense that is essential for their survival. However, such a defense is not complete, and superoxide plays a role during oxidative stress, as in postischemic reperfusion, organ transplantation, and various surgical interventions (Cadenas, 2004; Harman, 1993; Oberely, 1982; Rotelio, 1986). NO is formed enzymatically from L-arginine and O_2 in endothelial cells or neurons; it relaxes vascular smooth muscle, functions as a neurotransmitter and a mediator of glutamate neurotoxicity, and inhibits platelet aggregation and protein synthesis (Bredt and Snyder, 1994; Moncada *et al.*, 1991). NO is formed at much higher concentrations by inducible NO-synthases (iNOS) or overactivation of endothelial NOS (eNOS) and neuronal NOS (nNOS) (Bredt and Snyder, 1994; Gross and Wolin, 1995; Moncada *et al.*, 1991). Although NO is a potent biological mediator at low concentrations, high steady-state levels of NO play critical roles in modulating inflammatory and degenerative diseases, which often involve the nitration of many cellular components, such as tyrosine residues (Dedon and Tannenbaum, 2004; Ischiropoulos, 1998; Kubes and McCafferty, 2000; Patel *et al.*, 1999).

Despite their free radical nature, NO and O_2^- are in general not too reactive. Part of their toxicity is associated with their combination at a near diffusion-controlled rate (Goldstein and Czapski, 1995b; Huie and Padmaja, 1993; Nauser and Koppenol, 2002) to form peroxynitrite ion ($ONOO^-$). The importance of the coupling of NO with O_2^- to yield $ONOO^-$ in biological systems was first suggested by Beckman *et al.* (1990), and this reaction is currently accepted as the main biological source of peroxynitrite ($ONOOH/ONOO^-$). The reactivity of peroxynitrite toward biological molecules and its very high toxicity toward cells are assumed to be the potential cause of a number of diseases (Dedon and Tannenbaum, 2004; Greenacre and Ischiropoulos, 2001; Ischiropoulos, 1998; Kharitonov and Barnes, 2003; Radi, 2004; Turko and Murad, 2002).

Different methods have been utilized to expose biochemical or cellular systems to peroxynitrite, including addition or infusion of authentic peroxynitrite as well as fluxes of the radical precursors NO and O_2^-. The present paper describes in detail the chemistry of peroxynitrite in aqueous solutions and under physiological conditions and demonstrates the difference in the reactivity of peroxynitrite generated by these methods.

2. Mechanism of Peroxynitrite Decomposition in Aqueous Solutions

ONOOH is a rather strong acid and its pK_a for deprotonation to form $ONOO^-$ was found to be 6.5 to 6.8 (Goldstein and Czapski, 1995b; Kissner *et al.*, 1997; Logager and Sehested, 1993b; Pryor and Squadrito, 1995). ONOOH decays almost entirely into nitrate with $k_d = 1.25 \pm 0.05\ s^{-1}$ at 25° (Kissner *et al.*, 1997; Merényi *et al.*, 1999; Pryor and Squadrito, 1995), whereas $ONOO^-$ is essentially stable. The mechanism of the decomposition of ONOOH has become a matter of controversy during the last decade, despite the early and seminal work of Mahoney (Mahoney, 1970) that unambiguously proved that ONOOH homolyzes into free NO_2 and OH radicals. Nevertheless, homolysis was eventually confirmed in several studies (Goldstein *et al.*, 2005) and consensus has settled around a free radical yield of 28 ± 4% (Gerasimov and Lymar, 1999; Goldstein *et al.*, 2005; Hodges and Ingold, 1999).

The mechanism of the decomposition of ONOOH into nitrate is described in Scheme 1. The decomposition of ONOOH yields nitrate because OH reacts much faster, by about two orders of magnitude, with NO_2^- than with ONOOH.

When the pH is raised, O_2 and NO_2^- in a 1:2 proportion are formed at the expense of NO_3^-, reaching approximately 80% NO_2^- at pH 9 to 10 (Kirsch *et al.*, 2003; Kissner and Koppenol, 2002; Lymar *et al.*, 2003; Pfeiffer *et al.*, 1997). The rate constants for the reaction of OH with NO_2^- and $ONOO^-$ have been determined to be $k_4 = 5.3 \times 10^9\ M^{-1}s^{-1}$ (Merényi *et al.*, 1999) and $k_5 = 4.8 \times 10^9\ M^{-1}s^{-1}$ (Goldstein *et al.*, 1998), respectively. Because these rate constants are very similar, the reaction pattern of peroxynitrite changes when the pH approaches and exceeds pK_a(ONOOH) = 6.5 to 6.8. The results follow from the homolysis of ONOOH into NO_2 and OH, the homolytic equilibrium between $ONOO^-$ and NO + O_2^-, as well as other subsequent reactions of these reactive radicals, as illustrated in Scheme 2.

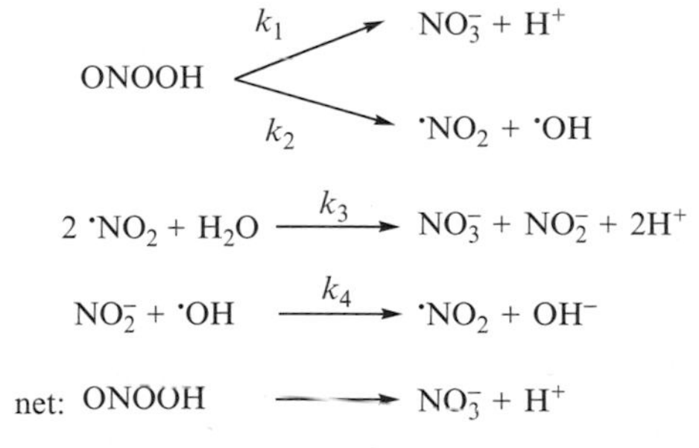

$$ONOOH \xrightarrow{k_1} NO_3^- + H^+$$

$$ONOOH \xrightarrow{k_2} {}^{\bullet}NO_2 + {}^{\bullet}OH$$

$$2\ {}^{\bullet}NO_2 + H_2O \xrightarrow{k_3} NO_3^- + NO_2^- + 2H^+$$

$$NO_2^- + {}^{\bullet}OH \xrightarrow{k_4} {}^{\bullet}NO_2 + OH^-$$

$$\text{net: } ONOOH \longrightarrow NO_3^- + H^+$$

Scheme 1 Decomposition of ONOOH.

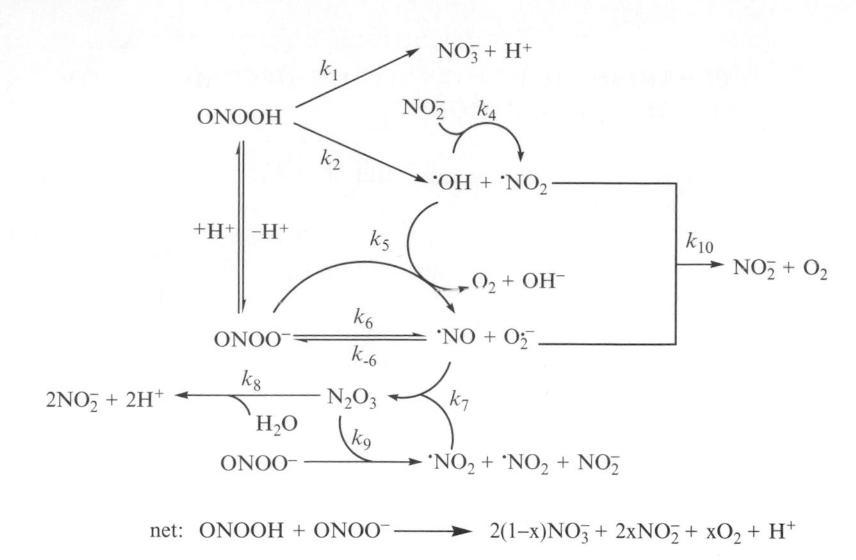

Scheme 2 Decomposition of peroxynitrite.

Scheme 2 also demonstrates the formation of the powerful electrophile N_2O_3. The reactions and the rate constants involved in the decomposition mechanism of peroxynitrite are summarized in Table 4.1.

3. Reactivity of Peroxynitrite

Peroxynitrite reactions can be classified into direct reactions and those occurring via the highly reactive radicals formed during its decomposition. In fact, peroxynitrite indirectly oxidizes any substrate due to its decomposition into NO_2 and OH radicals, as the latter radical is highly oxidizing.

Peroxynitrite reacts directly with sulfhydryls (Radi *et al.*, 1991), metal complexes (Goldstein and Czapski, 1995a), ebselen (Masumoto *et al.*, 1996), porphyrins (Ferrer-Sueta *et al.*, 2003; Lee *et al.*, 1998), heme proteins (Bourassa *et al.*, 2001; Exner and Herold, 2000; Thomson *et al.*, 1995), and CO_2 (Lymar and Hurst, 1995). The last reaction is most probably the predominant pathway for peroxynitrite disappearance in biological systems because of its high rate constant for this reaction (i.e., $k_{11} = [2.9 \pm 0.3] \times 10^4$ $M^{-1}s^{-1}$ at 24°) (Lymar and Hurst, 1995) and high levels of bicarbonate in intracellular (12 mM) and interstitial (30 m*M*) fluids (Carola *et al.*, 1990). The experimental evidence supports the nucleophilic addition of $ONOO^-$ to CO_2 to form $ONOOC(O)O^-$ (Goldstein *et al.*, 2002; Squadrito and Pryor, 2002), which decomposes via homolysis to form about 33% NO_2 and CO_3^- in the bulk of the solution (Scheme 3) (Goldstein *et al.*, 2001a).

Table 4.1 Summary of the reactions and their rate constants involved in the decomposition mechanism of peroxynitrite at 25°

No.	Reaction	*k*	Ref.
1	$ONOOH \rightarrow NO_3^- + H^+$	$0.90 \pm 0.05\ s^{-1}$	*
2	$ONOOH \rightarrow NO_2 + OH$	$0.35 \pm 0.03\ s^{-1}$	*
3	$2\ NO_2 + H_2O \rightarrow NO_2^- + NO_3^- + 2H^+$	$2k = 1.3 \times 10^8\ M^{-1}s^{-1}$	Gratzel *et al.*, 1969
4	$OH + NO_2^- \rightarrow OH^- + NO_2$	$5.3 \times 10^9\ M^{-1}s^{-1}$	Merenyi *et al.*, 1999
5	$OH + ONOO^- \rightarrow NO + O_2 + OH^-$	$4.8 \times 10^9\ M^{-1}s^{-1}$	Goldstein *et al.*, 1998
6	$NO + O_2^- \rightarrow ONOO^-$	$(5 \pm 1) \times 10^9\ M^{-1}s^{-1}$	Goldstein *et al.*, 2005
−6	$ONOO^- \rightarrow NO + O_2^-$	$0.020 \pm 0.003\ s^{-1}$	Goldstein *et al.*, 2001b; Merenyi and Lind, 1998
7	$NO_2 + NO$ **{ReversReact}** N_2O_3	$k_7 = 1.1 \times 10^9\ M^{-1}s^{-1}$ $k_{-7} = 8.4 \times 10^4\ s^{-1}$	Gratzel *et al.*, 1970
8	$N_2O_3 + H_2O \rightarrow 2\ NO_2^- + 2H^+$	$2 \times 10^3 + 10^8[OH^-] + (6.4 - 9.4) \times 10^5[\text{phosphate}]\ s^{-1}$	Goldstein and Czapski, 1996; Treinin and Hayon, 1970
9	$ONOO^- + N_2O_3 \rightarrow 2\ NO_2 + NO_2^-$	$3.1 \times 10^8\ M^{-1}s^{-1}$	Goldstein *et al.*, 1999
10	$NO_2 + O_2^- \rightarrow NO_2^- + O_2$	$4.5 \times 10^8\ M^{-1}s^{-1}$	Logager and Sehested, 1993a

* ONOOH decomposes with $k_d = 1.25\ s^{-1}$, yielding 72% $NO_3^- + H^+$ and 28% $NO_2 + OH$.

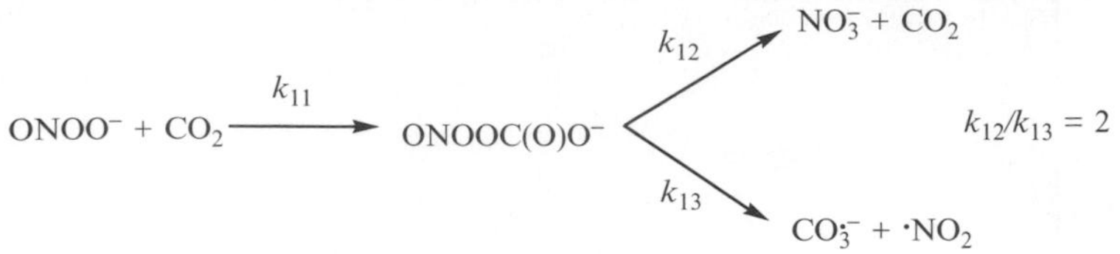

$NO_3^- + CO_2$
k_{12}
k_{11}
$ONOO^- + CO_2$
$ONOOC(O)O^-$
$k_{12}/k_{13} = 2$
k_{13}
$CO_3^{\cdot-} + {}^{\cdot}NO_2$

Scheme 3 Decomposition of peroxynitrite in the presence of excess of CO_2.

4. The Radical Model *In Vivo*

The formation and decomposition of peroxynitrite under physiological conditions, assuming the target molecule (RH) does not react directly with peroxynitrite, is given in Scheme 4.

The reaction of RH with OH and in some cases with CO_3^- forms less reactive secondary radicals, R. In most cases, R readily reacts with O_2 to form the corresponding peroxyl radical, ROO, which recombines quickly with NO_2 (Goldstein *et al.*, 2004) or decomposes into R^+ and O_2^- (von Sonntag and Schuchmann, 1997). If R is unreactive toward O_2, as in the case with phenoxyl radicals, it most probably scavenges NO_2 to form RNO_2. The dosage rate of peroxynitrite should affect the yields of $RNO_2/ROONO_2$ as follows. Upon bolus addition of authentic peroxynitrite, about 30% of it yields OH/CO_3^- and NO_2. Because the reaction of OH/CO_3^- with RH is relatively fast, about 30% of peroxynitrite forms R/ROO and NO_2. Thus, the initial concentrations of these radicals are relatively high; therefore, most of them will cross-recombine to yield the coupling products $RNO_2/ROONO_2$. Under continuous generation

RH
O_2
28%
ONOOH
$\cdot NO_2 + \cdot OH$
$H_2O + R\cdot$
$ROO^{\cdot}$
$R^+ + O_2^{\cdot-}$
H^+
$-H^+$
$ROONO_2$
$\cdot NO + O_2^{\cdot-}$
$ONOO^-$
$NO_3^- + H^+$
$NO_3^- + O_2$
CO_2
RNO_2
$ONOOCO_2^-$
33%
$\cdot NO_2 + CO_3^{\cdot-}$
$HCO_3^- + R\cdot$
RH

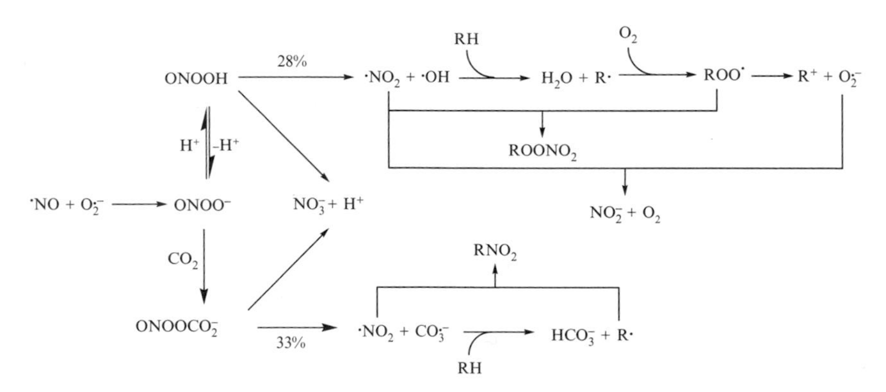

Scheme 4 Formation and decomposition of peroxynitrite under physiological conditions*.

* The mechanism in the presence of CO_2 represents the case where the secondary radical, $R^{\cdot}$, does not react with O_2.

of peroxynitrite, there is a decrease in the steady-state concentration of R/ROO and NO_2. As a result, any potential reaction between a radical and a nonradical substrate will be favored over radical–radical reactions. Consequently, such a competition depends on the flux of peroxynitrite and on the concentration of the target molecule, and influences the yields and distribution of the final products (Batthyany *et al.*, 2005; Goldstein and Czapski, 2000; Goldstein *et al.*, 2000; Niles *et al.*, 2006; Pfeiffer *et al.*, 2000; Schrammel *et al.*, 2003). Under nonequal fluxes of NO and O_2^-, the system is further complicated by the reactions of the radical being in excess with R or ROO, thus lowering the oxidation/nitration products (Daiber *et al.*, 2002; Demicheli *et al.*, 2007; Hodges *et al.*, 2000; Jourd'heuil *et al.*, 1999; Miles *et al.*, 1996; Quijano *et al.*, 2005; Sawa *et al.*, 2000; Thomas *et al.*, 2006; Trostchansky *et al.*, 2001; Wink *et al.*, 1997). The following section describes in detail the mechanism of tyrosine nitration and shows the difference in the nitration yield during addition of authentic peroxynitrite as opposed to its continuous generation by NO and O_2^- radicals.

5. Nitration of Tyrosine

Peroxynitrite has been shown to nitrate tyrosine and tyrosine residues in proteins under physiological conditions (Batthyany *et al.*, 2005; Beckman *et al.*, 1992; Berlett *et al.*, 1998; Demicheli *et al.*, 2007; Goldstein and Czapski, 2000; Goldstein *et al.*, 2000; Hodges *et al.*, 2000; Kong *et al.*, 1996; Lymar *et al.*, 1996; Pfeiffer *et al.*, 2000; Reiter *et al.*, 2000; Santos *et al.*, 2000; Sawa *et al.*, 2000; Van der Vliet *et al.*, 1995), and it therefore has been suggested that it can operate as a nitrating agent *in vivo*. In many cases the evidence that implicates $ONOO^-$ in a large variety of diseases is the detection of 3-nitrotyrosine in the injured tissues (Ischiropoulos, 1998). However, the discrepancies among the literature values of the nitration yield are enormous and have initiated continuous discussions and speculations on the involvement of peroxynitrite in this process. Our proposed model for nitration of tyrosine (Goldstein *et al.*, 2000), which is described below, is in line with Scheme 4. This model shows that most of the discrepancies in the literature result from the different methods used for peroxynitrite generation and tyrosine concentration.

The mechanism of tyrosine nitration by authentic peroxynitrite at pH 7.5 and presence of excess CO_2 is given in Scheme 5 (Goldstein *et al.*, 2000). All the rate constants in this reaction scheme are known (Jin *et al.*, 1993; Mallard *et al.*, 1998; Prutz *et al.*, 1985), and under such conditions the hydrolysis of NO_2 is relatively slow and can be ignored. Because about 33% of peroxynitrite yields CO_3^- and NO_2, and because the reaction of CO_3^- with tyrosine is relatively fast, about 33% of peroxynitrite forms

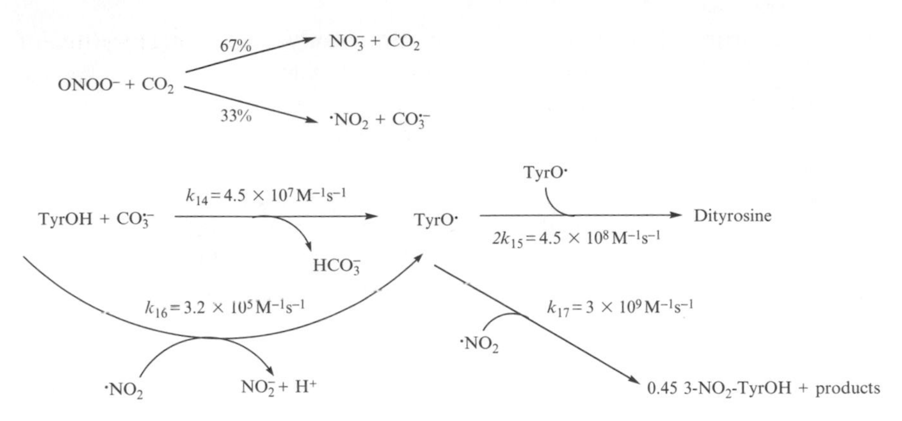

Scheme 5 Mechanism of nitration and oxidation of tyrosine by authentic peroxynitrite in the presence of excess of CO_2 at pH 7.5.

tyrosyl radical (TyrO) and NO_2, even in the presence of relatively low concentrations of tyrosine. Thus, the initial concentrations of these radicals are relatively high; therefore, most of them will cross-recombine to yield 3-nitrotyrosine.

It has been shown (Goldstein *et al.*, 2000) that the reaction of TyrO with NO_2 forms about 45% 3-nitrotyrosine. Thus, the nitration yield of tyrosine should be about 15% peroxynitrite, in agreement with experimental results (Goldstein *et al.*, 2000; Lymar *et al.*, 1996; Reiter *et al.*, 2000; Santos *et al.*, 2000). The yield of dityrosine was found to increase at the expense of 3-nitrotyrosine upon substantial lowering of the initial peroxynitrite concentration, or when peroxynitrite was generated continuously (Goldstein *et al.*, 2000; Pfeiffer *et al.*, 2000). As explained in the previous section, this observation is expected because of the decreases in the steady-state concentration of TyrO and NO_2. As a result, the reaction of tyrosine with NO_2 will come into efficient competition with the coupling of TyrO with NO_2. The same efficient competition takes place if the concentration of tyrosine is increased at a constant dosage rate of peroxynitrite (i.e., the relative steady-state concentration of NO_2 will decrease and the yield of dityrosine will increase at the expense of 3-nitrotyrosine). To conclude, the nitration yield depends on the dosage rate of peroxynitrite and on the concentration of tyrosine.

Upon excess production of O_2^-, the steady-state concentration of TyrO is significantly reduced due to reaction 16 (Eq. 18; Jin *et al.*, 1993).

$$\mathrm{TyrO} + \mathrm{O_2^-} \rightarrow \mathrm{Products}\ k_{16} = 1.5 \times 10^9\ \mathrm{M^{-1}s^{-1}} \qquad (4.18)$$

Thus, while reaction 16 is still the major consumption route of NO_2, reaction 17 is suppressed in favor of reaction 18, and nitration is completely inhibited in the presence of excess generation of O_2^-. At excess production of NO, the steady-state concentration of TyrO is unaffected due to the reversible binding of NO to TyrO (Goldstein *et al.*, 2000). However, the - steady-state concentration of NO_2 is reduced due to its reaction with NO to form N_2O_3, which hydrolyzes fast under physiological conditions (Goldstein and Czapski, 1996; Treinin and Hayon, 1970). Theoretically (Goldstein *et al.*, 2000), the maximum nitration yield should occur at equal rates of O_2^- and NO production and should decrease under unequal production rates, resulting in a bell-shaped curve. This was confirmed by Hodges *et al.* (2000). Other theoretical reports, however, showed that this bell-shaped profile was completely lost, due to the incorporation of biologically relevant events such as nitric oxide diffusion (Lancaster, 2006) and superoxide-catalyzed dismutation (Quijano *et al.*, 2005).

6. Summary

Most of peroxynitrite chemistry is initiated by or involves O–O bond homolysis forming OH and NO_2 radicals. In the presence of bicarbonate, OH is replaced by the less reactive but more selective CO_3^-. The biologically toxic effect of peroxynitrite is contingent on the simultaneous generation of OH/CO_3^- and NO_2. The radicals OH/CO_3^- usually react with the target molecule to form a secondary radical, and the less reactive NO_2 couples with this radical to produce a nitrated product. Overall, it transpires that, largely due to radical generation, peroxynitrite is able to attack most molecules of biological importance. The rate of radical production, which in turn parallels peroxynitrite dosage, is shown strongly to affect the outcome and yield of such reactions.

ACKNOWLEDGMENTS

This research was supported by the Israel Science Foundation, founded by the Israel Academy of Sciences and Humanities, and by the Swedish Research Council.

REFERENCES

Batthyany, C., Souza, J. M., Duran, R., Cassina, A., Cervenansky, C., and Radi, R. (2005). Time course and site(s) of cytochrome c tyrosine nitration by peroxynitrite. *Biochemistry* **44,** 8038–8046.

Beckman, J. S., Beckman, T. W., Chen, J., Marshall, P. A., and Freeman, B. A. (1990). Apparent hydroxyl radical production by peroxynitrite: Implications for endothelial injury from nitric oxide and superoxide. *Proc. Natl. Acad. Sci. USA* **87,** 1620–1624.

Beckman, J. S., Ischiropoulos, H., Zhu, L., van der Wored, M., Smith, C., Chen, J., Harrison, J., Martin, J. C., and Tsai, J.-H. M. (1992). Kinetics of superoxide dismutase- and iron-catalyzed nitration of phenolic by peroxynitrite. *Arch. Biochem. Biophys.* **298,** 438–445.

Berlett, B. S., Levine, R. L., and Stadman, E. R. (1998). Carbon dioxide stimulates peroxynitrite-mediated nitration of tyrosine residues and inhibits oxidation of methionine residues of glutamine synthetase: Both modifications mimic effects of adenylylation. *Proc. Natl. Acad. Sci. USA* **95,** 2784–2789.

Bourassa, J. L., Ives, E. P., Marqueling, A. L., Shimanovich, R., and Groves, J. T. (2001). Myoglobin catalyzes its own nitration. *J. Am. Chem. Soc.* **123,** 5142–5143.

Bredt, D. S., and Snyder, S. H. (1994). Nitric oxide: A physiologic messenger molecule. *Annu. Rev. Biochem.* **63,** 175–195.

Cadenas, E. (2004). Mitochondrial free radical production and cell signaling. *Mol. Asp. Med.* **25,** 17–26.

Carola, R., Harely, J. P., and Noback, C. R. (1990). "Human anatomy & physiology." McGraw-Hill, New York.

Daiber, A., Frein, D., Namgaladze, D., and Ullrich, V. (2002). Oxidation and nitrosation in the nitrogen monoxide/superoxide system. *J. Biol. Chem.* **277,** 11882–11888.

Dedon, P. C., and Tannenbaum, S. R. (2004). Reactive nitrogen species in the chemical biology of inflammation. *Arch. Biochem. Biophys.* **423,** 12–22.

Demicheli, V., Quijano, C., Alvarez, B., and Radi, R. (2007). Inactivation and nitration of human superoxide dismutase (SOD) by fluxes of nitric oxide and superoxide. *Free Rad. Biol. Med.* **42,** 1359–1368.

Exner, M., and Herold, S. (2000). Kinetic and mechanistic studies of the peroxynitrite-mediated oxidation of oxymyoglobin and oxyhemoglobin. *Chem. Res. Toxicol.* **13,** 287–293.

Ferrer-Sueta, G., Vitturi, D., Batinic-Haberle, I., Fridovich, I., Goldstein, S., Czapski, G., and Radi, R. (2003). Reactions of manganese porphyrins with peroxynitrite and carbonate radical anion. *J. Biol. Chem.* **278,** 27432–27438.

Gerasimov, O. V., and Lymar, S. V. (1999). The yield of hydroxyl radical from the decomposition of peroxynitrous acid. *Inorg. Chem.* **38,** 4317–4321.

Goldstein, S., and Czapski, G. (1995a). Direct and indirect oxidations by peroxynitrite. *Inorg. Chem.* **34,** 4041–4048.

Goldstein, S., and Czapski, G. (1995b). The reaction of $^{\cdot}NO$ with $O_2^{\cdot-}$ and $HO_2^{\cdot}$: A pulse-radiolysis study. *Free Rad. Biol. Med.* **19,** 505–510.

Goldstein, S., and Czapski, G. (1996). Mechanism of the nitrosation of thiols and amines by oxygenated center dot NO solutions: The nature of the nitrosating intermediates. *J. Am. Chem. Soc.* **118,** 3419–3425.

Goldstein, S., and Czapski, G. (2000). Reactivity of peroxynitrite versus simultaneous generation of $^{\cdot}NO$ and $O_2^{\cdot-}$ toward NADH. *Chem. Res. Toxicol.* **13,** 736–741.

Goldstein, S., Lind, J., and Merényi, G. (2002). The reaction of $ONOO^-$ with carbonyls: Estimation of the half-lives of $ONOOC(O)O^-$ and $O_2NOOC(O)O^-$. *J. Chem. Soc. Dalton Trans.* 808–810.

Goldstein, S., Lind, J., and Merényi, G. (2004). The reaction of organic peroxyl radicals with NO2 and NO in aqueous solution: The intermediacy of organic peroxynitrate and peroxynitrite species. *J. Phys. Chem. A.* **108,** 1719–1725.

Goldstein, S., Lind, J., and Merényi, G. (2005). The chemistry of peroxynitrites as compared to peroxynitrates. *Chem. Rev.* **105,** 2457–2470.

Goldstein, S., Czapski, G., Lind, J., and Merényi, G. (1999). Effect of $^{\cdot}NO$ on the decomposition of peroxynitrite: Reaction of N_2O_3 with $ONOO^-$. *Chem. Res. Toxicol.* **12,** 132–136.

Goldstein, S., Czapski, G., Lind, J., and Merényi, G. (2000). Tyrosine nitration by simultaneous generation of ·NO and $O_2^{\cdot -}$ under physiological conditions. How the radicals do the job? *J. Biol. Chem.* **275,** 3031–3036.

Goldstein, S., Czapski, G., Lind, J., and Merényi, G. (2001a). Carbonate radical ion is the only observable intermediate in the reaction of peroxynitrite with CO_2. *Chem. Res. Toxicol.* **14,** 1273–1276.

Goldstein, S., Czapski, G., Lind, J., and Merényi, G. (2001b). Gibbs energy of formation of peroxynitrate-order restored. *Chem. Res. Toxicol.* **14,** 657–660.

Goldstein, S., Saha, A., Lymar, S. V., and Czapski, G. (1998). Oxidation of peroxynitrite by inorganic radicals: A pulse radiolysis study. *J. Am. Chem. Soc.* **120,** 5549–5554.

Gratzel, M., Taniguch, S., and Henglein, A. (1970). Study with pulse radiolysis of NO-oxidation and equilibrium $N_2O_3 = NO + NO_2$ in a water solution. *Berichte Bunsen-Gesellschaft Fur Physikalische Chemie* **74,** 488–492.

Gratzel, M., Henglein, A., Lilie, J., and Beck, G. (1969). Pulse radiolysis of some elementary oxidation-reduction processes of nitrite ion. *Berichte Der Bunsen-Gesellschaft Fur Physikalische Chemie* **73,** 646.

Greenacre, S. A., and Ischiropoulos, H. (2001). Tyrosine nitration: Localisation, quantification, consequences for protein function and signal transduction. *Free Rad. Res.* **34,** 541–581.

Gross, S. S., and Wolin, M. S. (1995). Nitric oxide: Pathophysiological mechanisms. *Annu. Rev. Physiol.* **57,** 737–769.

Harman, D. (1993). *In* "Free Radicals: From Basic Science to Medicine" (G. Poli, E. Albani, and M. O. Dizani, eds.) pp. 124–143. Birkhauser Verlag, Basel.

Hodges, G. R., and Ingold, K. U. (1999). Cage-escape of geminate radical pairs can produce peroxynitrate from peroxynitrite under a wide variety of experimental conditions. *J. Am. Chem. Soc.* **121,** 10695–10701.

Hodges, G. R., Marwaha, J., Paul, T., and Ingold, K. U. (2000). A novel procedure for generating both nitric oxide and superoxide *in situ* from chemical sources at any chosen mole of ratio. First application: tyrosine oxidation and a comparison with performed peroxynitrite. *Chem. Res. Toxicol.* **13,** 1287–1293.

Huie, R. E., and Padmaja, S. (1993). The reaction of NO with superoxide. *Free Radic. Res. Commun.* **18,** 195–199.

Ischiropoulos, H. (1998). Biological tyrosine nitration: A pathophysiological function of nitric oxide and reactive oxygen species. *Arch. Biochem. Biophys.* **356,** 1–11.

Jin, F. M., Leitich, J., and von Sonntag, C. (1993). The superoxide radical reacts with tyrosine-derived phenoxyl radicals by addition rather than by electron-transfer. *J. Chem. Soc. Perkin Trans.* **2,** 1583–1588.

Jourd'heuil, D., Miranda, K. M., Kim, S. M., Espey, M. G., Vodovotz, Y., Laroux, S., Mai, C. T., Miles, A. M., Grisham, M. B., and Wink, D. A. (1999). The oxidative and nitrosative chemistry of the nitric oxide/superoxide reaction in the presence of bicarbonate. *Arch. Bioch. Biophys.* **365,** 92–100.

Kharitonov, S. A., and Barnes, P. J. (2003). Nitric oxide, nitrotyrosine, and nitric oxide modulators in asthma and chronic obstructive pulmonary disease. *Curr. Allergy Asthma Rep.* **3,** 121–129.

Kirsch, M., Korth, H.-G., Wensing, A., Sustmann, R., and de Groot, H. (2003). Product formation and kinetic simulations in the pH range 1–14 account for a free-radical mechanism of peroxynitrite decomposition. *Arch. Biochem. Biophys.* **418,** 133–150.

Kissner, R., and Koppenol, W. H. (2002). Product distribution of peroxynitrite decay as a function of pH, temperature, and concentration. *J. Am. Chem. Soc.* **124,** 234–239.

Kissner, R., Nauser, T., Bugnon, P., Lye, P. G., and Koppenol, W. H. (1997). Formation and properties of peroxynitrite as studied by laser flash photolysis, high-pressure stopped-flow technique, and pulse radiolysis. *Chem. Res. Toxicol.* **10,** 1285–1292.

Kong, S.-K., Yim, M. B., Stadman, E. R., and Chock, P. B. (1996). *Proc. Natl. Acad. Sci. USA* **93,** 3377–3382.

Kubes, P., and McCafferty, D. M. (2000). Nitric oxide and intestinal inflammation. *Am. J. Med.* **109,** 150–158.

Lancaster, J. R., Jr. (2006). Nitroxidative, nitrosative, and nitrative stress: Kinetic predictions of reactive nitrogen species chemistry under biological conditions. *Chem. Res. Toxicol.* **19,** 1160–1174.

Lee, J. B., Hunt, J. A., and Groves, J. T. (1998). Manganese porphyrins as redox-coupled peroxynitrite reductases. *J. Am. Chem. Soc.* **120,** 6053–6061.

Logager, T., and Sehested, K. (1993a). Formation and decay of peroxynitric acid: A pulse-radiolysis study. *J. Phys. Chem.* **97,** 10047–10052.

Logager, T., and Sehested, K. (1993b). Formation and decay of peroxynitrous acid: A pulse-radiolysis study. *J. Phy. Chem.* **97,** 6664–6669.

Lymar, S. V., and Hurst, J. K. (1995). Rapid reaction between peroxynitrite ion and carbon-dioxide: Implications for biological-activity. *J. Am. Chem. Soc.* **117,** 8867–8868.

Lymar, S. V., Jiang, Q., and Hurst, J. K. (1996). Mechanism of carbon dioxide-catalyzed oxidation of tyrosine by peroxynitrite. *Biochemistry* **35,** 7855–7861.

Lymar, S. V., Khairutdinov, R. F., and Hurst, J. K. (2003). Hydroxyl radical formation by O-O bond homolysis in peroxynitrous acid. *Inorg. Chem.* **42,** 5259–5266.

Mahoney, L. R. (1970). Evidence for the formation of hydroxyl radicals in the isomerisation of pernitrous acid to nitric acid in aqueous solution. *J. Am. Chem. Soc.* **92,** 5262–5263.

Mallard, W. G., Ross, A. B., and Helman, W. P. (1998). "NIST Standard References Database 40, Version 3.0."

Masumoto, H., Kissner, R., Koppenol, W. H., and Sies, H. (1996). Kinetic study of the reaction of ebselen with peroxynitrite. *FEBS Lett* **398,** 179–182.

Merényi, G., and Lind, J. (1998). Free radical formation in the peroxynitrous acid (ONOOH)/peroxynitrite($ONOO^-$) system. *Chem. Res. Toxicol.* **11,** 243–248.

Merényi, G., Lind, J., Goldstein, S., and Czapski, G. (1999). Mechanism and thermochemistry of peroxynitrite decomposition in water. *J. Phys. Chem. A* **103,** 5685–5691.

Miles, A. M., Bohle, D. S., Glassbrenner, P. A., Hansert, B., Wink, D. A., and Grisham, M. B. (1996). Modulation of superoxide-dependent oxidation and hydroxylation reactions by nitric oxide. *J. Biol. Chem.* **271,** 40–47.

Moncada, S., Palmer, R. M. J., and Higgs, E. A. (1991). Nitric oxide: Physiology, pathophysiology, and pharmacology. *Pharmacol. Rev.* **43,** 109–142.

Nauser, T., and Koppenol, W. H. (2002). The rate constant of the reaction of superoxide with nitrogen monoxide: Approaching the diffusion limit. *J. Phys. Chem. A* **106,** 4084–4086.

Niles, J. C., Wishnok, J. S., and Tannenbaum, S. R. (2006). Peroxynitrite-induced oxidation and nitration products of guanine and 8-oxoguanine: Structures and mechanisms of product formation. *Nitric Oxide* **14,** 109–121.

Oberely, L. W. (1982). "Superoxidedismutase," Vols. 1 and 2. CRC Press, Boca Raton, FL.

Patel, R. P., McAndrew, J., Sellak, H., White, C. R., Jo, H. J., Freeman, B. A., and Darley-Usmar, V. M. (1999). Biological aspects of reactive nitrogen species. *Biochem. Biophys. Acta* **1411,** 385–400.

Pfeiffer, S., Schmidt, K., and Mayer, B. (2000). Dityrosine formation outcompetes tyrosine nitration at low steady-state concentrations of peroxynitrite: Implications for tyrosine modification by nitric oxide/superoxide *in vivo*. *J. Biol. Chem.* **275,** 6346–6352.

Pfeiffer, S., Gorren, A. C. F., Schmidt, K., Werner, E. R., Hansert, B., Bohle, D. S., and Mayer, B. (1997). Metabolic fate of peroxynitrite in aqueous solution. *J. Biol. Chem.* **272,** 3465–3470.

Prutz, W. A., Monig, H., Butler, J., and Land, E. J. (1985). Reactions of nitrogen-dioxide in aqueous model systems: Oxidation of tyrosine units in peptides and proteins. *Arch. Biochem. Biophys.* **243,** 125–134.

Pryor, W. A., and Squadrito, G. L. (1995). The chemistry of peroxynitrite: A product from the reaction of nitric oxide with superoxide. *Am. J. Physiol.* **268,** L699–L722.

Quijano, C., Romero, N., and Radi, R. (2005). Tyrosine nitration by superoxide and nitric oxide fluxes in biological systems: Modeling the impact of superoxide dismutase and nitric oxide diffusion. *Free Rad. Biol. Med.* **39,** 728–741.

Radi, R. (2004). Nitric oxide, oxidants, and protein tyrosine nitration. *Proc. Natl. Acad. Sci. USA* **101,** 4003–4008.

Radi, R., Beckman, J. S., Bush, K. M., and Freeman, B. A. (1991). *J. Biol. Chem.* **266,** 4244–4250.

Reiter, D. C., Teng, R.-J., and Beckman, J. S. (2000). Superoxide radical reacts with nitric oxide to nitrate tyrosine at physiological pH via peroxynitrite. *J. Biol. Chem.* **275,** 32460–32466.

Rotelio, G. (1986). "Superoxide and superoxide dismutase in chemistry, biology and medicine." Elsevier Science, Amsterdam.

Santos, C. X. C., Boinini, G. M., and Augusto, O. (2000). Role of the carbonate radical anion in tyrosine nitration and hydroxylation by peroxynitrite. *Arch. Bioch. Biophys.* **377,** 146–152.

Sawa, T., Akaike, T., and Maeda, H. (2000). Tyrosine nitration by peroxynitrite formed from nitric oxide and superoxide generated by xanthine oxidase. *J. Biol. Chem.* **275,** 32467–32474.

Schrammel, A., Gorren, A. C. F., Schmidt, K., Pfeiffer, S., and Mayer, B. (2003). S-nitrosation of glutathione by nitric oxide, peroxynitrite, and $NO/O_2^{\cdot -}$. *Free Rad. Biol. Med.* **34,** 1078–1088.

Squadrito, G. L., and Pryor, W. A. (2002). Mapping the reaction of peroxynitrite with CO_2: Energetics, reactive species, and biological implications. *Chem. Res. Toxicol.* **15,** 885–895.

Thomas, D. D., Ridnour, L. A., Espey, M. G., Donzelli, S., Ambs, S., Hussain, S. P., Harris, C. C., DeGraff, W., Roberts, D. D., Mitchell, J. B., and Wink, D. A. (2006). Superoxide fluxes limit nitric oxide-induced signaling. *J. Biol. Chem.* **281,** 25984–25993.

Thomson, L., Trujillo, M., Telleri, R., and Radi, R. (1995). *Arch. Biochem. Biophys.* **319,** 491–497.

Treinin, A., and Hayon, E. (1970). Absorption spectra and reaction kinetics of NO_2, N_2O_3, and N_2O_4 in aqueous solution. *J. Am. Chem. Soc.* **92,** 5821.

Trostchansky, A., Batthyany, C., Botti, H., Radi, R., Denicola, A., and Rubbo, H. (2001). Formation of lipid-protein adducts in low-density lipoprotein by fluxes of peroxynitrite and its inhibition by nitric oxide. *Arch. Biochem. Biophys.* **395,** 225–232.

Turko, I. V., and Murad, F. (2002). Protein nitration in cardiovascular diseases. *Pharmacol. Rev.* **54,** 619–634.

Van der Vliet, A., Eiserich, J. P., O'Neill, C. A., Halliwell, B., and Cross, C. E. (1995). *Arch. Biochem. Biophys.* **319,** 341–349.

von Sonntag, C., and Schuchmann, H. P. (1997). Peroxyl radicals in aqueous solution. *In* "Peroxyl Radicals" (Z. B. Alfassi, ed.). John Wiley & Sons, Chichester, UK.

Wink, D. A., Cook, J. A., Kim, S. Y., Vodovotz, Y., Pacelli, R., Krishna, M. C., Russo, A., Mitchell, J. B., Jourd'heuil, D., Miles, A. M., and Grisham, M. B. (1997). Superoxide modulates the oxidation and nitrosation of thiols by nitric oxide-derived reactive intermediates: Chemical aspects involved in the balance between oxidative and nitrosative stress. *J. Biol. Chem.* **272,** 11147–11151.

CHAPTER FIVE

Nitric Oxide Selective Electrodes

Ian R. Davies* *and* Xueji Zhang†

Contents

Abstract

Since nitric oxide (NO) was identified as the endothelial-derived relaxing factor in the late 1980s, many approaches have attempted to provide an adequate means for measuring physiological levels of NO. Although several techniques have been successful in achieving this aim, the electrochemical method has

* World Precision Instruments Limited, Aston, United Kingdom
† World Precision Instruments Inc., Sarasota, Florida

Reprinted from Zhang *et al.* Electrochemical Sensors, Biosensors and Their Biomedical Applications, "Nitric Oxide (NO) electrochemical sensors," pages 1–29, Elsevier, 2008.

Methods in Enzymology, Volume 436
ISSN 0076-6879, DOI: 10.1016/S0076-6879(08)36005-4

proved the only technique that can reliably measure physiological levels of NO *in vitro*, *in vivo*, and in real time.

We describe here the development of electrochemical sensors for NO, including the fabrication of sensors, the detection principle, calibration, detection limits, selectivity, and response time. Furthermore, we look at the many experimental applications where NO selective electrodes have been successfully used.

1. Significance of NO in Life Science

NO is reported to have been first prepared by the Belgian scientist Jan Baptist van Helmont in about 1620. The chemical properties of NO were first characterized by Joseph Priestly in 1772 (Schofield, 1966). However, until the mid-1980s, NO was regarded as an atmospheric pollutant and bacterial metabolite. NO is a hydrophobic, highly labile free radical that is catalytically produced in biological systems from the reduction of L-arginine by NO synthase (NOS) to form L-citrulline, which produces NO in the process. In biological systems, NO has long been known to play various roles in physiology, pathology, and pharmacology (Moncada *et al.*, 1991). In 1987 NO was identified as responsible for the physiological actions of endothelium-derived relaxing factor (EDRF) (Ignarro *et al.*, 1987). Since that discovery, NO has been shown to be involved in numerous biological processes, such as vasodilatation and molecular messaging (Ignarro *et al.*, 1987), penile erection (Ignarro, 1990), neurotransmission (Bredt *et al.*, 1991; Feldman *et al.*, 1993), inhibition of platelet aggregation (Radomski *et al.*, 1990), blood pressure regulation (Moncada *et al.*, 1988), immune response (Akaike *et al.*, 1993), and as a mediator in a wide range of both antitumor and antimicrobial activities (Anbar, 1994; Hibbs *et al.*, 1988). In addition, NO has been implicated in a number of diseases, including diabetes (Schmidt *et al.*, 1992), Parkinson's disease, and Alzheimer's disease (Moncada *et al.*, 1989). The importance of NO was confirmed in 1992 when *Science* magazine declared NO "Molecule of the Year," and in 1998, when F. Furchgott, Louis J. Ignarro, and Ferid Murad were awarded the Nobel Prize in Physiology or Medicine for unraveling the complex nature of this simple molecule. Despite the obvious importance of NO in so many biological processes, less than 10% of the thousands of scientific publications over the last decade dedicated to the field of NO research involve its direct measurement.

2. Methods of NO Measurement in Physiology

As stated previously, NO plays a significant role in a variety of biological processes, where its spatial and temporal concentration is of extreme importance. However, the measurement of NO is quite difficult because of its short half-life (approximately 5 s) and high reactivity with other biological components such as superoxide, oxygen, thiols, and others. To date, several techniques have been developed to measure NO, including chemiluminescence (Beckman and Congert, 1995; Robinson *et al.*, 1999), the Griess method (Green *et al.*, 1982), paramagnetic resonance spectrometry (Wennmalm *et al.*, 1990), paramagnetic resonance spectrophotometry (Bredt and Snyder, 1989), and bioassay (Wallace and Woodman, 1995). Each of these techniques has certain benefits associated with it but suffers from poor sensitivity and the need for complex and often-expensive experimental apparatus. In addition, these NO-sensing techniques are limited when it comes to continuous monitoring of NO concentration in real-time and, most important, *in vivo*.

3. Advantages of Electrochemical Sensors for Determination of NO

To date, electrochemical (amperometric) detection of NO is the only available technique sensitive enough to detect relevant concentrations of NO in real time and *in vivo* and suffers minimally from potential interfering species such as nitrite, nitrate, dopamine, ascorbate, and L-arginine. Also, because electrodes can be made on the micro- and nanoscales, these techniques also have the benefit of being able to measure NO concentrations in living systems without any significant effects from electrode insertion.

The first amperometric NO electrode used for direct measurement was described in 1990 (Shibuki, 1990). In 1992 the first commercial NO-sensor system was developed. Over subsequent years a range of highly specialized and sensitive NO electrodes have been developed that offer NO detection limits ranging from less than 1 n*M* up to 100 μ*M* (Zhang and Broderick, 2000). Most recently, a unique range of high-sensitivity NO sensors based on a membrane-coated activated carbon microelectrode with diameters ranging from 200 μm to 100 nm have been developed. These electrodes exhibit superior performance during NO measurement and feature a detection limit of less than 0.5 n*M* NO. Very recently, a new NO electrode has been developed with a detection limit of 50 p*M* (Zhang, 2007).

4. Principles of Determination of NO by Electrochemical Sensors

NO can be oxidized or reduced on an electrode surface. Because the reduction potential of NO is close to that of oxygen, which can lead to interference by oxygen when trying to measure NO, oxidation of NO is generally used to measure NO. NO oxidation on solid electrodes proceeds via an EC-mechanism electrochemical reaction (Trevin *et al.*, 1996) followed by a chemical reaction (Malinski *et al.*, 1993b). First, 1-electron transfers from the NO molecule to the electrode, resulting in the formation of a cation:

$$NO - e^- \rightarrow NO^+ \tag{5.1}$$

NO^+ is immediately and irreversibly converted into nitrite in the presence of OH^- because it is a relatively strong Lewis acid:

$$NO^+ + OH^- \rightarrow HNO_2 \tag{5.2}$$

According to Eq. 2, the rate of the chemical reaction increases with an increase in pH.

Among the several electrochemical techniques that have been shown to be useful for the measurement of NO, the most popular is amperometry. This technique uses the model set forth by Clark and Lyons (1962) for continuous gas monitoring during cardiovascular surgery. Generally, this technique involves applying a fixed (poise voltage) potential to a working electrode, versus a reference electrode, and monitoring the very low redox current produced (e.g., pA's) by the oxidation of NO. This technique has proved very useful for NO detection due to its fast response time, which is less than a few seconds, and its high sensitivity. As a result, it is possible to monitor changes in NO concentration on biologically relevant time scales and concentrations, which are typically in the n*M* range. A multitude of other electrochemical techniques have been used to detect NO, including differential pulse voltammetry (DPV), differential normal pulse voltammetry (DNPV), linear scanning voltammetry (LSV), square wave voltammetry (SWV), and fast scan voltammetry (FSV). These methods typically employ a classical three-electrode configuration consisting of a working electrode, a reference electrode, and a counter electrode. Scanning techniques, with the exception of FSV, require approximately 10 s for the voltammogram to be recorded, which precludes its use in most NO research applications. Moreover, because scanning voltammetry-based NO instrumentation is not currently available, NO researchers typically prefer to use the two-electrode amperometric technique. Because the amperometric method is so widely used, the subsequent discussion will focus on these techniques.

5. Fabrication of Electrodes for NO Determination

5.1. Clark-type NO electrodes

The first described electrochemical NO sensor was based on a classical Clark electrode design, where NO was directly oxidized on the working electrode surface (Shibuki, 1990). The NO sensor was composed of a fine platinum wire and a separate silver wire, which were then inserted into a glass micropipette. The micropipette was then filled with 30 m*M* NaCl and 0.3 m*M* HCl and sealed at the tip with a chloroprene rubber membrane. The platinum (working) electrode was positioned close to the surface of the membrane. The silver wire was then used as the reference/counter electrode. Although such electrodes could be used to measure NO, their inherent low sensitivity, narrow linear concentration measurement range, and fragility rendered them unsuitable for most research applications. In 1992, utilizing the Clark-type design, the first commercial electrochemical NO sensor and NO meter were produced. The NO sensor consisted of a platinum-wire disc working electrode and an Ag/AgCl reference electrode. Both electrodes were encased within a protective faraday-shielded stainless-steel sleeve. The tip of the sleeve was covered with a NO-selective membrane and the sleeve itself contained an electrolyte. The rugged design of this sensor made it extremely convenient in many research applications, and the sensor became widely used and established in numerous NO-measurement research applications. The basic design of this type of NO sensor is illustrated in Fig. 5.1.

In principle the NO sensor works as follows. The sensor is immersed in a solution containing NO and a positive potential of ≈860 mV (vs. Ag/AgCl reference electrode) is applied. NO diffuses across the gas-permeable, NO-selective membrane and is oxidized at the working electrode surface, producing a redox current. This oxidation proceeds via an electrochemical reaction followed by a chemical reaction. The electrochemical reaction is a 1-electron transfer from the NO molecule to the electrode, resulting in the formation of the nitrosonium cation:

$$NO - e^- \rightarrow NO^+ \tag{5.3}$$

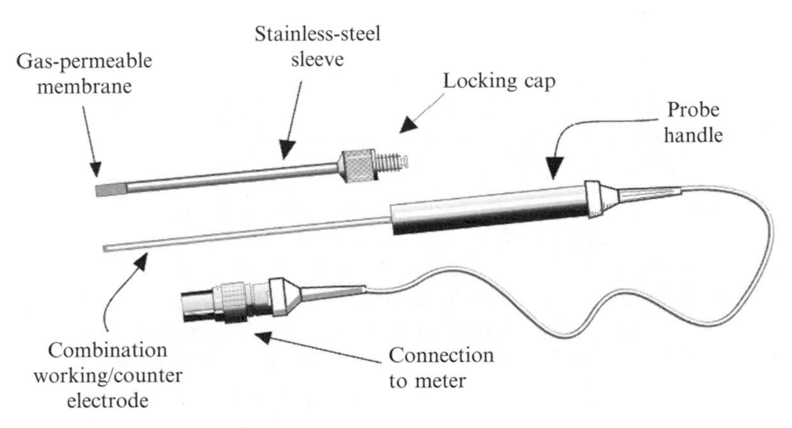

Figure 5.1 Illustration of WPI's ISO-NOP NO sensor. Reprinted with permission from Frontiers in Bioscience. (Reprinted from Zang, 2008.)

NO^+ is a relatively strong Lewis acid and in the presence of OH^- it is converted into nitrite (NO_2^-):

$$NO^+ + OH^- \rightarrow HNO_2 \rightarrow H^+ + NO_2^- \qquad (5.4)$$

Nitrite can then be further oxidized into nitrate. The amount of NO oxidized is thus proportional to the current flow between the working and the reference electrodes, which is measured by an NO meter.

The amount of redox current that is typically generated by the oxidation of NO in biological systems is extremely small, typically on the order of 1 to 10 pA. As a result of these extremely small currents, the design of an amperometric-based electrode NO-detection system requires very sensitive electronics and ultra-low noise-amplification circuitry. These measurement limitations have been overcome with the commercial development of low-noise and isolated-circuit electronic devices. These devices employ a unique electrically isolated low-noise circuit that permits measurement of redox currents as small as 0.1 pA. The design of the instrument also allows for measurement of NO to be performed without special electrical screening, such as a faraday cage. Although the Clark-type electrode has enjoyed great success for NO detection and, recently, NO reaction kinetics (Liu *et al.*, 2005), there have been other significant advances for NO detection that deserve mention.

5.2. Modified carbon-fiber NO microelectrodes

Surface-modified NO sensors incorporate an electrode surface that has been modified or treated in some way so as to increase the selectivity of the sensor for NO and to promote catalytic oxidation of NO. An early example of such a sensor was presented by Malinski and Taha (1992). In their publication an approximate 500-nm-diameter carbon-fiber electrode was coated with tetrakis(3-methoxy-4-hydroxyphenyl) porphyrin, via oxidative polymerization, and Nafion. This electrode was shown to have a detection limit of approximately 10 n*M* for NO and great selectivity against common interferences. However, it has been shown that this electrode suffers severe interference from H_2O_2 (Nagase *et al.*, 1997). Following this publication, other studies showed that pyrrole-functionalized porphyrins, containing metals such as Ni, Pd, and Mn, can be immobilized on carbon microelectrode surfaces via oxidative polymerization and can be used for NO detection (Diab and Schuhmann, 2001; Diab *et al.*, 2003). Other researchers have shown that carbon fibers coated with a variety of porphyrins, such as iron porphyrin (Hayon *et al.*, 1994) and cobalt porphyrin (Bedioui *et al.*, 1996; Brunet *et al.*, 2000; Kashevskii *et al.*, 2002; Malinski *et al.*, 1993b; Malinski and Taha, 1992), are also effective for NO detection. Although metal porphyrin–coated electrodes were successfully used to some extent for various applications (Malinski *et al.*, 1993a,b; Zhang *et al.*, 1995), subsequent studies have shown that carbon fibers modified with unmetallated porphyrins as well as bare carbon fibers can detect NO with similar sensitivity (Bedioui *et al.*, 1994; Lantoine *et al.*, 1995). The sensitivity and selectivity of porphyrinic NO sensors vary significantly from electrode to electrode and depend not only on the potential at which NO oxidizes but also on the effects of axial ligation to the central metal in the porphyrin, the modification/treatment procedure, and other experimental variations. Furthermore, because the surface of the electrode remains in direct contact with the measurement medium, a variety of biological species have been shown to interfere (i.e., give false responses) with the NO measurement. Adding a Nafion layer to the porphyrin-coated fibers could minimize these interferences. Other practical problems, such as easy porphyrin removal and degradation, have limited the usefulness of porphyrin for most applications (Yokoyama *et al.*, 1995). Phthalocyanines, with a similar structure to porphyrins, containing metals such as Fe, Ni, and Co, have also been used to modify electrode surfaces for NO sensing (Kim *et al.*, 2005; Vilakazi and Nyokong, 2001). Phthalocyanine-modified electrodes have comparable detection limits and selectivity to porphyrin-modified electrodes, with the added benefit of being more stable to degradation.

5.3. Integrated NO microelectrodes

During the mid- to late 1990s, a new range of combination NO sensors with tip diameters between 7 to 200 μm were developed (Zhang *et al.*, 2001). These sensors combine a carbon-fiber working electrode with a separate integrated Ag/AgCl reference electrode. The resulting combination sensor is then coated with a proprietary gas-permeable, NO-selective membrane. A high-performance faraday-shielded layer is then added to the sensor to minimize susceptibility to environmental noise. This electrode is then operated exactly as outlined previously for the Clark-type NO sensors. The use of these proprietary diffusion membranes and the novel design allows for NO measurement in small volumes and confined spaces, with great selectivity against a wide range of interferences, such as ascorbic acid, nitrite, and dopamine. This sensor design was elaborated upon by creating *L*-shaped sensors designed specifically for tissue bath studies. The design was further elaborated upon by creating flexible, virtually unbreakable NO sensors designed specifically for use in measuring NO concentrations in arteries and microvessels. This electrode combines a Pt/Ir wire with a separate integrated Ag/AgCl reference electrode (Dickson *et al.*, 2004). The resulting combination sensor is again coated with a proprietary gas-permeable,

NO-selective membrane and a high-performance faraday-shielded layer and operated as described previously.

Fairly recently, a novel combination NO nanosensor, which had a tip diameter of just 100 nm, was developed (Zhang *et al.*, 2002a). The design of this sensor can be seen in Fig. 5.2. This sensor was constructed using a 7-μm carbon fiber that was etched with an Ar ion beam to result in fibers with diameters in the 100-nm range. These 100-nm fibers were then used to construct combination electrodes, as described previously. These sensors are capable of making NO measurements at the cellular level. Later in 2002, a unique "microchip" combination sensor was developed (Zhang *et al.*, 2002b). This sensor is quite unique in its design, but its operation principle is the same as that of all other combination electrodes presented in this section. Briefly, a 5000-Å thick carbon film was deposited on a Si wafer by RF sputtering. This carbon layer was then covered with an insulation layer. Carbon electrodes (2 μm diameter) were then exposed by a dry etching process in a 50 $\times$ 50 array on the wafer surface, where each carbon electrode is 20 μm away from its nearest neighbor. An Ag/AgCl reference electrode and a Pt counter electrode were added to the wafer surface. The surface was then modified with a proprietary gas-permeable, NO-selective membrane and was ready for use. The resulting sensor exhibited an extremely low NO detection limit on the order of 300 p*M*; great selectivity against ascorbic acid, nitrite, and dopamine; and a superior response time.

5.4. Other NO electrodes

Various other types of carbon-fiber NO sensors that utilize a variety of different coatings have been described. Coatings used for these sensors include conducting and nonconducting polymers (Fabre *et al.*, 1997; Ferreira *et al.*, 2005; Friedemann *et al.*, 1996; Kato *et al.*, 2005; Park *et al.*, 1998), multiple membranes (Ichimori *et al.*, 1994; Pontie *et al.*, 1999), ruthenium (Allen *et al.*, 2000), iridium and palladium (Xian *et al.*, 1999),

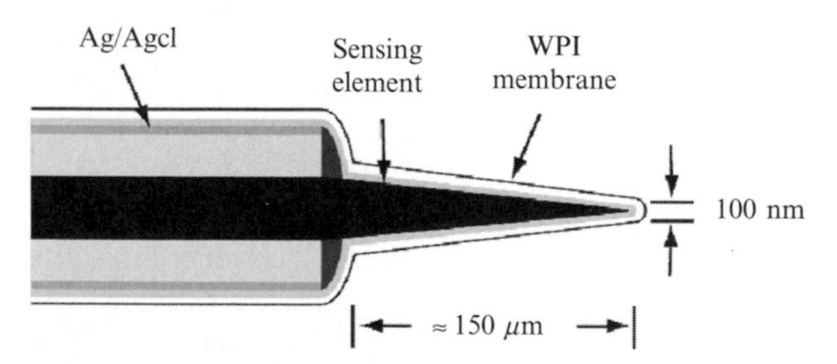

Figure 5.2 Illustration of a 100-nm combination NO nanosensor. Reprinted with permission from Elsevier Publishing (Zhang *et al.*, 2002a).

heated-denatured cytochrome c (Haruyama *et al.*, 1998), Nafion-Co (II)-1,10-phenanthroline (He and Mo, 2000), ferrioxamine (Smith and Thorp, 1998), a microcoaxial microelectrode for *in vivo* NO measurement (Kitamura *et al.*, 2000), siloxane polymer (Mizutani *et al.*, 2001), Nafion and cellulose (Katrlik and Zalesakova, 2002), Hb/phosphatidylcholine films (Fan *et al.*, 2002), hemoglobin-DNA film (Fan *et al.*, 2000), and ionic polymers and α-cyclodextrin (Kitajima *et al.*, 2001). Recently, researchers reported the coating of a 10-μm Pt disc electrode with a cross-linked sol-gel Langmuir-Blodgett film of siloxane polymer to render the electrode permselective for NO (Kato *et al.*, 2005). Others recently reported the addition of a sol-gel film to a Pt electrode for NO detection (Shin *et al.*, 2005). Still other researchers described an improved planar amperometric NO sensor based on a platinized anode (Lee *et al.*, 2004a,b) and its application for measurement of NO release from NO donors. Kroning *et al.* (2004) explored using a myoglobin-clay modified electrode for NO detection. Kamei *et al.* (2004) fabricated a NO-sensing device for drug screening using a polyelectrolyte film. Indium hexacyanoferrate film-modified electrodes were used for NO detection by Casero *et al.* (2003). Pailleret *et al.* (2003) developed a device for both *in situ* formation and scanning electrochemical microscopy–assisted positioning of NO sensors above human umbilical vein endothelial cells for the detection of NO release.

Unfortunately, despite the novelty of these approaches, none of the sensors has withstood the test of time, mostly due to various practical difficulties and/or poor sensitivity or selectivity. Furthermore, the lack of any published data describing the use of these sensors in any biological research applications limits any conclusions that can be made on their individual performance.

6. Calibration of NO Electrodes

Routine calibration of an NO sensor is essential in order to ensure accurate experimental results. One of three calibration techniques is generally used, depending on the sensor type, and will be described in the following section. Each of these methods has already been the subject of several reviews (Archer, 1993; Kiechle and Malinski, 1993; Magazine, 1995; Malinski *et al.*, 1993b) and will therefore only be summarized herein. NO sensors are typically sensitive to temperature; therefore, calibration is usually best performed at the temperature at which the measurements will be made.

6.1. Calibration using an NO standard solution

This calibration technique involves the production of an NO stock solution using a supply of compressed NO gas. One advantage of this method is that it allows an NO sensor to be calibrated in an environment similar to that in which the experimental measurements will be made. However, the major drawback is that it requires a source of compressed or other NO gas, and because NO gas is toxic, the whole procedure must be performed in a fume hood. Methods for preparing NO are described in this volume by Aga and Hughes (Chapter).

This method is summarized as follows. A Vacutainer is first filled with 10 mL deionized water and agitated ultrasonically for 10 min. Purified argon is then passed through an alkaline pyrogallol solution (5% w/v) to scavenge any traces of oxygen before being purged through a deionized water solution for 30 min. NO stock solution is prepared by bubbling compressed NO gas through the argon-treated water for 30 min. The NO gas is first purified by passing it through 5% pyrogallol solution in saturated potassium hydroxide (to remove oxygen) and then 10% (w/v) potassium hydroxide to remove all other nitrogen oxides. The resultant concentration of saturated NO in the water is 2 m*M* at 22°. This can be confirmed further by a photometric method based on the conversion of oxyhemoglobin to methemoglobin in the presence of NO (Kelm and Schrader, 1990). NO standard solutions can then be freshly prepared by serial dilution of saturated NO solution with oxygen-free deionized water before each experiment.

6.2. Calibration based on decomposition of SNAP

For this method, S-nitroso-N-acetyl-D, L-penicillamine (SNAP, FW = 220.3) is decomposed to NO in solution in the presence of a Cu (I) catalyst (Zhang *et al.*, 2000). The resultant NO can then be used to calibrate the sensor. The reaction proceeds in accordance with the following reaction:

$$2\,\mathrm{RSNO} \rightarrow 2\,\mathrm{NO} + \mathrm{RS} - \mathrm{SR} \qquad (5.5)$$

The stoichiometry of the reaction dictates that the final generated NO concentration will be equal to the concentration of SNAP in the solution. The method can be summarized as follows. Saturated cuprous chloride solution is first prepared by adding 150 mg CuCl to 500 ml distilled water. This solution is then deoxygenated by purging with pure nitrogen or argon gas for 15 min. The final saturated CuCl solution will have a concentration of approximately 2.4 m*M* at room temperature. The solution is light sensitive and must therefore be kept in the dark prior to use.

The SNAP solution is then prepared separately as follows. EDTA (5 mg) is dissolved in 250 mL of DI water (HPLC grade, Sigma or > 18 MΩ) and then adjusted to pH 9.0 using 0.1 *M* NaOH. The solution is then deoxygenated using the method described previously. Then 5.6 mg of SNAP is added to the solution to result in a SNAP concentration of approximately 0.1 m*M*. SNAP solution is also extremely sensitive to light and temperature and must therefore be stored refrigerated and in the dark until it is used. Under these conditions, and in the presence of the chelating reagent (EDTA), the decomposition of SNAP occurs extremely slowly. This allows the solution to be used to calibrate NO probes throughout the day. In practice, calibration is performed by placing an NO sensor into a vial containing a measured amount of the CuCl solution, and known volumes of the SNAP stock solution are then injected into the vial. The final concentration of NO can be calculated using dilution factors.

The concentration of SNAP in the stock solution is calculated as follows:

$$[C] = [A \times W/(M \times V)] \times 1000, \quad (5.6)$$

where C = concentration of SNAP (μM), A = purity of SNAP, W = weight of SNAP (mg), M = MW of SNAP (mg/mmole), and V = volume of the solution in liters (L).

If SNAP purity, for example, is 98.5%, then the concentration of SNAP is calculated as follows:

$$[C] = [98.5\% \times 5.6/(220.3 \times 0.25)] \times 1000 = 100.1\mu M \quad (5.7)$$

Figure 5.3 shows a typical calibration curve generated using an NO microsensor and the SNAP method described.

6.3. Calibration based on chemical generation of NO

This method of calibration generates known concentrations of NO from the reaction of nitrite with iodide in acid according to the following equation:

$$2\ KNO_2 + 2\ KI + 2\ H_2SO_4 \rightarrow 2\ NO + I_2 + 2\ H_2O + 2\ K_2SO_4 \quad (5.8)$$

The NO generated from the reaction is then used to calibrate the sensor. Because the conversion of NO_2^- to NO is stoichiometric (and KI and H_2SO_4 are present in excess), the final concentration of NO generated is equal to the concentration of KNO_2 in the solution. Thus, the

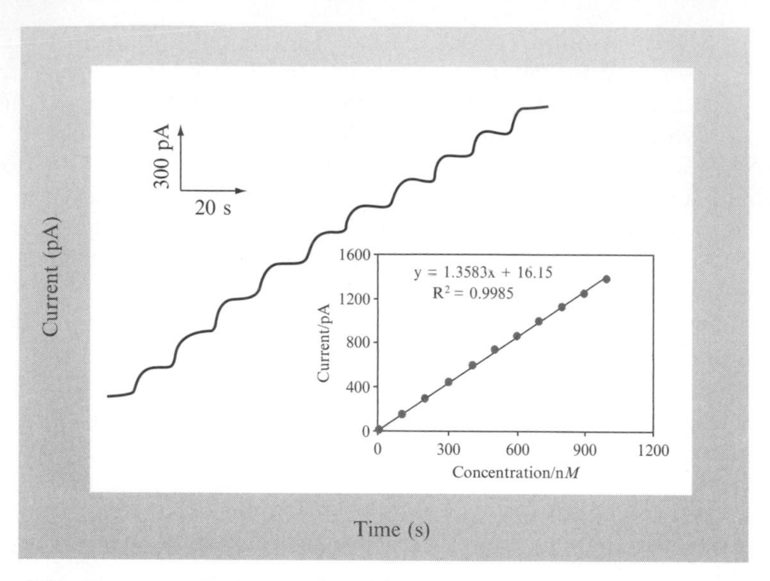

Figure 5.3 Response of a 7-μm carbon fiber NO sensor to increasing concentration of NO produced by introduction of SNAP to a solution of CuCl. Reprinted with permission from Wiley Publishing (Zhang *et al.*, 2000).

concentration of NO can be easily calculated by simple dilution factors. Experiments have demonstrated that NO generated from this reaction will persist sufficiently long enough to calibrate an NO sensor. However, because this technique involves the use of a strong acid, which can damage the delicate selective membranes of most NO microsensors, it is only suitable for use with Clark-type stainless steel–encased NO sensors. Figure 5.4 illustrates the amperometric response of a 2-mm sensor following exposure to increasing concentrations of NO. The sensor responds rapidly to NO and reaches steady-state current within a few seconds. The data generated from Fig. 5.4 are then used to construct a final calibration curve (Fig. 5.4, inset). The calibration curve illustrates the good linearity that exists between NO concentration and the current produced by its oxidation.

7. Characterization of NO Electrodes

NO sensors can be characterized in terms of sensitivity, detection limit, selectivity, response time, stability, linear range, lifetime, reproducibility, and biocompatibility. Sensor stability is important especially when measuring low NO concentrations. For example, when measuring low NO

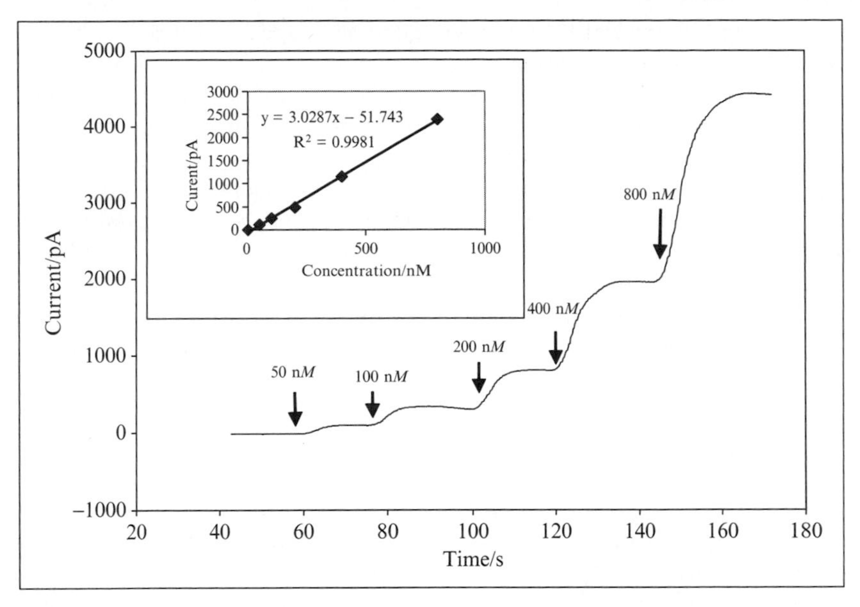

Figure 5.4 Response of WPI's ISO-NOP NO sensor to increasing concentration of chemically generated NO; inset shows the resulting calibration curve. Reprinted with permission from Frontiers in Bioscience.

concentrations it must be the case that the noise is lower than the anticipated current change upon NO addition. It is important to know the linear range of an NO sensor before performing any measurement. For example, the concentration of NO that is being measured must be in the linear range of the sensor in order for the measurement to be accurate. Typical commercially available NO sensors have a wide linear range of 1 n*M* to 10 μM. It is important to know sensor lifetime and reproducibility when using a sensor; these can be determined by frequent calibration, as described previously. Biocompatibility is also extremely important when making NO measurements *in vivo*. Frequently, NO sensors are impaled into living tissue so the reactivity of the tissue toward the NO-selective membrane must be minimal and cause a minimal amount of irritation in the tissue. In most applications, detection limit, sensitivity, selectivity, and response time are usually the most important requirements and will be described herein in further detail.

7.1. Sensitivity and detection limit

The sensitivity of an NO sensor depends largely on the reactive surface area of the sensor and the electrode materials used in the design and can range from 0.03 to 100 pA/n*M* NO. Generally speaking, this sensitivity is directly proportional to the electrode size and surface specifications, where an electrode with a small surface area will generally have a lower sensitivity than one with a larger surface area. Although a sensor's sensitivity is clearly important, its detection limit is often more important to the investigator. High sensitivity of a sensor does not necessarily equate to a low detection limit. For example, a highly sensitive NO sensor may have a high background-noise level, which may not be a problem at a high NO concentration. However, at lower NO concentrations, measurement can be hindered by excessive noise. Accordingly, in evaluating the performance of an NO sensor, the ultimate detection limit is usually more critical than sensitivity. Fortunately, most commercial NO sensors can detect NO at levels of 1 n*M* or less and are therefore well suited for the majority of research applications.

7.2. Selectivity

An NO sensor is practically useless unless it is immune to interference from other species likely to be present in the measurement environment. Selectivity is usually controlled by both the voltage applied between the working and reference electrode (poise voltage) and the selective membrane used to coat the sensor. Many species present in a biological matrix are easily oxidized at the poise voltage employed to detect NO (i.e., +860 mV vs Ag/AgCl). For example, monoamines such as dopamine (DA), 5-hydroxytryptamine (5-HT), and norepinephrine (NE), as well as their primary metabolites, can be oxidized at 0.3 V (and higher) versus Ag/AgCl. Ascorbic acid can be oxidized at 0.4 V (and higher). A Clark-type NO sensor is covered with a gas-permeable membrane; thus, the selectivity of such sensors in biological samples is extremely good. With other types of NO sensors, selectivity is usually achieved by coating the sensor surface with Nafion and other gas-permeable membranes. Nafion is widely used to eliminate interference caused by anions, such as ascorbate and nitrite, during measurement of catecholamine species. When used for NO detection, the negatively charged Nafion layer can stabilize NO^+ formed upon the oxidation of NO and can prevent a complicated pattern of reactions that could lead to the formation of nitrite and nitrate. However, the main drawback of Nafion is that it does not eliminate interference from cationic molecules such as dopamine, serotonin, epinephrine, and other catecholamines. Consequently, selectivity of the conventional Nafion-coated NO sensors is very poor. Nafion-coated NO sensors also exhibit other undesirable characteristics including unstable background current, continuous drift in the baseline, and extended polarization requirements. These problems significantly limit the use of Nafion-coated carbon-fiber electrodes for measurement of NO. During the late 1990s a unique multilayered proprietary membrane configuration was developed. NO sensors coated with this membrane exhibited increased selectivity and sensitivity for NO; moreover, they were shown to be immune from interference caused from a wide range of potentially interfering species (Zhang *et al.*, 2001).

7.3. Response time

Response and recovery times of NO sensors are extremely important for their use *in vivo*. Theoretically, because the rate of mass transport at a microelectrode is very high, it should have a response time on the order of μs; however, the addition of NO-selective membranes to the electrode surface decreases this response time significantly. With a membrane present the response time is then dependent on the diffusion rate of NO across the membrane, which is highly dependent on the nature of the membrane as well as its thickness. The response time depends not only on the electrode being used but also on the electronics being used to read out the current. For example, because the current being read out is typically on the order of pA's, an electronic filter is usually applied, which also slows the system response. Because the half-life of NO is from a few seconds to minutes in biological systems, a sensor response on the order of 3 to 4 s will work fine. Figure 5.5 is a typical response of a NO microsensor to addition of NO in PBS solution. It shows the 90% response of the sensor to a step from 0 to 100 n*M* NO and subsequently injection of NO scavenger 1 micromolar of oxyhemoglobin. As can be seen, the response times to both additions of NO and oxyhemoglobin are less than 4 s. This indicates that even with a heavy filter, the response is within few seconds, which is within the time frame required for measuring NO. To decrease the

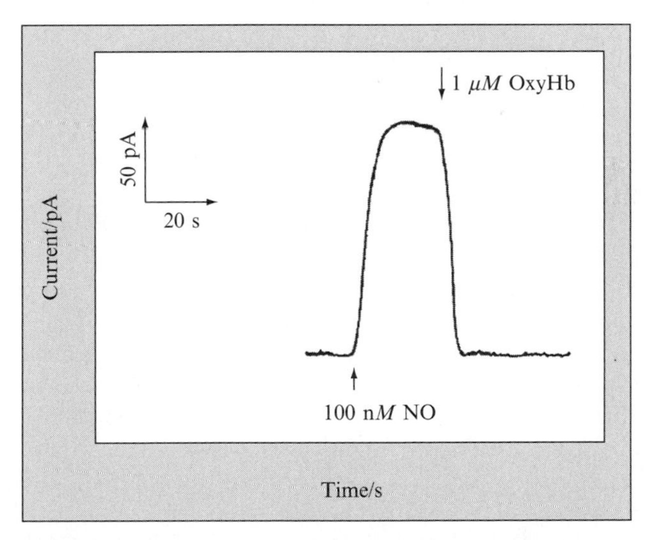

Figure 5.5 Amperometric response of a NO nanosensor upon addition of 100 n*M* NO and 1 μ*M* oxyhemoglobin to a stirred 0.1 *M* phosphate buffer solution (pH 7.4).

response time of NO sensors, less filtering can be applied, which will sacrifice the detection limit of the NO sensor.

7.4. Effect of temperature and pH on NO electrodes

The background current of all NO electrodes is sensitive to changes of temperature and pH. The magnitude of the effect depends upon the type of electrode. Clark-type NO electrodes are very sensitive to temperature changes. The temperature-induced response of this type of NO sensor is about 50 to 100 n*M* of NO/°C, while temperature-induced responses of carbon-fiber NO microelectrode are less than 10 n*M* of NO/°C. Thus, when measuring NO, the temperature should be simultaneously monitored. As far as pH-induced responses are concerned, changes of 1 pH unit can cause a 50- to 100-pA current change on Clark-type NO sensors.

8. Selected Applications of NO Electrodes

Several hundred research papers have been published over the last decade describing the amperometric detection of NO in biological systems. Because there has been such an explosion in the development of NO microsensors, these measurements can now be made in a variety of biological tissues and organs, as well as on the cellular level, without significant damage to the system. This section will point out several examples in which NO microsensors were used to determine the biological effects of NO (Tristani-Firouzi *et al.*, 1998).

NO release, NO consumption, and levels of applied NO have been measured in a variety of biological systems, including eyes (Akeo *et al.*, 2000; Amaki *et al.*, 2001; Millar, 2003; Sekaran *et al.*, 2005), gastrointestinal tract (Asanuma *et al.*, 2005; Iijima *et al.*, 2002; Stefano *et al.*, 2004), brain tissue (Buerk *et al.*, 2003; Cherian *et al.*, 2000; Fabre *et al.*, 1997; Ferreira *et al.*, 2005; Meulemans, 2002; Rocchitta *et al.*, 2004), kidney and kidney tubule fluid (Arregui *et al.*, 2004; Levine *et al.*, 2001, 2004; Levine and Iacovitti, 2003, 2006; Saitoand and Miyagawa, 2000; Thorup *et al.*, 1998), rat and guinea pig isolated and intact hearts (Fujita *et al.*, 1998; Novalija *et al.*, 1999), rat spinal cord (Schulte and Millar, 2003), human monocyte cells (Stefano *et al.*, 1999), human endothelial cells (Stefano *et al.*, 2000), mitochondria (Beltran *et al.*, 2002; Shiva *et al.*, 2001), rat penis corpus cavernosum (Mas *et al.*, 2002), granulocytes (Kedziora-Kornatowska *et al.*, 1998), invertebrate ganglia and immunocytes (Stefano *et al.*, 2002), choroidal endothelial cells (Uhlmann *et al.*, 2001), cancer cells (Bal-Price *et al.*, 2006; Kashiwagi *et al.*, 2005; Tsatmali *et al.*, 2000), peripheral blood (Rysz *et al.*, 1997), human blood (Rievaj *et al.*, 2004), human leukocytes

(Larfars *et al.*, 1999), platelets (De La Cruz *et al.*, 2003; Freedman *et al.*, 1997; Kasuya *et al.*, 2002), ears (Jiang *et al.*, 2004; Shi *et al.*, 2002), plants (Grande *et al.*, 2004; Sakihama *et al.*, 2002; Yamasaki *et al.*, 1999; Yamasaki and Sakihama, 2000), and pteropod mollusks (Moroz *et al.*, 2000).

Levine and colleagues first reported on the real-time profiling of kidney tubular fluid NO concentration *in vivo* (Levine *et al.*, 2001, 2004; Levine and Iacovitti, 2006). In the 2001 publication, a modified version of a 7-μm combination NO electrode was successfully used to measure NO concentration profiles along the length of a single nephron of a rat kidney tubular segment. Because it was shown that the electrode was sensitive to NO in the rat tubule, the electrode was used to detect NO concentration differences in rat kidney tubules before and after 5/6 nephrectomy. The results clearly showed that the NO concentration was much higher in nephrectomized rats than in unnephrectomized rats.

In a recent publication, investigators used a specially customized 2-mm sensor to monitor, in real time, NO production in the stomach and esophagus of human patients (Iijima *et al.*, 2002). A patient first swallowed two NO electrodes (Fig. 5.6), which were then withdrawn slowly at 1-cm increments every 2 min. The investigators were then able to establish a profile of NO concentration in the upper gastrointestinal tract.

Simonsen's group has performed some elegant work over the years on the characteristics of NO release from rat superior mesenteric artery. Initially, the group simultaneously monitored artery relaxation and NO concentration in the artery using a NO microsensor in response to various drugs (Simonsen *et al.*, 1999; Wadsworth *et al.*, 2006). NO concentration was monitored via a 30-μm electrode that was inserted into the artery lumen using a micromanipulator. The results of this work are shown in Fig. 5.7, which shows the force (upper traces) and NO concentration (lower traces) in an endothelium intact (+E) segment of the rat superior mesenteric artery and the same segment after mechanical endothelial cell removal (−E). As can be seen from the traces, if endothelial cells are present the artery is capable of relaxation because the endothelial cells release NO in response to acetylcholine, but if the endothelial cells are removed the artery is insensitive to acetylcholine injection but relaxes upon the introduction of the NO-releasing molecule SNAP. In a subsequent publication, Hernanz *et al.* (2004) again monitored artery relaxation and NO concentration in the rat superior mesenteric artery to monitor its hyporeactivity to various endotoxins. In this study, a 30-μm electrode was inserted into the lumen of the artery and the NO concentration, as well as artery relaxation, was measured as a function of endotoxin introduction to determine what effects lipopolysaccharide had on the artery function. The results nicely showed that lipopolysaccharide resulted in induction of iNOS and SOD associated with noradrenaline hyporeactivity, while increased NO concentration is measured only when L-arginine is present. Simonsen's group has also

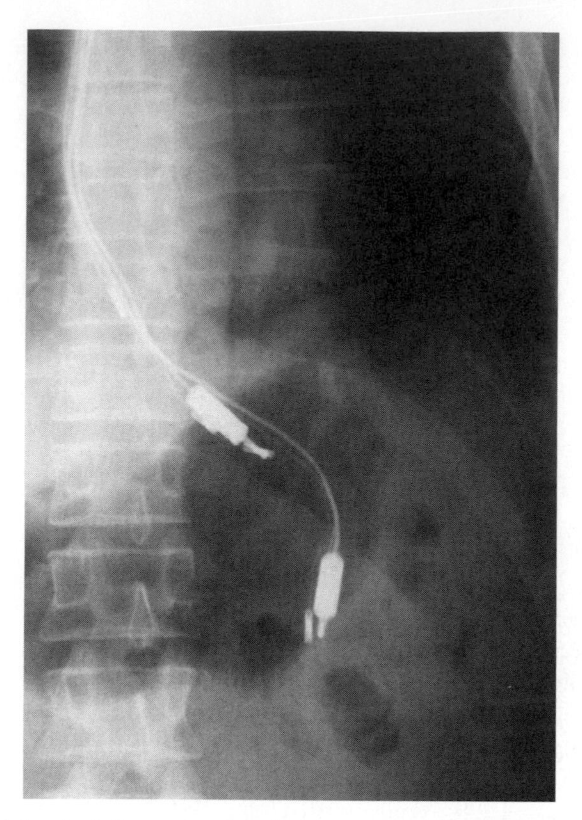

Figure 5.6 Abdominal X-ray showing the apparatus consisting of two NO sensors, a 4-channel pH catheter, and a Teflon nasogastric tube. Reprinted with permission from the American Gastroenterological Association (Iijima *et al.*, 2002).

measured NO concentration in the artery of a hypertensive rat (Stankevicius *et al.*, 2002) and in isolated human small arteries (Buus *et al.*, 2002, 2006) and has investigated the effects of Ca^{2+}-activated K^{+} channel blockers on acetylcholine-evoked NO release in rat mesenteric artery (Stankevicius *et al.*, 2006).

Isik *et al.* (2005) have recently published interesting results describing the measurement of NO release from human umbilical vein endothelial cells (HUVEC) using a unique array of microelectrodes. Figure 5.8 (top) shows an SEM image of the microelectrode array. Following microelectrode array construction, the surface of the electrodes were modified with nickel tetrasulfonate phthalocyanine tetrasodium salt using electrochemically induced deposition. Following this deposition, the HUVEC cells were allowed to grow in the interstitial spaces between the electrodes (see Fig. 5.8, bottom) and NO release from the cells was monitored as a function of growth and stimulation from bradykinin. This work is unique because the cells are

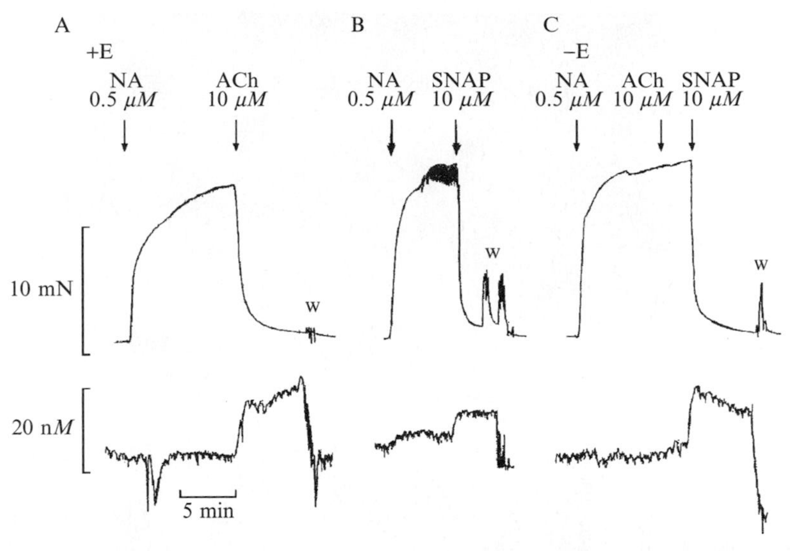

Figure 5.7 Simultaneous measurements of force (upper traces) and NO concentration (lower traces) in an endothelium intact (+E) segment of rat superior mesenteric artery contracted with 0.5 μM noradrenaline (NA) and relaxed with either 10 μM acetylcholine (ACh) (A), or 10 μM SNAP (B). Panel C shows a similar measurement in the rat superior mesenteric artery after mechanical endothelial cell removal. As can be seen in C, ACh addition does not cause NO production from the artery but shows a NO increase upon SNAP addition causing artery relaxation. W = washout. Reprinted with permission from Blackwell Publishing (Simonsen *et al.*, 1999).

actually being grown on the Si_3N_4-insulating layer so they are not being affected by the microelectrode working potential (which typically causes cell death), which allows for NO concentration to be monitored.

A publication by Millar (2003) has shown that NO concentrations can be measured in bovine eyes' trabecular meshwork *in situ* using a 200-μm electrode. For this study, the tip of the electrode was inserted into the region of the trabecular meshwork and NO monitored as a function of epinephrine concentration. The study found that NO generation increased as a function of epinephrine addition, resulting in reduced intraocular pressure. This finding is important as it sheds light on treatments for patients with primary open-angle glaucoma. A Pt/Ir electrode, coated with NO-selective membranes, has also been used to determine the effects of L-DOPA on eyes (Akeo *et al.*, 2000; Amaki *et al.*, 2001).

As NO is a diffusible messenger molecule in the brain, the measurement of NO concentration *in vitro* is important in understanding its action. In order to accomplish this general goal, Ferreira *et al.* (2005) recently developed an 8-μm diameter carbon-fiber electrode coated with Nafion and *o*-phenylenediamine. Nafion was added to the electrode surface by dipping the electrode into a 5% solution followed by drying for 10 min

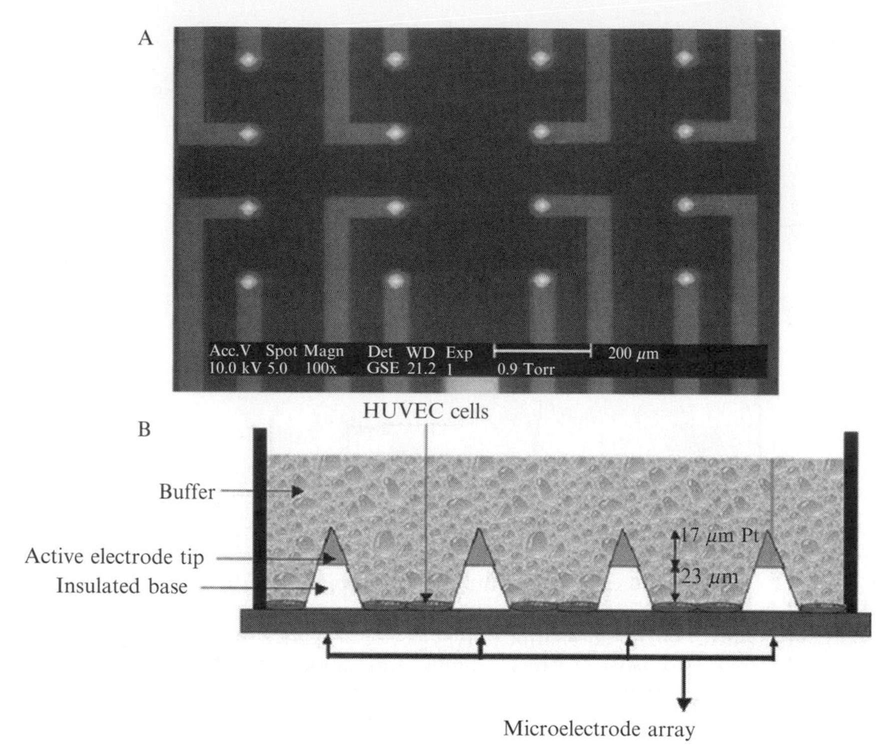

Figure 5.8 A representative SEM image of the microelectrode array (A) and a schematic representation of the experimental setup (B). Reprinted with permission from Elsevier Publishing (Isik *et al.*, 2005).

at 170°, followed by *o*-phenylenediamine electropolymerization on the electrode by holding the potential at +0.9 V vs Ag/AgCl reference electrode. These electrodes were then inserted into the CA1 region of the hippocampal slices from a Wistar rat, and the NO concentration monitored as a function of L-glutamate and *N*-methyl-D-aspartate introduction. The results can be seen in Fig. 5.9. The main finding of this study was that the electrodes were sufficiently sensitive to monitor brain NO concentrations with a sensitivity of 954 ± 271 pA/μM and a limit of detection of 6 ± 2 nM. These measurements can also be made with minimal interference from common interferences such as ascorbate, nitrite, and H_2O_2. These electrodes can be useful for unraveling pathways for memory and learning processes in the hippocampus.

Mas *et al.* (2002) published results on the measurement of NO release from the corpus cavernosum of the penis and its relation to penile erection. For this study, a 30-μm-diameter carbon fiber was coated with Nafion and nickel tetrakis (3-methoxy-4-hydroxyphenyl) porphyrin via electrodeposition using differential pulse voltammetry. This electrode was then inserted into

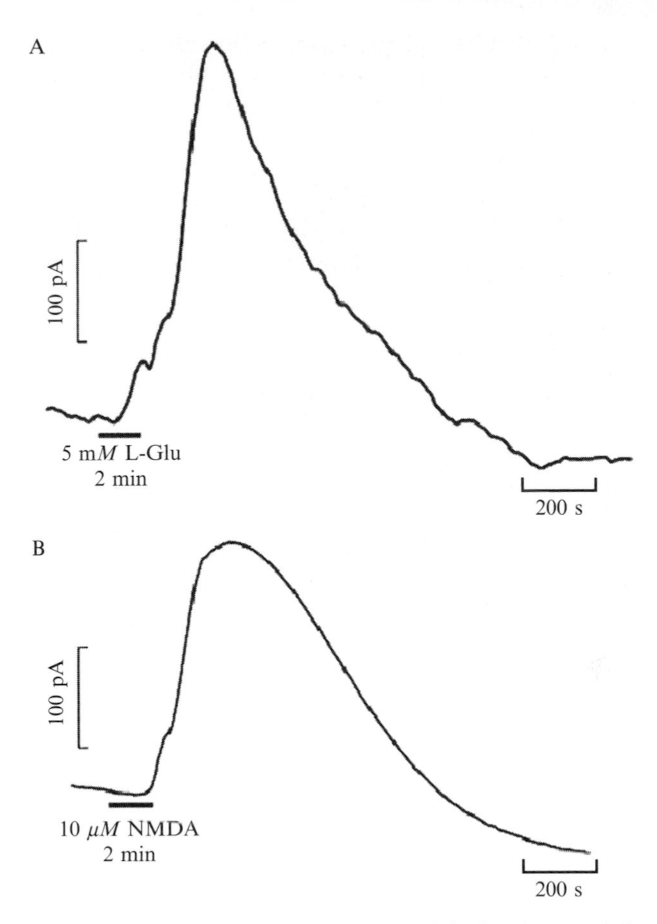

Figure 5.9 NO production from the CA1 region of the hippocampal slice following addition of 5 m*M* L-glutamate (A) and 10 μ*M* *N*-methyl-D-aspartate (B). Reprinted with permission from Elsevier Publishing (Ferreira *et al.*, 2005).

the cavernous bodies of a rat penis. The results showed an increase in NO concentration and intracavernous pressure upon cavernous nerve stimulation and subsequently decreased upon introduction of NO synthase isoenzymes. This study was important to facilitate further measurements of NO concentrations *in vivo* in the penis.

NO measurements have also been made in the ears of a guinea pig. The first of these studies was performed by Shi *et al.* (2002). For this investigation, a 30-μm carbon fiber was inserted into the perilymph of the basal turn of the guinea pig ear to measure changes in NO concentration as a result of noise stimulation. This study showed that guinea pigs exposed to broadband noise for 3 h/day at 120 dBA for 3 days exhibited an increase in NO concentration in the perilymph. This result is important in order to understand the role that NO plays in hearing loss. A subsequent publication by Jiang *et al.* (2004) used a 200-μm carbon-fiber electrode to measure NO concentration in the spiral modiolar artery (SMA) of a guinea pig. To perform this study a 3-mm section of the SMA was added to a bath solution and the complete tip of the 200-μm electrode was inserted into the bath, parallel to the SMA, to measure basal NO concentration as well as drug-induced NO release and how that relates to cochlear blood-flow regulation. The key findings of this study were that the SMA continuously releases NO and that a blockage of this release by L-NAME causes a decrease in NO production and a vasoconstriction. This study is important to gain a better understanding of how NO potentially regulates cochlear blood flow to set a benchmark for pharmacological and pathological evaluation.

An interesting study was performed by Kashiwagi *et al.* (2005) on the role that NO plays in tumor vessel morphogenesis and maturation. For this study, B16 tumor cells were injected into mice and tumor tissue removed from the mice when it reached ≈8 mm in diameter. The tip of a Nafion polymer–coated Au microelectrode was subsequently inserted into the tumor to monitor NO production. The results of this study showed that NO mediates mural cell coverage and vessel branching/longitudinal extension but does not play a part in the growth of tumor blood vessels. The investigators also used a NOS inhibitor to show that NOS from endothelial cells in tumors is the primary source of NO and mediates tumor growth.

Kasuya *et al.* (2002) used a commercial NO-selective microelectrode to monitor the effect that exercise has on platelet-derived NO. This study used 23 healthy male nonsmokers who underwent treadmill exercise. Blood samples were taken from the subjects before and directly following exercise and blood platelets isolated. The study showed that NO concentration and platelet levels were increased following exercise. This increase in NO concentration is thought to play a role in the prevention of exercise-induced platelet activation in humans.

Kellogg *et al.* (2003) reported on the measurement of NO under the human skin in response to heat stress. For this study, a flexible 200-μm NO microelectrode was inserted into the cutaneous interstitial space of the forearm of nine human patients to measure NO concentration while the subjects were at low (34°) and high (39°) temperature. Laser-Doppler flowmetry (LDF) was used to monitor skin blood flow (SkBF). This publication demonstrated that NO concentration, as well as SkBF, increased in the cutaneous interstitial space during heat stress in humans. Figure 5.10 shows the key results for these experiments. As the temperature is increased at around 10 min, the data show that the blood flow (top plot) and NO concentration (bottom plot) increased as a function of temperature. Also, as the temperature was decreased again, at around 45 min, the blood flow and NO concentration returned to their original value. At approximately 130 min, while the subjects were at low temperature, the investigators

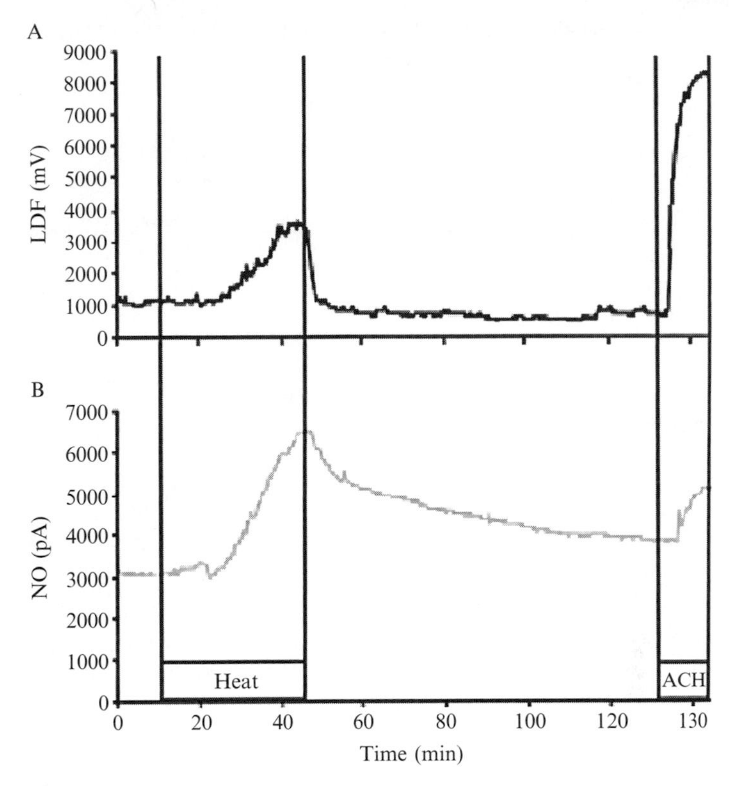

Figure 5.10 Laser-Doppler flowmetry (A) and NO measurement (B) from one subject. Upon heating the subject to 39°, at ≈10 min, NO production and skin blood flow increased, which instantly returned to normal upon cooling the subject at ≈45 min. After heat stress and cooling, ACh was administered by intradermal microdialysis to confirm the ability of the microelectrode to measure NO concentrations. Reprinted with permission from the American Physiological Society (Kellogg *et al.*, 2003)

injected acetylcholine to show that NO was indeed being detected in the subjects. The same group used a similar experimental design to monitor NO concentration, as well as SkBF, during reactive hyperemia under the human skin (Zhao *et al.*, 2004).

Thom *et al.* (2003) published results on the stimulation of perivascular NO synthesis by oxygen. To perform this study, a 200-μm diameter electrode was placed between the aorta and vena cava of anesthetized rats and mice and then the rodents were placed inside a hyperbaric chamber. Inside the hyperbaric chamber the partial pressure of O_2 was regulated/changed as NO concentration was monitored. Figure 5.11 shows that NO concentration increased as a function of O_2 partial pressure. This experiment is important for understanding how NO synthesis, by NOS, is altered and regulated by O_2.

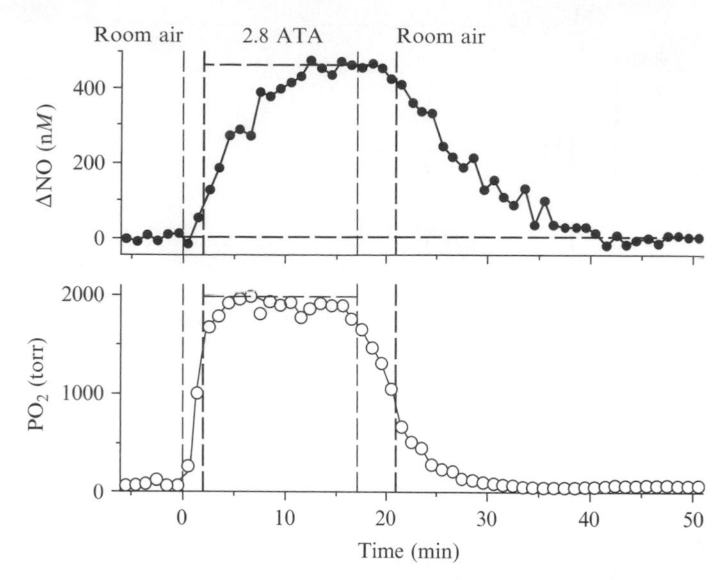

Figure 5.11 NO concentration (top) and O_2 concentration (bottom) as a function of O_2 pressure. As can be seen, NO and O_2 concentrations increase significantly when a rat is exposed to pressurized O_2 atmospheres (2.8 atmospheres absolute, ATA). Reprinted with permission from the American Physiological Society (Thom *et al.*, 2003).

Studies have also been conducted on the NO release from plants and the effects of a plant diet on atherosclerosis and endothelial cell dysfunction. Grande *et al.* (2004) studied the effects that a diet of wild artichoke and thyme had on the release of NO from porcine aortic endothelial cells and cerebral cell membranes. For this study, a rat brain was homogenized and cell membranes isolated by ultracentrifugation and NO release monitored using a 2-mm NO sensor. The cell membranes were then exposed to wild artichoke and thyme extracts. This study showed that NO release was significantly increased following wild artichoke and thyme extract addition. These results suggest that eating a diet rich in phenolic compounds, such as wild artichoke and thyme, contributes to maintenance of a healthy cardiovascular system. Using a 2-mm sensor, Yamasaki and Sakihama (2000) were able to study the ability of plant nitrate reductase to produce NO. Figures 5.12 and 5.13 show that plant nitrate reductase produces NO *in vitro* as well as its toxic derivative peroxynitrite.

Recently it was reported that pioglitazone treatment improves nitrosative stress in type 2 diabetes (Vinik *et al.*, 2006). In this research, NO was measured *in vivo* using an electrochemical NO microsensor inserted directly into the skin. The authors found that the NO production was significantly decreased in the pioglitazone-treated group in the basal condition.

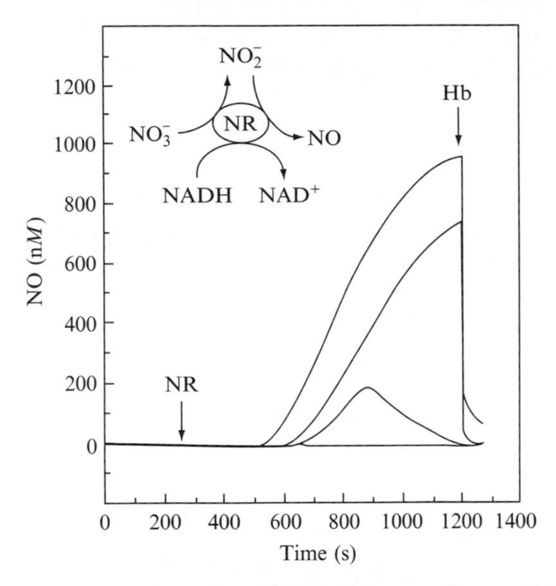

Figure 5.12 NO production from the addition of nitrate reductase (NR = 15 mU/mL) to a solution containing 50 μM sodium nitrate and various concentrations of NADH. The curves from top to bottom were obtained in solutions containing 100, 50, 40, and 0 μM NADH. Hemoglobin (Hb) was introduced to quench the production of NO. Reprinted with permission from Elsevier Publishing (Yamasaki and Sakihama, 2000).

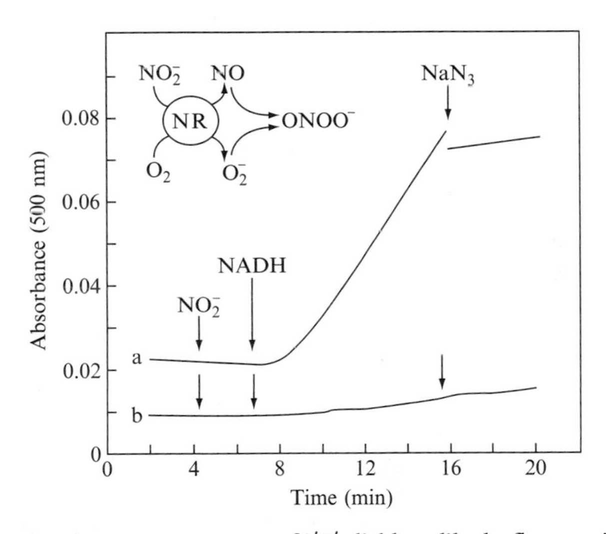

Figure 5.13 Absorbance measurement of 2′,7′-dichlorodihydrofluorescein (DCDHF), a peroxynitrite-sensitive dye, as a function of nitrite (1 mM) and NADH (1 mM) introduction to a solution containing 100 μM DCDHF and 30mU/mL NR under ambient (A) and nitrogen-saturated conditions (B). As can be seen, the absorbance of DCDHF increases upon nitrite and NADH introduction only under an oxygen atmosphere, indicative of peroxynitrite production. Reprinted with permission from Elsevier Publishing (Yamasaki and Sakihama, 2000).

9. Concluding Remarks and Other Directions

The use of the electrochemical NO microsensor provides an elegant and convenient way to detect NO in real time and in biological samples. Currently, such sensors provide the only means by which to measure NO continuously, accurately, and directly within living tissue without significant damage. The increasing acceptance of such sensors and their diversity of use in many NO research applications will help further the current understanding of the various clinical roles of this interesting and ubiquitous molecule. Continual improvements being made to NO microsensor design and technology will facilitate these studies in the foreseeable future.

ACKNOWLEDGMENTS

This research was supported by NIH grants (1 R43 GM62077-01, 2R44 GM62077-02, and 5R GM62077-3) to Xueji Zhang and WPI R&D priority research funds.

REFERENCES

Akaike, T., Yoshida, M., Miyamoto, Y., Sato, K., Kohno, M., Sasamoto, K., Miyazaki, K., Ueda, S., and Maeda, H. (1993). Antagonistic action of imidazolineoxyl N-oxides against endothelium-derived relaxing factor: NO through a radical reaction. *Biochemistry* **32,** 827–832.

Akeo, K., Amaki, S., Suzuki, T., and Hiramitsu, T. (2000). Melanin granules prevent the cytotoxic effects of L-DOPA on retinal pigment epithelial cells *in vitro* by regulation of NO and superoxide radicals. *Pigm. Cell. Res.* **13,** 80–88.

Allen, B. W., Piantadosi, C. A., and Coury, L. A. (2000). Electrode materials for nitric oxide detection. *Nitric Oxide-Biol. Chem.* **4,** 75–84.

Amaki, S. K., Oguchi, Y., Ogata, T., Suzuki, T., Akeo, K., and Hiramitsu, T. (2001). L-DOPA produced nitric oxide in the vitreous and caused greater vasodilation in the choroid and the ciliary body of melanotic rats than in those of amelanotic rats. *Pigm. Cell Res.* **14,** 256–263.

Anbar, M. (1994). Hyperthermia of the cancerous breast: Analysis of mechanism. *Cancer Lett.* **84,** 23–29.

Archer, S. (1993). Measurement of nitric-oxide in biological models. *FASEB J.* **7,** 349–360.

Arregui, B., Lopez, B., Salom, M. G., Valero, F., Navarro, C., and Fenoy, F. J. (2004). Acute renal hemodynamic effects of dimanganese decacarbonyl and cobalt protoporphyrin. *Kidney Int.* **65,** 564–574.

Asanuma, K., Iijima, H., Sugata, S., Ohara, T., Shimosegawa, T., and Yoshimura, T. (2005). Diffusion of cytotoxic concentrations of nitric oxide generated luminally at the gastro-oesophageal junction of rats. *Gut* **54,** 1072–1077.

Bal-Price, A., Gartlon, J., and Brown, G. C. (2006). Nitric oxide stimulates PC 12 cell proliferation via cGMP and inhibits at higher concentrations mainly via energy depletion. *Nitric Oxide* **14,** 238–246.

Beckman, J. S., and Congert, K. A. (1995). Direct measurement of dilute nitric oxide in solution with an ozone chemiluminescent detector. *Methods* **7,** 35–38.

Bedioui, F., Trevin, S., and Devynck, J. (1994). The use of gold electrodes in the electrochemical detection of nitric-oxide in aqueous-solution. *J. Electroanal. Chem.* **377,** 295.

Bedioui, F., Trevin, S., and Devynck, J. (1996). Chemically modified microelectrodes designed for the electrochemical determination of nitric oxide in biological systems. *Electroanalysis* **8,** 1085–1091.

Beltran, B., Quintero, M., Garcia-Zaragoza, E., O'Connor, E., Esplugues, J. V., and Moncada, S. (2002). Inhibition of mitochondrial respiration by endogenous nitric oxide: A critical step in Fas signaling. *Proc. Natl. Acad. Sci. USA* **99,** 8892–8897.

Bredt, D. S., and Snyder, S. H. (1989). Nitric-oxide mediates glutamate-linked enhancement of Cgmp levels in the cerebellum. *Proc. Natl. Acad. Sci. USA* **86,** 9030–9033.

Bredt, D. S., Hwang, P. M., Glatt, C. E., Lowenstein, C., Reed, R. R., and Snyder, S. H. (1991). Cloned and expressed nitric-oxide synthase structurally resembles cytochrome-P-450 reductase. *Nature* **351,** 714–718.

Brunet, A., Privat, C., Stepien, O., David-Dufilho, M., Devynck, J., and Devynck, M. A. (2000). Advantages and limits of the electrochemical method using Nafion and Ni-porphyrin-coated microelectrode to monitor NO release from cultured vascular cells. *Analysis* **28,** 469–474.

Buerk, D. G., Ances, B. M., Greenberg, J. H., and Detre, J. A. (2003). Temporal dynamics of brain tissue nitric oxide during functional forepaw stimulation in rats. *Neuroimage* **18,** 1–9.

Buus, N. H., Simonsen, U., Pilegaard, H. K., and Mulvaney, M. J. (2002). Nitric oxide, prostanoid and non-NO, non-prostanoid involvement in acetylcholine relaxation of isolated human small arteries. *Br. J. Pharmacol.* **129,** 184–192.

Buus, N. H., Simonsen, U., Pilegaard, H. K., and Mulvany, M. J. (2006). Intracellular smooth muscle[Ca^{2+}] in acetylcholine and nitric oxide mediated relaxation of human small arteries. *Eur. J. Pharmacol.* **535,** 243–247.

Casero, E., Pariente, F., and Lorenzo, E. (2003). Electrocatalytic oxidation of nitric oxide at indium hexacyanoferrate film-modified electrodes. *Anal. Bioanal. Chem.* **375,** 294–299.

Cherian, L., Goodman, J. C., and Robertson, C. S. (2000). Brain nitric oxide changes after controlled cortical impact injury in rats. *J. Neurophysiol.* **83,** 2171–2178.

Clark, L. C., and Lyons, C. (1962). Electrode systems for continuous monitoring in cardiovascular surgery. *Ann. N.Y. Acad. Sci.* **102,** 29–45.

De La Cruz, J. P., Gonzalez-Correa, J. A., Guerrero, A., Marquez, E., Martos, F., and De la Cuesta, F. S. (2003). Differences in the effects of extended-release aspirin and plain-formulated aspirin on prostanoids and nitric oxide in healthy volunteers. *Fundam. Clin. Pharmacol.* **17,** 363–372.

Diab, N., and Schuhmann, W. (2001). Electropolymerized manganese porphyrin/polypyrrole films as catalytic surfaces for the oxidation of nitric oxide. *Electrochim. Acta* **47,** 265–273.

Diab, N., Oni, J., Schulte, A., Radtke, I., Blochl, A., and Schuhmann, W. (2003). Pyrrole functionalised metalloporphyrins as electrocatalysts for the oxidation of nitric oxide. *Talanta* **61,** 43–51.

Dickson, A., Lin, J., Sun, J., Broderick, M., Fein, H., and Zhang, X. J. (2004). Construction and characterization of a new flexible and nonbreakable nitric oxide microsensor. *Electroanalysis* **16,** 640–643.

Fabre, B., Burlet, S., Cespuglio, R., and Bidan, G. (1997). Voltammetric detection of NO in the rat brain with an electronic conducting polymer and Nafion(R) bilayer-coated carbon fibre electrode. *J. Electroanal. Chem.* **426,** 75–83.

Fan, C. H., Li, G. X., Zhu, J. Q., and Zhu, D. X. (2000). A reagentless nitric oxide biosensor based on hemoglobin-DNA films. *Anal. Chim. Acta* **423,** 95–100.

Fan, C. H., Pang, J. T., Shen, P. P., Li, G. X., and Zhu, D. X. (2002). Nitric oxide biosensors based on Hb/phosphatidylcholine films. *Anal. Sci.* **18,** 129–132.

Feldman, P. L., Griffith, O. W., and Stuehr, D. J. (1993). The surprising life of nitric-oxide. *Chem. Eng. News* **71,** 26–37.

Ferreira, N. R., Ledo, A., Frade, J. G., Gerhardt, G. A., Laranjinha, J., and Barbosa, R. M. (2005). Electrochemical measurement of endogenously produced nitric oxide in brain slices using Nafion/o-phenylenediamine modified carbon fiber microelectrodes. *Anal. Chim. Acta* **535,** 1–7.

Freedman, J. E., Loscalzo, J., Barnard, M. R., Alpert, C., Keaney, J. F., and Michelson, A. D. (1997). Nitric oxide released from activated platelets inhibits platelet recruitment. *J. Clin. Invest.* **100,** 350–356.

Friedemann, M. N., Robinson, S. W., and Gerhardt, G. A. (1996). o-Phenylenediamine-modified carbon fiber electrodes for the detection of nitric oxide. *Anal. Chem.* **68,** 2621–2628.

Fujita, S., Roerig, D. L., Bosnjak, Z. J., and Stowe, D. F. (1998). Effects of vasodilators and perfusion pressure on coronary flow and simultaneous release of nitric oxide from guinea pig isolated hearts. *Cardiovasc. Res.* **38,** 655–667.

Grande, S., Bogani, P., De Saizieu, A., Schueler, G., Galli, C., and Visioli, F. (2004). Vasomodulating potential of Mediterranean wild plant extracts. *J. Agric. Food Chem.* **52,** 5021–5026.

Green, L. C., Wagner, D. A., Glogowski, J., Skipper, P. L., Wishnok, J. S., and Tannenbaum, S. R. (1982). Analysis of nitrate, nitrite, and N-15-labeled nitrate in biological fluids. *Anal. Biochem.* **126,** 131–138.

Haruyama, T., Shiino, S., Yanagida, Y., Kobatake, E., and Aizawa, M. (1998). Two types of electrochemical nitric oxide (NO) sensing systems with heat-denatured Cyt C and radical scavenger PTIO. *Biosens. Bioelectron.* **13,** 763–769.

Hayon, J., Ozer, D., Rishpon, J., and Bettelheim, A. (1994). Spectroscopic and electrochemical response to nitrogen monoxide of a cationic iron porphyrin immobilized in Nafion-coated electrodes or membranes. *J. Chem. Soc.-Chem. Commun.* 619–620.

He, X. C., and Mo, J. Y. (2000). Electrocatalytic oxidation of NO at electrode modified with Nafion-Co-II-1,10-phenanthroline film and its application to NO detection. *Analyst* **125,** 793–795.

Hernanz, R., Alonso, M. J., Zibrandtsen, H., Alvarez, Y., Salaices, M., and Simonsen, U. (2004). Measurements of nitric oxide concentration and hyporeactivity in rat superior mesenteric artery exposed to endotoxin. *Cardiovasc. Res.* **62,** 202–211.

Hibbs, J. B., Taintor, R. R., Vavrin, Z., and Rachlin, E. M. (1988). Nitric-oxide: A cyto-toxic activated macrophage effector molecule. *Biochem. Biophys. Res. Commun.* **157,** 87–94.

Ichimori, K., Ishida, H., Fukahori, M., Nakazawa, H., and Murakami, E. (1994). Practical nitric-oxide measurement employing a nitric oxide-selective electrode. *Rev. Sci. Instrum.* **65,** 2714–2718.

Ignarro, L. J. (1990). Nitric-oxide: A novel signal transduction mechanism for transcellular communication. *Hypertension* **16,** 477–483.

Ignarro, L. J., Buga, G. M., Wood, K. S., Byrns, R. E., and Chaudhuri, G. (1987). Endothelium-derived relaxing factor produced and released from artery and vein is nitric-oxide. *Proc. Natl. Acad. Sci. USA* **84,** 9265–9269.

Iijima, K., Henry, E., Moriya, A., Wirz, A., Kelman, A. W., and McColl, K. E. L. (2002). Dietary nitrate generates potentially mutagenic concentrations of nitric oxide at the gastroesophageal junction. *Gastroenterology* **122,** 1248–1257.

Isik, S., Berdondini, L., Oni, J., Blochl, A., Koudelka-Hep, M., and Schuhmann, W. (2005). Cell-compatible array of three-dimensional tip electrodes for the detection of nitric oxide release. *Biosens. Bioelectron.* **20,** 1566–1572.

Jiang, Z. G., Shi, X. R., Zhao, H., Si, J. Q., and Nuttall, A. L. (2004). Basal nitric oxide production contributes to membrane potential and vasotone regulation of guinea pig *in vitro* spiral modiolar artery. *Hear. Res.* **189,** 92–100.

Kamei, K., Haruyama, T., Mie, M., Yanagida, Y., Aizawa, M., and Kobatake, E. (2004). The construction of endothelial cellular biosensing system for the control of blood pressure drugs. *Biosens. Bioelectron.* **19,** 1121–1124.

Kashevskii, A. V., Lei, J., Safronov, A. Y., and Ikeda, O. (2002). Electrocatalytic properties of meso-tetraphenylporphyrin cobalt for nitric oxide oxidation in methanolic solution and in Nafion(R) film. *J. Electroanal. Chem.* **531,** 71–79.

Kashiwagi, S., Izumi, Y., Gohongi, T., Demou, Z. N., Xu, L., Huang, P. L., Buerk, D. G., Munn, L. L., Jain, R. K., and Fukumura, D. (2005). NO mediates mural cell recruitment and vessel morphogenesis in murine melanomas and tissue-engineered blood vessels. *J. Clin. Invest.* **115,** 1816–1827.

Kasuya, N., Kishi, Y., Sakita, S., Numano, F., and Isobe, M. (2002). Acute vigorous exercised primes enhanced NO release in human platelets. *Atherosclerosis* **161,** 225–232.

Kato, D., Kunitake, M., Nishizawa, M., Matsue, T., and Mizutani, F. (2005). Amperometric nitric oxide microsensor using two-dimensional cross-linked Langmuir-Blodgett films of polysiloxane copolymer. *Sens. Actuator B-Chem.* **108,** 384–388.

Katrlik, J., and Zalesakova, P. (2002). Nitric oxide determination by amperometric carbon fiber microelectrode. *Bioelectrochemistry* **56,** 73–76.

Kedziora-Kornatowska, K. Z., Luciak, M., Blaszczyk, J., and Pawlak, W. (1998). Effect of aminoguanidine on the generation of superoxide anion and nitric oxide by peripheral blood granulocytes of rats with streptozotocin-induced diabetes. *Clin. Chim. Acta* **278,** 45–53.

Kellogg, D. L., Zhao, J. L., Friel, C., and Roman, I. J. (2003). Nitric oxide concentration increases in the cutaneous interstitial space during heat stress in humans. *J. Appl. Physiol.* **94,** 1971–1977.

Kelm, J., and Schrader, J. (1990). Control of coronary vascular tone by nitric-oxide. *Circ. Res.* **66,** 1561–1575.

Kiechle, F. L., and Malinski, T. (1993). Nitric-oxide: Biochemistry, pathophysiology, and detection. *Am. J. Clin. Pathol.* **100,** 567–575.

Kim, K. I., Chung, H. Y., Oh, G. S., Bae, H. O., Kim, S. H., and Chun, H. J. (2005). Integrated gold-disk microelectrode modified with iron(II)-phthalocyanine for nitric oxide detection in macrophages. *Microchem. J.* **80,** 219–226.

Kitajima, A., Teranishi, T., and Miyake, M. (2001). Detection of nitric oxide on carbon electrode modified with ionic polymers and alpha-cyclodextrin. *Electrochemistry* **69,** 16–20.

Kitamura, Y., Uzawa, T., Oka, K., Komai, Y., Ogawa, H., Takizawa, N., Kobayashi, H., and Tanishita, K. (2000). Microcoaxial electrode for *in vivo* nitric oxide measurement. *Anal. Chem.* **72,** 2957–2962.

Kroning, S., Scheller, F. W., Wollenberger, U., and Lisdat, F. (2004). Myoglobin-clay electrode for nitric oxide (NO) detection in solution. *Electroanalysis* **16,** 253–259.

Lantoine, F., Trevin, S., Bedioui, F., and Devynck, J. (1995). Selective and sensitive electrochemical measurement of nitric-oxide in aqueous-solution: Discussion and new results. *J. Electroanal. Chem.* **392,** 85–89.

Larfars, G., Lantoine, F., Devynck, M. A., and Gyllenhammar, H. (1999). Electrochemical detection of nitric oxide production in human polymorphonuclear neutrophil leukocytes. *Scand. J. Clin. Lab. Invest.* **59,** 361–368.

Lee, Y., Oh, B. K., and Meyerhoff, M. E. (2004a). Improved planar amperometric nitric oxide sensor based on platinized platinum anode. 1. Experimental results and theory when applied for monitoring NO release from diazeniumdiolate-doped polymeric films. *Anal. Chem.* **76,** 536–544.

Lee, Y., Yang, J., Rudich, S. M., Schreiner, R. J., and Meyerhoff, M. E. (2004b). Improved planar amperometric nitric oxide sensor based on platinized platinum anode. 2. Direct real-time measurement of NO generated from porcine kidney slices in the presence of L-arginine, L-arginine polymers, and protamine. *Anal. Chem.* **76,** 545–551.

Levine, D. Z., and Iacovitti, M. (2003). Real time microelectrode measurement of nitric oxide in kidney tubular fluid *in vivo*. *Sensors* **3,** 314–320.

Levine, D. Z., and Iacovitti, M. (2006). Real time measurement of kidney tubule fluid nitric oxide concentrations in early diabetes: Disparate changes in different rodent models. *Nitric Oxide* **15,** 87–92.

Levine, D. Z., Burns, K. D., Jaffey, J., and Iacovitti, M. (2004). Short-term modulation of distal tubule fluid nitric oxide *in vivo* by loop NaCl reabsorption. *Kidney Int.* **65,** 184–189.

Levine, D. Z., Iacovitti, M., Burns, K. D., and Zhang, X. J. (2001). Real-time profiling of kidney tubular fluid nitric oxide concentrations *in vivo*. *Am. J. Physiol.-Renal Physiol.* **281,** F189–F194.

Liu, X. P., Liu, Q. H., Gupta, E., Zorko, N., Brownlee, E., and Zweier, J. L. (2005). Quantitative measurement of NO reaction kinetics with Clark-type electrode. *Nitric Oxide* **13,** 68–77.

Magazine, H. I. (1995). Detection of endothelial cell-derived nitric oxide: Current trends and future directions. *Adv. Neuroimmunol.* **5,** 479–490.

Malinski, T., and Taha, Z. (1992). Nitric-oxide release from a single cell measured *in situ* by a porphyrinic-based microsensor. *Nature* **358,** 676–678.

Malinski, T., Bailey, F., Zhang, Z. G., and Chopp, M. (1993b). Nitric-oxide measured by a porphyrinic microsensor in rat-brain after transient middle cerebral-artery occlusion. *J. Cereb. Blood Flow Metab.* **13,** 355–358.

Malinski, T., Taha, Z., Grunfeld, S., Burewicz, A., Tomboulian, P., and Kiechle, F. (1993a). Measurement of nitric oxide in biological materials using a porphyrinic microsensor. *Anal. Chim. Acta* **279,** 135–140.

Mas, M., Escrig, A., and Gonzalez-Mora, J. L. (2002). *In vivo* electrochemical measurement of nitric oxide in corpus cavernosum penis. *J. Neurosci. Methods* **119,** 143–150.

Meulemans, A. (2002). A brain nitric oxide synthase study in the rat: Production of a nitroso-corn pound NA and absence of nitric oxide synthesis. *Neurosci. Lett.* **321,** 115–119.

Millar, J. C. (2003). Real-time direct measurement of nitric oxide in bovine perfused eye trabecular meshwork using a Clark-type electrode. *J. Ocular Pharmacol. Ther.* **19,** 299–313.

Mizutani, F., Yabuki, S., Sawaguchi, T., Hirata, Y., Sato, Y., and Iijima, S. (2001). Use of a siloxane polymer for the preparation of amperometric sensors: O-2 and NO sensors and enzyme sensors. *Sens. Actuator B-Chem.* **76,** 489–493.

Moncada, S., Palmer, R. M. J., and Higgs, E. A. (1989). Biosynthesis of nitric-oxide from L-arginine: A pathway for the regulation of cell-function and communication. *Biochem. Pharmacol.* **38,** 1709–1715.

Moncada, S., Palmer, R. M. J., and Higgs, E. A. (1991). Nitric-oxide: Physiology, pathophysiology, and pharmacology. *Pharmacol. Rev.* **43,** 109–142.

Moncada, S., Radomski, M. W., and Palmer, R. M. J. (1988). Endothelium-derived relaxing factor: Identification as nitric-oxide and role in the control of vascular tone and platelet-function. *Biochem. Pharmacol.* **37,** 2495–2501.

Moroz, L. L., Norekian, T. P., Pirtle, T. J., Robertson, K. J., and Satterlie, R. A. (2000). Distribution of NADPH-diaphorase reactivity and effects of nitric oxide on feeding and locomotory circuitry in the pteropod mollusc, *Clione limacina*. *J. Comp. Neurol.* **427,** 274–284.

Nagase, S., Ohkoshi, N., Ueda, A., Aoyagi, K., and Koyama, A. (1997). Hydrogen peroxide interferes with detection of nitric oxide by an electrochemical method. *Clin. Chem.* **43,** 1246.

Novalija, E., Fujita, S., Kampine, J. P., and Stowe, D. F. (1999). Sevoflurane mimics ischemic preconditioning effects on coronary flow and nitric oxide release in isolated hearts. *Anesthesiology* **91,** 701–712.

Pailleret, A., Oni, J., Reiter, S., Isik, S., Etienne, M., Bedioui, F., and Schuhmann, W. (2003). *In situ* formation and scanning electrochemical microscopy assisted positioning of NO-sensors above human umbilical vein endothelial cells for the detection of nitric oxide release. *Electrochem. Commun.* **5,** 847–852.

Park, J. K., Tran, P. H., Chao, J. K. T., Ghodadra, R., Rangarajan, R., and Thakor, N. V. (1998). *In vivo* nitric oxide sensor using nonconducting polymer-modified carbon fiber. *Biosens. Bioelectron.* **13,** 1187–1195.

Pontie, M., Bedioui, F., and Devynck, J. (1999). New composite modified carbon microfibers for sensitive and selective determination of physiologically relevant concentrations of nitric oxide in solution. *Electroanalysis* **11,** 845–850.

Radomski, M. W., Palmer, R. M. J., and Moncada, S. (1990). An L-arginine nitric-oxide pathway present in human platelets regulates aggregation. *Proc. Natl. Acad. Sci. USA* **87,** 5193–5197.

Rievaj, M., Lietava, J., and Bustin, D. (2004). Electrochemical determination of nitric oxide in blood samples. *Chem. Pap.-Chem. Zvesti* **58,** 306–310.

Robinson, J. K., Bollinger, M. J., and Birks, J. W. (1999). Luminol/H_2O_2 chemiluminescence detector for the analysis of nitric oxide in exhaled breath. *Anal. Chem.* **71,** 5131–5136.

Rocchitta, G., Migheli, R., Mura, M. P., Esposito, G., Desole, M. S., Miele, E., Miele, M., and Serra, P. A. (2004). Signalling pathways in the nitric oxide donor-induced dopamine release in the striatum of freely moving rats: Evidence that exogenous nitric oxide promotes Ca2+ entry through store-operated channels. *Brain Res.* **1023,** 243–252.

Rysz, J., Luciak, M., Kedziora, J., Blaszczyk, J., and Sibinska, E. (1997). Nitric oxide release in the peripheral blood during hemodialysis. *Kidney Int.* **51,** 294–300.

Saitoand, M., and Miyagawa, I. (2000). Real-time monitoring of nitric oxide in ischemia-reperfusion rat kidney. *Urol. Res.* **28,** 141–146.

Sakihama, Y., Nakamura, S., and Yamasaki, H. (2002). Nitric oxide production mediated by nitrate reductase in the green alga *Chlamydomonas reinhardtii*: An alternative NO production pathway in photosynthetic organisms. *Plant Cell Physiol.* **43,** 290–297.

Schmidt, H., Warner, T. D., Ishii, K., Sheng, H., and Murad, F. (1992). Insulin-secretion from pancreatic B-cells caused by L-arginine derived nitrogen-oxides. *Science* **255,** 721–723.

Schofield, R. E. (1966). "A scientific autobiography of Joseph Priestley (1733–1804): Selected scientific correspondence." MIT Press, Cambridge, MA.

Schulte, D., and Millar, J. (2003). The effects of high- and low-intensity percutaneous stimulation on nitric oxide levels and spike activity in the superficial laminae of the spinal cord. *Pain* **103,** 139–150.

Sekaran, S., Cunningham, J., Neal, M. J., Hartell, N. A., and Djamgoz, M. B. A. (2005). Nitric oxide release is induced by dopamine during illumination of the carp retine: Serial neurochemical control of light adaption. *Eur. J. Neurosci.* **21,** 2199–2208.

Shi, X. R., Ren, T. Y., and Nuttall, A. L. (2002). The electrochemical and fluorescence detection of nitric oxide in the cochlea and its increase following loud sound. *Hear. Res.* **164,** 49–58.

Shibuki, K. (1990). An electrochemical microprobe for detecting nitric-oxide release in brain-tissue. *Neurosci. Res.* **9,** 69–76.

Shin, J. H., Weinman, S. W., and Schoenfisch, M. H. (2005). Sol-gel derived amperometric nitric oxide microsensor. *Anal. Chem.* **77,** 3494–3501.

Shiva, S., Brookes, P. S., Patel, P., Anderson, P. G., and Darley-Usmar, V. M. (2001). Nitric oxide partitioning into mitochondrial membranes and the control of respiration at cytochrome c oxidase. *Proc. Natl. Acad. Sci. USA* **98,** 7212–7217.

Simonsen, U., Wadsworth, R. W., Buus, N. H., and Mulvany, M. J. (1999). *In vitro* simultaneous measurements of relaxation and nitric oxide concentration in rat superior mesenteric artery. *J. Physiol.* **516,** 271–282.

Smith, S. R., and Thorp, H. H. (1998). Application of the electrocatalytic reduction of nitric oxide mediated by ferrioxamine B to the determination of nitric oxide concentrations in solution. *Inorg. Chim. Acta* **273,** 316–319.

Stankevicius, E., Martinez, A. C., Mulvany, M. J., and Simonsen, U. (2002). Blunted acetylcholine relaxation and nitric oxide release in arteries from renal hypertensive rats. *J. Hypertens.* **20,** 1571–1579.

Stankevicius, E., Lopez-Valverde, V., Rivera, L., Hughes, A. D., Mulvany, M. J., and Simonsen, U. (2006). Combination of Ca($^{2+}$)-activated K^+ channel blockers inhibits acetylcholine-evoked nitric oxide release in rat superior mesenteric artery. *Br. J. Pharmacol.* **149,** 560–572.

Stefano, G. B., Salzet, M., and Magazine, H. I. (2002). Cyclic nitric oxide release by human granulocytes, and invertebrate ganglia and immunocytes: Nano-technological enhancement of amperometric nitric oxide determination. *Med. Sci. Monit.* **8,** BR199–BR204.

Stefano, G. B., Zhu, W., Cadet, P., Bilfinger, T. V., and Mantione, K. (2004). Morphine enhances nitric oxide release in the mammalian gastrointestinal tract via the mu 3 opiate receptor subtype: A hormonal role for endogenous morphine. *J. Physiol. Pharmacol.* **55,** 279–288.

Stefano, G. B., Prevot, V., Beauvillain, J. C., Cadet, P., Fimiani, C., Welters, I., Fricchione, G. L., Breton, C., Lassalle, P., Salzet, M., and Bilfinger, T. V. (2000). Cell-surface estrogen receptors mediate calcium-dependent nitric oxide release in human endothelia. *Circulation* **101,** 1594–1597.

Stefano, G. B., Prevot, V., Beauvillain, J. C., Fimiani, C., Welters, I., Cadet, P., Breton, C., Pestel, J., Salzet, M., and Bilfinger, T. V. (1999). Estradiol coupling to human monocyte nitric oxide release is dependent on intracellular calcium transients: Evidence for an estrogen surface receptor. *J. Immunol.* **163,** 3758–3763.

Thom, S. R., Fisher, D., Zhang, J., Bhopale, V. M., Ohnishi, S. T., Kotake, Y., Ohnishi, T., and Buerk, D. G. (2003). Stimulation of perivascular nitric oxide synthesis by oxygen. *Am. J. Physiol.-Heart Circul. Physiol.* **284,** H1230–H1239.

Thorup, C., Kornfeld, M., Winaver, J. M., Goligorsky, M. S., and Moore, L. C. (1998). Angiotensin-II stimulates nitric oxide release in isolated perfused renal resistance arteries. *Pflugers Arch.* **435,** 432–434.

Trevin, S., Bedioui, F., and Devynck, F. (1996). Electrochemical and spectrophotometric study of the behavior of electropolymerized nickel porphyrin films in the determination of nitric oxide in solution. *Talanta* **43,** 303–311.

Tristani-Firouzi, M., DeMaster, E. G., Quast, B. J., Nelson, D. P., and Archer, S. L. (1998). Utility of a nitric oxide electrode for monitoring the administration of nitric oxide in biologic systems. *J. Lab. Clin. Med.* **131,** 281–285.

Tsatmali, M., Graham, A., Szatkowski, D., Ancans, J., Manning, P., McNeil, C. J., Graham, A. M., and Thody, A. J. (2000). Alpha-melanocyte-stimulating hormone modulates nitric oxide production in melanocytes. *J. Invest. Dermatol.* **114,** 520–526.

Uhlmann, S., Friedrichs, U., Eichler, W., Hoffmann, S., and Wiedemann, P. (2001). Direct measurement of VEGF-induced nitric oxide production by choroidal endothelial cells. *Microvasc. Res.* **62,** 179–189.

Vilakazi, S. L., and Nyokong, T. (2001). Voltammetric determination of nitric oxide on cobalt phthalocyanine modified microelectrodes. *J. Electroanal. Chem.* **512,** 56–63.

Vinik, A. I., Barlow, P. M. M., Ullal, J., Casellini, C. M., and Parson, H. K. (2006). Pioglitazone treatment improves nitrosative stress in type 2 diabetes. *Diabetes Care* **29,** 869–876.

Wadsworth, R., Stankevicius, E., and Simonsen, U. (2006). Physiologically relevant measurement of nitric oxide in cardiovascular research using electrochemical microsensors. *J. Vasc Res.* **43,** 70–85.

Wallace, J. L., and Woodman, R. C. (1995). Detection of nitric oxide by bioassay. *Methods* **7,** 55–78.

Wennmalm, A., Lanne, B., and Petersson, A. S. (1990). Detection of endothelial-derived relaxing factor in human plasma in the basal state and following ischemia using electron-paramagnetic resonance spectrometry. *Anal. Biochem.* **187,** 359–363.

Xian, Y. Z., Sun, W. L., Xue, J. A., Luo, M., and Jin, L. T. (1999). Iridium oxide and palladium modified nitric oxide microsensor. *Anal. Chim. Acta* **381,** 191–196.

Yamasaki, H., and Sakihama, Y. (2000). Simultaneous production of nitric oxide and peroxynitrite by plant nitrate reductase: *In vitro* evidence for the NR-dependent formation of active nitrogen species. *FEBS Lett.* **468,** 89–92.

Yamasaki, H., Sakihama, Y., and Takahashi, S. (1999). An alternative pathway for nitric oxide production in plants: New features of an old enzyme. *Trends Plant Sci.* **4,** 128–129.

Yokoyama, H., Mori, N., Kasai, N., Matsue, T., Uchida, I., Kobayashi, N., Tsuchihashi, N., Yoshimura, T., Hiramatsu, M., and Niwa, S. I. (1995). Direct and continuous monitoring of intrahippocampal nitric oxide (NO) by an NO sensor in freely moving rat after N-methyl-D-aspartic acid injection. *Denki Kagaku* **63,** 1167–1170.

Zhang, X. (2004). Real time and *in vivo* monitoring of nitric oxide by electrochemical sensors: From dream to reality. *Front. Biosci.* **9,** 3434–3446.

Zhang, X., and Broderick, M. (2000). Amperometric detection of nitric oxide. *Mod. Asp. Immunobiol.* **1,** 160–165.

Zhang, X. J. (2007). NO electrochemical sensor with pM detection limit: Fact or fiction. Gordon Conference, Nitric Oxide, Ventura, CA.

Zhang, X. J., Cardoso, L., Broderick, M., Fein, H., and Davies, I. R. (2000). Novel calibration method for nitric oxide microsensors by stoichiometrical generation of nitric oxide from SNAP. *Electroanalysis* **12,** 425–428.

Zhang, X. J., Cardoso, L., Broderick, M., Fein, H., and Lin, J. (2001). An integrated nitric oxide sensor based on carbon fiber coated with selective membranes. *Electroanalysis* **12,** 1113–1117.

Zhang, X. J., Lin, J., Cardoso, L., Broderick, M., and Darley-Usmar, V. (2002b). A novel microchip nitric oxide sensor with sub-n*M* detection limit. *Electroanalysis* **14,** 697–703.

Zhang, X. J., Kislyak, Y., Lin, J., Dickson, A., Cardoso, L., Broderick, M., and Fein, H. (2002a). Nanometer size electrode for nitric oxide and S-nitrosothiols measurement. *Electrochem. Commun.* **4,** 11–16.

Zhang, Z. G., Chopp, M., Bailey, F., and Malinski, T. (1995). Nitric-oxide changes in the rat-brain after transient middle cerebral-artery occlusion. *J. Neurol. Sci.* **128,** 22–27.

Zhao, J. L., Pergola, P. E., Roman, L. J., and Kellogg, D. L. (2004). Bioactive nitric oxide concentration does not increase during reactive hyperemia in human skin. *J. Appl. Physiol.* **96,** 628–632.

CHAPTER SIX

NO, N_2O, AND O_2 REACTION KINETICS: SCOPE AND LIMITATIONS OF THE CLARK ELECTRODE

L. A. M. Pouvreau,* M. J. F. Strampraad,* S. Van Berloo,† J. H. Kattenberg,* *and* S. de Vries*

Contents

Abstract

The Clark electrode, which has been commercially available for more than 50 years, is a robust first-generation sensor originally used to determine the concentration of dissolved oxygen. This paper describes a simple experimental setup employing the Clark electrode to measure low concentrations of aqueous solutions of dissolved nitric oxide (NO) (>5 n*M*) and nitrous oxide (N_2O) (>50 n*M*) in addition to oxygen (>5 n*M*). The Clark electrode is connected to a low-noise (home-built) amplifier interfaced via a 16-bit AD converter to a computer providing increased signal-to-noise performance.

Owing to the robustness of the Clark electrode, experiments can be performed routinely and repeatedly even to 90–95°, aiding, for example, in enzyme purification. The low noise enables determination of K_M values for O_2, NO, or N_2O from a single trace. Analyses can be conveniently performed on pure

* Section of Enzymology, Department of Biotechnology, Delft University of Technology, Delft, The Netherlands
† DEMO, Delft University of Technology, Delft, The Netherlands

Methods in Enzymology, Volume 436
ISSN 0076-6879, DOI: 10.1016/S0076-6879(08)36006-6

enzymes or on membranes from psychrophilic, mesophilic, and (hyper)thermophilic microorganisms. The sensitivity for O_2 and NO of the current apparatus approaches that of commercially available microelectrodes, while that for N_2O is superior.

1. Introduction

Clark-type electrodes have been successfully used in recording reaction kinetics of oxygen (O_2), hydrogen (H_2) (Wang *et al.*, 1971), and NO in various systems (Carr *et al.*, 1989; Girsch and de Vries, 1997; Liu *et al.*, 2005).

The Clark electrode consists of a working electrode (usually a platinum cathode) and a counter/reference Ag/AgCl anode immersed in an electrolyte solution (e.g., 3 *M* KCl). Both electrodes and the electrolyte are covered with a gas-permeable membrane (e.g., Teflon of 5 to 25 μm thickness), effectively separating the electrodes and electrolyte from the outer solution with dissolved gases, the concentrations of which are measured. Due to the membrane, the chemical and the physical properties of the outer solution have little effect on the measurement, a clear advantage. However, even a membrane of 5-μm thickness yields a minimal response time for the Clark-type electrode of 3 to 10 s. As a consequence, the concentration curves recorded by the electrode will deviate from true concentrations (Liu *et al.*, 2005). Because of this problem, the Clark electrode is most useful for the determination of steady-state rates. A further disadvantage of the presence of a membrane is the resultant unstirred layer, limiting the sensitivity in dynamic measurements of, for example, K_M values to a value calculated at $\approx$100 n*M*. (Lundsgaard *et al.*, 1978; see also below). To determine lower K_M values, alternative methods must be applied (D'Mello *et al.*, 1995; Lundsgaard *et al.*, 1978; Preisig *et al.*, 1996; Rice and Hempfling, 1978).

Despite these disadvantages, the Clark electrode is robust and flexible. By changing the polarization voltage (V_{pol}) some differential selectivity can be achieved with respect to dissolved O_2 vs. NO. For example, at $V_{pol} = 0.5$ to 0.6 V, the electrode is much more sensitive to O_2 than to NO; while at $V_{pol} = 0.8$V, both dissolved gases display the same sensitivity (see below). More important, by changing the polarity and thus making the Pt electrode the anode, H_2 (being oxidized to $2H^+ + 2e$) can be measured and NO (oxidized at the electrode to nitrite) without interference of the signal due to O_2.

The Clark electrode also responds to N_2O gases, but the sensitivity is about 25 times less than that for NO. N_2O synthesis or reduction was therefore studied with a modified Clark-type electrode in which the working electrode was made of Ag in place of Pt (Alefounder and Ferguson, 1982).

However, the sensitivity was still too low for quantitative measurements of N_2O reaction kinetics.

In this article, we will describe a home-made setup that enables us to amplify the signal response of the Clark electrode up to 10,000 times while suppressing the electronic noise. This new setup makes it possible to record traces of O_2, NO, and N_2O consumption down to the 5–50 n*M* range. K_M values for these dissolved gases can easily be determined from a single experimental trace.

2. Materials

The O_2, NO, N_2O, and argon gases were purchased from Hoekloos BV (The Netherlands). Stock aqueous solutions contained 1.3 m*M* O_2, 0.1 or 2 m*M* NO, and 25 m*M* N_2O. To get rid of higher nitrogen oxides, NO solutions were prepared by scrubbing over 2 *M* NaOH and then over 0.1 *M* phosphate buffer, pH 7.5, previously deaerated by sparging with 100% argon.

Enzyme sources, assay buffers, and reductants used for activity measurements are listed in Table 6.1. Each experiment was performed at least in duplicate, and the data shown were reproducible.

3. The Clark Electrode Holder

The Clark electrode (YSI Life Sciences, Yellow Springs, OH) is covered by a high-sensitivity membrane (type 5331; YSI Life Sciences). The following electrolytes and polarization voltages (V_{pol}) of the electrode were used. For O_2 and NO measurements, the electrolyte was 3 *M* KCl, and for N_2O measurements it was 3 *M* KCl + 100 m*M* KOH. The polarization voltage is set to 0.7V, 0.85V, and 1.0V (with the Pt electrode serving as the cathode) for O_2, NO, and N_2O, respectively. The sensitivity of the Clark electrode for each of these gases is depicted in Table 6.2. When expressed as mV per μM electrons (mV/μM e), the Clark electrode displays similar sensitivities toward O_2 and NO, while for N_2O the sensitivity is approximately 10-fold lower. NO could also be detected in the reverse polarization mode (same electrolyte as previously, but $V_{pol} = -0.8V$); the sensitivity is then approximately twofold less. The advantage of the reverse polarization is that oxygen is not detected under such conditions. Commercially available NO-sensors from, for example, World Precision Instruments (NOP200), work in the reversed polarized mode (see the chapter by Davies and Zhang; Chapter 5).

Table 6.1 Enzyme sources and experimental conditions

Studied enzyme	Origin	Buffer	Electron donors	References
qCu_ANor*	*B. azotoformans*	50 m*M* KP_i, pH 7.0	10 m*M* ascorbate + 100 μM PES	(Suharti *et al.*, 2001)
Nor	*P. aerophilum*	25 m*M* KP_i, pH 6.0	10 m*M* formate	(de Vries and Schröder, 2002; de Vries *et al.*, 2003)
N_2OR	*B. azotoformans*	20 m*M* KP_i, pH 7.0	10 m*M* ascorbate + 100 μM PES	(Suharti and de Vries, 2005)
cbb$_3$ oxidase	*B. japonicum*	50 m*M* KP_i, pH 7.0	3 m*M* TMPD	(Delgado *et al.*, 1995; Preisig *et al.*, 1996)
aa$_3$ oxidase*	*P. denitrificans*	50 m*M* KP_i, pH 7.0	3 m*M* TMPD	(Ludwig and Schatz, 1980)
bo$_3$ and *bd* oxidase	*E. coli*	25 m*M* KP_i, pH 7.0	10 m*M* ascorbate + 5 μM PES	(D'Mello *et al.*, 1995)

* Purified protein was used in the assays; for all other assays purified membranes were used. PES: phenazine ethosulfate; TMPD: N,N,N′,N′-tetramethyl-1,4-phenyl-ene-diammonium dichloride.

Table 6.2 Sensitivity of the Clark electrode for O_2, NO, and N_2O

	O_2	NO	N_2O
Concentration (μM)	250	100	100
Signal (V)*	10 ± 2	1.8 ± 0.5	0.4 ± 0.1
Sensitivity (mV/μM)	40 ± 8	18 ± 5	4 ± 1
Absolute sensitivity (mV/μM e)	20 ± 4	18 ± 5	2 ± 0.5

* Voltage at a 1000-fold amplification of the signal.

The cylindrical reaction vessel is made of Plexiglas (PMDA) or polycarbonate for experiments at temperatures >75° and can be used for volumes between 1.5 and 3.0 ml. The Clark electrode is fitted from the bottom, and stirring occurs from the side with a magnetic stirring bar (Fig. 6.1).

The setup is thermostatted (allowing measurements up to ≈95°) via a water jacket connected to a thermostat. Since the rate of O_2 influx at >50° via the inlet port of the cylindrical stopper (see No. 1, Fig. 6.1) is too high for routine measurements, a special stopper was designed (see Nos. 11 and 12, Fig. 6.1). When closed with a black (not gray) butyl rubber stopper, the rate of O_2 influx could be sufficiently reduced.

4. Schematics of the Clark-Electrode Setup

Figure 6.2A describes schematically the home-built setup. The Clark electrode is connected to an amplifier working on two 9V batteries to prevent 50/60 Hz interference from the net. The amplifier is connected to a home-made channel selector, in which five channel sensitivities are selected: ±10V, ±5V, ±1, ±0.325V, and ±0.0065V. This channel selector is subsequently connected to a 16-bit analog digital converter (ADC) (National Instruments, the Netherlands) in which the channels are preselected. The ADC is further connected to a computer for data recording and storage. Signal recording is via a home-written program (Labview) that enables a data-transfer time of 200 ms.

To suppress electric or electrostatic interferences during measurements, the Clark-electrode holder, the rotating magnet, and the amplifier are contained in a Faraday cage. The magnet is connected via a shielded and grounded cable to its power supply, mounted outside the Faraday cage.

The amplifier was designed and built in-house and has three main purposes: (1) Transforming current to voltage and amplification of the voltage by a factor of 10 up to 10000, (2) zeroing the instrument, and (3) changing the polarization voltage of the electrode. The full schematic diagram of the amplifier is given in Fig. 6.2B, while Fig. 6.2C clarifies the

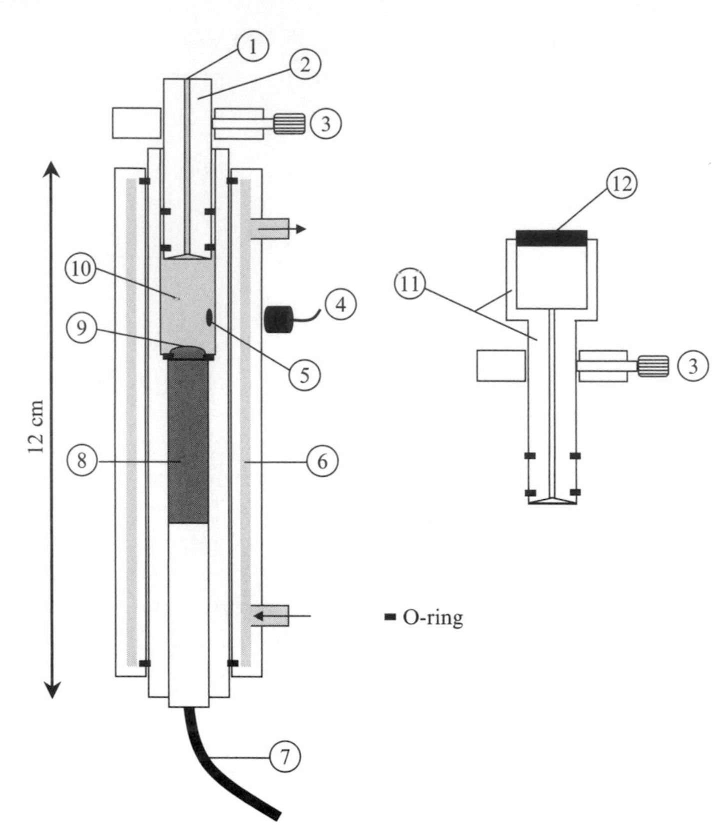

Figure 6.1 Clark-electrode set-up, placed in a Faraday cage. (1) Movable cylindrical stopper, (2) hole for addition with a Hamilton syringe, (3) stop-screw for setting the desired volume, (4) rotating magnet connected to motor, (5) stirring bar, (6) water jacket, (7) Clark electrode, (8) cable to amplifier, (9) high-sensitivity membrane, (10) reactant volume (1.4–4.2 ml; 1–3 cm long × 1.35 cm inner diameter), (11) stopper used for high temperature measurement under strictly anaerobic conditions, (12) black butyl rubber septum.

operation of the amplifier. The V_{pol} is set by the voltage source, which is connected to the + port of the op-amp (Fig. 6.2C); the Clark electrode (probe) is connected to the − port of the op-amp. During the measurement, the current changes because of the changing concentration of the dissolved gas. This change in current is compensated by the current source, amplified by the differential amplifier, and led outside to the channel selector. The critical element in the circuit is the IL300 used in the current source (Fig. 6.2B). The IL300 is a linear optocoupler with a high-gain stability yielding low noise, yet at high amplifications.

In practice, the Clark electrode is calibrated with atmospheric oxygen (250 μM) or any of the stock NO or N_2O solutions, and the amplification is

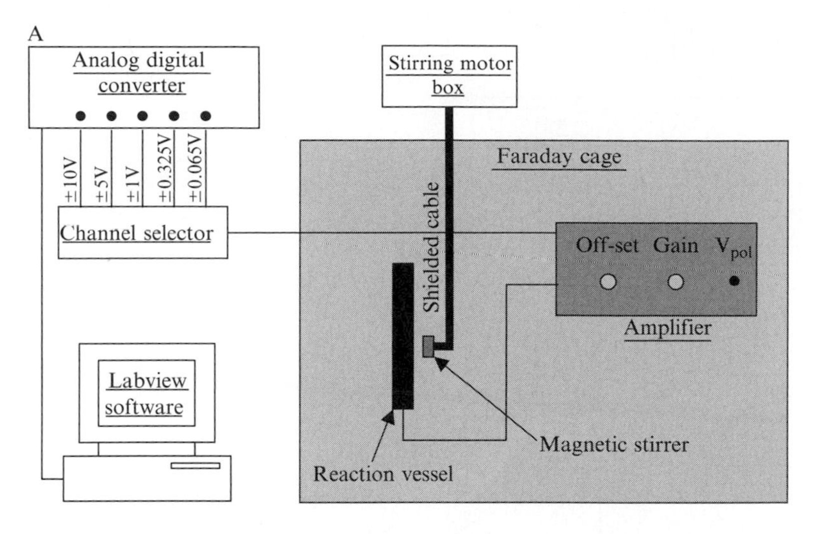

Figure 6.2 (*continued*)

set to 100. For measurements at much lower concentrations of the dissolved gases (e.g., by injection from the stock solutions), the amplification might be set at 10,000. However, in general, an amplification of 1,000 (yielding $\approx$10V for 250 μM O_2) was used (cf. Table 6.2).

5. Performance of the Apparatus

5.1. Determination of NO reduction activity of Nor

NO reductases (Nor) are found in various denitrifying organisms (Carr *et al.*, 1989; de Vries *et al.*, 2003; de Vries and Schröder, 2002; Girsch and de Vries, 1997; Suharti *et al.*, 2001, 2004; Suharti and de Vries, 2005; Wasser *et al.*, 2002; Zumft, 1997) and catalyze the reduction of NO to N_2O according to Eq. (6.1):

$$2\ NO + 2H^+ + 2e^- \rightarrow N_2O + H_2O \tag{6.1}$$

Nors from Gram-positive, Gram-negative bacteria and Archaea have been shown to display similar steady-state kinetics, in which the enzyme is inhibited at high ($\approx$10 μM) concentrations of NO (Fig. 6.3A) (de Vries *et al.*, 2003; Girsch and de Vries, 1997; Suharti *et al.*, 2001). The inhibition by NO yields sigmoid activity traces (Fig. 6.3A). The NO substrate inhibition is

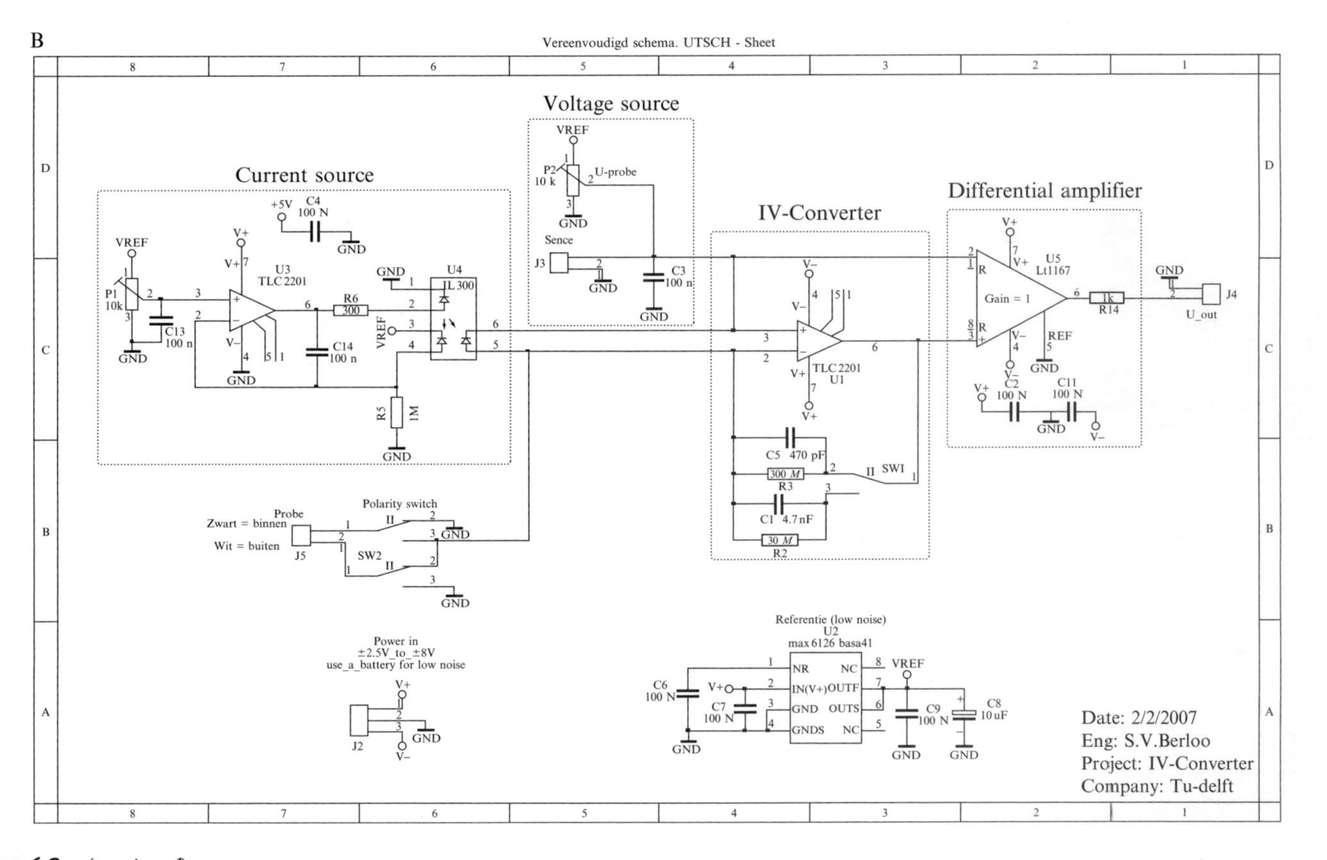

Figure 6.2 (*continued*)

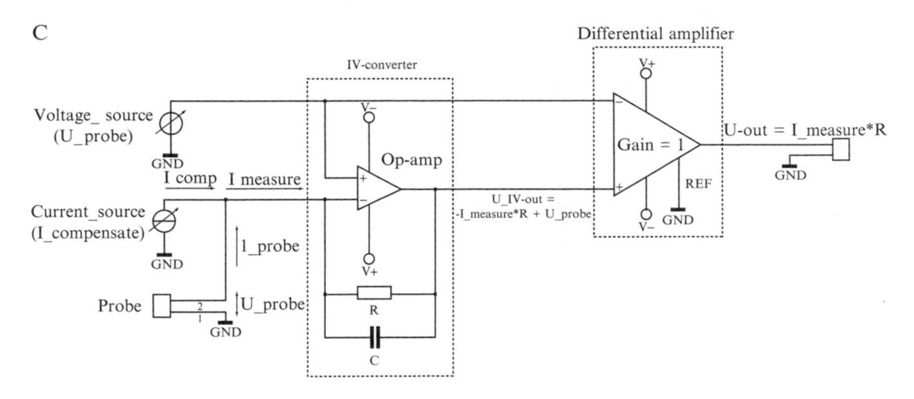

Figure 6.2 The Clark-electrode set-up (A), the schematics of the low-noise amplifier (B), and the schematic drawing of the mode of operation of the amplifier (C).

suggested to be due to binding of NO ($K_d = K_i \sim 10\ \mu M$) to the oxidized form of the enzyme (de Vries *et al.*, 2003). On the basis of this assumption, a simple (non-Michaelis-Menten) steady-state rate equation (Eq. [6.2]) has been derived (Girsch and de Vries, 1997) in which K_1 represents the binding affinity for the first NO and K_2 for the second:

$$V = k_{cat} \times e/(1 + K_2(1/[NO] + K_1/[NO]^2) + [NO]/K_i) \quad (6.2)$$

In Fig. 6.3, 7 μM NO from a 2-mM solution was injected into a buffer containing *B. azotoformans* qCu_ANor. Given the response time, the trace represents true NO consumption kinetics at [NO] < 6 μM. The three parameters of Eq. (2) can be determined directly by fitting the rate vs. [NO] curve (Fig. 6.3B) obtained from the kinetic trace in Fig. 6.3A by differentiation and smoothing (20-fold). The fitting parameters (Table 6.3) for the *B. azotoformans* enzyme are close to those of the *Paracoccus denitrificans* Nor (Girsch and de Vries, 1997).

Figures 6.4A and 6.4B show NO reduction by membranes of *Pyrobaculum aerophilum* at 75 °C (de Vries and Schröder, 2002; de Vries *et al.*, 2003). Data analysis was similar as for the *B. azotoformans*, yielding similar NO affinity constants, but the inhibition constant could not be determined owing to the low [NO] (see Table 6.3). The traces indicate that the Clark electrode also performs well at this elevated temperature. In fact, owing to the larger diffusion rate of NO (or other dissolved gases) at 75°, both sensitivity and response time are somewhat better than at room temperature.

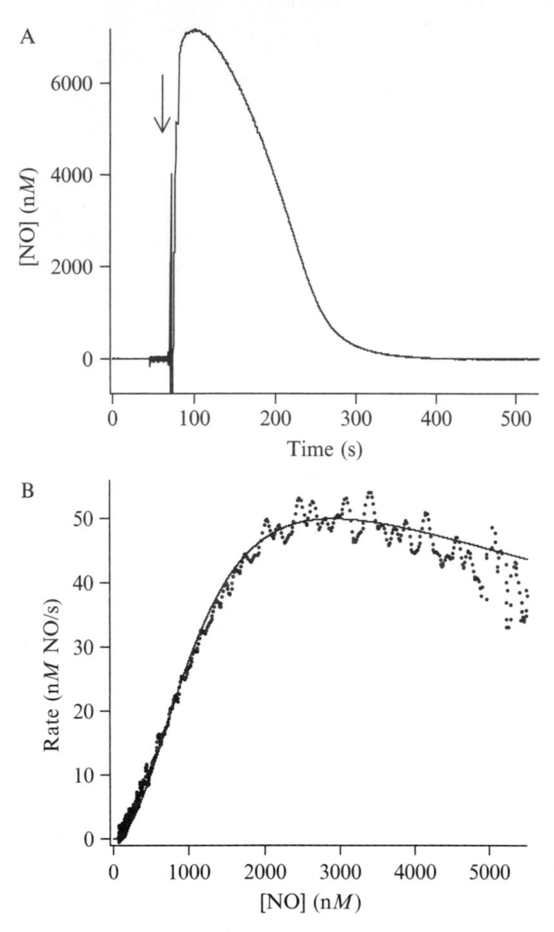

Figure 6.3 (A) Steady-state reduction of NO by purified *B. azotoformans* qCu_ANor. At the arrow, 7 μM NO was added from a 2-m*M* NO stock solution. Temperature: 21°. (B) Rate vs. [NO] plot. The rate was calculated from the slope of the trace in (A) and smoothed 20-fold. Dots: experimental rate. Line: fit to Eq. (6.2) with fit parameters listed in Table 6.3.

5.2. Determination of N_2O reduction activity of N_2OR

N_2O reductase (N_2OR) is found in various denitrifiers and performs the last step of denitrification (Wasser *et al.*, 2002; Zumft, 1997) given by Eq. (6.3):

$$N_2O + 2H^+ + 2e^- \rightarrow N_2 + H_2O \tag{6.3}$$

N_2O reductase activity was determined in membranes from *B. azotoformans*. For this measurement, the amplification is set at 1000, yielding 0.4V at 100 μM N_2O. The enzyme follows Michaelis-Menten kinetics, as indicated

Table 6.3 K_M values determined for NO–, N_2O–, and O_2-reducing enzymes as determined with the Clark electrode

Enzyme	K_1 (n*M*)	K_2 (n*M*)	K_i (n*M*)
qCu_ANor	10,000	180	6,000
Nor *P. aerophilum*	2,200	180	Omitted from calculation
	K_M (n*M*)	K_M (n*M*) (literature)	References
N_2OR	3,400	2,000–25,000	(Zumft, 1997)
Cytochrome *cbb*$_3$ oxidase	75–125	7	(Preisig *et al.*, 1996)
Cytochrome *bd* oxidase	75–125	20–40	(D'Mello *et al.*, 1995; Rice and Hempfling, 1978)
Cytochrome *bo*$_3$ oxidase	150–250	150–300	(D'Mello *et al.*, 1995; Rice and Hempfling, 1978)
Cytochrome *aa*$_3$ oxidase	250–350	—	—

in Fig. 6.4B. Figure 6.5A shows a typical measurement of reduction of N_2O to N_2 by N_2OR. The K_M for N_2O is 3.4 μM (see Table 6.3), in the range of values determined for N_2OR from Gram-negative denitrifiers (Zumft, 1997). The increased sensitivity of the set-up thus allows determination of kinetic parameters for N_2O reduction in a single trace, rather than sampling in time followed by determination of N_2O by gas chromatography (Frunzke and Zumft, 1984).

5.3. O_2 measurements

According to Preisig *et al.* (1996), the *cbb*$_3$-type cytochrome oxidase from *B. japonicum* has a K_M of 7 n*M* for O_2, a value obtained by a spectrophotometric method with oxygenated soybean leghemoglobin as the O_2 delivery and analysis system. The K_M for the *E. coli* cytochrome *bd* oxidase has been determined at 20–40 n*M* (D'Mello *et al.*, 1995; Rice and Hempfling, 1978). Both K_M values are close to the noise level ($\approx$5–10 n*M* O_2) of the Clark-electrode set-up. Activity traces for the *cbb*$_3$ and *bd* oxidases are straight lines down to $\approx$50–100 n*M* O_2 (data not shown). Experimental K_M values for the cytochrome *cbb*$_3$ and *bd* oxidases obtained with the current set-up using Michealis-Menten or Eadie-Hofstee analyses ranged between 75 and

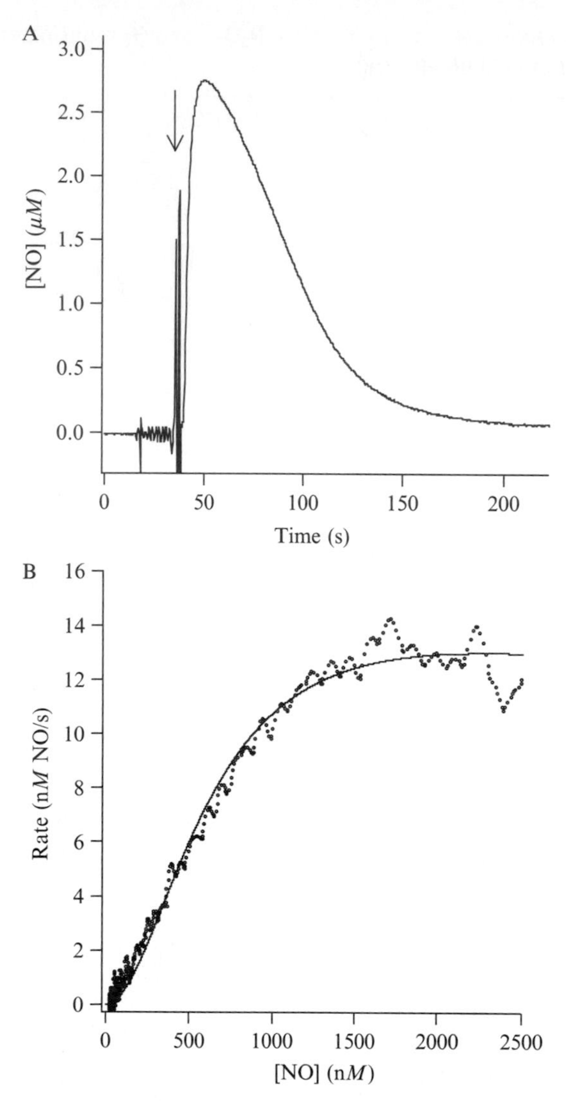

Figure 6.4 (A) Steady-state reduction of NO by membranes of *P. aerophilum*. At the arrow, 3 μM NO was added from a 2-mM NO stock solution. Temperature: 75°. (B) Rate vs. [NO] plot. The rate was calculated from the slope of the trace in (A), and smoothed 20-fold. Dots: experimental rate. Line: fit to Eq. (6.2) with fit parameters listed in Table 6.3.

125 nM, much higher than the true K_M values (Table 6.3). K_M values obtained for the *E. coli* cytochrome bo_3 oxidase and the *P. denitrificans* cytochrome aa_3 oxidase were 200 ± 50 μM and 300 ± 50 μM, respectively (data not shown, but see Table 6.3), in good agreement with the literature (D'Mello *et al.*, 1995; Rice and Hempfling, 1978).

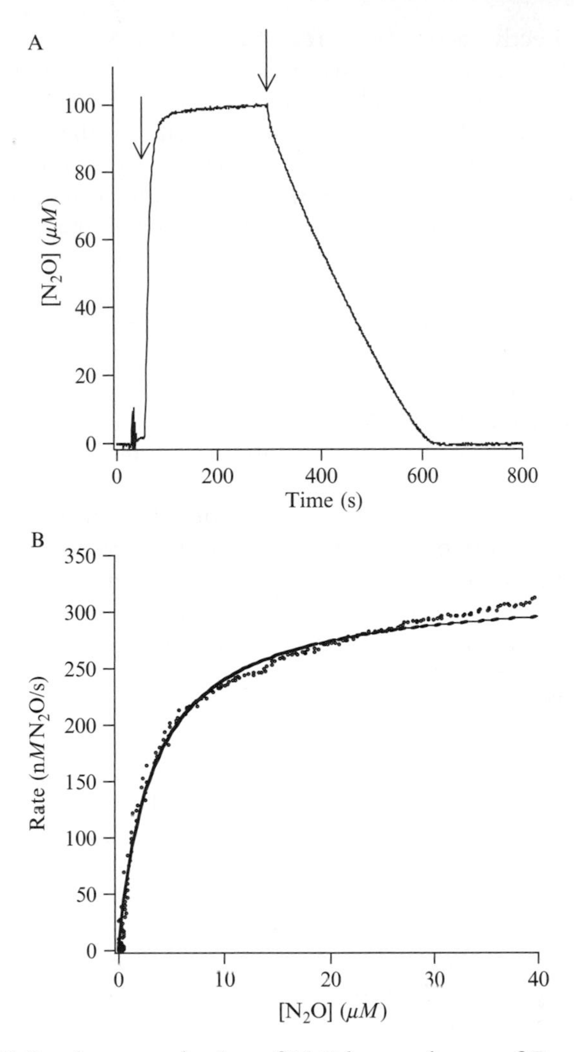

Figure 6.5 (A) Steady-state reduction of N_2O by membranes of *B. azotoformans*. At the left arrow, 100 μM N_2O was added from a 25-mM N_2O stock solution. At the right arrow, the membranes were added. Temperature: 21°. (B) Rate vs. [N_2O] plot. The rate was calculated from the slope of the trace in (A) and smoothed 20-fold. Dots: experimental rate. Line: fit to Eq. (6.2) with fit parameters listed in Table 6.3.

6. Scope and Limitations of the Clark Electrode

The noise level of the current set-up, ≈0.2–0.3 mV, corresponds to NO and O_2 levels of 5–10 nM and approaches that of commercially available microelectrodes. For N_2O the noise level of 50–100 nM made it possible to directly record N_2O reductase activities without the need for sampling and gas chromatography analysis (Frunzke and Zumft, 1984).

In the present set-up the electronic noise level is only 20 to 30 μV, ten times lower than the practical value; we surmise that the magnetic stirrer is the main source of noise. Thus, there might be room for an approximate 10-fold improvement to achieve n*M* to subnanomolar sensitivities by using alternative stirring devices. However, this might well make the total set-up less convenient for routine measurements and should probably be limited to special applications in which low amounts of the gases are being produced.

Owing to the low noise level and the great number of data points recorded (5 per s), experimental traces can be differentiated and smoothed without losing essential information, enabling, for example, the determination of K_M values (and V_{max}) from a single trace and thus eliminating errors due to, for example, differences in enzyme concentration in different experiments.

Work on the cytochrome oxidases clearly shows the limit of the Clark electrode. K_M values below 100 n*M* O_2 (or NO) cannot be determined. While this is in part due to the noise level of the current set-up, it is mainly due to presence of the unstirred layer, the Teflon membrane, with the dissolved gases, which do not equilibrate with the bulk concentration of the dissolved gases (Lundsgaard *et al.*, 1978). Calculations for a 12.5-μm Teflon membrane indicated that K_M values around 10 n*M* cannot be determined. K_m values of 100 n*M* can be determined with 50% accuracy, K_m values of 1 μM with 90% accuracy (Lundsgaard *et al.*, 1978). However, the data shown in Table 6.3 obtained with a 5-μM Teflon membrane, which might yield slightly better results than calculated (Lundsgaard *et al.*, 1978), are in very good agreement with the calculations regarding the lower limit of ≈100 n*M* for determination of the K_M.

Commercial (micro)electrodes for detection of NO or O_2 are more sensitive than the Clark-electrode set-up described here and are also significantly smaller, enabling measurements on cultivated cells or on smaller volumes. However, with these electrodes, the same problem of the unstirred layer is encountered as is with the Clark electrode, and it is not possible to use them for low K_M determinations. In our experience, the Clark electrode is much more robust than the microelectrodes. In addition, the Clark electrode can be used for various gases by simply changing the V_{pol} and has a stable response and low background signal even at 95°. This makes the Clark electrode an excellent tool for biochemical research on enzyme systems over a great temperature range including the hyperthermophilic enzymes.

ACKNOWLEDGMENTS

We greatly appreciate the gift of *P. denitrificans* cytochrome aa_3 oxidase from Dr. B. Ludwig, of *p. aerophilum* cells from Dr. I. Schröder, and of *B. japonicum* cells from Dr. M.J. Delgado to measure the cbb_3 oxidase activity. We also thank Bsc. H. Wolleswinkel for performing some of the NO reductase measurements.

REFERENCES

Alefounder, P. R., and Ferguson, S. J. (1982). Electron transport-linked nitrous oxide synthesis and reduction by *Paracoccus denitrificans* monitored with an electrode. *Biochem. Biophys. Res. Commun.* **104,** 1149–1155.

Carr, G. J., Page, M. D., and Ferguson, S. J. (1989). The energy-conserving nitric-oxide-reductase system in *Paracoccus denitrificans*: Distinction from the nitrite reductase that catalyses synthesis of nitric oxide and evidence from trapping experiments for nitric oxide as a free intermediate during denitrification. *Eur. J. Biochem.* **179,** 683–692.

D'Mello, R., Hill, S., and Poole, R. K. (1995). The oxygen affinity of cytochrome bo′ in *Escherichia coli* determined by the deoxygenation of oxyleghemoglobin and oxymyoglobin: K_M values for oxygen are in the submicromolar range. *J. Bacteriol.* **177,** 867–870.

de Vries, S., and Schröder, I. (2002). Comparison between the nitric oxide reductase family and its aerobic relatives, the cytochrome oxidases. *Biochem. Soc. Trans.* **30,** 662–667.

de Vries, S., Strampraad, M. J., Lu, S., Moënne-Loccoz, P., and Schröder, I. (2003). Purification and characterization of the MQH_2: NO oxidoreductase (qNOR) from the hyperthermophilic Archaeon *Pyrobaculum aerophilum. J. Biol. Chem.* **278,** 35861–35868.

Delgado, M. J., Yeoman, K. H., Wu, G., Vargas, C., Davies, A. E., Poole, R. K., Johnston, A. W., and Downie, J. A. (1995). Characterization of the cycHJKL genes involved in cytochrome *c* biogenesis and symbiotic nitrogen fixation in *Rhizobium leguminosarum. J. Bacteriol.* **177,** 4927–4934.

Frunzke, K., and Zumft, W. G. (1984). Rapid, single sample analysis of H_2, O_2, N_2, NO, CO, N_2O and CO_2 by isothermal gas chromatography: Applications to the study of bacterial denitrification. *J. Chromatogr.* **299,** 477–483.

Girsch, P., and de Vries, S. (1997). Purification and initial kinetic and spectroscopic characterization of NO reductase from *Paracoccus denitrificans. Biochim. Biophys. Acta* **1318,** 202–216.

Liu, X., Liu, Q., Gupta, E., Zorko, N., Brownlee, E., and Zweier, J. L. (2005). Quantitative measurements of NO reaction kinetics with a Clark-type electrode. *Nitric Oxide* **13,** 68–77.

Ludwig, B., and Schatz, G. (1980). A two-subunit cytochrome c oxidase (cytochrome aa_3) from *Paracoccus dentrificans. Proc. Natl. Acad. Sci. USA* **77,** 196–200.

Lundsgaard, J. S., Gronlund, J., and Degn, H. (1978). Error in oxygen measurements in open systems owing to oxygen consumption in unstirred layer. *Biotechnol. and Bioeng.* **20,** 809–819.

Preisig, O., Zufferey, R., Thony-Meyer, L., Appleby, C. A., and Hennecke, H. (1996). A high-affinity cbb_3-type cytochrome oxidase terminates the symbiosis-specific respiratory chain of *Bradyrhizobium japonicum. J. Bacteriol.* **178,** 1532–1538.

Rice, C. W., and Hempfling, W. P. (1978). Oxygen-limited continuous culture and respiratory energy conservation in *Escherichia coli. J. Bacteriol.* **134,** 115–124.

Suharti, W. P., and deVries, S. (2005). Membrane-bound denitrification in the Gram-positive bacterium *Bacillus azotoformans. Biochem. Soc. Trans.* **33,** 130–133.

Suharti, S., Heering, H. A., and de Vries, S. (2004). NO reductase from *Bacillus azotoformans* is a bifunctional enzyme accepting electrons from menaquinol and a specific endogenous membrane-bound cytochrome c551. *Biochemistry* **43,** 13487–13495.

Suharti, Strampraad, M. J., Schröder, I., and de Vries, S. (2001). A novel copper A containing menaquinol NO reductase from *Bacillus azotoformans. Biochemistry* **40,** 2632–2639.

Wang, R., Healey, F. P., and Myers, J. (1971). Amperometric measurement of hydrogen evolution in *Chlamydomonas. Plant Physiol.* **48,** 108–110.

Wasser, I. M., de Vries, S., Moënne-Loccoz, P., Schröder, I., and Karlin, K. D. (2002). Nitric oxide in biological denitrification: Fe/Cu metalloenzyme and metal complex NO(x) redox chemistry. *Chem. Rev.* **102,** 1201–1234.

Zumft, W. G. (1997). Cell biology and molecular basis of denitrification. *Microbiol. Mol. Biol. Rev.* **61,** 533–616.

CHAPTER SEVEN

Chemiluminescence Quantification of NO and Its Derivatives in Liquid Samples

Jay R. Laver, Tânia M. Stevanin, *and* Robert C. Read

Contents

Abstract

Nitric oxide (NO) is a ubiquitous gas with potent biological effects, including vasodilation, neuronal signaling, and antimicrobial activity. NO is a free radical and can readily react with other molecules, in particular, iron centers and oxygen. At physiological concentrations in aqueous solutions, even in the presence of oxygen, NO is reasonably stable. Under these conditions, NO is oxidized almost exclusively to nitrite (NO_2^-). In cell lysates and tissue extracts with iron-containing proteins, however, NO is postulated to have a very short half-life, with the major oxidation product being nitrate (NO_3^-). In mammalian cells, NO is generated via the action of the NO synthases (NOS), of which there

Academic Unit of Infection and Immunity, School of Medicine and Biomedical Sciences, University of Sheffield, Sheffield, United Kingdom

Methods in Enzymology, Volume 436
ISSN 0076-6879, DOI: 10.1016/S0076-6879(08)36007-8

are three known isotypes. NO can also be generated from the chemical decomposition of *S*-nitrosothiols, and there is some indication that naturally occurring *S*-nitrosothiols, such as *S*-nitrosoalbumin, may be natural reservoirs of NO *in vivo*. Here we describe a methodology to measure variations in NO in liquid samples using chemiluminescence. The protocols described allow us to distinguish between various products of NO chemistry, thus providing a sensitive method of measurement of NO concentration within a sample. They also allow us to distinguish between the various products that may be generated when NO reacts with molecules in complex biological samples such as cell lysates and supernatants.

1. Introduction

This chapter discusses the use of chemiluminescence for accurate and sensitive measurement of the oxidation products of NO: nitrites (NO_2^-), nitrates (NO_3^-), and *S*-nitrosothiols (RSNOs), using the Sievers NO analyzer.

The Griess reagent has long been used to measure changes in NO/NO_2^- content in cell lines and murine samples; however, it is relatively insensitive and therefore unsuitable for the measurement of nitrite in primary human cells, due to the low levels of NO normally present in the samples and the difficulties in inducing expression of iNOS in human cells.

1.1. The myriad roles of NO

NO is a free radical that has been shown to play an important role in a variety of biological processes. NO is involved in the control of vascular tone through the stimulation of guanylate cyclase, having been identified as the endothelium-derived relaxing factor (EDRF) (Ignarro *et al.*, 1987). The reactions of NO with hemoglobin, myoglobin, and the metal centers of other proteins have been well characterized (reviewed in Cooper, 1999), as has its role as a signaling molecule in neurons (reviewed in Guix *et al.*, 2005). There is cumulative evidence implicating NO in mechanisms such as apoptosis (Hara *et al.*, 2005), intracellular signaling cascades (Lander *et al.*, 1995) and gene expression (Reynaert *et al.*, 2004). NO plays a very important role in innate immunity, where it is synthesized by the constitutively active inducible NO synthase (iNOS). In combination with the superoxide radical, which is also produced in response to infection, NO forms the toxic peroxynitrite ($ONOO^-$). $ONOO^-$ is capable of nitrosating the tyrosine residues of proteins (Ischiropoulos *et al.*, 1992), causing detrimental conformational changes and disrupting function (Beckman and Koppenol, 1996).

The observed antimicrobial action of NO and its reaction products (Huang *et al.*, 2002; Stevanin *et al.*, 2005) is undoubtedly beneficial, but it should also be noted that nitrogen oxides (especially $ONOO^-$) are also responsible for pathophysiological conditions, such as ischemia (Eliasson *et al.*, 1999), chronic inflammatory conditions (Stichtenoth and Frolich, 1998), and septic shock syndrome (Kirkebøen and Strand, 1999).

1.2. The chemistry of NO

NO is poorly soluble in water, does not undergo any form of hydration reaction, is freely diffusible in biological systems and capable of traversing cellular boundaries. In aqueous solution in the presence of oxygen, NO is oxidized almost entirely to nitrite (NO_2^-), with nitrate (NO_3^-) levels below the limit of detection (Ignarro *et al.*, 1993). Given that the half-life of NO is inversely proportional to the square of its concentration (Beckman and Koppenol, 1996), biologically active concentrations of NO (5 n*M*–4 μM) in pure aqueous solution have a half-life in excess of 500 s. In biological tissues, however, the observed half-life of NO is only 3–5 sec (Ignarro *et al.*, 1990) owing to the presence of oxyhemoproteins and the abundance of superoxide, the latter reacting with NO at the near-diffusion-limited rate of 6.7×10^9 M^{-1} s^{-1}, to produce $ONOO^-$. NO_3^- is the predominant oxidation product of NO in biological samples such as tissue homogenates and impure protein extracts.

Another important aspect of NO chemistry is the formation of *S*-nitrosothiols. *S*-nitrosothiols are the products of the reaction between an intermediate in NO oxidation (e.g., N_2O_3) and the free thiol of cysteine residues (Zhang and Hogg, 2004). *S*-nitrosothiol formation is an important mechanism of post-translational modification of proteins by NO, having been shown to play a role in a multitude of molecular processes. For example, it is involved in the regulation of caspases (Li *et al.*, 1997), the activation of NF-κB (Reynaert *et al.*, 2004), signaling mediated through the $p^{ras}21$ pathway (Lander *et al.*, 1995; Williams *et al.*, 2003), the nuclear localization of GAPDH during apoptosis (Hara *et al.*, 2005), and the activity of inducible NO synthase (iNOS) (Mitchell *et al.*, 2005).

1.3. The NO-ozone reaction and the chemiluminescence analyzer

Quantification of nitrogen oxides in biological samples is achieved through the combinatorial use of reducing agents, which reduce the different nitrogen oxide species to NO, and gas-phase chemiluminescent detection of NO using ozone.

Reaction 7.1 shows that some of the NO_2 produced from the reaction of ozone and NO is in an excited state (NO_2★), which releases energy (*hv*) in the red and near-infrared region of the spectrum. The release of energy is increased in low-pressure environments, is heavily dependent on temperature (Gorimar, 1985), and is directly proportional to the NO content of the sample. Chemiluminescent detectors of NO such as the Sievers 280i NOA used in our laboratory are equipped with "a thermoelectrically cooled, red-sensitive photomultiplier tube" (Sievers nitric oxide analyzer NOA^{TM} 280i: Operation and Maintenance Manual, 1995–2000) to detect this release of energy.

$$NO + O_3 \rightarrow NO_2{}^* + O_2 \rightarrow NO_2 + hv \qquad (7.1)$$

NO has a high partition coefficient (>20), which means that it readily diffuses into the gas phase of the reaction cell. Since the reaction of NO with ozone takes place in the gas phase, it is necessary to assist this diffusion and strip the NO from liquid samples. An inert gas, such as helium or nitrogen, is appropriate for this process and helps maintain an oxygen-depleted environment, where NO is unable to oxidize and form equimolar concentrations of NO_2^- and NO_3^-.

Generating NO from the nitrogen oxides present in liquid samples is achieved with the use of reducing agents. The reduction step takes place in the purge vessel, where samples are injected into reducing solution, which is itself bubbled (or sparged) with the inert gas. The purge vessel is sheathed in a water jacket, allowing the operator to stabilize the temperature at which the different reduction reactions take place. The outlet from the purge vessel contains a water-cooled condenser, to help prevent hot acidic vapors from entering and corroding the plastic tubing. Note that although subsequent protocols refer to the use of N_2, any other inert gas can be substituted in its place.

Reactions 7.2 and 7.3 illustrate how NO is evolved in the purge vessel from nitrite and nitrate, respectively. Reaction 2 takes place at room temperature and requires the use of a 1% solution of sodium iodide (NaI) in glacial acetic acid (Section 2.1). Much stronger reducing conditions are required to convert nitrate to NO, however, and reaction 3 takes place in a saturated acidified vanadium (III) chloride (VCl_3) solution at 90° (see Section 2.2). The strength of this reducing agent is sufficient to also convert nitrite and *S*-nitrosocompounds into NO. Other nitrocompounds, such as nitroarginine, are slowly reduced over time in the presence of acidified VCl_3.

$$I^- + NO_2^- + 2H^+ \rightarrow NO + \{1/2\}I_2 + H_2O \qquad (7.2)$$

$$2NO_3^- + 3V^{3+} + 2H_2O \rightarrow 2NO + 3VO_2^+ + 4H^+ \qquad (7.3)$$

2. Methods

2.1. Determination of NO_2^- concentration in liquid samples

Reagents

NaI (MW: 149.89) (Fisher Scientific)
Glacial acetic acid (BDH Laboratory Supplies)
$NaNO_2$ (MW: 69.00, minimum 99.5%) (Sigma-Aldrich)
Antifoaming agent (Analytix Ltd.)
dH_2O

Reagents preparation

1. Dissolve 50 mg NaI into 500 μl of dH_2O. Aliquots of this solution can be kept at room temperature in the dark but should be used within the day (aliquots of NaI in salt form can be stored for longer in the dark). Immediately prior to analysis, add 4.5 ml glacial acetic acid. The solution will turn a yellow color.
2. Dissolve 0.345 *g* $NaNO_2$ into 5 ml dH_2O to make up a 1 *M* stock solution. This stock solution can be stored for up to a month if protected from light and kept at 4°.

Procedures

1. Using the 1 *M* $NaNO_2$ solution, prepare a range of standards by 10-fold serial dilution from 10 m*M* to 10 n*M*. Store at room temperature and protect the standards from light when not in use. Dilute standards are unstable and should be prepared fresh on the day of the experiment.
2. Connect the feed line from the condenser to the NO analyzer (NOA). Turn on the NOA and allow it to cool. The photomultiplier and detection apparatus operate below freezing, because the emission of light from the ozone-NO reaction is strongly influenced by temperature (Gorimar, 1985).
3. Turn on the supplies of O_2 (used to generate ozone in the NOA) and N_2.
4. Start the "analysis" function of the NOA. *Never* activate this function without first supplying the NOA with oxygen. Upon activation of the pump, the pressure of the cell should drop. If the system is open to the atmosphere, the cell pressure (measured in Torr) registered by the NOA is the current atmospheric pressure. Make a note of this value, as the pressure of the closed cell will be adjusted to match it.
5. Add 4 ml of the NaI in glacial acetic acid (hereafter referred to as "reaction mixture") and an adequate amount of antifoaming agent to

the purge vessel. Antifoaming agent is added because on addition of proteinaceous samples, the sparging reaction mixture will otherwise have a tendency to foam. Aspiration of this foam by the NOA will coat the inside of the tubing, thus affecting the results. The amount of antifoaming agent added will vary according to the concentration of the particular antifoaming agent, as well as to the volume of reaction mixture. It is important to note that too much antifoaming may affect the release of NO from the solution creating broader and flatter peaks and may affect sensitivity.

6. Adjust the flow of N_2 so that the cell pressure matches the previously recorded atmospheric pressure.
7. Allow approximately 5–10 min for the reaction mixture to equilibrate. An initial spike in luminescence is detected following addition to the purge vessel, but this soon drops off as any NO/NO_2^- present in the reaction mixture is removed. In a properly maintained NOA, the baseline signal using this reaction mixture is low and flat (approximately 4–6 mV in our system).
8. Use a glass Hamilton microsyringe (Analytix, United Kingdom) to inject known volumes of the NO_2^- standards into the purge vessel. When performing injections, be sure to leave at least a 10-μl headspace of air before taking the sample into the syringe. This ensures that when the plunger is completely depressed, the entire sample will be delivered into the purge vessel and none will be left in the needle. Injections will produce a distinctive peak compared to baseline, as the NO derived from the NO_2^- reacts photolytically with ozone. Especially when injecting very small amounts of sample, ensure that the injected liquids are delivered directly into the reaction mixture and are not splashed on the wall of the purge vessel. Delivery of material as a bolus will produce a more distinctive peak, which is easier to define for the purposes of integration.
9. Integrate the area under each peak. By plotting the amount of NO_2^- injected along the x-axis and the corresponding area along the y-axis, a standard curve is generated for determining the NO_2^- concentration of samples. Use of peak height to produce the standard curve is inaccurate, as the height is determined by the rate at which NO is purged from solution and not by the concentration of NO in solution. A number of factors influence this rate, such as the presence of protein in the reaction mixture, which itself changes throughout the course of a single experiment.
10. Injection of standards and samples should be performed preferably in duplicate. This is important because even small variations in the volume of sample added can produce large differences in output, especially if the concentration of NO_2^- is high. Duplicate sampling, therefore, helps mitigate operator error. The maximum signal output (measured in mV)

of the luminometer is 1000 mV. Thus, it will quickly become apparent if samples require dilution. It is possible to perform a number of injections into a given aliquot of reaction mixture, but for more accurate measurements, the maximum volume of sample added should not exceed 1 ml prior to changing the reaction solution (see below). The volume of sample added per aliquot of reaction mixture will also depend on the protein content of the sample. When performing measurements, ensure that the samples are flanked by injections of a known NO_2^- standard. This enables the determination of interrun accuracy when using a standard curve.

11. To discard used reaction mixture, first stop the flow of N_2, then release the cell pressure by unscrewing the cap of the purge vessel. Use the tap located at the bottom of the purge vessel to flush out the used solution. Close the tap, wash the reaction chamber thoroughly with dH_2O, and apply another 4 ml of reaction mixture and antifoaming agent. Screw on the cap and restart the flow of N_2 to allow the fresh solution to equilibrate and the cell pressure to return to the desired level.

2.2. Determination of NO_2^-, NO_3^-, and *S*-nitrosothiol concentrations in liquid samples

Reagents

VCl_3 (Acros Organics)
HCl (BDH Laboratory Supplies)
$NaNO_3$ (minimum 99.5%) (Sigma-Aldrich)
NaOH (minimum 99.5%) (Sigma-Aldrich)
Antifoaming agent (Analytix Ltd.)
dH_2O

Reagents preparation

1. Dissolve 0.8 *g* VCl_3 (MW: 157.30) and slowly add 100 ml 1 *M* HCl. Close the container and invert several times to mix. Note that not all the solid will dissolve, which requires the solution to be filtered after preparation. The solution should turn bright blue. Also note that the glassware will get hot during this stage, due to the exothermic reaction of solid VCl_3 with water. It is possible to store the solution for up to 1 month at 4° in a container sealed from light. Vanadium (III) will react slowly with the oxygen in air; therefore, aliquot and filter sterilize only the volume needed for the day to avoid continuous exposure of the stock solution to air.
2. To produce a calibration curve, make up a 1 *M* stock solution of $NaNO_3$ by dissolving 0.425 *g* $NaNO_3$ (MW: 84.99) into 5 ml dH_2O.

Protected from light and kept at 4°, the stock solution can be kept for at least 1 month.

3. Dissolve 40 *g* of NaOH (MW: 40.00) to 1L of dH_2O to make up a 1*M* stock solution.

Procedure

1. Using the 1 *M* $NaNO_3$ solution, prepare a range of standards by 10-fold serial dilution from 10 m*M* to 100 n*M*. Store at room temperature and protect the standards from light when not in use. Diluted standards are unstable and should be prepared fresh on the day of the experiment.
2. Connect the feed line from the condenser to an acid vapor trap containing 15 ml NaOH (1 *M*) solution. This will prevent the corrosion of the NOA by hot acid vapors produced during the analysis. Such a trap is unnecessary when using 1% NaI, as acetic acid is less corrosive than mineral acid, and because the reduction of NO_2^- takes place at room temperature. A fresh aliquot of NaOH solution must be added on the day of the experiment.
3. Refer to Section 2.1 (Procedure, Steps 3–11), bearing in mind the following particularities:
   a. Add 4 ml of saturated VCl_3.
   b. The heating jacket of the purge vessel should be connected to a water bath and the temperature in the water bath set to 90° to ensure conversion of NO_3^- to NO. As the VCl_3 approaches this temperature, it will turn from blue to turquoise. If this does not occur, then the solution should be discarded and a fresh solution prepared. Note that the bubbling of the VCl_3 solution is more violent than that of 1% NaI due to the increased temperature. Be careful when injecting standards and/or samples in subsequent stages, the glassware is very hot!
   c. Ensure that there is cold water running through the condenser, in order to prevent hot acidic vapors from entering and corroding the plastic tubing.
   d. Allow at least 30 min for the VCl_3 to stabilize and contaminating nitrogen oxides present in the solution to be purged. Under good conditions, in a properly maintained NOA, the baseline is low and flat (signal varies between 4 and 10 mV in our system).
   e. The calibration curve is obtained using NO_3^- standards. Note that the generation of peaks using VCl_3 solution is slower than 1% NaI, so more time is required between injections.
   f. It is possible to perform a number of injections into a given aliquot of VCl_3 solution, but this solution is not as resilient as the 1% NaI solution, and therefore must be changed more frequently.

2.3. Determination of SNO concentration in liquid samples

Measurement of *S*-nitrosothiols (SNO) in a given biological sample requires the construction of a standard curve. Use of a small molecular weight SNO, such as *S*-nitrosoglutathione (GSNO), is recommended. Below is methodology for in-house synthesis of GSNO.

2.3.1. Synthesis of GSNO

Reagents

L-glutathione (minimum 99%) (Sigma-Aldrich)
Ice-cold 2N hydrochloric acid (BDH Laboratory Supplies)
$NaNO_2$ (minimum 99.5%) (Sigma-Aldrich)
Ice-cold acetone (Sigma-Aldrich, St. Louis, MO)
Ice-cold glacial distilled H_2O

Procedure

1. Fill a large beaker with ice and place it on a stirring plate. Place a smaller beaker containing 0.76 *g* of L-glutathione into the ice along with a magnetic stirrer.
2. Add 4 ml of ice-cold glacial dH_2O and turn on the stirrer.
3. When the L-glutathione has dissolved, add 1.25 ml ice cold 2N HCl and transfer the apparatus to a fume cupboard. This is a precautionary measure because the addition of $NaNO_2$ in the next step may generate noxious NO fumes.
4. Add between 0.17 and 0.2 *g* of $NaNO_2$ to the glutathione solution, ensuring it is being stirred. The mixture should turn a vibrant pink color.
5. Transfer the apparatus to the cold room or a refrigerator (4°) and continue to stir for 40 min.
6. To precipitate GSNO, add 10 ml ice-cold acetone and stir for a further 10 min.
7. Harvesting of GSNO is performed with vacuum flask and filter paper. Ensure that the solid GSNO, which is pink, is pulled onto the filter paper with a gentle vacuum.
8. Wash the solid with a further 10 ml of ice-cold acetone to remove remaining reaction substrates and the acid.
9. Carefully transfer the filter paper and the freshly synthesized GSNO to a vacuum desiccator for drying. Ensure that the desiccator is shielded from light, as GSNO is light sensitive. Following overnight desiccation, aliquot small amounts of GSNO into light-protected microfuge tubes and store at −80°.

2.3.2. Preparation of GSNO standards

Due to the light-sensitive nature of SNO, a fresh set of standards needs to be prepared on the day of the experiment.

Reagents

GSNO
Na_2HPO_4 (1 *M*) (Sigma-Aldrich)
NaH_2PO_4 (1 *M*) (BDH Laboratory Supplies)
diethylenetriamine pentaacetic acid (DTPA) (Sigma-Aldrich)
dH_2O

Reagent preparation To prepare 100 ml of 50 m*M* phosphate buffer, mix 3.87 ml of 1 *M* Na_2HPO_4 with 1.13 ml of 1 *M* NaH_2PO_4 and top up to 100 ml with dH_2O. Adjust pH to 7.4. To this, add 39.3 mg DTPA, a Cu(I) chelator that will prevent Cu-mediated degradation of the SNO. This buffer is subsequently referred to as PB + DTPA.

Procedure

1. Weigh out between 5 and 10 mg of GSNO in a clean, dry microfuge tube and add 100 μl of PB + DTPA and vortex to dissolve. The solution should turn pink. Note that not all of the GSNO will dissolve into the buffer. Allow the undissolved GSNO to settle, taking care not to disturb it in subsequent steps.
2. Dilute the resulting solution 100-fold in ice-cold PB + DTPA by vortexing.
3. To calculate the concentration of GSNO in the diluted GSNO solution, transfer an aliquot of the solution to a quartz cuvette and measure the absorbance at OD_{336nm}, using the PB + DTPA buffer as the blank. Divide the absorbance by 0.77, which is the extinction coefficient of a 1 m*M* solution of GSNO at OD_{336nm}. This will yield the actual concentration of the diluted solution, which should be in the region of 0.9 to 1.1 m*M*.
4. Using the PB + DTPA buffer, serially dilute the GSNO solution to produce a range of standards of known concentration, typically from 20 to 1 μM GSNO.
5. Store the standards on ice and in the dark until required. Standards kept in this way remain stable and generate reproducible peaks on the NOA for at least 3 to 4 h.

NB: SNOs are photolabile and will degrade during prolonged exposure to light. At each step of the following protocol, ensure that handling times are kept to a minimum and that lysates are stored as directed.

2.3.3. Production of a GSNO standard curve

Chemiluminescence detection of SNO in a liquid sample necessitates the reduction of the thiol–ester bond, which subsequently liberates NO. Triiodide (I_3) reacts with SNOs producing nitrous acid, which is then reduced by potassium iodide to form NO (Samouilov and Zweier, 1998).

Reagents

KI (MW: 166.00) (Fluka Biochemika)
I_2 (MW: 126.904) (Sigma-Aldrich)
Glacial acetic acid (BDH Laboratory Supplies)
Sulfanilamide (MW: 172.21) (Sigma-Aldrich)
2N HCl (BDH Laboratory Supplies)
NaOH (minimum 99.5%) (Sigma-Aldrich)
Antifoaming agent (Analytix Ltd.)
GSNO
dH_2O

Reagents preparation

1. To produce 100 ml triiodide reaction mixture, dissolve 525 m*g* KI and 336 m*g* I_2 into 21 ml glass-distilled H_2O and 73.5 ml glacial acetic acid. Stir with a magnetic stirrer for at least 15 minutes, or until the I_2 has dissolved. The final solution should be a rich dark-brown color. Prepare fresh at the start of each day and store away from bright light between analyses.
2. The sulfanilamide solution is prepared by dissolving 1.72 *g* sulfanilamide in 100 ml HCl.
3. 1 *M* NaOH is prepared, as in Section 2.2.
4. GSNO: Standards (and samples) require pretreatment with 10% (w/v) acidified sulfanilamide, to complex any free nitrite and render it incapable of distorting the NO signal. The coupling reaction between NO_2^- and sulfanilamide produces a purple diazonium compound under acidic conditions, which does not register a signal in the triiodide assay. Pretreatment of GSNO standards is achieved by addition of 180 μl of the GSNO standard to a fresh, clean microfuge tube containing 20 μl of 100 m*M* sulfanilamide in 2N HCl. Store on ice in the dark for 10 min to allow the coupling reaction to take place.

Procedure

1. Connect the condenser to the acid vapor trap, as shown in Section 2.2 (Procedure, Step 2).
2. Refer to Section 2.1 (Procedure, Steps 3–11), but with the following modifications.

a. Ensure that the purge vessel is warmed to 30° and that there is cold water running through the condenser.
b. Add 4 ml triiodide reaction mixture and antifoaming agent to the purge vessel.
c. Allow approximately 5–10 min for the reaction mixture to equilibrate. In a properly maintained NOA, the baseline signal using triiodide reaction mixture is low and flat (approximately 2–6 mV in our system).

3. Following approximately three injections (addition of 150 μl) of a cellular lysate or highly proteinaceous solution to the purge vessel, the triiodide reaction mixture and antifoaming agent will need to be replaced; otherwise a decrease in sensitivity is observed. Rinsing out the triiodide reaction mixture is performed in exactly the same way as described in Section 2.1 (Step 11).

3. Applications

3.1. Measurement of nitrate/nitrite in cell supernatants

The output of NO from murine macrophages and murine macrophage cell lines (e.g., RAW264.7, J774.2) far exceeds that of human macrophages, although both have been shown to express iNOS under stimulating conditions (Gao *et al.*, 2005; Marriott *et al.*, 2004). The precise reason for this has yet to be elucidated. Supernatants from cultures of LPS-stimulated murine macrophages contain high levels of detectable nitrate and nitrite, which are the original markers of the biosynthesis of reactive nitrogen intermediates (Ding *et al.*, 1988; Stuehr *et al.*, 1989; Stuehr and Marletta, 1985). Work on murine cells has mostly utilized the Griess reagent (*N*-[1-napthyl]-ethylenediamine hydrochloride); however, the limit of detection using this assay is only $\approx$2.5 μM (Grand *et al.*, 2001), making it unsuitable for measuring more subtle changes in biological levels of NO and its derivatives. Using the chemiluminescence techniques described previously, the accumulated NO_2^- output of primary human peripheral blood monocyte-derived macrophages (MDM) following 6 h of incubation in fresh medium was determined to be $\approx$0.5 μM. Further illustrating the sensitivity of the technique, a 48-h pretreatment of MDM with the iNOS inhibitor NG-monomethyl-L-arginine (L-NMMA) caused a significant reduction in this level (Stevanin *et al.*, 2002) and was sufficient to increase the intracellular survival of both wild-type and the *hmp* mutant of *Salmonella*, thus demonstrating that the *Salmonella* flavohemoglobin Hmp protects this bacterium from macrophage killing (Stevanin *et al.*, 2002). The methods described here have also been key in demonstrating that *N. meningitidis* is capable of depleting NO in MDM cultures, even in the presence of an NO donor (Stevanin *et al.*, 2005).

3.2. Measurement of SNO in murine macrophage cell lysates

A number of groups have been using the triiodide chemiluminescence method to measure the intracellular levels of SNO in complex biological samples (Marley, *et al.*, 2000; Wang, *et al.*, 2006; Zhang & Hogg, 2004), and is still regarded as the gold-standard method for determining SNO concentrations. Zhang and Hogg (2004) studied the formation of SNOs in LPS-treated murine macrophages and showed it to be dependent on the endogenous formation of NO by iNOS and not on the presence of nitrite. Measurements of SNO concentration in plasma and whole blood have also been made using a variant of the triiodide assay, which also includes a source of Cu(I) in the reaction mixture. This modified assay detected SNO concentrations as low as 5 n*M* in plasma samples (Marley *et al.*, 2000). The work of Wang *et al.* (2006) addressed several criticisms of using acidified sulfanilamide pretreatment to remove contaminating nitrite from biological samples. They showed that using acidified sulfanilamide does not lead to the mooted degradation of SNO, and that this technique does not underestimate their concentration.

4. Final Remarks

NO produced from the three isoforms of NOS has many different fates; and cellular lysates contain a wide variety of NO-adducted compounds. In addition to broad detection range, one of the greatest advantages of the methodologies described here is that they can be employed sequentially on a given sample, to determine the relative abundance of several nitrogen species, including SNOs, and provide the researcher with a more complete picture of the fate of NO in biological samples.

When compared with most other available methods, most notably the Griess reagent, the advantages of using chemiluminescence to assay for NO and NO metabolites in biological samples are clear. The technique is not flawless, however, and there are some limitations such as the incapability of obtaining real-time measurements of NO *in situ* or the difficulty in obtaining reliable measurements of nitrate due to the high background levels of nitrate in buffers, tissue culture medium, and so on.

In addition to the great sensitivity and reproducibility afforded by the ozone-based chemiluminescence method, the protocols described previously are only some of the methods available to the researcher for measuring NO and its derivatives in biological samples using the NOA. The NOA can also be adapted to measure lung- and nose-exhaled NO. Work currently taking place in our laboratory seeks to modify the above methodology to study alterations in *S*-nitrosylation in bacteria-infected macrophages.

ACKNOWLEDGMENTS

Our work is supported by the Wellcome Trust (Project Grant 069791).

REFERENCES

Beckman, J. S., and Koppenol, W. H. (1996). Nitric oxide, superoxide and peroxynitrite: The good, the bad, and the ugly. *Am. J. Physiol.* **271,** C1424–C1437.

Cooper, C. E. (1999). Nitric oxide and iron proteins. *Biochim. Biophys. Acta* **1411,** 290–309.

Ding, A. H., Nathan, C. F., and Stuehr, D. J. (1988). Release of reactive nitrogen intermediates and reactive oxygen intermediates from mouse peritoneal macrophages: Comparison of activating cytokines and evidence for independent production. *J. Immunol.* **141,** 2407–2412.

Eliasson, M. J. L., Huang, Z., Ferrante, R. J., Sasamata, M., Molliver, M. E., Snyder, S. H., and Moskowitz, M. A. (1999). Neuronal nitric oxide synthase activation and peroxynitrite formation in ischemic stroke linked to neural damage. *J. Neurosci.* **19,** 5910–5918.

Gao, C., Guo, H., Wei, J., Mi, Z., Wai, P. Y., and Kuo, P. C. (2005). Identification of *S*-nitrosylated proteins in endotoxin-stimulated RAW264.7 murine macrophages. *Nitric Oxide* **12,** 121–126.

Gorimar, T. S. (1985). Total nitrogen determination by chemiluminescence. *In* "Bioluminescence and chemiluminescence: Instruments and applications." (K. Van Dyke, ed.), pp. 77–93. CRC Press, Boca Raton, FL.

Grand, F., Guitton, J., and Goudable, J. (2001). Optimisation of the measurement of nitrite and nitrate in serum by the Griess reaction. *Ann. Biol. Clin. (Paris)* **59,** 559–565.

Guix, F. X., Uribesalgo, I., Coma, M., and Muñoz, F. J. (2005). The physiology and pathophysiology of nitric oxide in the brain. *Prog. Neurobiol.* **76,** 126–152.

Hara, M. R., Agrawal, N., Kim, S. F., Cascio, M. B., Fujimuro, M., Ozeki, Y., Takahashi, M., Cheah, J. H., Tankou, S. K., Hester, L. D., Ferris, C. D., Hayward, S. D., *et al.* (2005). *S*-nitrosylated GAPDH initiates apoptotic cell death by nuclear translocation following Siah1 binding. *Nat. Cell Biol.* **7,** 665–674.

Huang, J., DeGraves, F. J., Lenz, S. D., Gao, D., Feng, P., Li, D., Schlapp, T., and Kaltenboeck, B. (2002). The quantity of nitric oxide released by macrophages regulates Chlamydia-induced disease. *Proc. Natl. Acad. Sci. USA* **99,** 3914–3919.

Ignarro, L. J. (1990). Biosynthesis and metabolism of endothelium-derived nitric oxide. *Annu. Rev. Pharmacol. Toxicol.* **30,** 535–560.

Ignarro, L. J., Buga, G. M., Wood, K. S., Byrns, R. E., and Chaudhuri, G. (1987). Endothelium-derived relaxing factor produced and released from artery and vein is nitric oxide. *Proc. Natl. Acad. Sci. USA* **84,** 9265–9269.

Ignarro, L. J., Fukuto, J. M., Griscavage, J. M., Rogers, N. E., and Byrns, R. E. (1993). Oxidation of nitric oxide in aqueous solution to nitrite but not nitrate: Comparison with enzymatically formed nitric oxide from L-arginine. *Proc. Natl. Acad. Sci. USA* **90,** 8103–8107.

Ischiropoulos, H., Zhu, L., Chen, J., Tsai, M., Smith, C. D., Martin, J. C., and Beckman, J. S. (1992). Peroxynitrite-mediated tyrosine nitration catalyzed by superoxide dismutase. *Arch. Biochem. Biophys.* **298,** 431–437.

Kirkebøen, K. A., and Strand, Ø. A. (1999). The role of nitric oxide in sepsis: An overview. *Acta Anaesthesiol. Scand.* **43,** 275–288.

Lander, H. M., Ogiste, J. S., Pearce, S. F., Levi, R., and Novogrodsky, A. (1995). Nitric oxide-stimulated guanine nucleotide exchange on p21ras. *J. Biol. Chem.* **270,** 7017–7020.

Li, J., Billiar, T. R., Talanian, R. V., and Kim, Y. M. (1997). Nitric oxide reversibly inhibits seven members of the caspase family via S-nitrosylation. *Biochem. Biophys. Res. Commun.* **240,** 419–424.

Marley, R., Feelisch, M., Holt, S., and Moore, K. (2000). A chemiluminescense-based assay for *S*-nitrosoalbumin and other plasma S-nitrosothiols. *Free Radic. Res.* **32,** 1–9.

Marriott, H. M., Ali, F., Read, R. C., Mitchell, T. J., Whyte, M. K., and Dockrell, D. H. (2004). Nitric oxide levels regulate macrophage commitment to apoptosis or necrosis during pneumococcal infection. *FASEB J.* **18,** 1126–1128.

Mitchell, D. A., Erwin, P. A., Michel, T., and Marletta, M. A. (2005). *S*-nitrosation and regulation of inducible nitric oxide synthase. *Biochemistry* **44,** 4636–4647.

Reynaert, N. L., Ckless, K., Korn, S. H., Vos, N., Guala, A. S., Wouters, E. F., van der Vliet, A., and Janssen-Heininger, Y. M. (2004). Nitric oxide represses inhibitory κB kinase through *S*-nitrosylation. *Proc. Natl. Acad. Sci. USA* **101,** 8945–8950.

Samouilov, A., and Zweier, J. L. (1998). Development of chemiluminescence-based methods for specific quantitation of nitrosylated thiols. *Anal. Biochem.* **258,** 322–330.

"Sievers nitric oxide analyzer NOA™ 280i (1995–2000): Operation and Maintenance manual." Sievers Instruments, Boulder, CO.

Stevanin, T. M., Moir, J. W. B., and Read, R. C. (2005). Nitric oxide detoxification systems enhance survival of *Neisseria meningitidis* in human macrophages and in nasopharyngeal mucosa. *Infect. Immun.* **73,** 3322–3329.

Stevanin, T. M., Poole, R. K., Demoncheaux, E. A., and Read, R. C. (2002). Flavohemoglobin Hmp protects *Salmonella enterica* serovar typhimurium from nitric oxide-related killing by human macrophages. *Infect. Immun.* **70,** 4399–4405.

Stichtenoth, D. O., and Frolich, J. C. (1998). Nitric oxide and inflammatory joint diseases. *Br. J. Rheumatol.* **37,** 246–257.

Stuehr, D. J., and Marletta, M. A. (1985). Mammalian nitrate biosynthesis: Mouse macrophages produce nitrite and nitrate in response to *Escherichia coli* lipopolysaccharide. *Proc. Natl. Acad. Sci. USA* **82,** 7738–7742.

Stuehr, D. J., Gross, S. S., Sakuma, I., Levi, R., and Nathan, C. F. (1989). Activated murine macrophages secrete a metabolite of arginine with the bioactivity of endothelium-derived relaxing factor and the chemical reactivity of nitric oxide. *J. Exp. Med.* **169,** 1011–1020.

Wang, X., Bryan, N. S., MacArthur, P. H., Rodriguez, J., Gladwin, M. T., and Feelisch, M. (2006). Measurement of nitric oxide levels in the red cell: Validation of tri-iodide-based chemiluminescence with acid-sulfanilamide pretreatment. *J. Biol. Chem.* **281,** 26994–27002.

Williams, J. G., Pappu, K., and Campbell, S. L. (2003). Structural and biochemical studies of p21Ras S-nitrosylation and nitric oxide-mediated guanine nucleotide exchange. *Proc. Natl. Acad. Sci. USA* **100,** 6376–6381.

Zhang, Y., and Hogg, N. (2004). Formation and stability of *S*-nitrosothiols in RAW 264.7 cells. *AJP: Lung* **287,** 467–474.

SECTION TWO

BACTERIAL AND ARCHAEAL HEMOGLOBINS

CHAPTER EIGHT

Interactions of NO with Hemoglobin: From Microbes to Man

Michael Angelo,* Alfred Hausladen,* David J. Singel,* *and* Jonathan S. Stamler*,†

Contents

* Department of Chemistry and Biochemistry, Montana State University, Bozeman, Montana
† Department of Medicine and Department of Biochemistry, Duke University Medical Center, Durham, North Carolina

Methods in Enzymology, Volume 436
ISSN 0076-6879, DOI: 10.1016/S0076-6879(08)36008-X

Abstract

Hemoglobins are found in organisms from every major phylum and subserve life-sustaining respiratory functions across a broad continuum. Sustainable aerobic respiration in mammals and birds relies on the regulated delivery of oxygen (O_2) and nitric oxide (NO) bioactivity by hemoglobin, through reversible binding of NO and O_2 to hemes as well as *S*-nitrosylation of cysteine thiols (SNO synthase activity). In contrast, bacterial and yeast flavohemoglobins function *in vivo* as denitrosylases (O_2 nitroxylases), and some multimeric, invertebrate hemoglobins function as deoxygenases (Cys-dependent NO dioxygenases), which efficiently consume rather than deliver NO and O_2, respectively. Analogous mechanisms may operate in plants. Bacteria and fungi deficient in flavohemoglobin show compromised virulence in animals that results from impaired resistance to NO, whereas animals and humans deficient in *S*-nitrosylated Hb exhibit altered vasoactivity. NO-related functions of hemoglobins center on reactions with ferric (FeIII) heme iron, which is exploited in enzymatic reactions that address organismal requirements for delivery or detoxification of NO and O_2. Delivery versus detoxification of NO/O_2 is largely achieved through structural changes and amino acid rearrangements within the heme pockets, thereby influencing the propensity for heme/cysteine thiol redox coupling. Additionally, the behavior exhibited by hemoglobin *in vivo* may be profoundly dependent both on the abundance of NO and O_2 and on the allosteric effects of heterotropic ligands. Here we review well-documented examples of redox interactions between NO and hemoglobin, with an emphasis on biochemical mechanisms and physiological significance.

1. Introduction

Hemoglobins (Hbs) comprise a heterogeneous group of proteins found within all phyla of living organisms (Vinogradov *et al.*, 2006; Weber and Vinogradov, 2001). Hbs are composed, wholly or in part, of structurally homologous monomers consisting of an iron porphyrin ring folded within 6–8 helices (Vinogradov *et al.*, 2006; Weber and Vinogradov, 2001). The functional properties of microbial and invertebrate Hbs reflect adaptive mechanisms that have evolved within a given lineage to counter metabolic inhibition by oxidative and nitrosative stresses, whereas hemoglobins of mammals and birds subserve metabolic coupling of O_2 delivery with tissue demand (Poole, 2005; Singel and Stamler, 2005; Sonveaux *et al.*, 2007). NO and O_2 are employed in various enzymatic reactions that, in microbes and invertebrates (and probably plants), serve to eliminate (i.e., detoxify) each other and that, in mammals, conspire to preserve each other's bioavailability. Notably, ferric (FeIII) heme iron is pivotal in all NO-based functions of hemoglobins, and cysteine thiol/heme-iron redox coupling is a central feature of O_2 processing: both detoxification and delivery of O_2 involve heme-catalyzed *S*-nitrosylation of cysteine thiols. In contrast, while detoxification of NO does not require cysteine thiols, delivery of NO groups does (Foster *et al.*, 2005; Gow *et al.*, 1999; Gow and Stamler, 1998; Hess *et al.*, 2005; McMahon *et al.*, 2002b; Singel and Stamler, 2005).

The physiological role of Hbs varies considerably among taxa, and even within a taxon expression levels can differ by 1000-fold (Perutz, 1987; Vinogradov *et al.*, 2006; Weber and Vinogradov, 2001). Hbs are found in both intracelluar and extracellular compartments, and encounter widely varying levels of NO and O_2 (Antonini *et al.*, 1984; Chiancone and Gibson, 1989; Gow and Stamler, 1998; Hausladen *et al.*, 2001; Liu *et al.*, 2000; Minning *et al.*, 1999; Vinogradov *et al.*, 2006; Weber and Vinogradov, 2001). Enzymatic reduction of oxidized Hb is obligatory for ongoing function in some organisms but seemingly unimportant in others (Gardner *et al.*, 1998; Hausladen *et al.*, 2001; Liu *et al.*, 2000; Singel and Stamler, 2005). Consistent with the functional diversity exhibited *in vivo*, the biochemical mechanisms of Hb action are manifold and cannot be understood without taking adequate account of the redox environment *in situ*. In general, experimental analysis that fails to recapitulate the salient features of that environment will generate artifactual reactions and erroneous conclusions about Hb function in the physiological milieu.

Previous work on mammalian Hb has indicated large differences in how O_2 and NO are handled (Antonini *et al.*, 1984; Carver *et al.*, 1990; Dewilde *et al.*, 2001; Gow *et al.*, 1999; Gow and Stamler, 1998; Jia *et al.*, 1996; Luchsinger *et al.*, 2003; Moore and Gibson, 1976; Sharma *et al.*, 1983; Sharma *et al.*, 1987; Singel and Stamler, 2005; Stamler *et al.*, 1997; Taketa *et al.*, 1978; Van Doorslaer *et al.*, 2003). It is important to emphasize at the outset that, unlike O_2, the *in vivo* concentration of NO is often far exceeded by that of Hb (Gow *et al.*, 1999; Hausladen *et al.*, 2001), and the impact of NO and O_2 concentration and other redox parameters on reactions with Hb is profound. We also note that room-air concentrations of O_2 typically far exceed those *in vivo* (Eu *et al.*, 2000; Hausladen *et al.*, 2001; Singel and Stamler, 2005). With this in mind, we provide here an overview of some of the phylogenetic, biochemical, and functional concepts of particular relevance to these reactions, with an emphasis on well-characterized examples, and suggest some general guidelines for experimental design and data analysis.

2. Microbes, Plants, and Invertebrates

The Hbs and flavohemoglobins (FHbs) of bacteria, fungi, worms, and plants have been the subject of intensive analysis (Appleby, 1984; Das *et al.*, 2000; Farres *et al.*, 2005; Giangiacomo *et al.*, 2001; Hargrove *et al.*, 1997;

Membrillo-Hernandez *et al.*, 1999; Moore *et al.*, 2004; Mukhopadhyay *et al.*, 2004; Nakano, 2002; Poole, 1994; Poole *et al.*, 1996; Richardson *et al.*, 2006; Sebbane *et al.*, 2006; Stevanin *et al.*, 2000; Thorsteinsson *et al.*, 1999; Zhu and Riggs, 1992). Originally described in only a few microbes and the root nodules of legumes, their function was poorly defined and assumed to be primarily oxygen storage by analogy to mammalian Hbs. With the discovery of the FHb of *E. coli*, it became clearer that Hbs also catalyze redox reactions (Vasudevan *et al.*, 1991). However, the function of FHb would remain a mystery until 1998, when three groups independently established its role in protecting against nitrosative stress (Crawford and Goldberg, 1998; Gardner *et al.*, 1998; Hausladen *et al.*, 1998). An enzymatic function of FHb in detoxifying NO aerobically (Gardner *et al.*, 1998; Hausladen *et al.*, 1998) and anaerobically (Hausladen *et al.*, 1998; Kim *et al.*, 1999) was proposed, and a unifying enzymatic mechanism was ultimately revealed (Hausladen *et al.*, 2001). Yeast flavoHb was subsequently shown to exhibit a similar function in protecting against NO both aerobically and anaerobically (Liu *et al.*, 2000). Both *Cryptococcus* (De Jesus-Berrios *et al.*, 2003) and *Salmonella* (Bang *et al.*, 2006) deficient in FHb show impaired virulence in mice that is attributable entirely to impaired NO consumption (i.e., virulence is fully restored in iNOS mutant mice), establishing a primary function of FHb in NO detoxification by both genetic and physiological criteria. In addition, a novel catalytic role for NO in detoxifying O_2 was identified with *Ascaris* Hb (Minning *et al.*, 1999). Elimination of the worm's Hb resulted in hazardous accumulations of O_2 within its tissues (Minning *et al.*, 1999). In contrast, NO turnover by the worm Hb is relatively inefficient. It is now well established that Hbs occur in all organisms, and ubiquitous NO-related roles seem likely. It has even been proposed that the principal ancestral function of Hb involved NO chemistry (Freitas *et al.*, 2004; Hausladen *et al.*, 1998; Minning *et al.*, 1999; Vinogradov *et al.*, 2005; Vinogradov *et al.*, 2006).

2.1. FHbs

FHb protects against NO both aerobically and anaerobically (Crawford and Goldberg, 1998; Hausladen *et al.*, 1998; Justino *et al.*, 2005; Liu *et al.*, 2000; Poole, 2005) and is required for virulence in mice where tissue pO_2 is low (Bang *et al.*, 2006). Consumption of NO (Hausladen *et al.*, 2001) and resultant protection (Goncalves *et al.*, 2006) is more efficient in the presence of O_2 than its absence. The FHb binds NO at the heme iron, effectively reducing it to nitroxyl anion (NO^-/HNO). Heme-bound nitroxyl then reacts with O_2 to form nitrate:

2.1.1. O_2 nitroxylase mechanism (denitrosylase)

$$\text{FHb-[Fe(II)]} \xrightarrow{\text{NO}} \text{FHb-[Fe(II)NO]} \tag{8.1}$$

$$\begin{aligned} &\text{FHb-[Fe(II)NO]} \leftrightarrow \text{FHb-[Fe(III)NO}^-] \\ &\text{FHb-[Fe(III)NO}^-] \xrightarrow{O_2} \text{FHb-[Fe(III)]} + \text{NO}_3^- \end{aligned} \tag{8.2}$$

During steady-state turnover, the enzyme can be found in the ferric (FeIII) state (Hausladen *et al.*, 2001). The ferric heme produced in this reaction is reduced by the reductase domain to the ferrous form via NADH and the FAD prosthetic group (Gardner *et al.*, 1998; Hausladen *et al.*, 1998).

2.1.2. Enzymatic reduction of ferric hemes

$$\begin{aligned} &\text{FAD} + \text{NADH} \xrightarrow{H^+} \text{FADH}_2 + \text{NAD}^+ \\ &\text{FHb-[Fe(III)]} + \text{FADH}_2 \rightarrow \text{FHb-[Fe(II)]} + \text{FADH} + \text{H}^+ \\ &\text{FHb-[Fe(III)]} + \text{FADH} \rightarrow \text{FHb-[Fe(II)]} + \text{FAD} + \text{H}^+ \end{aligned} \tag{8.3}$$

This scheme, which operates very efficiently even under microaerobic conditions that predominate *in vivo* (Goncalves *et al.*, 2006; Hausladen *et al.*, 2001), is consistent with a peroxidase-like function that activates ligands by a polarizing push-pull mechanism (Mukai *et al.*, 2001). This mechanism may not be completely unique to FHbs, as Gow and Stamler (1998) have reported that nitroxyl is liberated by mammalian Hb in T-state. Herold and Rock (2005) have noted, however, that the reactions of nitrosylated human Hb and nitrosyl neuroglobin (Herold *et al.*, 2004) with O_2 do not occur readily.

Under complete anaerobiosis, FHb has a slow NO reductase activity (Gardner *et al.*, 2000b; Hausladen *et al.*, 1998; Hausladen *et al.*, 2001; Kim *et al.*, 1999; Mills *et al.*, 2001), in which NO is effectively converted to N_2O (Hausladen *et al.*, 2001; Hausladen *et al.*, 1998; Kim *et al.*, 1999).

2.1.3. NO reductase mechanism (denitrosylase)

$$\begin{aligned} &\text{2FHb-[Fe(II)NO]} \leftrightarrow \text{2FHb-[Fe(III)NO}^-] \\ &\text{2FHb-[Fe(III)NO}^-] \xrightarrow{2H^+} \text{2FHb-[Fe(III)]} + \text{N}_2\text{O} + \text{H}_2\text{O} \end{aligned} \tag{8.4}$$

NO reductase activity of FHb might not be sufficient to protect against NO in all circumstances (Pathania *et al.*, 2002). Nonetheless, it should be pointed out that *Hmp* (FHb) is among the genes induced most strongly by anaerobic NO exposure (Justino *et al.*, 2005; Pullan *et al.*, 2007) and protection by *Hmp* (as well as flavorubredoxin) against 50–150 μM NO was observed under anaerobic conditions (Justino *et al.*, 2005). In addition, flavoHb protected yeast against NO under stringent anaerobic conditions (Liu *et al.*, 2000).

It was proposed initially that *E. coli* FHb mediated the dioxygenation of NO, in which heme-bound O_2 has partial superoxide character and reacts with NO via a heme-bound peroxynitrite intermediate, which is then released as nitrate (Fig. 8.1).

2.1.4. NO dioxygenase mechanism

$$\text{FHb-[Fe(II)]} \xrightarrow{O_2} \text{FHb-[Fe(II)O}_2] \tag{8.5}$$

$$\begin{aligned} &\text{FHb-[Fe(II)O}_2] \leftrightarrow \text{FHb-[Fe(III)O}_2^-] \\ &\text{FHb-[Fe(III)O}_2^-] \xrightarrow{NO} \text{FHb-[Fe(III)]} + NO_3^- \end{aligned} \tag{8.6}$$

However, it was subsequently found that FHb has relatively low O_2 affinity, and it was therefore expected that at low physiologic O_2 concentrations, the ability of FHb (Hmp) to detoxify NO would be compromised (Gardner *et al.*, 2000a; Mukai *et al.*, 2001). It was suggested that the very high NO affinity of FHb would lead to the accumulation of an inactive, NO-bound heme (Gardner *et al.*, 2000a). But when anoxic FHb-NO was exposed to air, rapid O_2 consumption occurred, and no inhibition was observed even at very high NO concentrations (10–20 μM) (Frey *et al.*, 2002; Hausladen *et al.*, 2001; Mills *et al.*, 2001) or very low O_2 concentrations. Furthermore, FHb was able to support microaerobic growth with NO ($\approx$5 μM O_2, 0.5 μM NO) in a flavorubredoxin mutant (*norV*, a major anaerobic NO reductase in *E. coli*), whereas a *hmp/norV* double mutant was severely compromised under the same conditions (Pathania *et al.*, 2002). In addition, although NO indeed outcompetes O_2 for heme binding at low O_2 concentrations, nitrosyl-FHb is by no means inactive: it consumes NADH with NO bound to the heme and contributes a partial ferric character to the Hmp spectrum (Hausladen *et al.*, 2001). Therefore, FHb/Hmp functions as an NO dioxygenase only at very high O_2 concentrations and very low NO concentrations—a situation encountered rarely *in vivo*. In contrast, at most O_2 concentrations and across a wide range of NO concentrations—recapitulating the physiological situation—FHb functions as an O_2 nitroxylase, catalyzing the addition of heme-bound nitroxyl (NO^-) to O_2 to form nitrate (see Fig. 8.1).

2.2. Monomeric Hbs

Monomeric Hbs are highly homologous to the globin domain of FHbs, but lack a reductase domain. Like FHbs, they form the typical globin fold of 6–8 helices that binds the heme group. These O_2-binding monomeric Hbs, best studied in the bacterium *Vitreoscilla* (VHb), are thought to aid in respiration under microaerobic conditions, and their expression has been manipulated

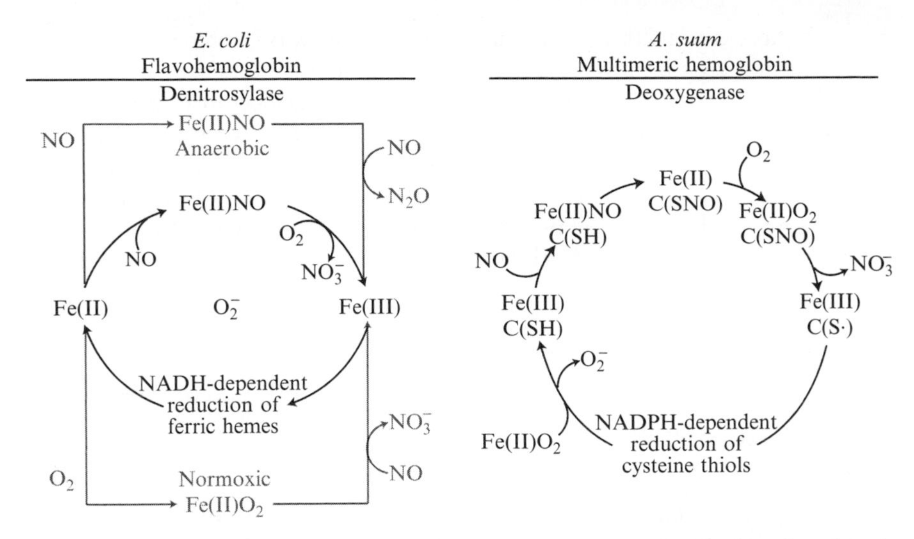

Figure 8.1 Protection against NO (FHb) and O_2 (AHb). Although both FHb and AHb consume NO and O_2, the enzymatic mechanisms employed differ significantly. Under physiological conditions, FHb functions as a denitrosylase (O_2 nitroxylase) to protect against NO. (left): ferrous hemes rapidly bind and eliminate NO by catalyzing the addition of nitroxyl (NO^-) to dioxygen, yielding an equivalent of nitrate for each oxidized heme. NADH-dependent reduction of ferric hemes by FHb-bound FAD is independent of NO binding to heme and occurs rapidly. In contrast, NO binds to ferric rather than ferrous hemes in AHb (right), and subsequent reduction of ferric hemes is obligatorily coupled to *S*-nitrosylation of C72 (CysNO) within the distal heme cavity. The juxtaposition of Cys(NO) and heme-bound O_2 enables formation of a peroxynitrite intermediate, which rapidly isomerizes to nitrate, leaving an oxidized heme and cysteinyl radical. This radical is reduced by NADPH, rather than heme iron. In AHb, therefore, *S*-nitrosylation of a critical Cys residue mediates NO-regulated heme reduction, and thereby governs the binding and elimination of molecular oxygen (deoxygenase activity; formally a Cys-dependent NO dioxygenase). Thus, AHb protects against O_2.

to improve growth yields of bacteria. Monomeric Hbs are also found in *Aquifex aeolicus*, *Campylobacter*, and *Clostridium* (Frey and Kallio, 2003).

It is interesting that the same residues that give Hmp its peroxidase character are present in the heme pocket of VHb (Mukai *et al.*, 2001), raising the possibility that VHb supports similar chemistry. Whereas the flavin domain of FHbs serves to reduce ferric heme after turnover with NO/O_2 (Eq. 8.3), the single-domain Hbs need a reductase partner in order to sustain such a reaction. VHb contains a methemoglobin reductase that copurifies with VHb (Gemzik *et al.*, 1992), but no studies of NO consumption by this bacterium have been performed. However, a chimeric VHb carrying a reductase domain from *Ralstonia* Fhp has been expressed in *E. coli* and was shown to consume NO and protect respiration and growth from NO inhibition (Kaur *et al.*, 2002). Similarly, several bacterial Hbs and FHbs have been expressed in *E. coli* and shown to protect from growth inhibition by NO (Frey *et al.*, 2002). Equivalent protection was conferred by Hbs or FHbs, indicating that monomeric Hbs are efficiently reduced in the *E. coli* cytoplasm. In contrast, the Hbs did not exhibit NO-consuming activity in the presence of NADH when assayed in cell-free extracts, whereas the FHbs consumed NO at high rates (Frey *et al.*, 2002). Taken together, these observations indicate that the beneficial effect of VHb expression on microaerobic growth is likely due not to improved O_2 delivery but to protection of the respiratory chain and other sensitive targets from NO formed under these conditions (Frey and Kallio, 2005).

The single-domain Hb in *Campylobacter jejuni* (Cgb) scavenges NO, protects from nitrosative stress, and is induced by *S*-nitroso-glutathione (GSNO) in bacteria cultured in the presence of nitrite or nitrate (Elvers *et al.*, 2004; Elvers *et al.*, 2005; Pittman *et al.*, 2007). Although a reductase partner for Cgb has not been identified, its existence is indicated by the sustained consumption of NO observed in both the native organism and in genetically modified *E. coli*. Conversely, Cgb in cell-free extracts does not consume NO (Frey *et al.*, 2002).

2.3. Truncated Hbs

Truncated hemoglobins (trHbs) with a four-helix globin fold are found in bacteria, unicellular eukaryotes, and plants (Pesce *et al.*, 2000). Several trHbs have been implicated in tolerance to nitrosative stress, and others appear to be involved in respiration.

The trHb in *C. jejuni* (Ctb) is up-regulated by nitrosative stress (Elvers *et al.*, 2005). However, Ctb knockout strains are not sensitized to GSNO (Elvers *et al.*, 2005). *Mycobacterium leprae* trHb (GlbO) oxidizes NO via a heme-bound peroxynitrite (Ascenzi *et al.*, 2006) and rescues *E. coli* Hmp mutants from NO sensitivity (Fabozzi *et al.*, 2006). The trHb in *M. tuberculosis* (HbN) protects from nitrosative stress (Ouellet *et al.*, 2002) and scavenges NO when expressed in heterologous hosts (Pathania *et al.*, 2002; Pawaria *et al.*, 2007). HbN processes NO in a novel way, in which two "channels" in the protein allow ligand migration to the heme. It has been proposed that this feature allows access of both O_2 and NO to the active site (Bidon-Chanal *et al.*, 2006). HbN also protects a *Salmonella* Hmp mutant from nitrosative stress and killing by mouse macrophages (Pawaria *et al.*, 2007). It is interesting that the homologous HbO, while consuming NO when expressed in *S. typhimurium*, did not protect from killing by macrophages, which suggests a different role for this trHb (Pawaria *et al.*, 2007).

Although truncated Hbs may in some cases protect against NO, the mechanisms remain unclear. Neither steady-state turnover of NO nor mass balance accounting for product has been demonstrated convincingly (unlike the FHb) (Hausladen *et al.*, 2001), and complementary reductases that would be necessary to support the catalytic cycle have not been identified.

Studies of the reaction of NO with truncated Hbs *in vitro* during a single turnover are not informative in this regard. It has been reported that truncated Hb of *C. jejuni* can support electron transfer in the respiratory transport chain and thereby bypass an NO-induced block on respiration (Lu *et al.*, 2007); protective mechanisms such as this merit further consideration. *Bacillus subtilis* Hbs are a case in point. Although *B. subtilis* FHb is one of the genes most highly induced by nitrosative stress and contributes significantly to resistance, *B. subtilis* trHb (YjbI) is also involved in protection, as mutants lacking YjbH show a 100-fold increase in sensitivity to sodium nitroprusside. However, YjbI has not been shown to be up-regulated by nitrosative stress, and the mechanism for this heightened sensitivity remains to be established (Rogstam *et al.*, 2007).

3. Yeast and Fungi

3.1. FHbs

Yeast FHbs (YHbs) are highly homologous to bacterial FHbs, perhaps due to lateral gene transfer between bacteria and eukaryotes (Gardner *et al.*, 2000b; Moens *et al.*, 1996; Vinogradov *et al.*, 2005, 2006; Zhu and Riggs, 1992). While YHbs have been shown to consume NO and protect against nitrosative stress *in vitro* (Liu *et al.*, 2000) and *in vivo* (De Jesus-Berrios *et al.*, 2003), other functions are still being explored (Cassanova *et al.*, 2005). *Saccharomyces cerevisiae* cells or cell-free extracts consume NO, and deletion of the *yhb1* gene or heat inactivation abolishes this activity (Liu *et al.*, 2000). The *yhb1* mutation also leads to strong growth inhibition and a marked increase in nitrosylated proteins when exposed to NO donor compounds (Liu *et al.*, 2000). Greater sensitivity to NO donor compounds was also seen under anaerobic conditions, although as in *E. coli*, anaerobic NO reduction was much slower than the aerobic reaction (Liu *et al.*, 2000). Much like *S. cervisiae*, *Cryptococcus neoformans* contains a FHb that consumes NO and protects from NO toxicity. Remarkably, it also promotes survival in macrophages and enhances fungal virulence in mice by consuming NO derived from the host iNOS (De Jesus-Berrios *et al.*, 2003).

YHb expression and inducibility is more variable than for FHb. Unlike bacterial FHbs, *S. cerevisiae* YHb is constitutively expressed and not induced by nitrosative stress (Landfear *et al.*, 2004; Liu *et al.*, 2000). In contrast, *Candida albicans* contains an NO-consuming YHb1 that is strongly induced by NO and nitrite and confers resistance to growth inhibition by NO donors. Mutants in the *C. albicans* YHb1 also showed attenuated virulence in mice (Landfear *et al.*, 2004). However, *C. albicans* contains two additional YHb genes (inferred by sequence homology), the products of which neither consume NO nor contribute to NO resistance (Landfear *et al.*, 2004), indicating that these YHbs may have functions other than NO detoxification.

4. Plants

Three types of Hbs are found in plants: the symbiotic leghemoglobins (Elvers *et al.*, 2004) in the root nodules of legumes (Lbs), nonsymbiotic Hbs (nsHb) that are ubiquitously expressed, and trHbs that are also widely expressed. With the discovery of NO biosynthesis in plants, and the relatively recent descriptions of nsHbs and trHbs, a new area of research is emerging on the interactions between plant Hbs and NO.

4.1. Lbs

Lbs have been studied in detail for decades. They bind O_2 with high affinity and are thought to fulfill two functions: (1) to limit O_2 concentration in the root nodule to a level at which the O_2-sensitive nitrogenase can function, and (2) to deliver O_2 to the respiring bacteroids to meet the high ATP demands of nitrogen fixation (Kundu *et al.*, 2003).

The structural features that allow Lbs to fulfill O_2-related functions have recently been reviewed (Kundu *et al.*, 2003). Despite a high affinity for O_2, only about 20–30% of Lb is in the oxy form (Denison and Harter, 1995), with the remainder believed to be deoxy Lb. Like all Hbs, Lb can also bind NO with high affinity (Appleby, 1984; Fleming *et al.*, 1987; Gibson *et al.*, 1989; Harutyunyan *et al.*, 1996; Wittenberg *et al.*, 1986). Indeed, naturally occurring nitrosyl-Lb (Lb-NO) has been found in Lb preparations from soybean and cowpea (Maskall *et al.*, 1977), and Lb-NO found in intact, nitrate-free soybean nodules has been proposed to arise from a nodular NO synthase (Baudouin *et al.*, 2006; Cueto *et al.*, 1996; Mathieu *et al.*, 1998). Nitrate has long been known to inhibit nitrogen fixation (Streeter, 1988), but the mechanism by which this occurs remains unclear. One proposal is that the nitrate reductases of either plant or bacteroids accumulate nitrite, which is subsequently converted to Lb-NO, thereby disabling the O_2-binding function. Thus, when soybeans were supplied with nitrate, N_2 fixation decreased and Lb-NO could be isolated from the nodules (Kanayama *et al.*, 1990; Kanayama and Yamamoto, 1990a,b). Recognizing that NO production in roots is increased under hypoxic conditions (Dordas *et al.*, 2003b,c), a function for Lb in protecting against NO has been suggested. A recent study revealed that up to 70% of total Lb is converted to Lb-NO when soybean roots are incubated with nitrate under hypoxic conditions, compared to 20% Lb-NO under normoxia

Table 8.1 Previous work on the mechanism and function of bacterial flavohemoglobins

	Enzymatic activity	*In vivo* analysis		Proteomics
	NO consumption	Tolerance to nitrosative stress	Inducible expression	Microarray analysis
E. coli	Gardner *et al.*, 1998	Gardner *et al.*, 1998	Poole *et al.*, 1996	Mukhopadhyay *et al.*, 2004
	Hausladen *et al.*, 1998	Hausladen *et al.*, 1998	Gardner *et al.*, 1998	Moore *et al.*, 2004
	Frey *et al.*, 2002	Membrillo-Hernadez *et al.*, 1999	Hausladen *et al.*, 1998	Justino *et al.*, 2005
	Stevanin *et al.*, 2000	Stevanin *et al.*, 2000	Stevanin *et al.*, 2000	Pullan *et al.*, 2007
	Mills *et al.*, 2001	Frey *et al.*, 2002	Nakano, 2002	
S. typhimurium	Bang *et al.*, 2006	Crawford *et al.*, 1998	Crawford *et al.*, 1999	–
		Stevanin *et al.*, 2002	Bang *et al.*, 2006	
		Bang *et al.*, 2006		
S. enterica	Frey *et al.*, 2002	–	–	–
B. subtillis	Frey *et al.*, 2002	Frey *et al.*, 2002	Rogstam *et al.*, 2007	Moore *et al.*, 2004
		Rogstam *et al.*, 2007		
A. eutrophus	Frey *et al.*, 2002	Frey *et al.*, 2002		
P. aeroginosa	Frey *et al.*, 2002	Frey *et al.*, 2002	Arai *et al.*, 2005	Firoved *et al.*, 2004
	Firoved *et al.*, 2004	Arai *et al.*, 2005	Firoved *et al.*, 2002	
	Arai *et al.*, 2005			
D. radiodurans	Frey *et al.*, 2002	Frey *et al.*, 2002	–	–
C. jejuni	Frey *et al.*, 2002	Frey *et al.*, 2002	–	–
K. pneumoniae	Frey *et al.*, 2002	Frey *et al.*, 2002	–	–
Y. pestis	–	Sebbane *et al.*, 2006	–	–
S. aureus	Richardson *et al.*, 2006	Richardson *et al.*, 2006	Richardson *et al.*, 2006	Richardson *et al.*, 2002
		Goncalves *et al.*, 2006	Goncalves *et al.*, 2006	

(Meakin *et al.*, 2007). Lb-NO is supposedly stable in the presence of O_2 (Kanayama and Yamamoto, 1990a; Maskall *et al.*, 1977), however, so the means to sustain NO turnover is unclear. Catalytic mechanisms of NO consumption might include the reaction of Lb-O_2 with either NO ($k = 8.2 \times 10^7\ M^{-1}s^{-1}$) or peroxynitrite/$CO_2$ ($k = 5.5 \times 10^4 - 8.8 \times 10^5\ M^{-1}s^{-1}$) (Herold and Puppo, 2005a), and/or of Lb-NO with peroxynitrite ($k = 8.8 \times 10^3 - 1.2 \times 10^5\ M^{-1}s^{-1}$) (Herold and Puppo, 2005b). A catalytic cycle involving Lb^{3+} (by analogy to FHb) might then be supported by a methemoglobin reductase, which is present in root nodules (Ji *et al.*, 1994). It is interesting to note that this reductase is also expressed in leaf tissue (Ji *et al.*, 1994), where it could function to maintain nsHbs in the reduced state.

4.2. nsHbs

Whereas decades of research on Lb have yielded few publications on NO, the more recently discovered nonsymbiotic plants Hbs (nsHb) were quickly identified with a role in NO metabolism, no doubt reflecting the description in plants of NO synthesis (Dordas *et al.*, 2003a,b,c; Perazzolli *et al.*, 2006).

The class I nsHbs were found to have an extremely high O_2 affinity, making them unlikely candidates for O_2 delivery (Duff *et al.*, 1997). However, overexpressed nsHb improved energy status in hypoxic maize cells (Sowa *et al.*, 1998) and promoted early growth and survival under hypoxia in *Arabidopsis* (Hunt *et al.*, 2002). It was soon realized that hypoxic conditions increased NO formation from nitrate in plants, and nsHb induction functioned to protect from this nitrosative stress (Dordas *et al.*, 2003a,b,c; Perazzolli *et al.*, 2004).

Energy production during hypoxic growth is mainly achieved through glycolysis. In classic fermentation pathways, NADH generated in glycolysis is reoxidized (to NAD^+) by the formation of ethanol or lactate, which accumulate as end products. In hypoxic plants, an alternative mechanism of NADH oxidation involving nsHb and nitrate has been proposed: nitrate reductase generates nitrite, which is subsequently converted to NO. The nsHb, with an O_2 affinity higher than cytochrome reductase, binds O_2 in hypoxic conditions and promotes reaction with NO, regenerating nitrate. Reduction of nitrate and nitrite and the regeneration of ferrous heme all consume NADH, making NAD^+ available for glycolysis (Abir *et al.*, 2004; Igamberdiev and Hill, 2004). As attractive as this idea may be, the exact mechanism of nitrate formation under physiologically relevant conditions has not been elucidated; particularly, an O_2-nitroxylase mechanism by analogy to flavoHb has not been excluded. More generally, analysis of the enzyme during steady-state turnover of NO/O_2 and the exact basis of the reductase function have not been adequately explored (Table 8.2).

Table 8.2 Rate constants for the association (k'), dissociation (k), and dissociation equilbirium (K_d) of O_2 and NO from ferrous and ferric hemes of hemoglobin from bacteria, yeast, and plants

	Fe(II)			Fe(II)			Fe(III)		
	O_2			NO			NO		
	K'	k	K_d	k'	k	K_d	k'	k	K_d
	$\mu M^{-1}s^{-1}$	s^{-1}	μM	$\mu M^{-1}s^{-1}$	s^{-1}	μM	$\mu M^{-1}s^{-1}$	s^{-1}	μM
Bacteria									
E. coli[a] FHb	38	0.44	.012	26 (4)†	2.0×10^{-4}	8.0×10^{-6}	44	4×10^3	100
B. subtilis[b] FHb	44 (2.8)	7 (0.3)†	0.16	38	2.1	0.055	2.4	–	–
A. eutrophus[c] FHb	50	0.2	4.0×10^{-3}	10–20	–	–	2.4	1.2×10^3	500
S. enterica[b] FHb	6	1.5 (0.08)†	0.25	42	3.4	0.081	48	–	–
C. jejuni[b] Hb	150	1.1	7.3×10^{-3}	–	1.6	–	130 (4)†	–	–
C. perfringens[b] Hb	210	1.8	8.6×10^{-3}	–	6	–	310 (26)†	–	–
Vitreoscilla[d] Hb	200	4.2 (0.15)†	2.1×10^{-2}	–	–	–	–	–	–
P. caudatum[e] Hb	30	25	0.83	–	–	–	–	–	–
M. tuberculosis[f] trHbO	0.11	0.0014	0.013	0.18	–	–	–	–	–
N. commune[g] GlbN	390	79	0.20	600	2.2×10^{-4}	3.7×10^{-7}			
Yeast[c]									
S. cerevesial YHb	17	0.6	0.035	10–20	–	–	230	~200	0.87
Plants[h]									
Soybean[i,j] Lb	130	5.6	0.047	170	2×10^{-5}	1.6×10^{-7}	0.14	3	21

(continued)

Table 8.2 *(continued)*

	Fe(II)			Fe(II)			Fe(III)		
	O_2			NO			NO		
	K'	k	K_d	k'	k	K_d	k'	k	K_d
	$\mu M^{-1} s^{-1}$	s^{-1}	μM	$\mu M^{-1} s^{-1}$	s^{-1}	μM	$\mu M^{-1} s^{-1}$	s^{-1}	μM
Kideny bean Lb	130	6.2	0.048	240	–	–	–	–	–
Cowpea II Lb	140	5.5	0.039	190	–	–	–	–	–
Sesbania II Lb	210	7.5	0.036	–	–	–	–	–	–
Green pea I Lb	250	16	0.064	250	–	–	–	–	–
Green pea IV Lb	260	16	0.062	180	–	–	–	–	–
Broad bean V Lb	260	19	0.073	200	–	–	–	–	–
Lupin I Lb	540	20	0.037	200	–	–	–	–	–
Lupin II Lb	320	25	0.078	–	–	–	–	–	–
Parasponia[k] I Hb	170	15	0.091	240	–	–	–	–	–
Casuarina[l] II Hb	41	6	0.135	–	–	–	–	–	–
Rice[m] nsHb	68	–	–	–	–	–	–	–	–

[a] Gardner *et al.*, 2000a.
[b] Farres *et al.*, 2005.
[c] Gardner *et al.*, 2000b.
[d] Giangiacomo *et al.*, 2001.
[e] Das *et al.*, 2000.
[f] Ouellet *et al.*, 2002.
[g] Thorsteinsson *et al.*, 1999.
[h] Gibson *et al.*, 1989.
[i] Hargrove *et al.*, 1997.
[j] Herold and Puppo, 1997.
[k] Wittenberg *et al.*, 1986.
[l] Fleming *et al.*, 1987.
[m] Appleby *et al.*, 1984.
† Biphasic kinetics, slow rate constant enclosed in parenthesis.

In this regard, a protective function of nsHb against nitrosative stress has been suggested based on *in vitro* analysis, which demonstrated that nsHb scavenges NO; the ferric heme can be reduced by NADH and then rebinds O_2 to complete the catalytic cycle (NO-dioxygenase or O_2-nitrosylase, see Eqs. 8.1–8.6). Direct reduction of *Arabidopsis* nsHb by NADPH has also been implicated in NO turnover (Perazzolli *et al.*, 2004) in a manner reminiscent of an invertebrate Hb (Minning *et al.*, 1999). Enzymatic catalysis involving a reductase remains an open question. A candidate is methemoglobin reductase from alfalfa root extracts that is dependent on NADH, and flavin copurifies with the nsHb from barley or alfalfa (Abir *et al.*, 2004; Seregelyes *et al.*, 2004). Additionally, a monodehydroascorbate reductase has been shown to catalyze the reduction of ferric nsHb in barley (Igamberdiev *et al.*, 2006). Inasmuch as the reduction of ferric heme is crucial in catalytic NO removal, it is interesting to note that hexacoordination of the heme, as in nsHbs, has been shown to facilitate heme reduction (Weiland *et al.*, 2004). Ferric Hb may also be important in production of plant *S*-nitrosohemoglobin, which has been isolated from *Arabidopsis* and implicated in the mechanism by which its nsHb processes NO: drawing directly on analogy to *Ascaris* Hb (see the subsequent section), a cysteine residue in the distal heme pocket (Fig. 8.2) was shown to subserve degradation of *S*-nitrosothiols and NO (Perazzolli *et al.*, 2004, 2006).

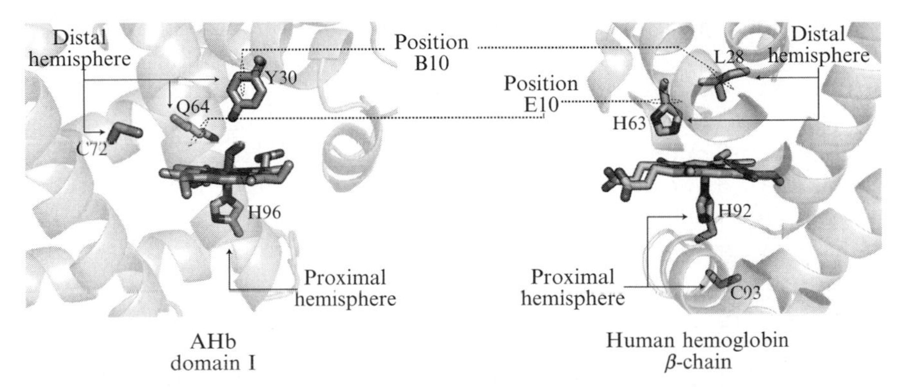

Figure 8.2 The positioning of β93C in the proximal heme cavity of human Hb (PDB ID: 2DN1) (at right) hinders reactions between β93C(NO) and heme-bound dioxygen, and thereby promotes the preservation and delivery of NO bioactivity. In contrast (at left), NO scavenging is favored by the close proximity of Cys(NO) and heme-bound dioxygen in *Ascaris* Hb (PDB ID: 1ASH). The stability of heme-bound oxygen is increased through electrophilic interaction with the hydroxyl side chain of Y30, and Q64 (position E10) enhances the rate of binding of NO to ferric hemes and thereby facilitates S-nitrosylation of C72 (see Eqs. 8.7–8.12). In human Hb versus AHb, L28 is found at position B10 in place of Y30, and facilitates O_2 delivery by increasing k_{off} (see Table 8.3). Position E10 is occupied by H63 in human Hb and in all mammals and birds. β93C and H92, which are located in the proximal hemisphere of the heme cavity, are similarly conserved.

It remains to be shown how rates of NO turnover by nsHbs compare to those of bacterial and yeast FHbs, and whether nsHbs play a physiologically relevant role in protection from NO. Although *Arabidopsis* nsHb is apparently not involved in pathogen-defense signaling (Perazzolli *et al.*, 2004), alfalfa nsHb appears to modulate pathogenesis-related gene expression (Seregelyes *et al.*, 2004).

Little is known about class II nsHbs, although they have a lower O_2 affinity than do class 1 nsHbs and may therefore store O_2 (Trevaskis *et al.*, 1997). No function has been proposed for the truncated, "2-on-2" plant hemoglobins (Vieweg *et al.*, 2004; Watts *et al.*, 2001).

In summary, plant nsHbs appear to catalyze NO-related detoxification chemistry in a manner similar to FHbs and to form *S*-nitrosohemoglobin in a manner reminiscent of worm and human Hbs (as discussed in the subsequent section). The physiological roles of plant Hbs in NO signaling or protection remain to be established.

5. Invertebrates

5.1. Nematode worms

Ascaris lumbricodes is a parasitic worm infecting nearly 25% of the world's population, and is a major cause of bowel obstruction in children younger than 15 years of age (Leder and Weller, 2006). It is the largest human pathogenic worm, reaching 40 cm in length and 6 mm in diameter (Leder and Weller, 2006). The coelom that transverses the length of *Ascaris* holds extracellular hemolymph that contains a polymeric Hb (De Baere *et al.*, 1992,1994). *Ascaris* Hb (AHb) is a multimer large enough to prevent extravasation into the inner lumen, so that it remains confined to the coelomic cavity (Vinogradov *et al.*, 2006; Weber and Vinogradov, 2001).

5.2. AHb: Structural and biochemical properties

The overall structure of AHb is a hexagonal bilayer composed of 12 dodecamers, each of which contain 3 trimers, 3 monomers, and 3 linker regions, for a total of 144 heme cofactors (Vinogradov *et al.*, 2006; Weber and Vinogradov, 2001). The heme pocket exhibits three distinguishing differences from mammalian Hb with respect to redox function. First, tyrosine substitutes for leucine B10 (Minning *et al.*, 1999). The electrophilic character of B10Y stabilizes heme-bound O_2 relative to leucine, thereby decreasing the rate of disassociation (k_{off}, off rate) by 500-fold (De Baere *et al.*, 1992, 1994; Gibson and Smith, 1965). Because the enhanced affinity that is conferred by B10Y relies on the polarizable character of the Fe(II)–O_2 bond, the affinity for carbon monoxide and other nonpolar ligands differs little from most other Hbs (De Baere *et al.*, 1992, 1994).

In addition, the distal position in the heme pocket that is occupied by histidine in most vertebrate Hbs is replaced by glutamine (E10Q) (Minning *et al.*, 1999). Analogous mutations in mammalian myoglobin result in a 1000-fold increase in the rate of NO binding (k_{on}; rate of association) to ferric hemes and thereby, remarkably, eliminates the disparity between the rate of NO binding to oxidized versus reduced hemes found in the native protein (De Baere *et al.*, 1992, 1994; Springer *et al.*, 1989). Finally, cysteine E15 (E15C) extends into the distal hemisphere of the heme cavity (see Fig. 8.2), such that its thiol is within 9 Å of the heme-dioxygen complex (PDB ID: 1ASH).

5.3. Physiological role and biochemical mechanism

Ascaris thrives in anaerobic environments; in room air, the worm's movement is slowed. This, in conjunction with the high O_2 affinity of AHb, has prompted the suggestion that O_2 is toxic (De Baere *et al.*, 1992, 1994; Minning *et al.*, 1999). Consistent with this interpretation, it was shown that AHb functions *in vivo* as an NADPH-dependent enzyme that sustains tissue hypoxia (Minning *et al.*, 1999). Like FHb, AHb catalyzes nitrate formation from O_2 and NO. However, the role of AHb is to mitigate the toxic effects of O_2 rather than NO, evidenced by the accumulation of O_2 *in situ* upon removal of Hb (Minning *et al.*, 1999). Whereas microbes are exposed to high concentrations of exogenous NO, AHb produces its own, which can be detected within the coelom as either Hb-[FeNO] or *S*-nitrosohemoglobin, reflecting the unique mechanism of NO turnover (Minning *et al.*, 1999). That is, AHb deviates mechanistically from FHb in using cysteines both as a redox couple and as an active site for the formation of transient intermediates in the form of *S*-nitrosothiols (see Fig. 8.1).

5.3.1. Cysteine-mediated dioxygenase mechanism (NO-dependent deoxygenase)

$$\text{AHb-[Fe(II)O}_2\text{||C(SH)]} \leftrightarrow \text{AHb-[Fe(III)O}_2^-\text{||C(SH)]} \quad (8.7)$$

$$\text{AHb-[Fe(III)O}_2^-\text{||C(SH)]} \xrightarrow{\text{NO}} \text{AHb-[Fe(III)||C(SH)]} + \text{NO}_3^- \quad (8.8)$$

$$\text{AHb-[Fe(III)||C(SH)]} \xrightarrow{\text{NO}} \text{AHb-[Fe(III)NO||C(SH)]}$$
$$\text{AHb-[Fe(III)NO||C(SH)]} \leftrightarrow \text{AHb-[Fe(II)||C(SNO)]} + \text{H}^+ \quad (8.9)$$

$$\text{AHb-[Fe(II)||C(SNO)]} \xrightarrow{O_2} \text{AHb-[Fe(II)O}_2\text{||C(SNO)]}$$
$$\text{AHb-[Fe(II)O}_2\text{||C(SNO)]} \rightarrow \text{AHb-[Fe(III)||C(S}^\bullet\text{)]} + \text{NO}_3^- \tag{8.10}$$

$$\begin{aligned}&\text{AHb-[Fe(III)||C(S}^\bullet\text{)]} + \text{NADPH}\\ &\rightarrow \text{AHb-[Fe(III)||C(SH)]} + \text{NADP}^\bullet\end{aligned} \tag{8.11}$$

$$\text{NADP}^\bullet \xrightarrow{O_2} \text{NADP}^+ + O_2^- \tag{8.12}$$

In contrast to the O_2-nitroxylase function of FHb, AHb functions as a deoxygenase (formally an NO dioxygenase) through a sequential mechanism that does not involve Fe(II)NO but that hinges on an *S*-nitrosothiol intermediate that is formed following binding of NO to Fe(III). Specifically, AHb couples heme reduction by NO to formation of an *S*-nitrosothiol at C72 (AHb-SNO); ferric heme reduction to Fe(II) allows O_2 to bind. Therefore, O_2 binding is necessarily preceded by the formation of AHb-SNO, with which it reacts rapidly (Eqs. 8.9 and 8.10).

Heme-bound O_2 and SNO are thought to form a peroxynitrite intermediate that dissociates and subsequently dismutates to nitrate (Minning *et al.*, 1999). Two electrons are required to sustain a catalytic cycle. NADPH-dependent reduction of a heme/cysteine redox couple thus completes a single enzymatic turnover (Eqs. 8.11 and 8.12). Mutational analysis suggests that the reaction is initiated by either NO binding to Fe(III) followed by transnitrosylation of the adjacent E15C or by *S*-nitrosylation of A7C located near the protein surface followed by NO transfer to E15C or heme iron located within the central region of the protein (Minning *et al.*, 1999).

In the phylogeny of Hbs, nematode Hb sits at an evolutionary divide between primordial flavoHb, which functions in NO detoxification, and mammalian Hb, which functions in NO/SNO delivery (see the subsequent section). The introduction of an *S*-nitrosothiol intermediate thus signifies a significant divergence in the physiological role of Hb (see Fig. 8.1) and suggests that the evolution of Hbs may be rationalized by NO-related functions. More specifically, thiols have been adapted to provide functionality other than heme-based with which to handle NO. *Ascaris* employs thiol-bound NO to control O_2 concentrations by fixing O_2 as nitrate. In doing so, AHb exemplifies a key feature of mammalian and avian hemoglobins, namely, heme/Cys redox coupling. But whereas AHb subserves NO-dependent deoxygenation, the mammalian and avian hemoglobins use SNO to enhance oxygenation. These alternative functions of deoxygenation and oxygenation, respectively, are enabled by distinct positioning of the thiol in the distal pocket of AHb versus the proximal pocket of mammalian Hb, as discussed subsequently.

In the following section, the evolutionary divergence of Hb function is explored further in mammalian and avian Hbs. In higher organisms that undergo aerobic respiration, allosteric regulation of O_2 binding is combined with regulation by NO of heme redox-dependent *S*-nitrosylation to enable O_2 delivery through vasodilation. The positioning of a highly conserved Cys (β93C) on the opposite side of the iron-dioxygen complex, namely in the proximal hemisphere of the heme cavity (Minning *et al.*, 1999), prevents SNO-β93C from interacting directly with O_2; the lifetimes of SNO-β93C and Fe(II)O_2 are thereby extended and their bioactivities preserved. Additionally, the essential nature of the mechanism of SNO formation in AHb has become more sophisticated in tetrameric human Hb, where small concentrations of NO are targeted to a minority of molecules (micropopulations) that are capable of preserving and delivering NO bioactivity (Angelo *et al.*, 2006; Gow *et al.*, 1999; Gow and Stamler, 1998; Luchsinger *et al.*, 2003; Singel and Stamler, 2005). The physiological role of Hb is thus altered a third time, to serve in coupled delivery of NO and oxygen.

6. Mammals and Birds

6.1. Tetrameric hemoglobins

Tetrameric mammalian and avian Hbs are intracellular proteins found in millimolar concentrations within erythrocytes (red blood cells, RBCs) (Perutz, 1987). Tetramers are composed of identical heterodimers, each containing an α-chain and β-chain. Low-level expression of β-subunits has also been observed within activated macrophages (Liu *et al.*, 2000), alveolar cells, and Clara cells in humans and mice (Bhaskaran *et al.*, 2005; Newton *et al.*, 2006). The affinity of Hb for ligand is pO_2 dependent (cooperative), which enables loading and unloading of O_2 within the lungs and hypoxic tissue, respectively (Perutz, 1987). In the absence of heterotropic ligands, the O_2-binding isotherm of tetrameric Hbs is predominantly hyperbolic, differing little from myoglobin (Imai, 1982). The sigmoidal-binding isotherm that is characteristic of tetrameric Hbs emerges prominently under conditions such as those in erythrocytes, where the concentration of tetramer is matched by the concentration of 2,3-bisphosphoglycerate (2,3-DPG) (Perutz, 1987). Endogenous allosteric effectors typically enhance cooperation by perturbing hydrogen bonding and ionic interactions between dimers that preferentially stabilize conformations with low ligand affinities, collectively referred to as T(ense)-state (Imai, 1982; Perutz, 1987). In contrast, structures with high ligand affinities are referred to as R(elaxed)-state.

Unlike oxyhemoglobin (HbO_2), concentrations of NO-bound hemoglobin (HbNO) *in vivo* are limited by the availability of NO, not Hb (Gow *et al.*, 1999; Gow and Stamler, 1998; Singel and Stamler, 2005).

Therefore, Hb molecules that contain NO comprise a micropopulation whose concentration rarely exceeds 0.1–0.01% of the total Hb concentration (Singel and Stamler, 2005). The HbNO micropopulation *in vivo* is found in the form of *S*-nitrosothiol (Hb-SNO) or iron nitrosyl (Hb-[FeNO]) (Doctor *et al.*, 2005; Gow and Stamler, 1998; McMahon *et al.*, 2005, 2002a; Pawloski *et al.*, 2001,2005; Pawloski and Stamler, 2002). The equilibrium *in vivo* between Hb-SNO and Hb-[FeNO] shifts in response to minor variations in ligand concentration, redox status, and heterotropic effectors, which should therefore be carefully considered in the design of experimental protocols.

6.2. Tetrameric Hb: Structural and biochemical properties

The proximal and distal positions within the heme pockets of both α- and β-chains are occupied by histidines, while the B10 position (a tyrosine in AHb and FHb) is occupied by leucine (Imai, 1982; Perutz, 1987) (see Fig. 8.2). β-chains are larger than α-chains (five amino acids) and contain a conserved cysteine at position 93 (β93C) near the $\alpha_1\beta_2$ dimer interface (Imai, 1982; Perutz, 1987; Reischl *et al.*, 2006) (abbreviated as β, β and α^+, β^+ for ferrous and ferric forms, respectively; Fe is used in cases where subunit identity is unknown). Consequently, the reactivity of the β93C thiol is dependent on quaternary structure, with solvent accessibility greater in T-state than in R-state (Gow and Stamler, 1998; Imai, 1982).

The biochemical features of each subunit differ significantly, with respect to both ligand binding and redox chemistry (Table 8.3) (Cassoly and Gibson, 1975; Doyle *et al.*, 1981, 1985; Edelstein and Gibson, 1975; Moore and Gibson, 1976; Perrella and Cera, 1999; Perutz, 1987; Sharma *et al.*, 1983, 1987; Sharma and Ranney, 1978; Taketa *et al.*, 1978; Tsuruga *et al.*, 1998). The off rate of O_2 and NO from ferrous hemes in β-chains equals or exceeds that of α-chains (Cassoly and Gibson, 1975; Moore and Gibson, 1976; Sharma and Ranney, 1978; Taketa *et al.*, 1978). Consequently, the ligand affinity of $\alpha \geq \beta$, even in the case of T-state Hb, where ${}^{O_2}k_{on}$ for β-chains is fourfold faster than α-chains. $\beta(O_2)$ formation is kinetically favored, and may transiently exceed concentrations of $\alpha(O_2)$, but the trend at equilibrium is reversed. In the case of NO and ferric hemes, ${}^{NO}k_{on}$ and ${}^{NO}k_{off}$ at pH 7.0 of β^+ meets or exceeds that of α^+ (Sharma *et al.*, 1987). Notably, in this case the ligand affinity of $\beta^+ > \alpha^+$ as well, making β^+(NO) the only species for which formation is more favorable, both kinetically and thermodynamically, than its α–chain equivalent (α^+[NO]).

In R-state, β-chains are more readily reduced than α-chains (Brunori *et al.*, 1967). This disparity widens in T-state, and is largely attributed to the functional and thermodynamic linkage between β93C and β-heme iron (Antholine *et al.*, 1984; Bonaventura *et al.*, 1999; Lau and Asakura, 1979; Manoharan *et al.*, 1990; Steinhoff, 1990). The transition to R-state

Table 8.3 Rate constants for the association (K), dissociation (k), and dissociation equilibrium (K_d) of O_2 and NO from ferrous and ferric hemes of hemoglobin from vertebrates and invertebrates

				Fe(II)			Fe(II)			Fe(III)		
				O_2			NO			NO		
				K'	k	K_d	K'	k	K_d	K'	k	K_d
				$\mu M^{-1} s^{-1}$	s^{-1}	μM	$\mu M^{-1} s^{-1}$	s^{-1}	μM	$\mu M^{-1} s^{-1}$	s^{-1}	μM
Vertebrates												
Tetrameric	Carp[a]	Fully oxidized		–	–	–	–	–	–	0.014	13	930
										$(5 \times 10^{-3})^{\dagger}$	$(3.2)^{\dagger}$	
	Opposum[b]	α		–	–	–	–	–	–	1.3	13	10
		β		–	–	–	–	–	–	0.017	2.9	170
		R-state[c,d]	α	33	13	0.39	14	4.6×10^{-5}	3.3×10^{-6}	–	–	–
			β	33	21	0.64	14	2.2×10^{-5}	1.6×10^{-6}	–	–	–
		T-state[e,f]	α	2.9	180	62	24	3.3×10^{-4}	1.4×10^{-5}	–	–	–
			β	12	2.5×10^{3}	210	24	1.3×10^{-3}	5.4×10^{-5}	–	–	–
	Human	Fully[a] oxidized	α	–	–	–	–	–	–	1.7×10^{-3}	0.65	380
			β	–	–	–	–	–	–	6.4×10^{-3}	1.5	230
		Mixed valency[a]	$\alpha^{+}\beta(CO)$	–	–	–	–	–	–	1.2×10^{-3}	1.4	1.2×10^{3}
			$\alpha(CO)\beta^{+}$	–	–	–	–	–	–	7.3×10^{-3}	1.4	190
Monomeric			Ngb[g,h]				150	2.0×10^{-4}	$1.1 \times 10^{-3\ddagger}$	–	–	–
			Cgb[i]	30	0.35	0.012	–	–	–	–	–	–
			Mb[j,b]	14	12	0.86	20	–	–	0.053	14	260
	Elephant[b]		Mb	–	–	–	–	–	–	22	40	1.8
	Sperm whale[a,k]		Mb	15	14	0.93	20	–	–	0.053	14	260

(continued)

Table 8.3 (*continued*)

	Fe(II)			Fe(II)			Fe(III)		
	O_2			NO			NO		
	K'	k	K_d	K'	k	K_d	K'	k	K_d
	$\mu M^{-1} s^{-1}$	s^{-1}	μM	$\mu M^{-1} s^{-1}$	s^{-1}	μM	$\mu M^{-1} s^{-1}$	s^{-1}	μM
Invertebrates									
Ascaris summ[l]	1.5	4.0×10^{-3}	2.7	–	–	–	–	–	–
Scapharca inequivalvas[m,n]	16	580 (50)[†]	3.6×10^{5} (3.1)	16	–	–	–	–	–

[a] Sharmaet *al.*, 1987.
[b] Sharma *et al.*, 1983.
[c] Cassoly and Gibson, 1975.
[d] Moore and Gibson, 1976.
[e] Sharma and Ranney., 1978.
[f] Taketa *et al.*, 1978.
[g] Van Doorsalaer *et al.*, 2003.
[h] Dewilde *et al.*, 2001.
[i] Trent and Hargrove., 2002.
[j] Springer *et al.*, 1989.
[k] Carver *et al.*, 1990.
[l] Gibson and Smith., 1965.
[m] Antonini *et al.*, 1984.
[n] Chiancone and Gibson, 1989.
[†] Biphasic kinetics, slow rate constant enclosed in parenthesis.
[‡] Disassociation constant accounts for competitive binding by distal histidine.

significantly alters the local environment around β93C: the orientation of β93C relative to β145Y undergoes a rotational translation of $\approx 90°$, while the movement of β146H toward the central cavity doubles the distance between the imidazole nitrogen and thiol sulfur, and solvent accessibility to β93C is hindered by steric influences from several nearby side chains (PDB ID: 2DN1, assumed biological unit). The pK_a of the proximal histidine (β92H) also increases. These changes alter redox coupling of β93C thiol and heme with resultant alterations of heme redox potential and spin state (Antholine *et al.*, 1985; Chan *et al.*, 1998; Imai, 1982; Mawjood *et al.*, 2000; Nothig-Laslo, 1977; Nothig-Laslo and Maricic, 1982).

The thermodynamic linkage between β-chain hemes and β93C is manifested *in vivo* as conformation-dependent redox coupling that facilitates NO processing within the β-subunit (NO is exchanged between hemes and thiols to regulate NO bioavailability)(Gow and Stamler, 1998; Hess *et al.*, 2005; McMahon *et al.*, 2002b, 2005; Pawloski *et al.*, 2001, 2005; Pawloski and Stamler, 2002; Singel and Stamler, 2005). As in AHb, biologically relevant interactions of Hb and NO involve both Hb-[FeNO] and *S*-nitrosothiol (Hb-SNO). However, whereas Hb-SNO is a transient intermediate in a mechanism for NO/O_2 elimination in the nematode, it functions in mammals to preserve and deliver NO in a pO_2-dependent fashion. Specifically, in mammals (and presumably in birds) *S*-nitrosylation/denitrosylation of Hb subserves RBC-mediated hypoxic vasodilation that matches blood flow to metabolic demand (Singel and Stamler, 2005; Sonveaux *et al.*, 2007).

6.3. Tetrameric Hb: Physiological role and biochemical mechanisms

Recent work indicates that functional differences between α- and β-subunits are physiologically relevant in Hb–SNO formation and the delivery of NO-related bioactivity. It has been shown that, when substrate concentrations and reaction times recapitulate the *in vivo* conditions, both NO and nitrite can be used by β-subunits of Hb to produce Hb-SNO (Angelo *et al.*, 2006; Gow *et al.*, 1999; Gow and Stamler, 1998; Luchsinger *et al.*, 2003, 2005). Thus, under aerobic conditions, with NO:heme ratios less than 0.01, Hb-SNO is a major product (Gow *et al.*, 1999):

$$HbO_2\text{-}[C(SH)] \xrightarrow{NO} HbO_2\text{-}[C(SNO)] + H^+ + e^- \qquad (8.13)$$

Analogous reactions (low physiological NO concentrations) carried out under anaerobic conditions produce Hb-[FeNO], which is converted to Hb-SNO if promptly oxygenated after NO addition (Eq. 8.14) (Gow and Stamler, 1998; Pawloski *et al.*, 2005). A similar scheme—involving a reactive SNO precursor with properties of $Fe(II)NO^+/Fe(III)NO$—rationalizes the reaction between deoxyhemoglobin (deoxyHb) and nitrite (NO_2^-) under physiological conditions (Angelo *et al.*, 2006) (Eq. 8.15):

$$\begin{aligned}&\text{Hb-[Fe||C(SH)]} \xrightarrow{\text{NO}} \text{Hb-[FeNO||C(SH)]}\\&\text{Hb-[FeNO||C(SH)]} \xrightarrow{\text{O}_2} \text{HbO}_2\text{-[Fe||C(SNO)]} + \text{H}^+ + \text{e}^-\end{aligned} \quad (8.14)$$

$$\begin{aligned}&\text{Hb-[Fe||C(SH)]}+\text{NO}_2^- \xrightarrow{\text{H}^+} \text{Hb-[Fe--NO}^+\text{||C(SH)]}+\text{OH}^-\\&\text{Hb-[Fe--NO}^+\text{||C(SH)]} \xrightarrow{\text{O}_2} \text{HbO}_2\text{-[Fe||C(SNO)]}+\text{H}^+\end{aligned} \quad (8.15)$$

In all three reactions, yields of Hb-SNO approach unity (actual yields 30–50%) with low substrate:protein ratios but progressively diminish as the relative NO concentration exceeds micromolar and the ratios exceed 1% (Angelo *et al.*, 2006; Gow *et al.*, 1999; Gow and Stamler, 1998). The amount of time between substrate addition and sample oxygenation is inversely related to Hb-SNO yields in reactions initiated under anaerobic conditions, such that product (Hb-SNO) yields in samples oxygenated 30 min after substrate addition are 10% the value of those oxygenated after 30 s (Angelo *et al.*, 2006; Gow and Stamler, 1998; McMahon *et al.*, 2002b; Singel and Stamler, 2005). Thus, HbNO micropopulations that serve as SNO precursors (e.g., valency hybrids containing β^+NO) in T-state are evidently short lived. *In vivo*, however, circulation times are on the order of 1 min, and Hb resides in the veins (deoxygenated conditions) for no more than 30 s.

6.4. Tetrameric Hb micropopulations

The maximum yield of Hb-SNO that is attainable under physiological conditions appears to be fixed at $\approx 1\mu$M, and has been suggested to indicate the saturation point of a micropopulation(s) involved in redox processing under limiting conditions (presumably reflecting the physiological milieu of avian and mammalian RBCs) (Singel and Stamler, 2005). Here, a micropopulation refers to a specific form of Hb, found in relatively small concentrations whose redox, ligation, or allosteric properties could enable it to exhibit unique biochemistry. Examples of micropopulations include subsaturated tetramers (e.g., $2[\text{Fe(II)O}_2\bullet\text{Fe(II)}]$) and mixed-valency species (e.g., $2[\text{Fe(II)} \bullet \text{Fe(III)}]$). Micropopulations may redox couple with cysteine β93.

Prolonged incubation under subsaturating conditions potentiates heme oxidation and NO reduction (assayed as hydroxylamine [in the presence of thiol] or as N_2O formation) (Eq. 8.16), leading to loss of bioactive NO. However, it has also been observed in some cases that time-dependent attenuation of Hb-SNO yields is inversely related to Hb-[FeNO] concentration. These results are difficult to rationalize exclusively within the context of NO-mediated heme oxidation but can be understood in conjunction with kinetic disparities between α- and β-chains in the rate of ligand binding (Angelo *et al.*, 2006; Cassoly and Gibson, 1975; Edelstein and Gibson, 1975; Moore and Gibson, 1976; Sharma *et al.*,

1983, 1987; Sharma and Ranney, 1978; Singel and Stamler, 2005). Because β(NO) may be favored kinetically but not thermodynamically, and/or may be favored in some (e.g., $\alpha \rightarrow \beta$) but not other micropopulations (Eq. 8.17), long incubation times may shift the relative partitioning of NO toward α-chains (Cassoly and Gibson, 1975; Edelstein and Gibson, 1975; Henry *et al.*, 1993; Moore and Gibson, 1976), altering the redox chemistry sustainable by the system upon allosteric transition (McMahon *et al.*, 2002b, 2005; Singel and Stamler, 2005). Moreover, elevated concentrations of NO and lower pH favor formation of high-affinity tetramers containing 5-coordinate α(NO) (Eq. 8.18), and thus dramatically alter allosteric properties (Singel and Stamler, 2005):

$$\begin{array}{c} \text{Hb-[Fe(II)NO]} \leftrightarrow \text{Hb-[Fe(III)NO}^-] \\ \text{Hb-[Fe(III)NO}^-] \xrightarrow{H^+} \text{Hb-[Fe(III)]} + \text{HNO} \\ 2\text{HNO} + \text{RSH} \rightarrow 2\text{NH}_2\text{OH} + \text{RSSR} \end{array} \tag{8.16}$$

$$\begin{array}{c} \text{Hb-}[\alpha\beta(\text{NO})] \rightarrow \text{Hb-}[\alpha(\text{NO})\beta] \\ \text{Hb-}[\alpha_6(\text{NO})\beta] \leftrightarrow \text{Hb-}[\alpha_5(\text{NO})\beta] \end{array} \tag{8.17}$$

Taken together, these experimental observations suggest the importance of a mixed-valency micropopulation that kinetically favors NO processing within β-chains, (Angelo *et al.*, 2006; Gow *et al.*, 1999; Gow and Stamler, 1998; Luchsinger *et al.*, 2003). Furthermore, they emphasize that the stability of intermediates formed en route to Hb-SNO likely depends on the allosteric and redox environment, and they may therefore not be found in significant quantities at equilibrium (Angelo *et al.*, 2006; Gow and Stamler, 1998; McMahon *et al.*, 2002b; Singel and Stamler, 2005). It is intriguing that the kinetic favorability of redox processing within β-chains may underlie a critical mechanism of Hb-SNO synthesis, whose operation is dependent on transient redox-binding intermediates that are formed and activated by cyclic perturbations in ligand concentration, much like those encountered by circulating RBCs. It is therefore of interest that the majority of ferric hemes occurring in native human erythrocytes are found within mixed-valency tetramers, the intracellular concentrations of which have been shown in mice to increase after intravenous infusions of NO-donating vasodilators (Kruszyna *et al.*, 1988).

Recent work has explored the role in Hb-SNO synthesis of NO-liganded, mixed valency micropopulations produced by partial oxidation of deoxyhemoglobin by nitrite (NO_2^-) (Angelo *et al.*, 2006)

$$\begin{array}{l} \text{Hb-[2Fe(II)]} + \text{NO}_2^- \xrightarrow{H^+} \text{Hb-[Fe(III)NO||Fe(II)]} + \text{OH}^- \\ \text{Hb-[Fe(III)NO||Fe(II)]} \rightleftharpoons \text{Hb-[Fe(III)||Fe(II)NO]} \end{array} \tag{8.18}$$

Analysis by spectral deconvolution revealed a product resembling a ferric nitrosyl whose equilibrium concentration was proportional to starting concentrations of nitrite in reactions using nitrite: Hb (tetramer) ratios ≤ 1. Higher ratios were associated with proportional decreases in yield from the maximum value observed at 1:1 ratios, and the product was undetectable with a twofold excess of nitrite. The disappearance of ferric nitrosyl was associated with an equivalent increase in Hb-SNO (the sum of these components was constant), and the yield of Hb-SNO was greatest with a twofold excess of nitrite over Hb tetramer. Additionally, Hb-SNO yields equal to those determined spectrally for ferric nitrosyl were observed in samples that were oxygenated at equilibrium:

$$\begin{aligned} &\text{Hb-[Fe(III)NO||C(SH)]} \leftrightarrow \text{Hb-[Fe(II)||C(SNO)]+H}^+ \\ &\text{Hb-[Fe(II)||C(SNO)]} \overset{O_2}{\rightarrow} \text{Hb-[Fe(II)O}_2\text{||C(SNO)]} \end{aligned} \tag{8.19}$$

Notably, the ligand-dependent features of Hb-[Fe(III)NO] formation are similar to those previously characterized for equilibrium concentrations of CO-binding intermediates containing two ligands. This suggests that tetramers containing Fe(III)NO may be doubly liganded as well, and therefore similar to micropopulations found *in vivo*.

6.5. Tetrameric Hb oxidation and Hb-SNO formation

The loss of one electron is formally required in forming *S*-nitrosothiols from NO and reduced cysteine (Jia *et al.*, 1996; Singel and Stamler, 2005). Some *in vitro* results suggest that methemoglobin, preferentially formed within the β subunit, assumes a major role in the formation of Hb-SNO (Luchsinger *et al.*, 2003):

$$\begin{aligned} &\text{Hb-[Fe(III)||C(SH)]} \xrightarrow{\text{NO}} \text{Hb-[Fe(III)NO||C(SH)]} \\ &\text{Hb-[Fe(III)NO||C(SH)]} \xrightarrow{\text{NO}} \text{Hb-[Fe(III)NO||C(SNO)]} + \text{H}^+ \end{aligned} \tag{8.20}$$

$$\text{Hb-[Fe(III)]} \xrightarrow{\text{2NO, OH}^-} \text{Hb-[Fe(II)NO]} + \text{NO}_2^- \tag{8.21}$$

More specifically, the product ratio [β(NO)]:[α(NO)] was highest under limiting conditions of ligand, and exhibited a sevenfold preference for β-chains. As the NO:heme ratio approached unity, subunit specificity was lost (β(NO)$\sim\alpha$(NO)). The ratio of [β(NO)]:[α(NO)] was directly proportional to the ratio of [Hb-SNO]:[NO_2^-] such that preferential formation of Hb-SNO versus NO_2^- and β(NO) versus α(NO) occurred concomitantly.

It is important to note that published values for the rate of the initial step in the reaction described by Eq. 8.20 (binding of NO by Hb Fe[III]) are approximately $10^4 M^{-1}s^{-1}$ (Sharma *et al.*, 1983, 1987), nearly 1000-fold less than published rate constants for the NO dioxygenase reaction between HbO_2 and NO (Eich *et al.*, 1996; Herold *et al.*, 2001):

$$\begin{aligned} &\text{Hb-[Fe(II)O}_2] \xrightarrow{\text{NO}} \text{Hb-[Fe(III)OONO}^-] \\ &\text{Hb-[Fe(III)OONO}^-] \rightarrow \text{Hb-[Fe(III)]} + \text{NO}_3^- \end{aligned} \tag{8.22}$$

Therefore, it would seem at first glance that the rate of NO scavenging *in vivo* precludes the binding of NO by Hb-[Fe(III)]. However, it is noteworthy that reactions of limiting NO with *Ascaris* Hb-[Fe(III)] outcompete the reaction with oxyHb (Minning *et al.*, 1999) and direct binding is not the only route to HbNO: Fe(II)NO can be oxidized to Fe (III)NO within both mammalian (Luchsinger *et al.*, 2003) and bacterial (Hausladen *et al.*, 2001) Hbs. Moreover, rates of redox processing of NO catalyzed by nonmammalian Hbs exceed those of the NO dioxygenase reaction by 10-fold or greater (Hausladen *et al.*, 2001; Liu *et al.*, 2000; Minning *et al.*, 1999). The kinetics of the reactions that describe the complex chemistry of oxyHb and NO in mammals under physiologically relevant conditions have not been explored.

7. Experimental Design

The concentration dependence of the reactions with NO illustrates the challenges of adequate experimental analysis of the redox chemistry of Hb. Generally, protein and reagent concentrations are critically important (Angelo *et al.*, 2006; Gow *et al.*, 1999; Gow and Stamler, 1998; Hausladen *et al.*, 2001; Luchsinger *et al.*, 2003; Singel and Stamler, 2005), as is the relative abundance of redox and allosteric effectors (Gow *et al.*, 1999). Hb concentration is directly related to the lifetime of transient intermediates (Nothig-Laslo, 1977), equilibrium concentrations of micropopulations (Zhang *et al.*, 1991), the buffering capacity of Hb solutions (Gary-Bobo and Solomon, 1968), and the minimum amount of time required for mixing in stopped-flow experiments (Liau *et al.*, 2005). Relevant concentrations of NO and O_2 depend on the physiological situation. High micromolar concentrations of NO that are produced, for example, to kill pathogens are relevant to the detoxification function of flavoHb but will suppress the heme-catalyzed SNO-synthase activity of mammalian Hbs. Moreover, whereas bacterial Hbs may experience *in vivo* concentrations of NO approaching those of both Hb and O_2, the NO:HbO_2 ratio will rarely exceed 1:1000 in mammals. Reaction channels, enzymatic mechanisms, and functions of Hbs may change depending on reactant conditions.

The functions of Hb have evolved within different intracellular environments. The high concentration of Hb within RBCs (300–350 g/L) greatly amplifies the effects of protein crowding (Minton, 2001). Similar considerations may apply to AHb, which is concentrated within the perienteric cavity (Weber and Vinogradov, 2001), and to neuroglobins in the central nervous system of mammals and invertebrates (Hankeln *et al.*, 2005). Thus, the effective concentration of RBC Hb, as judged by thermodynamic activity, may exceed actual values by nearly 100-fold (Ellis, 2001a,b; Minton, 2001). Consequently, protein–protein interactions are transition state limited, not diffusion limited (Zimmerman and Minton, 1993). Although biochemical behavior is effectively predicted by actual concentration in solutions less than $\approx$250 μM tetramer (generally greater than the concentrations employed in analysis *in vitro*), the magnitude of system nonideality increases exponentially with increasing concentrations of Hb (Ellis, 2001a,b; Minton, 2001; Zimmerman and Minton, 1993). Hb concentration dependence has been observed for reactions of NO (Gow and Stamler, 1998), likely reflecting both dimer-tetramer equilibria (3 nM [deoxy] to 3 μM [oxy]) and effects of protein crowding (reactions proceeding more or less efficiently at $\approx$250 μM). More generally, *in vitro* analyses to date have poorly replicated the physiological milieu, and heme redox chemistry has not been adequately considered.

We have observed that UV/visible spectra of Hb are sensitive to small changes in temperature, particularly between 15 and 40°. Use of a Peltier stage or water-jacketed cuvettes is strongly recommended, particularly when using a double-beam spectrophotometer for differential spectroscopy. Spectral changes induced by pressure can be problematic in stopped-flow kinetic analysis but can usually be identified when comparing experimental data with adequate controls. Additionally, photolytic effects on photolabile NO species, which may be exerted by the intense light sources that are used in most diode-array stopped-flow systems, should be considered carefully.

Contamination of solutions by metals, nitrite, and nitrate should be assessed. We have found that Tris buffers contain lower levels of contaminants than commercial phosphate-buffered saline. In most cases, submillimolar concentrations of chelators will effectively sequester and neutralize metals at the concentrations encountered. For example, diethylaminetriamine pentacetate (DTPA; 100 μM) or ethylenediaminetetraacetate (EDTA; 100 μM) may be employed to chelate monovalent and di- or trivalent metals, respectively. Avoiding contamination by nitrogen oxides can be particularly difficult when using aqueous NO solutions. However, nitrite and nitrate are ubiquitous in all physiological systems and small amounts should not be viewed as artifactual. A recent review treats this subject in detail (Lim *et al.*, 2005).

8. Concluding Remarks

The differences in mechanism and function among Hbs found in bacteria, nematodes, plants, and vertebrates highlight important parameters that should be considered in analyzing reactions between NO and Hb in living organisms and raise the idea that the evolution of Hb has been greatly influenced by NO-related functions. New *in vivo* roles for Hb are likely to be discovered. It is of note in this regard that recent work has revealed the expression of β-globin in macrophages (Liu *et al.*, 1999) and epithelium (Bhaskaran *et al.*, 2005; Newton *et al.*, 2006), as well as the existence of two monomeric, hexacoordinate Hbs, neuroglobin (Ngb) and cytoglobin (Cgb), which are ubiquitously expressed in all major classes of living organisms (Burmester *et al.*, 2000, 2002; Dewilde *et al.*, 2001; Trent and Hargrove, 2002; Van Doorslaer *et al.*, 2003). Currently, the functions of these globins are as yet unclear but may include roles in pO_2 sensing, NO processing, and hypoxic signaling (Hankeln *et al.*, 2005; Liu *et al.*, 1999). Future work should further widen the understanding of the continuum of evolutionary adaptations that have equipped Hbs to respond to tissue hypoxia on the one hand and oxidative and nitrosative stresses on the other.

REFERENCES

Abir, U. I., Csaba, S., Nathalie, M., and Robert, D. H. (2004). NADH-dependent metabolism of nitric oxide in alfalfa root cultures expressing barley hemoglobin. *Planta* **219,** 95–102.

Angelo, M., Singel, D. J., and Stamler, J. S. (2006). An *S*-nitrosothiol (SNO) synthase function of hemoglobin that utilizes nitrite as a substrate. *Proc. Natl. Acad. Sci. USA* **103,** 8366–8371.

Antholine, W. E., Basosi, R., Hyde, J. S., and Taketa, F. (1984). Interaction between low-affinity cupric ion and human methemoglobin. *J. Inorg. Biochem.* **21,** 125–136.

Antholine, W. E., Taketa, F., Wang, J. T., Manoharan, P. T., and Rifkind, J. M. (1985). Interaction between bound cupric ion and spin-labeled cysteine β93 in human and horse hemoglobins. *J. Inorg. Biochem.* **25,** 95–108.

Antonini, E., Ascoli, F., Brunori, M., Chiancone, E., Verzili, D., Morris, R. J., and Gibson, Q. H. (1984). Kinetics of ligand binding and quaternary conformational change in the homodimeric hemoglobin from *Scapharca inaequivalvis*. *J. Biol. Chem.* **259,** 6730–6738.

Appleby, C. (1984). Leghemoglobin and rhizobium respiration. *Ann. Rev. Plant Physiol.* **35,** 443–478.

Ascenzi, P., Bocedi, A., Bolognesi, M., Fabozzi, G., Milani, M., and Visca, P. (2006). Nitric oxide scavenging by *Mycobacterium leprae* GlbO involves the formation of the ferric heme-bound peroxynitrite intermediate. *Biochem. Biophys. Res. Commun.* **339,** 450–456.

Bang, I. S., Liu, L., Vazquez-Torres, A., Crouch, M. L., Stamler, J. S., and Fang, F. C. (2006). Maintenance of nitric oxide and redox homeostasis by the salmonella flavohemoglobin hmp. *J. Biol. Chem.* **281,** 28039–28047.

Baudouin, E., Pieuchot, L., Engler, G., Pauly, N., and Puppo, A. (2006). Nitric oxide is formed in *Medicago truncatula-Sinorhizobium meliloti* functional nodules. *Mol. Plant Microbe. Interact.* **19,** 970–975.

Bhaskaran, M., Chen, H., Chen, Z., and Liu, L. (2005). Hemoglobin is expressed in alveolar epithelial type II cells. *Biochem. Biophys. Res. Commun.* **333,** 1348–1352.

Bidon-Chanal, A., Marti, M. A., Crespo, A., Milani, M., Orozco, M., Bolognesi, M., Luque, F. J., and Estrin, D. A. (2006). Ligand-induced dynamical regulation of NO conversion in *Mycobacterium tuberculosis* truncated hemoglobin-N. *Proteins* **64,** 457–464.

Bonaventura, C., Godette, G., Tesh, S., Holm, D. E., Bonaventura, J., Crumbliss, A. L., Pearce, L. L., and Peterson, J. (1999). Internal electron transfer between hemes and Cu(II) bound at cysteine β93 promotes methemoglobin reduction by carbon monoxide. *J. Biol. Chem.* **274,** 5499–5507.

Brunori, M., Taylor, J. F., Antonini, E., Wyman, J., and Rossi-Fanelli, A. (1967). Studies on the oxidation-reduction potentials of heme proteins. VI. Human hemoglobin treated with various sulfhydryl reagents. *J. Biol. Chem.* **242,** 2295–2300.

Burmester, T., Ebner, B., Weich, B., and Hankeln, T. (2002). Cytoglobin: A novel globin type ubiquitously expressed in vertebrate tissues. *Mol. Biol. Evol.* **19,** 416–421.

Burmester, T., Weich, B., Reinhardt, S., and Hankeln, T. (2000). A vertebrate globin expressed in the brain. *Nature* **407,** 520–523.

Carver, T. E., Rohlfs, R. J., Olson, J. S., Gibson, Q. H., Blackmore, R. S., Springer, B. A., and Sligar, S. G. (1990). Analysis of the kinetic barriers for ligand binding to sperm whale myoglobin using site-directed mutagenesis and laser photolysis techniques. *J. Biol. Chem.* **265,** 20007–20020.

Cassanova, N., O'Brien, K. M., Stahl, B. T., McClure, T., and Poyton, R. O. (2005). Yeast flavohemoglobin, a nitric oxide oxidoreductase, is located in both the cytosol and the mitochondrial matrix: Effects of respiration, anoxia, and the mitochondrial genome on its intracellular level and distribution. *J. Biol. Chem.* **280,** 7645–7653.

Cassoly, R., and Gibson, Q. (1975). Conformation, co-operativity and ligand binding in human hemoglobin. *J. Mol. Biol.* **91,** 301–313.

Chan, N. L., Rogers, P. H., and Arnone, A. (1998). Crystal structure of the *S*-nitroso form of liganded human hemoglobin. *Biochemistry* **37,** 16459–16464.

Chiancone, E., and Gibson, Q. H. (1989). Ligand binding to the dimeric hemoglobin from *Scapharca inaequivalvis*, a hemoglobin with a novel mechanism for cooperativity. *J. Biol. Chem.* **264,** 21062–21065.

Crawford, M. J., and Goldberg, D. E. (1998). Role for the *Salmonella* flavohemoglobin in protection from nitric oxide. *J. Biol. Chem.* **273,** 12543–12547.

Cueto, M., Hernandez-Perera, O., Martin, R., Bentura, M. L., Rodrigo, J., Lamas, S., and Golvano, M. P. (1996). Presence of nitric oxide synthase activity in roots and nodules of *Lupinus albus*. *FEBS Lett.* **398,** 159–164.

Das, T. K., Weber, R. E., Dewilde, S., Wittenberg, J. B., Wittenberg, B. A., Yamauchi, K., Van Hauwaert, M. L., Moens, L., and Rousseau, D. L. (2000). Ligand binding in the ferric and ferrous states of *Paramecium* hemoglobin. *Biochemistry* **39,** 14330–14340.

De Baere, I., Perutz, M. F., Kiger, L., Marden, M. C., and Poyart, C. (1994). Formation of two hydrogen bonds from the globin to the heme-linked oxygen molecule in *Ascaris* hemoglobin. *Proc. Natl. Acad. Sci. USA* **91,** 1594–1597.

De Baere, I., Liu, L., Moens, L., Van Beeumen, J., Gielens, C., Richelle, J., Trotman, C., Finch, J., Gerstein, M., and Perutz, M. (1992). Polar zipper sequence in the high-affinity hemoglobin of *Ascaris suum*: Amino acid sequence and structural interpretation. *Proc. Natl. Acad. Sci. USA* **89,** 4638–4642.

De Jesus-Berrios, M., Liu, L., Nussbaum, J. C., Cox, G. M., Stamler, J. S., and Heitman, J. (2003). Enzymes that counteract nitrosative stress promote fungal virulence. *Curr. Biol.* **13,** 1963–1968.

Denison, R. F., and Harter, B. L. (1995). Nitrate effects on nodule oxygen permeability and leghemoglobin (nodule oximetry and computer modeling). *Plant Physiol.* **107,** 1355–1364.

Dewilde, S., Kiger, L., Burmester, T., Hankeln, T., Baudin-Creuza, V., Aerts, T., Marden, M. C., Caubergs, R., and Moens, L. (2001). Biochemical characterization and ligand binding properties of neuroglobin, a novel member of the globin family. *J. Biol. Chem.* **276,** 38949–38955.

Doctor, A., Platt, R., Sheram, M. L., Eischeid, A., McMahon, T., Maxey, T., Doherty, J., Axelrod, M., Kline, J., Gurka, M., Gow, A., and Gaston, B. (2005). Hemoglobin conformation couples erythrocyte *S*-nitrosothiol content to O_2 gradients. *Proc. Natl. Acad. Sci. USA* **102,** 5709–5714.

Dordas, C., Rivoal, J., and Hill, R. D. (2003a). Plant haemoglobins, nitric oxide and hypoxic stress. *Ann. Bot. (Lond)* **91,** 173–178.

Dordas, C., Hasinoff, B. B., Rivoal, J., and Hill, R. D. (2003b). Class-1 hemoglobins, nitrate and NO levels in anoxic maize cell-suspension cultures. *Planta* **219,** 66–72.

Dordas, C., Hasinoff, B. B., Igamberdiev, A. U., Manah, N., Rivoal, J., and Hill, R. D. (2003c). Expression of a stress-induced hemoglobin affects NO levels produced by alfalfa root cultures under hypoxic stress. *Plant J.* **35,** 763–770.

Doyle, M. P., Herman, J. G., and Dykstra, R. L. (1985). Autocatalytic oxidation of hemoglobin induced by nitrite: Activation and chemical inhibition. *J. Free. Radic. Biol. Med.* **1,** 145–153.

Doyle, M. P., Pickering, R. A., DeWeert, T. M., Hoekstra, J. W., and Pater, D. (1981). Kinetics and mechanism of the oxidation of human deoxyhemoglobin by nitrites. *J. Biol. Chem.* **256,** 12393–12398.

Duff, S. M., Wittenberg, J. B., and Hill, R. D. (1997). Expression, purification, and properties of recombinant barley (*Hordeum* sp.) hemoglobin: Optical spectra and reactions with gaseous ligands. *J. Biol. Chem.* **272,** 16746–16752.

Edelstein, S. J., and Gibson, W. H. (1975). The effect of functional differences in the alpha and beta chains on the cooperativity of the oxidation reduction reaction of hemoglobin. *J. Biol. Chem.* **250,** 961–965.

Eich, R. F., Li, T., Lemon, D. D., Doherty, D. H., Curry, S. R., Aitken, J. F., Mathews, A. J., Johnson, K. A., Smith, R. D., Phillips, G. N., Jr., and Olson, J. S. (1996). Mechanism of NO-induced oxidation of myoglobin and hemoglobin. *Biochemistry* **35,** 6976–6983.

Ellis, R. J. (2001a). Macromolecular crowding: An important but neglected aspect of the intracellular environment. *Curr. Opin. Struct. Biol.* **11,** 114–119.

Ellis, R. J. (2001b). Macromolecular crowding: Obvious but underappreciated. *Trends Biochem. Sci.* **26,** 597–604.

Elvers, K. T., Wu, G., Gilberthorpe, N. J., Poole, R. K., and Park, S. F. (2004). Role of an inducible single-domain hemoglobin in mediating resistance to nitric oxide and nitrosative stress in *Campylobacter jejuni* and *Campylobacter coli*. *J. Bacteriol.* **186,** 5332–5341.

Elvers, K. T., Turner, S. M., Wainwright, L. M., Marsden, G., Hinds, J., Cole, J. A., Poole, R. K., Penn, C. W., and Park, S. F. (2005). NssR, a member of the Crp-Fnr superfamily from *Campylobacter jejuni*, regulates a nitrosative stress-responsive regulon that includes both a single-domain and a truncated haemoglobin. *Mol. Microbiol.* **57,** 735–750.

Eu, J. P., Sun, J., Xu, L., Stamler, J. S., and Meissner, G. (2000). The skeletal muscle calcium release channel: Coupled O2 sensor and NO signaling functions. *Cell* **102,** 499–509.

Fabozzi, G., Ascenzi, P., Renzi, S. D., and Visca, P. (2006). Truncated hemoglobin GlbO from *Mycobacterium leprae* alleviates nitric oxide toxicity. *Microb. Pathog.* **40,** 211–220.

Farres, J., Rechsteiner, M. P., Herold, S., Frey, A. D., and Kallio, P. T. (2005). Ligand binding properties of bacterial hemoglobins and flavohemoglobins. *Biochemistry* **44,** 4125–4134.

Fleming, A. I., Wittenberg, J. B., Wittenberg, B. A., Dudman, W. F., and Appleby, C. (1987). The purification, characterization and ligand binding kinetics of hemoglobins from root nodules of the non-legume Camarina glauca-Frankza symbiosis. *Biochim. Biophys. Acta* **911,** 209–220.

Foster, M. W., Pawloski, J. R., Singel, D. J., and Stamler, J. S. (2005). Role of circulating *S*-nitrosothiols in control of blood pressure. *Hypertension* **45,** 15–17.

Freitas, T. A., Hou, S., Dioum, E. M., Saito, J. A., Newhouse, J., Gonzalez, G., Gilles-Gonzalez, M. A., and Alam, M. (2004). Ancestral hemoglobins in Archaea. *Proc. Natl. Acad. Sci. USA* **101,** 6675–6680.

Frey, A. D., Farres, J., Bollinger, C. J., and Kallio, P. T. (2002). Bacterial hemoglobins and flavohemoglobins for alleviation of nitrosative stress in *Escherichia coli. Appl. Environ. Microbiol.* **68,** 4835–4840.

Frey, A. D., and Kallio, P. T. (2003). Bacterial hemoglobins and flavohemoglobins: Versatile proteins and their impact on microbiology and biotechnology. *FEMS Microbiol. Rev.* **27,** 525–545.

Frey, A. D., and Kallio, P. T. (2005). Nitric oxide detoxification: A new era for bacterial globins in biotechnology? *Trends Biotechnol.* **23,** 69–73.

Gardner, A. M., Martin, L. A., Gardner, P. R., Dou, Y., and Olson, J. S. (2000a). Steady-state and transient kinetics of *Escherichia coli* nitric-oxide dioxygenase (flavohemoglobin): The B10 tyrosine hydroxyl is essential for dioxygen binding and catalysis. *J. Biol. Chem.* **275,** 12581–12589.

Gardner, P. R., Gardner, A. M., Martin, L. A., and Salzman, A. L. (1998). Nitric oxide dioxygenase: An enzymic function for flavohemoglobin. *Proc. Natl. Acad. Sci. USA* **95,** 10378–10383.

Gardner, P. R., Gardner, A. M., Martin, L. A., Dou, Y., Li, T., Olson, J. S., Zhu, H., and Riggs, A. F. (2000b). Nitric oxide dioxygenase activity and function of flavohemoglobins: Sensitivity to nitric oxide and carbon monoxide inhibition. *J. Biol. Chem.* **275,** 31581–31587.

Gary-Bobo, C. M., and Solomon, A. K. (1968). Properties of hemoglobin solutions in red cells. *J. Gen. Physiol.* **52,** 825–853.

Giangiacomo, L., Mattu, M., Arcovito, A., Bellenchi, G., Bolognesi, M., Ascenzi, P., and Boffi, A. (2001). Monomer-dimer equilibrium and oxygen binding properties of ferrous *Vitreoscilla* hemoglobin. *Biochemistry* **40,** 9311–9316.

Gibson, Q. H., and Smith, M. H. (1965). Rates of reaction of *Ascaris* haemoglobins with ligands. *Proc. R. Soc. Lond. B. Biol. Sci.* **163,** 206–214.

Gibson, Q. H., Wittenberg, J. B., Wittenberg, B. A., Bogusz, D., and Appleby, C. A. (1989). The kinetics of ligand binding to plant hemoglobins: Structural implications. *J. Biol. Chem.* **264,** 100–107.

Goncalves, V. L., Nobre, L. S., Vicente, J. B., Teixeira, M., and Saraiva, L. M. (2006). Flavohemoglobin requires microaerophilic conditions for nitrosative protection of *Staphylococcus aureus. FEBS Lett.* **580,** 1817–1821.

Gow, A. J., Luchsinger, B. P., Pawloski, J. R., Singel, D. J., and Stamler, J. S. (1999). The oxyhemoglobin reaction of nitric oxide. *Proc. Natl. Acad. Sci. USA* **96,** 9027–9032.

Gow, A. J., and Stamler, J. S. (1998). Reactions between nitric oxide and haemoglobin under physiological conditions. *Nature* **391,** 169–173.

Hankeln, T., Ebner, B., Fuchs, C., Gerlach, F., Haberkamp, M., Laufs, T. L., Roesner, A., Schmidt, M., Weich, B., Wystub, S., Saaler-Reinhardt, S., Reuss, S., *et al.* (2005). Neuroglobin and cytoglobin in search of their role in the vertebrate globin family. *J. Inorg. Biochem.* **99,** 110–119.

Hargrove, M. S., Barry, J. K., Brucker, E. A., Berry, M. B., Phillips, G. N., Jr., Olson, J. S., Arredondo-Peter, R., Dean, J. M., Klucas, R. V., and Sarath, G. (1997).

Characterization of recombinant soybean leghemoglobin a and apolar distal histidine mutants. *J. Mol. Biol.* **266,** 1032–1042.

Harutyunyan, E. H., Safonova, T. N., Kuranova, I. P., Popov, A. N., Teplyakov, A. V., Obmolova, G. V., Valnshtein, B. K., Dodson, G. G., and Wilson, J. C. (1996). The binding of carbon monoxide and nitric oxide to leghaemoglobin in comparison with other haemoglobins. *J. Mol. Biol.* **264,** 152–161.

Hausladen, A., Gow, A. J., and Stamler, J. S. (1998). Nitrosative stress: Metabolic pathway involving the flavohemoglobin. *Proc. Natl. Acad. Sci. USA.* **95,** 14100–14105.

Hausladen, A., Gow, A., and Stamler, J. S. (2001). Flavohemoglobin denitrosylase catalyzes the reaction of a nitroxyl equivalent with molecular oxygen. *Proc. Natl. Acad. Sci. USA* **98,** 10108–10112.

Henry, Y., Lepoivre, M., Drapier, J. C., Ducrocq, C., Boucher, J. L., and Guissani, A. (1993). EPR characterization of molecular targets for NO in mammalian cells and organelles. *FASEB J.* **7,** 1124–1134.

Herold, S., and Puppo, A. (2005a). Kinetics and mechanistic studies of the reactions of metleghemoglobin, ferrylleghemoglobin, and nitrosylleghemoglobin with reactive nitrogen species. *J. Biol. Inorg. Chem.* **10,** 946–957.

Herold, S., and Puppo, A. (2005b). Oxyleghemoglobin scavenges nitrogen monoxide and peroxynitrite: A possible role in functioning nodules? *J. Biol. Inorg. Chem.* **10,** 935–945.

Herold, S., and Rock, G. (2005). Mechanistic studies of the oxygen-mediated oxidation of nitrosylhemoglobin. *Biochemistry* **44,** 6223–6231.

Herold, S., Exner, M., and Nauser, T. (2001). Kinetic and mechanistic studies of the NO*-mediated oxidation of oxymyoglobin and oxyhemoglobin. *Biochemistry* **40,** 3385–3395.

Herold, S., Fago, A., Weber, R. E., Dewilde, S., and Moens, L. (2004). Reactivity studies of the Fe(III) and Fe(II)NO forms of human neuroglobin reveal a potential role against oxidative stress. *J. Biol. Chem.* **279,** 22841–22847.

Hess, D. T., Matsumoto, A., Kim, S. O., Marshall, H. E., and Stamler, J. S. (2005). Protein *S*-nitrosylation: Purview and parameters. *Nat. Rev. Mol. Cell Biol.* **6,** 150–166.

Hunt, P. W., Klok, E. J., Trevaskis, B., Watts, R. A., Ellis, M. H., Peacock, W. J., and Dennis, E. S. (2002). Increased level of hemoglobin 1 enhances survival of hypoxic stress and promotes early growth in *Arabidopsis thaliana*. *Proc. Natl. Acad. Sci. USA* **99,** 17197–17202.

Igamberdiev, A., and Hill, R. (2004). Nitrate, NO and haemoglobin in plant adaptation to hypoxia: An alternative to classic fermentation pathways. *J. Exp. Bot.* **55,** 2473–2482.

Igamberdiev, A. U., Bykova, N. V., and Hill, R. D. (2006). Nitric oxide scavenging by barley hemoglobin is facilitated by a monodehydroascorbate reductase-mediated ascorbate reduction of methemoglobin. *Planta* **223,** 1033–1040.

Imai, K. (1982). "Allosteric Effects in Haemoglobin." Cambridge University Press, Cambridge.

Jakob, W., Webster, D. A., and Kroneck, P. M. H. (1992). NADH-dependent methemoglobin reductase from the obligate aerobe Vitreoscilla: Improved method of purification and reexamination of prosthetic groups. *Arch Biochem Biophys.* **292,** 29–33.

Ji, L., Becana, M., Sarath, G., and Klucas, R. V. (1994). Cloning and sequence analysis of a cDNA encoding ferric leghemoglobin reductase from soybean nodules. *Plant Physiol.* **104,** 453–459.

Jia, L., Bonaventura, C., Bonaventura, J., and Stamler, J. S. (1996). *S*-nitrosohaemoglobin: A dynamic activity of blood involved in vascular control. *Nature* **380,** 221–226.

Justino, M. C., Vicente, J. B., Teixeira, M., and Saraiva, L. M. (2005). New genes implicated in the protection of anaerobically grown *Escherichia coli* against nitric oxide. *J. Biol. Chem.* **280,** 2636–2643.

Kanayama, Y., and Yamamoto, Y. (1990a). Inhibition of nitrogen fixation in soybean plants supplied with nitrate II: Accumulation and properties of nitrosylleghemoglobin in nodules. *Plant Cell Physiol.* **31,** 207–214.

Kanayama, Y., and Yamamoto, Y. (1990b). Inhibition of nitrogen fixation in soybean plants supplied with nitrate III: Kinetics of the formation of nitrosylleghemoglobin and of the inhibition of formation of oxyleghemoglobin. *Plant Cell Physiol.* **31,** 603–608.

Kanayama, Y., Watanabe, I., and Yamamoto, Y. (1990). Inhibition of nitrogen fixation in soybean plants supplied with nitrate I: Nitrite accumulation and formation of nitrosylleghemoglobin in nodules. *Plant Cell Physiol.* **31,** 341–346.

Kaur, R., Pathania, R., Sharma, V., Mande, S. C., and Dikshit, K. L. (2002). Chimeric *Vitreoscilla* hemoglobin (VHb) carrying a flavoreductase domain relieves nitrosative stress in *Escherichia coli*: New insight into the functional role of VHb. *Appl. Environ. Microbiol.* **68,** 152–160.

Kim, S. O., Orii, Y., Lloyd, D., Hughes, M. N., and Poole, R. K. (1999). Anoxic function for the *Escherichia coli* flavohaemoglobin (Hmp): Reversible binding of nitric oxide and reduction to nitrous oxide. *FEBS Lett.* **445,** 389–394.

Kruszyna, R., Kruszyna, H., Smith, R. P., and Wilcox, D. E. (1988). Generation of valency hybrids and nitrosylated species of hemoglobin in mice by nitric oxide vasodilators. *Toxicol. Appl. Pharmacol.* **94,** 458–465.

Kundu, S., Trent, J. T., III, and Hargrove, M. S. (2003). Plants, humans and hemoglobins. *Trends Plant Sci.* **8,** 387–393.

Lau, P. W., and Asakura, T. (1979). Use of heme spin-labeling to probe heme environments of alpha and beta chains of hemoglobin. *J. Biol. Chem.* **254,** 2595–2599.

Leder, K., and Weller, P. (2006). *Ascariasis* UpToDate.

Liau, A., Karnik, R., Majumdar, A., and Cate, J. H. (2005). Mixing crowded biological solutions in milliseconds. *Anal. Chem.* **77,** 7618–7625.

Lim, M. D., Lorkovic, I. M., and Ford, P. C. (2005). The preparation of anaerobic nitric oxide solutions for the study of heme model systems in aqueous and nonaqueous media: Some consequences of NO_x impurities. *Methods Enzymol.* **396,** 3–17.

Liu, L., Zeng, M., and Stamler, J. S. (1999). Hemoglobin induction in mouse macrophages. *Proc. Natl. Acad. Sci. USA* **96,** 6643–6647.

Liu, L., Zeng, M., Hausladen, A., Heitman, J., and Stamler, J. S. (2000). Protection from nitrosative stress by yeast flavohemoglobin. *Proc. Natl. Acad. Sci. USA* **97,** 4672–4676.

Lu, C., Egawa, T., Wainwright, L., Poole, R., and Yeh, S. (2007). Structural and functional properties of a truncated hemoglobin from a food-borne pathogen *Campylobacter jejuni*. *J. Biol. Chem.* **282,** 13627–13636.

Luchsinger, B. P., Rich, E. N., Gow, A. J., Williams, E. M., Stamler, J. S., and Singel, D. J. (2003). Routes to *S*-nitroso-hemoglobin formation with heme redox and preferential reactivity in the beta subunits. *Proc. Natl. Acad. Sci. USA* **100,** 461–466.

Luchsinger, B. P., Rich, E. N., Yan, Y., Williams, E. M., Stamler, J. S., and Singel, D. J. (2005). Assessments of the chemistry and vasodilatory activity of nitrite with hemoglobin under physiologically relevant conditions. *J. Inorg. Biochem.* **99,** 912–921.

Manoharan, P. T., Wang, J. T., Alston, K., and Rifkind, J. M. (1990). Spin label probes of the environment of cysteine beta-93 in hemoglobin. *Hemoglobin* **14,** 41–67.

Maskall, C. S., Gibson, J. F., and Dart, P. J. (1977). Electron-paramagnetic-resonance studies of leghaemoglobins from soya-bean and cowpea root nodules: Identification of nitrosyl-leghaemoglobin in crude leghaemoglobin preparations. *Biochem. J.* **167,** 435–445.

Mathieu, C., Moreau, S., Frendo, P., Puppo, A., and Davies, M. J. (1998). Direct detection of radicals in intact soybean nodules: Presence of nitric oxide-leghemoglobin complexes. *Free. Radic. Biol. Med.* **24,** 1242–1249.

Mawjood, A. H., Miyazaki, G., Kaneko, R., Wada, Y., and Imai, K. (2000). Site-directed mutagenesis in hemoglobin: Test of functional homology of the F9 amino acid residues of hemoglobin alpha and beta chains. *Protein Eng.* **13,** 113–120.

McMahon, T. J., Moon, R. E., Luschinger, B. P., Carraway, M. S., Stone, A. E., Stolp, B. W., Gow, A. J., Pawloski, J. R., Watke, P., Singel, D. J., Piantadosi, C. A., and Stamler, J. S. (2002a). Nitric oxide in the human respiratory cycle. *Nat. Med.* **8,** 711–717.

McMahon, T. J., Moon, R. E., Luschinger, B. P., Carraway, M. S., Stone, A. E., Stolp, B. W., Gow, A. J., Pawloski, J. R., Watke, P., Singel, D. J., Piantadosi, C. A., and Stamler, J. S. (2002b). Nitric oxide in the human respiratory cycle. *Nat. Med.* **8,** 711–717.

McMahon, T. J., Ahearn, G. S., Moya, M. P., Gow, A. J., Huang, Y. C., Luchsinger, B. P., Nudelman, R., Yan, Y., Krichman, A. D., Bashore, T. M., Califf, R. M., Singel, D. J., *et al.* (2005). A nitric oxide processing defect of red blood cells created by hypoxia: Deficiency of *S*-nitrosohemoglobin in pulmonary hypertension. *Proc. Natl. Acad. Sci. USA* **102,** 14801–14806.

Meakin, G. E., Bueno, E., Jepson, B., Bedmar, E. J., Richardson, D. J., and Delgado, M. J. (2007). The contribution of bacteroidal nitrate and nitrite reduction to the formation of nitrosylleghaemoglobin complexes in soybean root nodules. *Microbiology* **153,** 411–419.

Membrillo-Hernandez, J., Coopamah, M. D., Anjum, M. F., Stevanin, T. M., Kelly, A., Hughes, M. N., and Poole, R. K. (1999). The flavohemoglobin of *Escherichia coli* confers resistance to a nitrosating agent, a "Nitric oxide Releaser," and paraquat and is essential for transcriptional responses to oxidative stress. *J. Biol. Chem.* **274,** 748–754.

Mills, C. E., Sedelnikova, S., Soballe, B., Hughes, M. N., and Poole, R. K. (2001). *Escherichia coli* flavohaemoglobin (Hmp) with equistoichiometric FAD and haem contents has a low affinity for dioxygen in the absence or presence of nitric oxide. *Biochem. J.* **353,** 207–213.

Minning, D. M., Gow, A. J., Bonaventura, J., Braun, R., Dewhirst, M., Goldberg, D. E., and Stamler, J. S. (1999). *Ascaris* haemoglobin is a nitric oxide-activated "deoxygenase." *Nature* **401,** 497–502.

Minton, A. P. (2001). The influence of macromolecular crowding and macromolecular confinement on biochemical reactions in physiological media. *J. Biol. Chem.* **276,** 10577–10580.

Moens, L., Vanfleteren, J., Van de Peer, Y., Peeters, K., Kapp, O., Czeluzniak, J., Goodman, M., Blaxter, M., and Vinogradov, S. (1996). Globins in nonvertebrate species: Dispersal by horizontal gene transfer and evolution of the structure-function relationships. *Mol. Biol. Evol.* **13,** 324–333.

Moore, C. M., Nakano, M. M., Wang, T., Ye, R. W., and Helmann, J. D. (2004). Response of *Bacillus subtilis* to nitric oxide and the nitrosating agent sodium nitroprusside. *J. Bacteriol.* **186,** 4655–4664.

Moore, E. G., and Gibson, Q. H. (1976). Cooperativity in the dissociation of nitric oxide from hemoglobin. *J. Biol. Chem.* **251,** 2788–2794.

Mukai, M., Mills, C. E., Poole, R. K., and Yeh, S. R. (2001). Flavohemoglobin, a globin with a peroxidase-like catalytic site. *J. Biol. Chem.* **276,** 7272–7277.

Mukhopadhyay, P., Zheng, M., Bedzyk, L. A., LaRossa, R. A., and Storz, G. (2004). Prominent roles of the NorR and Fur regulators in the *Escherichia coli* transcriptional response to reactive nitrogen species. *Proc. Natl. Acad. Sci. USA* **101,** 745–750.

Nakano, M. M. (2002). Induction of ResDE-dependent gene expression in *Bacillus subtilis* in response to nitric oxide and nitrosative stress. *J. Bacteriol.* **184,** 1783–1787.

Newton, D. A., Rao, K. M., Dluhy, R. A., and Baatz, J. E. (2006). Hemoglobin is expressed by alveolar epithelial cells. *J. Biol. Chem.* **281,** 5668–5676.

Nothig-Laslo, V. (1977). Collective behaviour of hemoglobin in dense solutions. *Biophys. Chem.* **7,** 71–75.

Nothig-Laslo, V., and Maricic, S. (1982). Temperature-dependent conformation change in spin-labeled hemoglobin. *Biophys. Chem.* **15,** 217–221.

Ouellet, H., Ouellet, Y., Richard, C., Labarre, M., Wittenberg, B., Wittenberg, J., and Guertin, M. (2002). Truncated hemoglobin HbN protects *Mycobacterium bovis* from nitric oxide. *Proc. Natl. Acad. Sci. USA* **99,** 5902–5907.

Pathania, R., Navani, N. K., Gardner, A. M., Gardner, P. R., and Dikshit, K. L. (2002). Nitric oxide scavenging and detoxification by the *Mycobacterium tuberculosis* haemoglobin, HbN in *Escherichia coli. Mol. Microbiol.* **45,** 1303–1314.

Pawaria, S., Rajamohan, G., Gambhir, V., Lama, A., Varshney, G. C., and Dikshit, K. L. (2007). Intracellular growth and survival of *Salmonella enterica serovar Typhimurium* carrying truncated hemoglobins of *Mycobacterium tuberculosis. Microb. Pathog.* **42,** 119–128.

Pawloski, J. R., and Stamler, J. S. (2002). Nitric oxide in RBCs. *Transfusion* **42,** 1603–1609.

Pawloski, J. R., Hess, D. T., and Stamler, J. S. (2001). Export by red blood cells of nitric oxide bioactivity. *Nature* **409,** 622–626.

Pawloski, J. R., Hess, D. T., and Stamler, J. S. (2005). Impaired vasodilation by red blood cells in sickle cell disease. *Proc. Natl. Acad. Sci. USA* **102,** 2531–2536.

Perazzolli, M., Dominici, P., Romero-Puertas, M. C., Zago, E., Zeier, J., Sonoda, M., Lamb, C., and Delledonne, M. (2004). *Arabidopsis* nonsymbiotic hemoglobin AHb1 modulates nitric oxide bioactivity. *Plant Cell* **16,** 2785–2794.

Perazzolli, M., Romero-Puertas, M. C., and Delledonne, M. (2006). Modulation of nitric oxide bioactivity by plant haemoglobins. *J. Exp. Bot.* **57,** 479–488.

Perrella, M., and Cera, E. D. (1999). CO ligation intermediates and the mechanism of hemoglobin cooperativity. *J. Biol. Chem.* **274,** 2605–2608.

Perutz, M. (1987). Molecular anatomy, physiology, and pathology of hemoglobin. *In* "Molecular Basis of Blood Diseases" (G. Stammatayanopoulos, ed.), pp. 127–177. Saunders.

Pesce, A., Couture, M., Dewilde, S., Guertin, M., Yamauchi, K., Ascenzi, P., Moens, L., and Bolognesi, M. (2000). A novel two-over-two α-helical sandwich fold is characteristic of the truncated hemoglobin family. *EMBO J.* **19,** 2424–2434.

Pittman, M. S., Elvers, K. T., Lee, L., Jones, M. A., Poole, R. K., Park, S. F., and Kelly, D. J. (2007). Growth of *Campylobacter jejuni* on nitrate and nitrite: Electron transport to NapA and NrfA via NrfH and distinct roles for NrfA and the globin Cgb in protection against nitrosative stress. *Mol. Microbiol.* **63,** 575–590.

Poole, R. K. (1994). Oxygen reactions with bacterial oxidases and globins: Binding, reduction and regulation. *Antonie Van Leeuwenhoek* **65,** 289–310.

Poole, R. K. (2005). Nitric oxide and nitrosative stress tolerance in bacteria. *Biochem. Soc. Trans.* **33,** 176–180.

Poole, R. K., Anjum, M. F., Membrillo-Hernandez, J., Kim, S. O., Hughes, M. N., and Stewart, V. (1996). Nitric oxide, nitrite, and Fnr regulation of hmp (flavohemoglobin) gene expression in *Escherichia coli* K-12. *J. Bacteriol.* **178,** 5487–5492.

Pullan, S. T., Gidley, M. D., Jones, R. A., Barrett, J., Stevanin, T. M., Read, R. C., Green, J., and Poole, R. K. (2007). Nitric oxide in chemostat-cultured *Escherichia coli* is sensed by Fnr and other global regulators: Unaltered methionine biosynthesis indicates lack of *S*-nitrosation. *J. Bacteriol.* **189,** 1845–1855.

Reischl, E., Dafre, A. L., Franco, J. L., and Wilhelm Filho, D. (2006). Distribution, adaptation and physiological meaning of thiols from vertebrate hemoglobins. *Comp. Biochem. Physiol. C. Toxicol. Pharmacol.*

Richardson, A. R., Dunman, P. M., and Fang, F. C. (2006). The nitrosative stress response of *Staphylococcus aureus* is required for resistance to innate immunity. *Mol. Microbiol.* **61,** 927–939.

Rogstam, A., Larsson, J. T., Kjelgaard, P., and von Wachenfeldt, C. (2007). Mechanisms of adaptation to nitrosative stress in *Bacillus subtilis. J. Bacteriol.* **189,** 3063–3071.

Sebbane, F., Lemaitre, N., Sturdevant, D. E., Rebeil, R., Virtaneva, K., Porcella, S. F., and Hinnebusch, B. J. (2006). Adaptive response of *Yersinia pestis* to extracellular effectors of innate immunity during bubonic plague. *Proc. Natl. Acad. Sci. USA* **103,** 11766–11771.

Seregelyes, C., Igamberdiev, A. U., Maassen, A., Hennig, J., Dudits, D., and Hill, R. D. (2004). NO degradation by alfalfa class 1 hemoglobin (Mhb1): A possible link to PR-1a gene expression in Mhb1-overproducing tobacco plants. *FEBS Lett.* **571,** 61–66.

Sharma, V. S., and Ranney, H. M. (1978). The dissociation of NO from nitrosylhemoglobin. *J. Biol. Chem.* **253,** 6467–6472.

Sharma, V. S., Traylor, T. G., Gardiner, R., and Mizukami, H. (1987). Reaction of nitric oxide with heme proteins and model compounds of hemoglobin. *Biochemistry* **26,** 3837–3843.

Sharma, V. S., Isaacson, R. A., John, M. E., Waterman, M. R., and Chevion, M. (1983). Reaction of nitric oxide with heme proteins: Studies on metmyoglobin, opossum methemoglobin, and microperoxidase. *Biochemistry* **22,** 3897–3902.

Singel, D. J., and Stamler, J. S. (2005). Chemical physiology of blood flow regulation by red blood cells: The role of nitric oxide and *S*-nitrosohemoglobin. *Annu. Rev. Physiol.* **67,** 99–145.

Sonveaux, P., Lobysheva, II, Feron, O., and McMahon, T. J. (2007). Transport and peripheral bioactivities of nitrogen oxides carried by red blood cell hemoglobin: Role in oxygen delivery. *Physiology (Bethesda)* **22,** 97–112.

Sowa, A. W., Duff, S. M., Guy, P. A., and Hill, R. D. (1998). Altering hemoglobin levels changes energy status in maize cells under hypoxia. *Proc. Natl. Acad. Sci. USA* **95,** 10317–10321.

Springer, B. A., Egeberg, K. D., Sligar, S. G., Rohlfs, R. J., Mathews, A. J., and Olson, J. S. (1989). Discrimination between oxygen and carbon monoxide and inhibition of autooxidation by myoglobin: Site-directed mutagenesis of the distal histidine. *J. Biol. Chem.* **264,** 3057–3060.

Stamler, J. S., Jia, L., Eu, J. P., McMahon, T. J., Demchenko, I. T., Bonaventura, J., Gernert, K., and Piantadosi, C. A. (1997). Blood flow regulation by *S*-nitrosohemoglobin in the physiological oxygen gradient. *Science* **276,** 2034–2037.

Steinhoff, H. J. (1990). Residual motion of hemoglobin-bound spin labels and protein dynamics: Viscosity dependence of the rotational correlation times. *Eur. Biophys. J.* **18,** 57–62.

Stevanin, T. M., Ioannidis, N., Mills, C. E., Kim, S. O., Hughes, M. N., and Poole, R. K. (2000). Flavohemoglobin Hmp affords inducible protection for *Escherichia coli* respiration, catalyzed by cytochromes bo′ or bd, from nitric oxide. *J. Biol. Chem.* **275,** 35868–35875.

Streeter, J. (1988). Inhibition of legume nodule formation and N_2 fixation by nitrate. *Crit. Rev. Plant Sci.* **7,** 1–23.

Taketa, F., Antholine, W. E., and Chen, J. Y. (1978). Chain nonequivalence in binding of nitric oxide to hemoglobin. *J. Biol. Chem.* **253,** 5448–5451.

Thorsteinsson, M. V., Bevan, D. R., Potts, M., Dou, Y., Eich, R. F., Hargrove, M. S., Gibson, Q. H., and Olson, J. S. (1999). A cyanobacterial hemoglobin with unusual ligand binding kinetics and stability properties. *Biochemistry* **38,** 2117–2126.

Trent, J. T., III, and Hargrove, M. S. (2002). A ubiquitously expressed human hexacoordinate hemoglobin. *J. Biol. Chem.* **277,** 19538–19545.

Trevaskis, B., Watts, R. A., Andersson, C. R., Llewellyn, D. J., Hargrove, M. S., Olson, J. S., Dennis, E. S., and Peacock, W. J. (1997). Two hemoglobin genes in *Arabidopsis thaliana*: The evolutionary origins of leghemoglobins. *Proc. Natl. Acad. Sci. USA* **94,** 12230–12234.

Tsuruga, M., Matsuoka, A., Hachimori, A., Sugawara, Y., and Shikama, K. (1998). The molecular mechanism of autoxidation for human oxyhemoglobin: Tilting of the distal histidine causes nonequivalent oxidation in the beta chain. *J. Biol. Chem.* **273,** 8607–8615.

Ullmann, B. D., Myers, H., Chiranand, W., Lazzell, A. L., Zhao, Q., Vega, L. A., Lopez-Ribot, J. L., Gardner, P. R., and Gustin, M. C. (2004). Inducible defense mechanism against nitric oxide in Candida albicans. *Eucaryotic Cell.* **3,** 715–723.

Van Doorslaer, S., Dewilde, S., Kiger, L., Nistor, S. V., Goovaerts, E., Marden, M. C., and Moens, L. (2003). Nitric oxide binding properties of neuroglobin: A characterization by EPR and flash photolysis. *J. Biol. Chem.* **278,** 4919–4925.

Vasudevan, S. G., Armarego, W. L., Shaw, D. C., Lilley, P. E., Dixon, N. E., and Poole, R. K. (1991). Isolation and nucleotide sequence of the *hmp* gene that encodes a haemoglobin-like protein in *Escherichia coli* K-12. *Mol. Gen. Genet.* **226,** 49–58.

Vieweg, M. F., Hohnjec, N., and Küster, H. (2004). Two genes encoding different truncated hemoglobins are regulated during root nodule and arbuscular mycorrhiza symbioses of *Medicago truncatula. Planta* **220,** 757–766.

Vinogradov, S. N., Hoogewijs, D., Bailly, X., Arredondo-Peter, R., Gough, J., Dewilde, S., Moens, L., and Vanfleteren, J. R. (2006). A phylogenomic profile of globins. *BMC. Evol. Biol.* **6,** 31.

Vinogradov, S. N., Hoogewijs, D., Bailly, X., Arredondo-Peter, R., Guertin, M., Gough, J., Dewilde, S., Moens, L., and Vanfleteren, J. R. (2005). Three globin lineages belonging to two structural classes in genomes from the three kingdoms of life. *Proc. Natl. Acad. Sci. USA* **102,** 11385–11389.

Watts, R. A., Hunt, P. W., Hvitved, A. N., Hargrove, M. S., Peacock, W. J., and Dennis, E. S. (2001). A hemoglobin from plants homologous to truncated hemoglobins of microorganisms. *Proc. Natl. Acad. Sci. USA* **98,** 10119–10124.

Weber, R. E., and Vinogradov, S. N. (2001). Nonvertebrate hemoglobins: Functions and molecular adaptations. *Physiol. Rev.* **81,** 569–628.

Weiland, T. R., Kundu, S., Trent, J. T., III, Hoy, J. A., and Hargrove, M. S. (2004). Bis-histidyl hexacoordination in hemoglobins facilitates heme reduction kinetics. *J. Am. Chem. Soc.* **126,** 11930–11935.

Wittenberg, J. B., Wittenberg, B. A., Gibson, Q. H., Trinick, M. J., and Appleby, C. A. (1986). The kinetics of the reactions of *Parasponia andersonii* hemoglobin with oxygen, carbon monoxide, and nitric oxide. *J. Biol. Chem.* **261,** 13624–13631.

Zhang, L., Levy, A., and Rifkind, J. M. (1991). Autoxidation of hemoglobin enhanced by dissociation into dimers. *J. Biol. Chem.* **266,** 24698–24701.

Zhu, H., and Riggs, A. F. (1992). Yeast flavohemoglobin is an ancient protein related to globins and a reductase family. *Proc. Natl. Acad. Sci. USA* **89,** 5015–5019.

Zimmerman, S. B., and Minton, A. P. (1993). Macromolecular crowding: Biochemical, biophysical, and physiological consequences. *Annu. Rev. Biophys. Biomol. Struct.* **22,** 27–65.

CHAPTER NINE

A Survey of Methods for the Purification of Microbial Flavohemoglobins

Megan E. S. Lewis,* Hazel A. Corker,† Bridget Gollan,‡ *and* Robert K Poole*

Contents

Abstract

Over the past decade, the flavohemoglobin Hmp has emerged as the most significant nitric oxide (NO)–detoxifying protein in many diverse organisms, including yeasts and fungi but particularly pathogenic bacteria. Flavohemoglobins—the best-characterized class of microbial globin—comprise two domains: a globin domain with a noncovalently bound heme B and a flavin domain with recognizable binding sites for FAD and NAD(P)H. Hmp was first identified in *Escherichia coli* and now has a clearly defined role in NO biology in that organism: its synthesis is markedly up-regulated by NO, and *hmp* knockout mutants of *E. coli* and *Salmonella typhimurium* are severely compromised for

* Department of Molecular Biology and Biotechnology, University of Sheffield, Sheffield, United Kingdom
† Syntopix Group Plc, Institute of Pharmaceutical Innovation, University of Bradford, Bradford, United Kingdom
‡ Centre for Molecular Microbiology and Infection, Department of Infectious Diseases, Imperial College of Science, London, United Kingdom

Methods in Enzymology, Volume 436
ISSN 0076-6879, DOI: 10.1016/S0076-6879(08)36009-1

survival in the presence of NO *in vitro* and in pathogenic lifestyles. In the presence of molecular O_2, Hmp catalyzes an oxygenase or denitrosylase reaction in which NO is stoichiometrically converted to nitrate ion, which is relatively innocuous. In this chapter, we present a survey of the methods used to express and purify the flavohemoglobins from diverse microorganisms and describe in more detail three methods developed and used in this laboratory for the *E. coli* protein. Particular problems are highlighted, particularly (a) the toxic consequences of Hmp overexpression that result from its ability to catalyze partial oxygen reduction and (b) the expression of protein with substoichiometric content of redox-active flavin and heme centers.

1. Introduction

The protein considered most important in the detoxification of NO by pathogenic bacteria, including *S. typhimurium*, *E. coli* and others, is the flavohemoglobin Hmp (Poole, 2005). Flavohemoglobins are the most well-characterized class of microbial globin. They comprise two domains, a globin domain with a noncovalently bound heme B and a flavin domain with recognizable binding sites for FAD and NAD(P)H (Wu *et al.*, 2003). Hmp was first identified in *E. coli* (Vasudevan *et al.*, 1991) and now has a clearly defined role in NO biology in that organism. Its synthesis is markedly up-regulated by NO (Poole *et al.*, 1996), and *hmp* knockout mutants of *E. coli* and *S. typhimurium* are severely compromised for survival in the presence of NO *in vitro* (Crawford and Goldberg, 1998b; Membrillo-Hernández *et al.*, 1999). *Salmonella* Hmp has also been implicated in response to NO in human (Stevanin *et al.*, 2002) and murine (Gilberthorpe *et al.*, 2007) macrophages. In the presence of molecular O_2, Hmp catalyzes an oxygenase (Gardner, 2005; Gardner *et al.*, 1998) or denitrosylase (Hausladen *et al.*, 1998) reaction in which NO is stoichiometrically converted to nitrate ion (Gardner, 2005; Gardner *et al.*, 1998), which is relatively innocuous. Extensive studies of the purified protein (briefly surveyed in Table 9.1) have revealed many details of the structure, function, and reaction mechanism of flavohemoglobin. A major incentive in studying bacterial flavohemoglobins is the rapidly accumulating evidence that these proteins are important in pathogenesis; in particular, flavohemoglobin synthesis is markedly increased in bacteria exposed to nitrosative stress in disease as in the case of *Yersinia pestis* recovered from rat buboes (Sebbane *et al.*, 2006) and *E. coli* recovered from urinary tract infections in humans (Roos and Klemm, 2006).

Regulation of Hmp levels in response to NO and related species in *E. coli* is complex. Control occurs predominantly at the transcriptional level and involves Fnr (Cruz-Ramos *et al.*, 2002; Poole *et al.*, 1996) and MetR (Membrillo-Hernández *et al.*, 1998). Computational and experimental

Table 9.1 Methods for the purification of flavohemoglobins from bacteria and eukaryotic microorganisms

Microorganism	Expression system	Purification protocol	Examples of properties/ exploitation of flavohemoglobin	Primary reference(s) to purification protocol
Bacteria				
E. coli	pPL304 in strain RSC419	DEAE-Sepharose CL-6B followed by gel filtration on Sephacryl S-200	Antibody generation (Vasudevan *et al.*, 1995)	(Vasudevan *et al.*, 1991)
	pBR322 (multicopy plasmid)	DEAE-Sepharose CL-6B followed by gel filtration on Sephacryl S-200	Determination of redox centers (Ioannidis *et al.*, 1992); anoxic NO binding and reduction (Kim *et al.*, 1999); Fe(III) and cytochrome *c* reduction (Membrillo-Hernández *et al.*, 1996; Poole *et al.*, 1997); cyanide titration and antibody preparation (Stevanin *et al.*, 2000)	(Ioannidis *et al.*, 1992)
	pPL757 in strain AN1459, a stable, high-copy-number λ promoter vector, temperature-inducible	MonoQ HR 10/10 then Q Sepharose FF and/or gel filtration on Superdex 200	Dioxygenase/denitrosylase assays	(Hausladen *et al.*, 1998, 2001)

(continued)

Table 9.1 (*continued*)

Microorganism	Expression system	Purification protocol	Examples of properties/ exploitation of flavohemoglobin	Primary reference(s) to purification protocol
	hmp gene cloned in pAlter (Promega)	Ammonium sulfate fractionation, DEAE Sepharose, Superdex 75, or Hi-Trap Mono Q	Steady-state and transient kinetics (Gardner *et al.*, 2000a); spectroscopic studies, kinetics studies of O_2, CO, NO binding, NO dioxygenase activity (Gardner *et al.*, 2000b)	(Gardner *et al.*, 1998, 2000a)
	pCL775, i.e., the *hmp* gene cloned in pPL452 (a stable, high-copy-number λ promoter vector, temperature-inducible)	Anion-exchange chromatography on DEAE-Sepharose Fast Flow at pH 8.0, followed by chromatography on DEAE-Toyopearl 650S at pH 6.5 and gel filtration on Superdex-200	Oxygen affinity, peroxide formation (Mills *et al.*, 2001); resonance Raman spectroscopy (Mukai *et al.*, 2001), cyanide sensitivity of dioxygenase activity (Stevanin *et al.*, 2000)	(Mills *et al.*, 2001)
	pEcHMP (subclone of pAlter in pUC19)	Ammonium sulfate fractionation, DEAE Sepharose, Superdex 75, or Hi-Trap Mono Q; followed by heme and flavin reconstitution	Assay of NOD activity and inhibition by imidazole antibiotics	(Helmick *et al.*, 2005)

	hmp gene cloned in pET11	DEAE 52, followed by microceramic hydroxyapatite to remove bound lipids, then DEAE-Toyopearl; additional step on lipid-avid hydroxyalkoyl-propyl dextran resin to remove physiologically bound lipids	FTIR spectroscopy (Bonamore *et al.*, 2001); crystallization (Ilari *et al.*, 2002); vibrational and X-ray spectroscopy (D'Angelo *et al.*, 2004); alkylhydroperoxide reductase assays (Bonamore *et al.*, 2003b); lipid interactions (Bonamore *et al.*, 2003a)	(Bonamore *et al.*, 2001; Bonamore *et al.*, 2003a)
	pBAD (arabinose-inducible vector)	Sepharose CL-6B, then Superdex-75	Site-directed mutagenesis, construction of proteins having only heme or flavin domains	This chapter
R. eutropha (formerly *A. eutrophus*)	No measures to induce overexpression	DEAE-Sepharose Fast-Flow (Pharmacia), then hydroxylapatite and Sephacryl S-11 (Pharmacia)	Crystallization and X-ray crystallography (Ermler *et al.*, 1995b) and structure determination (Ermler *et al.*, 1995a); electrochemical study of oxygen reduction (de Oliveira *et al.*, 2007)	(Ermler *et al.*, 1995b)
Staphylococcus aureus	*hmp* gene cloned into pET28a, induced with IPTG	Q-Sepharose High Performance column, then	Spectroscopic studies, redox titration, NADH oxidase	(Goncalves *et al.*, 2006; this volume)

(*continued*)

Table 9.1 (*continued*)

Microorganism	Expression system	Purification protocol	Examples of properties/ exploitation of flavohemoglobin	Primary reference(s) to purification protocol
		Superdex S-75, and reloaded onto Q-Sepharose High Performance column	and NO denitrosylase/ reductase assays	
Pseudomonas aeruginosa, *Deinococcus radiodurans*, *S. typhi*, and *Klebsiella pneumonia*	*hmp* genes cloned into pHIS (derivative of pKQV4, with His tag and P_{tac} promoter)	Sephadex HisTrap chelating column, then Sephadex G25	Spectroscopic studies, ligand association, and dissociation kinetics	(Bollinger *et al.*, 2001; Farres *et al.*, 2005)
Streptomyces antibioticus	No measures to induce overexpression	DEAE-cellulose (DE52), then DEAE Toyopearl 650M, then Phenyl Sepharose CL-4B, then DEAE Toyopearl 650M	Spectroscopic studies, analysis of prosthetic groups, and NO consumption	(Sasaki *et al.*, 2004)
Yeasts and Fungi				
C. norvegensis	No measures to induce overexpression	Ammonium sulfate fractionation, then Butyl-Toyopearl, Sephadex G-75 and DEAE cellulose	Sequence, spectroscopic, and autoxidation studies	(Kobayashi *et al.*, 2002)

	Recombinant heme domain: cloning into pMAL-c2 and expression in *E. coli*; induction with IPTG	Ammonium sulfate fractionation, then Sephadex G-75 and DEAE cellulose; cleavage from MBP with Xa proteinase and CM cellulose chromatography under CO	Autoxidation studies	(Kobayashi *et al.*, 2002)
Candida albicans	pCaYHB1 (*yhb* gene cloned in pUC19 derivative)	Ammonium sulfate fractionation, DEAE Sepharose, Superdex 75, or Hi-Trap Mono Q; followed by heme and flavin reconstitution	Assay of NOD activity and inhibition by imidazole antibiotics	(Helmick *et al.*, 2005)
Saccharomyces cerevisiae	Antimycin treatment of "respiration-proficient" strain induces Yhb expression	Ammonium sulfate fractionation, then Sephadex G-100 and Synchropak RP300 HPLC	Protein sequence determination	(Zhu and Riggs, 1992)
	Antimycin treatment	DEAE-Sepharose CL-6B, concentration on YM10 membrane, then gel	Spectroscopic studies, kinetic studies of O_2, CO, NO binding, and NO dioxygenase activity	(Gardner *et al.*, 2000b)

(*continued*)

Table 9.1 (*continued*)

Microorganism	Expression system	Purification protocol	Examples of properties/ exploitation of flavohemoglobin	Primary reference(s) to purification protocol
		electrophoresis, followed by electroelution		
Fusarium oxysporum	No measures to induce overexpression	DEAE-cellulose (DE52), then Hi-Trap SP (Pharmacia) then FPLC on MonoQ HR5/5, then Superdex 200HR (Pharmacia)	Spectroscopic studies, analysis of prosthetic groups	(Takaya *et al.*, 1997)

studies have also implicated NsrR (product of the *yjeB* [*nsrR*] gene) in *hmp* regulation (Bodenmiller and Spiro, 2006; Rodionov *et al.*, 2005). NsrR is an NO-sensitive transcriptional regulator of *hmp* and other genes known to be involved in nitrosative stress tolerance. It is a member of the Rrf2 family of transcriptional regulators, which also includes the IscR regulator involved in regulation of genes involved in [Fe-S] cluster biogenesis (Schwartz *et al.*, 2001). On the basis of the similarity of NsrR to IscR, which contains an [Fe-S] cluster (Schwartz *et al.*, 2001), and other members of the Rrf2 family (Rodionov *et al.*, 2005), it has been suggested that NsrR contains an NO-sensitive [Fe-S] cluster. A similar conclusion has been reached for NsrR in *Bacillus subtilis* (Nakano *et al.*, 2006), and a role for NsrR in *S. typhimurium hmp* regulation has been proposed (Bang *et al.*, 2006). In *Salmonella*, the response of *hmp* transcription to O_2^- (generated by addition to cells of paraquat) is mediated by a further regulator RamA (Hernandez-Urzua *et al.*, 2007). There has been some confusion over the possible role of the ferric uptake regulation (Fur) protein in *hmp* regulation. It was originally proposed (Crawford and Goldberg, 1998a) that the iron-responsive regulator Fur represses *hmp* transcription and that this is lifted by NO on inactivation of Fur. Although these results have been retracted (Crawford and Goldberg, 2006), and others have suggested that Fur is not involved in *Salmonella hmp* regulation (Bang *et al.*, 2006), it is clear that certain promoters, including *hmp*, are indeed controlled by nitrosylation of the Fur iron (D'Autreaux *et al.*, 2002). Furthermore, we have recently published evidence, based on newly constructed *hmp-lacZ* fusions and immunoblotting, that Fur is a repressor of *hmp* transcription in both *E. coli* and *Salmonella*, albeit a weak one (Hernandez-Urzua *et al.*, 2007).

Thus, flavohemoglobin expression in *E. coli* and *Salmonella* is complex involving Fnr, Fur, MetR, RamA, and NsrR (Spiro, 2007); repression of transcription particularly by NsrR and (anoxically) Fnr results in Hmp levels being extremely low in normally growing cells. Indeed, the presence of this hemoglobin in *E. coli* was unsuspected throughout the histories of biochemistry and microbiology until 1991 (Vasudevan *et al.*, 1991). Biochemical studies of Hmp may now provide a rational explanation of the suppression of Hmp synthesis: Hmp is a potent generator of the products of partial oxygen reduction (i.e., superoxide and peroxide). Such accumulation of oxygen radicals is evident in kinetic studies of purified Hmp (Orii *et al.*, 1992; Poole *et al.*, 1997; Wu *et al.*, 2004); in the ability of Hmp to reduce cytochrome *c*, Fe(III), these reductive activities are in part mediated via superoxide. Accumulation of radicals by Hmp can lead to self-destruction of the protein (Wu *et al.*, 2004). The superoxide-generating activity of Hmp is also detectable *in vivo* by the up-regulation of a *sodA-lacZ* fusion in an *E. coli* strain carrying the multicopy plasmid pPL341 (Membrillo-Hernández *et al.*, 1996) and is thought to explain the sensitivity of an *nsrR* mutant of *Salmonella* (i.e., which over-expresses Hmp) to oxidative stress *in vitro* and in murine macrophages (Gilberthorpe *et al.*, 2007). As described in the subsequent protocols, uncontrolled *hmp* expression (e.g., from a multicopy plasmid) may lead to cell damage and even promote spontaneous mutations in the *hmp* gene.

2. The "BC" Era of Flavohemoglobin Purification and Characterization

Before cloning ("BC"), attempts to purify flavohemoglobins were hampered by the expression levels attainable. Nevertheless, two landmark studies described the properties of flavohemoglobin from the bacterium *Alcaligenes eutrophus* (now *Ralstonia eutropha*) and the budding yeast *Candida mycoderma*. In the early 1970s, Oshino *et al.* (1972) described the purification of a "yeast hemoglobin-reductase complex" from *C. mycoderma* in which heme and flavin were associated with a single polypeptide reducible by NAD(P)H. Subsequent characterization revealed its oxygen affinity (Oshino *et al.*, 1971, 1973a,b), but no clear physiological function emerged. A similar protein was purified from *R. eutropha* (Probst and Schlegel, 1976; Probst *et al.*, 1979), and this protein was later to become the first flavohemoglobin for which a crystal structure would be solved (Ermler *et al.*, 1995a).

2.1. An overview of recent methods for flavohemoglobin expression and purification

Table 9.1 compiles a large number of methods used for expression and purification of flavohemoglobins from bacteria and eukaryotic microorganisms. Such proteins are not known to exist in higher animals and plants. With few exceptions, mostly from very early work (see previous section), investigators have used standard molecular cloning tools to express the flavohemoglobin in *E. coli* under the control of a variety of inducible promoters. The flavohemoglobins are soluble and cytoplasmic and are readily extracted by common cell-disruption techniques (e.g., ultrasonication, French pressure cell) followed by centrifugal clarification. Purification protocols generally use ion-exchange chromatography followed by gel-filtration chromatography. The color of the protein considerably aids monitoring of the purification process. Some protocols employ additional steps to improve purity or to remove bound lipids. Early reports found that ammonium sulfate fractionation should be avoided, but in other studies this step is routinely included. The references in Table 9.1 should be consulted for further details. Note that spectroscopic characterization of purified flavohemoglobins has often revealed a substoichiometric content of heme b and FAD (i.e., less than the 1 mole of each per mole of flavohemoglobin); in some studies, reconstitution of the purified protein has been performed to reinstate one or both cofactors.

3. Methods for Expression and Purification of *E. coli* Hmp

As Table 9.1 illustrates, many groups have devised methods for flavohemoglobin expression and purification over the past 10 years. Here, we provide further information on two published methods and describe an alternative that has proved useful more recently as the basis for separate expression of the flavin and heme domains and for site-directed mutagenesis.

3.1. Expression and purification of *E. coli* Hmp in pBR322, a multicopy vector

Our earliest preparations used a simple method: pPL341 was constructed in the laboratory of Dr. Nick Dixon and contained the entire *hmp* gene under the control of its own promoter cloned in pBR322 (Ioannidis *et al.*, 1992; Vasudevan *et al.*, 1991). Cells transformed with pPL341 (strain RSC521, which is RSC49 [*recA srlA*::Tn*10*]/pPL341) are deep brown due to the high expression levels attained during growth in Luria-Bertani broth supplemented with ampicillin (35 μg ml^{-1}). Cells are harvested from 10 l culture by continuous-flow centrifugation (yield approximately 5 *g*. (wet weight) per l), washed and disrupted by two passages through a French pressure cell at 138 MPa. The orange-red cytoplasmic fraction containing Hmp was obtained by centrifugation of the lysed cells at 132,000 *g* for 1.5 h. Hmp is readily purified in two chromatographic steps, namely anion exchange on DEAE-Sepharose CL-6B followed by gel filtration on Sephacryl S-200.

Prosthetic groups are determined as follows. Flavin is extracted by heating the protein at 97° for 3 min followed by centrifugation and flavin analysis on silica-gel thin-layer chromatography plates. The heme is assayed using the classical pyridine hemochrome method. Such analyses, together with electronic absorbance and EPR spectroscopy, reveal the protein to be purified in the ferric state but readily reducible *in vitro* by NADH or NADPH to give the oxyferrous form. A slow decay to the ferric form is the result of robust diaphorase activity and substrate depletion (Ioannidis *et al.*, 1992; Poole *et al.*, 1994). A nitrosyl species is readily formed on adding nitrite and dithionite (Ioannidis *et al.*, 1992).

Potential problems with this method are (1) the substoichiometry of redox centers, although many workers report such incidents with numerous and diverse preparation protocols, and (2) a propensity for irregular and unexplained poor expression. We have obtained some evidence (Membrillo-Hernández and Poole, unpublished data) for the accumulation of spontaneous mutations in the *hmp* gene cloned in pBR322, which might, in retrospect, be explained by the ability of Hmp to generate superoxide and peroxide in the presence of reductant and NAD(P)H. The accumulation of superoxide and peroxide may also explain the growth advantage recently reported in cocultivation experiments where an *hmp* mutant outcompetes a wild-type strain (Stevanin *et al.*, 2007). Thus, in a strain that constitutively synthesizes Hmp, a strain acquiring mutations in *hmp* would be at a selective advantage. In view of these difficulties, there are some advantages in placing *hmp* expression under the control of an inducible promoter, and this has been exploited in our laboratory in two subsequent protocols.

3.2. Expression and purification of *E. coli* Hmp from the vector pPL452 and thermal induction

A purification procedure for flavohemoglobin Hmp that gives high yields of protein with equistoichiometric heme and FAD contents was described by Mills *et al.* (2001). *E. coli* strain RSC2057 (AN1459 harboring the plasmid pCL775; Love *et al.*, 1996) was used for purification of Hmp and for assay of peroxide production in Hmp-rich cell extracts. Plasmid pCL775 contains the *hmp* gene cloned in pPL452, a stable, high-copy-number λ promoter vector conferring ampicillin resistance (Love *et al.*, 1996). All strains were grown in LB medium, supplemented where appropriate with kanamycin or ampicillin (each at 50 μg ml^{-1}).

E. coli strain RSC2057 is grown aerobically in LB supplemented with $FeCl_3$, δ-aminolevulinic acid and riboflavin at final concentrations, respectively, of 3, 50, and 100 μM. Thermal induction of gene expression is achieved by shifting cultures from 30 to 42° for 4 h. Cells are disrupted by sonication and the crude extract obtained after centrifugation used to purify Hmp by anion-exchange chromatography on DEAE-Sepharose Fast Flow at pH 8.0, followed by chromatography on DEAE-Toyopearl 650S at pH 6.5 and gel filtration on Superdex-200. Colored fractions at each step are combined, and the ratio A_{410}/A_{280} is determined. Purified Hmp (typically about 6 mg from 0.4 l culture) is a monomer in solution, has an A_{410}/A_{280} ratio of 0.9, migrates on SDS PAGE gels with an apparent molecular mass of 44 kDa, and is estimated to be 95% pure.

The yield of Hmp from aerobic cultures is typically about threefold higher than reported using anaerobic expression under nitrate-inducing conditions from the natural promoter of *hmp* cloned in pAlter (Gardner *et al.*, 2000a). Furthermore, typical preparations of Hmp have equimolar contents of heme and FAD as predicted by sequence and structural information on Hmp (Andrews *et al.*, 1992; Ilari *et al.*, 2002; Vasudevan *et al.*, 1991). Concentrations of redox centers, given as nmol (mg protein)$^{-1}$ with (standard deviations [SD] and numbers of preparations) are typically as follows: FAD (13.3, 1.9, 4), heme assayed as the pyridine hemochrome (13.2, 1.2, 3), and as the CO complex (13.0, 2.3, 5). We have found this to be a reliable method for purifying *E. coli* Hmp over many years and are adapting it for the purification of eukaryotic flavohemoglobins. The Hmp purification used here yields protein with a prosthetic group content and flavin:heme ratio superior to those in previous reports. Compared to our previous preparations (Ioannidis *et al.*, 1992), the FAD content is raised 1.4- to 2.1-fold and the heme is raised 2.7- to 3.6-fold in protein that has the expected 1:1 FAD: heme ratio. The prosthetic group content is 0.57 mol heme per mol protein and 0.58 mol FAD per mol protein and compares very favorably with the values of 0.1 and 0.01 mol heme and FAD, respectively, reported in azide-treated preparations (Gardner *et al.*, 1998). Despite the relatively high FAD content of the Hmp preparations used here, exogenous FAD markedly increased the apparent affinity of our preparation for oxygen in the absence of NO.

3.3. Expression of *E. coli hmp* in pBAD

We have more recently purified Hmp from *E. coli* strains transformed with a pBAD vector into which is cloned the native *hmp* gene of *E. coli*. Details of the purification process are also changed. In brief, the method involves extraction of genomic DNA from strain VJS676 (Poole *et al.*, 1996) using the Promega Wizard purification kit (product no. A1125) following the manufacturer's protocol for Gram-negative bacteria. The following primers were designed and checked in MacVector software and synthesized by Sigma-Genosys:

RP275(forward) : GAA GAC C*C/C ATG* GTT GAC GCT CAA ACC ATC GCT AC

RP276(reverse) : GTG ACG GTA A*A/A GCT T*CG TCT TTG ACG TGG CAC G

RP275, hybridizing upstream of *hmp*, contains an *Nco*I site (italicized and |); the reverse primer RP276 contains a *Hin*dIII site (italicized and |). Following optimization of PCR conditions, these primers were used to synthesize an *hmp* gene fragment containing both an *Nco*I site and a *Hin*dIII site. Digestion with both enzymes yielded a cohesive-ended fragment whose size was verified by electrophoresis. Vector pBAD/HisC (Invitrogen) was cleaved with the same enzymes (thus removing the His tag sequence), verified by electrophoresis, treated with alkaline phosphatase, and used in a ligation reaction with the *hmp* fragment at 16° for 16.5 h. The product was used to transform *E. coli* TOP10 cells, made competent using the one-step protocol of Chung (Chung *et al.*, 1989). Transformants were selected on ampicillin plates and screened by preparing plasmid DNA and digestion with both enzymes, revealing the vector (4.1 kb) and insert (1.2 kb) fragments. Selected clones were verified by DNA sequencing. The strain harboring this plasmid is deposited as RKP3040. Similar protocols have been devised for expression of the separate flavin and heme domains (not described here) and site-directed mutagenesis using the Quikchange method (Stratagene).

Hmp production in RKP3040 was induced with arabinose; optimization trials showed maximal levels of protein (judged by SDS PAGE) after treatment of exponentially growing cultures for 4 h with 0.2% (w/v) arabinose. The medium is supplemented with 3.1 μM $FeCl_3$, 100 μM riboflavin, and 50 μM δ-aminolevulinic acid.

Typically, Hmp purification is done on cells harvested from LB broth (4 l) grown to an OD_{600} of 0.5 before arabinose addition. Cells are harvested by centrifugation (10,000 *g*, 15 min), washed, and resuspended in a buffer containing 40 m*M* Tris-HCl, pH 8.0 to a volume of about 10 ml. This concentrated slurry is subjected to sonication to break cells (4 × 20 s, 40 s cooling intervals on ice). The sonicate is clarified by centrifugation at 50,000 rpm for 60 min and the supernatant, which is brown-red, is used to purify Hmp using DEAE Sepharose (CL-6B) followed by gel filtration on Superdex 75 or Superdex 200, essentially as described previously (Ioannidis *et al.*, 1992).

ACKNOWLEDGMENTS

Work from the Poole laboratory was supported generously by the UK Biotechnology and Biological Sciences Research Council. The Leverhulme Trust funds Megan E. S. Lewis.

REFERENCES

Andrews, S. C., Shipley, D., Keen, J. N., Findlay, J. B. C., Harrison, P. M., and Guest, J. R. (1992). The haemoglobin-like protein (HMP) of *Escherichia coli* has ferrisiderophore reductase activity and its C-terminal domain shares homology with ferredoxin $NADP^+$ reductases. *FEBS Lett.* **302,** 247–252.

Bang, I. S., Liu, L. M., Vazquez-Torres, A., Crouch, M. L., Stamler, J. S., and Fang, F. C. (2006). Maintenance of nitric oxide and redox homeostasis by the *Salmonella* flavohemoglobin Hmp. *J. Biol. Chem.* **281,** 28039–28047.

Bodenmiller, D. M., and Spiro, S. (2006). The *yjeB* (*nsrR)* gene of *Escherichia coli* encodes a nitric oxide-sensitive transcriptional regulator. *J. Bacteriol.* **188,** 874–881.

Bollinger, C. J. T., Bailey, J. E., and Kallio, P. T. (2001). Novel hemoglobins to enhance microaerobic growth and substrate utilization in *Escherichia coli. Biotechnol. Prog.* **17,** 798–808.

Bonamore, A., Chiancone, E., and Boffi, A. (2001). The distal heme pocket of *Escherichia coli* flavohemoglobin probed by infrared spectroscopy. *Biochim. Biophys. Acta* **1549,** 174–178.

Bonamore, A., Gentili, P., Ilari, A., Schinina, M. E., and Boffi, A. (2003a). *Escherichia coli* flavohemoglobin is an efficient alkylhydroperoxide reductase. *J. Biol. Chem.* **278,** 22272–22277.

Bonamore, A., Farina, A., Gattoni, M., Schinina, M. E., Bellelli, A., and Boffi, A. (2003b). Interaction with membrane lipids and heme ligand binding properties of *Escherichia coli* flavohemoglobin. *Biochemistry* **42,** 5792–5801.

Chung, C. T., Niemela, S. L., and Miller, R. H. (1989). One-step preparation of competent *Escherichia coli*: Transformation and storage of bacterial cells in the same solution. *Proc. Natl. Acad. Sci. USA* **86,** 2172–2175.

Crawford, M. J., and Goldberg, D. E. (1998a). Regulation of the *Salmonella typhimurium* flavohemoglobin gene: A new pathway for bacterial gene expression in response to nitric oxide. *J. Biol. Chem.* **273,** 34028–34032.

Crawford, M. J., and Goldberg, D. E. (1998b). Role for the *Salmonella* flavohemoglobin in protection from nitric oxide. *J. Biol. Chem.* **273,** 12543–12547.

Crawford, M. J., and Goldberg, D. E. (2006). Regulation of the *Salmonella typhimurium* flavohemoglobin gene: A new pathway for bacterial gene expression in response to nitric oxide. *J. Biol. Chem.* **281,** 3752.

Cruz-Ramos, H., Crack, J., Wu, G., Hughes, M. N., Scott, C., Thomson, A. J., Green, J., and Poole, R. K. (2002). NO sensing by FNR: Regulation of the *Escherichia coli* NO-detoxifying flavohaemoglobin, Hmp. *EMBO J.* **21,** 3235–3244.

D'Angelo, P., Lucarelli, D., della Longa, S., Benfatto, M., Hazemann, J. L., Feis, A., Smulevich, G., Ilari, A., Bonamore, A., and Boffi, A. (2004). Unusual heme iron-lipid acyl chain coordination in *Escherichia coli* flavohemoglobin. *Biophys. J.* **86,** 3882–3892.

D'Autreaux, B., Touati, D., Bersch, B., Latour, J. M., and Michaud-Soret, I. (2002). Direct inhibition by nitric oxide of the transcriptional ferric uptake regulation protein via nitrosylation of the iron. *Proc. Natl. Acad. Sci. USA* **99,** 16619–16624.

de Oliveira, P., Ranjbari, A., Baciou, L., Bizouarn, T., Ollesch, G., Ermler, U., Sebban, P., Keita, B., and Nadjo, L. (2007). Preliminary electrochemical studies of the flavohaemoprotein from *Ralstonia eutropha* entrapped in a film of methyl cellulose: Activation of the reduction of dioxygen. *Bioelectrochemistry* **70,** 185–191.

Ermler, U., Siddiqui, R. A., Cramm, R., and Friedrich, B. (1995a). Crystal structure of the flavohemoglobin from *Alcaligenes eutrophus* at 1.75 angstrom resolution. *EMBO J.* **14,** 6067–6077.

Ermler, U., Siddiqui, R. A., Cramm, R., Schroder, D., and Friedrich, B. (1995b). Crystallization and preliminary X-ray diffraction studies of a bacterial flavohemoglobin protein. *Proteins* **21,** 351–353.

Farres, J., Rechsteiner, M. P., Herold, S., Frey, A. D., and Kallio, P. T. (2005). Ligand binding properties of bacterial hemoglobins and flavohemoglobins. *Biochemistry* **44,** 4125–4134.

Gardner, A. M., Martin, L. A., Gardner, P. R., Dou, Y., and Olson, J. S. (2000a). Steady-state and transient kinetics of *Escherichia coli* nitric-oxide dioxygenase (flavohemoglobin): The B10 tyrosine hydroxyl is essential for dioxygen binding and catalysis. *J. Biol. Chem.* **275,** 12581–12589.

Gardner, P. R. (2005). Nitric oxide dioxygenase function and mechanism of flavohemoglobin, hemoglobin, myoglobin and their associated reductases. *J. Inorg. Biochem.* **99,** 247–266.

Gardner, P. R., Gardner, A. M., Martin, L. A., and Salzman, A. L. (1998). Nitric oxide dioxygenase: An enzymic function for flavohemoglobin. *Proc. Natl. Acad. Sci. USA* **95,** 10378–10383.

Gardner, P. R., Gardner, A. M., Martin, L. A., Dou, Y., Li, T. S., Olson, J. S., Zhu, H., and Riggs, A. F. (2000b). Nitric-oxide dioxygenase activity and function of flavohemoglobins: Sensitivity to nitric oxide and carbon monoxide inhibition. *J. Biol. Chem.* **275,** 31581–31587.

Gilberthorpe, N. J., Lee, M. E., Stevanin, T. A., Read, R. C., and Poole, R. K. (2007). NsrR: A key regulator circumventing *Salmonella enterica* serovar Typhimurium oxidative and nitrosative stress *in vitro* and in IFN-g-stimulated J774.2 macrophages. *Microbiology* **153,** 1756–1771.

Goncalves, V. L., Nobre, L. S., Vicente, J. B., Teixeira, M., and Saraiva, L. M. (2006). Flavohemoglobin requires microaerophilic conditions for nitrosative protection of *Staphylococcus aureus*. *FEBS Lett.* **580,** 1817–1821.

Hausladen, A., Gow, A. J., and Stamler, J. S. (1998). Nitrosative stress: Metabolic pathway involving the flavohemoglobin. *Proc. Natl. Acad. Sci. USA* **95,** 14100–14105.

Hausladen, A., Gow, A., and Stamler, J. S. (2001). Flavohemoglobin denitrosylase catalyzes the reaction of a nitroxyl equivalent with molecular oxygen. *Proc. Natl. Acad. Sci. USA* **98,** 10108–10112.

Helmick, R. A., Fletcher, A. E., Gardner, A. M., Gessner, C. R., Hvitved, A. N., Gustin, M. C., and Gardner, P. R. (2005). Imidazole antibiotics inhibit the nitric oxide dioxygenase function of microbial flavohemoglobin. *Antimicrob. Agents Chemo.* **49,** 1837–1843.

Hernandez-Urzua, E., Zamorano-Sanchez, D. S., Ponce-Coria, J., Morett, E., Grogan, S., Poole, R. K., and Membrillo-Hernandez, J. (2007). Multiple regulators of the flavohaemoglobin (hmp) gene of *Salmonella enterica* serovar Typhimurium include RamA, a transcriptional regulator conferring the multidrug resistance phenotype. *Arch. Microbiol.* **187,** 67–77.

Ilari, A., Bonamore, A., Farina, A., Johnson, K. A., and Boffi, A. (2002). The X-ray structure of ferric *Escherichia coli* flavohemoglobin reveals an unexpected geometry of the distal heme pocket. *J. Biol. Chem.* **277,** 23725–23732.

Ioannidis, N., Cooper, C. E., and Poole, R. K. (1992). Spectroscopic studies on an oxygen-binding haemoglobin-like flavohaemoprotein from *Escherichia coli*. *Biochem. J.* **288,** 649–655.

Kim, S. O., Orii, Y., Lloyd, D., Hughes, M. N., and Poole, R. K. (1999). Anoxic function for the *Escherichia coli* flavohaemoglobin (Hmp): Reversible binding of nitric oxide and reduction to nitrous oxide. *FEBS Lett.* **445,** 389–394.

Kobayashi, G., Nakamura, T., Ohmachi, H., Matsuoka, A., Ochiai, T., and Shikama, K. (2002). Yeast flavohemoglobin from *Candida norvegensis*: Its structural, spectral, and stability properties. *J. Biol. Chem.* **277,** 42540–42548.

Love, C. A., Lilley, P. E., and Dixon, N. E. (1996). Stable high-copy-number bacteriophage lambda promoter vectors for overproduction of proteins in *Escherichia coli*. *Gene* **176,** 49–53.

Membrillo-Hernández, J., Ioannidis, N., and Poole, R. K. (1996). The flavohaemoglobin (HMP) of *Escherichia coli* generates superoxide *in vitro* and causes oxidative stress *in vivo*. *FEBS Lett.* **382,** 141–144.

Membrillo-Hernández, J., Coopamah, M. D., Channa, A., Hughes, M. N., and Poole, R. K. (1998). A novel mechanism for upregulation of the *Escherichia coli* K-12 *hmp* (flavohaemoglobin) gene by the "NO releaser," *S*-nitrosoglutathione: Nitrosation of homocysteine and modulation of MetR binding to the *glyA-hmp* intergenic region. *Mol. Microbiol.* **29,** 1101–1112.

Membrillo-Hernández, J., Coopamah, M. D., Anjum, M. F., Stevanin, T. M., Kelly, A., Hughes, M. N., and Poole, R. K. (1999). The flavohemoglobin of *Escherichia coli* confers resistance to a nitrosating agent, a "nitric oxide releaser," and paraquat and is essential for transcriptional responses to oxidative stress. *J. Biol. Chem.* **274,** 748–754.

Mills, C. E., Sedelnikova, S., Søballe, B., Hughes, M. N., and Poole, R. K. (2001). *Escherichia coli* flavohaemoglobin (Hmp) with equistoichiometric FAD and haem contents has a low affinity for dioxygen in the absence or presence of nitric oxide. *Biochem. J.* **353,** 207–213.

Mukai, M., Mills, C. E., Poole, R. K., and Yeh, S. R. (2001). Flavohemoglobin, a globin with a peroxidase-like catalytic site. *J. Biol. Chem.* **276,** 7272–7277.

Nakano, M. M., Geng, H., Nakano, S., and Kobayashi, K. (2006). The nitric oxide-responsive regulator NsrR controls ResDE-dependent gene expression. *J. Bacteriol.* **188,** 5878–5887.

Orii, Y., Ioannidis, N., and Poole, R. K. (1992). The oxygenated flavohaemoglobin from *Escherichia coli*: Evidence from photodissociation and rapid-scan studies for two kinetic and spectral forms. *Biochem. Biophys. Res. Commun.* **187,** 94–100.

Oshino, R., Oshino, N., and Chance, B. (1971). The oxygen equilibrium of yeast hemoglobin. *FEBS Lett.* **19,** 96–100.

Oshino, R., Oshino, N., Chance, B., and Hagihara, B. (1973b). Studies on yeast hemoglobin: The properties of yeast hemoglobin and its physiological function in the cell. *Eur. J. Biochem.* **35,** 23–33.

Oshino, R., Asakura, T., Tamura, M., Oshino, N., and Chance, B. (1972). Yeast hemoglobin-reductase complex. *Biochem. Biophys. Res. Commun.* **46,** 1055–1060.

Oshino, R., Asakura, T., Takio, K., Oshino, N., Chance, B., and Hagihara, B. (1973a). Purification and molecular properties of yeast hemoglobin. *Eur. J. Biochem.* **39,** 581–590.

Poole, R. K. (2005). Nitric oxide and nitrosative stress tolerance in bacteria. *Biochem. Soc. Trans.* **33,** 176–180.

Poole, R. K., Anjum, M. F., Membrillo-Hernández, J., Kim, S. O., Hughes, M. N., and Stewart, V. (1996). Nitric oxide, nitrite, and Fnr regulation of *hmp* (flavohemoglobin) gene expression in *Escherichia coli* K-12. *J. Bacteriol.* **178,** 5487–5492.

Poole, R. K., Ioannidis, N., and Orii, Y. (1994). Reactions of the *Escherichia coli* flavohaemoglobin (Hmp) with oxygen and reduced nicotinamide adenine dinucleotide: Evidence for oxygen switching of flavin oxidoreduction and a mechanism for oxygen sensing. *Proc. Biol. Sci.* **255,** 251–258.

Poole, R. K., Rogers, N. J., D'mello, R. A. M., Hughes, M. N., and Orii, Y. (1997). *Escherichia coli* flavohaemoglobin (Hmp) reduces cytochrome *c* and Fe(III)-hydroxamate K by electron transfer from NADH via FAD: Sensitivity of oxidoreductase activity to haem-bound dioxygen. *Microbiology* **143,** 1557–1565.

Probst, I., and Schlegel, H. G. (1976). Respiratory components and oxidase activities in *Alcaligenes eutrophus*. *Biochim. Biophys. Acta* **440,** 412–428.

Probst, I., Wolf, G., and Schlegel, H. G. (1979). An oxygen-binding flavohemoprotein from *Alcaligenes eutrophus*. *Biochim. Biophys. Acta* **576,** 471–478.

Rodionov, D. A., Dubchak, I. L., Arkin, A. P., Alm, E. J., and Gelfand, M. S. (2005). Dissimilatory metabolism of nitrogen oxides in bacteria: Comparative reconstruction of transcriptional networks. *PLoS Comput. Bio.* **1,** 415–431.

Roos, V., and Klemm, P. (2006). Global gene expression profiling of the asymptomatic bacteriuria *Escherichia coli* strain 83972 in the human urinary tract. *Infect. Immun.* **74,** 3565–3575.

Sasaki, Y., Takaya, N., Nakamura, A., and Shoun, H. (2004). Isolation of flavohemoglobin from the actinomycete *Streptomyces antibioticus* grown without external nitric oxide stress. *Biosci. Biotech. and Biochem.* **68,** 1106–1112.

Schwartz, C. J., Giel, J. L., Patschkowski, T., Luther, C., Ruzicka, F. J., Beinert, H., and Kiley, P. J. (2001). IscR, an Fe-S cluster-containing transcription factor, represses expression of *Escherichia coli* genes encoding Fe-S cluster assembly proteins. *Proc. Natl. Acad. Sci. USA* **98,** 14895–14900.

Sebbane, F., Lemaitre, N., Sturdevant, D. E., Rebeil, R., Virtaneva, K., Porcella, S. F., and Hinnebusch, B. J. (2006). Adaptive response of *Yersinia pestis* to extracellular effectors of innate immunity during bubonic plague. *Proc. Natl. Acad. Sci. USA* **103,** 11766–11771.

Spiro, S. (2007). Regulators of bacterial responses to nitric oxide. *FEMS Microbiol. Rev.* **31,** 193–211.

Stevanin, T. A., Read, R. C., and Poole, R. K. (2007). The *hmp* gene encoding the NO-inducible flavohaemoglobin in *Escherichia coli* confers a protective advantage in resisting killing within macrophages, but not *in vitro*: Links with swarming motility. *Gene* **398,** 62–68.

Stevanin, T. M., Poole, R. K., Demoncheaux, E. A. G., and Read, R. C. (2002). Flavohemoglobin Hmp protects *Salmonella enterica* serovar Typhimurium from nitric oxide-related killing by human macrophages. *Infect. Immun.* **70,** 4399–4405.

Stevanin, T. M., Ioannidis, N., Mills, C. E., Kim, S. O., Hughes, M. N., and Poole, R. K. (2000). Flavohemoglobin Hmp affords inducible protection for *Escherichia coli* respiration, catalyzed by cytochromes *bo'* or *bd*, from nitric oxide. *J. Biol. Chem.* **275,** 35868–35875.

Takaya, N., Suzuki, S., Matsuo, M., and Shoun, H. (1997). Purification and characterization of a flavohemoglobin from the denitrifying fungus *Fusarium oxysporum*. *FEBS Lett.* **414,** 545–548.

Vasudevan, S. G., Tang, P., Dixon, N. E., and Poole, R. K. (1995). Distribution of the flavohaemoglobin, HMP, between periplasm and cytoplasm in *Escherichia coli*. *FEMS Microbiol. Lett.* **125,** 219–224.

Vasudevan, S. G., Armarego, W. L. F., Shaw, D. C., Lilley, P. E., Dixon, N. E., and Poole, R. K. (1991). Isolation and nucleotide sequence of the *hmp* gene that encodes a haemoglobin-like protein in *Escherichia coli* K-12. *Mol. Gen. Genet.* **226,** 49–58.

Wu, G., Wainwright, L. M., and Poole, R. K. (2003). Microbial globins. *Adv. Microb. Physiol.* **47,** 255–310.

Wu, G., Corker, H., Orii, Y., and Poole, R. K. (2004). *Escherichia coli* Hmp, an "oxygen-binding flavohaemoprotein," produces superoxide anion and self-destructs. *Arch. Microbiol.* **182,** 193–203.

Zhu, H., and Riggs, A. F. (1992). Yeast flavohemoglobin is an ancient protein related to globins and a reductase family. *Proc. Natl. Acad. Sci. USA* **89,** 5015–5019.

CHAPTER TEN

Structural Studies on Flavohemoglobins

Andrea Ilari* *and* Alberto Boffi†

Contents

Abstract

The key three-dimensional features of flavohemoglobins have been unveiled by X-ray crystallographic investigations carried out on the *Alcaligenes eutrophus* and *Escherichia coli* proteins. Flavohemoglobins are made of a globin domain fused with a ferredoxin reductase-like FAD binding module and display highly conserved sequences in the active sites of both the heme-binding domain and the flavin-binding domain. Structural studies are discussed and methodological approaches to the solution of the crystal structures and to the analysis of the relevant stereochemical properties of the active sites are

* CNR Institute of Molecular Biology and Pathology, University of Rome La Sapienza, Rome, Italy
† Department of Biochemical Sciences, University of Rome La Sapienza, Rome, Italy

Methods in Enzymology, Volume 436
ISSN 0076-6879, DOI: 10.1016/S0076-6879(08)36010-8

presented. The understanding of the structural properties of flavohemoglobins serves as a guide for testing biological hypotheses and allows for a rational evaluation of structure-based alignments within the flavohemoglobin family.

1. Introduction

Flavohemoglobin genes are widespread within prokaryotes and eukaryotes, although their distribution among different species is apparently hazardous and does not follow a discernible evolutionary pattern. Sequence alignments show that the flavohemoglobin family is a very homogeneous group of proteins that share highly conserved active sites in both the heme and the flavin-binding domains. The conserved amino acids within the heme domain include the residues lining the heme pocket of the globin domain on both the proximal site and the distal site, thus indicating that there must be a strong regio- and/or stereochemical requirement for ligand binding and/or for gaseous ligand diffusion. In parallel, the amino acid residues responsible for flavin binding are conserved and conform to the typical architecture of flavodoxin-reductase proteins, thus indicating that the flavin moiety serves as an electron-transfer module from the NADH to the heme. In turn, sequence alignments of separate domains with homologous proteins diverge rapidly, thus suggesting that flavohemoglobins originated from the fusion of a protoglobin ancestor and a flavin-binding domain.

The first X-ray structure of a flavohemoglobin, obtained for the *A. eutrophus* protein (FHP, Ermler *et al.*, 1995) showed that the protein was made of a C-terminal NAD- and FAD-binding domain, a common arrangement among the members of the FdR-like family, and an N-terminal globin domain with a typical six-helix globin fold. The structure of the active site in the globin domain of FHP revealed that the distal heme pocket was occupied by a phospholipid molecule (Ollesch *et al.*, 1999). The presence of this bulky ligand was apparently unrelated to possible functional roles of FHP and impaired the assignment of the geometry of the heme active site according to the criteria established for vertebrate hemoglobins. The crystallization of the highly homologous *E. coli* flavohemoglobin (HMP) in its lipid-free form disclosed the architecture of the active site (Ilari *et al.*, 2002), brought about analogies and differences with the structure of FHP, and established a high degree of structural similarity with the single-chain hemoglobin from *Vitreoscilla* sp. (Tarricone *et al.*, 1997).

2. X-ray Structure of *A. Eutrophus* Flavohemoglobin

2.1. Crystallization

The flavohemoglobin has been purified as a native protein from *A. eutrophus* cells. A two-step purification procedure was employed that entails an anionic exchange chromatographic step on a DEAE-Sepharose column (Pharmacia LKB, Freiburg, Germany) and an additional step on a hydroxyapatite column equilibrated with 20 m*M* Tris-HCl, pH 7.5, containing 0.05 *M* KCl. The purified protein has been concentrated via Centricon-10 up to 25 mg/ml and stored in 20 m*M* TRIS-HCl buffer at pH = 7.5 and 2 m*M* DTT to maintain the protein in a reducing environment. Crystallization was carried out using the hanging-drop vapor-diffusion method. A volume of 2 μl of protein solution was mixed with an equal amount of reservoir solution containing 14% PEG 3350, 0.2 *M* NaCl 50 m*M* sodium citrate pH = 5.0. The droplets were equilibrated against 1 ml precipitant solution at 18°. Yellow-brown crystals grew in three days to a size of $0.5 \times 0.2 \times 0.1$ mm^3.

2.2. Data collection, data analysis, and structure solution

X-ray quality crystals were measured at the Max-Planck beam line of the Deutches Electronensynchrotron (DESY) in Hamburg at a wavelength of 1 Å. The data were collected at 4°. Reflection intensity was integrated using the program MOSFLM (Leslie, 1992) and scaled using SCALA (Evans, 1993). Two native data sets were collected on two different crystals and merged. The processed data containing 81,190 unique reflections were 90.5% complete with a R_{sym} of 5.2 % (see previous section) in the resolution range 20–1.75Å. The phase problems were solved using the multiple isomorphous replacement (MIR) method. Heavy atom derivatives were generated by soaking crystals in reservoir solution in the presence of 2 m*M* mersalylic acid for 48 hours (Hg1), 2 m*M* CH_3-Hg(CH_3COO) for 48 h (Hg2), Hg$(CH_3COO)_2$ for 48 h (Hg3), 0.2 m*M* K_2PtCl_4 for 12 h (Pt1), 0.2 m*M* $KAuCl_4$ for 6 h. (Au). The two Hg sites were located by interpretation of the difference Patterson maps. The heavy atom sites have been refined and the phases calculated using the program MLPHARE (Otwinowski, 1991). The statistics of the native and heavy atom derivatives data sets are reported in Table 10.1.

The MIRAS maps, calculated using the phase information obtained with the Hg derivatives, were extended and improved using solvent flattening by the program SQUASH (Zhang, 1993). The O program (Jones *et al.*, 1991) allowed the identification of the two molecules present in the asymmetric unit

Table 10.1 Data collections, structure solutions, and refinement statistics for FHP and HMP

FHP	Native	Mersalylic acid	$CH_3Hg(COOCH_3)$	$Hg(COOCH_3)_2$	K_2PtCl_4	$KAuCl_4$
Resolution (Å)	1.75	2.9	4.0	4.5	4.5	4.5
R_{sym} (%)	5.2	5.8	9.7	7.6	8.5	10.6
Completeness (%)	90.5	90.2	91.3	93.5	91.6	82.2
N° of sites		2	2	2	4	6
R_{cullis} (%) (ce)		0.7	0.68	0.81	0.82	0.81
Phasing power (ce)		1.07	1.11	1.14	0.98	1.06
R-factor (%)	19.6					
r.m.s angle (°)	1.5					
r.m.s. length (Å)	0.015					
HMP						
Resolution (Å)	2.19			2.9	3.0	
R_{sym} (%)	9.0			7.9	8.9	
Completeness (%)	99.5			98.8	90.3	
N° of sites				2	4	
R_{cullis} (%) (ace/ce)				0.70/0.69	0.72/0.76	
Phasing power (ace/ce)				1.59/1.17	1.52/0.93	
R-factor (%)	18.8					
r.m.s angle (°)	1.8					
r.m.s. length (Å)	0.008					

[a] $R_{sym}=\Sigma_{hkl}\,\Sigma_i\,|I_i(hkl)-\langle I(hkl)\rangle|/\langle I(hkl)\rangle$, where $I_i(hkl)$ is the intensity for the i_{th} measurement per reflection hkl and $\langle I(hkl)\rangle$ is the average intensity for the reflection hkl.

[b] $R_{cullis}=\Sigma_{hkl}\,(|F_{ph}\,(obs)\,|-|F_{ph}(calc)|)/\,\Sigma_{hkl}\,(|F_{ph}\,(obs)\,|-|F_p(obs)|)$.

[c] Phasing power$=(\Sigma_{hkl}\,(|F_h\,(calc)\,|^2\,/\,|_{hkl}\,(|\,|F_{ph}\,(obs)\,|-|F_{ph}(calc)||)^2).^{\{1/2\}}$.

[d] $R_{factor}=\Sigma_{hkl}\,(|F_p\,(obs)\,|-|F_p(calc)|)/\,(\Sigma_{hkl}\,(|\,F_p\,(obs)|)$.

and the calculation of a map mask. The molecular twofold averaging performed with the program RAVES (Kleywegt and Jones, 1994) using the symmetry operators derived from the heavy atom sites allowed for the calculation of interpretable maps.

2.3. Model building and refinement

The interpretable map has been used to build a partial polyalanine model containing 345 residues and the FAD and heme molecules. At this point, the R-factor between the F_{obs} and the F_{calc} was only 50.8% in a resolution range between 2.7 and 10 Å.

The use of the energy minimization and molecular dynamics options of the program XPLOR (Adams, *et al.* 1997) allowed the R factor to decrease to a value of 37%. The density modification performed with a more accurate map mask allowed for the calculation of more interpretable maps. After many cycles of alternate refinement and model building, the whole model has been built. The final R factor was 19.6% in the resolution range 1.75–10 Å. The deposited model contains 403 residues, one FAD, one heme, one 1-(glycerolylphosphonyl)-2-(8-(2-hexyl-cyclopropyl)-(octanal-1-yl)-3-(hexadecanal-1-yl)-glycerol), and 236 molecules for the first molecule and 251 solvent molecule for the second molecule in the asymmetric unit.

2.4. Lipid analysis

The shape of the electronic density map provided preliminary indications about the phospholipid nature of the ligand accommodated in the active site of the protein. In order to assess the nature of the phospholipid molecule, lipids were extracted from purified FHP fractions according to the method of Bligh and Dyer (1959) and analyzed by TLC. The two bands of the thin layer plate were ascribed to phosphatidylethanolamine (major band) and phosphatidylglycerol (minor band). The fatty acid methyl esters were then identified using a GC apparatus (Varian GC 3300) online-coupled to a mass selective detector (Finnigan MAT ion-trap detector ITS40). The major peaks of the GC spectra corresponded to hexadecanoic acid (palmitic acid) and cis-9,10-methylene hexadecanoic acid. The phosphorus content of the protein was also determined using the method described by Ames (1966). This method allowed the determination of a phosphorus/protein molar ratio of 2.9/1. On this basis, the phospholipid molecule was assigned to a phosphatidyl glycerol esterified with 9-cyclopropyl octadecanoic (SN1 position) acid and hexadecanoic acid (SN2 position), a common phospholipid of the *E. coli* membrane (Ollesch *et al.*, 1999).

3. X-RAY STRUCTURE OF *E. COLI* FLAVOHEMOGLOBIN

3.1. Crystallization

The *E. coli* flavohemoglobin has been expressed and purified as reported by Bonamore *et al.* (2001). Particular care was taken to remove bound lipids by means of hydroxylapatite chromatography followed by a hydrophobic exchange alkoxypropyldextrane column (Sigma Aldrich Co.). Lipid removal was checked by electrospray mass spectrometry carried out on the protein sample. The protein has been concentrated up to 20 mg/ml with a Centricon-PM10. The spectrum is measured at 20° in a 1-cm quartz cuvette. Protein concentration was determined by measuring the absorbance at 420 nm of the CN-adduct using the extinction coefficient of 156,000 $M^{-1}cm^{-1}$ (Bonamore *et al.*, 2001).

The HMP crystallization has been achieved using the sitting-drop vapor-diffusion method. The crystallization screening was performed using Hampton research crystallization plates (Hampton Research, CA, U.S.A.) on sitting-drop microbridges. A volume of 4 μl of the protein solution was mixed with an equal amount of each reservoir solution. Positive crystallization conditions were identified in 0.1 *M* sodium acetate buffer pH 5.1–5.3, 0.2 *M* NaCl with polyethylene glycol 3350 at a concentration ranging from 21 to 26% w/v. The crystallization plates were maintained at a controlled temperature of 21°. The crystals grew in three weeks, and were very small (50 × 50 × 50 μm) and surrounded by amorphous aggregates of precipitated protein.

3.2. Data collection, analysis, and structure solution

Data were collected as 0.50 oscillation frames on the ELETTRA beam line at Basovizza (Trieste, Italy) at a wavelength of 1.0 Å and at 100 K using 26% polyethylene glycol 200 as cryoprotectant. Data analysis performed with DENZO (Otwinowski and Minor, 1997) indicated that the crystals were hexagonal P622 with cell dimensions of a = b = 164.86 Å, c = 53.46 Å, δ = B = 90°, and γ = 120°. The data scaling performed with SCALEPACK (Otwinowski and Minor, 1997) gave an R_{sym} of 9.0 for 22,448 unique reflections with a completeness of 99.5% at 2.19 Å resolution (see Table 10.1). In spite of many attempts to solve the structure by molecular replacement (MR) using *A. eutrophus* flavohemoglobin as a search model, no clear solutions have been found. The phase problem was then solved by MIR. Heavy atom derivatives were generated by soaking crystals in reservoir solution plus 5 m*M* $Hg(CH_3COO)_2$ for 36 h or 5 m*M* K_2PtCl_4 for 48 h (see Table 10.1). Coordinates of Pt and Hg were located by the interpretation of the difference in Patterson maps of the derivatives using the programs CNS

(Brünger *et al.*, 1998) and PHASES (Furey and Swaminathan, 1997). The data set of the two different derivatives were combined, atom parameters were refined, and the starting phases were calculated using isomorphous and anomalous differences with the maximum-likelihood method incorporated in MLPHARE (Otwinowski, 1991). The phases were further improved by solvent flattening using DM (density modification) (Cowtan and Zhang, 1999). The Hg atoms are covalently bound to the two cysteine residues present on the NAD-binding domain (Cys 362, Cys 389) and topologically adjacent in the structure.

3.3. Model building and refinement

The model was built using the program XTALVIEW (McRee, 1999). The refinement was carried out with the maximum-likelihood method incorporated in REFMAC (Murshudov *et al.*, 1999) to an R factor of 19.5% and an R_{free} of 27.2% at 2.19-Å resolution. The model was complete with the exception of residues 333–336 and 347–354. The residues within these regions corresponding to two external loops were not well defined. The refinement of the atomic coordinates and displacement parameters was carried out by simulated annealing and by least-squared minimization using the program X-PLOR (Brünger, 1992). Water molecules and ions were added manually. The final model includes all 396 residues and 180 water molecules. The final R-factor at 2.19-Å resolution is 18.8% with final R_{free} of 24.7%, bond lengths r.m.s. deviation of 0.008, and bond angles r.m.s. deviation of 1.8°.

The quality of the model was assessed using the program PROCHECK (Laskowski *et al.,* 1993). The most favored regions of the Ramachandran plot contain 90.9% of nonglycine residues. The overall G factor is 0.22, which is better than the expected values. The structure and refinement statistics are listed in Table 10.1.

4. Structural Comparison Between *A. Euthrophus* and *E. coli* Flavohemoglobins

4.1. Overall fold

The overall fold of flavohemoglobins (Fig. 10.1A) consists of a heart-shaped structure in which three different domains, namely the C-terminal NAD-binding, the FAD-binding, and the N-terminal globin domains are clearly distinguished.

Despite the 39% identity between FHP and HMP (see Fig. 1B), the superposition of the Cα traces of the two structure results in a high r.m.s. deviation (3.3 Å) that is ascribed mainly to the 30° rotation of the

NAD-binding domain in HMP with respect to the analogous domain in FHP. The rotation of the NAD-binding domain has been interpreted as a conformational rearrangement within a common three-dimensional fold, possibly due to the different redox/ligation states of HMP and FHP. Further differences between the two proteins involve the E-helix positioning that is placed at about 4 Å from the heme plane in the lipid-free HMP, whereas it is shifted far apart in FHP to allow for the accommodation of the bulky phospholipid molecule.

4.2. FAD-binding domain

The FAD-binding domain (151–258 in FHP and 149–252 in HMP) consists of a six-stranded antiparallel β barrel, a small α-helix capping the β barrel on the bottom, and a long loop connecting the Fβ2, Fβ3 β-sheets on the top. The structural motif is identical in FHP and HMP (the superposition between the two FAD-binding domains yields an r.m.s. deviation of 1.1 Å) with the exception of the loop connecting the Fβ5 to the α-helix, which in FHP is six residues longer than in HMP (Fig. 10.2).

The *si*-side geometry of the FAD isoalloxazine ring is similar in the two flavoprotein domains. As shown in Figs. 10.3A and 10.3B, the conformation of the two tyrosine (206, 188 in HMP and 190, 208 in FHP) serine (207 in HMP and 209 in FHP) and glutamine residues (205 in HMP and 207 in FHP) are strictly conserved. In contrast, the geometry of the *re*-side, surrounded by residues of the NAD-binding domain, display different topological relationships with the flavin moiety in the two proteins. In particular, the loop connecting Fβ5 to the α-helix is responsible for the different orientations of the adenosine molecule in space.

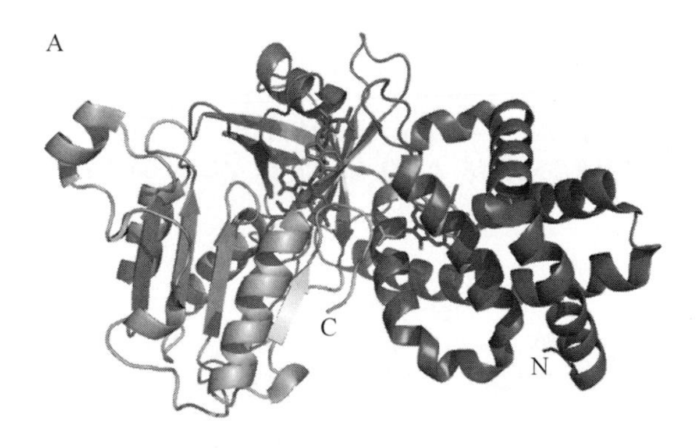

Figure 10.1 *(continued)*

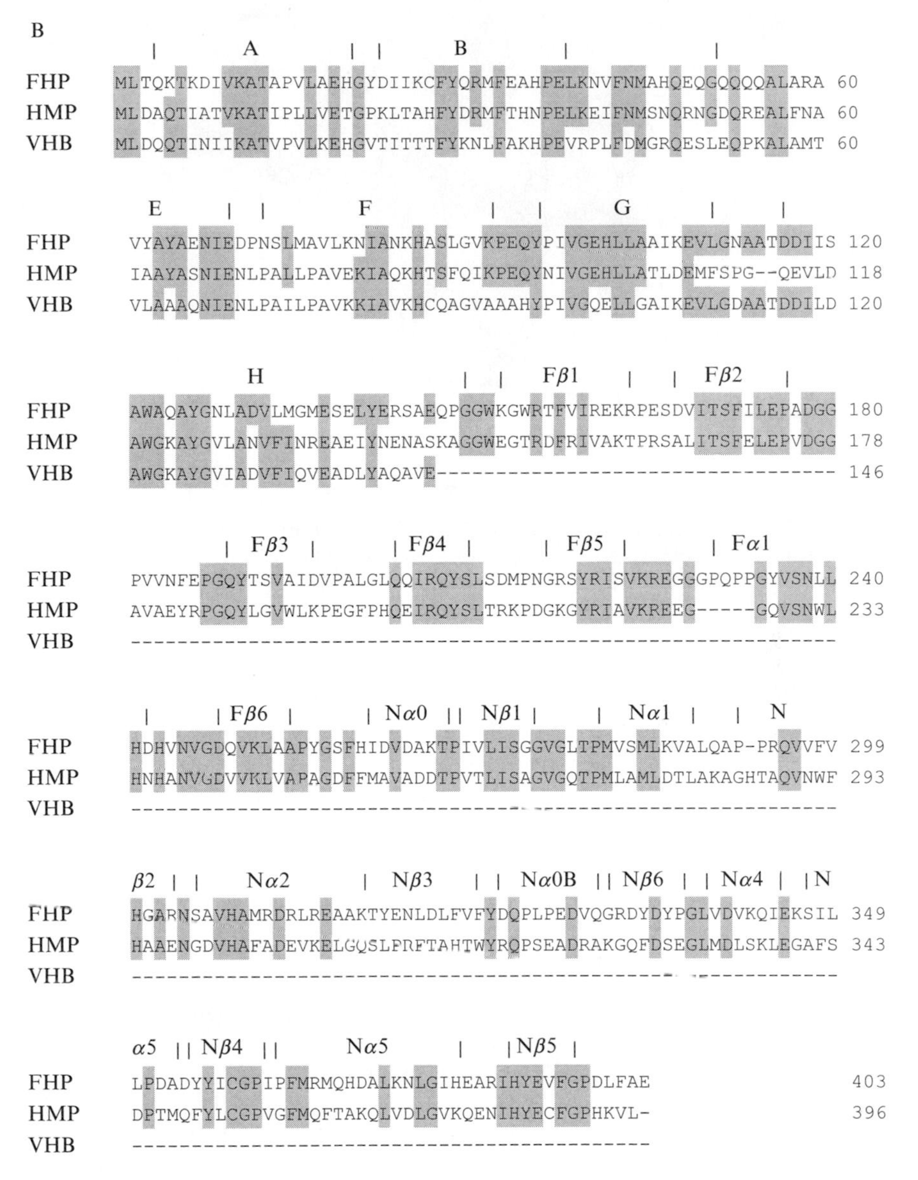

Figure 10.1 Overall fold of *E. coli* flavohemoglobin. (A) The heart-shaped structure is positioned with the flavin-binding domain at the upper apex (cyan), the globin domain on the lower-right side (red), and the NAD-binding domain on the lower-left side (green). The picture was depicted using PyMol (Delano, 1998). (B) Sequence alignment of *A. eutrophus* (FHP) and *E. coli* (HMP) flavohemoglobins, and *Vitreoscilla* sp. hemoglobin. Conserved residues are underlined. (See color insert.)

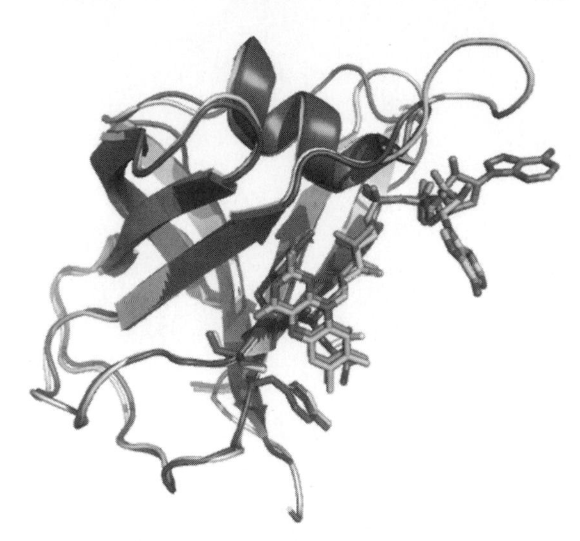

Figure 10.2 Structural overlay of the FAD-binding domain of *A. eutrophus* (FHP, yellow) and *E. coli* (HMP, red) flavohemoglobins. The FAD molecule is colored in green in HMP and in azure in FHP. The picture was generated with PyMol. (See color insert.)

In FHP, the longer loop determines a 130° rotation of the nucleotide phosphate with respect to the FAD adenosine in HMP (see Fig. 10.2). The additional five residues form a noose that anchors the nucleotide to the protein through two polar interactions (FAD AO2-O Pro 232 = 2.5 Å; Gln 231 OE1-AN3 FAD = 3.0 Å). No interactions between the protein and the adenosine nucleotide can be detected in FHP, thus allowing exposure of the adenine toward the solvent.

4.3. NAD-binding domain

The superposition between the two NAD-binding domains of FHP and HMP yields an r.m.s. deviation of 3.8 Å that is accounted for by marked structural differences in three regions (Fig. 10.4).

Nα1 is two helical turns longer in HMP than in FHP and rotated by about 10° around the conserved Met 277 (Met 286 in FHP). In FHP, the long segment connecting the two β-sheets Nβ2, Nβ3 is in a random coil conformation with the exception of a very short α-helix, whereas a supplementary α-helical segment (residues 328–335) is present in HMP. Last, the C-terminal region contacting the *re*-side of the isoalloxazine ring is turned in opposite directions in the two proteins. In particular, the aromatic ring of the conserved Phe390 (396 in FHP) is packed against the isoalloxazine central ring in HMP, whereas it protrudes toward the interdomain cleft in FHP. Moreover, the Glu 394 in HMP (400 in FHP) is closer than in FHP to

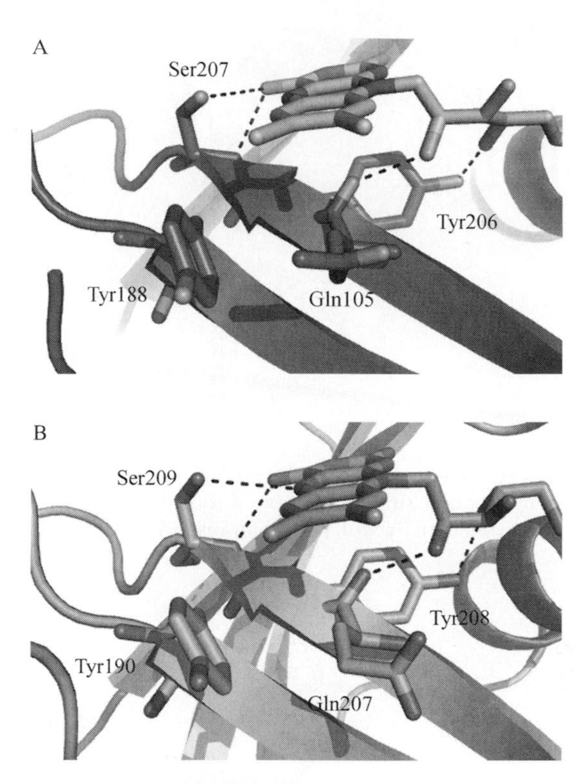

Figure 10.3 *Si*-side geometry of the isoalloxazine ring in *E. coli* (A) and *A. eutrophus* (B) flavohemoglobins. The residues within a distance of 5 Å from the ring are indicated in ball-and-stick: two tyrosine residues (206, 188 in HMP and 190, 208 in FHP) a serine residue (207 in HMP and 209 in FHP) and finally glutamine residues (205 in HMP and 207 in FHP). The images were generated with PyMol. (See color insert.)

the isoalloxazine ring (C7M-OE1 = 3.6 Å in HMP, C7M-OE1 = 4.33 Å). Considering the high sequence similarity between the NAD-binding domains, the differences in the *re*-side of the isoalloxazine ring may suggest a conformational change that accompanies the binding of the phospholipids molecule. Thus, it may be that the lipid-free species is stabilized by stacking interactions of the isoalloxazine ring with Phe390 in HMP, whereas, upon lipid binding, the whole C-terminal loop is shifted and rotated toward the ribose moiety with concomitant loss of stacking interaction.

4.4. Heme-binding domain

The architecture of the globin domain corresponds to a classical globin fold composed of six helices (segments C and D are considered loop regions) with an unusually long H helix and in which the D helix is substituted by a large loop region within the segment CD2-E6 (Asn44 to Asp52).

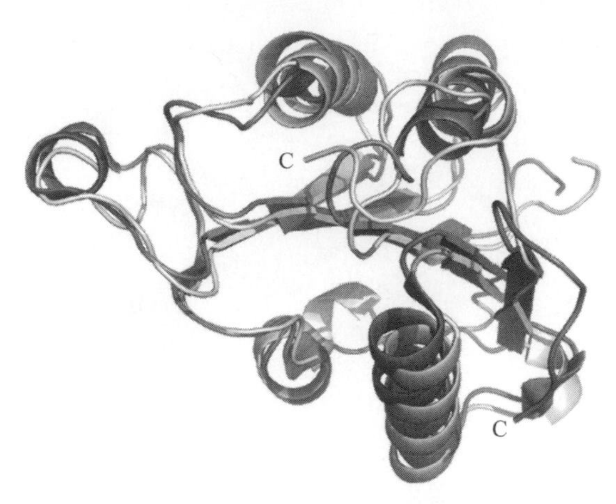

Figure 10.4 Structural overlay of the NAD-binding domain of *A. eutrophus* (yellow) and *E. coli* (red) flavohemoglobins. The C-termini of the two proteins are indicated. The picture was generated with PyMol. (See color insert.)

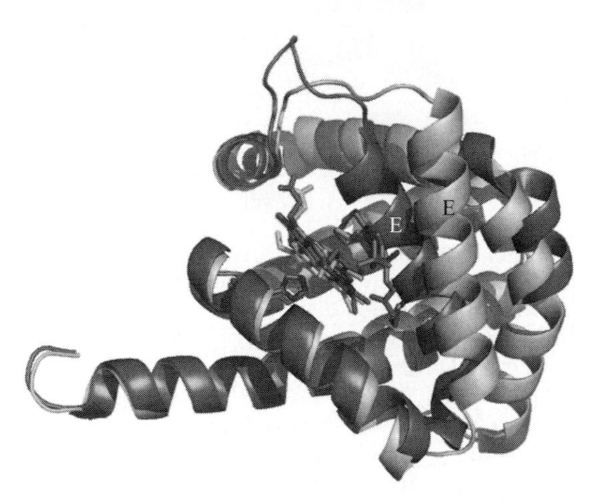

Figure 10.5 Structural overlay of the globin domain of *A. eutrophus* and *E. coli* flavohemoglobins. The heme rings and the proximal histidines are indicated in ball-and-stick. The FHP (yellow) bound dyacylglycerol-phosphatidic acid is indicated in ball-and-stick and colored cyan. The HMP (red) leucine 57 residue is indicated in ball-and-stick and in green. The E helices of both proteins are labeled. The picture was generated with PyMol. (See color insert.)

Apart from the invariant residues Phe CD1 and His F8, considered the hallmark of the globin structure, the heme pocket geometry shows little resemblance to that of vertebrate hemoglobins and myoglobins (Fig. 10.5).

Inspection of the distal pocket architecture in HMP reveals that the distal position, closest to the iron atom, is occupied by the Leu57 (E11) isobutyl side chain with the CG atom at 3.6 Å from the iron atom along the normal to the heme plane. Leu-E11 side chain together with the Phe-CD1 ring, the Ile61 (E15) and Val98 (G8) side chains, and the Gln53 (E7) backbone segment fill completely the first distal shell. In contrast, Tyr29 (B10) and Gln E7 side chains, previously proposed as the major candidates for iron-ligand distal stabilization, are shifted from the heme plane at a distance of more than 5.0 Å. The Tyr B10 phenol ring is confined to the second layer of distal pocket residues, with the phenol hydroxyl pointing toward the isobutyl side chain of Leu E11. The structural overlay of the Cα skeleton of HMP and FHP globin domains indicates that a major difference between the two proteins pertains to the positioning of the E helix that is shifted far from the heme plane in the latter protein and is determined by the accommodation of a bulky phospholipid molecule that alters the positioning of the relevant distal pocket residues (see Fig. 10.5). In particular, in FHP, the cyclopropane ring of a 9,10-methylene-hexadecanoic chain of a dyacylglycerol-phosphatidic acid (a common fatty acid component in bacterial membranes) sits on the top of the iron atom, apparently displacing the relevant distal site residues. In fact, in FHP, Leu-E11 accompanies the lipid-induced E-helix movement and is displaced by 4.6 Å from the heme pocket with respect to its positioning in HMP (Figs. 10.6A and 10.6B).

In FHP, the phospholipid acyl chains form a large number of van der Waal's contacts with nonpolar side chains (see Fig. 10.5A). The acyl chains are surrounded by the hydrophobic residues of Ile 25, Phe 43, Leu 57, and Leu 102, respectively (see Fig. 10.6A). Conversely, the glycerophosphate moiety, pointing toward the entrance of the channel, is linked to the protein matrix by polar interactions. The phosphate polar head establishes contacts with the C-terminal NAD-binding domain. In particular, the phosphate group is bound to the protein moiety by two salts bridges with Arg 375 (NH1-OP1 = 2.7 Å NE1-OP2 = 3.0 Å) and by a hydrogen bond with Tyr 393 (OH-OP3 = 2.9 Å), respectively. In HMP, the lower affinity for phospholipids is probably due to the absence of the Arg residue that binds the phosphate anion (Arg 375 in FHP) being replaced by an Asn residue. Conversely, the Tyr 393 and all hydrophobic residues forming van der Waal's contacts with phospholipids in FHP (Phe 28, 43; Leu107, 47) are conserved in HMP. Along this line, the major structural differences between HMP and FHP have been attributed to the different oxidation states (ferric heme and oxidized flavin in HMP; ferrous heme and reduced flavin in FHP) or to the presence of the phospholipid molecule in the active site of the latter protein.

An intriguing but unexplained feature of HMP is the presence of an anion-binding site next to the heme pocket. In fact, the loop between the C and E helices (CE loop) in the globin domain forms a distinct hydrophilic cavity. The 2fo-fc map showed the presence of a strong peak attributed to a chloride ion. The ion coordinating residues are Arg 54 on the E helix and Arg 49 on the CE loop (absent in FHP). This is intriguing because the

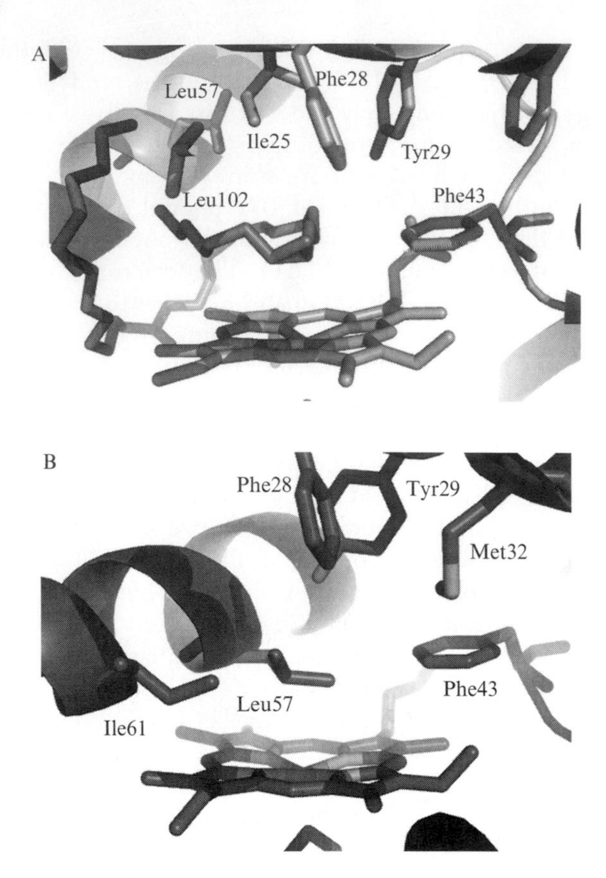

Figure 10.6 Structural details of the heme pocket in *A. eutrophus* and *E. coli* ferric unliganded flavohemoglobins. In the picture, all the hydrophobic residues lining the distal heme pocket in HMP (A) and interacting with the 9,10-methylene-hexadecanoic chain of the bound dyacylglycerol-phosphatidic acid in FHP (B) are indicated. The pictures were generated with PyMol. (See color insert.)

chloride ion resides next to the heme propionates (O2D $-$ Cl^- $=$ 4.38 Å) and the CE loop is in contact with the FAD cofactor (O Gln48 $-$ AO1 FAD $=$ 5.8 Å). Ilari *et al.* (2002) hypothesized that CE loop was be involved in modulating the heme iron–ligand binding. In particular, Arg 54, one of the anion-binding residues, is placed on the E helix and may have a pivotal role in the E-helix movement that opens and closes the heme pocket upon lipid binding.

The geometry of the heme pocket in the proximal region is highly conserved in both HMP and FHP and is characterized by the presence of the canonical proximal histidine (His F8) residue whose Nδ atom coordinates the ferric (in HMP) or ferrous (in FHP) heme iron. However, at variance with vertebrate hemoglobins, the proximal histidine is involved in a hydrogen-bonding network comprising His85 (F8), Tyr95 (G5), and Glu135 (H23) that imparts a rigid orientation to the imidazole histidine ring with respect to the heme plane. In particular, the OE2 atom of Glu135 is involved in a hydrogen-bonding interaction at 2.9 Å from the ND1 nitrogen of the proximal histidine. In turn, Glu135 is anchored to the phenol hydroxyl of Tyr95 by a hydrogen bond (OE2 at 2.7 Å from the Tyr oxygen atom).

In conclusion, the three-dimensional structures of HMP and FHP establish the key topological features that are common to the whole flavohemoglobin family and highlight differences that have been ascribed to ligand-linked or redox state–linked conformational rearrangements. Structural studies on flavohemoglobin derivatives with different ligands within each redox state will be necessary to shed light on the conformationally relevant motions that underline the ligand-binding properties and catalytic mechanisms of these proteins.

ACKNOWLEDGMENTS

Grants FIRB 2003 and PRIN 2004 to Alberto Boffi from the Ministero dell'Università e della Ricerca are gratefully acknowledged.

REFERENCES

Adams, P. D., Pannu, N. J., Read, R. J., and Brünger, A. T. (1997). Cross-validated maximum likelihood enhances crystallographic simulated annealing refinement. *Proc. Natl. Acad. Sci. USA* **94,** 5018–5023.

Ames, B. N. (1966). Assay of inorganic phosphate, total phosphate and phosphatases. *Meth. Enzymol.* **8,** 115–118.

Bligh, E. G., and Dyer, W. J. (1959). A rapid method of total lipid extraction and purification. *Can. J. Biochem. Phys.* **37,** 911–915.

Bonamore, A., Chiancone, E., and Boffi, A. (2001). The distal heme pocket of *Escherichia coli* flavohemoglobin probed by infrared spectroscopy. *Biochim. Biophys. Acta* **1549,** 174–178.

Brünger, A. T. (1992). A system for X-ray crystallography and NMR New Haven: X-PLOR Version 3.1. Yale University Press, New Haven CT.

Brünger, A. T., Adams, P. D., Clore, G. M., DeLano, W. L., Gros, P., Grosse-Kunstleve, R. W., Jiang, J. S., Kuszewski, J., Nilges, M., Pannu, N. S., Read, R. J., Rice, L. M., Simonson, T., and Warren, G. L. (1998). Crystallography & NMR system: A new software suite for macromolecular structure determination. *Acta Crystallogr. D* **54,** 905–921.

Cowtan, K. D., and Zhang, K. Y. (1999). Density modification for macromolecular phase improvement. *Prog. Biophys. Mol. Biol.* **72,** 245–270.

Delano, W. L. (1998). The *PyMol molecular graphics system.* San Carlos, LA: Delano Scientific LLC, San Carlos, LA.

Ermler, U., Siddiqui, R. A., Cramm, R., Schröder, D., and Friedrich, B. (1995). Crystallization and preliminary X-ray diffraction studies of a bacterial flavohemoglobin protein. *Proteins* **21,** 351–353.

Evans, P. R. (1993). *"Data reduction," Proceedings of CCP4 Study Weekend on Data Collection & Processing* 114–122.

Furey, W., and Swaminathan, S. (1997). PHASES-95: A program package for the processing and analysis of diffraction data from macromolecules. *Meth. Enzymol.* **277,** 590–620.

Ilari, A., Bonamore, A., Farina, A., Johnson, K. A., and Boffi, A. (2002). The X-ray structure of ferric *Escherichia coli* flavohemoglobin reveals an unexpected geometry of the distal heme pocket. *J. Biol. Chem.* **277,** 23725–23732.

Jones, T. A., Zou, J. Y., Cowan, S., and Kieldgaard, M. (1991). Improved methods for building protein models in electron density maps and the location of errors in these models. *Acta Crystallogr. A* **47,** 110–119.

Kleywegt, G. J., and Jones, T. A. (1994). *In* "From first map to final model" (S. Bailey, R. Hubbard, and D. Waller, eds.), pp. 59–66. SERC Daresbury Laboratory, Warrington, UK.

Laskowski, R. A., McArthur, M. W., Moss, D. S., and Thornton, J. M. (1993). PROCHECK: A program to check the stereochemical quality of protein structures. *J. Appl. Crystallogr.* **26,** 283–291.

Leslie, A. G. W. (1992). *Joint CCP4 + ESF-EAMCB newsletter on protein crystallography* **26.**

McRee, D. E. (1999). XtalView/Xfit: A versatile program for manipulating atomic coordinates and electron density. *J. Struct. Biol.* **125,** 156–165.

Murshudov, G. N., Lebedev, A., Vagin, A., Wilson, K. S., and Dodson, E. J. (1999). Efficient anisotropic refinement of macromolecular structures using FFT. *Acta Crystallogr. D* **55,** 247–255.

Ollesch, G., Kaunzinger, A., Juchelka, D., Schubert-Zsilavecz, M., and Ermler, U. (1999). Phospholipid bound to the flavohemoprotein from *Alcaligenes eutrophus. Eur. J. Biochem.* **262,** 396–405.

Otwinowski, Z. (1991). *In* "Isomorphous replacement and anomalous scattering" (W. Wolf, P. R. Evans, and A. G. W. Leslie, eds.), pp. 80–86. Science Engineering Research Council, Warrington, UK.

Otwinowski, Z., and Minor, W. (1997). Processing of X-ray diffraction data collected in oscillation mode. *Meth. Enzymol.* **276,** 307–326.

Tarricone, C., Galizzi, A., Coda, A., Ascenzi, P., and Bolognesi, M. (1997). Unusual structure of the oxygen-binding site in the dimeric bacterial hemoglobin from Vitreoscilla sp. *Structure* **5,** 497–507.

Zhang, K. Y. J. (1993). SQUASH: Combining constraints for macromolecular phase refinement and extension. *Acta Crystallogr. D* **49,** 429–439.

CHAPTER ELEVEN

Flavohemoglobin of *Staphylococcus aureus*

Lígia S. Nobre, Vera L. Gonçalves, *and* Lígia M. Saraiva

Contents

Abstract

Biotically, bacteria encounter nitrogen-reactive species in environments where denitrification occurs or when nitric oxide (NO) is generated by the mammal NO synthase, particularly during the infectious processes. In bacteria, flavohemoglobins have been shown to be one of the major systems responsible for the scavenging of these chemical species, either in aerobic or in anaerobic environments. *Staphylococcus aureus*, a pathogenic bacterium with high impact on human heath, also contains a gene encoding a homologue of a flavohemoprotein. The study of the recombinant protein and of the knockout mutant strain allowed us to conclude that, in spite of sharing similar physicochemical properties with the other known flavohemoglobins, the *S. aureus* flavohemoglobin requires microaerobic conditions to protect this bacterium from the nocive effects of the nitrosative stress.

Instituto de Tecnologia Química e Biológica, Universidade Nova de Lisboa, Oeiras, Portugal

Methods in Enzymology, Volume 436
ISSN 0076-6879, DOI: 10.1016/S0076-6879(08)36011-X

1. Introduction

Pathogens counteract the deleterious effects of NO produced by the activated macrophages through the induction of NO-metabolizing enzymes. Flavodiiron NO reductases and the flavohemoglobins (Hmps) (Frey and Kallio, 2003; Saraiva *et al.*, 2004) are examples of proteins that directly detoxify NO or S-nitrosothiols. Flavohemoglobins are widespread among bacteria and yeast and have a two-domain structure formed by a hemoglobin-like domain in the N-terminal (containing a single B-type heme) and a ferredoxin-$NADP^+$ reductase-like domain in the C-terminal (harboring a FAD moiety and a NAD[P]H binding motif).

For a long time, the function of microbial flavohemoglobins was obscure, and the first indication that Hmp might be involved in NO biochemistry was the up-regulation of the *hmp* gene by NO or nitrosating agents (Membrillo-Hernandez *et al.*, 1998; Poole *et al.*, 1996). It has been demonstrated for several microorganisms (Gardner and Gardner, 2002; Justino *et al.*, 2005; Wu *et al.*, 2004) that Hmps have an important role in protection against nitrosative stress. Indeed, Hmp is able to detoxify NO aerobically by catalyzing its oxidation to nitrate, with a range of activities that varies from 7.4 to 128 s^{-1}, at 20° (Gardner *et al.*, 1998; Hausladen *et al.*, 1998, 2001), and anaerobically by reducing NO to N_2O, with lower activity values between 0.14 to 0.5 s^{-1} (Frey and Kallio, 2003; Kim *et al.*, 1999). The enzymatic mechanism of oxidation and reduction are proposed to occur through a common NO-(nitroxyl anion-bound heme) intermediate (Gardner *et al.*, 1998; Hausladen *et al.*, 1998, 2001).

S. aureus is a pathogenic bacterium with serious impact on human health, being responsible for a variety of community- and hospital-acquired infections, causing from mild to life-threatening systemic diseases, as a result of its capacity to colonize different environmental niches and to survive a diverse range of stresses (Lowy, 1998; Lowy, 2003). The mechanisms by which the bacterium resists or avoids the effects of reactive nitrogen species remain almost unknown. *S. aureus* lacks a flavodiiron-like gene and has only a putative flavohemoprotein, making *S. aureus* a good system for studying the role of Hmp.

This chapter describes the cloning and purification of the recombinant *S. aureus* Hmp followed by its biochemical characterization. The methods used to disrupt the *hmp* gene and to perform the complementation studies are also described, and the physiological function of the protein is analyzed.

2. Preparation of Recombinant *S. aureus* Hmp

Hmp is purified by standard techniques following overproduction of the protein in an *Escherichia coli* host carrying an *hmp*-expressing plasmid.

2.1. Protein expression

To clone the gene encoding *S. aureus* flavohemoglobin, two oligonucleotide primers based on flaking sequences that generated *Nco*I and *Eco*RI restriction sites are used to amplify from the *S. aureus* NCTC8325 chromosomal DNA (Iandolo *et al.*, 2002), isolated as described in the work of Aires de Sousa *et al.* (1996), a 1.2-kb fragment by means of a polymerase chain reaction (PCR). Using standard techniques, the PCR product is cloned into the *Nco*I/*Eco*RI sites of the *E. coli* expression vector pET28a^{+} (Novagen, VWR, Portugal) to yield plasmid pETHmp, which is then sequenced to ensure the integrity of the gene.

Culture (100 ml) of *E. coli* BL21Gold(DE3) cells harboring pETHmp, a strain that efficiently produces recombinant proteins, is grown aerobically for 14 h in Luria Bertani (LB) medium with 30 μg/ml kanamycin. This culture is used as inoculum for growth in a 10-L fermentor (Braun) filled with LB, and supplemented with 3 μM $FeCl_3$, 100 μM riboflavin, and 30 μg/ml kanamycin. When cells reach an optimal density at 600 nm (OD_{600}) of 0.3, 500 μM isopropyl-1-thio-β-D-galactopyranoside (IPTG) and 50 μM aminolevulinic acid are added and growth continued for another 6 h. The fermentor is always maintained at 37° and stirred at 150 rpm using a 5 l/min oxygen supply. The cells are harvested by centrifugation (11,000 × *g* for 6 min, at 4°), resuspended in Tris-HCl 10 m*M* pH 7.5 buffer supplemented with 20 μg/ml DNase, disrupted in a French press (1,000 bar), and ultracentrifuged at 1,38,000 × *g* for 2 h. The *S. aureus* Hmp is isolated from the soluble extract.

2.2. Protein purification

All purification steps are performed under aerobic conditions at 4°, and to avoid flavin degradation the protein is always kept protected from light by using aluminum foil. The soluble extract is dialyzed overnight against buffer A (Tris-HCl 10 m*M* pH 7.6, glycerol 20% and 1 m*M* of phenylmethylsulfonyl fluoride [>99% GC, Sigma]) and applied on to a Q-Sepharose Fast Flow column, previously equilibrated with buffer A. Hmp is eluted at 230 m*M* NaCl and then loaded on to a Superdex 200 gel-filtration column equilibrated with buffer A and supplemented with 150 m NaCl. The protein is reloaded on to a Q-Sepharose High Performance column, and the protein eluted at 200 m*M* NaCl was found to be pure, as judged by SDS-PAGE. Protein concentration is assayed by the bicinchoninic acid (BCA) method with BSA as standard (Smith *et al.*, 1985). The purified protein has an A_{400}/A_{280} ratio of 1. The typical yield from this purification procedure is about 6.5 mg of purified *S. aureus* Hmp per gram of *E. coli* cells (wet weight).

2.3. Physicochemical properties of *S. aureus* Hmp

2.3.1. Characterization of Hmp

S. aureus Hmp contains 381 amino acids and the SDS-PAGE of cell extracts and of the purified Hmp reveals a strong band at ≈45 kDa that agrees with the molecular mass predicted by the gene-derived amino acid sequence and the prosthetic groups. Upon purification, the molecular mass of Hmp is determined in a Superdex 75 (10/300) GL (Tricorn) column equilibrated with Tris-HCl 20 m*M* pH 7.6 and 150 m*M* NaCl buffer, using as standards ovalbumin, bovine serum albumin, horse cytochrome *c*, and desulfoferrodoxin monomer and tetramer forms. The pure protein elutes from a gel-filtration column with an apparent molecular weight of 44 kDa, indicating that it is a monomeric protein. The purified Hmp contains 1 mol of flavin per monomer, as quantified after acid extraction with trichloroacetic acid (Susin *et al.*, 1993), showing that in this preparation the protein is fully loaded in flavin. The heme content is determined by the hemochromopyridine method (Berry and Trumpower, 1987), and the purified protein contains 1 mol heme per monomer.

At room temperature, the UV-visible absorption spectra of *S. aureus* Hmp, acquired in a Shimadzu UV-1603 spectrophotometer, displays the features typical of the canonical Hmps in the oxidized, reduced, and CO-bound states (Fig. 11.1A). In the oxidized state, recombinant *S. aureus* Hmp exhibits absorbance maxima at 646, 461, and 400 nm. After reduction with a saturated solution of sodium dithionite, buffered at pH 9, the spectrum displays maxima at 558 (shoulder at 590 nm) and 435 nm. The reaction of the reduced Hmp with CO yields a spectrum with maxima at 571, 541, and 423 nm. The spectroscopic features of the ferric, ferrous, and carbonmonoxyferrous states of *S. aureus* Hmp resemble those of *E. coli* Hmp (Ioannidis *et al.*, 1992).

The EPR spectrum is acquired on a Bruker ESP 380 spectrometer, equipped with an Oxford Instruments continuous-flow helium cryostat under the following conditions: 10 K, with a microwave frequency of 9.64 GHz, a microwave power of 2.4 mW, and a modulation amplitude of 1 mT. The EPR spectrum of the as-isolated protein displays $g_{y,x,z} = 6.3$, 5.5, 2.0, which are characteristic of high-spin ferric iron.

2.4. Redox titration and analysis

Redox titration of *S. aureus* Hmp (≈11 μM) is determined anaerobically in 50 m*M* Tris-HCl, pH 7.6, by the stepwise addition of buffered sodium dithionite (250 m*M* Tris-HCl at pH 8.0). The following compounds are required as redox mediators, with concentration typically between 0.25 and 0.5 μM: methylene blue (E'_o = 11 mV), indigo tetrasulfonate (E'_o = -30 mV), indigo trisulfonate (E'_o = −70 mV), indigo disulfate

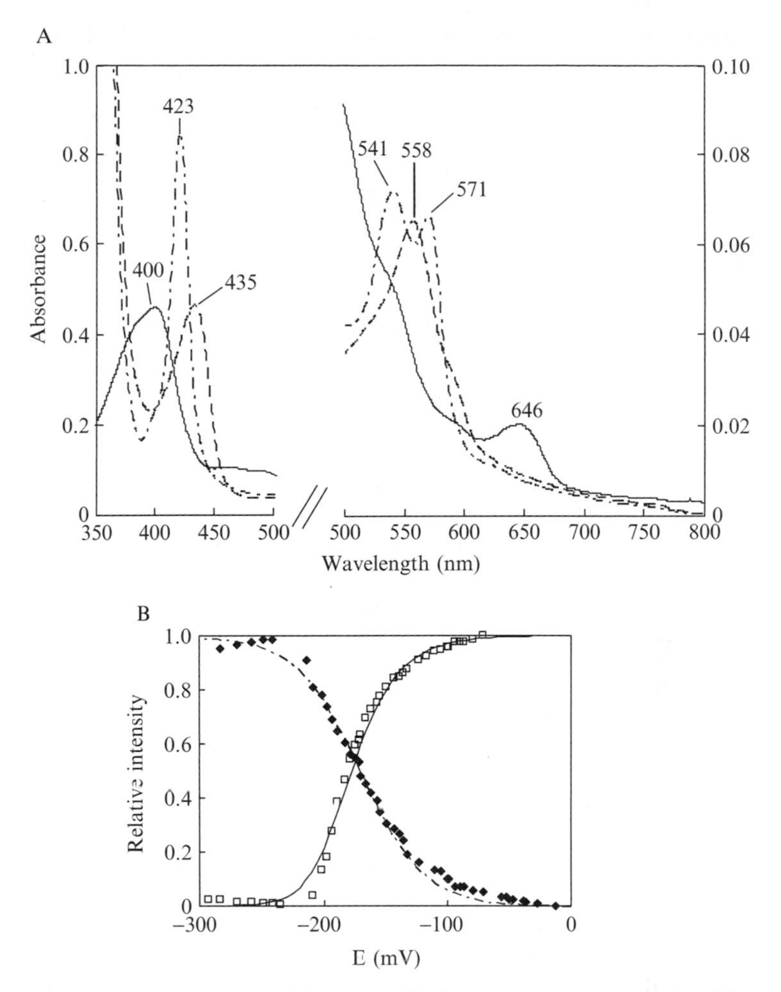

Figure 11.1 Characterization of the as-purified *S. aureus* Hmp. (A) UV/visible spectra of Hmp oxidized (1, —), reduced with sodium dithionite (2, - - -), and reduced and CO ligated (3, -. -. -). (B) Redox titration of the heme center (◆) and of the FAD center (□) of *S. aureus* Hmp. Full line is calculated with the Nernst equation for a monoelectronic (reduction potential of −170 mV) and two monoelectronic consecutive processes (identical reduction potentials of −190 mV), for the heme and FAD centers, respectively.

(E'_o = −182 mV), anthraquinone 2,7-disulfonate (E'_o = −182 mV), safranine (E'_o = −280 mV), neutral red (E'_o = −325 mV), benzyl viologen (E'_o = − 359 mV), and methyl viologen (E'_o = −446 mV). A silver–silver chloride electrode, calibrated against a saturated quinhydrone solution, is used and the reduction potentials are quoted against the standard hydrogen electrode. The experimental data are manipulated and analyzed using MATLAB (Mathworks, South Natick, MA) for Windows, which is also used for the optical deconvolution and the data-fitting calculations. The data are adjusted to monoelectronic Nernst equations, not taking into consideration possible homotropic (electron–electron) interactions between the redox centers. Hmp exhibits a reduction potential of −170 mV for the heme center and two identical reduction potentials of −190 mV for the FAD centers (Fig. 11.1B). These potentials are lower than those reported for *E. coli* Hmp (−120 mV for the heme center and −150 mV for the FAD centers) (Cooper *et al.*, 1994).

2.5. Catalytic properties of *S. aureus* Hmp

The activities of *S. aureus* Hmp are determined in kinetic experiments recorded in a Shimadzu UV-1603 spectrophotometer, equipped with a cell-stirring magnetic system and connected to a water bath that allows the control of temperature. All assays are performed at 25°, in 50 m*M* Tris-HCl, pH 7.5 buffer supplemented with 18% glycerol. Anaerobic conditions are generated by thoroughly degassing the reaction buffer, which is achieved by performing several argon/vacuum cycles. In addition, the assays are done under a flush of argon and in the presence of the oxygen-scavenging system composed by glucose oxidase (4 U/ml), catalase (130 U/ml) and glucose (3 m*M*) to ensure the anaerobicity of the system. Activities are reported in terms of NO consumption, using the proposed stoichiometry of 2 NO molecules per NADH molecule, for both the denitrosylase and the reductase reactions (Gardner, 2005).

The NADH oxidase activity of *S. aureus* Hmp is measured by monitoring the anaerobic NADH consumption, using potassium ferricyanide ($[Fe(CN)_6]^-$) as an artificial electron acceptor. To this end, NADH (200 μM) is anaerobically incubated with ferricyanide (500 μM) and Hmp (30 m*M*) is added to the reaction mixture; a fast zero-order consumption of $[Fe(CN)_6]^-$ is observed, as judged by the absorbance decrease at 420 nm ($\epsilon_{[Fe(CN)6]^-} = 1020\ M^{-1}.cm^{-1}$). Further additions of NADH results in the complete exhaustion of $Fe(CN)_6$. Under these conditions, the *S. aureus* Hmp displays a NADH oxidase activity of 34 s^{-1}.

The NADH:NO oxidoreductase activity is determined by the anaerobic incubation of NADH (200 μM) with Hmp (365 n*M*), and by monitoring the NADH oxidation at 340 nm ($\epsilon_{340\ nm} = 6200\ M^{-1}.cm^{-1}$) upon addition of aliquots of the 2 m*M* NO-saturated water solution, prepared as described in Justino *et al.* (2005). Hmp exhibits a NADH:NO oxidoreductase activity of 0.7 s^{-1}. The NO denitrosylase activity of Hmp (36 n*M*) is measured aerobically by following the NADH oxidation (200 μM), and a NO denitrosylase activity of 66 s^{-1} is measured upon addition of 20 μM of NO (saturated solution). The values of *S. aureus* Hmp activities are within the range usually reported for homologous enzymes (Gardner, 2005; Kim *et al.*, 1999; Poole and Hughes, 2000).

3. Nitrosative Protection of *S. aureus* by Hmp

3.1. RNA extraction and RT-PCR analysis

Total RNA is isolated from cells grown in LB medium under aerobic conditions (in flasks filled with 1/5 of its volume), microaerobically (in closed flasks filled with 1/2 of its volume), or anaerobically (in rubber seal–capped flasks that, once filled with media and closed, are extensively bubbled with nitrogen). *S. aureus* cells are grown in the absence or in the presence of 50 or 200 μM of S-nitrosoglutathione (GSNO) for 4 h. After that, 15 ml of each culture is collected by centrifugation at 7000 $\times$ *g* for 5 min. Cells are lysed by the addition of 0.5 mg/ml of lysozyme (Sigma) and 77 μg/ml of lysostaphin (Sigma) for 15 min at 37°. Then 1% of SDS is added to the cell suspension followed by 2 min of incubation at 64°, after which 0.1 V of 1 m*M* of sodium acetate (NaAc), pH 5.2, is added and the solution is mixed by inversion. The hot-phenol RNA extraction is performed by incubation of the cells at 64° for 6 min with 1 V of phenol (pH 4.5). After centrifugation at 13000 $\times$ *g* for 10 min, the aqueous phase is transferred to a new tube containing 1 V of trichloromethane. The aqueous phase is collected by centrifugation and frozen at −80° for 2 h in 2.5 V of ethanol and 0.1 V of 3 *M* NaAc plus 1 m*M* of EDTA. The total RNA is treated with DNaseI (Roche), and after confirming the absence of any residual DNA in PCR reactions, RT-PCR reactions are performed with 150 ng of RNA using the USB Reverse Transcriptase Kit. The forward and reverse primers used to create the disruption of the *hmp* gene are also utilized in the RT-PCR assays, and the 16S rRNA gene is used to guarantee comparison of equal amounts of RNA.

The transcription of *hmp* was analyzed under different oxygen-growth conditions, and the results reveal that *hmp* has a low level of expression under aerobic conditions, which is significantly enhanced by the decreased oxygen content in the growth conditions (Fig. 11.2). In contrast, it is only under aerobic conditions that we observed a further increase in the induction of *hmp* upon addition of GSNO. Therefore, the major trigger for *hmp* induction is oxygen limitation, in agreement with a possible physiological function under oxygen-limited conditions.

3.2. Construction of *S. aureus hmp* deletion strain

An internal fragment of the *S. aureus hmp* is amplified by PCR, using genomic DNA isolated from *S. aureus* NCTC8325 and two oligonucleotides that carry two restriction sites sequences (italic): SAHECO:5′-GAAAGGACA-*GAATTC*ACGTCAATC-3′ and SAHBAM: 5′-GTTGTCATGAT*GGA-TCC*CGATACT-3′. The ligation of the 883 bp fragment into pSP64D-E,

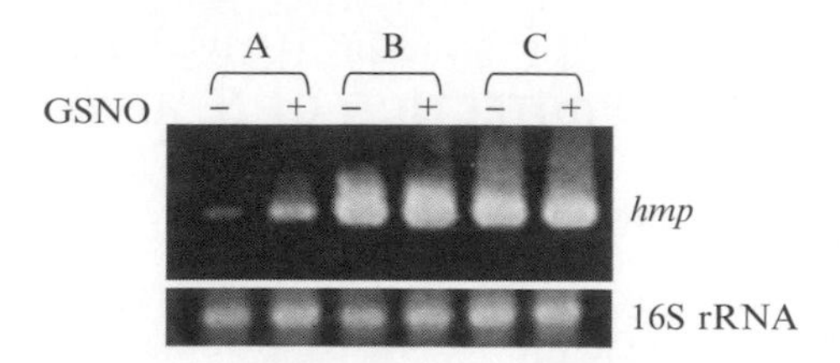

Figure 11.2 *S. aureus hmp* expression increases upon oxygen limitation. RNAs isolated from *S. aureus* grown aerobically (A), microaerobically (B), or anaerobically (C) in LB for 4 h in the absence (−) or presence (+) of 200 μM (A, B) or 50 μM of GSNO (C). The equal loading of total RNA is confirmed by the same intensity of the 16S rRNA band (lower panel). Data are representative of reactions performed with two independent RNA samples.

which contains an erythromycin (Erm) cassette cloned in *XbaI/Bam*HI restriction sites (de Lencastre *et al.*, 1994; Pinho *et al.*, 2000) and was previously digested with the *Eco*RI and *Bam*HI, generates the recombinant pSPHmp.

S. aureus RN4220 cells are prepared for electrotransformation (Kraemer and Iandolo, 1990; Novick *et al.*, 1990). Briefly, an overnight culture of *S. aureus* RN4220 is diluted 1/50 into 100 ml of Tryptic Soy Broth (TSB) and incubated at 37° and 150 rpm until an OD_{600} of 0.5 is reached. At this point cells are collected by centrifugation at 7000 × *g* for 10 min and washed twice with 1 V of 500 m*M* of sucrose (filter sterilized). The cell pellet is resuspended in 300 μl of sucrose (500 m*M*), and aliquots of 50 μl of that suspension are frozen in liquid nitrogen and stored at −80°. The vector pSPHmp (10 μg) is then incubated, on ice, with a 50-μl aliquot of the cell suspension (thawed at room temperature) in a 0.1-cm BioRad Gene Pulser cuvette for 30 min and electroporated in a BioRad MicroPulser using the following conditions: 2.5 KV and 100Ω during 2.5 msec. Immediately after the pulse, 950 μl of SMMP medium (2X strength Penassay broth, 250 m*M* of sucrose, 10 m*M* of maleate, and 10 m*M* of $MgCl_2$, pH 6.5) is added and the suspension is mixed by inversion. After 15 min on ice, the cell suspension is incubated at 37° and 200 rpm for 1 h. Selection of transformants is done on TSB-agar plates containing erythromycin (10 μg/ml). PCR analysis on chromosomal DNA isolated from single colonies is used to confirm the correct integration of pSPHmp into the chromosome of RN4220, using two oligonucleotides: 5′-TCACATTTTTATTATCATGTTTACTTTTTTCTAGGA-3′, located 200 bp upstream of the start codon of the *hmp* gene and 5′-GCTCTAGAACATTCCCTTTAGTAACGTG-3′, located downstream of the Erm cassette. One such colony was designated LMS800 and used in the subsequent studies.

3.3. Protection of *S. aureus* by Hmp in response to GSNO and NO

Evaluation of the role of Hmp in nitrosative-stress protection is done by analyzing the effect of various concentrations of GSNO and oxygen on the growth of *S. aureus hmp* mutant and wild-type strains. To this end, overnight cultures of wild-type *S. aureus* RN4220 and mutant strain grown in TSB or TSB plus 10 μg/ml of Erm, respectively, are used to inoculate LB medium. Cells are then grown aerobically, microaerobically, or anaerobically at 37° and 150 rpm (except the aerobically grown cultures, which are shaken at 180 rpm). The GSNO is added at the beginning of the growth ($OD_{600} \approx 0.1$).

S. aureus wild-type response to GSNO is shown to be dependent on the oxygen-related conditions. It is observed that concentrations of GSNO ranging from 50 to 200 μM cause little effect on microaerobic growth of wild-type *S. aureus* (Fig. 11.3C), while 200 μM GSNO induces a severe growth arrestment under aerobic conditions (Fig. 11.3A). Furthermore, under anaerobic conditions, and for all tested concentrations of GSNO, a significant growth inhibition is always observed (Fig. 11.3E).

S. aureus hmp mutant exhibits an oxygen-dependent growth behavior identical to the wild-type strain (Fig. 11.3B, D). Although under aerobic conditions no differences are observed in the rates and extents of growth between the wild-type and the mutant strain for all the concentrations of GSNO tested (Fig. 11.3B), under microaerobic conditions the exposure to 200 μM of GSNO causes an $\approx$50% decrease in the growth rate of Δ*hmp* mutant strain. On the contrary, the wild-type strain displays only a decrease of no more than $\approx$7% (Fig. 11.3D). A similar behavior is observed for the *S. aureus hmp* mutant in the presence of 50 μM of NO gas (not shown). Thus, Hmp seems to be able to protect *S. aureus* submitted to strong nitrosative stress only under microaerobic conditions.

3.4. Complementation analysis

For the complementation analysis, plasmids pETHmp and pET28a$^+$ are individually transformed into LMS2710 strain. This *E. coli* strain has a deletion on the flavorubredoxin gene (*norV*), and our previous work showed that it is a mutant strain with increased sensitivity to NO under anaerobic conditions (Justino *et al.*, 2005). Single colonies of LMS2710 strain carrying pETHmp or pET28a$^+$ and grown overnight aerobically in LB medium are used to inoculate minimal salt medium (Justino *et al.*, 2005) containing 25 μg/ml of chloramphenicol and 30 μg/ml of kanamycin. The growth is performed under anaerobic conditions, at 37° and 150 rpm, and monitored at 600 nm. Expression of *S. aureus* Hmp in the *E. coli norV* mutant, LMS2710, leads to a significant increase in the anaerobic GSNO resistance of the mutant strain (Fig. 11.3F), showing that, under these conditions, *S. aureus* Hmp has the ability to perform anaerobic nitrosative detoxification.

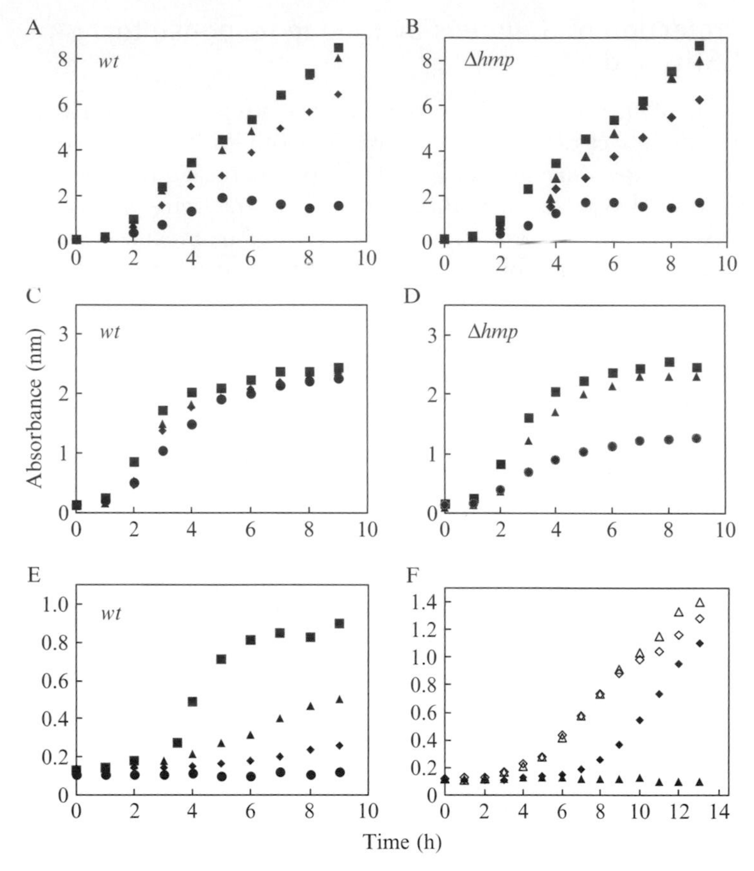

Figure 11.3 Hmp confers microaerobic GSNO-protection to *S. aureus*. *S. aureus* wild-type strain RN4220 (*wt*) and mutant strain LMS800 (Δ*hmp*), are grown aerobically (A, B), microaerobically (C, D), or anaerobically (E) in LB medium and left untreated (▪) or treated with 50 μM GSNO (▲), 100 μM GSNO (◆), and 200 μM GSNO (•). Each growth curve represents the average of at least three independent cultures. Panel (F): analysis of complementation of *E. coli* LMS2710 (Δ*norV*) with *S. aureus* Hmp. Growth curves are acquired in minimal medium under anaerobic conditions for the *E. coli* flavorubredoxin mutant strain LMS2710 harboring either vector alone (pET) (Δ) and with pETHmp expressing *S. aureus* Hmp (□) without addition of GSNO or in the presence of 50 μM GSNO: LMS2710 (pET) (π) and LMS2710 (pETHmp) (v).

4. Discussion

The amino acid sequence of the *S. aureus* Hmp exhibits 30%, 33%, and 40% sequence identity with Hmp of *E. coli*, *Alcaligenes eutrophus*, and *Saccharomyces cerevisiae*, respectively. Sequence comparison (Fig. 11.4) shows that among the several conserved residues is the proximal histidine

```
                 *    *       *         *                   ** *
Saureus     MLTEQEKDIIKQTVPLLKEKGTEITSIFYPKMFKAHPELLNMFNQTNQKRGMQSSALAQA :60
Ecoli       MLDAQTIATVKATIPLLVETGPKLTAHFYDRMFTHNPELKEIFNMSNQRNGDQREALFNA :60
Aeutropha   MLTQKTKDIVKATAPVLAEHGYDIIKCFYQRMFEAHPELKNVFNMAHQEQGQQQQALARA :60
Scerevisae  MLAEKTRSIIKATVPVLEQQGTVITRTFYKNMLTEHTELLNIFNRTNQKVGAQPNALATT :60
              *
Saureus     VMAAAVNIDNLSVIKPVIMPVAYKHCALQVYAEHYPIVGKNLLKAIQDVTGLEENDPVIQ :120
Ecoli       IAAYASNIENLPALLPAVEKIAQKHTSF-QKPEQYNIVGEHLLATLDEMFS--PGQEVLD :117
Aeutropha   VYAYAENIEDPNSLMAVLKNIANKHASLGVKPEQYPIVGEHLLAAIKEVLGNAATDDIIS :120
Scerevisae  VLAAAKNIDDLSVLMDHVKQIGHKHRALQIKPEHYPIVGEYLLKAIKEVLGDAATPEIIN :120

Saureus     AWAKAYGVIADVFIQIEKEIYDQMM-----WIGFKPFKITNIKQESEDIKSFTVET---- :171
Ecoli       AWGKAYGVLANVFINREAEIYNENASKAGGWEGTRDFRIVAKTPRSALITSFELEP---- :173
Aeutropha   AWAQAYGNLADVLMGMESELYERSAEQPGGWKG RTFVIREKRPESDVITSFILEP---- :174
Scerevisae  AWGEAYQAIADIFITVEKKMYEEAL-----WPG KPFDITAKEYVASDIVEFTVKPKFGS :175

Saureus     -EEYDFSEFTPGQYITVDVS--SDKLPYRAKRHYSIVSG-EKNHLTFGVKRDVTTEH--E :225
Ecoli       VDGGAVAEYRPGQYLGV LK--PEGFPHQEIRQYSLTRKPDGKGYRIAVKREEGGQ---- :227
Aeutropha   ADGGPVVNFEPGQYTSVAID--VPALGLQQIRQYSLSDMPNGRTYRISVKREGGGPQ-PP :231
Scerevisae  GIELESLPITPGQYITVNTHPIRQENQYDALRHYSLCSASTKNGLRFAVKMEAARENFPA :235

Saureus     GEVSTILHDEIKEGDMINLAAPVGGFVL-----ENTTEPQLFLGSGIGVTPLVAMYEAAS :280
Ecoli       --VSN LHNHANVGDVVKLVAPAGDFFMA----VADDTPVTLISAGVGQTPMLAMLDTLA :281
Aeutropha   GYVSNLLHDHVNVGDQVKLAAPYGSFHID----VDAKTPIVLISGGVGLTPMVSMLKVAL :287
Scerevisae  GLVSEYLHKDAKVGDEIKLSAPAGDFAINKELIHQNEVPLVLLSSGVGVTPLLAMLEEQV :295

Saureus     AKGLDT--QMVQVAENEQHLPFKDNFNSIASHHDNAKLYTHLK-------------DKQGY :326
Ecoli       KAGHTAQVN FHAAENGDVHAFADEVKELGQSLPRFTAHT YRQPSEADRAKGQFDSEGL :341
Aeutropha   QA-PPRQVVFVHGARNSAVHAMRDRLREAAKTYENLDLFVFYDQPLPEDVQGRDYDYPGL :346
Scerevisae  KCNPNRPIY IQSSYDEKTQAFKKHVDELLAECANVDKIIVHT-------------DTEPL :343

Saureus     IGAEELQ-VFLANKPEIYICGGTKFLQSMIEALKSLNYDMDRVHYETFIPRLSVAV:   381
Ecoli       MDLSKLEGAFSDPTMQFYLCGPVGFMQFTAKQLVDLGVKQENIHYECFGPHKVL:     396
Aeutropha   VDVKQIEKSILLPDADYYICGPIPFMRMQHDALKNLGIHEARIHYEVFGPDLFAE    :403
Scerevisae  INAAFLK-EKSPAHADVYTCGSLAFMQAMIGHLKELEHRDDMIHYEPFGPKMSTVQV  :399
```

Figure 11.4 Amino acid sequence comparison of flavohemoglobins of *S. aureus*, *E. coli (Ecoli)*, *A. eutropha (Aeutropha)*, and *S. cerevisiae (Scerevisiae)*. Asterisks refer to amino acid residues involved in ligation to the phospholipids in *A. eutropha* (Ermler *et al.*, 1995).

ligand His-85 (considering the residue numbering of the *S. aureus* sequence). In *E. coli*, the two distal pocket residues Tyr-29 and Gln-53 are proposed to determine the functional properties of most bacterial hemoglobins (Ilari *et al.*, 2002). In particular, the nature of the Tyr-29 residue is considered responsible for establishing an interaction with diatomic ligands, controlling the oxygen release kinetics of the protein. In *S. aureus* these residues, Tyr-29 and Gln-53, are also conserved, together with two other amino acid residues, Phe-43 and Leu-57, which are usually strictly conserved in bacterial Hmps (Ilari *et al.*, 2002).

The three-dimensional structure of the *E. coli* Hmp (Ilari *et al.*, 2002) was used to predict by molecular modeling techniques the structure of *S. aureus* Hmp. The results indicate that *S. aureus* Hmp has a similar overall folding, with no major rearrangements (not shown). In particular, the residues that in *E. coli* dominate the architecture of the proximal region of the heme pocket (His 85, Tyr95, and Glu139) (Ilari *et al.*, 2002) are also structurally conserved in the *S. aureus* protein. The residues that in *A. eutrophus* Hmp are in van der Waals contact with the phospholipids (Ermler *et al.*, 1995) are structurally and sequentially maintained in *S. aureus* Hmp (see Fig. 11.4), thus suggesting that binding of a lipid may also occur in the *S. aureus* protein. The similarity between *S. aureus* Hmps and the other hemoglobins suggest that the physicochemical properties of *S. aureus* should not differ substantially, which was confirmed by the spectroscopic and biochemical data acquired for *S. aureus* Hmp. In particular, the kinetic parameters of the NO activities are within the range of values measured for homologous proteins (Gardner, 2005; Kim *et al.*, 1999; Poole and Hughes, 2000).

Although it is now evident that many organisms employ Hmps to metabolize NO, the mechanism by which they eliminate NO and the conditions under which they confer protection are still controversial. Our work in *S. aureus* reveals that protection occurs exclusively under micro-aerobic conditions, and under those conditions, the NO dioxygenase activity does not operate. Thus, the data indicate that at least in *S. aureus* the *in vitro* dioxygenase mechanism is unrelated to the *in vivo* function, and Hmp seems to be only able to act as a denitrosylase. The fact that the *S. aureus* enzyme is not a dioxygenase under *in vivo* conditions is most probably related to the requirements of the NO dioxygenase reaction (Gardner *et al.*, 1998; Hausladen *et al.*, 1998) for very high (nonphysiological) O_2 concentrations and very low (nontoxic) NO concentrations, because transcription of *S. aureus hmp* is not induced under high O_2 conditions (see Fig. 11.2) and NO is not toxic to *S. aureus* unless it is present in high micromolar concentrations (see Fig. 11.3). Thus, the measured NO dioxygenase activity is an artifact of the *in vitro* conditions.

Anaerobic protection by Hmp was first seen in yeast (Liu *et al.*, 2000), followed by observations that Hmp protects the anaerobic growth of *Salmonella enterica* serovar Typhimurium against GSNO (Bang *et al.*, 2006) and is necessary for the anaerobic survival of *Bacillus subtilis* (Nakano, 2006). We also observed that elevated expression of *S. aureus* Hmp is able to protect *E. coli* against NO-mediated growth inhibition. Altogether these results suggest that the microbial function of Hmps as denitrosylase and anaerobic NO reductases cannot be ruled out and are of extreme importance, because the pathogenicity, in particular the production of virulence factors, for many organisms is highly dependent on the oxygen concentration and related to the oxygen levels in human tissues (Ross and Onderdonk, 2000; Yarwood *et al.*, 2001).

ACKNOWLEDGMENTS

We thank Prof. Hermínia de Lencastre (ITQB) and Prof. Ana M. Ludovice (ITQB) for providing the *S. aureus* strains and plasmid pSP64D-E and for technical advice, Dr. Claúdio Soares for the molecular modeling studies, and Prof. Miguel Teixeira for the EPR studies. Lígia S. Nobre and Vera L. Gonçalves are recipients of grants from FCT, respectively SFRH/BD/22425/2005 and SFRH/BD/29428/2006. This work was supported by FCT projects POCTI/2002/BME/44597 and POCI/SAU-IMI/56088/2004.

REFERENCES

Aires de Sousa, M., Sanches, I. S., van Belkum, A., van Leeuwen, W., Verbrugh, H., and de Lencastre, H. (1996). Characterization of methicillin-resistant *Staphylococcus aureus* isolates from Portuguese hospitals by multiple genotyping methods. *Microb. Drug Resist.* **2,** 331–341.

Bang, I. S., Liu, L., Vazquez-Torres, A., Crouch, M. L., Stamler, J. S., and Fang, F. C. (2006). Maintenance of nitric oxide and redox homeostasis by the *Salmonella* flavohemoglobin Hmp. *J. Biol. Chem.* **281,** 28039–28047.

Berry, E. A., and Trumpower, B. L. (1987). Simultaneous determination of hemes *a*, *b*, and *c* from pyridine hemochrome spectra. *Anal. Biochem.* **161,** 1–15.

Cooper, C. E., Ioannidis, N., D'mello, R., and Poole, R. K. (1994). Haem, flavin and oxygen interactions in Hmp, a flavohaemoglobin from *Escherichia coli*. *Biochem. Soc. Trans.* **22,** 709–713.

de Lencastre, H., Couto, I., Santos, I., Melo-Cristino, J., Torres-Pereira, A., and Tomasz, A. (1994). Methicillin-resistant *Staphylococcus aureus* disease in a Portuguese hospital: Characterization of clonal types by a combination of DNA typing methods. *Eur. J. Clin. Microbiol. Infect. Dis.* **13,** 64–73.

Ermler, U., Siddiqui, R. A., Cramm, R., and Friedrich, B. (1995). Crystal structure of the flavohemoglobin from *Alcaligenes eutrophus* at 1.75Å resolution. *EMBO J.* **14,** 6067–6077.

Frey, A. D., and Kallio, P. T. (2003). Bacterial hemoglobins and flavohemoglobins: Versatile proteins and their impact on microbiology and biotechnology. *FEMS Microbiol. Rev.* **27,** 525–545.

Gardner, A. M., and Gardner, P. R. (2002). Flavohemoglobin detoxifies nitric oxide in aerobic, but not anaerobic, *Escherichia coli*: Evidence for a novel inducible anaerobic nitric oxide-scavenging activity. *J. Biol. Chem.* **277,** 8166–8171.

Gardner, P. R. (2005). Nitric oxide dioxygenase function and mechanism of flavohemoglobin, hemoglobin, myoglobin and their associated reductases. *J. Inorg. Biochem.* **99,** 247–266.

Gardner, P. R., Gardner, A. M., Martin, L. A., and Salzman, A. L. (1998). Nitric oxide dioxygenase: An enzymic function for flavohemoglobin. *Proc. Natl. Acad. Sci. USA* **95,** 10378–10383.

Hausladen, A., Gow, A. J., and Stamler, J. S. (1998). Nitrosative stress: Metabolic pathway involving the flavohemoglobin. *Proc. Natl. Acad. Sci. USA* **95,** 14100–14105.

Hausladen, A., Gow, A., and Stamler, J. S. (2001). Flavohemoglobin denitrosylase catalyzes the reaction of a nitroxyl equivalent with molecular oxygen. *Proc. Natl. Acad. Sci. USA* **98,** 10108–10112.

Iandolo, J. J., Worrell, V., Groicher, K. H., Qian, Y., Tian, R., Kenton, S., Dorman, A., Ji, H., Lin, S., Loh, P., Qi, S., *et al.* (2002). Comparative analysis of the genomes of the temperate bacteriophages phi 11, phi 12 and phi 13 of *Staphylococcus aureus* 8325. *Gene* **289,** 109–118.

Ilari, A., Bonamore, A., Farina, A., Johnson, K. A., and Boffi, A. (2002). The X-ray structure of ferric *Escherichia coli* flavohemoglobin reveals an unexpected geometry of the distal heme pocket. *J. Biol. Chem.* **277,** 23725–23732.

Ioannidis, N., Cooper, C. E., and Poole, R. K. (1992). Spectroscopic studies on an oxygen-binding haemoglobin-like flavohaemoprotein from *Escherichia coli*. *Biochem. J.* **288,** 649–655.

Justino, M. C., Vicente, J. B., Teixeira, M., and Saraiva, L. M. (2005). New genes implicated in the protection of anaerobically grown *Escherichia coli* against nitric oxide. *J. Biol. Chem.* **280,** 2636–2643.

Kim, S. O., Orii, Y., Lloyd, D., Hughes, M. N., and Poole, R. K. (1999). Anoxic function for the *Escherichia coli* flavohaemoglobin (Hmp): Reversible binding of nitric oxide and reduction to nitrous oxide. *FEBS Lett.* **445,** 389–394.

Kraemer, G. R., and Iandolo, J. J. (1990). High-frequency transformation of *Staphylococcus aureus* by electroporation. *Curr. Microbiol.* **21,** 373–376.

Liu, L., Zeng, M., Hausladen, A., Heitman, J., and Stamler, J. S. (2000). Protection from nitrosative stress by yeast flavohemoglobin. *Proc. Natl. Acad. Sci. USA* **97,** 4672–4676.

Lowy, F. D. (1998). *Staphylococcus aureus* infections. *N. Engl. J. Med.* **339,** 520–532.

Lowy, F. D. (2003). Antimicrobial resistance: The example of *Staphylococcus aureus*. *J. Clin. Invest.* **111,** 1265–1273.

Membrillo-Hernandez, J., Coopamah, M. D., Channa, A., Hughes, M. N., and Poole, R. K. (1998). A novel mechanism for upregulation of the *Escherichia coli* K-12 *hmp* (flavohaemoglobin) gene by the "NO releaser," S-nitrosoglutathione: Nitrosation of homocysteine and modulation of MetR binding to the *glyA-hmp* intergenic region. *Mol. Microbiol.* **29,** 1101–1112.

Nakano, M. M. (2006). Essential role of flavohemoglobin in long-term anaerobic survival of *Bacillus subtilis*. *J. Bacteriol.* **188,** 6415–6418.

Novick, R., Kornblum, J., Kreiswirth, B., Projan, S., and Ross, H. (1990). Regulation of post-exponential-phase exoprotein synthesis in *Staphylococcus aureus*. *In* "Microbial determinants of virulence and host response" (T. J. Henry, ed.), pp. 3–18. American Society for Microbiology, Washington DC.

Pinho, M. G., de Lencastre, H., and Tomasz, A. (2000). Cloning, characterization, and inactivation of the gene *pbpC*, encoding penicillin-binding protein 3 of *Staphylococcus aureus*. *J. Bacteriol.* **182,** 1074–1079.

Poole, R. K., and Hughes, M. N. (2000). New functions for the ancient globin family: Bacterial responses to nitric oxide and nitrosative stress. *Mol. Microbiol.* **36,** 775–783.

Poole, R. K., Anjum, M. F., Membrillo-Hernandez, J., Kim, S. O., Hughes, M. N., and Stewart, V. (1996). Nitric oxide, nitrite, and Fnr regulation of *hmp* (flavohemoglobin) gene expression in *Escherichia coli* K-12. *J. Bacteriol.* **178,** 5487–5492.

Ross, R. A., and Onderdonk, A. B. (2000). Production of toxic shock syndrome toxin 1 by *Staphylococcus aureus* requires both oxygen and carbon dioxide. *Infect. Immun.* **68,** 5205–5209.

Saraiva, L. M., Vicente, J. B., and Teixeira, M. (2004). The role of flavodiiron proteins in nitric oxide detoxification. *Adv. Microb. Physiol.* **49,** 77–130.

Smith, P. K., Krohn, R. I., Hermanson, G. T., Mallia, A. K., Gartner, F. H., Provenzano, M. D., Fujimoto, E. K., Goeke, N. M., Olson, B. J., and Klenk, D. C. (1985). Measurement of protein using bicinchoninic acid. *Anal. Biochem.* **150,** 76–85.

Susin, S., Abian, J., Sanchez-Baeza, F., Peleato, M. L., Abadia, A., Gelpi, E., and Abadia, J. (1993). Riboflavin 3′- and 5′-sulfate, two novel flavins accumulating in the roots of iron-deficient sugar beet (*Beta vulgaris*). *J. Biol. Chem.* **268,** 20958–20965.

Wu, G., Wainwright, L. M., Membrillo-Hernandez, J., and Poole, R. K. (2004). Bacterial hemoglobins: Old proteins with new functions? Roles in respiratory and nitric oxide metabolism. *In* "Respiration in Archaea and bacteria diversity of prokaryotic electron transport carriers" (Zanoni Davide, ed.), pp. 251–284. Kluwer Academic Publishers, Dordecht, The Netherlands.

Yarwood, J. M., McCormick, J. K., and Schlievert, P. M. (2001). Identification of a novel two-component regulatory system that acts in global regulation of virulence factors of *Staphylococcus aureus*. *J. Bacteriol.* **183,** 1113–1123.

CHAPTER TWELVE

Assay and Characterization of the NO Dioxygenase Activity of Flavohemoglobins

Paul R. Gardner

Contents

Abstract

A variety of hemoglobins, including several microbial flavohemoglobins, enzymatically dioxygenate the free radical nitric oxide (•NO) to form nitrate. Many of these •NO dioxygenases have been shown to control •NO toxicity and signaling. Furthermore, •NO dioxygenation appears to be an ancient and intrinsic function for members of the hemoglobin superfamily found in Archaea, eukaryotes, and bacteria. Yet for many hemoglobins, a function remains to be elucidated.

Department of Chemistry, University of Dayton, Dayton, Ohio

Methods in Enzymology, Volume 436
ISSN 0076-6879, DOI: 10.1016/S0076-6879(08)36012-1

Methods for the assay and characterization of the •NO dioxygenase (EC 1.14.12.17) activity and function of flavohemoglobins are described. The methods may also be applied to the discovery and design of inhibitors for use as antibiotics or as modulators of •NO signaling.

1. Introduction

Evidence suggests that the earliest hemoglobin functioned not as an O_2 transport or storage protein, but as an enzyme with the capacity to dioxygenate and detoxify •NO (Gardner, 2005; Gardner *et al.*, 1998b; Wu *et al.*, 2003). Moreover, this primitive enzymatic activity and function appears to have been retained in the modern O_2 transport or storage hemoglobin and myoglobin during approximately 2 billion years of evolution (Miranda *et al.*, 2005; Vinogradov *et al.*, 2006). Thus, several of the hemoglobins show a capacity for •NO dioxygenation, albeit with varying efficiencies. Bimolecular rate constants for the reaction of •NO with oxyhemoglobins vary over a ≈50-fold range from a near diffusion-limited $\approx 2 \times 10^9\ M^{-1}s^{-1}$ for the microbial flavohemoglobins at 37° (Gardner *et al.*, 2000b) to $\approx 4 \times 10^7\ M^{-1}s^{-1}$ for sperm whale myoglobin at 20° (Eich *et al.*, 1996). Nitrate is produced in •NO reactions with oxyflavohemoglobins with both atoms of O incorporated from O_2 (Gardner *et al.*, 2006) [Eq. (12.1)].

$$\bullet NO + HbFe(II)O_2 \rightarrow HbFe(III) + NO_3- \qquad (12.1)$$

2. The •NO Dioxygenase Mechanism

The proposed dioxygenase mechanism for the enzymatic conversion of •NO to nitrate by hemoglobins and myoglobins was recently reviewed and will not be extensively discussed here (Gardner, 2005; Gardner *et al.*, 2006; Olson *et al.*, 2004). Here, I focus on the methods used for characterization of a •NO dioxygenase activity as well as the issues and difficulties one can expect to encounter when trying to assign a •NO dioxygenase function to various flavohemoglobins. Nevertheless, it is important to note that several defining attributes of hemoglobin appear critical for its ability to function as a •NO dioxygenase. These include (1) a high O_2 affinity [Eq. (12.2)], (2) a superoxide radical-like character of the heme-bound O_2 [Eq. (12.3)], (3) a protein "cage" or protected pocket for the high-fidelity iron-catalyzed isomerization of the peroxynitrite intermediate to nitrate [Eq. (12.4)], and (4) a mechanism for univalent reduction of the oxidized methemoglobin [Eq. (12.5)].

$$Fe(II) + O_2 \Leftrightarrow Fe(II)O_2 \tag{12.2}$$

$$Fe(II)O_2 \Leftrightarrow Fe(III)^{-}O_2\bullet \tag{12.3}$$

$$Fe(III)^{-}O_2\bullet + \bullet NO \rightarrow [Fe(III)^{-}OONO] \rightarrow Fe(III) + NO_3^- \tag{12.4}$$

$$Fe(III) + e^- \rightarrow Fe(II) \tag{12.5}$$

3. Susceptibility of the •NO Dioxygenase to •NO Inhibition

In order for a hemoglobin to function as an efficient •NO dioxygenase, it must avoid or limit the inhibition caused by the high affinity binding of •NO to the ferrous heme [Eq. (12.6)] in place of the substrate O_2 [Eq. (12.2)] (Gardner *et al.*, 2000a,b). Flavohemoglobins can avoid, but not totally escape, •NO inhibition by binding O_2 with high affinity or by reducing •NO to nitroxyl (NO^-) [Eq. (12.7)].

The importance of a high O_2 affinity is best illustrated by the effect of mutating the highly conserved distal B10 tyrosine to a phenylalanine in *E. coli* flavohemoglobin. The mutation increases the O_2 dissociation rate constant ≈80-fold, increases the dissociation equilibrium constant for O_2 ≈56-fold, and accordingly increases the susceptibility of the •NO dioxygenase activity to •NO inhibition (Gardner *et al.*, 2000a). Full activity is achieved by elevating O_2 and decreasing the •NO:O_2 concentration ratio.

Complicating matters, direct measurements of the O_2 and •NO affinities of flavohemoglobins using laser photolysis and CO or •NO competition can belie the ability of flavohemoglobins to function as •NO dioxygenases. These measurements may predict that •NO outcompetes O_2 and inhibits the •NO dioxygenase activity at far lower •NO:O_2 ratios than are actually observed.

For example, the measured equilibrium constants for the *E. coli* flavohemoglobin suggest an equal competition between •NO and O_2 for the ferrous heme at an •NO:O_2 concentration ratio of approximately 1:1500 (Gardner *et al.*, 2000a,b). However, inhibition of the •NO dioxygenase activity is apparent only at •NO:O_2 concentration ratios of >1:100. Below a ratio of 1:100 the flavohemoglobin shows high catalytic efficiency with a near diffusion-limited k_{cat}/K_M(•NO) value of $\approx 2.4 \times 10^9\ M^{-1}s^{-1}$.

In fact, these equilibrium data prompted others (Hausladen *et al.*, 2001) to suggest an O_2 nitrosylase or denitrosylase mechanism for •NO metabolism by flavohemoglobin in which •NO first binds the ferrous heme and O_2 then reacts with the Fe(II)•NO. However, this proposed revision of the mechanism is clearly inconsistent with the steady-state kinetic data. In an O_2-nitrosylase mechanism, CO would be expected to compete with •NO for the ferrous heme. However, CO clearly competes directly with O_2 with a K_i value of ≈ 1 μM (Gardner *et al.*, 2000b) (Fig. 12.1A), and shows uncompetitive inhibition with respect to •NO in double-reciprocal plots (Fig. 12.1B). Also, note the apparent inhibition by •NO as the •NO:O_2 concentration ratio exceeds 1:100 (0.2 μM •NO : 20 μM O_2) (see Fig 12.1B, *line 1*). Inhibition would not be expected at increasing •NO:O_2 concentration ratios for an O_2-nitrosylase mechanism. Moreover, mutation of B10 tyrosine to a phenylalanine impairs O_2 binding and •NO metabolic activity without affecting •NO affinity, which is best explained by a dioxygenase mechanism (Gardner *et al.*, 2000a). Nevertheless, the >15-fold discordance of the •NO and O_2 affinity measurements with steady-state dioxygenase activity measurements of flavohemoglobins, including the >10-fold discordant $K_M(O_2)$ values, clearly requires a resolution. For

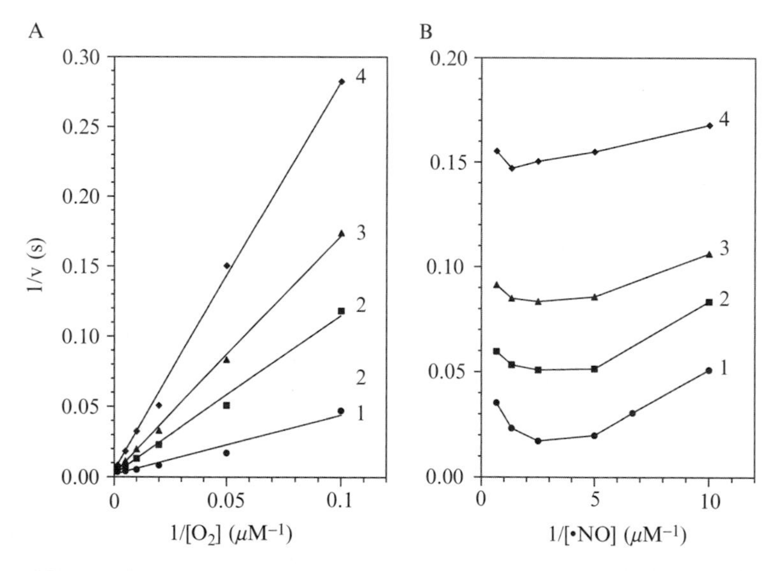

Figure 12.1 Effect of CO on the •NO dioxygenase activity of flavohemoglobin. (A) •NO dioxygenase activity of the *E. coli* flavohemoglobin was measured at 0.4 μM •NO with different O_2 concentrations in the presence of 0 μM CO (*line 1*), 0.5 μM CO (*line 2*), 1.0 μM CO (*line 3*), and 2.0 μM CO (*line 4*). (B) •NO dioxygenase activity was measured at various •NO concentrations with 20 μM O_2 and 0 μM CO (*line 1*), 0.5 μM CO (*line 2*), 1.0 μM CO (*line 3*), and 2.0 μM CO (*line 4*). Reactions were at 37° in 100 mM sodium phosphate buffer, pH 7.0, containing 0.3 mM EDTA with 100 μM NADH and 1 μM FAD.

mammalian myoglobins and hemoglobins, the ratio of •NO/O_2 affinities is even greater, ≈200,000, thus making the susceptibility to •NO inhibition even more substantial than that for the bacterial flavohemoglobins.

Flavohemoglobins also possess a unique capacity for •NO reduction, and this may partly account for a lower than anticipated •NO inhibition. •NO reduction generates nitroxyl anion [Eq. (12.7)], and two nitroxyl anions ultimately form N_2O (Kim *et al.,* 1999; Poole and Hughes, 2000). Measured •NO reductase turnover rates of flavohemoglobins vary between 0.02 and 0.24 •NO per heme per second (Gardner *et al.,* 2000b; Kim *et al.,* 1999). Although the maximal rates of •NO reduction are relatively low compared to maximal •NO dioxygenase turnover rates (670 s^{-1}), the activity may be important at low O_2 concentrations and high •NO concentrations, where •NO inhibition is more prominent, and •NO dioxygenase turnover rates are lower.

$$\mathrm{Fe(II)} + \bullet\mathrm{NO} \Leftrightarrow \mathrm{Fe(II)} \bullet \mathrm{NO} \quad (12.6)$$

$$\mathrm{Fe(II)} \bullet \mathrm{NO} \rightarrow \mathrm{Fe(III)} + \mathrm{NO}^- \quad (12.7)$$

Finally, when testing for a •NO-dioxygenase function of a candidate flavohemoglobin, it is critical to evaluate for •NO dioxygenase activity and function over the full range of physiologically relevant concentrations of •NO and O_2. For example, •NO acts as a signal molecule in the range of 1 n*M* but clearly acts as a toxin at >50 n*M* and can approach 1 μM under extreme nitrosative-stress conditions. Thus, a hemoglobin or myoglobin may function well at low nanomolar •NO concentrations where available O_2 adequately competes with •NO for binding to the ferrous heme and turnover rates are low, but the enzyme may be inhibited and fail at higher •NO concentrations. By scavenging •NO at low concentrations, a hemoglobin may modulate •NO signaling and reduce chronic •NO toxicity caused by its attack of critical and sensitive targets such as aconitase (Gardner *et al.,* 1997) or through its reaction with superoxide radical to form peroxynitrite [see Eq. (12.8)] (Beckman and Koppenol, 1996). In the latter reaction, $HbFe(III)^{-}O_2\bullet$ would directly compete with $^{-}O_2\bullet$ for •NO. Given that there are few paths for •NO consumption in cells at nanomolar •NO levels, one can estimate the flux of •NO going to hemoglobin versus $^{-}O_2\bullet$ from the second-order rate constants and the concentrations of reactants. •NO reacts with free $^{-}O_2\bullet$ with a second–order rate constant of $6.7 \times 10^9\ M^{-1}s^{-1}$ (Beckman and Koppenol, 1996; Huie and Padmaja, 1993). The cellular $^{-}O_2\bullet$ concentration is normally maintained in the range of 10^{-10} *M* (Fridovich, 1978; Gardner, 2002). Thus, for an oxyhemoglobin reaction [Eq. (12.4)] showing a second-order rate constant of $10^8\ M^{-1}s^{-1}$, 99% of the •NO flux to $^{-}O_2\bullet$ and toxic peroxynitrite formation [Eq. (12.8]) would be diverted to oxyhemoglobin and nitrate formation at an oxyhemoglobin concentration of only $\approx 0.7\ \mu M$.

$$\bullet NO + {}^{-}O_2\bullet \rightarrow {}^{-}OONO \qquad (12.8)$$

When evaluating a potential •NO dioxygenase function, it is also important to realize that multiple •NO dioxygenases (and other •NO-metabolizing enzymes) may be expressed within the same organism or cell. These enzymes may show different optima for [•NO] and [O_2], employ unique reducing systems, and localize to different subcellular compartments (Hallstrom *et al.*, 2004). Moreover, the •NO metabolic activities may be complementary and overlapping, like the •NO dioxygenase and •NO reductase in bacteria (Gardner *et al.*, 2002; Gardner and Gardner, 2002). It is noteworthy that such a scenario exists for the metabolism and detoxification of hydrogen peroxide by multiple peroxidases and catalase in cells.

4. Autooxidation of Hemoglobins and •NO Decomposition

In addition, mechanisms for limiting autooxidation and ${}^{-}O_2\bullet$ generation by hemoglobin [Eq. (12.9)] are critical for an •NO dioxygenase function. High rates of ${}^{-}O_2\bullet$ release necessarily increase cellular ${}^{-}O_2\bullet$ levels, which is counterproductive for organisms in the absence of sufficient superoxide dismutase to scavenge the excess toxic ${}^{-}O_2\bullet$ (Fridovich, 1995). •NO reacts with free ${}^{-}O_2\bullet$ with a second-order rate constant of 6.7×10^9 $M^{-1}s^{-1}$ to form highly reactive and toxic peroxynitrite [Eq. (12.8)] (Beckman and Koppenol, 1996; Huie and Padmaja, 1993). Thus, •NO can effectively outcompete superoxide dismutase, which efficiently dismutates ${}^{-}O_2\bullet$ with a second-order rate constant of $\approx 2 \times 10^9$ $M^{-1}s^{-1}$ [Eq. (12.10)] (Fridovich, 1995). The rapid reaction of •NO with free ${}^{-}O_2\bullet$ can be especially problematic in assays for •NO dioxygenase activity where ${}^{-}O_2\bullet$ is generated either *in vitro* or within cells or tissues. Free ${}^{-}O_2\bullet$ can consume •NO faster than the oxyhemoglobins.

$$Fe(II)O_2 \Leftrightarrow Fe(III){}^{-}O_2\bullet \rightarrow Fe(III) + {}^{-}O_2\bullet \qquad (12.9)$$

$$2{}^{-}O_2\bullet + 2H^+ \rightarrow O_2 + H_2O_2 \qquad (12.10)$$

Using the previous rate constants, one can calculate that for >90% of the ${}^{-}O_2\bullet$ to be removed via dismutation rather than reaction with •NO, superoxide dismutase needs to be present at a >34-fold molar excess over •NO. Within cells, reaction of •NO with free $^-O_2$• will be limited by the endogenous superoxide dismutases present at concentrations of 1–10 μM (Fridovich, 1978). However, even at 10 μM superoxide dismutase, >10% of the $^-O_2$• flux will go toward peroxynitrite formation when [•NO] exceeds 300 nM. Hemoglobins acting as efficient •NO-scavenging NO dioxygenases lower the steady-state [•NO] levels within cells and tissues and thus collaborate with the superoxide dismutases to avert toxic peroxynitrite formation. Averting peroxynitrite damage may be particularly important in sensitive tissues exposed to sustained fluxes of •NO and $^-O_2$•.

5. Methemoglobin Reduction

Perhaps no issue demands greater attention than the cellular reducing systems for the various single-domain hemoglobins and myoglobins acting as dioxygenases [Eq. (12.5)]. In the erythrocyte, hemoglobin is reduced by a FAD-containing methemoglobin reductase and cytochrome b_5 (Hultquist and Passon, 1971), but the reducing system for myoglobin acting as a •NO dioxygenase within myocytes remains poorly defined (Flögel *et al.*, 2004; Flögel *et al.*, 2001). Cytochrome b_5 can reduce metmyoglobin and is one possibility (Liang *et al.*, 2002). Neuroglobin efficiently dioxygenates •NO *in vitro* with a second-order rate constant of >70 $M^{-1}s^{-1}$ at 5° (Brunori *et al.*, 2005), yet a reducing system that would support enzymatic turnover of neuroglobin within neurons or *in vitro* remains to be identified. Cytochrome b_5 reduces ferric neuroglobin, albeit relatively slowly, with a second-order rate constant of 6×10^2 $M^{-1}s^{-1}$ (Fago *et al.*, 2006). For flavohemoglobins, heme reduction is readily achieved by an associated FAD-containing reductase domain. The isoalloxazine ring of FAD transfers electrons to the heme iron via the heme propionate substituent over a minimum distance of ≈6.3 Å, presumably through a bridging water molecule (Ermler *et al.*, 1995). The *E. coli* flavohemoglobin shows a first-order rate constant for electron transfer of ≈120 s^{-1} at 20° which is also the rate-limiting step for maximal turnover in the •NO dioxygenase mechanism (Gardner *et al.*, 2000a). Indeed, electron transfer [Eq. (12.5)] would appear to be the rate-limiting step for all hemoglobins functioning as •NO dioxygenases. Within legume root nodules, ascorbate and flavin-containing reductases participate in leghemoglobin reduction (Becana and Klucas, 1990; Moreau *et al.*, 1995). A similar role for ascorbate in nonsymbiotic plant hemoglobin reduction is also suggested (Igamberdiev *et al.*, 2006). Thus, a variety of reducing systems, including NADH cytochrome

b_5 reductase:b_5, cytochrome *b*-type NAD(P)H oxidoreductase (Zhu *et al.*, 1999), ascorbate, ferredoxin reductase (Hayashi *et al.*, 1973), and NADPH cytochrome P450 reductase, may contribute to methemoglobin, metmyoglobin, metneuroglobin, and metcytoglobin reduction within cells. It is also conceivable that within any cell there are multiple reducing systems for the single-domain hemoglobins, as described for leghemoglobin in root nodules (Becana and Klucas, 1990). Moreover, for abundant hemoglobins functioning primarily in the enzymatic removal of relatively low •NO fluxes, such as those associated with modulating •NO signaling or chronic •NO toxicity, a similar low rate of reduction should suffice.

6. Heme and Flavin Cofactors

It has become increasingly evident that the content of the heme and FAD cofactors in isolated flavohemoglobins, and heme in isolated hemoglobins, is more often than not substoichiometric. Although initially unexpected and troubling for the work on flavohemoglobins (Gardner *et al.*, 1998b), there is now ample evidence that heme and FAD can be substoichiometric within cells or in isolated flavohemoglobins (Kobayashi *et al.*, 2002). For *E. coli* flavohemoglobin, one can observe the separation of the bright yellow FAD from the flavohemoglobin protein during isolation, and FAD is generally found in a ratio of $\approx$0.4 FAD per isolated flavohemoglobin. Moreover, my lab has encountered several instances where added FAD is required during •NO dioxygenase assays to maintain the activity of the enzyme (Gardner *et al.*, 1998a, 2000b; Ullmann *et al.*, 2004). The dependence upon FAD appears greatest in assays of the activity within cell extracts. For example, both the *E. coli* and *Candida albicans* flavohemoglobin (YHB1) require added FAD for detectable •NO dioxygenase activity, whereas the *Saccharomyces cerevisiae* enzyme did not (Gardner *et al.*, 1998a; Ullmann *et al.*, 2004). In this regard, FAD (1 μM) is routinely included in assays of flavohemoglobins. The heme content of isolated flavohemoglobins is also routinely measured and is generally found in the range of 0.5 to 0.9 heme per flavohemoglobin. Fractionation with ammonium sulfate and mutations near the heme site can decrease the fraction of heme present in the isolated flavohemoglobin. In some instances, reconstitution of flavohemoglobins with heme may be required. Given the usual substoichiometric amounts of heme, •NO dioxygenase activities and turnover rates are routinely reported relative to the heme content of flavohemoglobins rather than total protein.

7. •NO Dioxygenase Assays

7.1. Reagents

100 m*M* potassium phosphate buffer, pH 7.0, containing 0.3 m*M* EDTA
•NO (2 m*M*) prepared in water and stored at 4° (see subsequent section)
O_2 (1.14 m*M*) prepared in phosphate buffer (see subsequent section)
CO (1 m*M*) prepared in water and stored at 4° (see subsequent section)
NADH or NADPH (10 m*M*) prepared fresh in water
Manganese-containing superoxide dismutase (*E. coli*) (100 mg/ml) in potassium phosphate buffer and stored at −80°
FAD (1 m*M*) prepared in water and stored at −20°
1 *M* glucose in water stored at −20°
Glucose oxidase (*Aspergillus niger*) (4000 units/mL)
Catalase (bovine liver) (260,000 units per mL)

Rates of •NO decomposition by cells or isolated flavohemoglobins are measured amperometrically with a 2-mm ISONOP •NO electrode and a •NO meter (World Precision Inst., Sarasota, FL) in a 2-mL thermostatted and magnetically stirred reaction vessel (Fig. 12.2). Reactions are conducted in a glass-stoppered zero-headspace volume, and reactants (e.g., •NO, CO, enzyme) are injected with a Hamilton syringe through the port in the glass stopper.

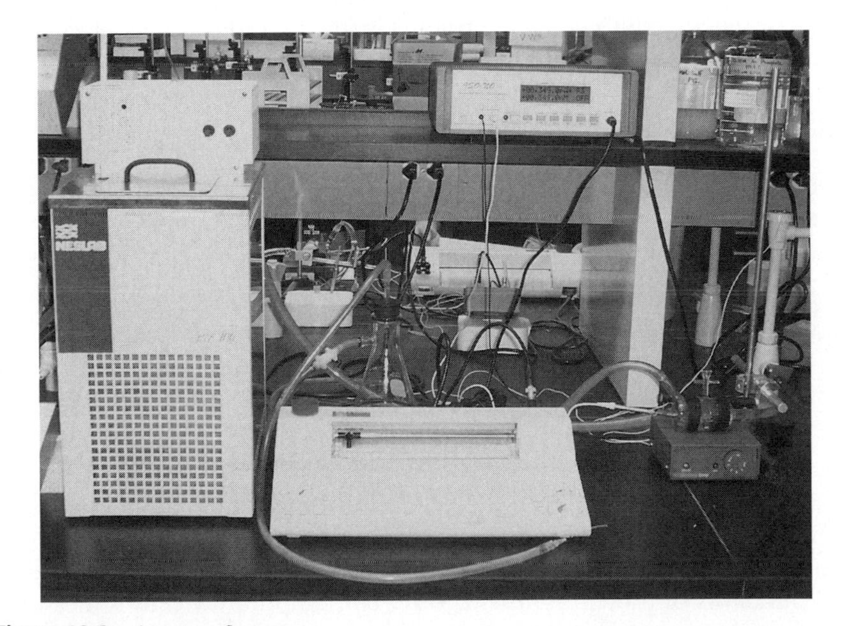

Figure 12.2 (*continued*)

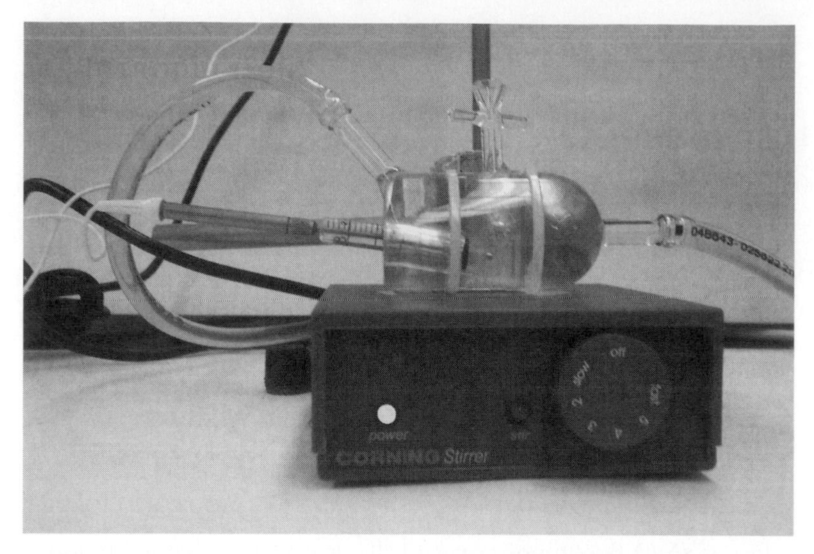

Figure 12.2 (*Top*) •NO measurement station equipped with ISO •NO meter, circulating water bath, chart recorder, thermostatted vessel, 2-mm ISONOP •NO electrode, magnetic, stirrer and aspirator. (*Bottom*) Close-up view of the •NO electrode, glass-stoppered and thermostatted reaction vessel. Note that the 2-mm ISNOP electrode has been fitted to an older O_2 electrode glass chamber (Gilson Inc.) using a custom-cut plastic pipette, epoxy, and an O-ring seal.

•NO decomposition rates are measured from the slope of chart recorder traces at specific •NO concentrations (Fig. 12.3). Calculations of rates are simplified by using a protractor to measure the angle (θ) of the slope and by using the equation relating the rate of •NO decomposition to the angle of the slope, chart speed, signal response to •NO and the total •NO in the 2-mL reaction [Eq. (12.11)]. Flavohemoglobin-catalyzed decomposition rates are routinely corrected for background rates of •NO decomposition at specific •NO concentrations under similar conditions to those of reactions containing the flavohemoglobin or catalyst (see Fig. 12.3, *bottom*).

$$\text{rate (nmol/min)} = \frac{\tan\theta \times \text{chart speed(mm/min)} \times \text{total} \bullet\text{NO(nmol)}}{\text{signal response height(mm)}} \quad (12.11)$$

For measurements of •NO metabolism by flavohemoglobins expressed within whole bacteria, fungi, and mammalian cells, cells are harvested and washed with an appropriate buffered salts medium containing glucose and inhibitors of protein synthesis. •NO consumption by cells is measured in the same glucose-containing medium at 37° (or the appropriate

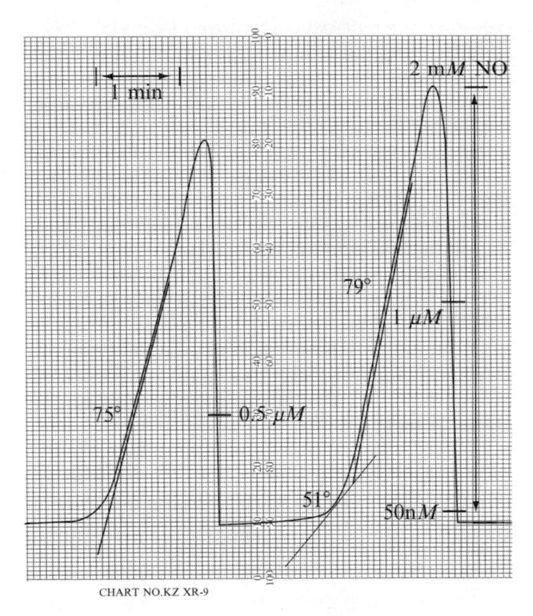

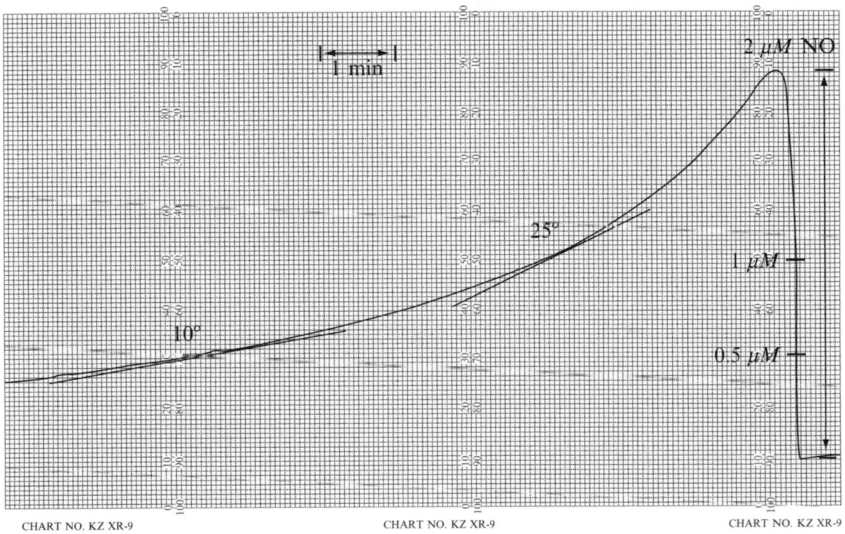

Figure 12.3 (*Top*) Chart recorder traces for flavohemoglobin-catalyzed •NO decomposition measured for rates at various •NO concentrations. Flavohemoglobin was added following the apex of the signal response to 2 μM •NO and •NO was completely metabolized. A second bolus of 2 μM •NO was then added. (*Bottom*) Background •NO decomposition rates measured from the angle of the slopes at 1 μM and 0.5 μM •NO.

temperature) in the 2-mL reaction chamber. The •NO (2 μL) from a 2-mM saturated solution is injected through the glass-stoppered port in the reaction vessel and the signal response is allowed to reach its apex. Cells are then injected, and the rate of •NO consumption is measured for various •NO concentrations. From the traces, a K_M(•NO) value for normoxia (200 μM O_2 at 37°) can be calculated.

Subsequently, 20 μL of manganese-containing superoxide dismutase (2 mg or $\approx$22 μM) is added to the reaction to determine the contribution of extracellular $\bullet O_2^-$ to cellular •NO metabolism. It should be noted that manganese-containing superoxide dismutase is preferred as a $\bullet O_2^-$ scavenger because the copper, zinc-containing superoxide dismutase shows interference with the assay, presumably by reacting with •NO. Intracellular superoxide dismutase levels normally suffice in inhibiting the reaction of intracellular $\bullet O_2^-$ with •NO that would otherwise be measured as cellular •NO metabolism.

The dependence of cellular •NO metabolism on O_2 and an *in vivo* $K_M(O_2)$ value for the •NO metabolic enzyme can be approximated by varying $[O_2]$ within the reaction chamber. The reaction chamber is first scrubbed with argon by bubbling the reaction mixture at 30 mL/min for 5 min through a fitted rubber stopper, then pierced with syringe needles for gas entry and exit, and then the rubber stopper carefully replaced with the ported glass stopper. While replacing the glass stopper, stirring is stopped and the reaction chamber is blown with argon gas to limit contamination by O_2 from the atmosphere. The O_2 is then delivered through the glass stopper port with a Hamilton syringe from an O_2-saturated buffer solution (1.14 mM) to achieve the desired O_2 concentration. Anaerobic conditions are achieved by adding 2 μL of glucose oxidase (4 U/μL), 10 μL of glucose (1 M), and 2 μL of catalase (2,60,000 U/mL) to the 2–mL reaction volume and preincubating the reaction medium for 5 min or longer before adding •NO and cells.

The sensitivity of whole-cell •NO metabolism to flavohemoglobin •NO dioxygenase inhibitors can be tested by injecting compounds directly into the 2-mL reaction chamber. •NO dioxygenases show sensitivity to cyanide (Gardner *et al.*, 1998a), CO (Gardner *et al.*, 2000b), and imidazoles (Helmick *et al.*, 2005), and these and other agents can be directly tested for their effects on •NO metabolism by intact cells. Furthermore, cyanide and CO sensitivity and O_2 dependence are characteristic of flavohemoglobin-catalyzed •NO metabolism.

Cells are injected at lower •NO concentrations (>50 nM) when a susceptibility to •NO inhibition is suspected. In this case, the signal response to the •NO electrode should be amplified by increasing the chart recorder output to the maximum permissible signal–to–noise ratio. If this fails, the ISNOP •NO electrode should be replaced with a more sensitive electrode.

For measurements of the •NO dioxygenase activity of flavohemoglobins in cell-free extracts (Gardner *et al.*, 1998a; Ullmann *et al.*, 2004), •NO dioxygenase activity is measured in a 2-mL reaction mixture containing 100 m*M* potassium phosphate buffer, pH 7.0, 0.3 m*M* EDTA, 1 μM FAD, 200 μM O_2, 2 mg manganese-containing superoxide dismutase, 100 μM NADH (or NADPH), and 2 μM •NO and thermostated at 37°.

For characterizations of the •NO dioxygenase activities of flavohemoglobins *in vitro*, purified flavohemoglobins are measured for •NO dioxygenase activity in a reaction mixture containing buffer, a reductant or reducing system, FAD (for flavohemoglobins), O_2, •NO, and superoxide dismutase (as required).

Flavohemoglobin reactions are typically carried out in 100 m*M* potassium phosphate buffer, pH 7.0, containing 0.3 m*M* EDTA, 100 μM NAD(P)H, with varying concentrations of •NO, and O_2 as required by the experiment. Apparent $K_M(O_2)$ and apparent K_M(•NO) values are determined by varying $[O_2]$ and measuring •NO decomposition rates at various •NO concentrations. Concentrations of $[O_2]$ are varied, as described previously, for whole-cell measurements. Apparent K_M values rather than actual K_M values are determined, because as saturating •NO levels are approached (>2 μM at 200 μM O_2) •NO inhibits the enzyme. Inhibitors such as CO, imidazoles, and cyanide are tested for mechanism of inhibition by varying substrates (see Fig. 12.1). Manganese-containing superoxide dismutase (2 mg) is added to test for the contribution of $•O_2^-$. However, in studies of isolated flavohemoglobins, free $•O_2^-$ has not been found to be significant. Nevertheless, it is critical to measure and rule out the contribution of free $•O_2^-$ by any •NO-decomposing catalyst, given the ease of $•O_2^-$ formation and the extreme reactivity of •NO with $•O_2^-$.

Studies of single-domain hemoglobins coupled to reducing systems can be carried out with the same precautions and variations as described for flavohemoglobins. Again, superoxide dismutase should always be included in reactions to assess the contribution of free $•O_2^-$ to •NO metabolism and to limit its reaction with •NO. Also, manganese-containing superoxide dismutase is recommended for these reactions over the copper, zinc-containing superoxide dismutase because of the marked interferences of the latter enzyme with the •NO assay at high enzyme concentrations. Manganese-containing superoxide dismutase has no appreciable effect on •NO decomposition at 1 mg per mL.

8. Preparation of •NO-, CO-, and O_2-Saturated Solutions

8.1. Reagents and materials

98.5% •NO gas lecture bottle
99.9% O_2 gas tank

99.5% CO gas lecture bottle
99.99% Ar or ultrapure N_2 gas tank
AG 1-X8 anion exchange resin (200 to 400 mesh acetate form) stored at 4°
Potassium permanganate (0.16 *M*) and 2.5% sulfuric acid prepared freshly in water
Potassium phosphate (100 m*M*) buffer, pH 7.0, containing 0.3 m*M* EDTA
Stainless-steel gas proportioner/flow regulator
Vacutainer tubes (10 mm × 100 mm)

Rubber septum-sealed glass tubes containing 3 mm × 10 mm micro stir bars and 1 ml pure water are used to prepare •NO and CO stock solutions (Fig. 12.4, *top*). Glass tubes are custom made from rubber septum-sealed 10 mm × 100 mm (7 ml) Vacutainer tubes (Becton-Dickinson) by cutting with a file to 45 mm and fire polishing the cut end. The rubber septum is reinserted in the tube for a gas-tight seal. To remove trace nitrate and nitrite formed during the preparation, storage, and use of saturated •NO stocks, AG 1-X8 anion exchange resin (BioRad Laboratories) (100 mg) is added to the tube.

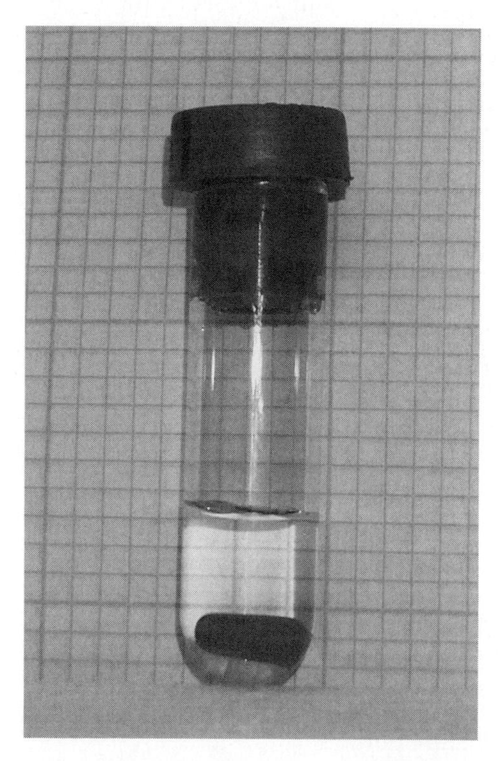

Figure 12.4 (*continued*)

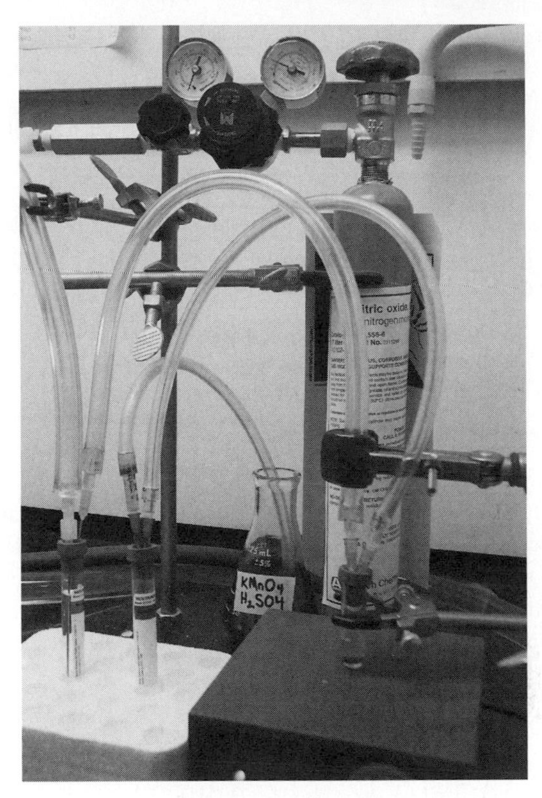

Figure 12.4 (*Top*) Septum-sealed tubes for preparation and delivery of gas-saturated solutions. Background reference squares are 2 mm × 2 mm. (*Bottom*) •NO gassing apparatus showing NaOH scrubbing tube (*far left*), exhaust tubing, and stirring saturated stock tube (no AG 1-X8 resin added).

For the preparation of •NO-saturated water solutions, septum-sealed tubes containing 1 mL of water (and anion exchange resin as required) are vigorously stirred under a stream of ultrapure N_2 or Ar delivered with a gas proportioner at a rate of 30 mL per minute for 8 min or longer. Gases are delivered through thick-walled low-gas-permeability Tygon tubing connected to 25 to 27 gauge syringe needles penetrating the rubber septum. •NO gas is first scrubbed of nitrogen oxides by bubbling slowly ($\approx$5 mL per minute) through a solution of 1 N NaOH and then passed over the 1 mL of argon-scrubbed water while stirring for $\approx$8 minutes. The •NO gas stream is ultimately passed through a 2.5% sulfuric acid solution containing 0.16 *M* potassium permanganate to decompose •NO exhausting from the system. In addition, •NO is handled in a fume hood to limit human exposure to toxic nitrogen oxides. Saturated •NO stocks are stored at 4° and are stable for months. The saturated •NO solution ($\approx$2 m*M* at 20°) is then conveniently removed using a 10 μL Hamilton syringe for •NO dioxygenase assays or assays of nitrate or nitrite formation. Typically, 100 μL of the 1 mL of saturated •NO stock can be used before discarding the remaining •NO stock because of decreased •NO saturation. •NO stocks are generally prepared in lots of 20 or more tubes and tested for saturation using the •NO electrode.

For the preparation of saturated CO stocks (1 mM at 20°), rubber septum-sealed tubes containing 1 mL of water are first stirred under a stream of ultrapure Ar or N_2 gas at a rate of 30 mL per minute for 8 min to remove O_2. The solution is then stirred under a stream of 99.5% CO gas delivered at a rate of ≈5 mL per minute for 8 min or longer. CO gas exhausting from the system is vented in a chemical hood to avoid the danger of CO poisoning. CO stocks are stored at 4° and are stable for months. The saturated CO solution is then conveniently removed using a 10–to 50–μL Hamilton syringe for •NO dioxygenase inhibition studies.

For the preparation of saturated O_2 solution (1.14 mM at 20°), 99.99% O_2 gas is passed over vigorously stirring phosphate buffer in a 10 mL septum-sealed vial at 30 mL per minute. The gas exhaust is vented to the atmosphere. Aliquots of O_2-saturated buffer are removed with a syringe as needed for studies of the O_2 dependence of the •NO dioxygenase activity.

9. Assay of Heme and FAD Content

9.1. Reagents

FAD standard (1.0 mM) prepared in water and stored at −20°
Pyridine (4.4 M) prepared in NaOH (0.2 M) and stored at room temperature
Dithionite (solid)
Potassium ferricyanide (100 mM) prepared in water and stored at −20°

The FAD content of flavohemoglobin is determined by first boiling the flavohemoglobin (10 to 20 nmol protein) in 1 mL of 100 mM potassium phosphate, pH 7.0, buffer containing 0.3 mM EDTA for 3 min. The denatured protein is then removed by centrifugation at 20,000 × g for 3 min at room temperature. A 0.8 mL fraction of the supernatant is added to 1.2 mL of phosphate buffer for the measurement of fluorescence and the quantification of FAD. The fluorescence of the FAD in the supernatant is measured at 520 nm using excitation at 460 nm. Commercial FAD is used to produce a linear standard curve for fluorescence. An extinction coefficient of 11.5 $mM^{-1}cm^{-1}$ at 450 nm is applied in the preparation of the FAD standard (≈1.0 mM) in water. The fraction of FAD in flavohemoglobin is calculated relative to protein concentration (Lowry *et al.*, 1951).

The heme content of flavohemoglobins is measured using the alkaline-pyridine method (Appleby, 1978). Briefly, a 300-μL sample of flavohemoglobin containing 1.2 to 12 nmol of heme is added to 300 μL of a 4.4 *M* pyridine/0.2 *M* NaOH solution and mixed. Absorbances are measured at 556 nm and 539 nm before and immediately after adding a small pinch ($\approx$2 mg) of dithionite and gently mixing the sample in a cuvette. A second pinch of dithionite is added and absorbances are recorded again to verify complete reduction of heme. Potassium ferricyanide (100 m*M*) is then added stepwise in 3-μL aliquots to the cuvette and absorbances at 556 nm and 539 nm are recorded. The reduced heme solution turns from pink to yellow when fully oxidized. Heme concentration in the sample is calculated from the absorbance differences of the reduced minus oxidized heme at 556 nm and 539 nm, where μM heme $= 46.7\ (\Delta A;_{556} - \Delta A_{539})$ (Appleby, 1978). Flavohemoglobin protein is assayed (Lowry *et al.*, 1951) to determine the stoichiometry of heme to protein.

10. Reconstitution of Flavohemoglobin with Heme

10.1. Reagents

20 m*M* hemin prepared in dimethyl sulfoxide and stored at $-20°$
100 m*M* dithiothreitol in water stored at $-20°$
Dithionite (solid)
Tris-Cl (50 m*M*), pH 8, containing EDTA (1 m*M*)
Catalase (bovine liver) (2,60,000 units per mL)

To reconstitute heme-deficient flavohemoglobin with heme, flavohemoglobin is incubated with heme under reducing conditions in the presence of catalase to scavenge potentially damaging peroxide. Briefly, flavohemoglobin (0.5 to 0.75 m*M*) is prepared in a volume of 1.5 mL containing 50 m*M* Tris-Cl, pH 8, 1 m*M* EDTA, 10 m*M* DTT, 3000 U of catalase, and hemin is then added slowly from a 20 m*M* stock in dimethyl sulfoxide to a final concentration of 0.5 m*M*. Dithionite ($\approx$2 mg freshly dissolved in 25 μL of water) is added after 15 min of incubation to fully reduce hemin and eliminate O_2. The reaction is incubated at 37° for 60 min. The flavohemoglobin is then immediately separated from free heme, catalase, and reductants by gel-filtration chromatography. For this purpose, a 1.5 cm $\times$ 60 cm Superdex 200 column equilibrated with N_2-sparged 50 m*M* Tris-Cl, pH 8, buffer containing 1 m*M* EDTA is used.

11. Assay for Nitrate and Nitrite Reaction Products

11.1. Reagents

Tris-Cl, pH 7.5 (50 m*M*)
NADPH (solid)
Nitrate reductase (*Aspergillus niger*) (30 mg/mL in Tris-Cl buffer) stored at −80°
Sodium nitrite standard (10 m*M*) prepared in water and stored at 4°
Sodium nitrate standard (10 m*M*) prepared in water and stored at 4°
Sulfanilamide (1% w/v) in phosphoric acid (5% v/v) and stored for weeks at 4°
N-(1-naphthyl-) ethylenediamine 2HCl (0.1% w/v) in water stored for weeks at 4°

•NO-saturated water prepared over AG 1-X8 anion exchange resin is used to measure nitrate and nitrite formation following the reaction of flavohemoglobins in the 2-ml thermostatted reaction chamber. Sufficient enzyme or oxyhemoglobin is used to produce a high rate of •NO conversion. In these reactions, it is important to minimize the reaction of •NO with O_2-forming $•NO_2$ and nitrite. This is achieved by minimizing the •NO concentration and the time for reaction. •NO (40 nmol; 20 μL) is injected at a slow rate such that •NO does not accumulate appreciably (<0.5 μ*M*) as measured by the •NO electrode. The reaction is allowed to go to completion and then the mixture is quantitatively removed and stored frozen at −80° until assay.

Nitrite and nitrate are measured spectrophotometrically using the Griess reaction and NADPH-dependent nitrate reductase (Green *et al.*, 1982; Titheradge, 1998). Amounts of nitrite and nitrate produced as a result of flavohemoglobin metabolism are corrected for any nitrite or nitrate introduced with the stock •NO solution and not removed by the AG 1-X8 resin. For this purpose, the remaining •NO stock solution is thoroughly scrubbed with ultrapure N_2 or argon and the sparged solution is assayed for nitrite and nitrate. With the use of the AG 1-X8 resin the contamination by nitrite and nitrate is generally inconsequential but should nevertheless be measured and accounted for.

Briefly, nitrate and nitrate are measured by incubating 100 μL of the 2-mL reaction mixture containing a theoretical maximum of 2 nmol of nitrate/nitrite with 100 μL of a solution containing 50 m*M* Tris-Cl, pH 7.5, buffer, and 400 μ*M* NADPH with or without added nitrate reductase (30 μg). Nitrate reductase reduces nitrate to nitrite, and the difference in nitrite measured in the two reactions allows for the calculation of nitrate.

The reaction is incubated at room temperature for 60 min or longer. Sodium nitrite (10 m*M*) and sodium nitrate (10 m*M*) are diluted to 0.5 m*M* in 50 m*M* Tris-Cl (50 m*M*) buffer and are used to prepare standard curves over the range of 0 to 5.0 nmol under similar reaction conditions and volumes as the test samples. Nitrate reductase is added to the nitrate standards but not to the nitrite standards. After the 60-min incubation, 300 μL of a freshly prepared 1:1 mixture of sulfanilamide (1% w/v) in phosphoric acid (5% v/v) with *N*-(1-naphthyl-) ethylenediamine 2HCl (0.1% w/v) is added to each of the reactions. The stable pink-colored product of the reaction with nitrite is measured spectrophotometrically at 540 nm and the yields of nitrate and nitrite are calculated from the standard curves. The standard curves for nitrate and nitrite should be nearly identical.

12. Conclusions

Hemoglobins and myoglobins bear an intrinsic capacity for high fidelity •NO dioxygenation. Moreover, •NO dioxygenation by hemoglobins may serve important roles in modulating •NO signaling and in limiting •NO toxicity. While evidence suggests that many flavohemoglobins from diverse sources function as •NO dioxygenases, many hemoglobins, including neuroglobin and cytoglobin, remain to be extensively tested for this function both *in vitro* and within cells. Reducing systems for single-domain hemoglobins also require identification. It is hoped that the methods and arguments outlined here will facilitate future investigations of hemoglobins and dioxygenase inhibitors that may be useful as therapeutics.

ACKNOWLEDGMENTS

I gratefully acknowledge the past support of Grant GM65090 from the National Institutes of Health. I especially thank my coworkers, many of whom are listed in the cited publications, for their valuable contributions.

REFERENCES

Appleby, C. A. (1978). Purification of *Rhizobium* cytochromes P-450. *Meth. Enzymol.* **52,** 157–166.

Becana, M., and Klucas, R. V. (1990). Enzymatic and nonenzymatic mechanisms for ferric leghemoglobin reduction in legume root nodules. *Proc. Natl. Acad. Sci. USA* **87,** 7295–7299.

Beckman, J. S., and Koppenol, W. H. (1996). Nitric oxide, superoxide, and peroxynitrite: The good, the bad, and the ugly. *Am. J. Physiol.* **271,** C1424–C1437.

Brunori, M., Giuffrè, A., Nienhaus, K., Nienhaus, G. U., Scandurra, F. M., and Vallone, B. (2005). Neuroglobin, nitric oxide, and oxygen: Functional pathways and conformational changes. *Proc. Natl. Acad. Sci. USA* **102,** 8483–8488.

Eich, R. F., Li, T., Lemon, D. D., Doherty, D. H., Curry, S. R., Aitken, J. F., Mathews, A. J., Johnson, K. A., Smith, R. D., Phillips, G. N., Jr., and Olson, J. S. (1996). Mechanism of NO-induced oxidation of myoglobin and hemoglobin. *Biochemistry* **35,** 6976–6983.

Ermler, U., Siddiqui, R. A., Cramm, R., and Friedrich, B. (1995). Crystal structure of the flavohemoglobin from *Alcaligenes eutrophus* at 1.75Å resolution. *EMBO J.* **14,** 6067–6077.

Fago, A., Mathews, A. J., Moens, L., Dewilde, S., and Brittain, T. (2006). The reaction of neuroglobin with potential redox protein partners cytochrome b_5 and cytochrome *c*. *FEBS Lett.* **580,** 4884–4888.

Flögel, U., Gödecke, A., Klotz, L.-O., and Schrader, J. (2004). Role of myoglobin in the antioxidant defense of the heart. *FASEB J.* **18,** 1156–1158.

Flögel, U., Merx, M. W., Gödecke, A., Decking, U. K. M., and Schrader, J. (2001). Myoglobin: A scavenger of bioactive NO. *Proc. Natl. Acad. Sci. USA* **98,** 735–740.

Fridovich, I. (1978). The biology of oxygen radicals. *Science* **201,** 875–880.

Fridovich, I. (1995). Superoxide radical and superoxide dismutases. *Annu. Rev. Biochem.* **64,** 97–112.

Gardner, A. M., and Gardner, P. R. (2002). Flavohemoglobin detoxifies nitric oxide in aerobic, but not anaerobic, *Escherichia coli*: Evidence for a novel inducible anaerobic nitric oxide scavenging activity. *J. Biol. Chem.* **277,** 8166–8171.

Gardner, A. M., Helmick, R. A., and Gardner, P. R. (2002). Flavorubredoxin: An inducible catalyst for nitric oxide reduction and detoxification in *Escherichia coli*. *J. Biol. Chem.* **277,** 8172–8177.

Gardner, A. M., Martin, L. A., Gardner, P. R., Dou, Y., and Olson, J. S. (2000a). Steady-state and transient kinetics of *Escherichia coli* nitric oxide dioxygenase (flavohemoglobin): The tyrosine B10 hydroxyl is essential for dioxygen binding and catalysis. *J. Biol. Chem.* **275,** 12581–12589.

Gardner, P. R. (2002). Aconitase: Sensitive target and measure of superoxide. *Meth. Enzymol.* **349,** 9–23.

Gardner, P. R. (2005). Nitric oxide dioxygenase function and mechanism of flavohemoglobin, hemoglobin, myoglobin and their associated reductases. *J. Inorg. Biochem.* **99,** 247–266.

Gardner, P. R., Costantino, G., and Salzman, A. L. (1998a). Constitutive and adaptive detoxification of nitric oxide in *Escherichia coli*. Role of nitric oxide dioxygenase in the protection of aconitase. *J. Biol. Chem.* **273,** 26528–26533.

Gardner, P. R., Costantino, G., Szabó, C., and Salzman, A. L. (1997). Nitric oxide sensitivity of the aconitases. *J. Biol. Chem.* **272,** 25071–25076.

Gardner, P. R., Gardner, A. M., Martin, L. A., and Salzman, A. L. (1998b). Nitric oxide dioxygenase: An enzymic function for flavohemoglobin. *Proc. Natl. Acad. Sci. USA* **95,** 10378–10383.

Gardner, P. R., Gardner, A. M., Brashear, W., Suzuki, T., Hvitved, A. N., Setchell, K. D. R., and Olson, J. S. (2006). Hemoglobins dioxygenate nitric oxide with high fidelity. *J. Inorg. Biochem.* **100,** 542–550.

Gardner, P. R., Gardner, A. M., Martin, L. A., Dou, Y., Li, T., Olson, J. S., Zhu, H., and Riggs, A. F. (2000b). Nitric oxide dioxygenase activity and function of flavohemoglobins: Sensitivity to nitric oxide and carbon monoxide inhibition. *J. Biol. Chem.* **275,** 31581–31587.

Green, L. C., Wagner, D. A., Glogowski, J., Skipper, P. L., Wishnok, J. S., and Tannenbaum, S. R. (1982). Analysis of nitrate, nitrite, and [^{15}N]nitrate in biological fluids. *Anal. Biochem.* **126,** 131–138.

Hallstrom, C. K., Gardner, A. M., and Gardner, P. R. (2004). Nitric oxide metabolism in mammalian cells: Substrate and inhibitor profiles of a NADPH-cytochrome P450 oxidoreductase-coupled microsomal nitric oxide dioxygenase. *Free Radic. Biol. Med.* **37,** 216–228.

Hausladen, A., Gow, A., and Stamler, J. S. (2001). Flavohemoglobin denitrosylase catalyzes the reaction of a nitroxyl equivalent with molecular oxygen. *Proc. Natl. Acad. Sci. USA* **98,** 10108–10112.

Hayashi, A., Suzuki, T., and Shin, M. (1973). An enzymic reduction system for metmyoglobin and methemoglobin, and its application to functional studies of oxygen carriers. *Biochim. Biophys. Acta* **310,** 309–316.

Helmick, R. A., Fletcher, A. E., Gardner, A. M., Gessner, C. R., Hvitved, A. N., Gustin, M. C., and Gardner, P. R. (2005). Imidazole antibiotics inhibit the nitric oxide dioxygenase function of microbial flavohemoglobin. *Antimicrob. Agents Chemother.* **49,** 1837–1843.

Huie, R. E., and Padmaja, S. (1993). The reaction of NO with superoxide. *Free Radic. Res. Commun.* **18,** 195–199.

Hultquist, D. E., and Passon, P. G. (1971). Catalysis of methaemoglobin reduction by erythrocyte cytochrome b_5 and cytochrome b_5 reductase. *Nat. New Biol.* **229,** 252–254.

Igamberdiev, A. U., Bykova, N. V., and Hill, R. D. (2006). Nitric oxide scavenging by barley hemoglobin is facilitated by a monodehydroascorbate reductase-mediated ascorbate reduction of methemoglobin. *Planta* **223,** 1033–1040.

Kim, S. O., Orii, Y., Lloyd, D., Hughes, M. N., and Poole, R. K. (1999). Anoxic function for the *Escherichia coli* flavohemoglobin (Hmp): Reversible binding of nitric oxide and reduction to nitrous oxide. *FEBS Lett.* **445,** 389–394.

Kobayashi, G., Nakamura, T., Ohmachi, H., Matsuoka, A., Ochiai, T., and Shikama, K. (2002). Yeast flavohemoglobin from *Candida norvegensis*: Its structural, spectral, and stability properties. *J. Biol. Chem.* **277,** 42540–42548.

Liang, Z.-X., Jiang, M., Ning, Q., and Hoffman, B. M. (2002). Dynamic docking and electron transfer between myoglobin and cytochrome b_5. *J. Biol. Inorg. Chem.* **7,** 580–588.

Lowry, O. H., Rosebrough, N. J., Farr, A. L., and Randall, R. J. (1951). Protein measurement with the Folin phenol reagent. *J. Biol. Chem.* **193,** 265–275.

Miranda, J. L., Maillett, D. H., Soman, J., and Olson, J. S. (2005). Thermoglobin, oxygen-avid hemoglobin in a bacterial hyperthermophile. *J. Biol. Chem.* **280,** 36754–36761.

Moreau, S., Puppo, A., and Davies, M. J. (1995). The reactivity of ascorbate with different redox states of leghaemoglobin. *Phytochemistry* **39,** 1281–1286.

Olson, J. S., Foley, E. W., Rogge, C., Tsai, A.-L., Doyle, M. P., and Lemon, D. D. (2004). NO scavenging and the hypertensive effect of hemoglobin-based blood substitutes. *Free Radic. Biol. Med.* **36,** 685–697.

Poole, R. K., and Hughes, M. N. (2000). New functions for the ancient globin family: Bacterial responses to nitric oxide and nitrosative stress. *Mol. Microbiol.* **36,** 775–783.

Titheradge, M. A. (1998). The enzymatic measurement of nitrate and nitrite. *In* "Nitric Oxide Protocols" (M. A. Titheradge, ed.), vol. 100, pp. 83–91. Humana Press, Brighton, UK.

Ullmann, B. D., Myers, H., Chiranand, W., Lazzell, A. L., Zhao, Q., Vega, L. A., Lopez-Ribot, J. L., Gardner, P. R., and Gustin, M. C. (2004). An inducible defense mechanism against nitric oxide in *Candida albicans*. *Eukaryot. Cell* **3,** 715–723.

Vinogradov, S. N., Hoogewijs, D., Bailly, X., Arredondo-Peter, R., Gough, J., Dewilde, S., Moens, L., and Vanfleteren, J. R. (2006). A phylogenomic profile of globins. *BMC Evol. Biol.* **6,** 31.

Wu, G., Wainwright, L. M., and Poole, R. K. (2003). Microbial globins. *Adv. Microb. Physiol.* **47,** 255–310.

Zhu, H., Qiu, H., Yoon, H.-W. P., Huang, S., and Bunn, H. F. (1999). Identification of a cytochrome *b*-type NAD(P)H oxidoreductase ubiquitously expressed in human cells. *Proc. Natl. Acad. Sci. USA* **96,** 14742–14747.

CHAPTER THIRTEEN

Globin Interactions with Lipids and Membranes

Antonio Di Giulio* *and* Alessandra Bonamore[†]

Contents

Abstract

Many bacterial globins have been demonstrated to interact with membrane lipids, and several hypotheses in support of a functional role for membrane localization have been set forth. Bacterial globins have been suggested to facilitate oxygen diffusion to terminal oxidases, to protect oxidases from nitric oxide or eventually to preserve the integrity of the membrane lipids through

* Department of Science and Biomedical Technology, University of L'Aquila, L'Aquila, Italy
† Department of Biochemical Sciences, University of Rome La Sapienza, Rome, Italy

Methods in Enzymology, Volume 436
ISSN 0076-6879, DOI: 10.1016/S0076-6879(08)36013-3

peroxide-reducing activities as a response to oxidative/nitrosative stress. In this framework, methodological approaches to the study of globin–membrane interactions need to be analyzed in depth in order to single out the relevant features of these interactions and to clearly distinguish the specific membrane and lipid binding process from trivial effects related to the possible partitioning of the lipid side chains to the hydrophobic heme pocket or to the presence of partially folded, insoluble protein aggregates within membranous pellets. Methods for qualitative lipid analysis, liposome-protein binding studies, and analysis of protein insertion into lipid monolayer are thus described with the aim of providing rapid and efficient screening of specific globin–membrane interactions.

1. Introduction

Studies on the intracellular localization of bacterial hemoglobins have shown that these proteins are not simply distributed in the cytoplasmatic compartment but frequently occur in the periplasmic space (Khosla and Bailey, 1989) and in many cases partition to the plasma membrane of the bacterial cell. Thus, as demonstrated by electron microscopy measurements, the intracellular localization of *Vitreoscilla* sp. Hb (VHb) is adjacent to the cell membrane (Ramandeep *et al.*, 2001), where it has been proposed to act as a facilitator of oxygen transfer to the terminal oxidases. *Mycobacterium tuberculosis* truncated hemoglobin (HbO) has been reported to be preferentially located in contact with the bacterial inner membrane *in vivo* and to be capable of reversible binding to liposomes *in vitro* (Liu *et al.*, 2004). *Escherichia coli* flavohemoglobin (HMP) has also been demonstrated to partition efficiently to lipid membrane and to be capable of reversible binding to liposomes. Last, *Alcaligenes eutrophus* flavohemoglobin (FHP) has been shown to bind directly to phospholipids within the heme active site, as observed in the X-ray crystal structure (Ermler *et al.*, 1995) and solution experiments (Ollesch *et al.*, 1999).

In this framework, it is of interest to study the interaction of bacterial hemoglobins with naturally occurring membrane lipids and with their corresponding fatty acids. The present work focuses on the investigation of recombinantly expressed HMP and VHb proteins in their interactions with *E. coli* total lipid extracts and with saturated, unsaturated, and cyclopropanated fatty acids. Complementary methodological approaches include solution-binding studies to liposome systems as well as the study of the interaction of globins with phospholipids at the air–water interface by means of lateral pressure measurements on monomolecular phospholipid films. This latter system has become increasingly popular as a tool for investigating the interactions of a wide range of proteins and peptides with biological membranes (Brockman, 1999; Maget-Dana, 1999; Zhao *et al.*, 2003) because it can monitor directly and with high sensitivity the protein–monolayer interaction.

2. Lipid Extraction

2.1. Overview

Extraction and characterization of lipid fractions either from whole bacterial cells or from purified bacterial hemoglobins is a prerequisite for the study of globin–lipid interactions. The methods hitherto developed are optimized for *E. coli* cells but can be conveniently carried over to most bacterial species. Lipid extraction from whole cells rests essentially on the method of Bligh and Dyer (1959), which allows for nonselective extraction of both polar and neutral lipid species. The method is not intended for high-yield lipid extracts but has the advantage of rigorously excluding proteins, nucleic acids, and metabolites from the nonaqueous solvent phase.

It is worth mentioning that the fatty acid composition of phospholipids and triglycerides may vary significantly in bacteria as a function of the growth phase, temperature, and pH of the medium. In particular, in standard *E. coli* strains (BL21 cells), higher content of unsaturated fatty acids (palmitoleic and cis-vaccenic acids) is observed at low temperature and the corresponding cyclopropanated derivatives become the predominant species under stationary phase conditions and low pH values (<5.0) (Cronan, 2002). Thus, growth conditions must be standardized in order to obtain reproducible lipid extracts of constant lipid composition.

2.2. Extraction method

E. coli cells recovered from 1 L of LB overgrown culture (shake flask, 48 h at 37° corresponding to 1.4 OD/cm at 600 nm) are resuspended in 20 ml of PBS buffer, sonicated for 20 min, and centrifuged at 13,000 rpm for 20 min. Then 1 g wet weight pellet is washed twice with 10 ml of distilled water and thereafter resuspended with 3 ml of a chloroform/methanol solution (1:2 v/v). The suspension is vortexed at room temperature until homogenous and stored at 4° for 24 h. Thereafter, the sample is centrifuged at 6,000 rpm and the supernatant is added to 1 ml of chloroform and 1 ml of water. The dilution procedure results in a biphasic system in which the chloroform layer contains the lipid fraction. The chloroform bottom phase can be quantitatively removed with a Pasteur pipette, evaporated to dryness under a nitrogen stream at 4°, and stored at −20°. Larger amounts of cells can be dealt with by scaling up the procedure proportionally. When the final volume is higher than 20 to 30 ml, a Rotavapor apparatus is necessary for evaporating chloroform.

This procedure allows for reproducible recovery of phospholipids (phosphatidyl glycerol and phosphatidyl ethanolamine) and triglycerides (cardiolipin) and currently yields 40 to 60 mg of total lipid fraction (TLE) from 1 g of *E. coli* cells wet pellet. It must be pointed out that the TLE fraction thus obtained contains significant amounts (1 to 5%) of fatty acid methyl esters generated during the methanol chloroform extraction. The amount of methyl ester significantly increases when the procedure is carried out at room temperature.

3. Qualitative Analysis of Total Lipid Extracts

3.1. Overview

Gas chromatography/mass spectrometry (GC/MS) is the technique of choice for the analysis of fatty acid methyl esters (FAME) obtained from TLE. Technical aspects of the FAME analysis have been extensively reviewed elsewhere (Lewis *et al.*, 2000) and will not be considered here. Special caution, however, should be given to the preparation of methyl ester extracts (see the subsequent section) because of the presence of possible artifacts when dealing with bacterial lipids. In particular, when TLE is subjected to acidic methanolysis, the proportion of cyclopropanated fatty acids can be underestimated and unusual methoxylated fatty acids of artifactual nature can be produced (Orgambide *et al.*, 1993). A convenient transesterification procedure is outlined subsequently.

3.2. Transesterification reaction and GC/MS analysis

TLE (1 to 2 mg) is directly resuspended in 1 ml of a transesterification mix composed of methanol:hydrochloric acid:chloroform, 10:1:1 v/v/v. The reaction is carried out for 1 h at 90° and then 300 μl of water is added to stop the reaction. FAME so obtained are extracted three times with 1 ml hexane: chloroform, 4:1, and analyzed by GC/MS using an Agilent 6,850A gas chromatograph coupled to a 5,973 N quadrupole mass selective detector (Agilent Technologies, Palo Alto, CA). Chromatographic separation is carried out on a Zebron ZB-WAX fused-silica capillary column (30 m × 0.25 mm i.d.) coated with polyethylene glycol (film thickness 0.25 μm) as stationary phase. No standard need be used for qualitative analysis because spectral libraries already contain fragmentation profiles of most FAMEs.

4. Hemoglobin-Liposome Binding Studies

4.1. Overview

Protein–membrane interactions can be conveniently studied by monitoring the protein binding to the surface of a liposome, provided that the lipid composition of the liposome reflects that of the native membrane.

Unfortunately, not all lipid mixtures give rise to liposomes stable enough to run reproducible experiments. For instance, phospholipids containing only phosphatidic acid or phosphatidylethanolamine derivatives will not produce a stable liposome phase. In turn, lipids obtained from bacterial membranes contain mixtures of phospholipids with different polar heads, and in most cases simple bacterial lipid extracts are suitable for liposome preparations. Thus, the preparation of liposomes can be carried out using the TLE described previously. As a general rule, it is advisable to use TLE extracts from cells grown at a temperature slightly higher than that of the programmed liposome experiments. TLE from cells grown at low temperature (<20°) will result in unstable, low-temperature-melting liposomes.

4.2. Liposome preparations

Depending on the conditions of lipid storage (freezing temperature and age of the sample), direct resuspension of the raw extracts with aqueous buffer is not recommended because amorphous aggregates, frequently occurring in aged extracts, are likely to yield poor liposome preparations. It is advisable to first resuspend the TLE aggregates in chloroform-methanol 1:2 v/v (10 mg/ml) under vigorous stirring and to then dry the sample under nitrogen stream on a glass tube by carefully spreading the solution on the wall. Thereafter, 50 m*M* phosphate buffer at pH 7.0 (1 ml/10 mg of lipid) is added, heated at 60° for 15 min, and sonicated in a Soniprep 150 sonicator for 30 min to disperse the lipids. The liposomes thus obtained can be stored for 4 h at 4°, pelleted by centrifugation, and resuspended in buffer before use. TLE liposomes thus prepared have limited stability and must be used within 24 h of preparation. Other methods can be used to yield more stable, homogeneous, small-unilamellar-vesicle (SUV) liposomes (Duzgunes, 2003). However, the lipid composition of SUV is highly constrained to limited lipid species and would not reproduce the lipid composition of the parent bacterial strain.

4.3. Gel-filtration experiments

Gel-permeation chromatography is a simple and direct method for measuring protein liposome interactions. In the case of globins, the method is especially useful and simple because of the high molar absorptivity of the heme in the visible region. The interaction of *E. coli* HMP with TLE liposomes has been measured in a gel-filtration experiment that demonstrated a strong interaction of the protein for the lipid layer (Bonamore *et al.*, 2003b). Liposome solutions (0.4 ml) are mixed with 0.1 ml of protein solution (25 μM) and then applied on a Sephadex G-75 (Pharmacia, Uppsala, Sweden) column (0.5 × 35 cm). This analysis is carried out in 50 mM phosphate buffer at pH 7.0 at a constant flow rate of 5 ml/h at 20°, and the elution profile is monitored in the

Soret region (400 to 420 nm) with a Jasco 7,800 spectrophotometer equipped with a thermostatted flow cell. The void volume of the column can be measured by dextran blue, and elution volumes of human hemoglobin (CO adduct) and myoglobin in both the presence and the absence of liposomes can be used as standards.

As shown in (Fig. 13.1), the elution profile of HMP in the presence of TLE liposomes displays two separate peaks at 8.3 and 11.7 ml, whose integrated surfaces account for 55 and 45% of the total area. The peak at 8.3 ml contains both the TLE liposome fraction and ferric HMP, whereas the peak at 11.7 corresponds to the elution volume of free HMP. The absorption spectra of the two fractions (see Fig. 13.1, panel 3) correspond to the spectrum of ferric, lipid-bound HMP. The gel-filtration data indicate clearly that HMP is capable not only of strong binding to liposomes

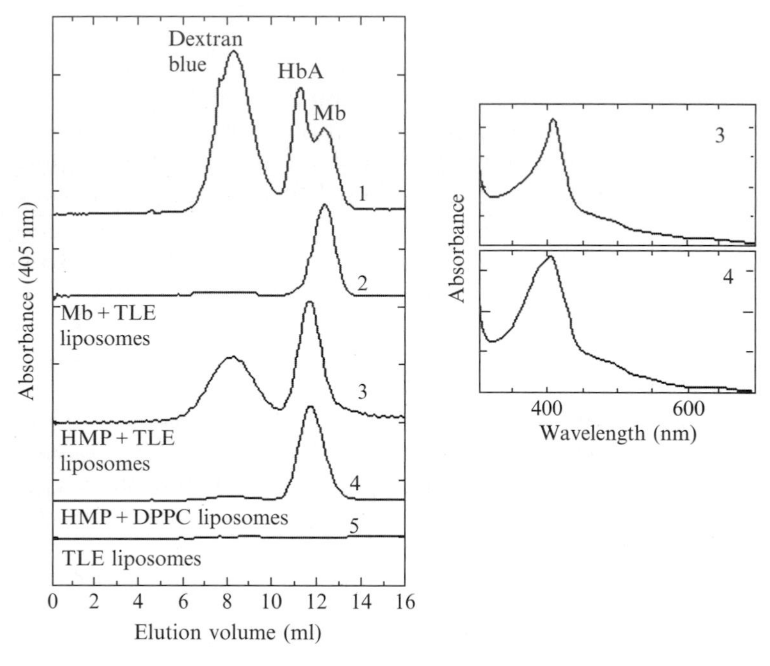

Figure 13.1 Gel-permeation profiles of HMP in the presence of liposomes. Profile 1: standard solution composed of blue dextrane (that corresponds to the void volume), human hemoglobin (HbA), and horse heart myoglobin (Mb). Profile 2: ferric Mb mixed with TLE from *E. coli* cells. Profile 3: purified ferric unliganded HMP mixed with liposomes (20:1 phospholipid:protein molar ratio) obtained from *E. coli* TLE. Profile 4: purified ferric unliganded HMP mixed with DPPC liposomes (20:1 phospholipid: protein molar ratio). Profile 5: TLE liposomes. On the right-hand side of the picture, the UV-visible absorption spectra of HMP from the eluates relative to profiles 3 and 4 are shown. Experiments were carried out on a G-75 Sephadex column equilibrated with 50 m*M* phosphate buffer at pH 7.0 and 20°.

obtained from *E. coli* lipid extracts (TLE) but also of abstracting phospholipids from the liposome fraction, although it cannot be excluded that free, monodisperse lipids might be present in the liposome preparation even after centrifugation. The absence of binding to pure DPPC liposomes (Fig. 13.1, profile 4 and panel 4) indicates that HMP can actually recognize the polar heads of naturally occurring phospholipids of *E. coli* membrane (DPPC is not synthesized in bacteria) and does not interact with the hydrophobic core of the bilayer.

5. Lipid Monolayers as Models for Globin-Membrane Interaction Studies

5.1. Overview

Monomolecular lipid films (MLFs) at the air–water interface are ideal systems for the study of protein–membrane interactions (Brockman, 1999). MLFs have advantages over liposome-based systems as a result of their homogeneity, stability, and planar geometry. In fact, in MLFs, the lipid molecules have a specific orientation in which phospholipids polar heads point toward the water subphase, whereas acyl chains are in contact with the air phase. Moreover, the two-dimensional molecular density of MLFs can be adjusted to the desired value, and the ionic conditions of the subphase can be easily varied (Maget-Dana, 1999; Zhao *et al.*, 2003).

Monolayer penetration can be described as the interaction of an insoluble monolayer spread at the air–water interface with a soluble active compound present in the aqueous subphase. The interaction of the active compound with MLF can be measured by monitoring the surface increase of the film under constant lateral pressure or by monitoring the surface pressure changes upon addition of the active compound to the subphase under constant surface area. A schematic representation of a protein/MLF penetration experiment is reported in Fig. 13.2.

The experimental device consists of a small dish equipped with a channel for protein solution injection (see Fig. 13.2, B) and of a thin Wilhelmy wire probe (Fig. 13.2, D) for the lateral pressure measurements. The lipid monolayer is formed at the required initial pressure by intermittent spreading of the lipid solution in the appropriate solvent mixture (chloroform and methanol) on the buffer subphase. The following parameters can be monitored in this type of experiment: (1) the initial pressure of the lipid film, π_0, which reflects the packing of the lipids in the monolayer; (2) the concentration, C, of the protein in the subphase; and (3) the change ($\Delta\pi$) in MLF surface pressure upon interaction with the peptide dissolved in the subphase. As far as possible, C must be low enough to minimize (and thus neglect) the surface pressure, $\Delta\pi_8\Delta^{W}$, induced by adsorption of the protein at a clean

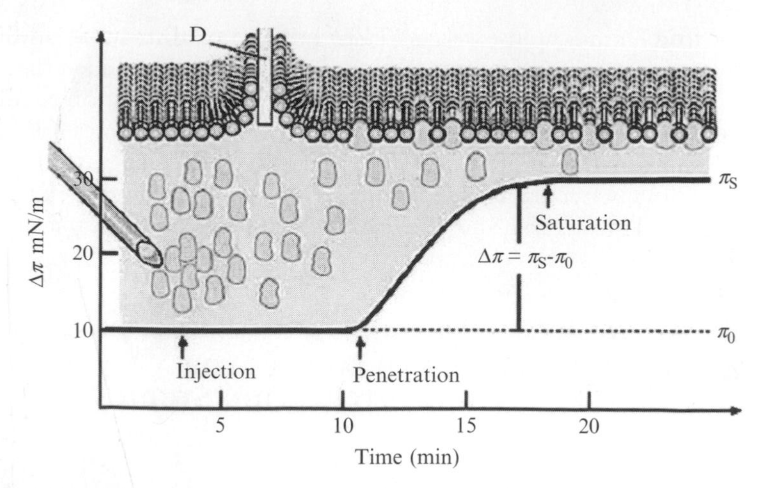

Figure 13.2 Schematic representation of a phospholipids monolayer experiment. The injection of the active compound and its penetration in the preformed film induces a surface pressure increase from π_0, the lateral pressure of the lipid monolayer, to π_S (i.e., $\pi_0 + \Delta\pi$), the lateral pressure of the monolayer after lipophilic compound insertion. A: intercalating agent (i.e., protein); B: injection tube; C: lipid monolayer (i.e., *E. coli* lipid extract), and D: Wilhelmy wire probe.

air–water interface under the same experimental conditions. The lipid film pressure at which the protein no longer penetrates the monolayer is called the exclusion pressure, π_s (an exclusion density, σ_{iex}, can also be defined; Maget-Dana, 1999).

5.2. Interaction of VHb with *E. coli* lipid monolayer

VHb interaction with MLF was performed using membrane lipids from *E. coli* (see previous section on *E. coli* TLE production). The film was obtained by spreading the TLE extract in a chloroform and methanol (2/1, v/v) mixture at an air–buffer (10 m*M* sodium phosphate at pH 7.2) interface. The surface pressure (π) was measured with a Wilhelmy wire probe attached to a microbalance (DeltaPi, Kibron Inc., Helsinki, Finland) connected to a personal computer in circular glass wells in which the subphase volume was 0.5 mL. After evaporation of lipid solvent and stabilization of monolayers at different initial surface pressures (π_0), various amount of the protein, from 0.05 to 1.5 μM, were injected into the subphase. The increments in surface pressure of the lipid film upon intercalation of the protein dissolved in the subphase were followed for the next 40 min at room temperature. The differences between the initial surface pressures and the values observed after the penetration of VHb into the film were taken as $\Delta\pi$.

Addition of a ferric VHb solution to the liquid phase underlying *E. coli* lipid monolayers caused a sudden increase in the film's surface pressure. Under the experimental conditions detailed earlier, the ?p shift was dependent on protein concentration, reaching a plateau around 0.5 μM VHb (Fig. 13.3, panel A). When data from similar measurements were analyzed in terms of $\Delta\pi$ versus π_0, the critical surface pressure corresponding to the lipid lateral-packing density preventing the intercalation of the protein into *E. coli* lipid films could be derived by extrapolating the $\Delta\pi - \pi_0$ slope to a $\Delta\pi$ of 0, giving a value of about 33 μN/m (Fig. 3, panel B).

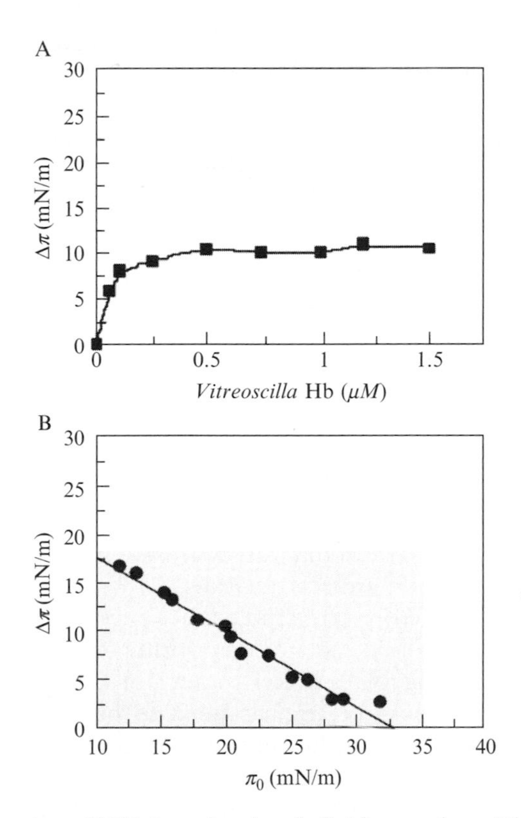

Figure 13.3 Insertion of VHb into the phospholipids monolayer. The increase in surface pressure is reported as a function of (A) VHb concentration in the subphase and (B) its initial value. *Panel A.* Various amounts of protein, to final concentrations ranging from 0.05 to 1.5 μM, were injected into the buffer (10 mM sodium phosphate at pH 7.2) subphase at an initial surface pressure varying between 19.8 and 20.2 mN/m. Data represent five independent measurements carried out on different protein and phospholipid preparations. The measurements were performed at room temperature. *Panel B.* Increments of surface pressure, $\Delta\pi$, of *E. coli* lipid films as a function of the initial surface pressure (π_0) at constant VHb concentration (0.5 μM). *X*-axis intercept is at 33 mN/m. Experiments were carried out under the same experimental conditions as in panel A.

The kinetics of protein insertion into the lipid monolayer were characterized by a rapid and marked increase in surface pressure that followed injection of the protein into the subphase; the lag phase for this process was too short to be measurable with our instrumentation. In a typical experiment, within 120 s of protein injection, π attained slightly more than 90% of that recorded at the end of the measurement. This initial peak was then followed by a slow increase in π for approximately the next 8 min, when a plateau was reached, and no more variation in π was detected up to 30 min. This kinetic pattern was independent of initial surface pressure and, to a lesser extent, protein concentration, because when low (0.05 to 0.25 μM) VHb concentrations were used, the initial spike in the increase of surface pressure was strongly diminished or negligible (not shown). As a whole, the data demonstrate that VHb efficiently penetrates the phospholipid monolayer. Control experiments demonstrated that the interaction of horse myoglobin with the membrane monolayer occurs but to a much lesser extent than with VHb and with very slow kinetics.

6. Lipid Binding Measurement by Absorption Spectroscopy

6.1. Overview

The high sensitivity of the heme-iron UV-visible absorption to iron ligand interactions or to changes in the dielectric properties of the solvent provides a most convenient way to monitor lipid–protein interaction, unique among all lipid-binding proteins. However, great care must be taken to single out possible artifactual phenomena that can be misinterpreted as binding to the heme. Possible source of artifacts include: (1) binding of cosolvents or impurities to the heme iron, (2) heme extraction within lipid micelles, (3) unspecific insertion of acyl side chains within the hydrophobic heme pocket, and (4) binding of carboxyl moiety of fatty acids to the ferric heme iron. All these possible artifacts need to be excluded by rigorous controls. Specific binding can be recognized only when (1) the spectral transition is fast (seconds or less after mixing) and complete, (2) the binding affinity is high (micromolar or below), and (3) only selected lipid species are recognized. Thus, specific binding of phospholipids can only be concluded after a complete screening process with adequate blank controls.

6.2. Lipid binding to VHb and HMP

Anomalous UV-visible spectral changes were first noticed during the VHb and HMP purification procedure. In particular, the visible absorption spectra of both the proteins before and after hydroxyapatite chromatography were different. Electrospray mass spectrometry, performed on HMP before the hydroxyapatite purification step, revealed the presence of heterogeneous protein-linked components attributed to phosphatidyl ethanolamine and phosphatidic acid derivatives, confirming previous observations in *A. eutrophus* flavohemoglobin (Ollesch *et al.*, 1999). Thereafter, simple UV-visible absorption experiments were performed to demonstrate that HMP and VHb are capable of interacting with bacterial lipid membranes and able to recognize specifically unsaturated and cyclopropanated fatty acids. These data added further focus to previous reports on VHb, *A. eutrophus* flavohemoglobin, and HMP that posited an unspecific interaction with membrane phospholipids (Ollesch *et al.*, 1999, Ramandeep *et al.*, 2001).

Simple mixing of TLE solutions with ferric VHb or HMP gives rise to a clear spectroscopic change in the visible spectrum of the ferric heme (Fig. 13.4). In fact, the absorption spectrum of the unliganded derivative in both proteins corresponds to a typical pentacoordinated heme adduct in which the proximal histidine is the fifth ligand.

The assignment is demonstrated by the presence of a peak at 645 nm and a broad Soret band, characterized by a low molar absorptivity (110 $mM^{-1}cm^{-1}$), and centered around 403 nm (Boffi *et al.*, 1999). In contrast, the spectrum of the TLE-saturated species displays a sharp peak at 407 nm and a shallower one at 625 nm and is similar to that of high-spin heme-iron adducts observed in hemoglobins and myoglobins in the presence of weak ligands such as fluoride (Antonini *et al.*, 1971, Giangiacomo *et al.*, 2001). Thus, the spectral profile of lipid-bound HMP or VHb is

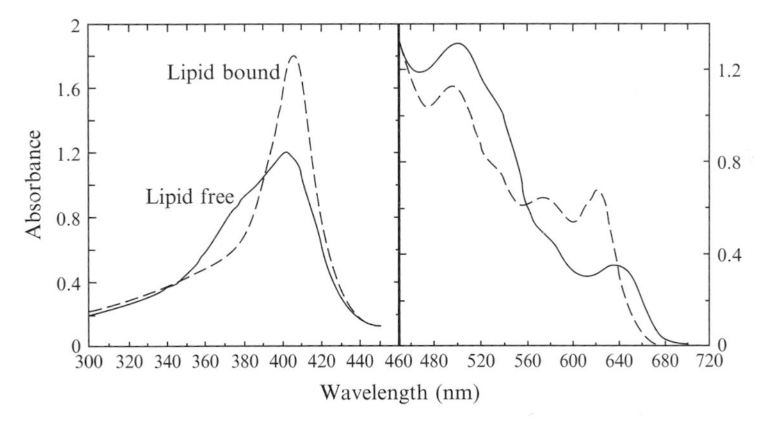

Figure 13.4 UV-visible absorption spectra of ferric HMP and VHb in their lipid-bound and lipid-free derivatives. The absorption spectra of ligand-free (continuous line) and ligand-bound (dashed line) HMP and VHb were measured on a 12-μM ferric protein solution in 0.1 *M* phosphate buffer at pH 7.0 and 20°. The spectrum of the lipid-bound species was obtained after addition of 10-fold molar excess of total *E. coli* lipid extract (dissolved in a 3:1 ethanol:chloroform mixture, 1 μl over a total volume of 3.5 ml).

diagnostic of the formation of a hexacoordinated species in which the sixth coordination position is occupied by a component of the fatty acid chain. The UV-visible spectra achieved by mixing lipid-free HMP or VHb with TLE or unsaturated fatty acids and cyclopropanated fatty acids were perfectly superimposable on that observed in the ferric proteins before hydroxyapatite chromatography, thus providing evidence that bound phospholipids are removed by this chromatographic step. No spectral change was detected by mixing the proteins with ethanol solutions of methyl ester derivatives of unsaturated and cyclopropanated fatty acids, saturated phospholipids, or saturated fatty acids, indicating that a free negative charge is essential for recognition and binding to the protein.

A number of phospholipids (e.g., dipalmitoleyl phosphatidil choline, dipalmit phosphatidil ethanolamine, dipalmit phosphatidic acid), saturated (e.g., palmitic and stearic), unsaturated (e.g., oleic, palmitoleic, linoleic, cis-vaccenic), and cyclopropanated (e.g., 9,10-methylen palmitoleic) fatty acids and their methyl esters have been screened for heme visible absorption spectral changes. The screening was performed on 10 to 12 μM ferric protein solution in 0.1 M phosphate buffer at pH 7.0 and 20°, containing 20% v/v ethanol by adding increasing amount of ligand (1 to 100 molar excess). All reagents were from Sigma Adrich Co., with the exception of cyclopropan palmitoleic acid (Larodan Fine Chemicals, Malmö, Sweden). Only unsaturated or cyclopropanated fatty acids or their phosphatidic acid esters were able to bind to the protein.

A further advantage of the spectrophotometric method concerns the possibility of investigating the kinetics of the hemoglobin–lipid binding process, thus allowing for an independent estimate of the affinity of bacterial hemoglobins. Fatty acid binding kinetics were performed by mixing 10 μM HMP or 10 μM VHb in 0.1 M phosphate buffer at pH 7 and 20°, containing 20% v/v ethanol with a solution containing increasing amounts of palmitoleic, cyclopropan palmitoleic, or linoleic acid for HMP and cyclopropan palmitoleic for VHb dissolved in the same buffer in an Applied Photophysics stopped-flow apparatus (Leatherhead, UK) at the selected wavelength of 410 nm.

Fatty acid concentrations higher than 100 μM could not be obtained due to low solubility, and a full pseudo-first-order plot could not be produced. The experimental conditions after mixing were protein concentration, 5 μM, and fatty acid concentrations, 50 and 25 μM in the presence of ethanol 20% (Fig. 13.5). The apparent second-order rate constant obtained for HMP with linoleic acid was around 2×10^6 $M^{-1}s^{-1}$.

Fatty acid release kinetics were estimated in competition essays with the lipid avid hydroxyalkoxypropyl-dextrane resin (Type X, Sigma Alrich Co.). A total of 5 ml of a gel suspension containing 20 mg of dried resin in 0.1 M phosphate buffer at pH 7 and 20% v/v ethanol were placed in a small beaker under stirring at 20°. The fiberoptic dip probe of a Varian Cary 50

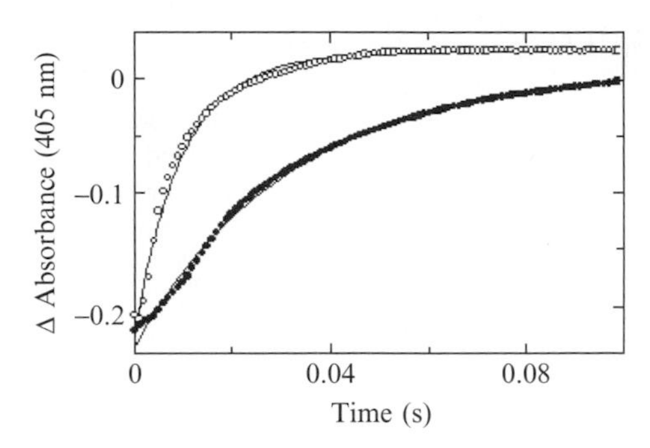

Figure 13.5 Kinetics of linoleic acid binding to ferric HMP. Time courses obtained in rapid mixing experiments at 405 nm are shown together with single exponential fitting curves (continuous lines). Apparent pseudo-first-order rates were 105 and 31 s^{-1}, respectively. Experimental conditions were: 5 μM protein concentration (after mixing); 5×10^{-5} M (Υ) and 2.5×10^{-5} M (λ) linoleic acid concentration (after mixing); 0.1 M phosphate buffer pH 7.0, containing 20% ethanol, at 20°.

spectrophotometer, equipped with a 150 W pulsed Xe lamp (Varian Instrument Co., Australia) was immersed in the suspension, and a baseline was recorded. A HMP solution (100 μl, 160 μM) saturated with oleic, cyclopropan-palmitoleic, or linoleic acid was then added to the suspension, and spectra were measured in sequential mode (9 s for each spectrum). Fatty acid release was monitored by following the spectral changes in the Soret region. Fatty acid release kinetics were measured by following the absorbance decreasing at 405 nm. The results obtained indicate that palmitoleic and cyclopropan-palmitoleic acids are rapidly released from HMP with a $k > 0.1\ s^{-1}$. The binding constant, estimated from ligand binding and release data, is about $4 \times 10^7 M^{-1}$ (or $k_d = 25$ nM), similar to that of true-type fatty acid binding proteins (Balendiran *et al.*, 2000).

7. Conclusions

The study of the interaction of bacterial globins with lipid membranes suggested that VHb and HMP most likely work at the interface between the cytosol and the bacterial inner membrane and are permanently saturated with a phospholipid esterified with either cyclopropanated or unsaturated fatty acids (Rinaldi *et al.*, 2006). Further investigations on the lipid binding properties of bacterial globins will be necessary to determine whether lipid binding is a general property of bacterial hemoglobins and flavohemoglobins and whether its physiological significance is related to nitric oxide dioxygenase or alkylhydroperoxide reductase activities (Bonamore *et al.*, 2003a).

ACKNOWLEDGMENTS

Grants FIRB 2003 and PRIN 2004 to Alessandra Bonamore from the Ministero dell'Università e della Ricerca are gratefully acknowledged. This work was partially supported by a grant from Instituto Pasteur, Fondazione Cenci Bolognetti.

REFERENCES

Antonini, E., and Brunori, M. (1971). Hemoglobin and myoglobin in their reactions with ligands. *In* "Specific aspects of the reactions of hemoglobin with ligands" (A Neuberger and EL Tatum, eds.), pp. 235–260. North Holland Publishing Co., Amsterdam.

Balendiran, G. K., Schnutgen, F., Scapin, G., Borchers, T., Xhong, N., Lim, K., Godbut, R., Spener, F., and Sacchettini, J. C. (2000). Crystal structure and thermodynamic analysis of human brain fatty acid-binding protein. *J. Biol. Chem.* **275,** 27045–27054.

Bligh, E. G., and Dyer, W. J. (1959). A rapid method of total lipid extraction and purification. *Can. J. Biochem. Physiol.* **37,** 911–917.

Boffi, A., Das, T. K., della Longa, S., Spagnolo, C., and Rousseau, D. L. (1999). Pentacoordinate hemin derivatives in sodium dodecyl sulfate micelles: Model systems for the assignment of the fifth ligand in ferric heme proteins. *Biophys. J.* **77,** 1143–1149.

Bonamore, A., Gentili, P., Ilari, A. S., Schinina, M. E., and Boffi, A. (2003a). *Escherichia coli* flavohemoglobin is an efficient alkylhydroperoxide reductase. *J. Biol. Chem.* **278,** 22272–22277.

Bonamore, A., Farina, A., Gattoni, M., Schininá, M. E., Bellelli, A., and Boffi, A. (2003b). Interaction with membrane lipids and heme ligand binding properties of *Escherichia coli* flavohemoglobin. *Biochemistry* **42,** 5792–5801.

Brockman, H. L. (1999). Lipid monolayers: Why use half a membrane to characterize protein-membrane interactions? *Curr. Opin. Struct. Biol.* **9,** 438–443.

Cronan, J. E., Jr. (2002). Phospholipid modifications in bacteria. *Curr. Opin. Microbiol.* **5,** 202–205.

Duzgunes, N. (2003). Preparation and quantitation of small unilamellar liposomes and large unilamellar reverse-phase evaporation liposomes. *Methods Enzymol.* **367,** 23–27.

Ermler, U., Siddiqui, R. A., Cramm, R., and Friedrich, B. (1995). Crystal structure of the flavohemoglobin from *Alcaligenes eutrophus* at 1.75Å resolution. *EMBO J.* **14,** 6067–6077.

Giangiacomo, L., Mattu, M., Arcovito, A., Bellenchi, G., Bolognesi, M., Ascenzi, P., and Boffi, A. (2001). Monomer-dimer equilibrium and oxygen binding properties of ferrous *Vitreoscilla* haemoglobin. *Biochemistry* **40,** 9311–9316.

Khosla, C., and Bailey, J. E. (1989). The *Vitreoscilla* hemoglobin gene: Molecular cloning, nucleotide sequence and genetic expression in *Escherichia coli. Mol. Gen. Genet.* **214,** 158–161.

Lewis, T., Nichols, P. D., and McMeekin, T. A. (2000). Evaluation of extraction methods for recovery of fatty acids from lipid-producing microheterotrophs. *J. Microbiol. Methods* **43,** 107–116.

Liu, C., He, Y., and Chang, Z. (2004). Truncated haemoglobin O of *Mycobacterium tuberculosis*: The oligomeric state change and its interaction with membrane components. *Biochem. Biophys. Res. Commun.* **316,** 1163–1172.

Maget-Dana, R. (1999). The monolayer technique: A potent tool for studying the interfacial properties of antimicrobial and membrane-lytic peptides and their interactions with lipid membranes. *Biochim. Biophys. Acta* **1462,** 109–140.

Ollesch, G., Kaunzinger, A., Juchelka, D., Schubert-Zsilavec, M., and Ermler, U. (1999). Phospholipid bound to the flavohemoprotein from *Alcaligenes eutrophus*. *Eur. J. Biochem.* **262,** 396–405.

Orgambide, G. G., Reusch, R. N., and Dazzo, F. B. (1993). Methoxylated fatty acids reported in *Rhizobium* isolates arise from chemical alterations of common fatty acids upon acid-catalyzed transesterification procedures. *J. Bacteriol.* **175,** 4922–4926.

Ramandeep, H. K. W., Raje, M., Kim, K. J., Stark, B. C., Dikshit, K. L., and Webster, D. A. (2001). *Vitreoscilla* hemoglobin: Intracellular localization and binding to membranes. *J. Biol. Chem.* **276,** 24781–24789.

Rinaldi, A. C., Bonomore, A., Macone, A., Boffi, A., Bozzi, A., and Di Giulio, A. (2006). Interaction of *Vitreoscilla* hemoglobin with membrane lipids. *Biochemistry.* **45,** 4069–4076.

Zhao, H. X., Rinaldi, A. C., Rufo, A., Bozzi, A., Kinnunen, P. K. J., and Di Giulio, A. (2003). Structural and charge requirements for antimicrobial peptide insertion into biological and model membranes. *In* "Pore-forming peptides and protein toxins" (G. Menestrina M. Della Serra, and P. Lazarovici, eds.) pp. 151–177. Taylor and Francis, London.

CHAPTER FOURTEEN

Assessment of Biotechnologically Relevant Characteristics of Heterologous Hemoglobins in *E. coli*

Pauli T. Kallio, Christian J. T. Bollinger, Taija Koskenkorva, *and* Alexander D. Frey

Contents

Abstract

The use of the heterologous bacterial hemoglobin (VHb) from *Vitreoscilla* to enhance growth and productivity of *Escherichia coli* under conditions of oxygen limitation has been one of the foremost examples of metabolic engineering. Although VHb has earned its merits during the last two decades by providing enhanced physiological enhancements to organisms from all kingdoms of life, it has been the candidate of choice primarily for historical reasons. Findings made

Institute of Microbiology, ETH Zürich, Zürich, Switzerland

Methods in Enzymology, Volume 436
ISSN 0076-6879, DOI: 10.1016/S0076-6879(08)36014-5

during the last years, however, suggest that hemoglobin and flavohemoglobin proteins from bacterial species other than *Vitreoscilla* or artificially generated mutant proteins or fusion variants of hemoglobins and flavohemoglobins may be better suited for use in biotechnological processes. This account provides guidelines for the assessment of biotechnologically relevant characteristics conferred by such novel heterologous hemoglobins and flavohemoglobins in *E. coli*.

1. Introduction

The commercial production of a variety of desirable biologicals uses the enzymatic capability of various microorganisms, plant, and cell cultures in bioreactors and typically has a high demand for oxygen. However, problems arise from the very low solubility of oxygen in water and are further accentuated during large-scale and high-cell-density production processes having very high growth rates, typical of industrial applications. Decreasing the oxygen transfer rate lowers oxygen concentration in the culture and, concomitantly, reduces the specific growth rate. This further leads to the shift from aerobic, exponential growth to microaerobic growth, which has major adverse effects on culture physiology. In *E. coli* cultivations this often leads to accumulation of several by-products, such as ethanol and acetate, and has negative effect on productivity, cell viability, and biomass concentration (Konz *et al.*, 1998). Usually, sterile oxygen is sparged directly into the bioreactor. However, this leads to a high oxygen partial pressure and potentially oxidizing conditions, which can cause damage to the product and decrease the fitness of cells. The high demand for oxygen can also be partially satisfied by improving process parameters such as mixing, by using high-efficiency dispersion systems to increase the rate of transport of the limiting substrate, or by modifying medium composition. In spite of all these improvements, limitations of desired cellular capabilities by oxygen remain a problem during various processes, and therefore more gentle methods for overcoming the oxygen deprivation and its negative effects have been sought.

Genetic approaches can be used to install systems, which are more efficient in utilizing the desired substrate, in this case oxygen, for useful metabolic activities. Therefore, bacterial flavohemoglobins and hemoglobins (collectively called *globins* in this chapter) have been of interest due to their proven potential in improving cell growth and productivity when expressed under low–oxygen conditions. Three classes of globin proteins have been identified in bacteria, namely, hemoglobins (Hbs); flavohemoglobins (flavoHbs), which comprise an amino-terminal globin and a carboxy-terminal reductase domain with binding sites for flavin adenine dinucleotide (FAD) and NAD(P)H; and truncated hemoglobin proteins, which lack the α-helical elements that are normally present in globin proteins (reviewed in Frey and Kallio, 2003; Frey and Kallio, 2005; Wu *et al.*, 2003). Truncated globin proteins are 20 to 40 amino acids shorter than normal hemoglobins and fold via a two-on-two helix structure into a globinlike conformation (Wittenberg *et al.*, 2002). The authentic globin proteins show the classical globin fold and are capable of reversibly binding oxygen. They all are also able to protect bacterial cells and heterologous hosts against nitrosative stress and NO *in vivo* (Frey *et al.*, 2002; Frey *et al.*, 2004; Gardner, 2005). Despite an almost random variation in amino acid sequence, globin proteins adopt a typical globin architecture having six to eight α-helical structural elements (Perutz, 1969). Three helices on each side of the oxygen-binding pocket, which contains the heme, form a sandwich-like assembly. In all known globin proteins, His-F8 (topological position of histidine in helix F) is conserved and required for coordinating the heme iron. The characteristics and biochemical properties of various bacterial globins have also been thoroughly described previously (Frey *et al.*, 2002; Frey *et al.*, 2004; Gardner, 2005; Wu *et al.*, 2003) and in various chapters of this volume; therefore, they are not discussed further in this chapter.

So far, only bacterial flavoHbs and Hbs have been used in biotechnological applications (Bollinger *et al.*, 2001; Frey *et al.*, 2000; Khosla and Bailey, 1988). The use of VHb, which was the first identified hemoglobin from bacteria, to enhance growth and metabolite production in heterologous *E. coli* under microaerobic conditions has been studied extensively (Frey and Kallio, 2003; Kallio *et al.*, 2001; Wu *et al.*, 2003). The natural habitats of the obligatory aerobic, Gram-negative *Vitreoscilla*, such as stagnant ponds and decaying vegetable matter, are extremely poorly oxygenated. In response to oxygen starvation, *Vitreoscilla* synthesizes significantly greater quantities of homodimeric hemoglobin capable of reversibly binding oxygen. This suggests that *Vitreoscilla* could employ VHb production as a natural physiological strategy to maintain its aerobic metabolism and growth under oxygen limitation. Motivated by this hypothesis, the *vhb* gene was isolated and expressed in *E. coli* under oxygen-limited conditions. The first experiments revealed that VHb provides biotechnologically advantageous functions to the host cells (Khosla *et al.*, 1990; Khosla and Bailey, 1988). The results obtained with heterologous expression of VHb show that this technology can generally be applied to improve oxygen-limited growth and metabolism of industrially important bacterial strains and to introduce beneficial properties into plants (Farrés and Kallio, 2002; Frey and Kallio, 2003; Häggman *et al.*, 2003; Wilhelmson *et al.*, 2005).

Although the exact biochemical mechanism of VHb expression on cell metabolism has to be analyzed more thoroughly, the recent experiments show that VHb-positive *E. coli* cells are able to generate 50% higher proton flux per reduced O_2 molecule and have fivefold higher amounts of energetically more favorable bo_3 complexes relative to the *vhb*-negative counterparts. The efficiency of proton flux is an important energetic determinant and leads to proton motive force, which is utilized for ADP phosphorylation during proton reentry from the periplasm to the cytoplasm. Consequently, these properties lead to elevated ATP levels and 65% higher ATP turnover rate (Kallio *et al.*, 1994; Tsai *et al.*, 1996b). Higher ATP synthesis could also increase the amount of available GTP, which is needed for efficient translation. This assumption is supported by the flow field-flow fractionation results of Nilsson *et al.* (1999) showing that VHb-positive *E. coli* cells have increased amounts of tRNA molecules, more than 2-fold higher numbers of translationally active 70S ribosome complexes, and a 1.6-fold increase of a cloned marker enzyme activity.

The methodology of heterologous expression of VHb seems to be especially effective in aerobic mycelium-forming microorganisms, such as the most common antibiotic producers, which are usually grown in a highly viscous culture environment. Therefore, oxygen delivery for such cellular processes remains a challenge in biotechnology. For example, VHb has been successfully expressed in *Streptomyces coelicolor* and *Streptomyces rimosus* and increases in the specific production rates of actinorhodin (10-fold) and oxytetracycline (2.2-fold), respectively, have been reported (Mangolo *et al.*, 1991). In *Acremonium chrysogenum*, production of cephalosporin C was increased 3.2-fold (DeModena *et al.*, 1993). The most notable example of heterologous VHb expression to improve antibiotic production of an industrial *Saccharopolyspora erythraea* strain has been described by Brünker *et al.* (1998) and Minas *et al.* (1998). A single copy of the *vhb* gene was integrated into the genome using the phage ϕ[CN1]C31 attachment site, and VHb production significantly increased erythromycin production in *S. erythraea* when industrial medium formulation and cultivation parameters were used. It should also be noted that the fermentation parameters were remotely monitored, and the dissolved oxygen levels in culture broth were controlled by increasing the stirrer and aeration rates if the dissolved oxygen level fell below 45% of air saturation. These results suggest that the cell pellets of the original production strain, having high respiration rates, may still suffer from oxygen limitation, and this limitation can be, at least partially, alleviated by VHb expression. The final reported erythromycin concentration was 7.4 g/l for VHb-expressing strain relative to control (4.0 g/l), and the space-time yield was enhanced approximately 100% in the VHb-positive strain (1.1 g of erythromycin/l/day versus 0.56 g of erythromycin/l/day in the controls). Unpublished experiments also suggest that the production of erythromycin can be enhanced further beyond 8 g/l by improving the process parameters for the VHb-expressing *S. erythraea* strain (P. T. Kallio and W. Minas, 2000, unpublished). The detailed description of the erythromycin process parameters for VHb-expressing *S. erythraea* strain can be found in the review by Minas (2005).

Generation of VHb mutants, construction of chimeric and fused hemoglobin proteins, and screening of new globin proteins have also been pursued to adapt hemoglobin technology to certain biotechnological applications (Andersson *et al.*, 2000; Bollinger *et al.*, 2001; Frey *et al.*, 2000; Kallio *et al.*, 2007; Roos *et al.*, 2002). We have successfully expanded this approach to significantly improved levels by using *Ralstonia eutropha* flavoHb (FHP), by constructing fusion proteins, and by generating VHb mutants (Andersson *et al.*, 2000; Frey *et al.*, 2000). The expression of novel globin constructions has been explored in microaerobic *E. coli* using fed-batch bioreactor cultivations.

Using the BLAST server at the National Center for Biotechnology Information, various microbial genomes have been screened for the presence of hemoglobin and flavohemoglobin genes, which could beneficially be used in biotechnological production processes to improve cell growth and metabolic properties under oxygen-limited conditions (Bollinger *et al.*, 2001; Kallio *et al.*, 2007). The effects of novel globins on growth behavior and by-product formation of *E. coli* have been studied using microaerobic fed-batch cultivations. Such screening experiments have revealed that it is possible to identify new bacterial globin proteins, which possess even better growth-promoting characteristics than VHb-expressing *E. coli*. For example, an approximate improvement of 52 and 34% in the yield of biomass on glucose was measured with *E. coli* cells expressing either *Deinococcus radiodurans* or *Bacillus halodurans* flavoHb, respectively, relative to VHb-producing control.

A similar fed-batch cultivation scheme, relative to cultivations utilizing wild-type globins, was also used to screen engineered hemoglobin-reductase fusion proteins and error-prone PCR generated *vhb* mutants for improved growth of *E. coli* (Andersson *et al.*, 2000; Frey *et al.*, 2000). One such candidate protein, a monomeric flavoHb (FHP) encountered in *R. eutropha*, is induced 20-fold by limited oxygen supply in its natural host. Frey *et al.* (2000) generated chimeric VHb-fusion proteins carrying the reductase domain of FHP at the C-terminus (VHb-Red). Similarly, the *fhp* gene was engineered to express various modules of flavoHb proteins. The microaerobic fed-batch cultivations showed that cells expressing the engineered VHb-Red protein were able to reach 75% higher final optical cell densities than the VHb-producing *E. coli* controls.

Alternative attempts to improve cell growth and/or protein productivity were based on the generation of VHb mutant proteins. Random mutants of the *vhb* gene carrying additionally an N-terminal extension (MTMITPSF) were generated and the selected mutants were assessed in terms of enhancing growth of *E. coli* during microaerobic fed-batch cultivations. The best VHb mutant protein had two mutations at positions His36Arg (C1) and Gln66Arg (E20) of the polypeptide and was able to elicit a 1.5-fold greater growth enhancement relative to wild-type VHb-expressing *E. coli*

MG1655 cells. In addition, the specific growth rate was almost 50% higher in the strain expressing mutated VHb ($0.49\ h^{-1}$) relative to the VHb-positive control ($0.33\ h^{-1}$; Andersson *et al.*, 2000).

Several attempts to downscale the fed-batch methods with respect to volume, time, and labor have been found to be nonapplicable. In small-scale cultivations (e.g., shake flasks and 96-deep well plates), it is difficult to reach and maintain cultivation conditions that would correspond to the industrial processes, especially in terms of aeration and cell density (T. Koskenkorva, A.D. Frey, and P.T. Kallio, 2007, unpublished). Thus, the previously mentioned positive effects of globin expression in heterologous hosts cannot be obtained in small-scale cultivations. However, the microaerobic fed-batch process developed in our laboratory has been simplified to the extent that it can be performed with very basic equipment using just manual control, if necessary at all.

2. Expression Strategy

Heterologous globin expression should be only an auxiliary measure in biotechnological processes, leaving as many cellular resources available as possible for the generation of the actual product. Thus, ensuring appropriate expression levels and accurate timing of expression to avoid metabolic burdening of the cell with the production of an extra protein in excessive amounts, or when it is not needed, is critical when adopting new globin proteins for biotechnological purposes.

2.1. Gene placement and plasmid copy number

First attempts to enhance biotechnological properties of *E. coli* under microaerobic conditions used amplification of the *vhb* gene by subcloning it into a derivative of pUC19. The high copy number of pUC-based plasmids, however, leads to excessive levels of intracellular hemoglobin and, consequently, its extrusion into the periplasm (Ramandeep *et al.*, 2001) or formation of inclusion bodies, especially when *E. coli* JM101 cells were used (Hart *et al.*, 1994).

An improved approach was the positioning of the *vhb* gene as a single copy on the chromosome. This solution provided good results with respect to the enhancement of productivity and activity for selected reporter proteins (Khosla *et al.*, 1990). Such an approach has the advantage of sparing the cell from the burden of plasmid replication, leaving metabolic capacities for the production of the actual product. Additionally, chromosomal integration provides greater genetic stability in *E. coli* and other microorganisms (e.g., *Bacillus subtilis*) than does a plasmid-based approach (Kallio *et al.*, 1987).

A good compromise with respect to copy number is the subcloning of a globin gene into a pBR322 derivative. Plasmids derived from pBR322 are maintained in the cells at a few dozen copies and not hundreds, as is the case with pUC derivatives. A dosage study for VHb carried out by Tsai *et al.* (1996a) indicates that the copy numbers provided by a pBR322 derivative are probably better suited for auxiliary flavoHb or Hb production in *E. coli* than chromosomal integration. VHb expression under the control of the IPTG-inducible P_{tac} promoter, from the pBR322 derivative pKTV1, was titrated in *E. coli* W3110 cells. The results revealed that the maximal growth-enhancing effect was 3.4 μmol/g (dry cell weight) when VHb production was induced with 0.5 m*M* IPTG. Considering that the chromosomal integration strategy provides a several-fold lower copy number, reaching an equally high expression level for VHb would, most likely, require prohibitively high amounts of IPTG.

2.2. Transcriptional regulation

At least in the case of VHb, it has been shown that its expression does not confer any enhancement to *E. coli* physiology under fully aerobic growth conditions. It seems reasonable to assume that this is also the case for other Hb and flavoHb proteins, which is why their expression should be restricted to microaerobic cultivation stages through the use of an inducible expression system. One such inducible expression system, used for VHb expression in the pioneering work of Khosla and Bailey (1988), employs the native *vhb* promoter, which is induced exclusively under microaerobic conditions. This scheme has the advantage that the transcriptional regulation is directly coupled to oxygen limitation; therefore, neither supervision of the culture nor external manipulations are required to keep the hemoglobin gene silent as long as oxygen is amply available. However, a distinct disadvantage of this system is that it relies on an intrinsic cultivation parameter (i.e., oxygen concentration) that cannot be adjusted to alter the induction strength without drastically affecting physiology. Furthermore, regulation by the FNR-dependent *vhb* promoter (Tsai *et al.*, 1995) excludes the possibility of preadapting the cell to microaerobic conditions by inducing hemoglobin expression before a critical drop in DO occurs. Khosla and Bailey (1989) have shown that the increase in active VHb trails the onset of microaerobiosis by a considerable amount of time when the *vhb* gene is subject to transcriptional regulation from its native promoter in *E. coli*.

Therefore, we recommend an induction scheme that relies on the timed addition of an external inducer, such as IPTG. Induction should be carried out shortly before the culture reaches microaerobic conditions. The optimal time point for induction is indicated by dissolved oxygen (DO) values falling in the range between 5 and 10% of oxygen saturation, which can be monitored using any standard, commercially available, polarographic DO electrode.

Giving general recommendations pertaining to the optimal expression level for Hb and flavoHb proteins in biotechnology is difficult. The most diligent, and probably the only way, to determine optimal level of globin expression is to carry out titration experiments by running a series of cultivations with increasing inducer concentrations until the desired production parameters reach an optimum. Such a study requires tremendous effort, and although it may be a valid optimization step, once a specific Hb or flavoHb protein has been chosen for a particular bioprocess, it is not practical for research purposes, particularly in comparative studies that involve more than one globin protein.

3. Considerations for the Choice of Growth Medium

3.1. Overview

The described cultivation protocol to assess the biotechnological suitability of heterologous globins attempts to mimic oxygen limitation as an *E. coli* culture might encounter it in a real-life bioreactor cultivation. All nutrients, besides oxygen, should therefore be supplied in sufficient amounts to ensure that oxygen remains the limiting nutrient throughout the entire cultivation.

In principle, a rich, complex media formulation would be sufficient to fulfill this criterion. In fact, Khosla and Bailey (1988) obtained an excellent degree of discrimination between the growth rate of a VHb-overexpressing strain and that of its VHb-free counterpart in a microaerobic fed-batch cultivation using PYA medium (0.25% peptone, 0.25% yeast extract, 200 mg/l Na-acetate, adjusted to pH 7.5 with 1 *M* NaOH) for both batch and feed phases.

Because, however, most bioprocesses utilizing *E. coli* do not involve complex media, but rather—for obvious reasons of reproducibility and cost—rely on a defined formulation with glucose as the major carbon source, we recommend the use of the following media formulation, which is a balanced mixture of common macronutrients supplemented with vitamins and trace elements typically found in complex media components such as yeast extract:.

1. *Glucose batch medium*: 4 g/l glucose, 1.5 g/l $(NH_4)_2SO_4$, 4.35 g/l K_2HPO_4, 1.5 g/l KH_2PO_4, 0.5 g/l NaCl, 10 ml/l vitamin mix, 1 ml/l trace element mix, 1 m*M* $MgSO_4$, 0.05 m*M* $CaCl_2$, 0.2 m*M* $FeCl_3$.
2. *Feed medium*: 250 g/l glucose, 110 g/l $(NH_4)_2SO_4$, 8 g/l $MgSO_4$, 10 ml/l vitamin mix, 1 ml/l trace metal mix, 0.05 m*M* $CaCl_2$, 0.2 m*M* $FeCl_3$.

3. *Vitamin mix*: 0.05 mg/l thiamine, 42 mg/l riboflavin, 540 mg/l Ca-panthotenate, 600 mg/l niacin, 140 mg/l pyridoxine•HCl, 6 mg/l biotin, 4 mg/l folic acid.
4. *Trace element mix*: 3.6 mg/l $(NH_4)_6Mo_7O_{24}$, 24.8 mg/l H_3BO_3, 4 mg/l $CoCl_2$, 2.4 mg/l $CuSO_45H_2O$, 16 mg/l $MnCl_2{\cdot}4H_2O$, 2.8 mg/l $ZnSO_47H_2O$.

Trace element mix and vitamin mix are prepared in distilled water and sterile filtered through 0.2-μm filters (Millipore). All chemicals can be obtained from commercial vendors. Stock solutions of the glucose batch medium components are autoclaved (at 121° for 20 min, and at 103 kPa above the atmospheric pressure [= 15 psi = 1.05 kg/cm^2]) except glucose (50% [w/v] stock solution), which is autoclaved at 121° for 15 min and 103 kPa.

For plasmid maintenance, both batch and feed medium are also supplemented with appropriate antibiotics (e.g., ampicillin 100 mg/l). In some cases, foaming can also occur. Foaming can be a serious problem when *E. coli* cells are grown in a rich medium (e.g., LB), and this can be prevented by using, for example, antifoams and/or specific foam brakers. To prevent foaming during bioreactor cultivations, we routinely add 1 ml/l of Structol J637 (Schill & Seilacher GmbH, Hamburg, Germany), diluted 1:20 in dH_2O, to the vessels before autoclaving. Antifoam can also be added (100 μl – 1 ml) to the reactors whenever needed during fed-batch cultivations.

4. Cofactor Imitation

Two metabolites that deserve special attention in globin biotechnology are the cofactors heme and FAD. Suboptimal concentrations of either Hb or flavoHb may result from cofactor limitation. The question as to what, if anything, can be done to adapt the cultivation medium to the increased biosynthetic needs caused by overexpression of a flavoHb is briefly addressed in the following paragraphs.

Saturated intracellular concentrations of active, soluble VHb have been shown to lie below 4 μmol/g (dry cell weight) at the very end of 30-h microaerobic fed-batch cultivations in a glucose minimal medium (Tsai *et al.*, 1996a). Although VHb does not contain flavin, it seems reasonable to assume that the maximal concentration of active flavoHb would lie in approximately the same range, since it has been shown that heme, and not protein synthesis, is the factor actually limiting the maximal level of active VHb (Hart *et al.*, 1994). In *E. coli* cells, VHb has been shown to accumulate at a fairly constant rate of 0.2 μmol/g (dry cell weight)/h during complex media shake-flask cultivations. Mulrooney (1997) showed that the production capacity of *E. coli* for FAD is approximately 0.6 μmol/g wet cell weight after 18 h of aerobic batch growth in 2xYT medium. This corresponds to around 6 μmol/g (dry cell weight), assuming a 90% H_2O content of the cell matter, and clearly exceeds the assumed maximal flavoHb concentration. Thus, the FAD productivity is likely to be sufficient for the synthesis of biotechnologically useful amounts of flavoHb.

To take into account the eventual case where FAD should nevertheless be limiting, our media formulation contains riboflavin, an immediate precursor for FAD. It should be noted, however, that *E. coli* is thought to be incapable of actively importing riboflavin (Bacher *et al.*, 1996). Thus, riboflavin uptake is governed by its diffusion rate through the cell membranes, a rate that is probably too slow to keep up with the synthesis rate of a flavoprotein.

Heme availability, as mentioned previously, is known to be limiting for the overexpression of bacterial hemoglobins in *E. coli*. Thus, if it would be possible to increase heme synthesis by supplementing the culture medium with appropriate precursors, an increase in maximal intracellular hemoglobin concentration would be feasible and, possibly, desirable. Attempts to overcome heme limitation in VHb-overexpressing *E. coli* have been made by Hart *et al.* (1994). They added extracellular δ-aminolevulinate (ALA), the product of the alleged rate-limiting enzymatic reaction in heme biosynthesis, but found that, in amounts ranging from 20 to 100 m*M*, ALA did have a negative effect on intracellular heme and holohemoglobin content. These findings are contrasted by the results obtained by Verderber *et al.* (1997), who found that, in an *E. coli* strain expressing recombinant human hemoglobin at moderate levels, intracellular heme concentration could be raised approximately seven-fold (free plus bound heme). Verderber *et al.* (1997) used ALA in submillimolar amounts and argue that the high concentrations used by Hart *et al.* (1994) may have been high enough to cause a deleterious effect on the general fitness of the cells. This seems unlikely because addition of ALA in the study of Hart *et al.* (1994) had a negative impact only on the relative portion of VHb containing heme, not on the total level of VHb, as would be expected upon impairment of general cellular fitness. On the basis of current information available about the addition of ALA to cultures overexpressing VHb, we advise against doing so, also for strains expressing flavoHbs.

5. Microaerobic Fed-Batch Cultivations

5.1. Overview

We have developed various fed-batch processes to assess the capabilities of globin proteins to support growth of a recombinant host organism under microaerobic conditions. In order to mimic the limited oxygen supply, often encountered in industrial large-scale processes, both oxygen supply and stirrer speed are reduced. This results in the generation of microaerobic growth conditions (concentration of $O_2 = 0.02$ mmol/l). For screening purposes, *E. coli* MG1655 cells are routinely used.

5.2. Small-scale bioreactor cultivations (300 ml)

Cultivations of *E. coli* MG1655 are performed under microaerobic conditions in a SixFors Bioreactor unit (Infors AG, Bottmingen, Switzerland) in fed-batch mode (Frey *et al.*, 2001). The SixFors Bioreactor unit allows six processes to run parallel at a small scale. One can control and adjust stirrer speed, temperature, pH, and DO levels. Furthermore, it contains three built-in pumps per reactor; two for pH regulation (addition of base and acid) and one for the supply of feed medium. The latter pump can also be used to control foaming by addition of antifoam. The bioreactors can be controlled either from the control unit or from associated IRIS-software, which can acquire and store process parameters and provides the necessary tools for graphics and analysis. Due to their small size, the assembled bioreactors can be sterilized (at 121° for 20 min, and 103 kPa) in tabletop autoclaves.

1. The vessels are filled with water, $(NH_4)_2SO_4$, K_2HPO_4, KH_2PO_4, NaCl, and $MgSO_4$ solutions (total working volume is 300 ml after all medium component additions) and sterilized at 121° for 20 min, and 103 kPa. The vessels are cooled down and the rest of the medium components (glucose, vitamin mix, trace element mix, 0.05 m*M* $CaCl_2$, 0.2 m*M* $FeCl_3$) are added in a sterile bench. Note that the pH probes should always be calibrated before sterilization.
2. The process parameters for an oxygen-limited cultivation are the following: working volume 300 ml, stirrer speed 300 rpm, temperature 37°, aeration rate 120 ml/min of air, and pH 7.0 ± 0.2, adjusted with addition of either 3 *M* NaOH or 3 *M* H_3PO_4.
3. DO is monitored using standard commercially available polarographic O_2 electrodes, which are calibrated by flushing the vessels with N_2 gas (0%) and then with air (100%). Air flow and stirrer speed are kept constant, resulting in the reduction of DO to microaerobic levels (concentration of $O_2 = 0.02$ mmol/l) within 5 to 7 h after inoculation. Air flow-in and exhaust gases are sterilized using 1 μm filters (e.g., Pall Bacterial Air Vent filters, Gelman Laboratory, Ann Arbor, MI), which can be autoclaved and reused and allow typical air flow rates of 40 l/min at 0.4 bar.
4. Inocula are grown in 100-ml shake flasks, which contain 10 ml of LB-media supplemented with ampicillin (100 mg/ml), 250 rpm at 37° for 14 h. Inoculation of bioreactors (usually 1:100) can be standardized to obtain a starting optical density at wavelength 600 nm (OD_{600}) of 0.2. The expression of the globin proteins is induced with addition of IPTG to a final concentration of 0.1 m*M* at $OD_{600} = 1$.

5. The fed-batch phase is divided into two modes with constant feed rates. Feeding is started at OD_{600} of 2 by supplementing a concentrated glucose minimal salt solution at a rate of 1 ml/h (mode 1). The feeding rate is increased to 2 ml/h (mode 2) when cultures reached an OD_{600} of 4.5 and is kept constant until the end of the cultivation. A typical process is depicted in Fig. 14.1.
6. The contents of the vessels are routinely autoclaved (at 121° for 20 min, and 103 kPa) at the end of the cultivations. Biosafety issues should also be taken into account, although such small-scale cultivations are usually classified as nontoxic, nonharmful level 1 fermentations.

5.3. Laboratory-scale bioreactor cultivations (1.3 l)

A SixFors bioreactor system cannot be found in all laboratories; therefore, a protocol for the assessments of physiological consequences of heterologous globin expression in *E. coli* using a regular laboratory-scale bioreactor,

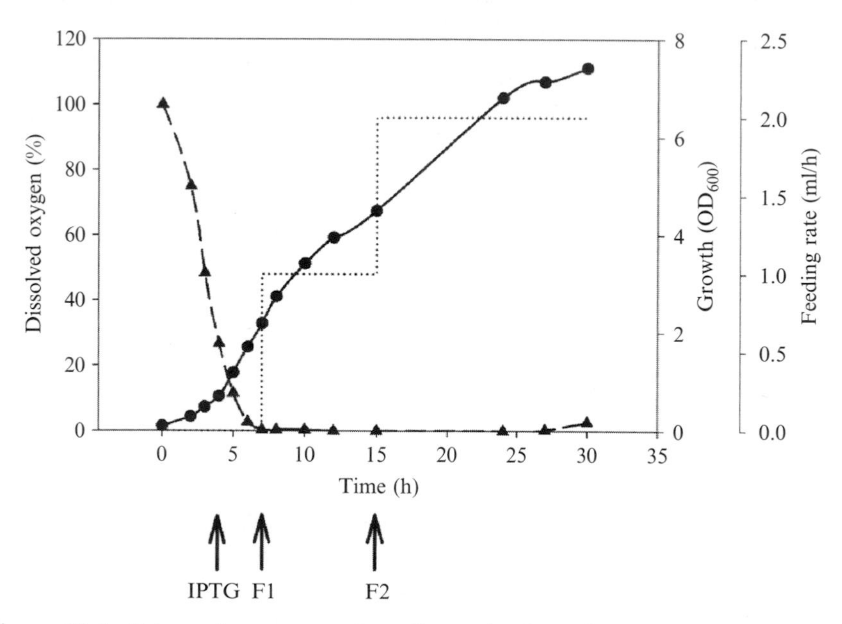

Figure 14.1 Schematic representation of growth of *E. coli* MG1655 cells expressing VHb. Microaerobic conditions are used to mimic growth of cells during high-cell-density fed-batch cultivations. Aeration conditions and stirring speed are set in order to limit oxygen transfer. Therefore, DO concentrations (black triangles) drop close to 0% of air saturation within the batch phase, where it remains for the whole process. Induction of VHb expression is started at an OD_{600} of 1 by addition of IPTG (cell growth is shown by black dots). Induction is performed at a time before dissolved oxygen reaches 0%. Feeding mode 1 (F1) is started at OD_{600} of 2 after the end of the exponential growth phase. Feed media is supplied at a constant rate of 1 ml/h and the feed rate is increased to 2 ml/h (F2) when cell density reaches an OD_{600} reading between 4.0 and 4.5.

working volume usually 1 to 2 l, is also given below. The protocol can easily be adapted (with slight modifications, if necessary) to other scales and vessels and is described here to give guidance. Usually, such reactors can also be controlled and the progress of the process monitored from external sources.

1. Microaerobic fed-batch cultivations can also be performed in any laboratory-scale fermentor such as a 2.5-l (total volume of the reactor vessel) KLF2000 bioreactor (Bioengineering AG, Wald, Switzerland) using starting volume of 1.3 l. The vessel can be autoclaved (at 121° for 20 min, and 103 kPa) *in situ* and the thermosensitive medium components (see previously) are added aseptically afterwards through membrane-covered ports with syringes and needles.
2. Cultivation parameters are: 37°, aeration rate 0.65 l/min, stirrer speed 800 rpm, and pH 7.0 ± 0.1, which is maintained by adding either 3 *M* NaOH or 3 *M* H_3PO_4. DO is monitored using an O_2 electrode, which is calibrated by flushing the vessels with N_2 gas (0%) and then with air (100%). We are using reusable 0.2 μm Acro 50 Vent filters (Pall, Gelman Laboratory, Ann Arbor, MI) to sterilize air flow-in and regular 1 μm Bacterial Air Vent filters (Pall) for exhaust gases.
3. Precultures are cultivated in 20 ml of either glucose batch medium or LB medium on a rotatory shaker 250 rpm, at 37° for 14 h, and 13 ml of cultures are used to inoculate the bioreactor (1:100 inoculum).
4. Feeding rate is 10 ml/h of feed medium (corresponding to 1 g of glucose/h) during the fed-batch mode 1, which is usually started 7 h post inoculation, when the OD_{600} is approximately 2.5. Fed-batch mode 2 (20 ml/h of feed medium corresponding to 2 g of glucose/h) is initiated 11.5 h post inoculation, when the OD_{600} value is approximately 7 ± 0.5, and the feeding rate is kept constant until the end of the microaerobic cultivations, which are usually run for 24 to 30 h.
5. The vessel can again be autoclaved (at 121° for 20 min, and 103 kPa) *in situ* at the end of the cultivation.

6. Physiological Analysis

6.1. DO, growth, carbon dioxide production, and oxygen uptake

During the cultivations the DO concentration of the cultures is monitored with a polarographic O_2 electrode (Mettler-Toledo, Switzerland), and growth of the cultures is followed spectrophotometrically (PerkinElmer) at a wavelength of 600 nm.

Furthermore, if desired, the consumption of O_2 and production of CO_2 can be measured. Off-gas (O_2 and CO_2) is, for example, measured online using a PRISMA600 mass spectrometer (FISONS Instr., Middlesex,

England). Mass fraction of consumed O_2 and produced CO_2 is normalized to the N_2-fraction in supply, and outlet air and can be used to calculate, for example, specific O_2 uptake [q_{O2}, mmol/g/min] and CO_2 evolution rates [q_{CO2}, mmol/g/min] (Bollinger *et al.*, 2001; Kallio *et al.*, 2007).

Furthermore, harvesting samples for the determination of cell dry weight is highly desired and can be done as follows. Culture (10 ml) is aliquoted into two to three preweighed glass tubes, and cells are harvested by centrifugation at 4,000 rpm, at 4° for 15 min using a cooled tabletop centrifuge. Cells are washed once with 5 ml of 0.9% (w/v) NaCl solution, harvested at 4,000 rpm, at 4° for 10 min, dried at 115° for 5 h, and allowed to cool for 12 h in a desiccator before measurements.

6.2. By-product formation

Samples for the measurement of glucose and by-product concentrations are collected throughout the cultivation. Cells are removed by centrifugation at 15,000 rpm, at 4° for 2 min, and supernatant is stored at −20°. Due to the oxygen-limited or partially anaerobic conditions, the metabolism of *E. coli* switches to mixed acids fermentation. Under these conditions, the tricarboxylic acid cycle functions in a branched fashion with the production of succinate and oxalacetate. Furthermore, acetate, ethanol, formate, and lactate are major products of the glucose metabolism under anaerobic conditions. Our previous results of ^{13}C glucose labeling of proteinogenic amino acids indicated that the change from aerobic to anaerobic metabolism occurs to varying extents, depending on the globin protein expressed (Frey *et al.*, 2001). The formation of formate from pyruvate is especially dependent on the intracellular oxygen levels. Pyruvate-formate lyase contains an oxygen sensitive radical that is critical for its activity. Exposure to oxygen abolishes the catalytic activity. Our previous data have shown that the carbon flux through the pyruvate-formate lyase catalyzed reaction differs among *E. coli* strains expressing different globin proteins (Frey *et al.*, 2001). Acetate is excreted into the medium as one of the major products of the fermentation of sugars. However, it is also produced and excreted in significant amounts during aerobic growth on glucose as a result of the overflow metabolism.

Various analytical methods such as GC, HPLC, MS, or enzymatic assays, most of which are also commercially available, can be used for the quantification of the metabolites and glucose concentration in the medium. Samples for glucose, acetate, and ethanol analyses are centrifuged at 15,000 rpm, at 4° for 15 min, and supernatants are stored at −20° until further analysis. We have routinely measured glucose, acetate, formate, and ethanol concentrations enzymatically using commercial kits in a Beckman Synchron CX5CE autoanalyzer (Frey *et al.*, 2000). Because the assays for acetate, formate, and ethanol rely on an NAD(P)H-dependent reaction, the quantification can easily be performed using a spectrophotometer capable of measuring the turnover of NAD(P)H at 340 nm. Therefore, self-defined enzyme assays based on standard protocols (Bergmeyer and Grassl, 1983) can also be used to measure, for example, D-lactate, succinate, formate, and pyruvate concentrations.

7. Verification of Expression of Globin Proteins

The expression of active globin proteins can be qualitatively verified by CO difference spectra (Webster and Liu, 1974). No quantitative data can be gained, because the extinction coefficients of the different globin proteins differ and are generally not known.

7.1. CO-difference spectrum

1. Culture (25 to 100 ml) is withdrawn at the end of microaerobic fed-batch cultivations and immediately stored on ice. Cells are harvested by centrifugation 5,000 rpm, at 4° for 15 min. Supernatant is discarded, pellet is resuspended to OD_{600} value of 100, or higher, in a lysis buffer (100 m*M* Tris-HCl, pH 7.5, 50 m*M* NaCl, 1 m*M* EDTA, 1 m*M* PMSF, 1 m*M* β-Mercaptoethanol), and cells are broken using a French pressure cell (AMINCO, SLM Instruments, Urbana, IL) at 7,582 kPa (=1,100 psi). The cell suspension is passed three times through the outlet. Cell debris is removed by centrifugation 15,000 rpm, at 4° for 15 min, using a tabletop centrifuge. The supernatant containing intracellular proteins is transferred to a fresh tube and centrifuged again 15,000 rpm, at 4° for 2 min.
2. Intracellular soluble protein extract containing heterologous globin proteins is reduced by addition of a few granules of $Na_2S_2O_4$ to reduce the heme iron to the ferrous state. The reduced protein extracts are aliquoted into two separate 1-ml cuvettes. CO is a stronger heme ligand than oxygen and forms a stable complex. Gaseous CO is bubbled into one of the cuvettes and the CO difference spectra (reduced + CO reduced) are recorded between 400 and 500 nm in a conventional spectrophotometer or using a 96-well plate reader (e.g., SpectraMAX Plus spectrophotometer; Molecular Devices, Sunnyvale, CA).

A typical CO difference spectrum for bacterial globin proteins displays an absorption peak approximately at 420 and an absorption minimum at 440 nm.

7.2. Western blotting

To compare the expression levels of different globin proteins, C-terminally tagged proteins, carrying, for example, a His-tag or a Myc-tag, can be constructed and their production levels can be analyzed using specific antibodies or commercially available conjugates. Such tags can be added during PCR either to the 5′- or 3′-end of the gene using specifically designed oligonucleotides or can already be present in suitable expression vectors (e.g., pHIS [Farrés *et al.*, 2005] or vectors from commercial sources [e.g., Qiagen, Basel, Switzerland]). For example, the biochemical properties of C-terminally His-tagged globin proteins have been shown to be identical to those of nontagged Hb and flavoHb proteins (Farrés *et al.*, 2005). This is an important observation for biotechnological applications, because the use of a His-tagged globin protein does not affect its biochemical properties in terms of ligand binding and release.

Therefore, the use of such tags is easy and highly recommended because cumbersome protein purification steps to raise antibodies against a specific protein can be avoided. However, the Western blotting method cannot discriminate between biologically active globins and globins lacking the prosthetic heme group. We are using standard SDS-PAGE and Western blotting protocols to detect heterologously expressed globin proteins and therefore, methods are not described in this chapter.

REFERENCES

Andersson, C. I. J., Holmberg, N., Farrés, J., Bailey, J. E., Bülow, L., and Kallio, P. T. (2000). Error-prone PCR of *Vitreoscilla* hemoglobin (VHb) to support the growth of microaerobic *Escherichia coli*. *Biotechnol. Bioeng.* **70,** 446–455.

Bacher, A., Eberhardt, S., and Richter, G. (1996). Biosynthesis of riboflavin. *In* "*Escherichia coli* and *Salmonella*: Cellular and molecular biology," 2d ed. (F. C., Neidhardt, I. R., Curtiss, J. L., Ingraham, E. C. C., Lin, K. B., Low, B., Magasanik, W. S., Reznikoff, M., Riley, M., Schaechter, and H. E., Umbarger, eds.), pp. 657–664. ASM Press, Washington, DC.

Bergmeyer, J., and Grassl, M. (1983). Methods of enzymatic analysis, Vol. 3. Enzymes 1: Oxidoreductases, transferases. 3d ed. Verlag-Chemie, Weinheim, Basel.

Bollinger, C. J. T., Bailey, J. E., and Kallio, P. T. (2001). Novel hemoglobins to enhance microaerobic growth and substrate utilization in *Escherichia coli*. *Biotechnol. Prog.* **17,** 798–808.

Brünker, P., Minas, W., Kallio, P. T., and Bailey, J. E. (1998). Genetic engineering of an industrial strain of *Saccharopolyspora erythraea* for stable expression of the *Vitreoscilla* haemoglobin gene (*vhb*). *Microbiology* **144,** 2441–2448.

DeModena, J. A., Gutiérrez, S., Velasco, J., Fernández, F. J., Fachini, R. A., Galazzo, J. L., Hughes, D. E., Mornon, J. P., and Martín, J. F. (1993). The production of cephalosporin C by *Acremonium chrysogenum* is improved by the intracellular expression of a bacterial hemoglobin. *Bio/Technol.* **11,** 926–929.

Farrés, J., and Kallio, P. T. (2002). Improved cell growth in tobacco suspension cultures expressing *Vitreoscilla* hemoglobin. *Biotechnol. Prog.* **18,** 229–233.

Farrés, J., Rechsteiner, M. P., Herold, S., Frey, A. D., and Kallio, P. T. (2005). Ligand binding properties of bacterial hemoglobins and flavohemoglobins. *Biochemistry* **44,** 4125–4134.

Frey, A. D., and Kallio, P. T. (2003). Bacterial hemoglobins and flavohemoglobins: Versatile proteins and their impact on microbiology and biotechnology. *FEMS Microbiol. Rev.* **27,** 525–545.

Frey, A. D., and Kallio, P. T. (2005). Nitric oxide detoxification: A new era for bacterial globins in biotechnology? *Trends Biotechnol.* **23,** 69–73.
Frey, A. D., Bailey, J. E., and Kallio, P. T. (2000). Expression of *Alcaligenes eutrophus* flavohemoprotein and engineered *Vitreoscilla* hemoglobin-reductase fusion protein for improved hypoxic growth of *Escherichia coli. Appl. Environ. Microbiol.* **66,** 98–104.
Frey, A. D., Farrés, J., Bollinger, C. J. T., and Kallio, P. T. (2002). Bacterial hemoglobins and flavohemoglobins for alleviation of nitrosative stress in *Escherichia coli. Appl. Environ. Microbiol.* **68,** 4835–4840.
Frey, A. D., Oberle, B. T., Farrés, J., and Kallio, P. T. (2004). Expression of *Vitreoscilla* haemoglobin in tobacco cell cultures relieves nitrosative stress *in vivo*, and protects from NO *in vitro*. *Plant Biotechnol. J.* **2,** 221–231.
Frey, A. D., Fiaux, J., Szyperski, T., Wüthrich, K., Bailey, J. E., and Kallio, P. T. (2001). Dissection of central carbon metabolism of hemoglobin-expressing *Escherichia coli* by ^{13}C nuclear magnetic resonance flux distribution analysis in microaerobic bioprocesses. *Appl. Environ. Microbiol.* **67,** 680–687.
Gardner, P. R. (2005). Nitric oxide dioxygenase function and mechanism of flavohemoglobin, hemoglobin and myoglobin and their associated reductases. *J. Inorg. Biochem.* **99,** 247–266.
Häggman, H., Frey, A. D., Aronen, T., Ryynänen, L., Julkunen-Tiitto, R., Tiimonen, H., Pihakaski-Maunsbach, K., Jokipii, S., Chen, X., and Kallio, P. T. (2003). Expression of *Vitreoscilla* hemoglobin in hybrid aspen (*Populus tremula* × *tremuloides*). *Plant Biotechnol. J.* **1,** 287–300.
Hart, R. A., Kallio, P. T., and Bailey, J. E. (1994). Effect of biosynthetic manipulation of heme on insolubility of *Vitreoscilla* hemoglobin in *Escherichia coli. Appl. Environ. Microbiol.* **60,** 2431–2437.
Kallio, P. T., Frey, A. D., and Bailey, J. E. (2001). From *Vitreoscilla* hemoglobin (VHb) to a novel class of growth stimulating hemoglobin proteins. *In* "Recombinant protein production with prokaryotic and eukaryotic cells: A comparative view on host physiology," (Merten, O.-W., Mattanovich, D., Lang, C., Larsson, G., Neubauer, P., Porro, D., Postma, P., Teixeira de Mattos, J., and Cole, J.A., eds.), pp. 75–87. Kluwer Academic Publishers, Dordrecht, The Netherlands.
Kallio, P., Palva, A., and Palva, I. (1987). Enhancement of α-amylase production by integrating and amplifying the α-amylase gene of *Bacillus amyloliquefaciens* in the genome of *Bacillus subtilis. Appl. Microbiol. Biotechnol.* **27,** 64–71.
Kallio, P. T., Kim, D.-J., Tsai, P. S., and Bailey, J. E. (1994). Intracellular expression of *Vitreoscilla* hemoglobin alters *Escherichia coli* energy metabolism under oxygen-limited conditions. *Eur. J. Biochem.* **219,** 201–208.
Kallio, P. T., Heidrich, J., Koskenkorva, T., Bollinger, C. J. T., Farrés, J., and Frey, A. D. (2007). Analysis of novel hemoglobins during microaerobic growth of HMP-negative *Escherichia coli. Enzyme Microb. Technol.* **40,** 329–336.
Khosla, C., and Bailey, J. E. (1988). Heterologous expression of a bacterial haemoglobin improves the growth properties of recombinant *Escherichia coli. Nature* **331,** 633–635.
Khosla, C., and Bailey, J. E. (1989). Characterization of the oxygen-dependent promoter of the *Vitreoscilla* hemoglobin gene in *Escherichia coli. J. Bacteriol.* **171,** 5995–6004.
Khosla, C., Curtis, J. E., DeModena, J., Rinas, U., and Bailey, J. E. (1990). Expression of intracellular hemoglobin improves protein synthesis in oxygen-limited *Escherichia coli. Bio/Technol.* **8,** 849–853.
Konz, J. O., King, J., and Cooney, C. L. (1998). Effects of oxygen on recombinant protein expression. *Biotechnol. Prog.* **14,** 393–409.
Mangolo, S. K., Leenutaphong, D. L., DeModena, J. A., Curtis, J. E., Bailey, J. E., Galazzo, J. L., and Hughes, D. E. (1991). Actinorhodin production by *Streptomyces coelicolor* and growth of *Streptomyces lividans* are improved by the expression of a bacterial hemoglobin. *Bio/Technol.* **9,** 473–476.

Minas, W. (2005). Production of erythromycin with *Saccharopolyspora erythraea*. *In* "Methods in biotechnology, microbial processes and products," Vol. 18 (Barredo, J. L., ed.), pp. 65–90. Humana Press Inc., Totowa, NJ.

Minas, W., Brünker, P., Kallio, P. T., and Bailey, J. E. (1998). Improved erythromycin production in a genetically engineered industrial strain of *Saccharopolyspora erythraea*. *Biotechnol. Prog.* **14,** 561–566.

Mulrooney, S. B. (1997). Application of a single-plasmid vector for mutagenesis and high-level expression of thioredoxin reductase and its use to examine flavin cofactor incorporation. *Protein Expr. Purif.* **9,** 372–378.

Nilsson, M., Kallio, P. T., Bailey, J. E., Bülow, L., and Wahlund, K.-G. (1999). Expression of *Vitreoscilla* hemoglobin in *Escherichia coli* enhances ribosome and tRNA levels: A flow field-flow fractionation study. *Biotechnol. Prog.* **15,** 158–163.

Perutz, M. F. (1969). Structure and function of hemoglobin. *Harvey Lect.* **63,** 213–261.

Ramandeep, H. K. W., Raje, M., Kim, K. J., Stark, B. C., Dikshit, K. L., and Webster, D. A. (2001). *Vitreoscilla* hemoglobin: Intracellular localization and binding to membranes. *J. Biol. Chem.* **276,** 24781–24789.

Roos, V., Andersson, C. I. J., Arfvidsson, C., Wahlund, K.-G., and Bülow, L. (2002). Expression of a double *Vitreoscilla* hemoglobin enhances growth and alters ribosome and tRNA levels in *Escherichia coli*. *Biotechnol. Prog.* **18,** 652–656.

Tsai, P. S., Hatzimanikatis, V., and Bailey, J. E. (1996a). Effect of *Vitreoscilla* hemoglobin dosage on microaerobic *Escherichia coli* carbon and energy metabolism. *Biotechnol. Bioeng.* **49,** 139–150.

Tsai, P. S., Kallio, P. T., and Bailey, J. E. (1995). Fnr, a global transcriptional regulator of *Escherichia coli*, activates the *Vitreoscilla* hemoglobin (VHb) promoter and intracellular VHb expression increases cytochrome *d* promoter activity. *Biotechnol. Prog.* **11,** 288–293.

Tsai, P. S., Nägeli, M., and Bailey, J. E. (1996b). Intracellular expression of *Vitreoscilla* hemoglobin modifies microaerobic *Escherichia coli* metabolism through elevated concentration and specific activity of cytochrome *o*. *Biotechnol. Bioeng.* **49,** 151–160.

Verderber, E., Lucast, L. J., Van Dehy, J. A., Cozart, P., Etter, J. B., and Best, E. A. (1997). Role of the *hemA* gene product and delta-aminolevulinic acid in regulation of *Escherichia coli* heme synthesis. *J. Bacteriol.* **179,** 4583–4590.

Webster, D. A., and Liu, C. Y. (1974). Reduced nicotinamide adenine dinucleotide cytochrome *o* reductase associated with cytochrome *o* purified from *Vitreoscilla*: Evidence for an intermediate oxygenated form of cytochrome *o*. *J. Biol. Chem.* **249,** 4257–4260.

Wilhelmson, A., Kallio, P. T., Oksman-Caldentey, K.-M., and Nuutila, A. M. (2005). Expression of *Vitreoscilla* hemoglobin enhances growth of *Hyoscyamus muticus* hairy root cultures. *Planta Medica* **71,** 48–53.

Wittenberg, J. B., Bolognesi, M., Wittenberg, B. A., and Guertin, M. (2002). Truncated hemoglobins: A new family of hemoglobins widely distributed in bacteria, unicellular eukaryotes and plants. *J. Biol. Chem.* **277,** 871–874.

Wu, G., Wainwright, L. M., and Poole, R. K. (2003). Microbial globins. *Adv. Microb. Physiol.* **47,** 255–309.

CHAPTER FIFTEEN

Applications of the VHb Gene *vgb* for Improved Microbial Fermentation Processes

Xiao-Xing Wei* *and* Guo-Qiang Chen†

Contents

Abstract

Dissolved oxygen (DO) plays an important role in cell growth, especially in industry-scale microbial production. To alleviate the defects of hypoxic conditions, *Vitreoscilla* hemoglobin (VHb) has been used to enhance respiration and energy metabolism by promoting oxygen delivery. Heterologous expression of VHb in a variety of hosts has been shown to improve cell growth, protein synthesis, metabolite productivity, and bioremediation under oxygen-restricted conditions. In this chapter, many well-studied areas are presented to illustrate the potential of VHb application in microbial metabolic engineering industry. Also, applications of the *vgb* promoter have been discussed.

* Department of Biological Sciences and Biotechnology, Tsinghua University, Beijing, China
† Multidisciplinary Research Center, Shantou University, Guangdong, China

Methods in Enzymology, Volume 436
ISSN 0076-6879, DOI: 10.1016/S0076-6879(08)36015-7

1. Introduction

Vitreoscilla is a filamentous Gram-negative bacterium genus belonging to the *Beggiatoa* genus. It is found in freshwater sediments and cow dung, where oxygen availability is limited (Khosla and Bailey, 1988b; Wakabayashi *et al.*, 1986). One species, *Vitreoscilla stercoraria*, is strictly aerobic, and to cope with hypoxic conditions, it has apparently evolved a strategy involving the synthesis of soluble hemoglobin (VHb). The globin has two identical subunits, each with a relative molecular mass of 15.8 kDa and two *b* hemes per molecule. VHb was initially believed to be a terminal oxidase and was called cytochrome *o* (Webster and Hackett, 1966; Webster and Liu, 1974). However, spectral, kinetic, and structural properties proved the hemoglobin character of this protein (Orii and Webster, 1986; Wakabayashi *et al.*, 1986). So far, VHb is the best-characterized member of the bacterial hemoglobin (Hb) proteins. Its coding gene (*vgb*) has been isolated and its three-dimensional structure recently solved (Dikshit and Webster, 1988; Khosla and Bailey, 1988b; Tarricone *et al.*, 1997).

Compared with other hemoglobins, VHb has a relatively low affinity for oxygen, with a K_D of 7.2 μ, almost equal to its affinity for CO ($K_D = 8\ \mu$) (Webster, 1988) (Table 15.1). This value is much greater than the dissociation constants for many other globins, which are in turn far lower than the oxygen concentrations expected to be encountered in bioreactors. VHb has an average oxygen association rate constant k_{on} of 78 $\mu M^{-1}s^{-1}$, while its oxygen dissociation rate constant k_{off} is 5600 s^{-1}, hundreds of times

Table 15.1 Kinetic constants for reactions of various hemoglobins with oxygen

Protein	k_{on} ($\mu M^{-1}s^{-1}$)	k_{off} (s^{-1})	K_D(nM)	References
VHb	78	5600	72,000	(Orii and Webster, 1986)
Barley Hb	9.5	0.0272	2.86	(Allocatelli *et al.*, 1994)
Soybean Lb	120	5.6	48	(Gibson *et al.*, 1989)
Parasponia Hb	165	15	89	(Bogusz *et al.*, 1988)
Sperm whale Mb	14	12	857	(Springer *et al.*, 1989)
Ascaris Hb	1.5	0.0041	2.7	(Gibson and Smith, 1965)

Notes: Abbreviations: k_{on}, oxygen association rate constant; k_{off}, oxygen dissociation rate constant; K_D, dissociation constant ($k_{off} \div k_{on}$).

greater than that of other globins. The unusually high k_{off} suggests that VHb releases oxygen rapidly and that it is designed to do so in the low range of oxygen concentrations encountered in the soil environment (Bulow *et al.*, 1999).

The functional investigation of VHb has suggested that VHb is especially suitable for engineering the energy metabolism of diverse organisms among all the currently known globins. However the mechanism of VHb action still remains unclear. On the basis of its biochemical characters and a great quantity of information from VHb-expressing *Escherichia coli*, Bailey *et al.* (1995) formulated a hypothesis concerning the mode of action of VHb in *E. coli*. The hypothesis states that the effective DO concentration in the cell is equal to the sum of the actual DO concentration and the concentration of oxygenated VHb when VHb is present (Kallio *et al.*, 1994). Thus, when VHb is present, the cells experience a more oxygenated environment than the external environment of that cell.

Under oxygen-limited conditions, heterologous expression of VHb has been reported to improve cell growth and protein synthesis of the hosts (Farres and Kallio, 2002; Khosla *et al.*, 1990; Khosla and Bailey, 1988a; Wei *et al.*, 1998). In this chapter, we give a short review of the application of VHb expression in the microbial fermentation process and discuss the industrial potential of this bacterial hemoglobin.

2. VHb Application in Microbial Metabolic Engineering

The industrial production of many desirable metabolites requires control and maintenance of growth conditions (e.g., temperature, pH value, DO concentration) for optimizing cell growth and production yields. When a bioprocess is scaled up from laboratory to industry, the key issue in any aerobic culture is the maintenance of DO concentration. This is due to the combination of low oxygen solubility in water and, with high cell density, high nutritional demand occurs. The importance of oxygen is related to both its primary use as a nutrient and its secondary effects on metabolism and physiology. Oxygen functions as a terminal electron acceptor in the electron transport chain. When electrons are transferred from reducing power (NADH and $FADH_2$) to oxygen through aerobic metabolism, more energy can be produced than by other electron acceptors through anaerobic or fermentative pathways (Ingledew and Poole, 1984). In contrast, oxygen influences metabolic changes, protein oxidation, DNA oxidation, and plasmid replication by its secondary effects (Konz *et al.*, 1998). During the cultivation of various microbial strains, insufficient oxygen supply can decrease the activity of cytochrome P-450 monooxygenases required for the processing of pathway intermediates into their final forms, resulting in overflow metabolism (i.e., the accumulation of intermediates as the primary products) (Urlacher *et al.*, 2004).

Webster (1988) and Bailey (1988) isolated the DNA fragment including the *vgb* gene, which encodes VHb protein and its native promoter respectively (Dikshit and Webster, 1988; Khosla and Bailey, 1988b). Southern-blot analysis indicated that only a single copy of this gene exists on the chromosome. Experimental data obtained by several studies showed that heterologous expression of VHb in *E. coli* resulted in enhanced growth and protein production under microaerobic conditions (Khosla *et al.*, 1990; Khosla and Bailey, 1988a; Khosravi *et al.*, 1990). Since then, VHb has been expressed in numerous microorganisms for the improvement of growth properties or metabolite production. The beneficial effects of VHb-expression technology on microbial metabolic engineering are well documented by Zhang *et al.* (2007) (Table 15.2, modified by the present authors). The following examples illustrate the application of VHb for microbial production of several products.

3. Effect of VHb on Protein Production

Expression of the *vgb* gene encoding VHb in various organisms was shown to improve microaerobic cell growth and enhance oxygen-dependent product formation. Khosla and Bailey (1998a) and Khosla *et al.* (1990) expressed the *vgb* gene in a multicopy plasmid in *E. coli* and found that the presence of VHb not only increased the final cell density but also improved protein synthesis in both total protein content and the activity of specific enzyme expressed under hypoxic conditions. Recently, some researchers coexpressed VHb and green fluorescent protein (GFP) in two typical industrial *E. coli* strains, BL21 and W3110, to observe the influence of VHb on foreign proteins expression in *E. coli* system (Kang *et al.*, 2002). VHb expression was under the oxygen-dependent promoter for self-tuning regulation via the natural changes in DO level during the cultivation. In shake-flask experiments, both VHb-positive *E. coli* strains showed a twofold enhancement of specific GFP fluorescence intensity; thus, there is linear correlation between fluorescence and GFP amount when compared to VHb negative strains in all cultures. Further research showed that coexpression of VHb could override the glucose-induced repression and result in steady expression of foreign proteins in strain W3110 (Kallio *et al.*, 1994).

In species other than *E. coli*, there are positive reports of VHb expression for elevated biomass yields and lower metabolite excretion during extended microaerobic, fed-batch cultivation. For *Bacillus subtilis*, an excellent system for protein production and secretion, 1.5-fold total protein was

Table 15.2 Effects of VHb expression in various bioprocesses

Categories	Organisms	Comments	References
Improved protein production	*Schwanniomyces occidentalis*	Increased alpha-amylase production and total protein secretion	(Suthar and Chattoo, 2006)
	P. pastoris	Enhanced growth and heterologous protein production	(Chien *et al.*, 2006)
	E. coli	Improved growth and alpha-amylase production	(Aydin *et al.*, 2000)
	B. subtilis	Enhanced total protein secretion and alpha-amylase production	(Kallio and Bailey, 1996)
	P. aeruginosa	Reduced glucose repression and increased L-asparaginase production	(Geckil *et al.*, 2006)
	E. coli	Increased total protein content	(Khosla *et al.*, 1990)
Elevated chemical production	*Gordonia amarae*	Improved growth and increased biosurfactant production	(Dogan *et al.*, 2006)
	Enterobacter aerogenes	Increased acetoin and butanediol production	(Geckil *et al.*, 2004a)
	E. coli	Improved growth and PHA accumulation	(Urlacher *et al.*, 2004; Yu *et al.*, 2002)
	A. hydrophila	Improved PHA accumulation and 3HHx content	(Ouyang *et al.*, 2005)
	Acetobacter xylinum	Improved yield and cellulose productivity	(Chien *et al.*, 2006)
Fortified antibiotic production	*A. chrysogenum*	Increased cephalosporin C production	(Khosla *et al.*, 1990)
	S. erythraea	Increased erythromycin production	(Brunker *et al.*, 1998; Minas *et al.*, 1998)

(*continued*)

Table 15.2 (*continued*)

Categories	Organisms	Comments	References
	S. cinnamonensis	Enhanced growth and monensin production	(Wen *et al.*, 2001)
	S. viridochromogenes	Increased level of bialaphos production	(Tabakov *et al.*, 2001)
Enhanced bioremediation	*P. aeruginosa*	Improved growth and degradation of 2,4-DNT	(Kallio *et al.*, 1994; So *et al.*, 2004)
	Xanthomonas maltophilia	Enhanced degradation of benzoic acid	(Webster and Liu, 1974)
	Burkholderia sp.	Improved growth and degradation of benzoic acid	(Kallio *et al.*, 1994)
	Burkholderia sp.	Improved growth and degradation of 2,4-DNT	(Lin *et al.*, 2003)
	B. cepacia	Enhanced degradation of 2-chlorobenzoate (2-CBA)	(Urgun-Demirtas *et al.*, 2006)
Physiological improvement	*Tremella fuciformis*	Enhanced growth	(Zhu *et al.*, 2006)
	E. coli	Increased copper uptake	(Khleifat, 2006)
	E. aerogenes	Increased sensitivity to mercury and cadmium	(Geckil *et al.*, 2004a,b)
	E. aerogenes	Improved growth and oxygen uptake rates	(Erenler *et al.*, 2004)
	P. pastoris	Increased beta-galactosidase activity	(Wu *et al.*, 2003)
	Yarrowia lipolytica	Improved growth and extracellular enzyme	(Bhave and Chattoo, 2003)
	E. coli	Enhanced resistance against the NO releaser sodium nitroprusside (SNP)	(Frey *et al.*, 2002)

secreted into its culture medium together with active VHb production (Kallio and Bailey, 1996). In addition, VHb-expressing *B. subtilis* cultures exhibited increases of approximately 30% and 5 to 17% in neutral protease activity and alpha-amylase activity, respectively, over the parental VHb-free strain (Kallio and Bailey, 1996). VHb had different effects in *Saccharomyces cerevisiae* and *Pichia pastoris*. In *S. cerevisiae*, results suggested that the action of VHb is likely linked to respiration and electron transfer, changing the carbon flux toward ethanol production (Chen *et al.*, 1994). Growth enhancement due to expression of VHb was observed only during the final stage of culture growth, when the acetaldehyde produced was used as a substrate. This effect was observed more clearly in an acetaldehyde fed-batch fermentation in which VHb-expressing cells grew to at least threefold higher in final cell density (Chen *et al.*, 1994). On the contrary, expression of VHb had no positive effect on cell growth but significantly enhanced the whole cell beta-galactosidase activity to a fourfold higher level under aerobic cultivation (Wu *et al.*, 2003).

In most of the previously mentioned experiments, the *vgb*-containing plasmids were constructed and expressed in the host cells in multiple copies. These experiments demonstrated the function and positive effect of VHb in microbial fermentations. However, the strains harboring plasmids with *vgb* are not suitable for industrial production in many cases because of the high cost of antibiotics to maintain the stability of the plasmids. With the maturation of DNA insertion techniques, several novel engineered strains were constructed to adapt to industrial scale fermentation via integration of the *vgb* gene into host chromosomes. In this way, a new *E. coli* strain G830 was obtained via homologous recombination of *vgb* into a specific location of the *E. coli* genome (Yang *et al.*, 2001). During high-cell-density cultivation, G830 entered the second exponential phase after the DO decreased to zero. This indicated that the VHb improved the efficiency of oxidation metabolism by participating in an oxygen-limited respiration chain, which offered an extra advantage in hypoxia growth. The final optical density (OD_{600}) and dry cell weight of G830 were enhanced by 55 and 47%. Chromosomal integration of the *vgb* gene was also achieved in other potential species and obtained good results (Chung *et al.*, 2001; Zhu *et al.*, 2006).

4. Effect of VHb on Polyhydroxyalkanoate Production

Poly(3-hydroxyalkanoates) (PHA), traditionally considered to be intracellular energy and carbon reserve materials that can be synthesized by many prokaryotes under unbalanced growth conditions, has been drawing significant interest in the areas of medicine, industry, and agriculture as a result of both its physical properties, similar to conventional plastics, and its other properties such as biodegradability and biocompatibility (Schubert *et al.*, 1988). The commercial applications of PHA have been hindered because of its higher cost of production than petrochemical-derived plastics. One of the major problems of low productivity is caused by insufficient oxygen supply. Traditional methods, such as enhancement of stirring and introduction of pure oxygen into the growth process, are usually characterized by high costs and low efficiency. Yu *et al.* (2002) constructed a recombinant *E. coli* VG1 (pTU14) for high-cell-density growth with the possibility of a low production cost of poly(3-hydroxybutyrate) (PHB), which is the simplest and most common PHA type. The *vgb* gene was integrated into the *E. coli* genome forming the VHb constitutively expressing strain VG1, while the PHB synthesis operon was strongly expressed on a plasmid. From the results in a 10-l fermentor, the presence of VHb had a positive effect in all the culture conditions. The enhancement level was dependent on the aeration level. Cell growth and PHB accumulation of VG1 (pTU14) were 3.4- and 4.2-fold higher than those of the VHb-negative control JM105 (pTU14), respectively, when the DO was maintained between 50 and 5% of air saturation. The superiority of VG1 (pTU14) increased to 1.3- and 1.8-fold, respectively, when the DO was above 50% of air saturation (Yu *et al.*, 2002).

Poly(3-hydroxybutyrate-co-3-hydroxyhexanoate) or PHBHHx is another type of PHA with better flexibility and biocompatibility in comparison with PHB and PHBV. The material properties are strongly influenced by the 3HHx fraction of PHBHHx (Favey *et al.*, 1995). As 3HHx content increases, the crystallinity and melting point decrease, while the flexibility and tractility increase. This laboratory has carried out many studies on the biological production of PHBHHx with different properties and investigated applications of PHBHHx in tissue engineering. VHb was used in *Aeromonas hydrophila* and recombinant *E. coli* to increase the PHBHHx yield and the 3HHx content (Ouyang *et al.*, 2005; Urlacher *et al.*, 2004). In *E. coli*, PHA content as high as 60.7% could be achieved via coexpression of the *vgb* and acyl-CoA dehydrogenase (*yafH*) genes together with the PHA synthesis operon. The VHb expression obviously enhanced the 3HHx monomer content from 12.3 mol% in the wild type to 28.5 mol % in the recombinant. *A. hydrophila* 4AK4 is a remarkable PHBHHx-producing strain in which PHA precursors can be synthesized from the β-oxidation pathway. In shake-flask experiments, VHb-expressing strain *A. hydrophila* 4AK4 (pVGAB) produced 20.5 g/L CDW containing 49.6% PHBHHx, while the wild type only produced 13.0 g/L CDW containing 31.5% PHBHHx. *A. hydrophila* 4AK4 (pVGAB) was further cultivated in a 6-L fermentor vessel with different carbon sources. CDW (54.0 g/L) containing 31.5% PHBHHx was obtained, and the 3HHx content could be controlled in the range from 7.4 mol% to 12.4 mol% (Ouyang *et al.*, 2005).

5. Effect of VHb on Antibiotic Production

Limited DO level significantly decreases the biosynthesis rate for antibiotic production. Tabakov *et al.* (2001) successfully modified antibiotic-producing *Streptomyces* strains by introducing VHb. Through intergeneric conjugation using conjugative-integrative plasmid vectors, the *vgb* gene was integrated into chromosomal DNAs of *Streptomyces* strains. Some exconjugants with stable inheritance were selected. Among these engineered strains, the maximal positive effect of *vgb* on antibiotic production was observed in *S. viridochromogenes* exconjugants in which the level of bialaphos production was five times higher than that of the control wild-type strain (Tabakov *et al.*, 2001). In another report, the industrial erythromycin production strain *Saccharopolyspora erythraea* displayed approximately a 60% increase in erythromycin productivity when chromosomally integrated with *vgb* (Brunker *et al.*, 1998). Other antibiotic-producing strains, such as *Streptomyces cinnamonensis* and *Acremonium chrysogenum*, showed similar improvements in antibiotic yield when engineered with VHb (Khosla *et al.*, 1990; Wen *et al.*, 2001). These results are encouraging because they point to the positive regulation of VHb in industrial antibiotic production.

6. Application of VHb in Bioremediation

Microorganisms have been widely used in the remediation of polluted soil and water. However, as oxygen was involved in some steps in the degradation pathways, limited oxygen in the subsurface of soil or water environments reduces the effectiveness of this microbial remediation method. VHb shows the ability to improve the oxygen usage for microorganisms and has been used to overcome the oxygen limitation problem in biodegradation process for contaminants under hypoxia conditions (Patel *et al.*, 2000; Urgun-Demirtas *et al.*, 2006) For example, oxygen was required for three reactions in the 2,4-dinitrotoluene (DNT) oxidative degradation pathway by a *Burkholderia* sp., formerly *Pseudomonas*. Expression of VHb was reported to enhance cell growth and DNT removal (Suen and Spain, 1993). Under oxygen-limited conditions (3.1 mg DO/L culture) with an influent DNT concentration of 214 mg/L, the effluent DNT concentration from the wild-type bioreactor reached more than 20 mg/L in 40 days, whereas it was up to only 1.7 mg/L for the bioreactor incubated with strains engineered with VHb in about 25 days (So *et al.*, 2004). In another study, *vgb* was integrated into the chromosomes of *Pseudomonas aeruginosa* and *Burkholderia cepacia* for the industrial application in benzoic acid and DNT degradation, respectively; the results were all positive (Chung *et al.*, 2001).

7. Application of *vgb* Promoter

The oxygen-dependent promoter of *vgb* (P_{vgb}) is induced under oxygen-limited conditions. P_{vgb} has been characterized in *E. coli* and has been shown to be functional in various heterologous hosts. P_{vgb} is regulated at the transcriptional level and is maximally induced under microaerobic conditions in both *Vitreoscilla* and *E. coli*, when the DO level is less than 2% of air saturation (Khosla and Bailey, 1988a, 1989). Transcriptional activity of P_{vgb} in *E. coli* is positively modulated by CRP (cAMP receptor protein) and FNR (fumarate nitrate reduction regulator) (Khosla and Bailey, 1989). The FNR protein is a global transcriptional regulator in *E. coli* that monitors the availability of oxygen in the environment. When the intracellular oxygen level changes, the FNR system controls the switch from aerobic to anaerobic metabolism by regulating the transcription of many oxygen-regulated genes (Unden and Schirawski, 1997). ArcA, another major oxygen sensor, was also found to be a positive regulator of P_{vgb}. To reach the maximal induction at low oxygen level, both ArcA and Fnr systems are needed (Yang *et al.*, 2005). Furthermore, induction of P_{vgb} can be repressed by the addition of a complex nitrogen source to the medium (e.g., yeast extract). This represents a third level of regulation of P_{vgb}, whose mechanism has not yet been studied in detail (Khosla *et al.*, 1990).

In bioreactor cultivations, approximately 10% of total cellular protein was expressed according to P_{vgb} under the maximally induced condition (Khosla *et al.*, 1990). Since high-cell-density batch cultures naturally progress toward an oxygen-limited regimen of growth, the P_{vgb} has a foreseeable prospect in industrial application. The induction of promoter activity can be achieved by lowering environmental DO levels without the need of expensive inducers such as IPTG. In our laboratory, several high-copy-number expression plasmids were constructed with P_{vgb} for highly efficient protein expression in *E. coli* under low oxygen conditions (Liu *et al.*, 2005). The GFP protein and toluene dioxygenase (TDO) were expressed under the control of P_{vgb}, without the addition of IPTG for induction. In shake-flask experiments, the GFP protein expression was enhanced six times by lowering aeration condition compared with that in strong aeration condition, while the TDO enzymatic activity was twofold stronger when produced under low aeration than in high aeration. In addition, the presence of VHb showed a positive effect on TDO productivity, probably because oxygen is required as the substrate of TDO (Liu *et al.*, 2005).

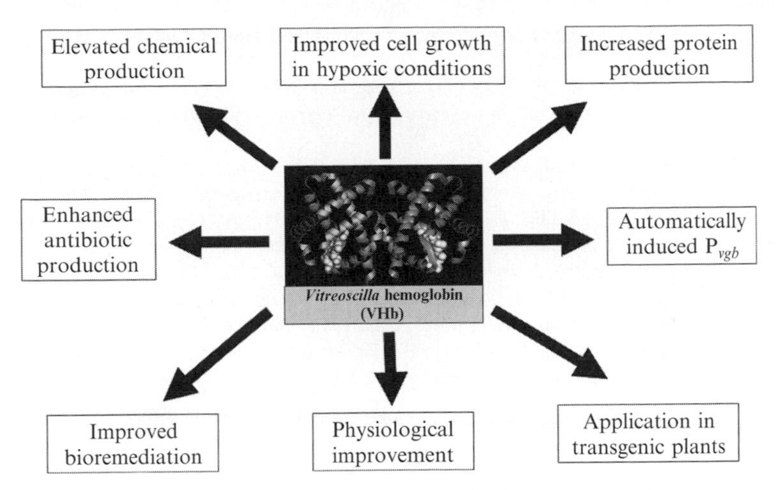

Figure 15.1 Potential applications of the *Vitreoscilla* hemoglobin (VHb) and its promoter. (See color insert.)

8. FUTURE PROSPECTS

VHb has not yet been widely used in biological engineering and industrial fermentation technology. In laboratory, VHb is a simple protein produced from bacteria with the natural function of transferring oxygen. Functional expression of VHb in a variety of hosts has led to various levels of improvement in industrial fermentation and metabolite production under oxygen-restricted conditions (Fig. 15.1). VHb has also been introduced in plants and even mammalian cells for improved oxygen uptake and better cell growth (Holmberg *et al.*, 1997; Pendse and Bailey, 1994). Although many encouraging results have been obtained in several areas, mechanisms of VHb action and regulation should be further elucidated.

Protein engineering gives a new way to improve the properties of VHb for biotechnological applications. By using molecular mutation methods, such as gene shuffling and rational or random evolution, VHb application is likely to be expanded and used on a wider scale.

REFERENCES

Allocatelli, C. T., Cutruzzola, F., Brancaccio, A., Vallone, B., and Brunori, M. (1994). Engineering Ascaris hemoglobin oxygen-affinity in sperm whale myoglobin: Role of tyrosine B10. *FEBS Lett.* **352,** 63–66.

Aydin, S., Webster, D. A., and Stark, B. C. (2000). Nitrite inhibition of Vitreoscilla hemoglobin (VHb) in recombinant *E. coli*: Direct evidence that VHb enhances recombinant protein production. *Biotechnol. Prog.* **16,** 917–921.

Bailey, J. E. (1995). Chemical engineering of cellular processes. *Chemical Engineering Science* **50,** 4091–4108.

Bhave, S. L., and Chattoo, B. B. (2003). Expression of Vitreoscilla hemoglobin improves growth and levels of extracellular enzyme in Yarrowia lipolytica. *Biotech. Bioeng.* **84,** 658–666.

Bogusz, D., Appleby, C. A., Landsmann, J., Dennis, E. S., Trinick, M. J., and Peacock, W. J. (1988). Functioning haemoglobin genes in non-nodulating plants. *Nature* **331,** 178–180.

Brunker, P., Minas, W., Kallio, P. T., and Bailey, J. E. (1998). Genetic engineering of an industrial strain of *Saccharopolyspora erythraea* for stable expression of the Vitreoscilla haemoglobin gene (vhb). *Microbiology-UK* **144,** 2441–2448.

Bulow, L., Holmberg, N., Lilius, G., and Bailey, J. E. (1999). The metabolic effects of native and transgenic hemoglobins on plants. *Trends Biotechnol.* **17,** 21–24.

Chen, W., Hughes, D. E., and Bailey, J. E. (1994). Intracellular expression of Vitreoscilla hemoglobin alters the aerobic metabolism of *Saccharomyces cerevisiae. Biotechnol. Prog.* **10,** 308–313.

Chien, L. J., Chen, H. T., Yang, P. F., and Lee, C. K. (2006). Enhancement of cellulose pellicle production by constitutively expressing Vitreoscilla hemoglobin in *Acetobacter xylinum. Biotechnol. Prog.* **22,** 1598–1603.

Chung, J. W., Webster, D. A., Pagilla, K. R., and Stark, B. C. (2001). Chromosomal integration of the Vitreoscilla hemoglobin gene in Burkholderia and Pseudomonas for the purpose of producing stable engineered strains with enhanced bioremediating ability. *J. Ind. Microbiol. Biotechnol.* **27,** 27–33.

Dikshit, K. L., and Webster, D. A. (1988). Cloning, characterization and expression of the bacterial globin gene from Vitreoscilla in *Escherichia coli. Gene* **70,** 377–386.

Dogan, I., Pagilla, K. R., Webster, D. A., and Stark, B. C. (2006). Expression of Vitreoscilla hemoglobin in Gordonia amarae enhances biosurfactant production. *J. Ind. Microbiol. Biotechnol.* **33,** 693–700.

Erenler, S. O., Gencer, S., Geckil, H., Stark, B. C., and Webster, D. A. (2004). Cloning and expression of the Vitreoscilla hemoglobin gene in *Enterobacter aerogenes*: Effect on cell growth and oxygen uptake. *Applied Biochemistry and Microbiology* **40,** 241–248.

Farres, J., and Kallio, P. T. (2002). Improved cell growth in tobacco suspension cultures expressing Vitreoscilla hemoglobin. *Biotechnol. Prog.* **18,** 229–233.

Favey, S., Labesse, G., Vouille, V., and Boccara, M. (1995). Flavohaemoglobin Hmpx: A new pathogenicity determinant in *Erwinia chrysanthemi* strain-3937. *Microbiology-UK* **141,** 863–871.

Frey, A. D., Farres, J., Bollinger, C. J. T., and Kallio, P. T. (2002). Bacterial hemoglobins and flavohemoglobins for alleviation of nitrosative stress in *Escherichia coli. Appl. Environ. Microbiol.* **68,** 4835–4840.

Geckil, H., Arman, A., Gencer, S., Ates, B., and Yilmaz, H. R. (2004a). Vitreoscilla hemoglobin renders *Enterobacter aerogenes* highly susceptible to heavy metals. *Biometals* **17,** 715–723.

Geckil, H., Barak, Z., Chipman, D. M., Erenler, S. O., Webster, D. A., and Stark, B. C. (2004b). Enhanced production of acetoin and butanediol in recombinant *Enterobacter aerogenes* carrying Vitreoscilla hemoglobin gene. *Bioprocess Biosyst. Eng.* **26,** 325–330.

Geckil, H., Gencer, S., Ates, B., Ozer, U., Uckun, M., and Yilmaz, I. (2006). Effect of Vitreoscilla hemoglobin on production of a chemotherapeutic enzyme, L-asparaginase, by *Pseudomonas aeruginosa. Biotechnol. J.* **1,** 203–208.

Gibson, Q. H., and Smith, M. H. (1965). Rates of reaction of Ascaris haemoglobins with ligands. *Proc. R. Soc. Lond. B Biol. Sci.* **163,** 206–214.

Gibson, Q. H., Wittenberg, J. B., Wittenberg, B. A., Bogusz, D., and Appleby, C. A. (1989). The kinetics of ligand binding to plant hemoglobins: Structural implications. *J. Biol. Chem.* **264,** 100–107.

Holmberg, N., Lilius, G., Bailey, J. E., and Bulow, L. (1997). Transgenic tobacco expressing Vitreoscilla hemoglobin exhibits enhanced growth and altered metabolite production. *Nature Biotechnol.* **15,** 244–247.

Ingledew, W. J., and Poole, R. K. (1984). The respiratory chains of *Escherichia coli. Microbiol. Rev.* **48,** 222–271.

Kallio, P. T., and Bailey, J. E. (1996). Intracellular expression of Vitreoscilla hemoglobin (VHb) enhances total protein secretion and improves the production of alpha-amylase and neutral protease in *Bacillus subtilis. Biotechnol. Prog.* **12,** 31–39.

Kallio, P. T., Kim, D. J., Tsai, P. S., and Bailey, J. E. (1994). Intracellular expression of Vitreoscilla hemoglobin alters *Escherichia coli* energy-metabolism under oxygen-limited conditions. *Eur. J. Biochem.* **219,** 201–208.

Kang, D. G., Kim, Y. K., and Cha, H. J. (2002). Comparison of green fluorescent protein expression in two industrial *Escherichia coli* strains, BL21 and W3110, under co-expression of bacterial hemoglobin. *Appl. Microbiol. Biotechnol.* **59,** 523–528.

Khleifat, K. M. (2006). Correlation between bacterial hemoglobin and carbon sources: Their effect on copper uptake by transformed *E. coli* strain alpha DH5. *Curr. Microbiol.* **52,** 64–68.

Khosla, C., and Bailey, J. E. (1988a). Heterologous expression of a bacterial haemoglobin improves the growth properties of recombinant *Escherichia coli. Nature* **331,** 633–635.

Khosla, C., and Bailey, J. E. (1988b). The Vitreoscilla hemoglobin gene: Molecular cloning, nucleotide sequence and genetic expression in *Escherichia coli. Mol. Gen. Genet.* **214,** 158–161.

Khosla, C., and Bailey, J. E. (1989). Characterization of the oxygen-dependent promoter of the Vitreoscilla hemoglobin gene in *Escherichia coli. J. Bacteriol.* **171,** 5995–6004.

Khosla, C., Curtis, J. E., DeModena, J., Rinas, U., and Bailey, J. E. (1990). Expression of intracellular hemoglobin improves protein synthesis in oxygen-limited *Escherichia coli. Biotechnology (NY)* **8,** 849–853.

Khosravi, M., Webster, D. A., and Stark, B. C. (1990). Presence of the bacterial hemoglobin gene improves alpha-amylase production of a recombinant *Escherichia coli* strain. *Plasmid* **24,** 190–194.

Konz, J. O., King, J., and Cooney, C. L. (1998). Effects of oxygen on recombinant protein expression. *Biotechnol. Prog.* **14,** 393–409.

Lin, J. M., Stark, B. C., and Webster, D. A. (2003). Effects of Vitreoscilla hemoglobin on the 2,4-dinitrotoluene (2,4-DNT) dioxygenase activity of Burkholderia and on 2,4-DNT degradation in two-phase bioreactors. *J. Ind. Microbiol. Biotechnol.* **30,** 362–368.

Liu, T., Chen, J. Y., Zheng, Z., Wang, T. H., and Chen, G. Q. (2005). Construction of highly efficient *E. coli* expression systems containing low oxygen induced promoter and partition region. *Appl. Microbiol. Biotechnol.* **68,** 346–354.

Minas, W., Brunker, P., Kallio, P. T., and Bailey, J. E. (1998). Improved erythromycin production in a genetically engineered industrial strain of *Saccharopolyspora erythraea. Biotechnol. Prog.* **14,** 561–566.

Orii, Y., and Webster, D. A. (1986). Photodissociation of oxygenated cytochrome o(s) (Vitreoscilla) and kinetic studies of reassociation. *J. Biol. Chem.* **261,** 3544–3547.

Ouyang, S. P., Han, J., Qiu, Y. Z., Qin, L. F., Chen, S., Wu, Q., Leski, M. L., and Chen, G. Q. (2005). Poly(3-hydroxybutyrate-co-3-hydroxyhexanoate) production in recombinant *Aeromonas hydrophila* 4AK4 harboring phbA, phbB and vgb genes. *Macromolecular Symposia* **224,** 21–34.

Patel, S. M., Stark, B. C., Hwang, K. W., Dikshit, K. L., and Webster, D. A. (2000). Cloning and expression of Vitreoscilla hemoglobin gene in Burkholderia sp strain DNT for enhancement of 2,4-dinitrotoluene degradation. *Biotechnol. Prog.* **16,** 26–30.

Pendse, G. J., and Bailey, J. E. (1994). Effect of Vitreoscilla hemoglobin expression on growth and specific tissue: Plasminogen activator productivity in recombinant Chinese-hamster ovary cells. *Biotechnol. Bioengin.* **44,** 1367–1370.

Schubert, P., Steinbuchel, A., and Schlegel, H. G. (1988). Cloning of the Alcaligenes eutrophus genes for synthesis of poly-beta-hydroxybutyric acid (PHB) and synthesis of PHB in *Escherichia coli*. *J. Bacteriol.* **170,** 5837–5847.

So, J. H., Webster, D. A., Stark, B. C., and Pagilla, K. R. (2004). Enhancement of 2,4-dinitrotoluene biodegradation by Burkholderia sp. in sand bioreactors using bacterial hemoglobin technology. *Biodegradation* **15,** 161–171.

Springer, B. A., Egeberg, K. D., Sligar, S. G., Rohlfs, R. J., Mathews, A. J., and Olson, J. S. (1989). Discrimination between oxygen and carbon monoxide and inhibition of auto-oxidation by myoglobin: Site-directed mutagenesis of the distal histidine. *J. Biol. Chem.* **264,** 3057–3060.

Suen, W. C., and Spain, J. C. (1993). Cloning and characterization of Pseudomonas sp. strain DNT genes for 2,4-dinitrotoluene degradation. *J. Bacteriol.* **175,** 1831–1837.

Suthar, D. H., and Chattoo, B. B. (2006). Expression of Vitreoscilla hemoglobin enhances growth and levels of alpha-amylase in *Schwanniomyces occidentalis*. *Appl. Microbiol. Biotechnol.* **72,** 94–102.

Tabakov, V. Y., Emelyanova, L. K., Antonova, S. V., and Voeikova, A. (2001). Effect of the bacterial hemoglobin vhb gene on the efficiency of intergeneric conjugation in *Escherichia coli*: Streptomyces and biosynthesis of antibiotics in Streptomyces. *Russian Journal of Genetics* **37,** 332–334.

Tarricone, C., Calogero, S., Galizzi, A., Coda, A., Ascenzi, P., and Bolognesi, M. (1997). Expression, purification, crystallization, and preliminary X-ray diffraction analysis of the homodimeric bacterial hemoglobin from Vitreoscilla stercoraria. *Proteins* **27,** 154–156.

Unden, G., and Schirawski, J. (1997). The oxygen-responsive transcriptional regulator FNR of *Escherichia coli*: The search for signals and reactions. *Mol. Microbiol.* **25,** 205–210.

Urgun-Demirtas, M., Stark, B., and Pagilla, K. (2006). Use of genetically engineered microorganisms (GEMs) for the bioremediation of contaminants. *Crit. Rev. Biotechnol.* **26,** 145–164.

Urlacher, V. B., Lutz-Wahl, S., and Schmid, R. D. (2004). Microbial P450 enzymes in biotechnology. *Appl. Microbiol. Biotechnol.* **64,** 317–325.

Wakabayashi, S., Matsubara, H., and Webster, D. A. (1986). Primary sequence of a dimeric bacterial haemoglobin from Vitreoscilla. *Nature* **322,** 481–483.

Webster, D. A. (1988). Structure and function of bacterial hemoglobin and related proteins. *Adv. Inorg. Biochem.* **7,** 245–265.

Webster, D. A., and Hackett, D. P. (1966). The purification and properties of cytochrome o from Vitreoscilla. *J. Biol. Chem.* **241,** 3308–3315.

Webster, D. A., and Liu, C. Y. (1974). Reduced nicotinamide adenine dinucleotide cytochrome o reductase associated with cytochrome o purified from Vitreoscilla: Evidence for an intermediate oxygenated form of cytochrome o. *J. Biol. Chem.* **249,** 4257–4260.

Wei, M. L., Webster, D. A., and Stark, B. C. (1998). Genetic engineering of *Serratia marcescens* with bacterial hemoglobin gene: Effects on growth, oxygen utilization, and cell size. *Biotechnol. Bioengin.* **57,** 477–483.

Wen, Y., Song, Y., and Li, J. L. (2001). [The effects of Vitreoscilla hemoglobin expression on growth and antibiotic production in *Streptomyces cinnamonensis*]. *Sheng Wu Gong Cheng Xue Bao* **17,** 24–28.

Wu, J. M., Hsu, T. A., and Lee, C. K. (2003). Expression of the gene coding for bacterial hemoglobin improves beta-galactosidase production in a recombinant Pichia pastoris. *Biotechnol. Lett.* **25,** 1457–1462.

Yang, J., Webster, D. A., and Stark, B. C. (2005). ArcA works with Fnr as a positive regulator of Vitreoscilla (bacterial) hemoglobin gene expression in *Escherichia coli*. *Microbiol. Res.* **160,** 405–415.

Yang, Y. G., Zhang, H. T., Tong, Q., Yang, Y. H., Wang, Y., Yang, S. L., and Gong, Y. (2001). Construction of a novel engineering strain *E-coli* G830, which is adoptable to high-cell-density fermentation. *Acta Biochim. Biophys.* **33,** 296–302.

Yu, H. M., Shi, Y., Zhang, Y. P., Yang, S. L., and Shen, Z. Y. (2002). Effect of Vitreoscilla hemoglobin biosynthesis in *Escherichia coli* on production of poly(beta-hydroxybutyrate) and fermentative parameters. *FEMS Microbiol. Lett.* **214,** 223–227.

Zhang, L., Li, Y., Wang, Z., Xia, Y., Chen, W., and Tang, K. (2007). Recent developments and future prospects of Vitreoscilla hemoglobin application in metabolic engineering. *Biotechnol. Adv.* **25,** 123–136.

Zhu, H., Wang, T. W., Sun, S. J., Shen, Y. L., and Wei, D. Z. (2006). Chromosomal integration of the Vitreoscilla hemoglobin gene and its physiological actions in Tremella fuciformis. *Appl. Microbiol. Biotechnol.* **72,** 770–776.

CHAPTER SIXTEEN

Expression and Purification of Cgb and Ctb, the NO-Inducible Globins of the Foodborne Bacterial Pathogen *C. jejuni*

James L. Pickford,* Laura Wainwright,† Guanghui Wu,‡ *and* Robert K. Poole*

Contents

Abstract

Campylobacter jejuni is a Gram-negative microaerophilic bacterium that occurs as a common gut commensal in many food-producing animals and birds. Contamination of meat during processing is an important route of transmission, and *C. jejuni* is now recognized as one of the most important causes of bacterial gastroenteritis worldwide. *C. jejuni* is notable, but not unique, in possessing two different hemoglobins. The first is termed Cgb and is a single-domain hemoglobin (i.e., having no other protein domain or cofactor) with clear structural similarities (3/3) with myoglobin, the heme domain of flavohemoglobins and *Vitreoscilla* hemoglobin. It is well established that Cgb plays a key role in providing resistance to *C. jejuni* in the face of NO and other reactive nitrogen

* Department of Molecular Biology and Biotechnology, University of Sheffield, Sheffield, United Kingdom
† Queen Alexandra Hospital, Cosham, Portsmouth, United Kingdom
‡ Food and Environmental Safety, Veterinary Laboratories Agency-Weybridge, New Haw, Addlestone, Surrey, United Kingdom

Methods in Enzymology, Volume 436
ISSN 0076-6879, DOI: 10.1016/S0076-6879(08)36016-9

species that might be encountered in its environments. The second globin is Ctb, a truncated globin (2/2 trHb) in class III, until recently the least well-understood class of these ubiquitous globins. In *C. jejuni*, both globin genes are members of a small regulon activated by the NssR protein, which acts as an NO sensor and transcriptional regulator. In this contribution, we describe the cloning of both the *cgb* and *ctb* genes from *C. jejuni* chromosomal DNA, construction of expression vectors in *E. coli*, and a simple purification procedure for each globin. A brief account of the spectroscopic characteristics of both globins is presented.

1. Introduction

Several classes of globin proteins are now recognized in addition to the classical vertebrate hemoglobins (Hbs), myoglobin, and the symbiotic Hbs of leguminous plants. These "new" globins, well represented in these two volumes, include the nonsymbiotic plant Hbs, chimeric flavohemoglobins (flavoHbs), and the trHbs. Globins may be classified functionally—into those that detoxify NO and those that are involved in the facilitation of O_2 transfer to respiratory oxidases (Wu *et al.*, 2005)—or structurally into three categories. The first structural group contains flavoHbs that possess a C-terminal ferredoxin-$NADP^+$ reductase-like domain and an N-terminal globin domain. Unlike many other microbial globins being intensively studied, there is sound genetic evidence for the function of these proteins; thus, bacterial (Crawford and Goldberg, 1998; Membrillo-Hernández *et al.*, 1999) and yeast (Liu *et al.*, 2000) flavoHbs function in the enzymic removal of NO to produce nitrate (Gardner *et al.*, 2006; Poole and Hughes, 2000), and this has been demonstrated by diminished tolerance of mutant bacteria or yeast to NO and nitrosative stress. The second functional group contains single-domain Hbs that are very similar to the N-terminal domain of the flavoHbs; however, their functions appear to be more varied than those of the flavoHbs. For example, Vgb, found in the obligate aerobe *Vitreoscilla*, is believed to function in the facilitation of O_2 transfer to the bacterial terminal oxidase cytochrome *bo′* (Park *et al.*, 2002), whereas the *C. jejuni* hemoglobin, Cgb, confers tolerance to NO (Elvers *et al.*, 2004) and is up-regulated by NO and its congeners by an NO-activated transcriptional regulator (Elvers *et al.*, 2005) (see the subsequent section). The third category comprises trHbs: members possess an altered structure whereby the globin fold is edited from the three-over-three α-helical sandwich (3/3) characteristic of vertebrate globins to a two-over-two arrangement (Pesce *et al.*, 2000). They are thus sometimes called 2/2 Hbs (Vinogradov *et al.*, 2005). This results in a considerably smaller globin composed typically of 110 to 130 amino acids.

C. jejuni is a Gram-negative microaerophilic bacterium that occurs as a common gut commensal in many food-producing animals and birds.

Contamination of meat during processing is an important method of transfer (Friedman *et al.*, 2000), and *C. jejuni* is now recognized as one of the most important causes of bacterial gastroenteritis worldwide (Friedman *et al.*, 2000). *C. jejuni* is notable, but not unique, in possessing two different Hbs: Cgb (a single-domain 3/3 hemoglobin; Elvers *et al.*, 2004) and Ctb (Wainwright *et al.*, 2005), which belongs to trHb group III. These proteins, although both up-regulated in response to NO and nitrosative stress (see the subsequent section), appear to have distinct functions.

We have established that Cgb, a single-domain globin, plays a key role in the defense of *C. jejuni* against NO and nitrosative stress (Elvers *et al.*, 2004). NO is a small molecule of diverse function and profound importance in biological systems. At low levels, it is a key signaling molecule in eukaryotes; at higher concentrations, it reacts with thiols and metal centers and so inhibits, for example, terminal oxidases (Mason *et al.*, 2006) and aconitases (Gardner *et al.*, 1997). Bacterial pathogens will be exposed to high fluxes of NO during their interaction with both macrophages and human colon epithelial cells, as NO is generated by the action of inducible NO synthases in both these cell types (Fang, 2004; Maresca *et al.*, 2005) and NO and/or its redox products (e.g. NO^+, NO^- and $ONOO^-$) are prominent agents in the bactericidal activity of host immune cells (Stevanin *et al.*, 2002). NO synthesis is markedly increased in patients with infective gastroenteritis, and particularly so following *C. jejuni* infection (Enocksson *et al.*, 2004). Thus, it is probable that *C. jejuni* encounters bactericidal levels of NO during infection. There are a number of possible NO-generating environments, including the following: (1) the chemical generation of NO in the stomach, as a consequence of microbial nitrite production in the mouth, may be a powerful defense against gut pathogens (Xu *et al.*, 2001), and (2) NO is also encountered as an intermediate in the global nitrogen cycle, namely in the process of denitrification, where it is formed by periplasmic nitrite reductases, as an intermediate during the reduction of nitrate to dinitrogen (Watmough *et al.*, 1999). Although *C. jejuni* is not a denitrifying bacterium, it may coexist with commensal anaerobes that produce NO in the gastrointestinal tract as a consequence of nitrite reduction (Salzman, 1995).

Consistent with its role, *cgb* expression is minimal in standard media but strongly and specifically induced following exposure to nitrosative stress (Elvers *et al.*, 2004, 2005) via the activity of a NO-responsive regulator (Elvers *et al.*, 2005). This regulator is a member of the Crp-Fnr superfamily (Cj0466), which we have termed NssR. Through microarray-based studies, we have shown that NssR controls a regulon of four genes (Elvers *et al.*, 2005). Intriguingly, the regulon includes Ctb, the second *C. jejuni* globin. The role of Ctb in the NO response is not known, but the high oxygen affinity of Ctb (Wainwright *et al.*, 2006) and the effect of a *ctb* mutation on microaerobic growth (Wainwright *et al.*, 2005) hint at a role in oxygen management, perhaps, for example, in the delivery of oxygen under conditions of NO stress. The NssR system probably represents the primary response of *C. jejuni* to NO, as, on the basis of examination of the gene sequence, there appear to be few, if any, alternative mechanisms of NO detoxification (Parkhill *et al.*, 2000). Indeed, only one other potential NO-detoxification pathway is present in *C. jejuni*, a nitrite reductase. In *E. coli*, periplasmic cytochrome *c* nitrite reductase (NrfA) (whose major function is respiratory reduction of nitrite) can also reduce NO *in vitro* and may play a role in NO management in oxygen-limited environments (Poock *et al.*, 2002). However, this protein confers only modest resistance to NO in *C. jejuni* (Pittman *et al.*, 2007), is not a member of the NssR regulon, and is not induced by NO.

Ctb belongs to trHb group III. Wittenberg *et al.* (2002) proposed a subdivision of the trHbs, based on the information available from more than 40 actual or putative trHb genes. Three distinct groups were identified (I, II, III), and four subgroups occur within group II. Few amino acids are strictly conserved throughout the trHb sequences; only the proximal HisF8 is invariant. Six different residues can occupy the distal E7, but it is always His in class III trHb, as in myoglobin. (Note that we use here I, II, and III as defined by Wittenberg *et al.* [2002] and not to distinguish between the truncated globins, myoglobin-like proteins, and flavohemoglobins [Egawa and Yeh, 2005].) Group III is by far the least well-understood family of Hbs, despite the fact that it embraces globins from the pathogens *Bordetella pertussis* and *Mycobacterium avium* and the obligately aerobic metal-leaching acidophilic bacterium *Thiobacillus ferrooxidans* (Wittenberg *et al.*, 2002). Recently, the three-dimensional structure of Ctb (in the cyano-met state) has been solved, showing that the 2-on-2 trHb fold is substantially conserved (Nardini *et al.*, 2006). Contrary to what has been observed in class I and II trHbs, no protein matrix tunnel/cavity system is evident in Ctb, and a gating movement of His (E7) may be involved in ligand entry to the heme distal site. Recent resonance Raman studies (Lu *et al.*, 2007) show that the oxy-derivative of the wild-type Ctb exhibits two ligand-related vibrational modes, $\nu_{Fe\text{-}O2}$ and $\nu_{O\text{-}O}$, at 542 and 1132 cm^{-1} respectively, suggesting an intertwined H-bonding network surrounding the heme-bound O_2, which is consistent with the positive electrostatic potential of the distal pocket revealed by the studies of the CO derivative. The oxygen association and dissociation kinetic measurements show an unusually high oxygen affinity, with moderate association rate (9.1×10^5 $M^{-1}s^{-1}$) and extremely slow dissociation rate (0.0041 s^{-1}). Mutagenesis studies with the three distal mutants, $B10_{YF}$, $E7_{HL}$, and $B10_{YF}/E7_{HL}$, indicate that the heme-bound dioxygen is stabilized by H bonds donated from the B10Tyr and G8Trp. An additional H bond between the B10Tyr and E7His further regulates these H-bonding interactions by restricting the conformational freedom of the phenolic side chain of the B10Tyr, thereby destabilizing the heme-bound O_2. We conclude that the intricate balance of the

H-bonding interactions determines the unique oxygen affinity of Ctb. While the physiological function of Ctb remains to be unveiled, the extremely high oxygen affinity suggests that it is unlikely that Ctb functions as an oxygen transporter. In contrast, the distal heme environment is surprisingly similar to that of cytochrome *c* peroxidase, suggesting a role for Ctb in performing peroxidase or P450-type of oxygen chemistry.

In this contribution, we describe methods for the cloning, expression and purification of both *C. jejuni* globins, the myoglobin-like hemoglobin Cgb, and the truncated globin, Ctb.

2. Methods for Cloning, Expression, and Purification of Cgb

2.1. Cloning and expression of *C. jejuni* Cgb in *E. coli*

C. jejuni strain NCTC 11168 was obtained from the National Collection of Type Cultures (London). *C. jejuni* strains were grown at 42° in Mueller-Hinton medium in a modular atmosphere controlled system (MACS) VA500 workstation (Don Whitley Scientific) with a constant gas supply of 10% oxygen, 10% carbon dioxide, and 80% nitrogen. Plates were incubated for 48 h; cells from these were inoculated into 50 ml of liquid culture in a 100-ml flask and grown for approximately 18 h. The apparent absorbance (OD_{600}) of the culture was adjusted to 0.5 and cultures reinoculated at 1.33% (v/v) into fresh medium (100 ml) in 250-ml baffled flasks, then grown with shaking at 115 rpm. Mueller-Hinton medium was supplemented with vancomycin (10 μg/ml).

The *cgb* gene (*cj1586)* was amplified from *C. jejuni* genomic DNA using the forward primer RP200, 5′ AAAAAGGAGAAA*CCATGG*CAAAA-GAACAAATTC 3′, containing an engineered *Nco*I restriction enzyme cut site (in italic), and the reverse primer RP201, 5′ AAATTGCTTTTTGG *CTCGAG*TTTTTAGAAC 3′, containing an engineered *Xho*I restriction enzyme cut site (in italic). The resulting product was cloned into pET16b (Qiagen) using standard techniques to yield plasmid pRKP1097. *E. coli* strain BL21 DE3 pLysS was transformed with pRKP1097 using standard techniques to yield strain RKP5341. This strain was spread from a frozen 25% glycerol stock onto nutrient agar plates containing ampicillin (150 μg/ml) and grown overnight at 37°. Despite the microaerophilic lifestyle of *C. jejuni*, *E. coli* strains expressing the *C. jejuni* globins are grown aerobically under normal laboratory conditions.

For expression, a single colony of this strain is used to inoculate 5 ml LB (10 g tryptone, 5 g yeast extract, and 10 g NaCl per 1 l dH_2O, pH adjusted to 7.0) containing ampicillin (150 μg/ml) in a universal bottle. This is grown overnight at 37° on a rotary shaker at 250 rpm. LB (500 ml) was prepared in a 2-l flask supplemented with 1 m*M* δ-aminolevulinic acid, 3 μ*M* $FeCl_3$, and ampicillin (150 μg/ml). To this medium, 500 μl of the overnight culture is added and grown at 37° on a rotary shaker at 250 rpm to an apparent absorbance (OD_{600}) of 0.6. At this point, 1 m*M* IPTG is added and the culture grown for a further 4 h. Cells are harvested by centrifugation at 5,500 rpm for 20 min and the brown-colored pellet is washed, then stored at −70° overnight.

The pellet is thawed and resuspended in approximately 8 ml of 50 m*M* Tris-HCl, pH 8.0. This simple freeze-thawing of the cell pellet is sufficient to lyse *E. coli* strain BL21 DE3 pLysS, which has been engineered to constitutively express T7 lysozyme from the pLysS plasmid (Stratagene, BL21 DE3 pLysS competent cell instruction manual, www.stratagene.com). The intended function of this was to inhibit RNA polymerase, and thus prevent leaky expression of pET derived genes, as T7 lysozyme is a natural inhibitor of T7 RNA polymerase (Studier, 1991). Damage to the inner membranes is caused by freeze-thawing, allowing the T7 lysozyme to leak out of the cytoplasm. This digests the cell wall, lysing the cell and releasing the T7 lysozyme into the surrounding media, which in turn causes a chain reaction with each lysed cell releasing lysozyme to digest nearby peptidoglycan. Using this simple method, we routinely purified Cgb in a ferric, nonligand-bound form, which was fully and readily reducible.

To ensure complete lysis, the suspension is briefly (10 s) vortexed. Cellular debris is removed by centrifugation at 12,000 rpm for 10 min and the supernatant is filtered through a filter having a pore size of 0.45 μm. Alternatively, because remaining cellular debris easily blocks the filters, this step could be replaced with an ultracentrifugation step (e.g., 200,000 g for 1 h) followed by retention of the supernatant. The supernatant fraction should appear deep red.

2.2. Purification of Cgb

A DEAE Sepharose Fast Flow 30-ml column (GE Healthcare Bio-Sciences, Amersham Biosciences UK Ltd) is equilibrated with 10 column volumes (CV) of 50 m*M* Tris-HCl, pH 8.0. Approximately 30 to 45 ml of supernatant from the broken cell suspension (previously) is applied to the column. The column is then washed with 50 ml of 50 m*M* Tris-HCl, pH 8.0, followed by 10 CV of a gradient of NaCl (0 to 0.25 *M*) in 50 m*M* Tris-HCl, pH 8.0. Fractions are monitored at both 280 and 400 nm. Fractions rich in Cgb (red in color and absorbing strongly at 400 nm, indicating heme, and 280 nm) elute early in the NaCl gradient. In our hands, alteration of this gradient, to allow Cgb fractions to elute later, has resulted in increased impurities in the Cgb fractions.

The fractions are pooled and concentrated in a Vivascience, Vivaspin 5,000 molecular weight cutoff (MWCO) tube at 7,000 rpm into a 1 ml final volume.

The collected fractions are then polished using either a Superdex-75 or a Superdex-200 column (GE Healthcare Bio-Sciences, Amersham Biosciences UK Ltd.) equilibrated with 5 CV 50 m*M* Tris-HCl, 0.1 M NaCl, pH 8.0. A volume (1 ml) of the concentrated protein solution is applied to the column and washed onto the column in 2 ml of buffer. Peak fractions (deep red in color) are collected and stored at 4°.

Collected samples are analyzed spectrophotometrically for the presence of Cgb (Fig. 16.1). Generally, Cgb has purified in our laboratory as the ferric form, exhibiting peaks at 398, 505, and 640 nm (Fig. 16.1). Upon addition of a few grains of sodium dithionite under aerobic conditions, Cgb was readily reducible to the oxy-ferrous form ($Fe^{2+}O_2$) with peaks at 411, 540, and 575 nm (see Fig. 16.1). Further addition of sodium dithionite resulted in the removal of oxygen, and peaks for reduced Cgb were seen at 434 and 555 nm (see Fig. 16.1). The spectrum of the reduced form of Cgb is similar to that of reduced VHb, which shows maxima at 432 and 555 nm.

2.3. Histidine-tagged Cgb: A cautionary note

A 6-histidine-tagged Cgb construct has also been made using standard techniques and purified down a 1-ml HisTrap Fast Flow column (GE Healthcare Bio-Sciences, Amersham Biosciences UK Ltd), according to the manufacturer's protocol (GE Healthcare). This protocol relied on a gradient of imidazole for elution of the target protein. However, Cgb was seen to elute from the column during the wash step before the imidazole gradient. The buffer used for binding (20 m*M* sodium phosphate, 0.5 *M*

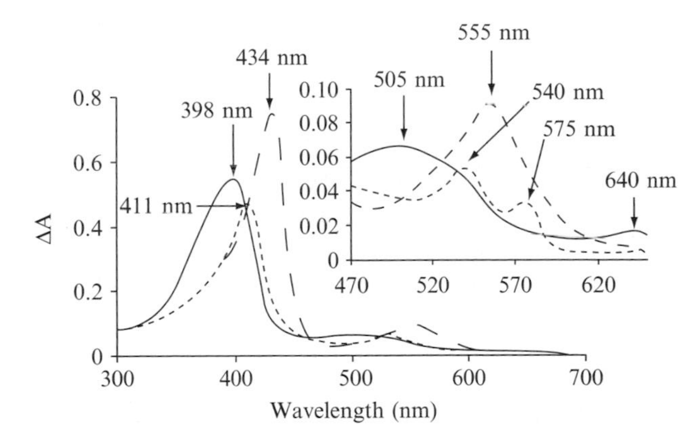

Figure 16.1 Electronic absorbance spectra for ferric (solid line), oxy-ferrous (dotted line), and ferrous (dashed line) forms of Cgb. Maxima for ferric Cgb seen at 398, 505, and 640 nm; oxy-ferrous Cgb observed at 411, 540, and 575 nm; and ferrous Cgb observed at 434 and 555 nm. The inset shows the α/β region on expanded wavelength and absorbance scales.

NaCl, pH 7.4) also contained 30 m*M* imidazole, which was intended to remove loosely bound impurities during the column-wash step. It is documented that imidazole is a ligand for heme-containing proteins, and we propose that the ligand maintains Cgb in solution rather than allowing it to bind to the column. Once the imidazole was removed from the binding buffer, Cgb was seen to bind well to the column and elute at 73% of elution buffer (20 m*M* sodium phosphate, 0.5 *M* NaCl, 365 m*M* imidazole, pH 7.4). The Cgb was then polished in the same manner as for untagged Cgb, using either a Superdex-75 or Superdex-200 column equilibrated with 50 m*M* Tris-HCl, pH 8.0.

The N-terminally His-tagged Cgb appeared to purify in the oxy-ferrous state, with absorbance maxima seen at 411, 540, and 575 nm, and the protein was readily reducible using a few grains of sodium dithionite, as shown in Fig. 16.1 for the untagged protein. Despite this, it was not possible to make the fully ferric form upon addition of potassium ferricyanide, suggesting the presence of a bound ligand affecting the redox capabilities of the protein. It is surmised that the bound ligand was imidazole. Even following attempted removal of this ligand by treatment of Cgb down a PD-10 column, the protein could not be fully oxidized.

Addition of a few grains of imidazole to untagged ferric Cgb results in an oxy-ferrous-type spectrum with maxima at 411, 540, and 575 nm, confirming imidazole as the bound ligand in the 6–histidine-tagged preparation. The imidazole-bound form of Cgb and the oxy-ferrous form can be distinguished by careful analysis of the spectra. The oxy-ferrous form has sharper peaks at 540 and 575 nm than the imidazole-bound form; however, these two forms cannot be distinguished by analysis of the Soret region (Fig. 16.2).

3. Methods for Cloning, Expression, and Purification of Ctb

These methods are essentially those described previously (Wainwright *et al.*, 2006). An alternative method was described subsequently by Nardini *et al.* (2006).

3.1. Cloning and expression of *C. jejuni* trHb in *E. coli*

C. jejuni NCTC 11168 is grown as described previously and genomic DNA isolated using guanidium thiocyanate (Pitcher *et al.*, 1989). The forward (RP268, 5′ AAAATTAACATTTAA*CCATGG*CTTATA TG AAATTTGAAAC-3′) and reverse (RP267, 5′ GAAAAGGTAAAAA*AAGC*

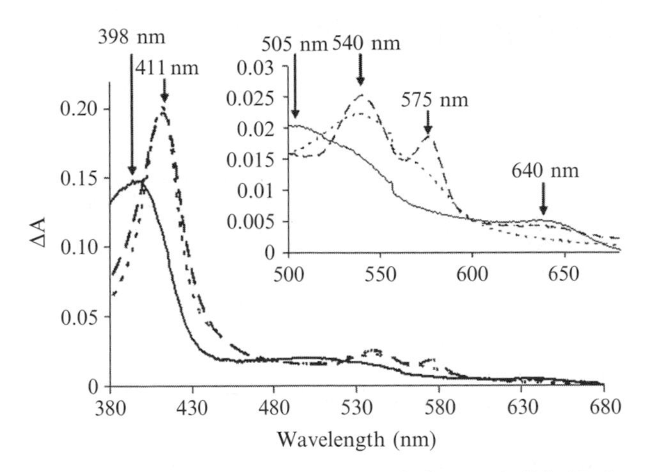

Figure 16.2 Electronic absorbance spectra of ferric (solid line), oxy-ferrous (dashed line), and imidazole-bound (dotted line) forms of Cgb. The oxy-ferrous and imidazole-bound forms can be distinguished by careful analysis of maxima at 540 and 575 nm; the peaks are of narrower bandwidth in the oxy-ferrous sample than in the imidazole-bound sample. The inset shows the α/β region on expanded wavelength and absorbance scales.

*TT*TGGCAAAAAAATTG 3′) primers described previously (Wainwright *et al.*, 2006) were designed based on the sequence of the *C. jejuni* genome and contained an *Nco*I and *Hin*dIII site, respectively (in italic). PCR products are viewed on a gel. The 0.38-kb fragment is recovered with a Qiaquick gel extraction kit (Qiagen), cloned between the *Nco*I and *Hin*dIII sites of pBAD/His (Invitrogen), and used to transform *E. coli* TOP10 using the method of Inoue *et al.* (Inoue *et al.*, 1990). The expressing strain is named RKP4979 (Wainwright *et al.*, 2006).

E. coli strains are grown in Luria-Bertani (LB) medium at 37° at 200 rpm, and on nutrient agar at the same temperature. The media are supplemented with ampicillin (50 μg/ml). The construct is checked by sequencing from just upstream of the promoter through to the end of the insert. For the overexpression, starter cultures grown overnight in LB supplemented with ampicillin are inoculated at 1% (v/v) into 500 ml LB in 2-L baffled flasks supplemented with ampicillin, 200 μM δ-aminolevulinic acid, and 12 μM $FeCl_3$. The cultures are shaken at 200 rpm until an OD_{600} of $\approx$0.5 is reached; protein expression is then induced by adding arabinose (0.02% w/v, final concentration) and the culture grown for an additional 4 h. The concentrations of δ-aminolevulinic acid, $FeCl_3$, and arabinose were selected after preliminary optimization studies. When an absorption spectrum in the visible region is taken of whole *E. coli* cells overexpressing Ctb, the globin is found to be in the oxy form (Wainwright *et al.*, 2006).

3.2. Purification of *C. jejuni* trHb

All buffers used in purification are made up using MilliQ water. Commonly, 4 L of culture are grown for a single purification. The cells are harvested by centrifugation at 5,000 × *g* for 10 min at 4° and resuspended in 80 ml of 50 m*M* Tris-HCl (pH 7.0). Cells are broken by ultrasonication and cell debris removed by centrifugation at 21,000 × *g* for 15 min at 4°. The clear supernatant obtained should be red-brown in color, and it is then loaded onto a 30-ml DEAE Sepharose Fast Flow (Pharmacia Biotech) column equilibrated with 50 m*M* Tris-HCl (pH 7.0) in an Åkta Purifier (GE Healthcare Bio-Sciences, Amersham Biosciences Ltd., UK). The column is washed with 40 ml of the same buffer and the trHb eluted with a NaCl gradient (from 0 to 0.5 *M*) in 50 m*M* Tris-HCl buffer (pH 7.0). Fractions to be carried forward for the next step are chosen on the basis of coincidence of the heme (412 nm) and protein (280 nm) peaks in the UV-visible absorption profile. The eluate is concentrated to ≈1.4 ml using a Vivaspin 20 concentrator (Vivascience) with a molecular mass cutoff of 5 kDa. This fraction is further purified by gel filtration. A Superdex-200 column (16 × 60 cm, GE Healthcare Bio-Sciences, Amersham Biosciences Ltd., UK) is equilibrated with 50 m*M* Tris-HCl (pH 7.0) containing 0.2 *M* NaCl. A 1-ml portion of the previous fraction is applied and eluted in the equilibration buffer at a flow rate of 1 ml/min. The globin elutes from the gel-filtration column as a monomer, and SDS-PAGE analysis gives an estimated molecular mass of 14 kDa. On the basis of the amino acid composition of the protein, this is consistent with the theoretical value of 14.06 kDa. Pure Ctb is best stored at 4°.

3.3. Visible absorbance spectroscopy

The absorption spectra shown here were recorded using a custom-built SDB4 dual wavelength scanning spectrophotometer (University of Pennsylvania School of Medicine Biomedical Instrumentation Group and Current Designs, Philadelphia, PA) as described previously (Kalnenieks *et al.*, 1998). Ctb may be reduced using a few grains of Na dithionite. CO binding is achieved by bubbling reduced samples for 2 min with the gas. Samples are oxygenated by passage of the reduced globin down a Sephadex G-25 column (Amersham Biosciences) in aerated 50 m*M* Tris-HCl pH 7.0. Ctb is oxidized using K ferricyanide or ammonium persulfate.

Typical absolute electronic absorbance spectra of Ctb are displayed in Fig. 16.3. As described before (Wainwright *et al.*, 2006), the trHb was found as a mixture of the ferric and oxy forms when isolated (native form). It is characterized by a Soret peak at 411 nm, a broad peak at around 512 nm, a charge-transfer band at 640 nm, and two peaks at 542 and 578 nm. In a typical Hb, the ferric heme iron in the resting state is axially coordinated by

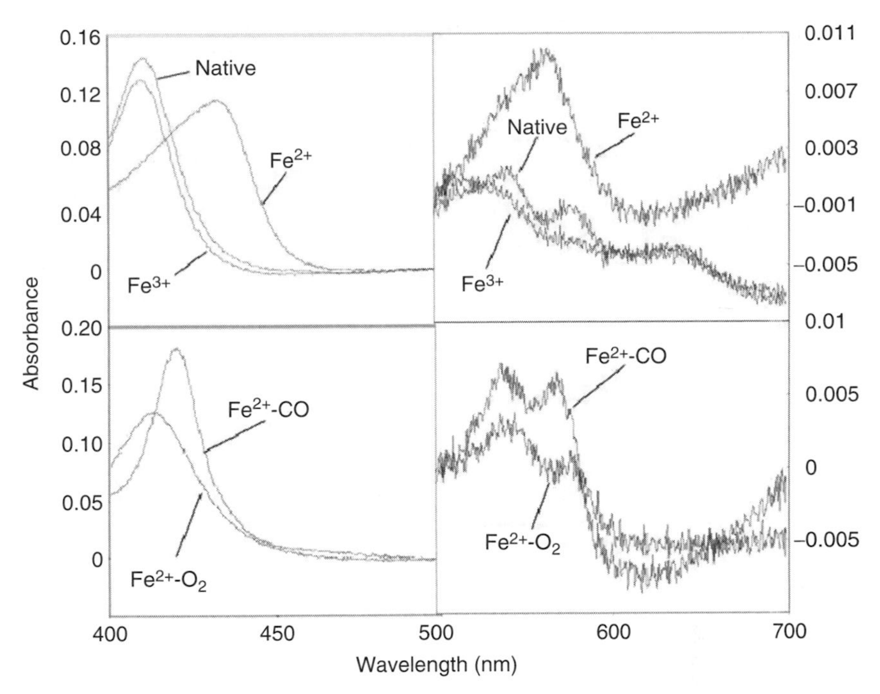

Figure 16.3 Electronic absorbance spectra of pure Ctb. All spectra were scanned against a 50 m*M* Tris-HCl pH 7.0 baseline and were normalized to heme concentration (0.001 n*M*). Native Ctb corresponds to the globin as it elutes from the gel-filtration column (peak maxima at 411, 512, 542, 578, and 640 nm). The other forms of Ctb were generated as described in the text. Peak maxima are as follows: Fe^{2+}, 432 and 565 nm; Fe^{3+}, 410, 512, 542, 582, and 640 nm; Fe^{2}-O, 421, 538, and 569 nm; Fe^{2+}-O_2, 414, 542, and 578 nm. (Reproduced from Wainwright *et al.*, 2006.)

an exogenous water ligand in the distal position. Because water is a weak field ligand, the heme iron typically exhibits a mixture of 6-coordinate high-spin and low-spin configurations. At high pH, the heme-bound water is partially deprotonated, and the contribution from the low-spin component increases, due to hydroxide being a strong field ligand for the heme iron. The 512 and 640 nm absorption bands found in Ctb are characteristic for a water-bound form with 6-coordinate high-spin configuration; the 542 and 578 nm bands signal the low-spin configuration. However, the latter two bands are similar to those of the oxy form (see the subsequent section). The intensity of these two bands decreases when native Ctb is exposes to an oxidant (K ferricyanide or ammonium persulfate). This oxy component is stable, decaying only gradually over a period of months at 4° to a predominantly ferric state.

We have obtained the pure ferric species by ferricyanide oxidation under anaerobic conditions and purification using a desalting column. The species exhibits a Soret band at 410 nm and the two pairs of bands at 512 and

640 nm, and 542 and 582 nm. The ferric species can be slowly reduced by dithionite over a period of ≈15 min. The reduced protein displays a Soret band at 432 nm and a broad $\alpha\beta$-band with a maximum at 565 nm. Addition of carbon monoxide to the reduced form readily results in formation of the carbonmonoxyferroheme. This displays a narrow Soret band at 421 nm and $\alpha-\beta$ bands at 538 and 569 nm, separated by a shallow trough. The oxygenated form is easily made by passage of ferrous Ctb down an aerated PD10 column. The species exhibits a Soret peak at 414 nm and $\alpha-\beta$ bands at 542 and 578 nm.

ACKNOWLEDGMENTS

We thank the UK Biotechnology and Biological Sciences Research Council (BBSRC) for funding this work and Syun-Ru Yeh for helpful suggestions and ongoing collaboration on these proteins.

REFERENCES

Crawford, M. J., and Goldberg, D. E. (1998). Role for the *Salmonella* flavohemoglobin in protection from nitric oxide. *J. Biol. Chem.* **273,** 12543–12547.

Egawa, T., and Yeh, S.-R. (2005). Structural and functional properties of hemoglobins from unicellular organisms as revealed by resonance Raman spectroscopy. *J. Inorg. Biochem.* **99,** 72–96.

Elvers, K. T., Wu, G., Gilberthorpe, N. J., Poole, R. K., and Park, S. F. (2004). Role of an inducible single-domain hemoglobin in mediating resistance to nitric oxide and nitrosative stress in *Campylobacter jejuni* and *Campylobacter coli*. *J. Bacteriol.* **186,** 5332–5341.

Elvers, K. T., Turner, S. M., Wainwright, L. M., Marsden, G., Hinds, J., Cole, J. A., Poole, R. K., Penn, C. W., and Park, S. F. (2005). NssR, a member of the Crp-Fnr superfamily from *Campylobacter jejuni*, regulates a nitrosative stress-responsive regulon that includes both a single-domain and a truncated haemoglobin. *Mol. Microbiol.* **57,** 735–750.

Enocksson, A., Lundberg, J., Weitzberg, E., Norrby-Teglund, A., and Svenungsson, B. (2004). Rectal nitric oxide gas and stool cytokine levels during the course of infectious gastroenteritis. *Clin. Diagn. Lab. Immunol.* **11,** 250–254.

Fang, F. C. (2004). Antimicrobial reactive oxygen and nitrogen species: Concepts and controversies. *Nat. Rev. Microbiol.* **2,** 820–832.

Friedman, C. R., Neimann, J., Wegener, H. C., and Tauxe, R. V. (2000). Epidemiology of *Campylobacter jejuni* infections in the United States and other industrialized nations. *In* "Campylobacter," 2d ed. (I. Nachamkin and M. J. Blaser, eds.), pp. 121–138. ASM Press, Washington, D.C.

Gardner, P. R., Costantino, G., Szabo, C., and Salzman, A. L. (1997). Nitric oxide sensitivity of the aconitases. *J. Biol. Chem.* **272,** 25071–25076.

Gardner, P. R., Gardner, A. M., Brashear, W. T., Suzuki, T., Hvitved, A. N., Setchell, K. D. R., and Olson, J. S. (2006). Hemoglobins dioxygenate nitric oxide with high fidelity. *J. Inorg. Biochem.* **100,** 542–550.

Inoue, H., Nojima, N., and Okayama, H. (1990). High efficiency transformation of *Escherichia coli* with plasmids. *Gene* **96,** 23–28.

Kalnenieks, U., Galinina, N., Bringer-Meyer, S., and Poole, R. K. (1998). Membrane D-lactate oxidase in *Zymomonas mobilis*: Evidence for a branched respiratory chain. *FEMS Microbiol. Lett.* **168,** 91–97.

Liu, L. M., Zeng, M., Hausladen, A., Heitman, J., and Stamler, J. S. (2000). Protection from nitrosative stress by yeast flavohemoglobin. *Proc. Natl. Acad. Sci. USA* **97,** 4672–4676.

Lu, C., Egawa, T., Wainwright, L. M., Poole, R. K., and Yeh, S.-R. (2007). Structural and functional properties of a truncated hemoglobin from a food-borne pathogen *Campylobacter Jejuni*. *J. Biol. Chem.* **282,** 13627–13636.

Maresca, M., Miller, D., Quitard, S., Dean, P., and Kenny, B. (2005). Enteropathogenic *Escherichia coli* (EPEC) effector-mediated suppression of antimicrobial nitric oxide production in a small intestinal epithelial model system. *Cell Microbiol.* **7,** 1749–1762.

Mason, M. G., Nicholls, P., Wilson, M. T., and Cooper, C. E. (2006). Nitric oxide inhibition of respiration involves both competitive (heme) and noncompetitive (copper) binding to cytochrome *c* oxidase. *Proc. Natl. Acad. Sci. USA* **103,** 708–713.

Membrillo-Hernández, J., Coopamah, M. D., Anjum, M. F., Stevanin, T. M., Kelly, A., Hughes, M. N., and Poole, R. K. (1999). The flavohemoglobin of *Escherichia coli* confers resistance to a nitrosating agent, a "nitric oxide releaser," and paraquat and is essential for transcriptional responses to oxidative stress. *J. Biol. Chem.* **274,** 748–754.

Nardini, M., Pesce, A., Labarre, M., Richard, C., Bolli, A., Ascenzi, P., Guertin, M., and Bolognesi, M. (2006). Structural determinants in the group III truncated hemoglobin from *Campylobacter jejuni*. *J. Biol. Chem.* **281,** 37803–37812.

Park, K. W., Kim, K. J., Howard, A. J., Stark, B. C., and Webster, D. A. (2002). *Vitreoscilla* hemoglobin binds to subunit I of cytochrome *bo* ubiquinol oxidases. *J. Biol. Chem.* **277,** 33334–33337.

Parkhill, J., Wren, B. W., Mungall, K., Ketley, J. M., Churcher, C., Basham, D., Chillingworth, T., Davies, R. M., Feltwell, T., Holroyd, S., Jagels, K., Karlyshev, A. V., *et al.* (2000). The genome sequence of the food-borne pathogen *Campylobacter jejuni* reveals hypervariable sequences. *Nature* **403,** 665–668.

Pesce, A., Couture, M., Dewilde, S., Guertin, M., Yamauchi, K., Ascenzi, P., Moens, L., and Bolognesi, M. (2000). A novel two-over-two alpha-helical sandwich fold is characteristic of the truncated hemoglobin family. *EMBO J.* **19,** 2424–2434.

Pitcher, D. G., Saunders, N. A., and Owen, R. J. (1989). Rapid extraction of bacterial genomic DNA with guanidium thiocyanate. *Lett. Appl. Microbiol.* **8,** 151–156.

Pittman, M. S., Elvers, K. T., Lee, L., Jones, M. A., Poole, R. K., Park, S. F., and Kelly, D. J. (2007). Growth of *Campylobacter jejuni* on nitrate and nitrite: Electron transport to NapA and NrfA via NrfH and distinct roles for NrfA and the globin Cgb in protection against nitrosative stress. *Mol. Microbiol.* **63,** 575–590.

Poock, S. R., Leach, E. R., Moir, J. W. B., Cole, J. A., and Richardson, D. J. (2002). Respiratory detoxification of nitric oxide by the cytochrome c nitrite reductase of *Escherichia coli*. *J. Biol. Chem.* **277,** 23664–23669.

Poole, R. K., and Hughes, M. N. (2000). New functions for the ancient globin family: Bacterial responses to nitric oxide and nitrosative stress. *Mol. Microbiol.* **36,** 775–783.

Salzman, A. L. (1995). Nitric oxide in the gut. *New Horiz.* **3,** 33–45.

Stevanin, T. M., Poole, R. K., Demoncheaux, E. A. G., and Read, R. C. (2002). Flavohemoglobin Hmp protects *Salmonella enterica* serovar Typhimurium from nitric oxide-related killing by human macrophages. *Infect. Immun.* **70,** 4399–4405.

Studier, F. W. (1991). Use of bacteriophage T7 lysozyme to improve an inducible T7 expression system. *J. Mol. Biol.* **219,** 37–44.

Vinogradov, S. N., Hoogewijs, D., Bailly, X., Arredondo-Peter, R., Guertin, M., Gough, J., Dewilde, S., Moens, L., and Vanfleteren, J. R. (2005). Three globin lineages belonging to two structural classes in genomes from the three kingdoms of life. *Proc. Natl. Acad. Sci. USA* **102,** 11385–11389.

Wainwright, L. M., Elvers, K. T., Park, S. F., and Poole, R. K. (2005). A truncated haemoglobin implicated in oxygen metabolism by the microaerophilic food-borne pathogen *Campylobacter jejuni*. *Microbiology* **151,** 4079–4091.

Wainwright, L. M., Wang, Y. H., Park, S. F., Yeh, S. R., and Poole, R. K. (2006). Purification and spectroscopic characterization of ctb, a group III truncated hemoglobin implicated in oxygen metabolism in the food-borne pathogen *Campylobacter jejuni*. *Biochemistry* **45,** 6003–6011.

Watmough, N. J., Butland, G., Cheesman, M. R., Moir, J. W. B., Richardson, D. J., and Spiro, S. (1999). Nitric oxide in bacteria: Synthesis and consumption. *Biochim. Biophys. Acta* **1411,** 456–474.

Wittenberg, J. B., Bolognesi, M., Wittenberg, B. A., and Guertin, M. (2002). Truncated hemoglobins: A new family of hemoglobins widely distributed in bacteria, unicellular eukaryotes, and plants. *J. Biol. Chem.* **277,** 871–874.

Wu, G., Wainwright, L. M., and Poole, R. K. (2005). Microbial globins. *Adv. Microb. Physiol.* **47,** 255–310.

Xu, J., Xu, X., and Verstraete, W. (2001). The bactericidal effect and chemical reactions of acidified nitrite under conditions simulating the stomach. *J. Appl. Microbiol.* **90,** 523–529.

CHAPTER SEVENTEEN

Mapping Heme-Ligand Tunnels in Group I Truncated(2/2) Hemoglobins

Alessandra Pesce,* Mario Milani,† Marco Nardini,† *and* Martino Bolognesi†

Contents

Abstract

Protein matrix cavities and extended tunnels can effectively channel substrates and products to and from active sites in enzymes. Substrate and product channeling can enhance catalytic efficiency by reducing the intramolecular diffusion times to and from reaction centers. Moreover, protected transfer between sites may prevent the release of reactive intermediates, thus promoting efficiency or regulation in a series of interconnected reactions. This overview concerns the characterization of matrix tunnels in 2/2 hemoglobins, the recently described family of small hemoproteins (found in bacteria, plants, and unicellular eukaryotes) that form a separate cluster within the hemoglobin superfamily. Crystallographic

* Department of Physics CNR-INFM, and Center for Excellence in Biomedical Research, University of Genoa, Genoa, Italy
† Department of Biomolecular Sciences and Biotechnology CNR-INFM, University of Milano, Milan, Italy

Methods in Enzymology, Volume 436
ISSN 0076-6879, DOI: 10.1016/S0076-6879(08)36017-0

investigations have shown that the 2/2 hemoglobin fold (a 2-on-2 α-helical sandwich) hosts a protein matrix tunnel system offering a potential path for ligand diffusion to and from the heme distal site. The tunnel topology is conserved in 2/2 hemoglobin group I, although with modulation of its size and/or structure. This article describes the methods that were adopted to characterize such matrix tunnels through analysis of the crystal structures and through binding of small apolar ligands to crystalline 2/2 hemoglobins. The methods are generally applicable and, in the case of 2/2 hemoglobins, underline the potential role of the tunnel system in supporting ligand diffusion to and from the heme, as well as ligand storage within the protein matrix.

1. Introduction

Despite the high-density packing of atoms and residues in their core regions, many proteins host inner cavities, or elongated core tunnels, that may play strategic functional roles. Small protein matrix cavities (10 to 20 Å^3 volume) may result from simple residue-packing defects, often linked to the presence of rigid-planar aromatic residue rings. Larger protein matrix cavities or tunnels that may hamper the thermodynamic stability of a folded protein should, in principle, be wiped out by molecular evolution. Their presence in several proteins suggests that cavities and tunnels offer an evolutionary, possibly functional, advantage to the hosting protein. Protein matrix tunnels in enzymes may play a relevant role in catalysis, providing preferred paths or intramolecular docking stations for the diffusion of substrates and products. Channeling of substrates or products can increase enzyme catalytic efficiency by reducing the time required for diffusion to and from active sites. Moreover, in the case of multifunctional enzymes or complexes, channeling between active centers prevents the release of substrates or of reactive intermediates that may be scavenged by competing enzymes or that may prove toxic to the cell. Substrate or ligand channeling may also serve to regulate a block of linked reactions within metabolic pathways or in a complex catalytic cycle (Huang *et al.*, 2001; Milani *et al.*, 2003; Raushel *et al.*, 2003; Weeks *et al.*, 2006).

A recently discovered example of proteins able to channel their ligands from the solvent phase to the active center through the protein matrix came from the analysis of the 2/2 hemoglobin family (2/2Hb), also known as truncated hemoglobins. 2/2Hbs are small oxygen-binding hemoproteins, identified in bacteria, higher plants, and in certain unicellular eukaryotes, that constitute a separate cluster within the hemoglobin (Hb) superfamily (Nardini *et al.*, 2007; Vuletich and Lecomte, 2006; Wittenberg *et al.*, 2002). 2/2Hbs display amino acid sequences that are 20 to 40 residues shorter than nonvertebrate Hbs, to which they are loosely related by sequence similarity (sequence identity to vertebrate Hbs falls well below 20%). On the basis of amino acid sequence analysis, three 2/2Hbs phylogenetic groups (groups I, II, and III, whose members are designated by the N, O, and P suffixes, respectively) were recognized. Sequence identity between 2/2Hbs from the three different groups is low ($\leq$20% overall identity) but may be greater than 80% within a given group. Some of the organisms hosting 2/2Hbs (in some cases 2/2Hbs from more than one group can coexist in the same host) are aggressive pathogenic bacteria; others perform photosynthesis, fix nitrogen, or display distinctive metabolic capabilities. Although very little is known about their *in vivo* role, possible functions of 2/2Hbs that are consistent with the observed biophysical properties include long-term ligand or substrate storage, nitric oxide (NO) detoxification, O_2/NO sensing, redox reactions, and O_2 delivery under hypoxic conditions (Nardini *et al.*, 2007; Vuletich and Lecomte, 2006; Wittenberg *et al.*, 2002).

So far, a number of three-dimensional structures belonging to all three groups have been characterized at atomic resolution by means of X-ray crystallography and NMR: four structures from group I 2/2Hbs (from *Chlamydomonas eugametos* [Pesce *et al.*, 2000], *Paramecium caudatum* [Pesce *et al.*, 2000], *Mycobacterium tuberculosis* [Milani *et al.*, 2001], and *Synechocystis* sp. [Falzone *et al.*, 2002; Hoy *et al.*, 2004; Trent *et al.*, 2004]), four structures from group II 2/2Hbs (from *M. tuberculosis* [Milani *et al.*, 2003], *Bacillus subtilis* [Giangiacomo *et al.*, 2005], *Thermobifida fusca* [Bonamore *et al.*, 2005], and *Geobacillus stearothermophilus* [Ilari *et al.*, 2007]), and one structure from group III 2/2Hbs (from *Campylobacter jejuni* [Nardini *et al.*, 2006]).

2. The 2/2 Hemoglobin Fold

Crystal structures of group I, group II, and group III 2/2Hbs show that their fold is based on a subset of the classical globin fold (the so-called 3-on-3 α-helical sandwich), typical of sperm whale myoglobin (Mb) (Holm and Sander, 1993; Perutz, 1979). Indeed, 2/2Hbs host the heme in a 2-on-2 α-helical sandwich (2/2 fold) based on four α-helices, corresponding to the B-, E-, G-, and H-helices of the classical globin fold (Milani *et al.*, 2003; Nardini *et al.*, 2007; Pesce *et al.*, 2000; Vinogradov *et al.*, 2006). The antiparallel helix pairs (B/E and G/H) are arranged in a sort of α-helical bundle, which surrounds and protects the heme group from the solvent phase (Fig. 17.1A). Residue deletions, insertions, and replacements relative to the classical vertebrate globin sequences have been shown to be distributed throughout the whole 2/2Hb chain. The most noticeable differences between the 2/2Hb and the full-length globin folds are the drastically shortened (or absent) A-helix, the absence of the D-helix, and a long polypeptide segment (pre-F) in extended conformation followed by a very short F-helix that supports the heme proximal HisF8 residue, enabling heme iron coordination. Local structural variations in the 2/2Hb fold are

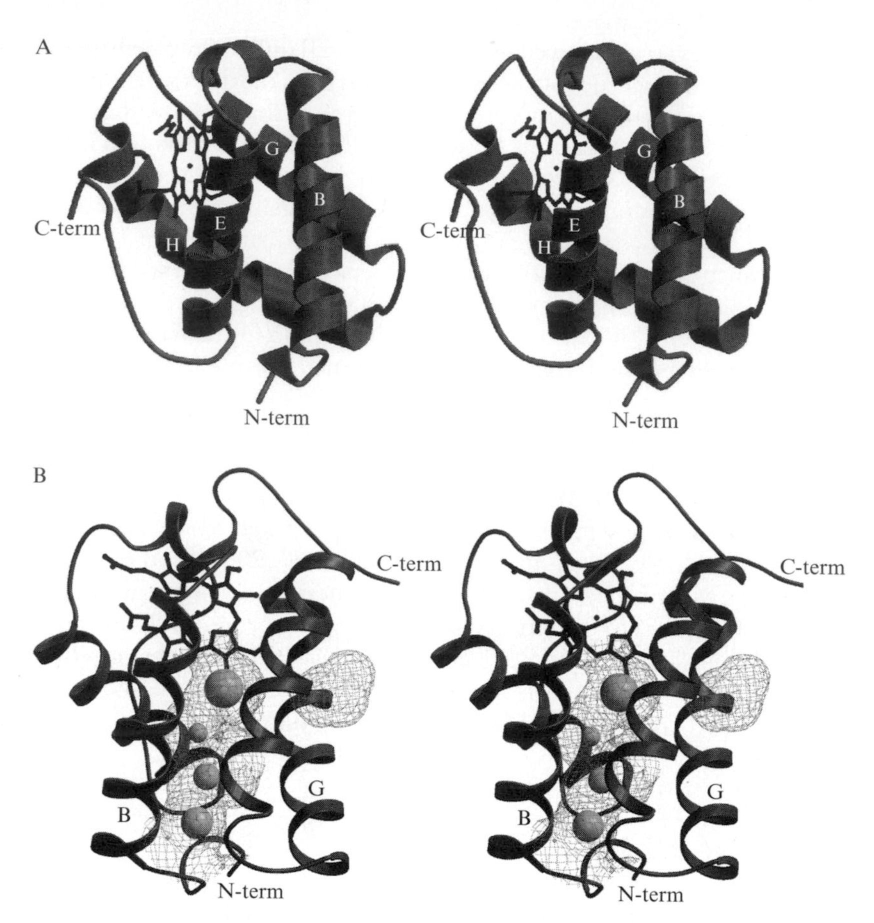

Figure 17.1 Panel A: Stereo view of the 2/2Hb fold, displaying the four main helices building up the protein scaffold and the heme group. Labels identify the B, E, G, and H helices and the protein termini. Panel B: Stereo view of *Ce*-2/2HbN displaying the protein matrix tunnel, shown as the gap-density grid provided by SURFNET. In addition, four Xe atoms identified through the crystallographic analyses described are shown as spheres, with a radius proportional to their occupancy in the crystal structure (Milani *et al.*, 2004). Figure drawn with MOLSCRIPT program (Kraulis, 1991).

evident in each of the three evolutionary groups (Milani *et al.*, 2003; Nardini *et al.*, 2006, 2007; Pesce *et al.*, 2000).

3. Protein Cavities/Tunnel and Ligand Entry

Detailed inspection of group I 2/2Hb three-dimensional structures has highlighted the presence of interconnected protein matrix apolar cavities, or a continuous tunnel, which connect the protein surface to an inner region merging with the heme distal site. Such a structural feature, never observed to such an extent in nonvertebrate Hbs, may have substantial implications for ligand diffusion, and thus heme-binding properties, in 2/2Hbs (Milani *et al.*, 2001; Pesce *et al.*, 2000; Fig. 17.1B). The tunnel/cavity network is topologically conserved in *C. eugametos* (*Ce*-2/2HbN), *P. caudatum* (*Pc*-2/2HbN), and *M. tuberculosis* (*Mt*-2/2HbN) three-dimensional structures, being built by almost invariant apolar residues (Milani *et al.*, 2001; Pesce *et al.*, 2000). In contrast, the crystal structure of hexacoordinated *Synechocystis* sp. 2/2HbN, where the sixth heme ligand is residue HisE10, did not show evidence of such a cavity network, likely related to the extensive conformational changes that are required in this protein distal site region to achieve heme hexacoordination (Hoy *et al.*, 2004).

In *Mt*-2/2HbN, the tunnel is composed of two orthogonal branches, yielding an *L*-shaped path through the protein matrix. The tunnel's short branch (about 8 Å long) connects the heme distal site to the outer solvent space, at a location between the central region of the G and H helices. The tunnel's long branch stretches for about 20 Å through the protein matrix, from the heme distal cavity to a solvent access site located between the interhelical AB and GH loops. Overall, the tunnel volume is about 265 $Å^3$ (Milani *et al.*, 2001). In *Ce*-2/2HbN a similar, but more open tunnel system displays a volume of about 400 $Å^3$, whereas in *Pc*-2/2HbN the tunnel is restricted to three cavities with an overall volume of 180 $Å^3$ (Pesce *et al.*, 2000). Residues lining the tunnel branches are hydrophobic and substantially conserved throughout group I (Vuletich and Lecomte, 2006). Small cavities have been observed in sperm whale Mb and are recognized as acting as temporary docking sites for small ligands such as O_2, NO, and CO (Brunori and Gibson, 2001; Schotte *et al.*, 2003; Scott *et al.*, 2001; Srajer *et al.*, 2001; Tilton *et al.*, 1984). However, the Mb cavities are much smaller (13 to 45 $Å^3$) than the cavities found in 2/2HbNs and are topologically unrelated.

The location and size of the hydrophobic tunnels in group I 2/2Hbs suggest important roles in controlling diffusion of small ligands to and from the heme distal pocket or in temporary ligand storage and/or accumulation (Milani *et al.*, 2001; Samuni *et al.*, 2003; Wittenberg *et al.*, 2002). In order to shed light on such stimulating ideas, we have analyzed crystals of *Mt*-2/2HbN, *Ce*-2/2HbN, and *Pc*-2/2HbN treated with xenon or with butyl-isocyanide, two distinct probes expected to diffuse differently through the 2/2HbN tunnel system, and we determined their three-dimensional structures. Our data show that diffusion to the heme distal site in the group I 2/2Hbs considered may follow a path matching the 2/2Hb hydrophobic tunnel system, in which Xe and butyl-isocyanide docking sites are experimentally identified.

4. Experimental Procedures

4.1. Use of atomic coordinates to unravel protein matrix tunnels

Over the years, several structural analysis packages have allowed for calculation of protein surface and cavities. Two widely diffused approaches among such software tools are provided by the VOIDOO and the SURFNET packages (Kleywegt and Jones, 1994; Laskowski, 1995). To explore the protein matrix cavities/tunnel in 2/2Hbs, our laboratory consistently used SURFNET (Laskowski, 1995). This program generates molecular surfaces and gaps from protein atom coordinates supplied in a PDB-format file. The gap regions, as defined by the program, may identify the empty space between two or more molecules, or the internal cavities or surface grooves within one protein molecule. In the case of 2/2Hbs, we were interested in locating cavities/tunnels within a single protein molecule. The gap regions are computed by positioning spheres (i.e., gap-spheres) between all pairs of protein atoms (as represented by van der Waals volumes). If a protein atom penetrates the gap-sphere, the gap-sphere radius is reduced until it just touches the protein atom. The process is repeated until all the neighboring atoms have been considered. If the radius of the gap-sphere falls below some predetermined minimum limit, it is rejected, otherwise the final gap-sphere is saved. The procedure is continued until all pairs of atoms have been considered and the gap region is filled with gap-spheres. In all our calculations, the minimum and the maximum sphere radii used were 1.4 Å (van der Waals radius of a water molecule) and 3.0 Å, respectively. Tunnels that reach the protein surface are terminated (at the solvent side) by a gap-sphere that is not limited by protein atoms while reaching the defined upper radius limit. All the saved gap-spheres are then used to define a grid that, if contoured at an arbitrary contour level of 100.0, portrays the surface of the saved gap-spheres and thus defines the extent of the gap region (i.e., the cavity or tunnel region). All output surfaces can be viewed interactively, along with the protein molecule, using conventional molecular modeling packages.

4.2. The use of Xe atoms to map cavities in 2/2HbN

The recent new developments in the generation of X-rays for macromolecular crystallography, with wavelengths in the range of 2 to 3 Å, suggested the use of xenon, and noble gases in general, for crystallographic phase determination (MIR or MAD methods) (Cohen *et al.*, 2001; Helliwell, 2004; Kim *et al.*, 2006; Mueller-Dieckmann *et al.*, 2004, 2005; Schiltz *et al.*, 2003; Weiss *et al.*, 2001). Because the applicability of this method may be limited by the lack of naturally occurring binding sites (i.e., cavities) in proteins, artificial noble-gas binding sites can be engineered in proteins by mutating large residues to alanine (Quillin and Matthews, 2003).

In the case of group I 2/2Hbs, Xe atoms were used as probes to map the naturally occurring protein matrix cavities/tunnels, with the aim of showing the suitability of such cavities for diffusion of small (diatomic) ligands from solvent to the heme distal site, and vice versa. In particular, to promote Xe diffusion within the protein matrix, crystals of selected group I 2/2Hbs (*Mt*-2/2HbN, *Ce*-2/2HbN, and *Pc*-2/2HbN), maintained in their mother liquor and cryo-protectant solutions, were treated in a high-pressure chamber (Xcell, Oxford Cryosystem, UK; Fig. 17.2). The crystals, hosted in a conventional crystallographic CryoLoop, could easily and quickly be transferred in and out of the high-pressure chamber (the chamber inner volume is about 10 cm^3) using a sliding sample holder. In the case of 2/2HbNs, the crystals were equilibrated with gaseous Xe to a max pressure of 25 bar for 20 to 30 min at room temperature.

Xe pressure and incubation times are two key parameters to be adjusted in order to obtain successful Xe binding to the protein and to keep a suitable diffraction pattern from the crystal. When the crystal is temporarily hosted in the cell and gaseous Xe is flowing into the chamber, a key factor is to prevent crystal drying. Indeed, when the chamber is pressurized, the gas is forced to flow through filter paper soaked with the crystal-stabilizing solution, which saturates the gas and the environment with aqueous vapor and prevents drying and collapse of the crystal under treatment. When the

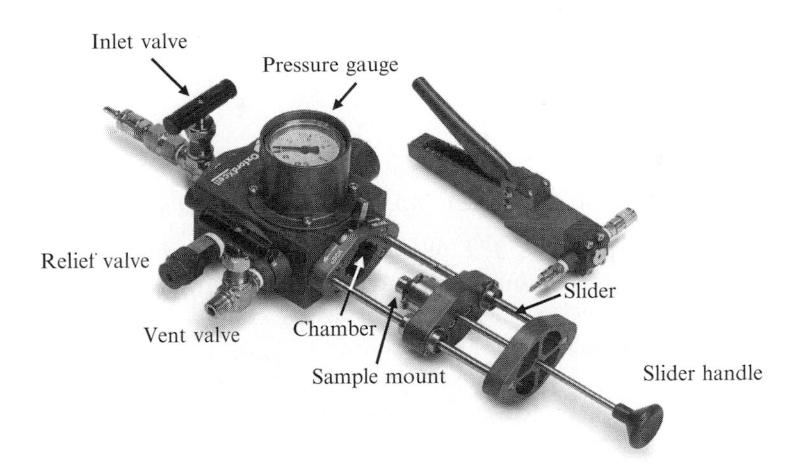

Figure 17.2 The pressure cell used for Xe diffusion in protein crystals (Xcell, from Oxford Cryosystem, UK). The picture identifies the main operative parts of the cell that are used for crystal loading and for Xe pressure treatment. (Courtesy of Oxford Cryosystem, UK, reproduced with permission.)

crystal is pulled out of the quickly depressurized chamber after treatment, it is immediately cryo-cooled to 100K, to avoid back diffusion of Xe atoms out of the crystallized protein molecules; in our laboratory, such an operation requires fewer than 15 s. The cryo-preserved crystal is then handled following common procedures and exposed to an X-ray source for data collection. Data reduction and protein structure analysis follows along conventional methods and according to the specific question addressed by the Xe diffusion experiments.

4.3. *n*-Butyl-isocyanide

Alkyl isocyanides have been extensively used to study equilibrium-binding properties of Hbs and Mbs (Mims *et al.*, 1983; Reisberg and Olson, 1980a,b). Furthermore, because of their apolar nature, alkyl isocyanides can be used as molecular probes to identify the location of apolar cavities. Trapping of alkyl isocyanides inside the protein matrix of heme-containing proteins may result in bond formation between the isocyanide carbon atom and the Fe atom of the heme group. In the case of group I 2/2Hbs, crystals of *Mt*-2/2HbN, *Ce*-2/2HbN, and *Pc*-2/2HbN were soaked in their stabilizing solution containing 20 to 40 m*M* butyl-isocyanide for up to 2 hours and then frozen and exposed to an X-ray source for data collection.

Alkyl isocyanides are able to bind to both ferric and ferrous iron of the heme in cytochrome P450; thus, this method has also been used for crystallographic and spectrophotometric characterizations of the *n*-butyl-isocyanide complex of cytochrome P450 (Lee *et al.*, 2001).

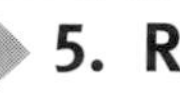

5. Results

5.1. Mapping of protein matrix tunnels in 2/2HbN

The computer output provided by SURFNET (Laskowski, 1995) consists of two pieces of information. First, each cavity is identified by its volume and by the coordinates of its center of gravity. Second, a wire model of the cavity (defined as gap density by SURFNET) is provided for display purposes. In addition, the whole protein surface is also mapped. No information is directly provided on the nature and sequence numbering of residues defining each cavity. Such description, if necessary, is available through VOIDOO. In the case of the 2/2HbNs tested, the protein matrix tunnel systems mapped displayed volumes in the range 180 to 400 Å^3 (Milani *et al.*, 2004).

5.2. 2/2HbN-Xe derivatives

Xe binding does not remarkably affect the crystal diffraction properties of 2/2HbNs tested relative to the native protein crystals; all data collections were performed on an in-house rotating anode X-ray source, at about 2.4 Å resolution.

Analysis of three-dimensional structures of *Mt*-2/2HbN, *Ce*-2/2HbN, and *Pc*-2/2HbN Xe-derivatives showed only minor overall structural readjustments of the proteins related to Xe binding. Inspection of the difference in Fourier maps for the Xe derivatives revealed several electron-density peaks along the branches of the protein matrix tunnel, unambiguously allowing for location of the bound Xe atoms (see Fig. 17.1B). The *Mt*-2/2HbN Xe-adduct showed eight Xe atoms with different occupancy levels distributed between the two protein molecules present in the asymmetrical crystal unit (A and B subunits). Three major Xe sites (Xe1, Xe2, and Xe4) are equally present and located in both *Mt*-2/2HbN A and *Mt*-2/2HbN B subunits. The Xe1 site (100% occupancy) is located in the tunnel's long branch, at 13.3 Å from the heme Fe atom, whereas site Xe2 (60% occupancy) falls in the short branch (at 6.3 Å from the heme Fe atom). The Xe4 site (50% occupancy) is located at the tunnel's short branch entrance, in a hydrophobic cleft between symmetrically related molecules. In contrast, the Xe3 site (50% occupancy) is present in subunit B only, being located between the Xe2 and Xe4 sites, at the interface between the tunnel short branch and the solvent. The Xe5 site, located at 9.0 Å from the heme Fe atom, between the sites Xe1 and Xe2 in subunit A, displays only 30% occupancy, likely representing a transition site bridging the Xe1 and Xe2 sites. The Xe2, Xe3, and Xe4 sites outline the path of the *Mt*-2/2HbN tunnel's short branch from the heme distal cavity to the solvent. In particular, the Xe3 site location on the protein surface suggests that access to the heme distal site may occur through an entry site between the G and H helices. Structural analysis also shows that the *Mt*-2/2HbN tunnel's long branch is accessible from the solvent, in both A and B subunits, through an aperture, as indicated by the location of a full occupancy Xe atom in the Xe1 cavity, at 5.5 Å from the long branch's solvent aperture (Milani *et al.*, 2004).

The *Ce*-2/2HbN protein structure displays the widest (about 5 Å in diameter) and most extended *L*-shaped tunnel observed so far in group I 2/2Hbs (see Fig. 17.1B). The tunnel's short branch of *Mt*-2/2HbN is more properly defined as a cavity in *Ce*-2/2HbN (about 90 Å^3). The four Xe atoms found in the *Ce*-2/2HbN-Xe adduct follow the long tunnel's branch, at mutual distances of 4 to 5 Å. In particular, three Xe atoms match the Xe1, Xe2, and Xe5 sites observed in *Mt*-2/2HbN (the *Mt*-2/2HbN-Xe site numbering scheme is kept for *Ce*-2/2HbN and *Pc*-2/2HbN to identify topologically equivalent Xe sites in the 2/2HbN fold). The Xe1 site (70% occupancy) falls 13.7 Å from the heme Fe atom, the Xe2 site (occupancy of 80%) falls 6.8 Å from the Fe atom, and the Xe site (40% occupancy) in *Ce*-2/2HbN is located at a position matching the Xe5 site of *Mt*-2/2HbN (10.8 Å from the heme Fe atom). Finally, the fourth Xe site (Xe6; 70% occupancy) does not match any Xe site in *Mt*-2/2HbN, as it is located at the protein-solvent interface at the exit of the tunnel's long branch (Milani *et al.*, 2004).

In *Pc*-2/2HbN, the tunnel is restricted and divided into three cavities (overall volume 180 Å^3) topologically distributed along the tunnel's long branch described previously. In agreement with this, only one Xe atom is bound by *Pc*-2/2HbN, at the Xe1 site (80% occupancy, at 13.8 Å from the heme Fe atom), in a 65 Å^3 cavity (Milani *et al.*, 2004).

5.3. 2/2HbN butyl-isocyanide derivatives

Under the experimental conditions applied, binding of butyl-isocyanide to *Mt*-2/2HbN occurs essentially only in the tunnel's long branch of the A subunit (with 100% occupancy). The B subunit displays electron density compatible only with very low occupancy binding. Butyl-isocyanide is completely buried within the tunnel's short branch and the heme distal cavity, stabilized by van der Waals contacts (Milani *et al.*, 2004). Additionally, binding of butyl-isocyanide to *Mt*-2/2HbN induces an important conformational change of residue PheE15(62) (subunit A), which acts as a gate at the heme distal region where the two tunnel branches merge (Milani *et al.*, 2001). It should be noted that coordination of the alkyl-isocyanide ligand to the heme Fe atom could not occur in the *Mt*-2/2HbN treated crystals, because the ligand diffusion experiment used crystals of the *Mt*-2/2HbN cyano-met form.

In the case of the *Ce*-2/2HbN butyl-isocyanide derivative, inspection of the different Fourier maps showed an elongated different electron-density peak in the long branch of the protein matrix tunnel, allowing for only a qualitative modeling of a bound butyl-isocyanide molecule as a result of low occupancy (<50%). In contrast, in the case of the butyl-isocyanide *Pc*-2/2HbN soaked crystals, no extra electron density could be located, indicating the lack of butyl-isocyanide binding to this 2/2HbN. All the previous structural data are in perfect agreement with the current view on the topology and size of matrix tunnels in 2/2HbNs (Milani *et al.*, 2004).

6. Conclusions

Protein matrix apolar and polar tunnels play an instrumental role in promoting and regulating substrate access to and product escape from enzyme active sites, as well as in intramolecular ligand migration and storage in different protein families (Huang *et al.*, 2001; Milani *et al.*, 2003; Raushel *et al.*, 2003; Weeks *et al.*, 2006). Intramolecular tunnel systems are a key structural and functional property not only of enzymes but also of proteins apparently devoid of enzymatic functions. The structural analyses on 2/2HbNs show that Xe binding to three group I 2/2Hbs occurs along the protein matrix tunnel path (Milani *et al.*, 2001; Pesce *et al.*, 2000) and follows an ordered pattern. These observations show that the Xe-binding sites outline experimentally, with perfectly comparable topology, the tunnel path previously proposed as a group I 2/2Hb conserved structural feature, and they highlight the general applicability of this experimental method for mapping protein matrix cavities.

Butyl-isocyanide binding to *Mt*-2/2HbN also occurs along the identified tunnel path. The fact that butyl-isocyanide can bind to a site located next to the heme while not exploiting Fe coordination, reveals both the potential role of the 2/2HbN apolar tunnel in supporting diffusion (and accumulation) of low polarity molecules and the suitability of this low-polarity small molecule in the study of elongated tunnels in proteins. It is worth noting that all the results here reported for 2/2HbNs show an excellent agreement between the two methods adopted, namely, the *in silico* localization of protein matrix cavities/tunnels and the crystallographic observation of bound Xe atoms and butyl-isocyanide molecules at the same sites. Extensive molecular dynamics simulations on *Mt*-2/2HbN also support the views here presented on the diffusion of small diatomic ligands through the 2/2HbN protein matrix tunnel system (Bidon-Chanal *et al.*, 2006; Milani *et al.*, 2004).

ACKNOWLEDGMENTS

This work was supported by grants from the Italian Ministry of University and Scientific Research (MIUR FIRB Project Biologia Strutturale). Mario Milani is recipient of a postdoctoral fellowship supported by NIH grant 1-R01-AI052258 (2004–2007). Martino Bolognesi is grateful to Fondazione Cariplo (Milan, Italy) and to CIMAINA (University of Milano) for continuous support.

REFERENCES

Bidon-Chanal, A., Martì, M. A., Crespo, A., Milani, M., Orozco, M., Bolognesi, M., Luque, F. J., and Estrin, D. A. (2006). Ligand-induced dynamical regulation of NO conversion in *Mycobacterium tuberculosis* truncated hemoglobin-N. *Proteins* **64,** 457–464.

Bonamore, A., Ilari, A., Giangiacomo, L., Bellelli, A., Morea, V., and Boffi, A. (2005). A novel thermostable hemoglobin from the actinobacterium *Thermobifida fusca. FEBS J.* **272,** 4189–4201.

Brunori, M., and Gibson, Q. H. (2001). Cavities and packing defects in the structural dynamics of myoglobin. *EMBO Rep.* **2,** 674–679.

Cohen, A., Ellis, P., Kresge, N., and Soltis, S. M. (2001). MAD phasing with krypton. *Acta Crystallogr. D* **57,** 233–238.

Falzone, C. J., Vu, B. C., Scott, N. L., and Lecomte, J. T. (2002). The solution structure of the recombinant hemoglobin from the cyanobacterium *Synechocystis* sp. PCC 6803 in its hemichrome state. *J. Mol. Biol.* **324,** 1015–1029.

Giangiacomo, L., Ilari, A., Boffi, A., Morea, V., and Chiancone, E. (2005). The truncated oxygen-avid hemoglobin from *Bacillus subtilis*: X-ray structure and ligand binding properties. *J. Biol. Chem.* **280,** 9192–9202.

Helliwell, J. R. (2004). Overview and new developments in softer X-ray (2 Å < lambda < 5 Å) protein crystallography. *J. Synchrotron Radiat.* **11,** 1–3.

Holm, L., and Sander, C. (1993). Structural alignment of globins, phycocyanins and colicin A. *FEBS Lett.* **315,** 301–306.

Hoy, J. A., Kundu, S., Trent, J. T., III, Ramaswamy, S., and Hargrove, M. S. (2004). The crystal structure of *Synechocystis* hemoglobin with a covalent heme linkage. *J. Biol. Chem.* **279,** 16535–16542.

Huang, X., Holden, H. M., and Raushel, F. M. (2001). Channeling of substrates and intermediates in enzyme-catalyzed reactions. *Annu. Rev. Biochem.* **70,** 149–180.

Ilari, A., Kjelgaard, P., von Wachenfeldt, C., Catacchio, B., Chiancone, E., and Boffi, A. (2007). Crystal structure and ligand binding properties of the truncated hemoglobin from *Geobacillus stearothermophilus*. *Arch. Biochem. Biophys.* **457,** 85–94.

Kim, C. U., Hao, Q., and Gruner, S. M. (2006). Solution of protein crystallographic structures by high-pressure cryocooling and noble-gas phasing. *Acta Crystallogr. D* **62,** 687–694.

Kleywegt, G. J., and Jones, T. A. (1994). Detection, delineation, measurement and display of cavities in macromolecular structures. *Acta Crystallogr. D* **50,** 178–185.

Kraulis, P. J. (1991). MOLSCRIPT: A program to produce both detailed and schematic plots of protein structures. *J. Appl. Crystallogr.* **24,** 946–950.

Laskowski, R. A. (1995). SURFNET: A program for visualizing molecular surfaces, cavities and intermolecular interactions. *J. Mol. Graph.* **13,** 323–330.

Lee, D. S., Park, S. Y., Yamane, K., Obayashi, E., Hori, H., and Shiro, Y. (2001). Structural characterization of n-butyl-isocyanide complexes of cytochrome P450nor and P450cam. *Biochemistry* **40,** 2669–2677.

Milani, M., Pesce, A., Bolognesi, M., Bocedi, A., and Ascenzi, P. (2003). Substrate channeling: Molecular bases. *Biochem. Mol. Biol. Edu.* **31,** 228–233.

Milani, M., Pesce, A., Ouellet, Y., Ascenzi, P., Guertin, M., and Bolognesi, M. (2001). Mycobacterium tuberculosis hemoglobin N displays a protein tunnel suited for O_2 diffusion to the heme. *EMBO J.* **20,** 3902–3909.

Milani, M., Pesce, A., Ouellet, Y., Dewilde, S., Friedman, J., Ascenzi, P., Guertin, M., and Bolognesi, M. (2004). Heme-ligand tunneling in group I truncated hemoglobins. *J. Biol. Chem.* **279,** 21520–21525.

Milani, M., Savard, P. Y., Ouellet, H., Ascenzi, P., Guertin, M., and Bolognesi, M. (2003). A TyrCD1/TrpG8 hydrogen bond network and a TyrB10-TyrCD1 covalent link shape the heme distal site of *Mycobacterium tuberculosis* hemoglobin O. *Proc. Natl. Acad. Sci. USA* **100,** 5766–5771.

Mims, M. P., Porras, A. G., Olson, J. S., Noble, R. W., and Peterson, J. A. (1983). Ligand binding to heme proteins: An evaluation of distal effects. *J. Biol. Chem.* **258,** 14219–14232.

Mueller-Dieckmann, C., Panjikar, S., Tucker, P. A., and Weiss, M. S. (2005). On the routine use of soft X-rays in macromolecular crystallography. Part III. The optimal data-collection wavelength. *Acta Crystallogr. D* **61,** 1263–1272.

Mueller-Dieckmann, C., Polentarutti, M., Djinovic-Carugo, K., Panjikar, S., Tucker, P. A., and Weiss, M. S. (2004). On the routine use of soft X-rays in macromolecular crystallography. Part II. Data-collection wavelength and scaling models. *Acta Crystallogr. D* **60,** 28–38.

Nardini, M., Pesce, A., Milani, M., and Bolognesi, M. (2007). Protein fold and structure in the truncated (2/2) globin family. *Gene* **398,** 2–11.

Nardini, M., Pesce, A., Labarre, M., Richard, C., Bolli, A., Ascenzi, P., Guertin, M., and Bolognesi, M. (2006). Structural determinants in the group III truncated hemoglobin from *Campylobacter jejuni*. *J. Biol. Chem.* **281,** 37803–37812.

Perutz, M. F. (1979). Regulation of oxygen affinity of hemoglobin: Influence of structure of the globin on the heme iron. *Annu. Rev. Biochem.* **48,** 327–386.

Pesce, A., Couture, M., Dewilde, S., Guertin, M., Yamauchi, K., Ascenzi, P., Moens, L., and Bolognesi, M. (2000). A novel two-over-two α-helical sandwich fold is characteristic of the truncated hemoglobin family. *EMBO J.* **19,** 2424–2434.

Quillin, M. L., and Matthews, B. W. (2003). Selling candles in a post-Edison world: Phasing with noble gases bound engineered sites. *Acta Crystallogr. D* **59,** 1930–1934.

Raushel, F. M., Thoden, J. B., and Holden, H. M. (2003). Enzymes with molecular tunnels. *Acc. Chem. Res.* **36,** 539–548.

Reisberg, P. I., and Olson, J. S. (1980a). Equilibrium binding of alkyl isocyanides to human hemoglobin. *J. Biol. Chem.* **255,** 4144–4150.

Reisberg, P. I., and Olson, J. S. (1980b). Rates of isonitrile binding to the isolated α and β subunits of human hemoglobin. *J. Biol. Chem.* **255,** 4151–4158.

Samuni, U., Dankster, D., Ray, A., Wittenberg, J. B., Wittenberg, B. A., Dewilde, S., Moens, L., Ouellet, Y., Guertin, M., and Friedman, J. (2003). Kinetics modulation in the carbonmonoxy derivatives of truncated hemoglobins: The role of distal heme pocket residues and extended apolar tunnel. *J. Biol. Chem.* **278,** 27241–27250.

Schiltz, M., Fourme, R., and Prange, T. (2003). Use of noble gases xenon and krypton as heavy atoms in protein structure determination. *Methods Enzymol.* **374,** 83–119.

Schotte, F., Lim, M., Jackson, T. A., Smirnov, A. V., Soman, J., Olson, J., Phillips, G. N., Jr., Wulff, M., and Anfinrud, P. A. (2003). Watching a protein as it functions with 150-ps time-resolved X-ray crystallography. *Science* **300,** 1944–1947.

Scott, E. E., Gibson, Q. H., and Olson, J. S. (2001). Mapping the pathways for O_2 entry into and exit from myoglobin. *J. Biol. Chem.* **276,** 5177–5188.

Srajer, V., Ren, Z., Teng, T. Y., Schmidt, M., Ursby, T., Bourgeois, D., Pradervand, C., Schildkamp, W., Wulff, M., and Moffat, K. (2001). Protein conformational relaxation and ligand migration in myoglobin: A nanosecond to millisecond molecular movie from time-resolved Laue X-ray diffraction. *Biochemistry* **40,** 13802–13815.

Tilton, R. F., Jr., Kuntz, I. D., Jr., and Petsko, G. A. (1984). Cavities in proteins: Structure of a metmyoglobin-xenon complex solved to 1.9 Å. *Biochemistry* **23,** 2849–2857.

Trent, J. T., III, Kundu, S., Hoy, J. A., and Hargrove, M. S. (2004). Crystallographic analysis of *Synechocystis* cyanoglobin reveals the structural changes accompanying ligand binding in a hexacoordinate hemoglobin. *J. Mol. Biol.* **341,** 1097–1108.

Vinogradov, S. N., Hoogewijs, D., Bailly, X., Arredondo-Peter, R., Gough, J., Dewilde, S., Moens, L., and Vanfleteren, J. R. (2006). A phylogenomic profile of globins. *BMC Evol. Biol.* **6,** 31–47.

Vuletich, D. A., and Lecomte, J. T. (2006). A phylogenetic and structural analysis of truncated hemoglobins. *J. Mol. Evol.* **62,** 196–210.

Weeks, A., Lund, L., and Raushel, F. M. (2006). Tunneling of intermediates in enzyme-catalyzed reactions. *Curr. Opin. Struct. Biol.* **10,** 465–472.

Weiss, M. S., Sicker, T., Djinovic-Carugo, K., and Hilgenfeld, R. (2001). On the routine use of soft X-rays in macromolecular crystallography. *Acta Crystallogr. D* **57,** 689–695.

Wittenberg, J. B., Bolognesi, M., Wittenberg, B. A., and Guertin, M. (2002). Truncated hemoglobins: A new family of hemoglobins widely distributed in bacteria, unicellular eukaryotes, and plants. *J. Biol. Chem.* **277,** 871–874.

CHAPTER EIGHTEEN

Scavenging of Reactive Nitrogen Species by Mycobacterial Truncated Hemoglobins

Paolo Ascenzi*,† *and* Paolo Visca*,†

Contents

Abstract

Tuberculosis and leprosy are among the most challenging infectious threats to human health. The ability of mycobacteria to persist *in vivo* in the presence of reactive nitrogen and oxygen species implies the presence in these bacteria of effective detoxification (pseudoenzymatic) systems. *Mycobacterium tuberculosis* and *Mycobacterium leprae* truncated hemoglobins (trHbs) belonging to group I (or N; trHbN) and group II (or O; trHbO) have recently been implicated in the scavenging of nitrogen monoxide (•NO) and peroxynitrite

* National Institute for Infectious Diseases IRCCS Lazzaro Spallanzani, Rome, Italy
† Department of Biology and Interdepartmental Laboratory for Electron Microscopy, University Roma Tre, Rome, Italy

Methods in Enzymology, Volume 436
ISSN 0076-6879, DOI: 10.1016/S0076-6879(08)36018-2

($ONOO^-/HOONO$). Furthermore, *M. leprae* trHbO was found to act as an efficient scavenger of the strong oxidant trioxocarbonate($^{\bullet}1^-$) ($CO_3^{\bullet-}$) following the reaction of peroxynitrite with carbon dioxide (CO_2). Here, mechanisms for scavenging of reactive nitrogen species by mycobacterial trHbs are reviewed, and detailed protocols for assessing pseudoenzymatic kinetics are provided.

1. Introduction

Tuberculosis and leprosy, caused by *M. tuberculosis* and *M. leprae*, respectively, have a global prevalence of ≈15 million cases. More than 14.5 million cases are tuberculosis, and there are ≈9 million new cases per year. Both diseases are curable, but despite antimicrobial strategies, ≈1.7 million people die every year from tuberculosis (Hussain, 2007).

Mycobacteria typically cause granulomatous disease, during which they face the toxic effects of reactive nitrogen and oxygen species, primarily nitrogen monoxide ($^{\bullet}NO$) and superoxide ($O_2^{\bullet-}$) produced by activated macrophages (see Adams *et al.*, 1997; Cooper *et al.*, 2002; Lau *et al.*, 1998; MacMicking *et al.*, 1997; Nathan and Shiloh, 2000; Ohno *et al.*, 2003; Ratledge and Dale, 1999; Schnappinger *et al.*, 2006; Visca *et al.*, 2002a; Voskuil *et al.*, 2003; Zahrt and Deretic, 2002). The reaction of $^{\bullet}NO$ with $O_2^{\bullet-}$ leads to the powerful oxidant peroxynitrite ($ONOO^-/HOONO$)[1], a more reactive species than its precursors $^{\bullet}NO$ and $O_2^{\bullet-}$ (see Ascenzi *et al.*, 2006b; Beckman *et al.*, 1990; Goldstein *et al.*, 2005). Peroxynitrite is implicated in atherosclerosis, inflammation, and neurodegenerative disorders (see Beckman and Koppenol, 1996; Clementi and Nisoli, 2005; Denicola and Radi, 2005; Ducrocq *et al.*, 1999; Ignarro, 2002). Peroxynitrite reacts with carbon dioxide (CO_2) to form 1-carboxylato-2-nitrosodioxidane ($ONOOC[O]O^-$), which decays by homolysis of the O–O bond to yield the reactive species nitrogen dioxide ($^{\bullet}NO_2$) and trioxocarbonate($^{\bullet}1-$) ($CO_3^{\bullet-}$); $CO_3^{\bullet-}$ is a stronger oxidant than $^{\bullet}NO_2$ and peroxynitrite (see Ascenzi *et al.*, 2006b; Goldstein *et al.*, 2005). Most reactions of $CO_3^{\bullet-}$ are one-electron oxidations with preference for Tyr and Trp residues; it is significant that high levels of NO_2-Tyr are detectable in *M. tuberculosis* and *M. leprae* granulomatous lesions (Choi *et al.*, 2002; Schon *et al.*, 2004; Visca *et al.*, 2002a).

The ability of mycobacteria to persist *in vivo* in the presence of reactive nitrogen and oxygen species implies the presence in these bacteria

[1] The recommended IUPAC nomenclature for peroxynitrite is oxoperoxonitrate(1−); for peroxynitrous acid, it is hydrogen oxoperoxonitrate. The term *peroxynitrite* is used in the text to refer generically to both $ONOO^-$ and its conjugate acid HOONO ($pK_a = 6.6$) (see Goldstein *et al.*, 2005).

of pseudoenzymatic[2] detoxification systems, including truncated hemoglobins (trHbs) (see Ascenzi *et al.*, 2006a,c, 2007b; Couture *et al.*, 1999; Fabozzi *et al.*, 2006; Lama *et al.*, 2006; Milani *et al.*, 2003a, 2005; Ouellet *et al.*, 2002, 2003; Pathania *et al.*, 2002a; Visca *et al.*, 2002a; Wittenberg *et al.*, 2002). The distinct features of the heme active site structure of •NO-responsive mycobacterial trHbs (see Bidon-Chanal *et al.*, 2006; Milani *et al.*, 2001, 2003b, 2004a,b, 2005; Mukai *et al.*, 2002, 2004; Ouellet *et al.*, 2006; Samuni *et al.*, 2004; Visca *et al.*, 2002a; Yeh *et al.*, 2000) and their ligand-binding properties (see Milani *et al.*, 2005) combined with co-occurrence of multiple trHb classes (I or N, II or O, and III or P) in individual mycobacterial species (see Ascenzi *et al.*, 2007b; Vinogradov *et al.*, 2006; Vuletich and Lecomte, 2006; Wittenberg *et al.*, 2002) and the temporal expression patterns of trHbs *in vivo* (see Fabozzi *et al.*, 2006; Lama *et al.*, 2006; Ouellet *et al.*, 2002, 2003) suggest that these globins also facilitate O_2 uptake or transport and cellular respiration. It is interesting that, having retained only one trHb, *M. leprae* trHbO has been proposed as representing the merging of multiple functions (see Ascenzi *et al.*, 2007b; Couture *et al.*, 1999; Liu *et al.*, 2004; Milani *et al.*, 2005; Pathania *et al.*, 2002b; Visca *et al.*, 2002a,b; Wittenberg *et al.*, 2002).

Here, mechanisms for the scavenging of reactive nitrogen species by mycobacterial trHbs are reviewed, and detailed protocols for assessing pseudoenzymatic kinetics are provided.

2. Materials

2.1. Mycobacterial trHbs

Recombinant *M. tuberculosis* trHbN was expressed in *Escherichia coli* BL21 cells and purified by a four-step chromatographic procedure (Couture *et al.*, 1999; Couture and Guertin, 1996). *M. tuberculosis* trHbN(II) was prepared by reducing the heme-Fe(III)-atom using ferredoxin and ferredoxin $NADP^+$ reductase or phenazine methosulfate, under anaerobic conditions. Then, trHbN(II)-O_2 was obtained by exposing the trHbN(II) solution to air (Couture *et al.*, 1999; Couture and Guertin, 1996). The *M. tuberculosis* trHbN(II)-O_2 concentration was determined by measuring the optical absorbance at 416 nm ($\epsilon_{416\ nm} = 94.4\ M^{-1}\ cm^{-1}$) (Ouellet *et al.*, 2002).

Recombinant *M. tuberculosis* trHbO was expressed in *E. coli* BL21 cells and purified by a three-step chromatographic procedure (Mukai *et al.*,

[2] Scavenging of reactive nitrogen and oxygen species by ferrous heme proteins might be considered pseudoenzymatic because the ferric-heme derivative produced by reaction with •NO and peroxynitrite must be reduced by a reductase before repeating a catalytic cycle (see Brunori, 2001).

2002). *M. tuberculosis* trHbO(II)-O_2 was prepared by reducing the heme-Fe(III)-atom with 10-molar excess of sodium dithionite under anaerobic conditions (i.e., after flushing the heme(III)-protein solution with N_2 for 20 min). The excess of dithionite and by-products were removed by passing the *M. tuberculosis* trHbO(II) solution through a Bio-Gel P6DG column (Bio-Rad Laboratories, Hercules, CA) equilibrated with 5.0×10^{-2} M phosphate buffer, pH 7.5, at 23.0°, under anaerobic conditions. Then, *M. tuberculosis* trHbO(II)-O_2 was obtained in a septum-sealed gas-tight syringe by injecting the solution of the heme(II)-protein into the appropriate volume of the O_2-saturated solution ($[O_2] = 1.2 \times 10^{-3}$ M in buffer) (Ouellet *et al.*, 2003). The *M. tuberculosis* trHbO(II)-O_2 concentration was determined by measuring the optical absorbance at 414 nm ($\epsilon_{414\ nm} = 1.12 \times 10^5\ M^{-1}\ cm^{-1}$) (Mukai *et al.*, 2002).

Recombinant *M. leprae* trHbO was expressed in the *E. coli* M15 strain as His_6-fusion protein and purified by a single-step affinity chromatographic procedure (Visca *et al.*, 2002b). *M. leprae* trHbO(II)-O_2 was prepared by reducing the heme-Fe-atom with sodium dithionite. The excess of dithionite and by-products were removed by passing the *M. leprae* trHbO (II) solution through a Sephadex G-25 gel-filtration column (Amersham Biosciences Europe GmbH, Freiburg, Germany) equilibrated in air with 1.0×10^{-1} M phosphate buffer, pH 7.3, at 20.0°, in the presence of 5.0×10^{-2} M EDTA. The *M. leprae* trHbO(II)-O_2 concentration was determined by measuring the optical absorbance at 412 nm ($\epsilon_{412\ nm} = 1.22 \times 10^5\ M^{-1}\ cm^{-1}$) (Ascenzi *et al.*, 2006c; Fabozzi *et al.*, 2006; Visca *et al.*, 2002b). *M. leprae* trHbO(II)-NO was prepared by reductive nitrosylation under anaerobic conditions. The *M. leprae* trHbO(III) solution was kept under •NO at 760 Torr overnight (pH 8.5, 5.0×10^{-3} M 2-amino-2-(hydroxymethyl)-1,3-propanediol buffer, at 20.0°). Then, the *M. leprae* trHbO(II)-NO solution was adjusted at pH 7.3 by adding degassed concentrated (3.0×10^{-1} M $-$ 5.0×10^{-1} M) phosphate buffer to attain the final phosphate buffer concentration of 1.0×10^{-1} M. The *M. leprae* trHbO(II)-NO concentration was determined by measuring the optical absorbance at 417 nm ($\epsilon_{417\ nm} = 1.31 \times 10^5\ M^{-1}\ cm^{-1}$) (Ascenzi *et al.*, 2006a).

2.2. Chemicals

Gaseous •NO was purchased from Aldrich Chemical Co. (Milwaukee, WI) and purified by flowing it through a NaOH column to remove acidic nitrogen oxides (Ascenzi *et al.*, 2006a,b; Fabozzi *et al.*, 2006). The •NO stock solution was prepared under anaerobic conditions by saturating 2.0×10^{-2} M phosphate buffer (pH 7.0, in the presence of 5.0×10^{-2} M EDTA) or 5.0×10^{-2} M phosphate buffer (pH 7.5, in the presence of 5.0×10^{-5} M EDTA) with •NO (Ascenzi *et al.*, 2006c; Fabozzi *et al.*, 2006; Ouellet *et al.*, 2002, 2003). The solubility of •NO in the aqueous buffered solution is 2.03 m*M* at

20.0° (Antonini and Brunori, 1971). The final NO concentration (1.5×10^{-3} M – 2.0×10^{-3} M) was measured with an NO electrode (World Precision Instruments, Sarasota, FL) (Ouellet *et al.*, 2002, 2003). The •NO stock solution was diluted with degassed 1.0×10^{-1} M phosphate buffer (pH 7.3, in the presence of 5.0×10^{-2} M EDTA) or 5.0×10^{-2} M phosphate buffer (pH 7.5, in the presence of 5.0×10^{-5} M EDTA) to reach the desired concentration (Ascenzi *et al.*, 2006c; Fabozzi *et al.*, 2006; Ouellet *et al.*, 2002, 2003).

Peroxynitrite was prepared from KO_2 and •NO, and from HNO_2 and H_2O_2, under anaerobic conditions. Peroxynitrite was purified by freeze-fractionation and stored at −80°. The peroxynitrite stock solution was diluted with degassed 1.0×10^{-2} M NaOH to reach the desired concentration. The peroxynitrite concentration was determined by measuring the optical absorbance at 302 nm ($\epsilon_{302\ nm} = 1.705 \times 10^3\ M^{-1}\ cm^{-1}$) (Ascenzi *et al.*, 2007a; Bohle *et al.*, 1996; Koppenol *et al.*, 1996). Decomposed peroxynitrite was obtained by acidification of the peroxynitrite solution with 1.0×10^{-1} M HCl, followed by neutralization with 3.0×10^{-1} M NaOH. Peroxynitrite decomposition was verified spectrophotometrically at 302 nm ($\epsilon_{302\ nm} = 1.705 \times 10^3\ M^{-1}\ cm^{-1}$) (Ascenzi *et al.*, 2006a,b, 2007a; Bohle *et al.*, 1996; Goldstein *et al.*, 2005; Koppenol *et al.*, 1996).

For the experiments carried out in the absence of CO_2, the 1.0×10^{-1} M phosphate buffer (pH 7.3, in the presence of 5.0×10^{-2} M EDTA) and 1.0×10^{-2} M NaOH solutions were prepared fresh daily and thoroughly degassed. Experiments in the presence of CO_2 ($=1.2 \times 10^{-3}$ M) were carried out by adding to the mycobacterial trHb solution the required amount from a freshly prepared 5.0×10^{-1} M sodium bicarbonate solution. The value for the constant of the hydration–dehydration equilibrium $CO_2 + H_2O \leftrightarrow H^+ + HCO_3^-$ at 20.0° is 8.41×10^{-7} M. The bicarbonate concentration present during the reactions was 1.08×10^{-2} M. The CO_2 concentration was expressed as the actual concentration in equilibrium with HCO_3^- (Ascenzi *et al.*, 2006a).

3. •NO Scavenging by *M. tuberculosis* trHbN(II)-O_2 and trHbO(II)-O_2 and by *M. leprae* trHbO(II)-O_2

3.1. Overview

Heme proteins share the ability to detoxify •NO by the rapid and irreversible reaction of their ferrous oxygenated derivative with •NO. This process is postulated to depend on the superoxide character of the heme-Fe(II)-bound O_2. Alternatively, O_2 can react with ferrous nitrosylated heme proteins, although this is a very slow process of questionable physiological relevance (see Ascenzi *et al.*, 2006c; Brunori, 2001; Brunori *et al.*, 2004, 2005; Brunori and Gibson, 2001; Fasano *et al.*, 2006; Flögel *et al.*, 2001; Frauenfelder *et al.*, 2003; Frey and Kallio, 2003; Frey and Kallio, 2005; Gardner *et al.*, 2000; Gow *et al.*, 1999; Herold, 1999, 2001; Herold and Fago, 2005; Milani *et al.*, 2004a; Ouellet *et al.*, 2002, 2003; Poole, 2005; Wu *et al.*, 2003).

•NO scavenging by *M. tuberculosis* trHbN(II)-O_2 and trHbO(II)-O_2 is consistent with the simple mechanism reported in Scheme 1 (Ouellet *et al.*, 2002, 2003).

$$\mathrm{Fe(II)\text{-}O_2 + {}^{\bullet}NO \xrightarrow{k_{on}} Fe(III) + NO_3^-}$$

Kinetics measured at 396 and 404 nm results in the appearance of trHb (III) without any evidence of intermediates. Over the whole heme-protein concentration range explored ($5.0 \times 10^{-7} - 1.8 \times 10^{-6}$ M), •NO scavenging by *M. tuberculosis* trHbN(II)-O_2 and trHbO(II)-O_2 exhibits a time course that can be fitted to a single exponential expression (Ouellet *et al.*, 2002, 2003). Values of the pseudo-first-order rate constant k for •NO scavenging by *M. tuberculosis* trHbN(II)-O_2 and trHbO(II)-O_2 depend linearly on [trHbN(II)-O_2] and [trHbO(II)-O_2], respectively. The slope corresponds to $k_{on} = 7.5 \times 10^8$ M^{-1} s^{-1} and 6.0×10^5 M^{-1} s^{-1}, respectively (Table 18.1), and the y-axis intercept is $a = 0$ s^{-1} (Ouellet *et al.*, 2002, 2003).

Table 18.1 Kinetic parameters for NO scavenging by ferrous oxygenated heme proteins

Heme protein	k_{on} ($M^{-1}s^{-1}$)	h (s^{-1})
M. tuberculosis trHbN[a]	7.5×10^8	—
M. tuberculosis trHbO[b]	6.0×10^5	—
M. leprae trHbO[c]	2.1×10^6	3.4
E. coli flavoHb[d]	$\geq 6 \times 10^8$	$\approx 2 \times 10^2$
Horse heart Mb[e]	4.4×10^7	$> 3.4 \times 10^2$
Murine Ngb[f]	$> 7.0 \times 10^7$	$\approx 3.0 \times 10^2$
Human Hb	8.9×10^7[g]	$> 5.8 \times 10^1$[h] $> 3.3 \times 10^1$[h]

[a] pH = 7.5 and 23°; from Ouellet *et al.* (2002).
[b] pH = 7.5 and 23°; from Ouellet *et al.* (2003).
[c] pH = 7.3 and 20°; from Ascenzi *et al.* (2006c).
[d] pH = 7.0 and 20°; from Gardner *et al.* (2000).
[e] pH = 7.0 and 20°; from Herold *et al.* (2001).
[f] pH = 7.0 and 20°; from Brunori *et al.* (2005).
[g] pH = 7.0 and 20°; from Herold *et al.* (2001).
[h] The two values represent the decay rates for Fe(III)OONO α− and β−Hb subunits, respectively. pH = 7.5 and 20°; from Herold (1999).

•NO scavenging by *M. leprae* trHbO(II)-O_2 is consistent with the two-step mechanism reported in Scheme 2 (Ascenzi *et al.*, 2006c).

$$Fe(II)\text{-}O_2 + {}^{\bullet}NO \xrightarrow{k_{on}} Fe(III)\text{-}OONO \xrightarrow{h} Fe(III) + NO_3^-$$

Mixing of the *M. leprae* trHbO(II)-O_2 and •NO solutions causes a shift of the optical absorption maximum of the Soret band from 412 (i.e., trHbO (II)-O_2) to 411 nm (i.e., trHbO(III)-OONO) and a change of the extinction coefficient from $\epsilon_{412\ nm} = 1.22 \times 10^5\ M^{-1}\ cm^{-1}$ (i.e., trHbO(II)-NO) to $\epsilon_{411\ nm} = 1.48 \times 10^5\ M^{-1}\ cm^{-1}$ (i.e., trHbO(III)-OONO). Then, the *M. leprae* trHbO(III)-OONO solution undergoes a shift of the optical absorption maximum of the Soret band from 411 (i.e., trHlbO(III)-OONO) to 409 nm (i.e., trHbO(III)) and a change of the extinction coefficient from $\epsilon_{411\ nm} = 1.48 \times 10^5\ M^{-1}\ cm^{-1}$ (i.e., trHbO(III)-OONO) to $\epsilon_{409\ nm} = 1.15 \times 10^5\ M^{-1}\ cm^{-1}$ (i.e., trHbO(III)) (Ascenzi *et al.*, 2006c).

Over the whole •NO concentration range explored (8.0×10^{-6} M – 4.0×10^{-5} M), the time course for •NO scavenging by *M. leprae* trHbO (II)-O_2 corresponds to a biphasic process between 360 and 450 nm. The first step (indicated by k_{on} in Scheme 2) is a bimolecular process. In contrast, the second step (indicated by h in Scheme 2) follows monomolecular behavior (Ascenzi *et al.*, 2006c).

Values of the pseudo-first-order rate constant k for the formation of the transient *M. leprae* trHbO(II)-OONO species are wavelength independent at fixed [•NO]. The plot of k versus [•NO] is linear; the slope corresponds to $k_{on} = 2.1 \times 10^6\ M^{-1}\ s^{-1}$ (see Table 18.1) and the y-axis intercept is $c = 0\ s^{-1}$. In contrast, values of the first-order rate constant h for the decay of the transient trHb(III)-OONO species are wavelength and [•NO] independent; the average value of h is $3.4\ s^{-1}$ (see Table 18.1) (Ascenzi *et al.*, 2006c).

•NO scavenging by *M. tuberculosis* trHbN(II)-O_2 and trHbO(II)-O_2 appears to be limited by the formation of the heme-Fe(III)-OONO intermediate (Ouellet *et al.*, 2002, 2003). In contrast, the dissociation of the *M. leprae* trHbO(III)-OONO species and peroxynitrite isomerization to nitrate (NO_3^-) is rate limiting (Ascenzi *et al.*, 2006c).

The catalytic parameters for •NO scavenging by *M. tuberculosis* trHbN (II)-O_2 are very favorable when compared with those of related heme proteins (see Table 18.1) (Ascenzi *et al.*, 2006c; Brunori *et al.*, 2005; Gardner *et al.*, 2000; Herold, 1999, 2001; Ouellet *et al.*, 2002, 2003). The fast rate of •NO scavenging by *M. tuberculosis* trHbN(II)-O_2 compared with that of *M. tuberculosis* and *M. leprae* trHbO(II)-O_2 (see Table 18.1) makes the •NO detoxification role of *M. tuberculosis* trHbN(II)-O_2 likely (Ouellet *et al.*, 2002). In contrast, *M. tuberculosis* and *M. leprae* trHbO(II)-O_2 are predicted to play only a marginal role in the protection of mycobacteria against •NO (Ascenzi *et al.*, 2006c; Ouellet *et al.*, 2003).

The involvement of *M. tuberculosis* trHbN in protecting against reactive nitrogen species, particularly •NO, has been documented *in vivo* using both reverse genetic approaches and homologous or heterologous expression systems (Lama *et al.*, 2006; Ouellet *et al.*, 2002; Pathania *et al.*, 2002a), and this has substantially been ascribed to the •NO-dioxygenase activity. In fact, an *M. bovis* mutant lacking trHbN does not oxidize •NO to NO_3^- and shows decreased respiration upon exposure to •NO (Ouellet *et al.*, 2002). A similar response can also be predicted for *M. tuberculosis*, given the close phylogenetic relationship between *M. bovis* and *M. tuberculosis* and the identity of trHbNs from these neighbor species (Ascenzi *et al.*, 2007b). Moreover, heterologous expression of *M. tuberculosis* trHbN significantly protects both *Mycobacterium smegmatis* and a flavohemoglobin (flavoHb) mutant of *E. coli* from •NO damage through an O_2-sustained detoxification mechanism (Pathania *et al.*, 2002a). Although to a lesser extent, a similar protective effect was also reported for *M. smegmatis* trHbN in the homologous and heterologous system (Lama *et al.*, 2006). Last, overexpression of *M. leprae* trHbO alleviates the growth inhibition of *E. coli hmp* (flavoHb, gene) mutants by NO donors, partly complementing the defect in flavoHb synthesis (Fabozzi *et al.*, 2006).

3.2. Assay protocol

Values of the second-order rate constant k_{on} for the •NO induced conversion of *M. tuberculosis* trHbN(II)-O_2 and trHbO(II)-O_2 to trHbN(III) and trHbO(III), respectively, as well as of *M. leprae* trHbO(II)-O_2 to trHbO(III)-OONO, and of the first-order rate constant *h* for the conversion of *M. leprae* trHbO(III)-OONO to trHbO(III), were determined as follows. *M. tuberculosis* trHbN(II)-O_2 and trHbO(II)-O_2 ($5.0 \times 10^{-7} - 1.8 \times 10^{-6}$ M) and the *M. leprae* trHbO(II)-O_2 (1.6×10^{-6} M) solutions (5.0×10^{-2} M or 1.0×10^{-1} M phosphate buffer, pH 7.5 or 7.3, in the presence of 5.0×10^{-5} M or 5.0×10^{-2} M or EDTA) were rapidly mixed with the •NO (1.0×10^{-7} M $- 4.0 \times 10^{-5}$ M) solution (5.0×10^{-2} M or 2.0×10^{-2} M phosphate buffer, pH 7.5 or 7.0, in the presence of 5.0×10^{-5} M or 5.0×10^{-2} M EDTA), at 23.0° or 20.0°, respectively. No gaseous phase was present. Kinetics of •NO scavenging by *M. tuberculosis* trHbN(II)-O_2 and trHbO(II)-O_2 was monitored at 404 and 396 nm, respectively. Kinetics of •NO scavenging by *M. leprae* trHbO(II)-O_2 was monitored between 360 and 450 nm. The dead time of the SX.18MV-R rapid-mixing stopped-flow apparatus (Applied Photophysics Ltd., Leatherhead, UK) was 0.9 or 1.6 ms (Ascenzi *et al.*, 2006c; Ouellet *et al.*, 2002, 2003).

The time course for •NO scavenging by *M. tuberculosis* trHbN(II)-O_2 and trHbO(II)-O_2 was fitted to a monoexponential process. Values of the pseudo-first-order rate constant k for the •NO induced conversion of *M. tuberculosis* trHbN(II)-O_2 and trHbO(II)-O_2 to trHbN(III) and trHbO(III), respectively, were determined from data analysis, according to [Eq. (18.1)], which treats the reaction as an irreversible first-order process (Ouellet *et al.*, 2002, 2003):

$$[\text{trHb(II)- O}_2]_t = [\text{trHb(II)- O}_2]_i \times e^{-k \times t} \tag{18.1}$$

Values of the second-order rate constant for •NO scavenging by *M. tuberculosis* trHbN(II)-O_2 and trHbO(II)-O_2 k_{on} were obtained from the linear dependence of k on the trHbN(II)-O_2 and trHbO(II)-O_2 concentration (i.e., [trHb(II)-O_2]) according to [Eq. (18.2)] (Ouellet *et al.*, 2002, 2003):

$$k = k_{on} \times [\text{trHb(II)- O}_2] + a \tag{18.2}$$

The time course for •NO scavenging by *M. leprae* trHbO(II)-O_2 was fitted to two consecutive monoexponential processes. Values of the •NO-dependent pseudo-first-order rate constant k for the formation and of the first-order rate constant h for the decay of the transient *M. leprae* trHbO(III)-OONO species have been determined from data analysis, according to [Eqs. (18.3) to (18.5)], which treat both reactions as irreversible first-order processes (Ascenzi *et al.*, 2006c):

$$[\text{trHbO(II)- O}_2]_t = [\text{trHbO(II)- O}_2]_i \times e^{-k \times t} \tag{18.3}$$

$$\begin{aligned}[\text{trHbO(III)- OONO}]_t = {} & [\text{trHbO(II)- O}_2]_i \\ & \times (k \times ([e^{-k \times t/(h-k)}] + [e^{-h \times t/(k-h)}]))\end{aligned} \tag{18.4}$$

$$\begin{aligned}[\text{trHbO(III)}]t = {} & [\text{trHbO(III)}-\text{OONO}]_i - ([\text{trHbO(II)}-\text{O}_2]_t \\ & + [\text{trHbO(III)}-\text{OONO}]_t)\end{aligned} \tag{18.5}$$

The value of the second-order rate constant k_{on} for •NO scavenging by *M. leprae* trHbO(II)-O_2 was obtained from the linear dependence of k on the •NO concentration (i.e., [•NO]) according to Eq. (18.6) (Ascenzi *et al.*, 2006c):

$$k = k_{on} \times [{}^{\bullet}\text{NO}] + c \tag{18.6}$$

Data were analyzed using the MATLAB program (The Math Works Inc., Natick, MA) (Ascenzi *et al.*, 2006b) or the SX.18MV-R rapid-mixing stopped-flow apparatus operating software (Applied Photophysics Ltd., Leatherhead, UK) (Ouellet *et al.*, 2002).

4. Peroxynitrite Scavenging by *M. leprae* trHbO(II)-O_2

4.1. Overview

Peroxynitrite scavenging by ferrous oxygenated heme proteins involves the transient formation of the ferryl-heme derivative (Ascenzi *et al.*, 2006a; Boccini and Herold, 2004; Herold *et al.*, 2003; Herold and Puppo, 2005b), which may induce chemical modifications of aromatic residues (e.g., Tyr and Trp) (Herold, 2004a; Ouellet *et al.*, 2007) and participate in sulfheme formation in the presence of hydrogen sulfide. Indeed, the biogenesis of sulfheme proteins is a still-obscure process (Bunn and Forget, 1986) in which peroxynitrite might take part.

In the absence and presence of CO_2 ($=1.2 \times 10^{-3}$ M), peroxynitrite scavenging by *M. leprae* trHbO(II)-O_2 is consistent with the two-step mechanism reported in Scheme 3 (Ascenzi *et al.*, 2006a):

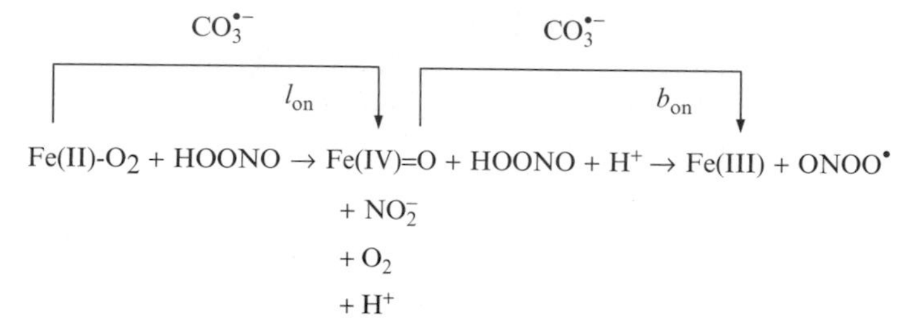

Mixing of the *M. leprae* trHbO(II)-O_2 and peroxynitrite solutions causes a shift of the optical absorption maximum of the Soret band from 412 (i.e., trHbO(II)-O_2) to 419 nm (i.e., trHbO(IV))=O) and a change of the extinction coefficient from $\epsilon_{412\ nm} = 1.22 \times 10^5\ M^{-1}\ cm^{-1}$ (i.e., trHbO(II)-NO) to $\epsilon_{419\ nm} = 1.18 \times 10^5\ M^{-1}\ cm^{-1}$ (i.e., trHbO(III)-NO). Then, the *M. leprae* trHbO(IV)=O solution undergoes a shift of the optical absorption maximum of the Soret band from 419 (i.e., trHbO(IV)=O) to 409 nm (i.e., trHbO(III)) and a change of the extinction coefficient from $\epsilon_{419\ nm} = 1.18 \times 10^5\ M^{-1}\ cm^{-1}$ (i.e., trHbO(IV)=O) to $\epsilon_{409\ nm} = 1.15 \times 10^5\ M^{-1}\ cm^{-1}$ (i.e., trHbO(III)) (Ascenzi *et al.*, 2006a).

Over the whole peroxynitrite concentration range explored (2.0×10^{-5} M – 2.5×10^{-4} M), the time course for peroxynitrite scavenging by *M. leprae* trHbO(II)-O_2 corresponds to a biphasic process between 360 and 460 nm, in the absence and the presence of CO_2. The first and the second steps (indicated by l_{on} and b_{on} in Scheme 3, respectively) are bimolecular processes (Ascenzi *et al.*, 2006a).

Values of the pseudo-first-order rate constant l for the formation and of the pseudo-first-order rate constant b for the decay of the transient trHbO-(IV)=O species are wavelength independent at fixed [peroxynitrite] and [CO_2]. The plots of l and b versus [peroxynitrite] are linear in the absence and the presence of CO_2. The slope of plots of l and b versus [peroxynitrite] corresponds to $l_{on} = 4.8 \times 10^4$ M^{-1} s^{-1} and $b_{on} = 1.3 \times 10^4$ M^{-1} s^{-1} in the absence of CO_2 and to $l_{on} = 6.3 \times 10^5$ M^{-1} s^{-1} and $b_{on} = 1.7 \times 10^4$ M^{-1} s^{-1} in the presence of CO_2 (=1.2×10^{-3} M) (Table 18.2). The y-axis intercept of plots of l and b versus [peroxynitrite] correspond to $m = n = 0$ s^{-1} in the absence of CO_2 and to $m = 1.3 \times 10^1$ s^{-1} and $n = 0$ s^{-1} in the presence of CO_2. Because peroxynitrite scavenging by *M. leprae* trHbO(II)-O_2 is not likely to be a reversible process, the value of $m = 1.3 \times 10^1$ s^{-1} is suggestive of a complex reaction mechanism (Ascenzi *et al.*, 2006a).

CO_2 facilitates the conversion of *M. leprae* trHbO(II)-O_2 to trHbO (IV)=O (see Table 18.2) by transient formation of the short-living species $CO_3^{\bullet -}$, without affecting the reduction of *M. leprae* trHbO(IV)=O (Ascenzi *et al.*, 2006a).

Under the experimental conditions tested, the reduction of *M. leprae* trHbO(IV)=O is rate limiting. Furthermore, decomposed peroxynitrite does not oxidize *M. leprae* trHbO(II)-O_2 (Ascenzi *et al.*, 2006a).

The catalytic parameters for peroxynitrite scavenging by the ferrous oxygenated heme proteins considered are similar (see Table 18.2) (Ascenzi *et al.*, 2006a; Boccini and Herold, 2004; Herold *et al.*, 2003; Herold and Puppo, 2005a). The unfavorable parameters for peroxynitrite scavenging by *M. leprae* trHbO(II)-O_2 when compared with those for $^{\bullet}$NO detoxification by trHbO(II)-O_2 and peroxynitrite scavenging by trHbO(II)-NO (for

Table 18.2 Kinetic parameters for peroxynitrite scavenging by ferrous oxygenated heme proteins

Heme protein	[CO_2] (M)	l_{on} ($M^{-1}s^{-1}$)	b_{on} ($M^{-1}s^{-1}$)
M. leprae trHbO[a]	—	4.8×10^4	1.3×10^4
	1.2×10^{-3}	6.3×10^5	1.7×10^4
Glycine max Lb[b]	—	5.5×10^4	2.1×10^4
	1.2×10^{-3}	8.8×10^5	3.6×10^5
Horse heart Mb[c]	—	5.4×10^4	2.2×10^4
	1.2×10^{-3}	4.1×10^5	3.2×10^4
Human Hb[d]	—	3.3×10^4	3.3×10^4
	1.2×10^{-3}	3.5×10^5	1.1×10^5

[a] pH 7.3 and 20°; from Ascenzi *et al.* (2006a).
[b] pH 7.3 and 20°; from Herold and Puppo (2005b).
[c] pH 7.5 and 20°; from Herold *et al.* (2003).
[d] pH 7.4 and 20°; from Boccini and Herold (2004).

comparison, see Tables 18.1 to 18.3) render this process unlikely *in vivo* (Ascenzi *et al.*, 2006a,c).

4.2. Assay protocol

Values of the second-order rate constant for peroxynitrite-mediated conversion of *M. leprae* trHbO(II)-O_2 to trHbO(IV)=O (indicated by l_{on} in Scheme 3) and of GlbO(IV)=O reduction (i.e., for the formation of GlbO-(III); indicated by b_{on} in Scheme 3) were determined by rapid-mixing the trHbO(II)-O_2 solution (2.5×10^{-6} M) with the peroxynitrite (2.0×10^{-5} M $- 2.4 \times 10^{-4}$ M) solution under anaerobic conditions, at pH 7.3 (1.0×10^{-1} M phosphate buffer, in the presence of 5.0×10^{-2} M EDTA) and 20°, in the absence and presence of CO_2 ($=1.2 \times 10^{-3}$ M). No gaseous phase was present. Kinetics were monitored between 360 and 460 nm. The dead time of the SX.18MV-R rapid-mixing stopped-flow apparatus (Applied Photophysics Ltd., Leatherhead, UK) was 1.6 ms (Ascenzi *et al.*, 2006a).

The time courses for peroxynitrite scavenging by *M. leprae* trHbO(II)-O_2 were fitted to two consecutive monoexponential processes. Values of the peroxynitrite-dependent pseudo-first-order rate constant l for the formation and of the pseudo-first-order rate constant b for the decay of the transient *M. leprae* GlbO(IV)=O species have been determined from data analysis, according to Eqs. 18.7 to 18.9 (Ascenzi *et al.*, 2006a):

$$[\mathrm{GlbO(II)\text{-}O_2}]_t = [\mathrm{GlbO(II)\text{-}O_2}]_i \times e^{-lt} \tag{18.7}$$

$$[\mathrm{GlbO(IV){=}O}]_t = [\mathrm{GlbO(II)\text{-}O_2}]_i \times \left(b \times \left[\left(e^{-lt/(b-l)}\right) + \left(e^{-bt/(l-b)}\right)\right]\right) \tag{18.8}$$

$$[\mathrm{GlbO(III)}]_t = [\mathrm{GlbO(II)\text{-}O_2}]_i - \left([\mathrm{GlbO(II)\text{-}O_2}]_t + [\mathrm{GlbO(IV){=}O}]_t\right) \tag{18.9}$$

Values of l_{on} and d_{on} were obtained from the linear dependence of l and b, respectively, on the peroxynitrite concentration (i.e., [peroxynitrite]) according to Eqs. 18.10 and 18.11) (Ascenzi *et al.*, 2006a):

$$l = l_{on} \times [\text{peroxynitrite}] + m \tag{18.10}$$

$$b = b_{on} \times [\text{peroxynitrite}] + n \tag{18.11}$$

Data were analyzed using the MATLAB program (The Math Works Inc., Natick, MA) (Ascenzi *et al.*, 2006a).

5. Peroxynitrite Scavenging by *M. leprae* trHbO(II)-NO

5.1. Overview

Peroxynitrite precursors, namely $^{\bullet}NO$ and O_2^{-}, are concomitantly produced by activated macrophages in order to limit mycobacterial proliferation *in vivo* (see Adams *et al.*, 1997; Cooper *et al.*, 2002; Lau *et al.*, 1998; MacMicking *et al.*, 1997; Nathan and Shiloh, 2000; Ohno *et al.*, 2003; Ratledge and Dale, 1999; Schnappinger *et al.*, 2006; Visca *et al.*, 2002a; Voskuil *et al.*, 2003; Zahrt and Deretic, 2002). Protection of host cells and infecting pathogens by $^{\bullet}NO$ and peroxynitrite may represent a prominent role of heme proteins (Ascenzi *et al.*, 2006a, 2007a; Ascenzi and Fasano, 2007; Herold, 2004b; Herold *et al.*, 2004; Herold and Boccini, 2006; Herold and Puppo, 2005a).

In the absence and presence of CO_2 ($=1.2 \times 10^{-3}$ M), peroxynitrite scavenging by *M. leprae* trHbO(II)-NO is consistent with the two-step mechanism reported in Scheme 4 (Ascenzi *et al.*, 2006a).

$$\begin{array}{c} CO_3^{\bullet-} \\ \xrightarrow{\quad d_{on} \quad} \quad f \\ \text{Fe(II)-NO} + \text{HOONO} \rightarrow \text{Fe(III)-NO} \rightarrow \text{Fe(III)} + {}^{\bullet}\text{NO} \\ + {}^{\bullet}\text{NO}_2 \\ + \text{OH}^- \end{array}$$

The first observable species after mixing the *M. leprae* trHbO(II)-NO ($\lambda_{max} = 417$ nm and $\epsilon = 1.31 \times 10^5$ M^{-1} cm^{-1}) and peroxynitrite solutions corresponds to that of the intermediate species trHbO(III)-NO ($\lambda_{max} =$ 421 nm and $\epsilon = 1.46 \times 10^5$ M^{-1} cm^{-1}), which is completely populated in the dead time of the rapid-mixing stopped-flow apparatus ($=1.6$ ms). Then, the *M. leprae* GlbO(III)-NO solution undergoes a shift of the optical absorption maximum from 421 (i.e., trHbO(III)-NO) to 409 nm (i.e., trHbO(III)) and a change of the extinction coefficient from $\epsilon_{421\ nm} =$ 1.46×10^5 M^{-1} cm^{-1} (i.e., trHbO(III)-NO) to $\epsilon_{409\ nm} = 1.15 \times$ 10^5 M^{-1} cm^{-1} (i.e., trHbO(III)) (Ascenzi *et al.*, 2006a).

Over the whole peroxynitrite concentration range explored (2.0×10^{-5} M $- 2.4 \times 10^{-4}$ M), the time course for the peroxynitrite-mediated oxidation of *M. leprae* trHbO(II)-NO corresponds to a monophasic process (indicated by *f* in Scheme 4) between 360 and 460 nm, in the absence and the presence of CO_2. A minimum estimate of the second-order rate constant in Scheme 4 yields $d_{on} > 1 \times 10^8$ M^{-1} s^{-1}. Values of the first-order rate constant *f* for the decay of the trHbO(III)-NO species are [peroxynitrite], [CO_2], and wavelength independent, the average value of *f* is $(2.5 \pm 0.4) \times 10^1$ s^{-1} (Table 18.3) (Ascenzi *et al.*, 2006a).

CO_2 may facilitate the conversion of *M. leprae* trHbO(II)-NO to trHbO-(III)-NO by transient formation of the short-living species $CO_3^{\bullet -}$. However, CO_2 does not affect the dissociation of $^\bullet$NO from *M. leprae* trHbO-(III)-NO (see Table 18.3) (Ascenzi *et al.*, 2006a).

Under the experimental conditions tested, *M. leprae* trHbO(III)-NO denitrosylation is rate limiting. Furthermore, decomposed peroxynitrite does not oxidize *M. leprae* trHbO(II)-NO (Ascenzi *et al.*, 2006a).

M. leprae trHbO(II)-NO displays the most favorable parameters for peroxynitrite scavenging both in the absence and presence of CO_2 (see Table 18.3), possibly contributing to bacterial persistence in the paucibacillary leprosy (Ascenzi *et al.*, 2006a). This form of the disease is characterized by a vigorous immune response of the host involving the release of reactive nitrogen and oxygen species by activated macrophages (e.g., $^\bullet$NO, $O_2^{\bullet -}$, and peroxynitrite), hypoxia, locally increased CO_2 production from respiratory burst concomitant with bacterial exposure to acidic pH resulting from phagolysosomal fusion (Britton and Lockwood, 2004; Scollard *et al.*, 2006; Visca *et al.*, 2002b).

Table 18.3 Kinetic parameters for peroxynitrite scavenging by ferrous nitrosylated heme proteins

Heme protein	[CO_2] (M)	d_{on} ($M^{-1}s^{-1}$)	f (s^{-1})
M. leprae trHbO[a]	—	$>1 \times 10^8$	2.6×10^1
	1.2×10^{-3}	$>1 \times 10^8$	2.4×10^1
Glycine max Lb[b]	—	8.8×10^3	2.0
	1.0×10^{-3}	1.2×10^5	2.5
Horse heart Mb	—	3.1×10^4[c]	$\approx 1.2 \times 10^1$[c]
	1.2×10^{-3}[d]	1.7×10^5[d]	1.1×10^1[d]
Human Ngb[e]	—	1.3×10^5	1.2×10^{-1}
Human Hb[f]	—	6.1×10^3	≈ 1
	1.2×10^{-3}	5.3×10^4	~ 1
Rabbit HPX-heme[g]	—	8.6×10^4	4.3×10^{-1}
	1.2×10^{-3}	1.2×10^6	4.3×10^{-1}
Human SA-heme[h]	—	6.5×10^3	1.9×10^{-1}
	1.2×10^{-3}	1.3×10^5	1.7×10^{-1}

[a] pH 7.3 and 20°; from Ascenzi *et al.* (2006a).
[b] pH 7.3 and 20°; from Herold and Puppo (2005a).
[c] pH 7.5 and 20°; from Herold and Boccini (2006).
[d] pH 7.0 and 20°; from Herold and Boccini (2006).
[e] pH 7.2 and 20°; from Herold *et al.* (2004).
[f] pH 7.2 and 20°; from Herold (2004b).
[g] pH 7.0 and 10°; from Ascenzi *et al.* (2007a).
[h] pH 7.0 and 10°; from Ascenzi and Fasano (2007).

5.2. Assay protocol

Values of the second-order rate constant d_{on} for peroxynitrite-mediated conversion of *M. leprae* GlbO(II)-NO to GlbO(III)-NO and of the first-order rate constant f for •NO dissociation from the GlbO(III)-NO complex (i.e., for GlbO(III) formation) were determined by rapid-mixing the GlbO(II)-NO solution (2.5×10^{-6} M) with the peroxynitrite (2.0×10^{-5} M – 2.0×10^{-4} M) solution under anaerobic conditions, at pH 7.3 (1.0×10^{-1} M phosphate buffer, in the presence of 5.0×10^{-2} M EDTA) and 20°, in the absence and the presence of CO_2 ($=1.2 \times 10^{-3}$ M). No gaseous phase was present. Kinetics were monitored between 360 and 460 nm. The dead time of the SX.18MV-R rapid-mixing stopped-flow apparatus (Applied Photophysics Ltd., Leatherhead, UK) was 1.6 ms (Ascenzi *et al.*, 2006a).

The peroxynitrite-mediated conversion of *M. leprae* trHbO(II)-NO to trHbO(III)-NO (indicated by d_{on} in Scheme 4) was undetectable even at the lowest peroxynitrite concentration investigated ($=2.0 \times 10^{-5}$ M), both in the absence and in the presence of CO_2 ($=1.2 \times 10^{-3}$ M). Values of d_{on} ($=d/[\text{peroxynitrite}] > 1 \times 10^{8}\ M^{-1}\ s^{-1}$) were estimated assuming that the peroxynitrite-mediated conversion of *M. leprae* trHbO(II)-NO to trHbO-(III)-NO was completed within the dead time of the rapid-mixing stopped-flow apparatus (=1.6 ms). Therefore, the value of the pseudo-first-order rate constant d for the peroxynitrite-mediated conversion of *M. leprae* trHbO(II)-NO to trHbO(III)-NO appears greater than $2 \times 10^{3}\ s^{-1}$ at [peroxynitrite] = 2.0×10^{-5} M (Ascenzi *et al.*, 2006a).

The time courses for •NO dissociation from *M. leprae* trHbO(III)-NO were fitted to a monoexponential process. Values of the first-order rate constant f for •NO dissociation from the transient trHbO(III)-NO complex (i.e., for the formation of trHbO(III)) were determined from data analysis, according to Eq. (18.12) (Ascenzi *et al.*, 2006a):

$$[\text{trHbO(III)-NO}]_t = [\text{trHbO(III)-NO}]_i \times e^{-ft} \tag{18.12}$$

Data were analyzed using the MATLAB program (The Math Works Inc., Natick, MA) (Ascenzi *et al.*, 2006a).

6. Structural Background

Some structural considerations are summarized in conclusion to rationalize the pseudoenzymatic activity of mycobacterial trHbs.

Despite its reduced size, *M. tuberculosis* trHbN displays a remarkable tunnel/cavity system through the protein matrix that may support ligand diffusion to and from the heme distal pocket, accumulation of heme ligands, and/or multiligand reactions. Such an inner tunnel/cavity system displays two branches of ≈20 Å and ≈8 Å in length, respectively; overall, the tunnel volume is ≈265 Å^3 (Bidon-Chanal *et al.*, 2006; Milani *et al.*, 2001, 2003a, 2004a, 2005; Ouellet *et al.*, 2006; Wittenberg *et al.*, 2002). In contrast, because of residue substitutions, the tunnel's long branch is restricted to two neighboring cavities (with an overall volume of ≈170 Å^3) in *M. tuberculosis* trHbO and in *M. leprae* trHbO (Ascenzi *et al.*, 2007b; Visca *et al.*, 2002b), and the tunnel's short branch is virtually absent (Milani *et al.*, 2003b).

The proximal HisF8 residue is observed in a staggered azimuthal orientation relative to the heme pyrrole N atoms in both *M. tuberculosis* trHbN and trHbO crystal structures (Milani *et al.*, 2001, 2003b). Such a feature indicates an unstrained proximal HisF8, supporting heme in-plane location of the Fe atom. This has been related to fast exogenous ligand association and electron donation to the heme-Fe bound distal ligand (see Milani *et al.*, 2005; Wittenberg *et al.*, 2002), which may facilitate pseudoenzymatic functions (see Ascenzi *et al.*, 2007a; Mukai *et al.*, 2001; Samuni *et al.*, 2004).

The mycobacterial trHb heme distal cavity is characterized by an array of unusual residues, if compared with classical nonvertebrate Hbs and Mbs. Residues at positions B10, CD1, E7, E11, E14, E15, and G8 are Tyr, Phe, Leu, Gln, Phe, Phe, and Val, respectively, in *M. tuberculosis* trHbN, and Tyr, Tyr, Ala, Leu, Phe, Leu, and Trp, respectively, in *M. tuberculosis* and *M. leprae* trHbO (Ascenzi *et al.*, 2007b; Vinogradov *et al.*, 2006; Vuletich and Lecompte, 2006). Such a selection of heme distal site residues, varying in size and polarity, is instrumental in achieving trHb family-specific hydrogen-bonded heme distal networks. They modulate the positioning and dynamics of heme distal residues that stabilize the heme-Fe-bound ligand and participate in the control of ligand access to and escape from the heme pocket (see Bidon-Chanal *et al.*, 2006; Milani *et al.*, 2004a, 2005).

Last, unlike *M. tuberculosis* Hmp, homologous to *E. coli* flavoHb (see Hu *et al.*, 1999), mycobacterial trHbs need a reductase partner to facilitate pseudoenzymatic scavenging of reactive nitrogen species. The identification of (1) an efficient trHb(III) reductase system to restore trHbO(II), (2) the structural bases for selective recognition of electron donor(s) by trHb(III)s, and (3) the electron transfer route(s) to the heme-Fe(III)-atom is a challenge for future research on trHbs.

ACKNOWLEDGMENTS

The authors wish to thank Prof. Martino Bolognesi and Prof. Robert K. Poole for critical reading of this article and for helpful advice. This work was partially supported by grants from the Ministry for University and Research of Italy (University Roma Tre, Rome, Italy, CLAR 2006 to both authors) and from the Ministry for Health of Italy (National Institute for Infectious Diseases IRCCS Lazzaro Spallanzani, Rome, Italy, Ricerca Corrente 2006 to both authors).

REFERENCES

Adams, L. B., Dinauer, M. C., Morgenstern, D. E., and Krahenbuhl, J. L. (1997). Comparison of the roles of reactive oxygen and nitrogen intermediates in the host response to *Mycobacterium tuberculosis* using transgenic mice. *Tuberc. Lung. Dis.* **78,** 237–246.

Antonini, E., and Brunori, M. (1971). "Hemoglobin and myoglobin in their reactions with ligands." Amsterdam: North Holland Publishing Co., Amsterdam.

Ascenzi, P., and Fasano, M. (2007). Abacavir modulates peroxynitrite-mediated oxidation of ferrous nitrosylated human serum heme-albumin. *Biochem. Biophys. Res. Commun.* **353,** 469–474.

Ascenzi, P., Milani, M., and Visca, P. (2006a). Peroxynitrite scavenging by ferrous truncated hemoglobin GlbO from *Mycobacterium leprae. Biochem. Biophys. Res. Commun.* **351,** 528–533.

Ascenzi, P., Bocedi, A., Antonini, G., Bolognesi, M., and Fasano, M. (2007a). Reductive nitrosylation and peroxynitrite-mediated oxidation of heme-hemopexin. *FEBS J.* **274,** 551–562.

Ascenzi, P., Bocedi, A., Visca, P., Minetti, M., and Clementi, E. (2006b). Does CO_2 modulate peroxynitrite specificity? *IUBMB Life* **58,** 611–613.

Ascenzi, P., Bolognesi, M., Milani, M., Guertin, M., and Visca, P. (2007b). Mycobacterial truncated hemoglobins: From genes to functions. *Gene* **398,** 42–51.

Ascenzi, P., Bocedi, A., Bolognesi, M., Fabozzi, G., Milani, M., and Visca, P. (2006c). Nitric oxide scavenging by *Mycobacterium leprae* GlbO involves the formation of the ferric heme-bound peroxynitrite intermediate. *Biochem. Biophys. Res. Commun.* **339,** 448–454.

Beckman, J. S., and Koppenol, W. H. (1996). Nitric oxide, superoxide, and peroxynitrite: The good, the bad, and ugly. *Am. J. Physiol.* **271,** C1424–C1437.

Beckman, J. S., Beckman, T. W., Chen, J., Marshall, P. A., and Freeman, B. A. (1990). Apparent hydroxyl radical production by peroxynitrite: Implications for endothelial injury from nitric oxide and superoxide. *Proc. Natl. Acad. Sci. USA* **87,** 1620–1624.

Bidon-Chanal, A., Martì, M. A., Crespo, A., Milani, M., Orozco, M., Bolognesi, M., Luque, F. J., and Estrin, D. A. (2006). Ligand-induced dynamical regulation of NO conversion in *Mycobacterium tuberculosis* truncated hemoglobin N. *Proteins* **64,** 457–464.

Boccini, F., and Herold, S. (2004). Mechanistic studies of the oxidation of oxyhemoglobin by peroxynitrite. *Biochemistry* **43,** 16393–16404.

Bohle, D. S., Glassbrenner, P. A., and Hansert, B. (1996). Syntheses of pure tetramethylammonium peroxynitrite. *Methods Enzymol.* **269,** 302–311.

Britton, W. J., and Lockwood, D. N. J. (2004). Leprosy. *Lancet* **363,** 1209–1219.

Brunori, M. (2001). Nitric oxide moves myoglobin centre stage. *Trends Biochem. Sci.* **26,** 209–210.

Brunori, M., and Gibson, Q. H. (2001). Cavities and packing defects in the structural dynamics of myoglobin. *EMBO Rep.* **2,** 674–679.

Brunori, M., Bourgeois, D., and Vallone, B. (2004). The structural dynamics of myoglobin. *J. Struct. Biol.* **147,** 223–234.

Brunori, M., Giuffrè, A., Nienhaus, K., Nienhaus, G. U., Scandurra, F. M., and Vallone, B. (2005). Neuroglobin, nitric oxide, and oxygen: Functional pathways and conformational changes. *Proc. Natl. Acad. Sci. USA* **102,** 8483–8488.

Bunn, H. F., and Forget, B. G. (1986). "Hemoglobin: Molecular, genetic, and clinical aspects." Philadelphia, PA: Sanders, Philadelphia, PA.

Choi, H. S., Rai, P. R., Chu. H. W., Cool, C., and Chan, E. D. (2002). Analysis of nitric oxide synthase and nitrotyrosine expression in human pulmonary tuberculosis. *Am. J. Respir. Crit. Care Med.* **166,** 178–186.

Clementi, E., and Nisoli, E. (2005). Nitric oxide and mitochondrial biogenesis: A key to long-term regulation of cellular metabolism. *Comp. Biochem. Physiol.* **142,** 102–110.

Cooper, A. M., Adams, L. B., Dalton, D. K., Appelberg, R., and Ehlers, S. (2002). IFN-γ and NO in mycobacterial disease: New jobs for old hands. *Trends Microbiol.* **10,** 221–226.
Couture, M., and Guertin, M. (1996). Purification and spectroscopic characterization of a recombinant chloroplastic hemoglobin from the green unicellular alga *Chlamydomonas eugametos*. *Eur. J. Biochem.* **242,** 779–787.
Couture, M., Yeh, S. R., Wittenberg, B. A., Wittenberg, J. B., Ouellet, Y., Rousseau, D. L., and Guertin, M. (1999). A cooperative oxygen-binding hemoglobin from *Mycobacterium tuberculosis*. *Proc. Natl. Acad. Sci. USA* **96,** 11223–11228.
Denicola, A., and Radi, R. (2005). Peroxynitrite and drug-dependent toxicity. *Toxicology* **208,** 273–288.
Ducrocq, C., Blanchard, B., Pignatelli, B., and Oshima, H. (1999). Peroxynitrite: An endogenous oxidizing and nitrating agent. *Cell Mol. Life Sci.* **55,** 1068–1077.
Fabozzi, G., Ascenzi, P., Di Renzi, S., and Visca, P. (2006). Truncated hemoglobin GlbO from *Mycobacterium leprae* alleviates nitric oxide toxicity. *Microb. Pathog.* **40,** 211–220.
Fasano, M., Antonini, G., and Ascenzi, P. (2006). O_2-mediated oxidation of hemopexin-heme(II)-NO. *Biochem. Biophys. Res. Commun.* **345,** 704–712.
Flögel, U., Merx, M. W., Gödecke, A., Decking, U. K., and Schrader, J. (2001). Myoglobin: A scavenger of bioactive NO. *Proc. Natl. Acad. Sci. USA* **98,** 735–740. *Erratum* in: *Proc. Natl. Acad. Sci. USA* **98,** 4276.
Frey, A. D., and Kallio, P. T. (2003). Bacterial hemoglobins and flavohemoglobins: Versatile proteins and their impact on microbiology and biotechnology. *FEMS Microbiol. Rev.* **27,** 525–545.
Frey, A. D., and Kallio, P. T. (2005). Nitric oxide detoxification: A new era for bacterial globins in biotechnology? *Trends Biotechnol.* **23,** 69–73.
Frauenfelder, H., McMahon, B. H., and Fenimore, P. W. (2003). Myoglobin: The hydrogen atom of biology and a paradigm of complexity. *Proc. Natl. Acad. Sci. USA* **100,** 8615–8617.
Gardner, A. M., Martin, L. A., Gardner, P. R., Dou, Y., and Olson, J. S. (2000). Steady-state and transient kinetics of *Escherichia coli* nitric-oxide dioxygenase (flavohemoglobin): The B10 tyrosine hydroxyl is essential for dioxygen binding and catalysis. *J. Biol. Chem.* **275,** 12581–12589.
Goldstein, S., Lind, J., and Merényi, G. (2005). Chemistry of peroxynitrites and peroxynitrates. *Chem. Rev.* **105,** 2457–2470.
Gow, A. J., Luchsinger, B. P., Pawloski, J. R., Singel, D. J., and Stamler, J. S. (1999). The oxyhemoglobin reaction of nitric oxide. *Proc. Natl. Acad. Sci. USA* **96,** 9027–9032.
Herold, S. (1999). Kinetic and spectroscopic characterization of an intermediate peroxynitrite complex in the nitrogen monoxide induced oxidation of oxyhemoglobin. *FEBS Lett.* **443,** 81–84.
Herold, S. (2004a). Nitrotyrosine, dityrosine, and nitrotryptophan formation from metmyoglobin, hydrogen peroxide, and nitrite. *Free Radic. Biol. Med.* **36,** 565–579.
Herold, S. (2004b). The outer-sphere oxidation of nitrosyliron(II)hemoglobin by peroxynitrite leads to the release of nitrogen monoxide. *Inorg. Chem.* **43,** 3783–3785.
Herold, S., and Boccini, F. (2006). NO• release from MbFe(II)NO and HbFe(II)NO after oxidation by peroxynitrite. *Inorg. Chem.* **45,** 6933–6943.
Herold, S., and Fago, A. (2005). Reactions of peroxynitrite with globin proteins and their possible physiological role. *Comp. Biochem. Physiol. A Mol. Integr. Physiol.* **142,** 124–129.
Herold, S., and Puppo, A. (2005a). Kinetics and mechanistic studies of the reactions of metleghemoglobin, ferrylleghemoglobin, and nitrosylleghemoglobin with reactive nitrogen species. *J. Biol. Inorg. Chem.* **10,** 946–957.
Herold, S., and Puppo, A. (2005b). Oxyleghemoglobin scavenges nitrogen monoxide and peroxynitrite: A possible role in functioning nodules? *J. Biol. Inorg. Chem.* **10,** 935–945.

Herold, S., Exner, M., and Boccini, F. (2003). The mechanism of the peroxynitrite-mediated oxidation of myoglobin in the absence and presence of carbon dioxide. *Chem. Res. Toxicol.* **16,** 390–402.
Herold, S., Exner, M., and Nauser, T. (2001). Kinetic and mechanistic studies of the NO•-mediated oxidation of oxymyoglobin and oxyhemoglobin. *Biochemistry* **40,** 3385–3395.
Herold, S., Fago, A., Weber, R. E., Dewilde, S., and Moens, L. (2004). Reactivity studies of the Fe(III) and Fe(II)NO forms of human neuroglobin reveal a potential role against oxidative stress. *J. Biol. Chem.* **279,** 22841–22847.
Hu, Y., Butcher, P. D., Mangan, J. A., Rajandream, M. A., and Coates, A. R. (1999). Regulation of hmp gene transcription in *Mycobacterium tuberculosis*: Effects of oxygen limitation and nitrosative and oxidative stress. *J. Bacteriol.* **181,** 3486–3493.
Hussain, T. (2007). Leprosy and tuberculosis: An insight-review. *Crit. Rev. Microbiol.* **33,** 15–66.
Ignarro, L. J. (2002). Nitric oxide as a unique signaling molecule in the vascular system: A historical overview. *J. Physiol. Pharmacol.* **53,** 503–514.
Koppenol, W. H., Kissner, R., and Beckman, J. S. (1996). Syntheses of peroxynitrite: To go with the flow or on solid grounds? *Methods Enzymol.* **269,** 296–302.
Lama, A., Pawaria, S., and Dikshit, K. L. (2006). Oxygen binding and NO scavenging properties of truncated hemoglobin, HbN, of *Mycobacterium smegmatis*. *FEBS Lett.* **580,** 4031–4041.
Lau, Y. L., Chan, G. C. F., Ha, S. Y., Hui, Y. F., and Yuen, K. Y. (1998). The role of phagocytic respiratory burst in host defense against *Mycobacterium tuberculosis*. *Clin. Infect. Dis.* **26,** 226–227.
Liu, C., He, Y., and Chang, Z. (2004). Truncated hemoglobin O of *Mycobacterium tuberculosis*: The oligomeric state change and the interaction with membrane components. *Biochem. Biophys. Res. Commun.* **316,** 1163–1172.
MacMicking, J., Xie, Q. W., and Nathan, C. (1997). Nitric oxide and macrophage function. *Annu. Rev. Immunol.* **15,** 323–350.
Milani, M., Pesce, A., Ouellet, H., Guertin, M., and Bolognesi, M. (2003a). Truncated hemoglobins and nitric oxide action. *IUBMB Life* **55,** 623–627.
Milani, M., Pesce, A., Ouellet, Y., Ascenzi, P., Guertin, M., and Bolognesi, M. (2001). *Mycobacterium tuberculosis* hemoglobin N displays a protein tunnel suited for O_2 diffusion to the heme. *EMBO J.* **20,** 3902–3909.
Milani, M., Savard, P. Y., Ouellet, H., Ascenzi, P., Guertin, M., and Bolognesi, M. (2003b). A TyrCD1/TrpG8 hydrogen bond network and a TyrB10TyrCD1 covalent link shape the heme distal site of *Mycobacterium tuberculosis* hemoglobin O. *Proc. Natl. Acad. Sci. USA* **100,** 5766–5771.
Milani, M., Pesce, A., Ouellet, Y., Dewilde, S., Friedman, J., Ascenzi, P., Guertin, M., and Bolognesi, M. (2004a). Heme-ligand tunneling in group I truncated hemoglobins. *J. Biol. Chem.* **279,** 21520–21525.
Milani, M., Ouellet, Y., Ouellet, H., Guertin, M., Boffi, A., Antonimi, G., Bocedi, A., Mattu, M., Bolognesi, M., and Ascenzi, P. (2004b). Cyanide binding to truncated hemoglobins: A crystallographic and kinetic study. *Biochemistry* **43,** 5213–5221.
Milani, M., Pesce, A., Nardini, M., Ouellet, H., Ouellet, Y., Dewilde, S., Bocedi, A., Ascenzi, P., Guertin, M., Moens, L., Friedman, J. M., and Wittenberg, J. B., *et al.* (2005). Structural bases for heme binding and diatomic ligand recognition in truncated hemoglobins. *J. Inorg. Biochem.* **99,** 97–109.
Mukai, M., Mills, C. E., Poole, R. K., and Yeh, S. R. (2001). Flavohemoglobin, a globin with a peroxidase-like catalytic site. *J. Biol. Chem.* **276,** 7272–7277.

Mukai, M., Ouellet, Y., Ouellet, H., Guertin, M., and Yeh, S. R. (2004). NO binding induced conformational changes in a truncated hemoglobin from *Mycobacterium tuberculosis*. *Biochemistry* **43,** 2764–2770.

Mukai, M., Savard, P. Y., Ouellet, H., Guertin, M., and Yeh, S. R. (2002). Unique ligand-protein interactions in a new truncated hemoglobin from *Mycobacterium tuberculosis*. *Biochemistry* **41,** 3897–3905.

Nathan, C., and Shiloh, M. U. (2000). Reactive oxygen and nitrogen intermediates in the relationship between mammalian hosts and microbial pathogens. *Proc. Natl. Acad. Sci. USA* **97,** 8841–8848.

Ohno, H., Zhu, G., Mohan, V. P., Chu, D., Kohno, S., Jacobs, W. R., Jr., and Chan, J. (2003). The effects of reactive nitrogen intermediates on gene expression in *Mycobacterium tuberculosis*. *Cell Microbiol.* **5,** 637–648.

Ouellet, H., Ouellet, Y., Richard, C., Labarre, M., Wittenberg, B., Wittenberg, J., and Guertin, M. (2002). Truncated hemoglobin HbN protects *Mycobacterium bovis* from nitric oxide. *Proc. Natl. Acad. Sci. USA* **99,** 5902–5907.

Ouellet, H., Ranguelova, K., Labarre, M., Wittenberg, J. B., Wittenberg, B. A., Magliozzo, R. S., and Guertin, M. (2007). Reaction of *Mycobacterium tuberculosis* truncated hemoglobin O with hydrogen peroxide: Evidence for peroxidatic activity and formation of protein-based radicals. *J. Biol. Chem.* in press.

Ouellet, H., Juszczak, L., Dantsker, D., Samuni, U., Ouellet, Y. H., Savard, P. Y., Wittenberg, J. B., Wittenberg, B. A., Friedman, J. M., and Guertin, M. (2003). Reactions of *Mycobacterium tuberculosis* truncated hemoglobin O with ligands reveal a novel ligand-inclusive hydrogen bond network. *Biochemistry* **42,** 5764–5774.

Ouellet, Y., Milani, M., Couture, M., Bolognesi, M., and Guertin, M. (2006). Ligand interactions in the distal heme pocket of *Mycobacterium tuberculosis* truncated hemoglobin N: Roles of TyrB10 and GlnE11 residues. *Biochemistry* **45,** 8770–8781.

Pathania, R., Navani, N. K., Rajamohan, G., and Dikshit, K. L. (2002b). *Mycobacterium tuberculosis* hemoglobin HbO associates with membranes and stimulates cellular respiration of recombinant *Escherichia coli*. *J. Biol. Chem.* **277,** 15293–15302.

Pathania, R., Navani, N. K., Gardner, A. M., Gardner, P. R., and Dikshit, K. L. (2002a). Nitric oxide scavenging and detoxification by the *Mycobacterium tuberculosis* haemoglobin, HbN in *Escherichia coli*. *Mol. Microbiol.* **45,** 1303–1314.

Poole, R. K. (2005). Nitric oxide and nitrosative stress tolerance in bacteria. *Biochem. Soc. Trans.* **33,** 176–180.

Ratledge, C., and Dale, J., (1999). "Mycobacteria, molecular biology and virulence." Blackwell Science, Oxford.

Samuni, U., Ouellet, Y., Guertin, M., Friedman, J. M., and Yeh, S. R. (2004). The absence of proximal strain in the truncated hemoglobins from *Mycobacterium tuberculosis*. *J. Am. Chem. Soc.* **126,** 2682–2683.

Schnappinger, D., Schoolnik, G. K., and Ehrt, S. (2006). Expression profiling of host pathogen interactions: How *Mycobacterium tuberculosis* and the macrophage adapt to one another. *Microbes Infect.* **8,** 1132–1140.

Schon, T., Hernandez-Pando, R., Baquera-Heredia, J., Negesse, Y., Becerril-Villanueva, L. E., Eon-Contreras, J. C., Sundqvist, T., and Britton, S. (2004). Nitrotyrosine localization to dermal nerves in borderline leprosy. *Br. J. Dermatol.* **150,** 570–574.

Scollard, D. M., Adams, L. B., Gillis, T. P., Krahenbuhl, J. L., Truman, R. W., and Williams, D. L. (2006). The continuing challenges of leprosy. *Clin. Microbiol. Rev.* **19,** 338–381.

Vinogradov, S. N., Hoogewijs, D., Bailly, X., Arredondo-Peter, R., Gough, J., Dewilde, S., Moens, L., and Vanfleteren, J. R. (2006). A phylogenomic profile of globins. *BMC Evol. Biol.* **6,** 31.

Visca, P., Fabozzi, G., Milani, M., Bolognesi, M., and Ascenzi, P. (2002a). Nitric oxide and *Mycobacterium leprae* pathogenicity. *IUBMB Life* **54,** 95–99.

Visca, P., Fabozzi, G., Petrucca, A., Ciaccio, C., Coletta, M., De Sanctis, G., Milani, M., Bolognesi, M., and Ascenzi, P. (2002b). The truncated hemoglobin from *Mycobacterium lepre. Biochem. Biophys. Res. Commun.* **294,** 1064–1070.

Voskuil, M. I., Schnappinger, D., Visconti, K. C., Harrell, M. I., Dolganov, G. M., Sherman, D. R., and Schoolnik, G. K. (2003). Inhibition of respiration by nitric oxide induces a *Mycobacterium tuberculosis* dormancy program. *J. Exp. Med.* **198,** 705–713.

Vuletich, D. A., and Lecomte, J. T. (2006). A phylogenetic and structural analysis of truncated hemoglobins. *J. Mol. Evol.* **62,** 196–210.

Wittenberg, J. B., Bolognesi, M., Wittenberg, B. A., and Guertin, M. (2002). Truncated hemoglobins: A new family of hemoglobins widely distributed in bacteria, unicellular eukaryotes, and plants. *J. Biol. Chem.* **277,** 871–874.

Wu, G., Wainwright, L. M., and Poole, R. K. (2003). Microbial globins. *Adv. Microb. Physiol.* **47,** 255–310.

Yeh, S. R., Couture, M., Ouellet, Y., Guertin, M., and Rousseau, D. L. (2000). A cooperative oxygen binding hemoglobin from *Mycobacterium tuberculosis*: Stabilization of heme ligands by a distal tyrosine residue. *J. Biol. Chem.* **275,** 1679–1684.

Zahrt, T. C., and Deretic, V. (2002). Reactive nitrogen and oxygen intermediates and bacterial defenses: Unusual adaptation in *Mycobacterium tuberculosis*. *Antioxid. Redox Signal.* **4,** 141–159.

SECTION THREE

OTHER HEMOGLOBINS

CHAPTER NINETEEN

Expression, Purification, and Crystallization of Neuro- and Cytoglobin

Sylvia Dewilde,* Kirsten Mees,* Laurent Kiger,† Christophe Lechauve,† Michael C. Marden,† Alessandra Pesce,‡ Martino Bolognesi,** *and* Luc Moens*

Contents

* Department of Biomedical Sciences, University of Antwerp, Antwerp, Belgium
† Le Kremlin-Bicetre, France
‡ Department of Physics CNR-INFM, and Center for Excellence in Biomedical Research, University of Genoa, Genoa, Italy
**Department of Biomolecular Sciences and Biotechnology CNR-INFM, University of Milan, Milan, Italy

Methods in Enzymology, Volume 436
ISSN 0076-6879, DOI: 10.1016/S0076-6879(08)36019-4

Abstract

Neuroglobin and cytoglobin, members of the globin family, are present in vertebrate cells at very low concentrations. As the function of both proteins is still a matter of debate, it is very important to be able to produce and purify these proteins, and in general all members of the globin family, to homogeneity. For this purpose, this chapter describes the expression of neuro- and cytoglobin by *E. coli* and its preparative purification. These proteins are then used in crystallization experiments. Also an analytical purification strategy is discussed in detail.

1. Introduction

Neuro- (Ngb) and cytoglobin (Cygb) were recently discovered vertebrate globins. Ngb is expressed mainly in the brain and other nervous tissues, such as the retina, whereas Cygb occurs in most body cells. Ngb is localized only in the cytoplasm, whereas Cygb is found in the nucleus and cytoplasm of most cells (Burmester *et al.*, 2000, 2002; Schmidt *et al.*, 2003; Trent and Hargrove, 2002; Geuens *et al.*, 2003). The primary structure of both proteins displays all determinants of the globin fold (Bashford *et al.*, 1987; Moens *et al.*, 1996). Cygb has extensions of ≈20 residues at the N- and C-termini. In contrast with hemo- (Hb) and myoglobin (Mb), in which the heme iron atom is pentacoordinated, Ngb and Cygb are hexacoordinated. This means that the sixth position of the heme iron atom is bound to an amino acid residue (the HisE7 distal residue) of the globin chain. This results in a competition of an external ligand (O_2) with the internal ligand for binding to the heme iron atom (Dewilde *et al.*, 2001b; Pesce *et al.*, 2002a). Both globins display the classical α-helical globin fold, although specific structural features in such a fold can be noted. For example, the presence of extended matrix cavities in both molecules and the presence of the terminal extensions in Cygb are noteworthy (de Sanctis *et al.*, 2004; Pesce *et al.*, 2003). In human Ngb, a disulfide bond occurs between Cys CD7 and Cys D5. The presence of this disulfide bond, which should be linked to the redox state of the cell, has a strong effect on the oxygen affinity (Hamdane *et al.*, 2003; Kiger *et al.*, 2004). The function of Ngb and Cygb is, as for other hexacoordinated Hbs, still a matter of debate. Ngb clearly has a neuroprotective function under hypoxia (Fordel *et al.*, 2004, 2006; Khan *et al.*, 2006; Sun *et al.*, 2001, 2003); how such protection occurs, however, is less clear.

As Ngb and Cygb are both present in the cell in very low concentrations, it is impossible to purify the proteins out of the tissue. For the biochemical characterization of the proteins (e.g., oxygen affinity, three-dimensional structure) and functional studies, expression cloning was the best alternative approach. Here we describe the expression cloning of both proteins and the subsequent purification and crystallization.

2. Expression and Preparative Purification of Human Wild-Type NGB

2.1. Cloning of NGB cDNA in the expression vector

The Ngb cDNA is cloned in the expression vector pET3a as follows. Total RNA was extracted using the TriZol method (Invitrogen). First-strand cDNA was synthesized using random primers (Promega, Madison, WI) and Superscript II RT-enzyme (Invitrogen) or by using the One-step RT-PCR kit from Qiagen. Full-length Ngb cDNA was amplified by PCR using specific primers containing the correct adaptors. The 5′ primer contains an *Nde*I restriction site that covers the initiating Met codon of the globin gene. The 3′ primer contains a *Bam*HI restriction site. The amplified product was cleaned and cut with *Nde*I and *Bam*HI and subsequently ligated into the equivalently cleaved expression vector pET3a. Recombinants obtained in the *E. coli* strain XL1-Bleu were tested by PCR and restriction digests. The complete sequence of the construct was checked by dideoxy sequencing; as such, it was verified that the construct was correctly positioned for expression. The recombinant expression plasmid was then successfully transformed in the *E. coli* strain BL21(DE3)pLysS (Invitrogen).

When necessary, mutations can be made on the recombinant Ngb pET3a vector using the QuickChange site-directed mutagenesis method (Stratagene).

2.2. Expression of NGB

The cells are grown overnight in an orbital shaker at 37° in 6 ml L-broth (10 g tryptone, 5 g yeast extract, and 0.5 g NaCl per l) containing 200 μg/ml ampicillin and 30 μg/ml chloramphenicol. The grown culture is transferred into a 1-l Erlenmeyer flask containing 250 ml TB medium (1.2% bactotryptone, 2.4% yeast extract, 0.4% glycerol, 17 m*M* KH_2PO_4, and 72 m*M* $K_2HPO_4.3H_20$) containing 200 μg/ml ampicillin, 30 μg/ml chloramphenicol, and 1 m*M* δ-aminolevulinic acid and shaken in an orbital shaker at 160 rpm at 25° (New Brunswick Scientific Innova 4230). The culture is induced at $A_{600} = 0.8$ by the addition of isopropyl-1-thio-D-galactopyranoside to a final concentration of 0.4 m*M*, and expression is continued overnight. The cells of a 250-ml culture are harvested (20 min at 3220 × *g*) and resuspended in 12 ml lysis buffer (50 m*M* Tris-HCl, pH 8.0, 1 m*M* EDTA, 0.5 m*M* DTT). The cells express Ngb in the cytoplasm, where it is properly folded, resulting in a bright red color as a result of overexpression.

Under our conditions, the metabolism of the *E. coli* cells will gradually shift from aerobic respiration to anaerobic fermentation, due to the complete consumption of the available oxygen by the fast-growing cells. This induces a nitrate consumption and a nitrite production. Thus, it is likely that NO is formed in these *E. coli* cells (Van Doorslaer *et al.*, 2003). Indeed, when studying the overexpression of Ngb in intact *E. coli* cells using UV-vis and EPR, it is clear that Ngb is predominantly present in its deoxy ferrous hexacoordinated form, where a small fraction of the nitrosyl ferrous form is found. When overexpressing Ngb E7 mutants, this equilibrium is completely shifted to the nitrosyl ferrous form (Trandafir *et al.*, 2004; Van Doorslaer *et al.*, 2003).

2.3. Preparative purification of recombinant Ngb

2.3.1. Preparation of a crude Ngb extract

The resuspended cells are exposed to three freeze-thaw steps and are sonicated (1 min at 60 Hz and 3-s pulses at 4°) until completely lysed. Sonication will also shear the released genomic DNA. The extract is clarified by low (10 min at 10,700 × *g*, 4°) and high (60 min at 1,05,000 × *g*, 4°) speed centrifugation and made up of 60% ammonium sulfate. After mixing for 10 min, the crude Ngb is collected by centrifugation (10 min at 10,700 × *g* at 4°). It was inefficient to use different steps of ammonium sulfate precipitation for Ngb (e.g., 40 and 90% ammonium sulfate precipitation) because Ngb has the tendency to start precipitating already at 35% ammonium sulfate. The ammonium sulfate pellet is dissolved in 5 m*M* Tris-HCl, pH 8.5, and dialyzed overnight against the same buffer. A sample was taken for analyses on SDS-PAGE, for determination of total protein content using the BCA Protein Assay kit (Pierce) and for determination of the heme content. The latter was done using the pyridine hemochromogen method (Riggs, 1981) (Table 19.1).

2.3.2. DEAE Sepharose Fast Flow chromatography

The dialyzed material is mixed in bulk with an excess DEAE Sepharose Fast-Flow (Amersham Biosciences) slurry in a 6.0 × 5.0-cm column. The unbound material is eluted with 1 l 5 m*M* Tris-HCl, pH 8.5, after which the crude Ngb is eluted with 200 m*M* NaCl in 5 m*M* Tris-HCl, pH 8.5. The eluted Ngb is concentrated to approximately 6 ml using an Amicon ultrafiltration apparatus under 2 bar air pressure, using a PM-10 membrane (cutoff 10,000 Da). A sample was taken for analyses on SDS-PAGE.

2.3.3. Sephacryl S-200 high-resolution chromatography

The concentrated material is loaded on a Sephacryl S-200 high-resolution (Amersham Biosciences) column (3 × 100 cm). The column is developed at 4° with the column-equilibrating buffer (50 m*M* Tris-HCl, pH 8.5, 150 m*M* NaCl, and 5 m*M* EDTA), at a flow rate of 99 ml/h (Fig. 19.1A). Ngb elutes as a symmetrical peak at the same position of sperm whale Mb. Analysis of the fractions by 15% SDS-PAGE in presence of β-mercapto-ethanol reveals a major monomeric Ngb band at an apparent Mr of ≈17,000, as expected. In the leading edge of the peak, a prominent band with an apparent Mr of ≈35,000 together with minor nonglobin impurities can be seen. The top fractions and trailing edge of the peak contains the purest Ngb (Fig. 19.1B). The band of Mr ≈35,000 represents a dimeric Ngb fraction in addition to a majority of *E. coli* β-lactamase, as shown by aminoterminal sequencing after blotting to a PVDF membrane. The percentage of this Ngb

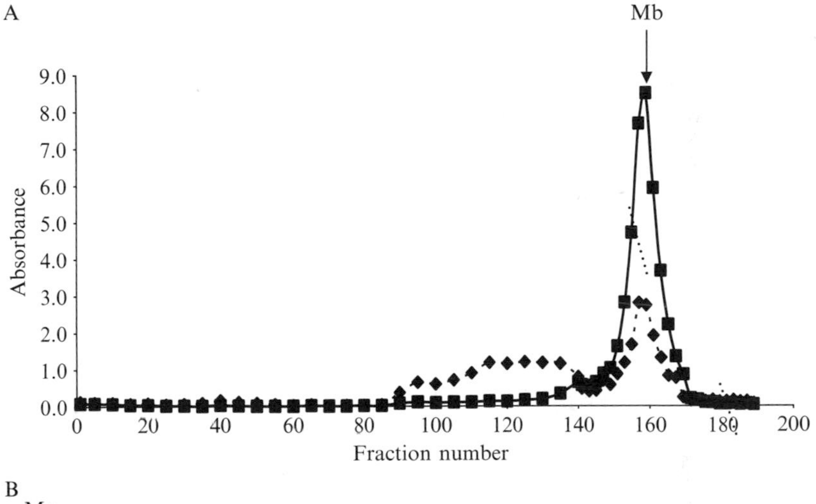

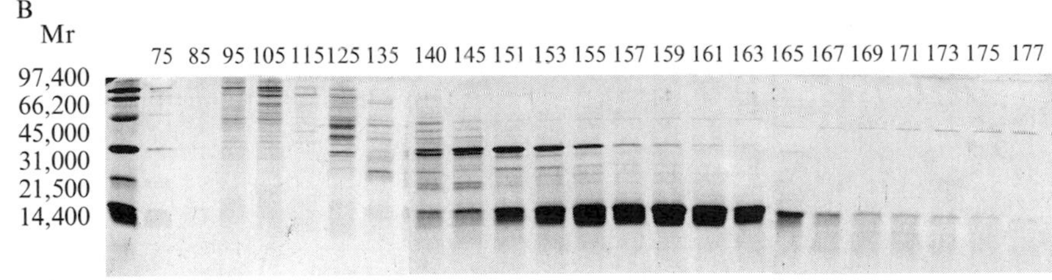

Figure 19.1 Preparative Sephacryl S-200 high-resolution chromatography of recombinant Ngb. Panel A: Sephacryl S-200 high-resolution chromatography as described. Detection is performed at 280 (◆) and 412 (■) nm. The elution position of Mb as marker is indicated. Panel B: Analysis of representative fractions by 15% SDS-PAGE containing 2-mercapto-ethanol.

dimer and the β-lactamase contaminant to the monomeric Ngb is variable from batch to batch.

Why a dimeric fraction of Ngb still persists, even after reduction with 1% β-mercapto-ethanol at 100° for 5 min, is unclear. This phenomenon, however, in our laboratory has been observed in many other globin samples as well and must therefore be considered nonspecific.

The fractions (Fig. 19.1A: 151 to 165) containing the purified Ngb are pooled, dialyzed against 5 m*M* Tris-HCl, pH 8.5, and concentrated. The final purity of the pooled Ngb fractions is checked by 15% SDS-PAGE (Fig. 19.1B) and mass spectrometry when needed.

An overview of the purification progress of the recombinant wild-type Ngb is given in Table 19.1 and Fig. 19.2A.

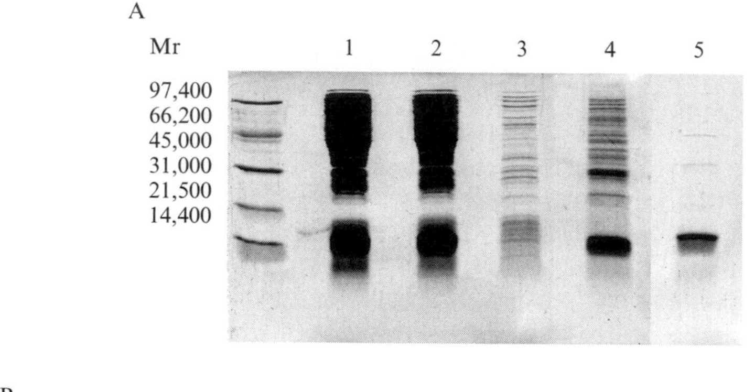

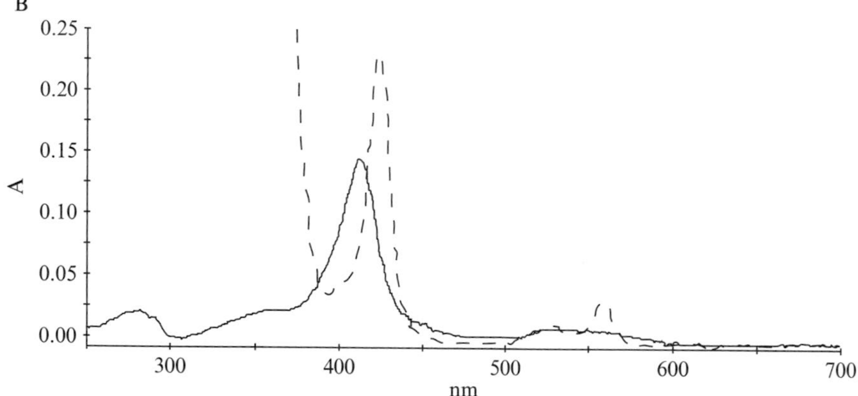

Figure 19.2 Purity of the obtained neuroglobin preparation. Panel A: SDS-PAGE of samples taken at different purifications steps. Lane 1: after ultracentrifugation; lane 2: after ammonium precipitation; lane 3: unbound fraction DEAE bulk; lane 4: bound fraction DEAE bulk; lane 5: after gel filtration. Panel B: UV/vis spectrum with (interrupted line) and without (full line) reduction of the Ngb preparation.

Table 19.1 Purification progress of neuro- and cytoglobin

Sample	Total Protein [mg]	Globin [mg]	Globin/total protein %
NGB after ultracentrifugation	126.5	6.0	4.8
NGB after $(NH_4)_2SO_4$ precipitation	121.8	14.9	12.3
NGB after DEAE	31.0	7.2	23.2
NGB after gel filtration	7.3	5.7	77.9
CYGB after refolding	8.7	13.0*	150.1*
CYGB after gel filtration fraction 1	6.4	4.6	72.6
CYGB after gel filtration fraction 2	3.3	2.9	86.7

* This value exceeds 100% due to the presence of an excess of free hemin from the refolding procedure. The excess of hemin is removed later by the gel filtration step

Total protein content was determined using the BCA Protein Assay kit (Pierce); amount of neuro- and cytoglobins was determined using the pyridine hemochromogen method (PHC) (Riggs, 1981).

3. Expression and Preparative Purification of Human Wild-Type CYGB

3.1. Cloning of CYGB cDNA in the expression vector and expression of CYGB

The cDNA of Cygb was cloned into the pET3a vector as described for Ngb. The growth of the transformed bacteria and the overexpression of Cygb was performed as for Ngb. After expression, the cells of a 250-ml culture are harvested (20 min at 3220 × *g*) and resuspended in 12 ml lysis buffer (50 m*M* Tris-HCl, pH 8.0, 5 m*M* EDTA, 1 m*M* PMSF). In contrast to Ngb, Cygb is expressed in inclusion bodies and has not incorporated the heme group (the cells are not red). As such, prior to the purification of the protein, the inclusion bodies are separated and the protein is refolded in presence of hemin (see the subsequent section).

3.2. Preparative purification of recombinant CYGB

3.2.1. Preparation of a crude Cygb extract

The cells are exposed to three freeze-thaw steps and sonicated (1 min at 60 Hz and 3-s pulses at 4°) until completely lysed. Inclusion bodies are isolated by centrifugation at 6,900 × *g* for 30 min at 4°. The pellet is washed two times with 50 m*M* Tris-HCl, pH 8.0, 5 m*M* EDTA, 2% deoxycholic acid, and finally with distilled water by resuspension and centrifugation after each step during 10 min at 10,700 × *g* at 4°. The resultant inclusion bodies are solubilized in 5 ml of 6 M guanidinium hydrochloride, 50 m*M* Tris-HCl, pH 7.5, 72 m*M* DTT, and boiled for 5 min. After elimination of the insoluble material by centrifugation (10 min at 10,700 × *g* at 4°), Cygb is refolded by adding 1.4 *M* excess of hemin (stock solution is 5 mg hemin dissolved in 100 μl 0.1 *M* NaOH and 900 μl 50 m*M* Tris-HCl, pH 7.5). The insoluble material is eliminated by centrifugation (10 min at 10,700 × *g* at 4°). The supernatant is dialyzed overnight against 5 m*M* Tris-HCl, pH 8.5. The refolded Cygb was concentrated to approximately 6 ml using an Amicon ultrafiltration apparatus under 2 bar air pressure, over a PM-10 membrane (cutoff 10,000 Da).

3.2.2. Sephacryl S-200 high-resolution chromatography

Because the isolation of inclusion bodies on itself is a strong purification step, Cygb is purified further only by gel-filtration chromatography, as described for Ngb. The elution profile displays two major peaks containing, respectively, fractions 75 to 100 and 110 to 135 (Fig. 19.3A). The second peak elutes at an apparent Mr of ≈45,000, which corresponds to the expected Mr of dimeric native Cygb (Hamdane *et al.*, 2003), whereas the first peak represents much higher Mr. SDS-PAGE analysis reveals a monomeric Cygb

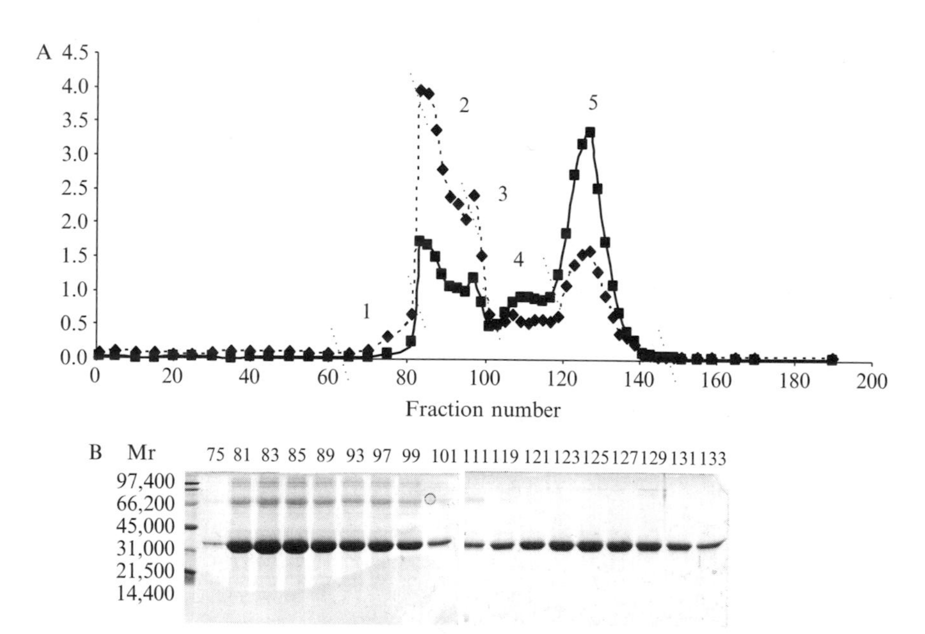

Figure 19.3 Preparative Sephacryl S-200 high-resolution chromatography of recombinant cytoglobin. Panel A: Preparative Sephacryl S-200 high-resolution chromatography of recombinant cytoglobin as described. Panel B: Analysis of representative fractions by 15% SDS-PAGE containing 2-mercapto-ethanol.

band of Mr ≈21000 for all fractions (75 to 133; Fig 19.3B). The fractions of the first peak mainly display the Cygb band and the nonglobin bands of higher M, whereas the second peak contains the Cygb band only. During chromatography, the first peak is clearly visible as a greenish-colored band, whereas the second peak is bright red. We assume that the first peak represents incompletely folded Cygb molecules that bind heme aspecifically.

The final purity of the pooled Cygb fractions (110 to 133) is checked by 15% SDS-PAGE (Fig. 19.4A, B) and mass spectrometry when needed.

An overview of the purification progress of the recombinant wild-type Cygb is given in Table 19.1 and Fig. 19.4.

4. Alternative Methods for the Preparative Purification of Recombinant Ngb and Cygb and Their Mutants

The physicochemical behavior of mutants of both molecules compared to their wild-type form may be different, and adaptations to the described methods must be applied.

The percentage of ammonium sulfate for precipitation of the crude Ngb or Cygb mutants must be tested individually for each molecule. For example, a progressive precipitation of the protein with 30, 40, 50, 60, and 90% ammonium sulfate can be used.

An alternative use for the denaturation of inclusion bodies and the refolding to the native molecule is as follows. Inclusion bodies are resuspended in 5 ml of 2 M urea, 100 m*M* Tris-HCl, pH 12, and incubated for 30 min at room temperature. The remaining insoluble material is removed by centrifugation at 10,000 × *g* for 10 min at 4°. The globin is refolded by adding 1.4 *M* excess of hemin and incubated for 10 min at room temperature. The insoluble material is eliminated by centrifugation (10 min at 10,700 × *g* at 4°). The pH of the supernatant is set at pH 8.5 after adding 1 *M* HCl.

After incubation at room temperature for 10 min, the globin is 5 times diluted with 100 m*M* Tris-HCl, pH 8.0, 0.2 *M* KCl, 0.4 *M* L-arginine, 5 m*M* DTT, and 2% glycine. After dialyzing overnight against the same buffer, the globin is concentrated to approximately 6 ml using an Amicon ultrafiltration apparatus under 2 bar air pressure. Further purification is obtained by gel-filtration chromatography.

When necessary, further preparative purification of Ngb and Cygb can be accomplished by chromatography on a DEAE column (under HPLC conditions) using gradient elution (buffer A: 5 m*M* Tris-HCl pH 8.5; buffer B: 5 m*M* Tris-HCl pH 8.5, 250 m*M* NaCl) as follows. Load sample onto the column (Waters, Protein Pak DEAE SPW 5×75 mm), carefully washed with

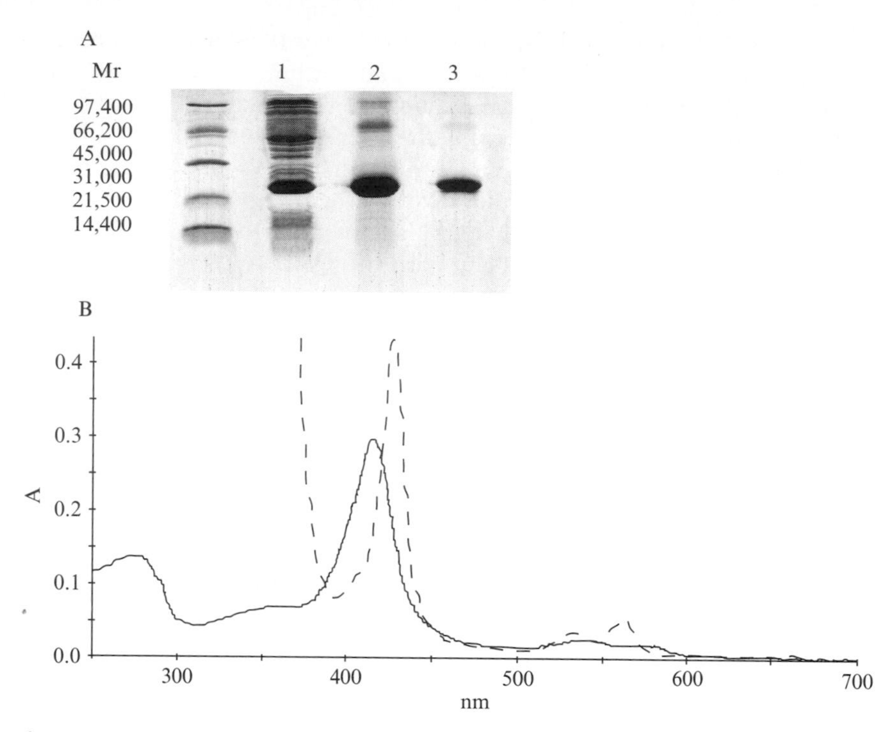

Figure 19.4 Purity of the cytoglobin preparation. Panel A: SDS-PAGE of samples taken at different purifications steps. Lane 1: inclusion bodies resuspended in 6M Gu-HCl, boiled, and centrifuged for 10 min; lane 2: after refolding before gel filtration; lane 3: after gel filtration. Panel B: UV/vis spectrum with (interrupted line) and without (full line) reduction of the Cygb preparation.

equilibration buffer (100% buffer A), and elute with a NaCl gradient (flow rate 1 ml/min; 100% A for 10 min, linear gradient in 30 min to 100% B).

5. Analytical Purification of Recombinant Ngb and Cygb

The Ngb and Cygb preparations obtained by preparative purification methods are free of nonglobin contaminants but are still heterogeneous (see Figs. 19.2 and 19.4). Indeed, the presence of three (CD7, D5, and G19) and two (B2 and E9) cysteines in Ngb and Cygb, respectively, results in the potential formation of inter- and intramolecular disulfide bridges during the purification procedure in the absence of reducing compounds. As a reducing agent, DTT is recommended at concentrations not exceeding the globin concentration. Low concentrations of DTT will not reduce the heme iron atom and will not produce ROS after heme reduction with dithionite in the presence of O_2. Hexacoordinated globins are particularly sensitive to such oxidation. As mentioned previously, the presence of an intramolecular disulfide bond in Ngb and Cygb has an effect on their O_2-binding affinity and, as such, has potential physiological significance. It is therefore important to know which conformation, with or without an intramolecular disulfide bond, must be obtained and consequently to use the appropriate procedure to obtain it. The purification of the different conformers will be achieved by analytical gel filtration and DEAE chromatography.

5.1. Gel filtration on Superose 12 column: Separation of aggregates

Gel filtration on Superose 12 (Amersham) column is used to eliminate undesirable high Mr aggregates due to intermolecular disulfide bridges. The column is equilibrated with 150 m*M* Tris-acetate buffer, pH 7.5, and the flow rate is 0.4 ml/mn for all the analytical experiments. The retention volumes are about 13.4 and 14.1 ml for Cygb and Ngb samples, respectively. A void volume of 7.9 ml and a total volume of the gel bed of 24 ml result in the molecular sieve coefficients of 0.34 and 0.38 for Cygb and Ngb, respectively.

Figure 19.5A shows the separation of Cygb with monomeric, dimeric, and tetrameric Hb species as markers. Cygb clearly behaves as a dimer in solution. No shift of the elution peak was observed at a low globin concentration (<0.1 μM) due to the strong interface stability between both subunits (>10 kcal/mol) and not to the presence of a covalent bond (Hamdane *et al.*, 2003). Indeed, the crystal structure of Cygb showed that the cysteine locations are far away from the monomer–monomer interface but inside each monomer close enough to form an intramolecular disulfide bond (de Sanctis *et al.*, 2004; Hamdane *et al.*, 2003).

After refolding from inclusion bodies as described, Cygb preparation contains a tetrameric fraction, as revealed by analytical gel filtration (Fig. 19.5B). For preparative purposes, this fraction can be eliminated by preparative chromatography on a Superose 12HR 16/50 and an anion exchange chromatography, as described subsequently.

Reduction of the Cygb with DTT results in a decrease of the tetramer and an increase of the dimeric fraction, which proves the presence of at least one covalent bond between two subunits inside the tetramer (Fig. 19.5C). A plausible explanation is that Cygb is able to form a dimer through the formation of two intermolecular disulfide bridges observed in a crystal structure (Makino *et al.*, 2006). It is interesting that the monomer–monomer interface found in another available crystal structure of dimeric Cygb (de Sanctis *et al.*, 2004) in which serine residues have replaced the two

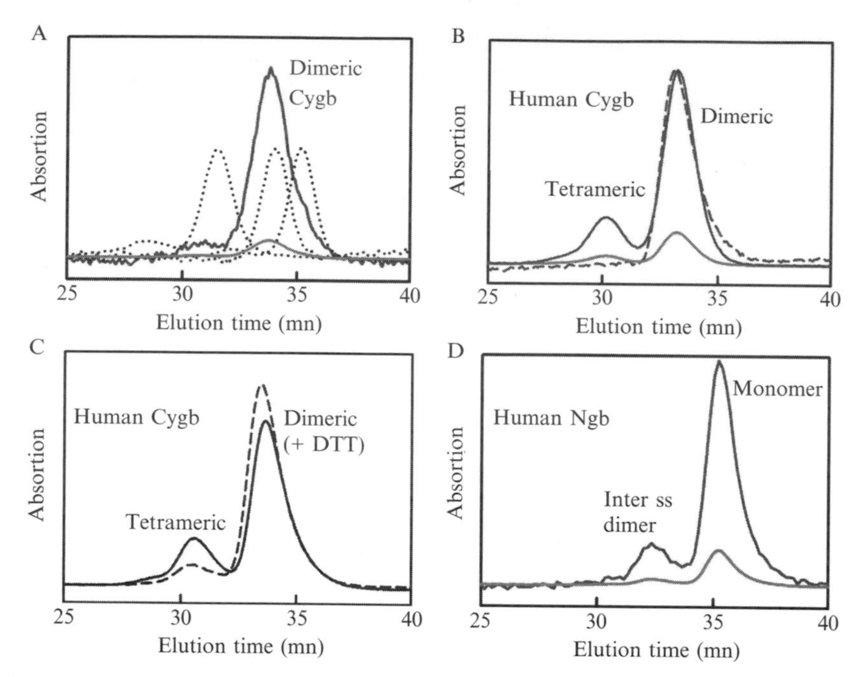

Figure 19.5 Analytical gel-filtration chromatogram on Superose 12 10/300 GL (Amersham Biosciences). (A) Cygb (415-nm red line, 280-nm blue line) compared to control samples for tetrameric (diaspirin cross-linked) Hb, dimeric Hb Rothschild (β 37 Arg) and monomeric horse heart Mb (dotted lines from left to right respectively). (B) Cygb (415-nm red line, 280-nm blue line) after refolding the globin with hemin. The elution profile after an extend purification consisting of preparative gel-filtration chromatography on Superose 12 HR 16/50 and anion-exchanged chromatography on HiTrap DEAE fast-flow 5 ml (Amersham Biosciences) is shown by the blue dotted line. For both preparative chromatographies, the flow rate was 1 ml/mn. (C) Cygb (415 nm). Incubation 1 h at 25° in the presence of 5 m*M* DDTdecreases the fraction of tetrameric species (dotted line), although the conversion to dimers is not complete even for 1 h incubation at 40°, which may indicate that tetramers are re-formed during the elution process in the absence of dithiotreitol. (D) Ngb (415-nm red line, 280-nm blue line). The higher MW species is an Ngb dimer due to the formation of an intersubunit disulfide bridge. Following the purification procedure, the dimer fraction does not exceed a low percentage but increases in solution at room temperature in the absence of a reducing agent. (See color insert.)

cysteines is fully exposed to the solvent and thus able to reform from two covalent S–S bound monomers. This minor tetrameric fraction can be removed by preparative gel-filtration and anion-exchange chromatography, as mentioned previously and as shown in Fig. 19.5B.

The same analytical gel filtration can be applied on Ngb to reveal a major monomeric and a minor dimeric fraction due to the formation of an intersubunit disulfide bridge. This dimeric fraction does not exceed a low percentage but increases in solution at room temperature in the absence of a reducing agent (Fig. 19.5D).

5.2. Anion-exchange chromatography on a HiTrap DEAE fast-flow column

Additional purification of Cygb and Ngb can be performed on a HiTrap DEAE fast-flow column. The elution profiles for both molecules are shown in Figs. 19.6A and 19.6B.

For Ngb, a major fraction is eluted at 50 m*M* NaCl and a minor one at a higher salt concentration (Fig. 19.6A). A similar profile is obtained for Cygb (Fig. 19.6B). The analysis of the different globin components of Ngb by 12% SDS-PAGE (Fig. 19.6A, inset) shows a slight difference in the location of the two protein fraction bands. This could be explained by the absence of an intradisulfide bond in the minor fraction due to different refolding or heme insertion, resulting in a change of the protein's hydrodynamic radius and/or isoelectric point (presence of a protein adduct may also be responsible for this gel shift). Gel-filtration chromatography also reveals a different retention

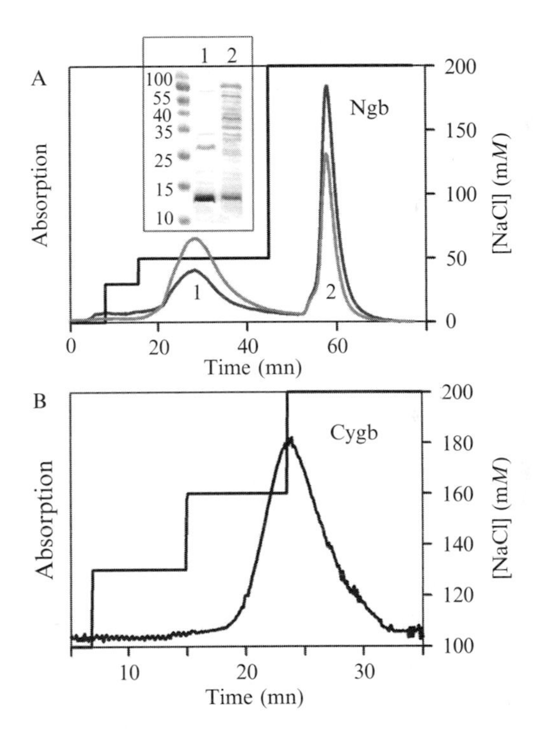

Figure 19.6 (A) Anion exchanged chromatography on HiTrap DEAE fast-flow 5 ml (Amersham Biosciences). The purity of the globin sample is controlled with 12% SDS-PAGE. Note always the presence of a small amount of Ngb dimers (lane 1) in the eluted fraction peak 1. The location of the minor globin fraction eluted at high salt concentration (lane 2) is slightly different from that of the major globin fraction. (B) The same column is used for the purification of Cygb, which is eluted at pH 8.5 by a step gradient of salt. (See color insert.)

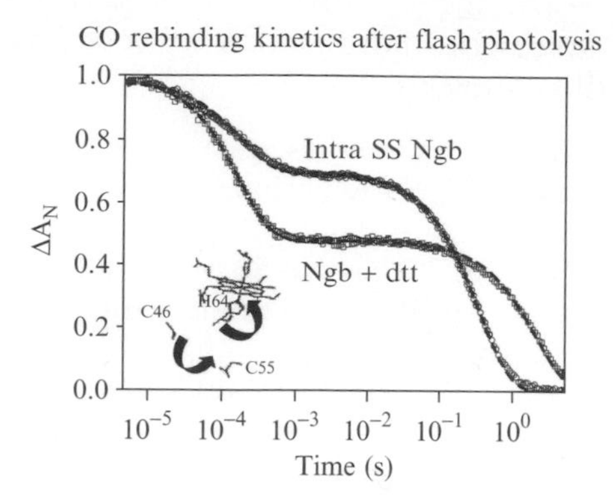

Figure 19.7 CO rebinding kinetics for the NGB purified by DEAE chromatography (main fraction 1, see Fig. 19.2B) for samples with 100 μM [CO]. There are two kinetic phases; the rapid phase is competitive binding of CO and the E7 His to the hemes; the second is the replacement of the protein residue by CO to return to the more stable ligand. As previously shown (Hamdane *et al.*, 2003), the presence of the Cys46 Cys55 S–S bond induces a change in the microscopic binding parameters, especially those for the histidine E7; breaking this bond by addition of DTT leads to a slower His to CO replacement reaction. Note that the minor fraction of Ngb eluted on the DEAE column exhibits more than two phases (not shown).

time (about 10% of the molecular sieve coefficient difference between monomer and dimer) between intra-SS Ngb WT and the Ngb mutant for which serine residues replaced the cysteines (not shown). To unambiguously determine which protein fraction is functionally simulated by a simple model of competition for heme binding between the internal E7 His and the external ligands, we measured the CO rebinding kinetics after flash photolysis (Fig. 19.7). Only the major globin fraction exhibits a biphasic kinetic pattern predicted by the model, while the kinetic for the other fraction exhibits more kinetic components (not shown). Note the effect of the reduction of the intradisulfide bond on the binding kinetics after the addition of DTT. As shown previously, the S–S bridge formation leads to an approximate 10-fold increase of the His affinity and the O_2 affinity compared to the other Ngb conformer in which the cysteines are reduced (Dewilde *et al.*, 2001a; Hamdane *et al.*, 2003). SDS-PAGE after an anion-exchanged column or gel-filtration chromatography also reveals the presence of a higher MW species (see Fig. 19.5D). This species is an Ngb dimer, not the monomeric form of Ngb as previously detected by mass spectroscopy (Hamdane *et al.*, 2003). Ngb has one more cysteine than Cygb, which is available to form an intermolecular disulfide bridge while the two others create an intramolecular disulfide. The dimer fraction is removed by preparative gel-filtration chromatography and the pure fraction readily stored frozen to avoid the re-formation of the dimeric molecules such as occurs at room temperature.

6. Crystallization of NGB and CYGB

The Ngb triple mutant (CysCD5Gly, CysD5Ser, and CysG19Ser; Ngb★) and the Cygb double mutant (Cys38Ser and Cys83Ser; Cygb★) were used for crystallization trials. The Cys residues were mutated to prevent aggregation and formation of insoluble precipitates that were observed during crystallization trials on both wild-type globins.

Ngb★ and Cygb★ crystals were grown using the hanging-drop vapor-diffusion method. Each crystallization droplet was prepared by mixing 1 μl of protein (at a given concentration) and 1 μl of reservoir solution. The most promising initial conditions that yielded crystals were optimized by varying protein concentration, precipitant, temperature, and pH. After optimization, the following conditions were adopted for reproducible production of Ngb★ and Cygb★ crystals.

6.1. Human NGB

Two crystal forms of Ngb★ (at a concentration of 33 mg/ml) were obtained.

6.1.1. Crystal form I

The best crystallization conditions were 20% polyethylene glycol (PEG) 4000, 20% isopropanol, 0.1 *M* sodium citrate, pH 5.6, at 277 K. Crystals of rhombic shape and typical dimensions of $0.4 \times 0.2 \times 0.1$ mm^3 grew in about one week. The crystals, however, proved very unstable, especially when transferred to the cryo-solution for data collection at 100 K, likely in relation to the high percentage of volatile isopropanol. Accordingly, the crystals diffracted to only a 7-Å resolution on a rotating anode X-ray source.

6.1.2. Crystal form II

The optimized growth conditions were 1.4 *M* ammonium sulfate, 3% isopropanol, 0.05 *M* sodium citrate, pH 6.5, at 277 K. In about 3 to 4 weeks, bunches of thin plates grew, each plate having the typical dimensions of $0.3 \times 0.3 \times 0.005$ mm^3. The crystals could be safely stored in a stabilizing solution containing 2.4 *M* ammonium sulfate, 3% isopropanol, 0.05 *M* sodium citrate, pH 7.0, at 27 K; they were transferred to the same solution supplemented with 20% (v/v) glycerol, for cryo-protection, immediately before X-ray data collection at 100 K (Pesce *et al.*, 2002b). The Ngb★ crystals grown under these conditions were characterized as belonging to the monoclinic space group $P2_1$, with unit-cell parameters as follows: $a = 39.6$ Å, $b = 94.9$ Å, $c = 67.5$ Å, $\beta = 94.4$ Å; four Ngb★ molecules were located in the crystal asymmetric unit (V_M 1.90 Å^3 Da^{-1}) (Pesce *et al.*, 2002b).

6.2. Human CYGB

Cygb* crystals were obtained at a protein concentration of 44 mg/ml, using a reservoir solution containing 20% (w/v) PEG 4000, 200 m*M* NaCl, 0.05 *M* sodium acetate, 0.01 *M* potassium ferricyanide, and 1 m*M* KCN, pH 4.0 to 5.0. Under these conditions, Cygb* crystals grew at 277 or 294 K, as orthorhombic prisms of considerable size in about 2 months.

After recovery from the growth chambers, Cygb* crystals were transferred into a stabilizing solution that had the same composition as the reservoir solution but contained 30% (w/v) PEG. For cryo-protection during X-ray data collection the same stabilizing solution was supplemented with 20% (v/v) glycerol. Cygb* crystals were characterized as belonging to the orthorhombic space group $P2_12_12_1$, with unit cell constants $a = 46.8$ Å, $b = 73.1$ Å, $c = 98.9$ Å; two Cygb* molecules were located in the crystal asymmetric unit ($V_M = 2.05$ Å^3/Da) (de Sanctis *et al.*, 2003).

ACKNOWLEDGMENTS

This study was supported by Inserm, University of Paris-XI, by EU Grant QLG3-CT-2002–01548 and by the Fund for Scientific Research of Flanders (FWO) Grant G.0468,03. Sylvia Dewilde is a postdoctoral fellow of the FWO.

REFERENCES

Bashford, D., Chothia, C., and Lesk, A. M. (1987). Determinants of a protein fold: Unique features of the globin amino acid sequences. *J. Mol. Biol.* **196,** 199–216.

Burmester, T., Ebner, B., Weich, B., and Hankeln, T. (2002). Cytoglobin: A novel globin type ubiquitously expressed in vertebrate tissues. *Mol. Biol. Evol.* **19,** 416–421.

Burmester, T., Weich, B., Reinhardt, S., and Hankeln, T. (2000). A vertebrate globin expressed in the brain. *Nature* **407,** 520–523.

de Sanctis, D., Dewilde, S., Pesce, A., Ascenzi, P., Burmester, T., Hankeln, T., Moens, L., and Bolognesi, M. (2003). New insight into the haemoglobin superfamily: Preliminary crystallographic characterization of human cytoglobin. *Acta Crystallogr. D* **59,** 1285–1287.

de Sanctis, D., Dewilde, S., Pesce, A., Moens, L., Ascenzi, P., Hankeln, T., Burmester, T., and Bolognesi, M. (2004). Crystal structure of cytoglobin: The fourth globin type discovered in man displays heme hexa-coordination. *J. Mol. Biol.* **336,** 917–927.

Dewilde, S., Van Hauwaert, M. L., Vinogradov, S., Vierstraete, A., Vanfleteren, J., and Moens, L. (2001a). Protein and gene structure of a chlorocruorin chain of Eudistylia vancouverii. *Biochem. Biophys. Res. Commun.* **281,** 18–24.

Dewilde, S., Kiger, L., Burmester, T., Hankeln, T., Baudin-Creuza, V., Aerts, T., Marden, M. C., Caubergs, R., and Moens, L. (2001b). Biochemical characterization and ligand binding properties of neuroglobin, a novel member of the globin family. *J. Biol. Chem.* **276,** 38949–38955.

Fordel, E., Geuens, E., Dewilde, S., Rottiers, P., Carmeliet, P., Grooten, J., and Moens, L. (2004). Cytoglobin expression is upregulated in all tissues upon hypoxia: An *in vitro* and *in vivo* study by quantitative real-time PCR. *Biochem. Biophys. Res. Commun.* **319,** 342–348.

Fordel, E., Thijs, L., Martinet, W., Lenjou, M., Laufs, T., Van Bockstaele, D., Moens, L., and Dewilde, S. (2006). Neuroglobin and cytoglobin overexpression protects human SH-SY5Y neuroblastoma cells against oxidative stress-induced cell death. *Neurosci. Lett.* **410,** 146–151.

Geuens, E., Brouns, I., Flamez, D., Dewilde, S., Timmermans, J. P., and Moens, L. (2003). A globin in the nucleus! *J. Biol. Chem.* **278,** 30417–30420.

Hamdane, D., Kiger, L., Dewilde, S., Green, B. N., Pesce, A., Uzan, J., Burmester, T., Hankeln, T., Bolognesi, M., Moens, L., and Marden, M. C. (2003). The redox state of the cell regulates the ligand binding affinity of human neuroglobin and cytoglobin. *J. Biol. Chem.* **278,** 51713–51721.

Khan, A. A., Wang, Y., Sun, Y., Mao, X. O., Xie, L., Miles, E., Graboski, J., Chen, S., Ellerby, L. M., Jin, K., and Greenberg, D. A. (2006). Neuroglobin-overexpressing transgenic mice are resistant to cerebral and myocardial ischemia. *Proc. Natl. Acad. Sci. USA* **103,** 17944–17948.

Kiger, L., Uzan, J., Dewilde, S., Burmester, T., Hankeln, T., Moens, L., Hamdane, D., Baudin-Creuza, V., and Marden, M. (2004). Neuroglobin ligand binding kinetics. *IUBMB Life* **56,** 709–719.

Makino, M., Sugimoto, H., Sawai, H., Kawada, N., Yoshizato, K., and Shiro, Y. (2006). High-resolution structure of human cytoglobin: Identification of extra N- and C-termini and a new dimerization mode. *Acta Crystallogr. D* **62,** 671–677.

Moens, L., Vanfleteren, J., Van de, P. Y., Peeters, K., Kapp, O., Czeluzniak, J., Goodman, M., Blaxter, M., and Vinogradov, S. (1996). Globins in nonvertebrate species: Dispersal by horizontal gene transfer and evolution of the structure-function relationships. *Mol. Biol. Evol.* **13,** 324–333.

Pesce, A., Bolognesi, M., Bocedi, A., Ascenzi, P., Dewilde, S., Moens, L., Hankeln, T., and Burmester, T. (2002a). Neuroglobin and cytoglobin: Fresh blood for the vertebrate globin family. *EMBO Rep.* **3,** 1146–1151.

Pesce, A., Dewilde, S., Nardini, M., Moens, L., Ascenzi, P., Hankeln, T., Burmester, T., and Bolognesi, M. (2003). Human brain neuroglobin structure reveals a distinct mode of controlling oxygen affinity. *Structure* **11,** 1087–1095.

Pesce, A., Nardini, M., Dewilde, S., Ascenzi, P., Burmester, T., Hankeln, T., Moens, L., and Bolognesi, M. (2002b). Human neuroglobin: Crystals and preliminary X-ray diffraction analysis. *Acta Crystallogr. D* **58,** 1848–1850.

Riggs, A. (1981). Preparation of blood hemoglobins of vertebrates. *Methods Enzymol.* **76,** 5–29.

Schmidt, M., Giessl, A., Laufs, T., Hankeln, T., Wolfrum, U., and Burmester, T. (2003). How does the eye breathe? Evidence for neuroglobin-mediated oxygen supply in the mammalian retina. *J. Biol. Chem.* **278,** 1932–1935.

Sun, Y., Jin, K., Mao, X. O., Zhu, Y., and Greenberg, D. A. (2001). Neuroglobin is up-regulated by and protects neurons from hypoxic-ischemic injury. *Proc. Natl. Acad. Sci. USA* **98,** 15306–15311.

Sun, Y., Jin, K., Peel, A., Mao, X. O., Xie, L., and Greenberg, D. A. (2003). Neuroglobin protects the brain from experimental stroke *in vivo*. *Proc. Natl. Acad. Sci. USA.*

Trandafir, F., Van Doorslaer, S., Dewilde, S., and Moens, L. (2004). Temperature dependence of NO binding modes in human neuroglobin. *Biochim. Biophys. Acta* **1702,** 153–161.

Trent, J. T., III, and Hargrove, M. S. (2002). A ubiquitously expressed human hexacoordinate hemoglobin. *J. Biol. Chem.* **277,** 19538–19545.

Van Doorslaer, S., Dewilde, S., Kiger, L., Nistor, S. V., Goovaerts, E., Marden, M. C., and Moens, L. (2003). Nitric oxide binding properties of neuroglobin: A characterization by EPR and flash photolysis. *J. Biol. Chem.* **278,** 4919–4925.

CHAPTER TWENTY

Measurement of Distal Histidine Coordination Equilibrium and Kinetics in Hexacoordinate Hemoglobins

Benoit J. Smagghe,* Puspita Halder,† *and* Mark S. Hargrove†

Contents

Abstract

The kinetics of ligand binding to hemoglobins has been measured for decades. Initially, these studies were confined to readily available pentacoordinate oxygen transport proteins like myoglobin, leghemoglobin, and red blood cell hemoglobin. Bimolecular ligand binding to these proteins is relatively simple, as ligand association is largely unimpeded at the heme iron. Although many techniques have been used to examine these reactions in the past, stopped-flow rapid mixing and flash photolysis are the most common ways to measure rate constants for ligand association and dissociation.

* Immune Disease Institute, Harvard Medical School, Boston, Massachusetts
† Department of Biochemistry, Biophysics, and Molecular Biology, Iowa State University, Ames, Iowa

Methods in Enzymology, Volume 436
ISSN 0076-6879, DOI: 10.1016/S0076-6879(08)36020-0

Expression of recombinant proteins has allowed for examination of many newly discovered hemoglobins. The hexacoordinate hemoglobins are one such group of proteins that exhibit more complex binding kinetics than pentacoordinate hemoglobins due to reversible intramolecular coordination by a histidine side chain. Here, we describe methods for characterizing the kinetics of ligand binding to hexacoordinate hemoglobins with a focus on measurement of histidine coordination and exogenous ligand binding in both the ferrous and the ferric oxidation states.

1. Introduction

In the last decade, the hemoglobin (Hb) superfamily has seen the addition of the new group of hexacoordinate hemoglobins (hxHbs). Members include neuroglobin (Ngb) and cytoglobin (Cgb) from vertebrates (Burmester *et al.*, 2000, 2002; Dewilde *et al.*, 2001; Trent and Hargrove, 2002; Trent *et al.*, 2001b), plant nonsymbiotic Hbs (nsHbs; Kundu *et al.*, 2003a), and cyanoglobin from *Synechocystis* (Hvitved *et al.*, 2001; Scott *et al.*, 2002). The histidine side chain binding to the sixth coordination site of the heme iron is the common structural feature from which the hxHb name is derived. Coordination is reversible and occurs in most cases in both the ferric and the ferrous forms (Hargrove, 2000; Hargrove *et al.*, 2000; Trent and Hargrove, 2002; Trent *et al.*, 2001b).

The two most common methods for measuring ligand-binding kinetics are flash photolysis and stopped-flow rapid mixing (Hargrove, 2000; Kundu *et al.*, 2003b; Olson, 1981). Both were implemented decades ago for investigations of pentacoordinate Hbs, such as Mb and red blood cell Hb. The benefit of stopped-flow mixing is its simplicity, but it is limited by mixing dead times (usually around 1 ms) that prevent its use with many exogenous ligands binding faster than $\approx 10\ \mu M^{-1}\ s^{-1}$ (Hargrove, 2005). This limits its usefulness for studies of association reactions with ferrous Hbs, where reactions with O_2 and CO are usually much faster, but it allows for measurement of many rate constants for ligand binding in the ferric oxidation state (Antonini and Brunori, 1971; Brancaccio *et al.*, 1994). Stopped-flow reactions have also been the standard for measuring ligand dissociation rate constants (Olson, 1981). Flash photolysis is functionally unlimited in time scale and is thus superior for measuring association rate constants in the ferrous oxidation state, but it has not been successful for studies of ferric ligand binding due to poor photolability for most ligands in this oxidation state.

When studying reactions with pentacoordinate Hbs that can be measured using both techniques, stopped-flow and flash photolysis provide the same rate constants. The reactions for which either technique is superior are largely complementary. Ferrous association rate constants are amenable to flash photolysis, while those in the ferric state are measurable by stopped flow. In concert, the two techniques can provide bimolecular and geminate association rate constants, and dissociation rate constants for a large variety of ligands (Gibson *et al.*, 1986; Olson *et al.*, 1988). In most bimolecular reactions with pentacoordinate Hbs, time courses are monophasic (i.e., they can be described by a single rate constant) and easily interpretable, and consistent results are garnered from kinetic and equilibrium measurements using different techniques. Such experiences have given researchers the confidence to use kinetic information to calculate equilibrium constants within the context of known reaction schemes. This is convenient for calculating equilibrium constants for high-affinity ligands for which direct equilibrium measurement might be difficult (Kundu *et al.*, 2003b).

As with other Hbs, hxHbs bind exogenous ligands in both oxidation states. However, measurements of ligand binding to hxHbs are somewhat more complex than those involving pentacoordinate Hbs because of the effects of reversible histidine coordination. Both stopped-flow and rapid mixing have been employed in the study of hxHbs, and both have shown multiphasic time courses for these reactions (Dewilde *et al.*, 2001; Fago *et al.*, 2006; Smagghe *et al.*, 2006; Trent *et al.*, 2001a,b). As with studies of pentacoordinate Hbs, the two techniques are largely complementary in studies of hxHbs. However, the complexity brought about by intramolecular histidine coordination and the resulting multiphasic time courses generate a greater degree of uncertainty (compared to pentacoordinate Hbs) in the rate constants measured for these proteins and the models used to describe their reactions. Together, uncertainty in these two properties makes the calculation of equilibrium constants from kinetic constants more difficult than it is with pentacoordinate Hbs, and efforts should be made to measure equilibrium constants independently in all reactions if possible.

Presented here are experimental procedures for analyzing ligand binding to hxHbs using stopped-flow experiments, absorbance spectrophotometry, and electrochemistry with the goal of providing equilibrium and rate constants for intramolecular histidine and exogenous ligand binding. The purpose is to outline the simplest way to characterize these reactions for newly discovered or genetically modified hxHbs.

2. Ligand Binding to hx

The reaction associated with exogenous ligand binding to hxHbs is shown in Scheme 1, where k_{-H} and k_H are the rate constants for binding of the endogenous ligand, and k'_L is the bimolecular rate constant for binding of the ligand to the pentacoordinate form of the hxHb (Hb_P).

$$Hb_H \underset{k_H}{\overset{k_{-H}}{\rightleftarrows}} Hb_P \xrightarrow{k'_L[L]} Hb_L$$

Scheme 1

By assuming a steady state equilibrium between Hb_H and Hb_P one can derive the following equation for the observed rate constant for the reaction (Hargrove, 2000):

$$k_{obs,L} = \frac{k_{-H}k'_L[L]}{k_H + k_{-H} + k'_L[L]} \tag{20.1}$$

From this equation, two situations can be observed. First, when $k'_L[L]$ becomes large compared to k_{-H} and k_H [Eq. (20.1)] predicts a time course with an observed rate (k_{obs}) approaching an asymptote equal to k_{-H}. Second, when $k'_L[L] << k_{-H}$ and k_H, [Eq. (20.1)] predicts a linear relationship between k_{obs} and L ($k_{obs} = k'_L[L]/(1 + K_H)$ where $K_H = k_H/k_{-H}$). Under both situations, single exponential time courses are predicted.

We have recently introduced a comprehensive equation to describe the multiphasic time courses observed for ligand binding (initiated by stopped flow) to hxHbs (Smagghe *et al.*, 2006). This takes into account the situation occurring when an appreciable fraction of Hb_P reacts rapidly with the exogenous ligand (i.e., when $k'_L[L] >> k_{-H}$ and k_H). Under these circumstances, two phases are expected in the reaction time course as described by [Eq. (20.2)].

$$\Delta A_{obs} = \Delta A_T(F_P e^{-k'_L[L]^* t} + F_H e^{-k'_{obs,L}[L]^* t}) \tag{20.2}$$

In Eq. 20.2, ΔA_{obs} is the observed time course for binding, $k'_{obs,L}$ is calculated from [Eq. (20.1)], the fraction of Hb_P and Hb_H are denoted F_P and F_H, $F_H = K_H/(1+ K_H)$, $F_P + F_H = 1$, and L is an exogenous ligand. ΔA_T is the total change in absorbance expected for the reaction (calculated independently from the ligand-free and ligand-bound forms of the protein). Equation (20.2) predicts biexponential behavior with a fraction of ΔA_T associated with ligand binding to F_P and a fraction associated with binding to the hexacoordinate species F_H. The ligand concentration dependence of the rate constant describing the reaction with F_P (k'_L) is linearly dependent on [L], while that of the reaction with F_H would follow [Eq. (20.1)] (as the second term of Eq. [20.2]).

Four kinetic scenarios are possible for ligand binding to hxHbs according to Eqs. (20.1) and (20.2), resulting from relative values of $k'[L]$, k_H, and k_{-H}. First we present examples and analysis for three that are common in the ferrous oxidation state of hxHbs, then an example of the fourth that is characteristic of reactions in the ferric oxidation state.

3. Reactions in the Ferrous Oxidation State

The value $k'_L[L]$ is usually greater for ferrous ligands than for ferric ligands, and K_H is generally smaller in the ferrous oxidation state than in the ferric state. This produces kinetic competition between exogenous and endogenous ligand binding and typically causes heterogeneous reactions. Due to the photolability of many ferrous ligands, flash photolysis has successfully measured these rate constants independently in several instances (Dewilde *et al.*, 2001; Ioanitescu *et al.*, 2005; Van Doorslaer *et al.*, 2003). However, large values of k_H compared to k_{-H} and $k'_L[L]$ can make measurements more difficult, and the resultant time courses can show phases over time scales spanning many orders of magnitude and can be confused with geminate recombination. However, flash photolysis of ferrous ligands is necessary for independent measurement of $k'_L[L]$, which can be unambiguously identified from the ligand dependence of the rebinding time courses following the flash. Specific flash photolysis methods vary and have been described elsewhere (Hargrove, 2000; Ioanitescu *et al.*, 2005; Sawicki and Morris, 1981).

3.1. Methods

The simplest reaction for evaluating ligand binding by a ferrous hxHb is that of carbon monoxide (CO) initiated by rapid mixing. In these reactions, an Hb solution of appropriate concentration for absorbance measurements (≈ 6 μM for a 1-cm cuvette and measuring in the Soret region) is prepared with ≈ 150 μM sodium dithionite in a nitrogen- or argon-purged gastight syringe. A second gastight syringe containing the same concentration of sodium dithionite is sparged with 100% CO. At room temperature and atmospheric pressure, this produces a solution that is 1 mM in CO. The syringe can be diluted into others sparged with N_2 to generate the [CO] desired for rapid mixing. Ideally, one should not lower ligand concentration below $10 \times$ [Hb] to ensure pseudo-first-order conditions in ligand (Espenson, 1995).

Prior to collecting kinetic time courses, CO and deoxy spectra of the Hb should be measured to calculate the expected change in absorbance for the reaction (ΔA_T). Ideally this is done inside the cuvette of the stopped-flow reactor, but adjustments for path length between the stopped-flow reactor and separate spectrophotometer can be used for this calculation. Once k'_{CO} (as measured by flash photolysis) and ΔA_T are measured, time courses for reactions at various [CO] can be used to assign one of the following kinetic scenarios to the hxHb.

In the following examples, the deoxy-ferrous spectrum was first recorded by mixing 150 μl of the sodium dithionite reduced protein

(≈5 μM after mixing) solution and 150 μl of the N_2 solution also containing sodium dithionite. The mixing time was set to 75 ms and the flow rate was equal to 4 ml/s with a dead time of 7.6 ms. The parameters were chosen because the reactions are not fast enough to justify a shorter dead time and the volumes are on the upper limit for the complete flushing of the mixing line and cuvette (Hargrove, 2005). Kinetic time courses were collected at different ligand concentrations by recording the change in absorbance at fixed wavelength (the Soret maximum) after mixing 150 μl reduced protein solution with 150 μl of the CO solution. At least three kinetic traces were collected and averaged. After the final collection, the final CO-bound spectrum of the sample was collected. Deoxy-ferrous and CO-bound spectra are recorded to calculate the total expected ΔAbs associated with the reaction (ΔA_T). This absorbance is compared to the change in absorbance associated with each kinetic trace to calculate any amplitude lost in the dead time (due to reactions with Hb_P). All rapid mixing experiments were conducted at 20° in 0.1 M potassium phosphate, pH 7.0, but buffering conditions varied in other experiments. In these examples, equations are fit to Eq. (20.2) using Igor Pro with the following macro:

```
Function Equation2_fit(w, x)
   Wave w; Variable x
variable A
A=w[0]*w[2]*w[3]/(w[0]+w[1]+w[2]*w[3])
variable B
B=1/(1+w[1]/w[0])
   Return w[5]+w[4]*B*exp(-w[2]*w[3]*x)+w[4]*(1-B)*exp(-A*x)
end
```

In this macro, which defines a fitting equation within Igor Pro, variable A is Eq. (20.1) (defining the rate constant for binding to the hexacoordinate complex) with w[0] = k_{-H}, w[1] = k_H, w[2] = k'_{CO}, and w[3] = [CO]. Variable B defines F_P as $1/(1 + K_H)$. The fitting equation following the "Return" command is Eq. (20.2), where B and A are as defined by the variable statements, w[4] is ΔA_T, and w[5] is an offset.

3.2. Analysis

3.2.1. Scenario 1: F_P is appreciable ($K_H < 10$) and $k'_L[L] >> k_{-H}$ and k_H

This scenario holds for hxHbs that exhibit a measurable contribution of Hb_P, like the plant nsHbs. From Eq. (20.2), two exponential phases are expected. The first, associated with CO binding to Hb_P, will be fast and linearly dependent on [CO] (as would be expected for CO reactions with pentacoordinate Hbs like Mb). The second, associated with CO binding following His dissociation, will obey Eq. (20.1). It will be slower, and at high [CO] the rate will approach k_{-H}. The fractional change in absorbance associated with the slow phase will be equal to F_H, facilitating calculation of K_H from $K_H = F_H/(1 - F_H)$. In this case, k'_{CO}[CO] is greater than the dead time of the mixing apparatus and ΔA_{obs} will be smaller than ΔA_T (due to the loss of the bimolecular phase in the mixing dead time), and $\Delta A_{obs}/\Delta A_T = F_H$. When working with a 1-cm cuvette, dead times are usually ≈3 ms. Because [CO] must be $\geq \approx 60\ \mu M$ to maintain pseudo-first-order conditions, k'_{CO} would have to be very low ($<0.5\ s^{-1}$) to avoid amplitude loss. Therefore, in most cases following this scenario, where k'_{CO} is usually $> 1\ s^{-1}$, amplitude loss is observed.

This scenario is demonstrated in Fig. 20.1. Figure 20.1A is a simulation of time courses for CO binding using the rate constants listed in the legend. The values of k_H and k_{-H} cause F_P to be ≈0.5, and the value of k'_{CO} ($2\ \mu M^{-1} s^{-1}$) is fast enough to cause amplitude loss over 3 ms at the [CO] used in the calculation. CO binding to riceHb1, which follows this scenario, is shown in Fig. 20.1B. The observed association rate constant (k_{obs}) increases as [CO] increases, with a maximum value (at high [CO]) corresponding to k_{-H} as predicted by Eq. (20.1) (or the second part of Eq. (20.2)). From Fig. 20.1B it is also clear that as [CO] increases, data are lost in the dead time (shaded part) due to the fast reaction with the appreciable fraction of Hb_P (first part of Eq. (20.2)). In this case, the fraction of absorbance associated with the observed time course ($\Delta A_{obs}/\Delta A_T$ at high [CO]) corresponds to F_H and can be used to calculate K_H. This correspondence is independent of the individual rate constants assigned to k_H and k_{-H}, thus providing a robust fit for K_H using Eq. (20.2).

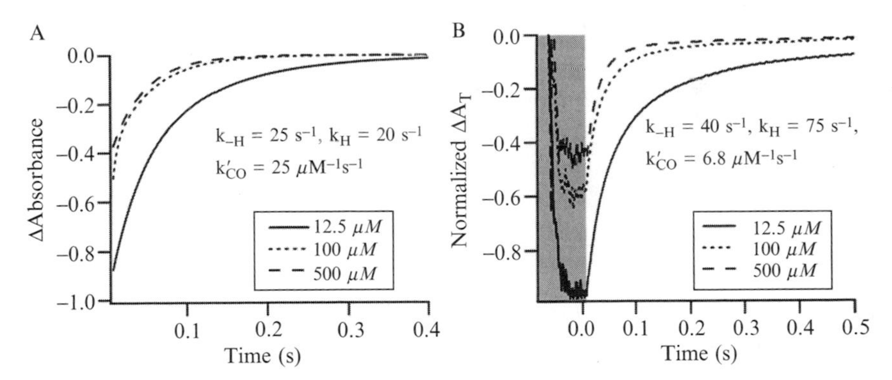

Figure 20.1 Ligand binding following scenario 1. A. Simulated time courses for ligand binding following rapid mixing at different ligand concentrations (12.5, 100, and 500 μM) using Eq. 20.2. B. Time courses for CO binding to RiceHb1 following rapid mixing at different CO concentrations (12.5, 100, and 500 μM). The gray-shaded region corresponds to the dead time (7.6 ms).

3.2.2. Scenario 2: F_P is not appreciable ($K_H > 10$) and $k'_L[L] >> k_{-H}$ and k_H

Under these conditions, the hxHb is fully hexacoordinate, and kinetics of binding are completely limited by k_{-H}. Only the F_H term in Eq. (20.2) (determined by Eq. (20.1)) is meaningful as $F_P \approx 0$. In this condition, a single exponential phase is expected with a rate constant equal to k_{-H} for any [CO], as demonstrated by the simulated time courses (Fig. 20.2A). In addition, ΔA_{obs} will be equal to ΔA_T for all [CO] (Fig. 20.2B), consistent with the fact that $F_H \approx 1$.

This specific scenario was observed for CO binding to the human hxHbs Ngb and Cgb (Smagghe *et al.*, 2006). In the example shown here, time courses for CO binding to Cgb are not dependent on [CO] (Fig. 20.2C). However, a double exponential is necessary to fit the data, which is not predicted by the model. The observed fast phase (70%) is independent

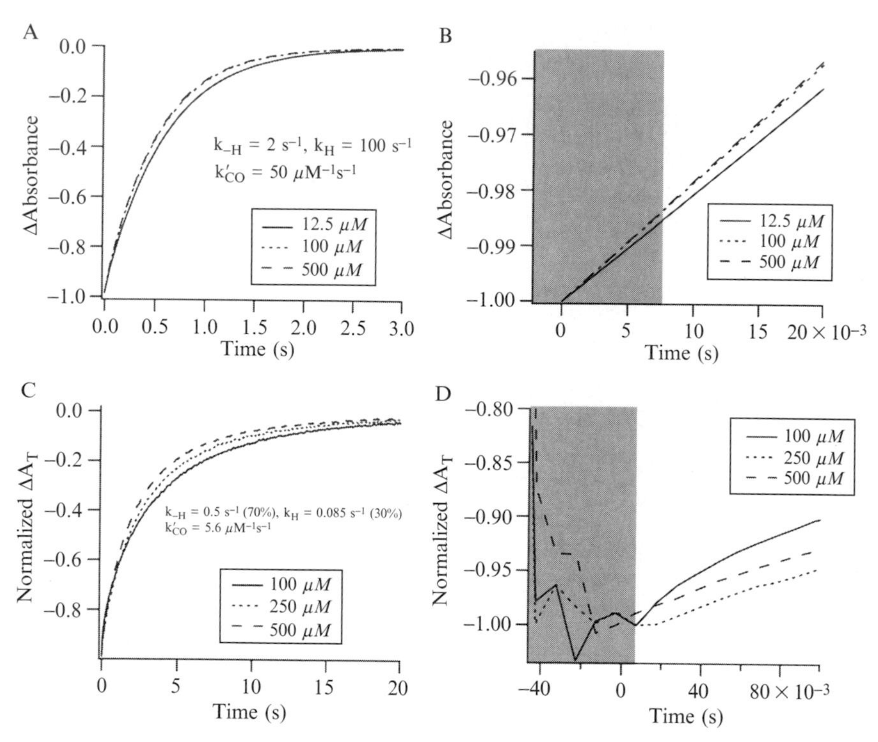

Figure 20.2 Ligand binding following scenario 2. A. Simulated time courses for ligand binding following rapid mixing at different ligand concentrations (12.5, 100, and 500 μM) using Eq. 20.2. B. The simulated data in A are expanded for the observation of the starting time point of the reaction. C. Time courses for CO binding to Cgb following rapid mixing at different CO concentrations (100, 250, and 500 μM). D. The data in C are expanded for the observation of the starting time point of the reaction. The gray-shaded region corresponds to the dead time (7.6 ms).

of [CO] and corresponds to k_{-H} (Fig. 20.2C). In addition, no significant portion of the data is lost in the dead time (Fig. 20.2D). This is an example of when stopped-flow analysis is useful; even though the time courses are complex, the lack of [CO] dependence or loss of amplitude allow for unambiguous assignment of the kinetic scenario. This provides an understanding that k_{-H} is the rate-limiting step and that k'_{CO}[CO] is faster than k_H and k_{-H}.

3.2.3. Scenario 3: F_P is not appreciable ($K_H > 10$) k'_L[L] is not >> k_{-H} and k_H

As in scenario 2, only the F_H term of Eq. (20.2) describes ligand binding because $F_P \approx 0$. However, in this case, the reaction is at rapid equilibrium with Hb_P, as the bimolecular reaction is not fast enough to sample the relative populations of Hb_P and Hb_H. Time courses are single exponential and dependent on [CO] as described for Eq. (20.1), and no significant loss in amplitude is observed (as simulated in Figs. 20.3A and 20.3B). This scenario is exemplified by CO binding to the cyanoglobin from *Synchocystis* (*Syn*Hb, Figs. 20.3C and 20.3D). As expected, time courses for CO binding are dependent on [CO], exhibit no loss in amplitude at higher [CO], and k_{obs} reaches a maximum corresponding to k_{-H} (see Fig. 20.3C).

4. Reactions in the Ferric Oxidation State

All ligand-binding reactions can be described by Eq. (20.2), and the differences between ferric and ferrous hxHbs that should be considered stem from three general features of these proteins: (1) K_H is much larger in the ferric oxidation state than the ferrous; (2) ligand association rate constants for binding to pentacoordinate Hbs are typically slower for ferric ligands than for ferrous ligands; (3) because flash photolysis is not generally applicable to reactions with ferric ligands, direct measurement of k'_L is usually precluded (Antonini and Brunori, 1971; Brancaccio *et al.*, 1994). The following example shows measurement of NO binding to several hxHbs in the ferric oxidation state.

4.1. Methods

A difficulty in working with NO is that it reacts with O_2, so all buffers must be O_2 free. In the following reactions, a glucose oxidase system (Brunori *et al.*, 2005) was used to catalytically remove O_2 from the syringe containing NO.

The NO solution stock (2 m*M*) was obtained by equilibrating a deoxygenated solution with NO gas that was first passed through 20% NaOH solution. NO solutions were obtained by dilution of the NO solution stock

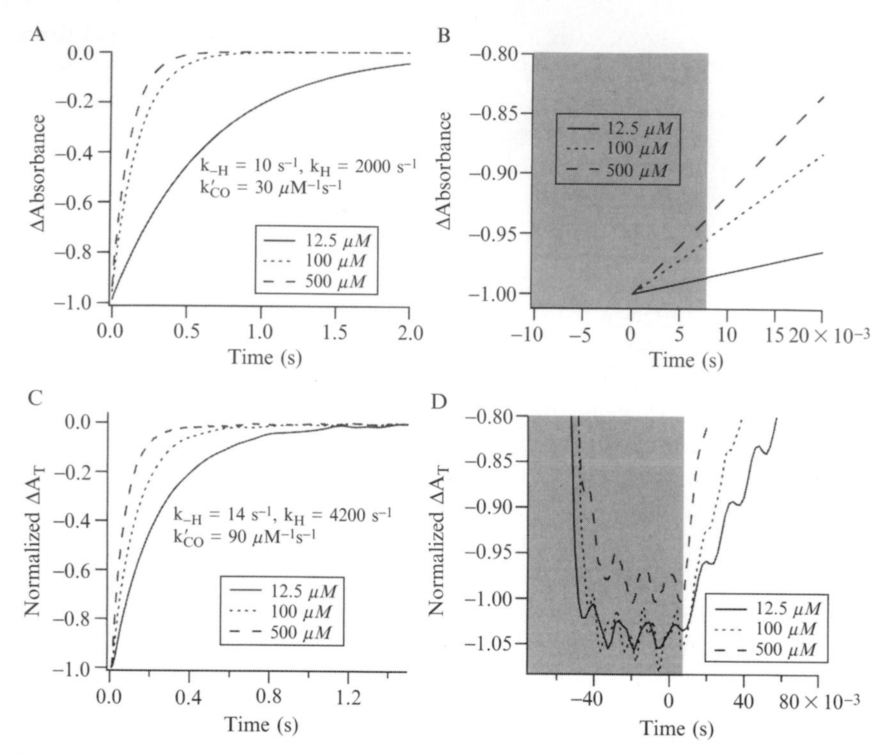

Figure 20.3 Ligand binding following scenario 3. A. Simulated time courses for ligand binding following rapid mixing at different ligand concentrations (12.5, 100, and 500 μM) using Eq. 20.2. B. The simulated data in A are expanded for the observation of the starting time point of the reaction. C. Time courses for CO binding to SynHb following rapid mixing at different CO concentrations (12.5, 100, and 500 μM). D. The data in C are expanded for the observation of the starting time point of the reaction. The gray-shaded region corresponds to the dead time (7.6 ms).

with N_2-equilibrated buffer containing 2 mM glucose, 8 units/ml glucose oxidase from *Aspergillus niger* (Sigma, St. Louis, MO), and 260 units/ml catalase from bovine liver (Sigma, St. Louis, MO).

Kinetic time courses were recorded by mixing 150 μl of the ferric protein solution (0.5 μM for Cgb E7L and 5 μM for other protein, after mixing) and 150 μl of a NO solution. The protein solution was not deoxygenated, as the consumption of NO by oxygen is very slow (seconds to minutes; Kharitonov *et al.*, 1994). The fast ferric-NO binding rate required a minimal dead time of 5 to 3 ms. Such dead-time values are on the limit of the apparatus used, and were obtained by changing the mixing time to 50 and 30 ms with a flow rate equal to 6 and 10 ml/s. Also, due to its fast binding rate (Cgb E7L), and to keep the pseudo-first-order conditions, low concentrations of NO and protein solutions were used. Those conditions were required to avoid losing the majority of the signal in the dead time. Kinetic time courses were collected at different ligand concentrations by recording the change in absorbance at fixed wavelength (the Soret maximum). Due to a small change in absorbance, at least 10 kinetic traces were collected and averaged.

4.2. Analysis

4.2.1. Scenario 4: F_P is not appreciable ($K_H > 10$) and $k'_L[L] << k_{-H}$ and k_H

This condition exists in the ferric oxidation state because of the combination of the large K_H and relatively slow k'_L for many ferric ligands. Under these conditions, Eq. (20.2) predicts a single exponential time course (associated with the F_H term) and a linear relationship between k_{obs} and L. This relationship, derived by setting $k_H/k_{-H} = K_H$ in Eq. (20.1), is

$$k_{obs} = \frac{k'_L[L]}{1 + K_H} \tag{20.3}$$

In this way, the slope of a plot of k_{obs} versus [L] will be linear and small ($k_{obsNO,(Fe3+)}$), deviating from that observed for the pentacoordinate form of the protein (if it could be measured) by dividing the slope by $1 + K_H$. Thus, hxHbs with large K_H values have very shallow slopes when plotting k_{obs} versus [L]. In contrast, pentacoordinate Hbs exhibit simple bimolecular reactions where the slope of a plot of k_{obs} versus [L] is equal to k'_L.

Figure 20.4A shows the NO dependence of the observed rate constant for NO binding to three hxHbs. For each, a linear dependence of the reaction with [NO] is observed. The linear fit gives the observed binding rate constant ($k_{obsNO,\ Fe3+}$) as described in Eq. (20.3) where $k_{obsNO,(Fe3+)} = k'_L/(1 + K_H)$. In these cases, the values are 0.073 (*Syn*Hb), 3.5 (riceHb1), and 13 $mM^{-1}s^{-1}$ (Cgb). Evaluation of either K_H or k'_L from this rate constant required independent knowledge of the other.

Figure 20.4B shows these reactions for two pentacoordinate Hbs, Mb and the E7L mutant of Cgb (in which the coordinating distal His is replaced with Leu). Removal of this coordinating side chain has the effect of increasing $k_{obsNO,(Fe3+)}$ from 13 to 3,100 $mM^{-1}s^{-1}$. The rate constant for Mb (72 $mM^{-1}s^{-1}$) is limited by water coordination of the binding site in the ferric form (Antonini and Brunori, 1971; Quillin *et al.*, 1993) which is not present in E7L Cgb. The linearity of these data for Mb indicate that water coordination, like intramolecular His coordination, is much faster than exogenous ligand binding and thus also obeys scenario 4.

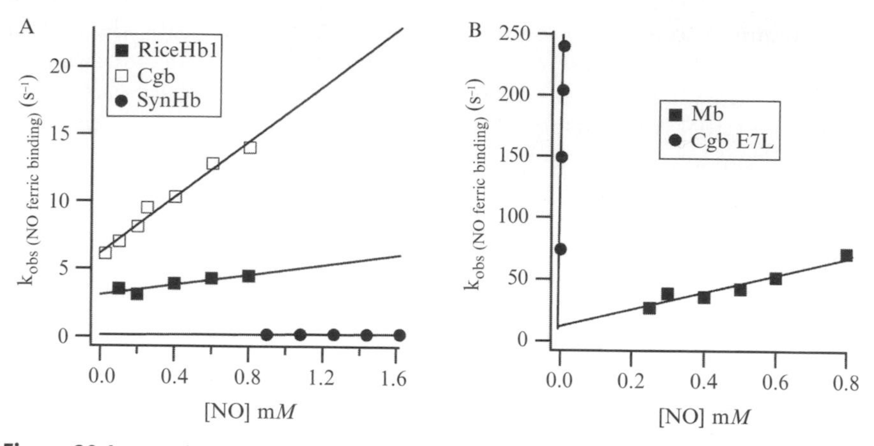

Figure 20.4 NO binding to ferric Hbs. A. Plots of k_{obs} versus [NO] for hxHb (riceHb1, Cgb, and SynHb). B. Plots of k_{obs} versus [NO] for pentacoordinate Hbs (Mb and E7L Cgb). Time courses giving rise to these values were measured at different [NO] and were fitted to a single exponential to extract the observed rate constants (k_{obs}). The linear fit to these data provides the observed ferric NO binding association rate constant ($k_{obs, NO(Fe3+)}$).

5. Measurement of K_H in the Ferrous Oxidation State Using Absorbance Spectroscopy

The visible absorbance spectra of hxHbs provide spin state and spectral shape information that are indicative of the nature and strength of coordination. In the ferrous oxidation state, the differences between penta- and hexacoordination is revealed in the α and β bands of *bis*-histidyl hxHbs and have been used to quantify the equilibrium constant for hexacoordination, K_H. By comparing the dithionite-reduced absorbance spectrum of a hxHb to reference spectra for $F_H = 0$ and $F_H = 1$, K_H can be measured.

The reference spectrum for $F_H = 0$ is H61A soybean leghemoglobin (Lba) (Kundu and Hargrove, 2003), which is completely pentacoordinate in the absence of exogenous ligands. However, it binds exogenous imidazole with high affinity, and when imidazole is saturating, the absorbance spectrum of H61A soybean leghemoglobin serves as the reference for $F_H = 1$ (Fig. 20.5A). Examination of these endpoint spectra and those at different fractional degrees of saturation revealed a good correlation between the ratio of the α band (555 nm) absorbance to that at the trough between the α and β bands (540 nm) and F_H (Smagghe *et al.*, 2006). An additional benefit of a ratio is that it is not dependent on the absolute protein concentration within the useful range of absorbance values. As shown in Fig. 20.5B, a linear correlation is observed between this ratio (A_{555}/A_{540}) and F_H at each imidazole (black dots).

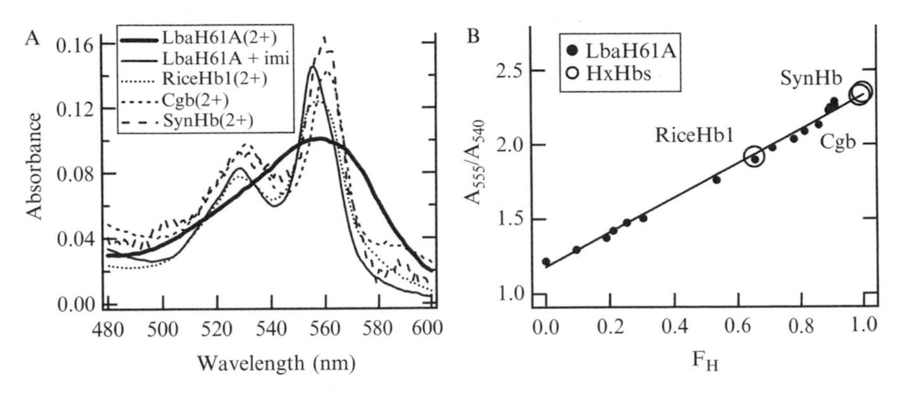

Figure 20.5 Evaluation of K_H using UV-vis spectroscopy. A. Visible absorbance spectra of deoxy LbaH61A (bold line), LbaH61A plus imidazole (1.6 m*M*, thin line), deoxy riceHb1 (dotted line), deoxy Cgb (short dashed line), and deoxy SynHb (long dashed line). B. For each imidazole concentration the ratio of A_{555}/A_{540} is plotted versus F_H (filled circles) and fit to a line (solid line). Each hxHb ratio is also plotted versus its F_H using the values published earlier (Smagghe *et al.*, 2006). Those data fall exactly along the LbaH61A fitted data confirming the good correlation between visible absorbance spectrum and F_H.

The same ratio has been calculated for different hxHbs and plotted versus their respective F_H (calculated from rapid mixing experiments). As shown in Fig. 20.5B, those data are near the linear fit for H61A, which confirms that this method is adequate for the estimation of K_H in the ferrous oxidation state. Therefore, if the dithionite-reduced absorbance spectrum of an hxHb gives a value of A_{555}/A_{540} near 2.2, it is fully hexacoordinate with F_H near 1. Lower values of this ratio indicate fractional saturation.

6. Measurement of K_H by Electrochemistry

The redox potential associated with the ferrous/ferric heme iron has contributions from many factors, including differential coordination in each oxidation state. For example, if the hexacoordinate His side chain binds more tightly in the ferric than in the ferrous oxidation state, the reduction potential will be more negative than if the reverse is true. Reduction potentials of several hxHbs are listed in Table 20.1, and for each the value is lower than those of pentacoordinate Hbs, like Mb and Lba. This suggests that K_H is higher in the ferric than in the ferrous oxidation state for hxHbs in general. In fact, electrochemistry can be used to measure these relative equilibrium constants.

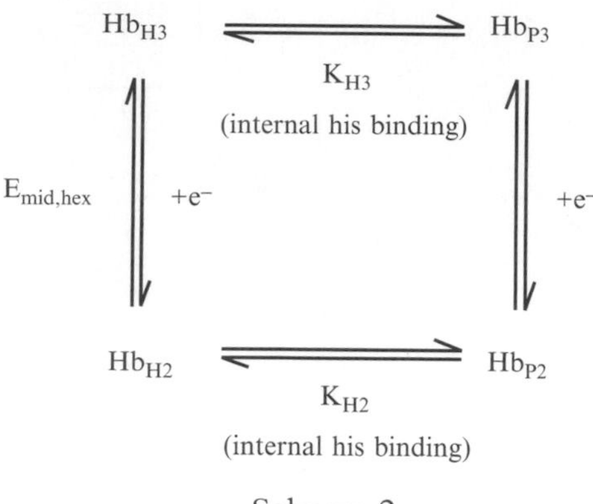

Scheme 2

Scheme 2 presents the different reactions occurring during hexacoordination in the ferrous and ferric oxidation states. Hb_{P2} and Hb_{P3} are the fraction of pentacoordinate Hb, and Hb_{H2} and Hb_{H3} are the fraction of hexacoordinate Hb. The numbers 2 and 3 represent the ferrous and the ferric oxidation states, respectively. The association equilibrium constants for hexacoordination are K_{H2} and K_{H3} for the ferrous and ferric oxidation states. Reduction of the ferric protein (Hb_3) can be described by a midpoint reduction potential (E_{mid}), which is thermodynamically linked to the reactions in Scheme 1.

Normally, the Nernst equation leads to Eq. (20.4) for simple redox reactions (Moore and Pettigrew, 1990), where $F_{reduced}$ is the observed fraction of protein in the reduced state, E_{obs} is the observed cell potential, and E_{mid} is the fitted midpoint potential (the potential at which $F_{reduced} = 0.5$).

Table 20.1 Reduction potentials

Protein	E_{Mid} mV	$E_{mid,\ pent}$ mV	$(1+K_{H3})/(1+K_{H2})$
RiceHb1[a]	−143	−30	81
Cgb[a]	−28	84	78
Ngb[a]	−115		
SynHb[a]	−195		
Sw Mb[b]	59		
Human Mb[c]	54		
Lba[d]	21		

[a] Halder *et al.* (2006).
[b] Van *et al.* (1996).
[c] Dou *et al.* (1995).
[d] Jones *et al.* (2002).

$$F_{reduced} = \frac{e^{-\left(\frac{nF(E_{obs}-E_{mid})}{RT}\right)}}{1 + e^{-\left(\frac{nF(E_{obs}-E_{mid})}{RT}\right)}} \quad (20.4)$$

The influence of hexacoordination in Scheme 1 leads to a modification of Eq. (20.4) Halder *et al.*, 2006)

$$F_{reduced} = \frac{e^{-\left(\frac{nF(E_{obs}-E_{mid,pent})}{RT}\right)}}{\left(\frac{1 + K_{H3}}{1 + K_{H2}}\right) + e^{-\left(\frac{nF(E_{obs}-E_{mid,pent})}{RT}\right)}} \quad (20.5)$$

Equation (20.5) allows for quantification of the influence of the equilibrium constants for reversible histidine coordination on reduction midpoint potential for the pentacoordinate form of the hxHb ($E_{mid,pent}$). Thus, if one can measure $E_{mid,pent}$ independently for the hexacoordinate form of the hxHb, the relative values of K_{H3} and K_{H2} can be determined by Eq. (20.6) (Halder *et al.*, 2006).

$$\left(\frac{1 + K_{H3}}{1 + K_{H2}}\right) = e^{-\left(\frac{nF(E_{obs}-E_{mid})}{RT}\right)} \quad (20.6)$$

Furthermore, independent measurement of either of these constants (K_{H3} and K_{H2}) allows for calculation of the other (Halder *et al.*, 2006).

6.1. Methods

Potentiometric titrations are performed using spectrophotometric analysis of a ferric Hb solution that is titrated with a reductant in the presence of redox mediators. The examples here were used an Ocean-Optics UV-Vis spectrophotometer (USB2000) coupled to an Oakton pH-mV meter (pH 1100 Series). This method was adapted from the work of Altuve *et al.* (2004) and Halder *et al.* (2006). A standard saturated calomel electrode (SCE) was used as a reference with a platinum working electrode in all experiments. Reduction potentials (E_{obs}) and midpoint potentials (E_{mid}) are reported with reference to a standard hydrogen electrode (SHE). A custom-made electrochemical cell was employed for all the measurements and is described in detail in Fig. 20.6. The lower portion of the cell consists of a quartz cuvette, and the upper glass portion contains three openings. The top opening is used for inserting the electrodes into the cell solution through an airtight rubber stopper. The side-arm openings of the cell are sealed with open-top plastic caps fitted with airtight Teflon septa. One of the side arms of the cell is used for the gas inlet and the other for the gas outlet and for

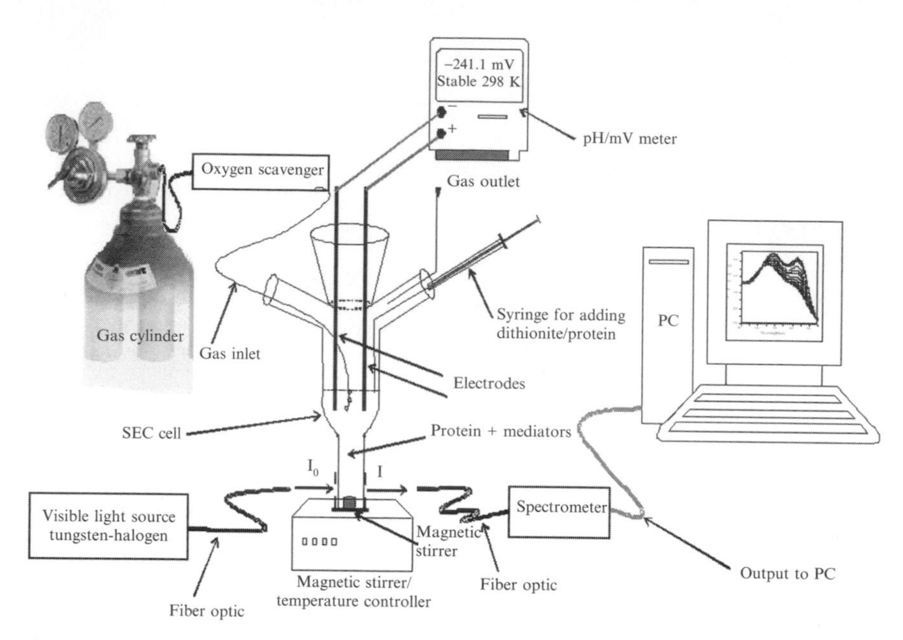

Figure 20.6 Spectroelectrochemistry apparatus for potentiometric titration of heme proteins.

addition of mediators, proteins, and the reductant (in these examples a dithionite stock solution) via an airtight glass syringe. Argon is passed through an oxygen-scavenging solution to remove traces of oxygen. Oxygen-scavenging solutions consist of two 25 ml saturated KOH solution containing ≈60 m*M* anthraquinone-2-sulfonic acid (sodium salt, Sigma) and ≈1 *M* sodium dithionite and one 25 ml solution of saturated lead acetate.

Titrations of 10 μM Hb are carried out at 25° in argon saturated 0.1 *M* potassium phosphate, pH 7.0. Ferric proteins are titrated stepwise with a solution of sodium dithionite (40 m*M*) previously sparged with argon. Reduction is monitored by recording the absorbance spectrum in the visible region (500 to 700 nm), and the corresponding cell potential was noted for every addition of dithionite after attainment of equilibrium.

A group of redox mediators were used to buffer the potential range from +160 to −440 mV. Their standard reduction potentials versus SHE are 1,2-naphthoquinone (E_{mid} = +157 mV), toluylene blue (E_{mid} = +115 mV), duroquinone (E_{mid} = +5 mV), hexaamineruthenium (III) chloride (E_{mid} = +50 mV), pentaaminechlororuthenium (III) chloride (E_{mid} = −40 mV), 5,8-dihydroxy-1,4-naphthoquinone (E_{mid} = −50 mV, Alfa Aesar/Avocado), 2,5-dihydroxy-1–4 benzoquinone (E_{mid} = −60 mV), 2-hydroxy-1,4-naphthoquinone (E_{mid} = −137), anthraquinone-1,5-disulfonic acid (E_{mid} = −175 mV), 9,10-anthraquinone-2,6-disulfonic acid (E_{mid} = −184 mV), and methyl viologen (E_{mid} = −440 mV).

6.2. Analysis

Figure 20.7 shows the plot of F_{red} as a function E_{obs} for riceHb1 (A) and Cgb (B). The data were fitted to Eq. (20.4) and the calculated E_{mid} value for each wild-type protein is reported in Table 20.1. As noted in Eq. (20.6), measurement of the ratio of hexacoordination affinity constants is only possible if $E_{mis,pent}$ can be determined, usually by the production of a mutant protein in which the hexacoordinating histidine side chain has been replaced by one incapable of heme iron coordination. For the hxHbs under investigation here (riceHb1 and Cgb) this is true, but this is not the case with all hxHbs (Halder *et al.*, 2006; Nienhaus *et al.*, 2004). Figure 20.7 also shows plots of F_{red} as a function E_{obs} for these respective pentacoordinate mutant proteins (riceHb1 E7L [A] and Cgb E7L [7B]). The data were fitted to Eq. (20.4) and the calculated E_{mid} value (which is $E_{mid,pent}$) for each mutant protein is reported in Table 20.1. The shift to higher midpoint potentials for E7L riceHb1 and E7L Cgb compared to their wild-type proteins indicates that the His^{E7} is responsible for a tighter coordination in the ferric form. Equation 20.6 can be used to calculate this ratio of affinity constants, which is provided in Table 20.1 for these two proteins.

Electrochemistry thus provides a means to measure relative values of K_H in different oxidation states, or to measure K_{H3} and K_{H2} directly if either is known independently. However, the thermodynamic linkage shown in Scheme 1 can also be extended to include binding of exogenous ligands (Halder *et al.*, 2006) and could be used to measure equilibrium affinity constants for these ligands as well, using the following equation:

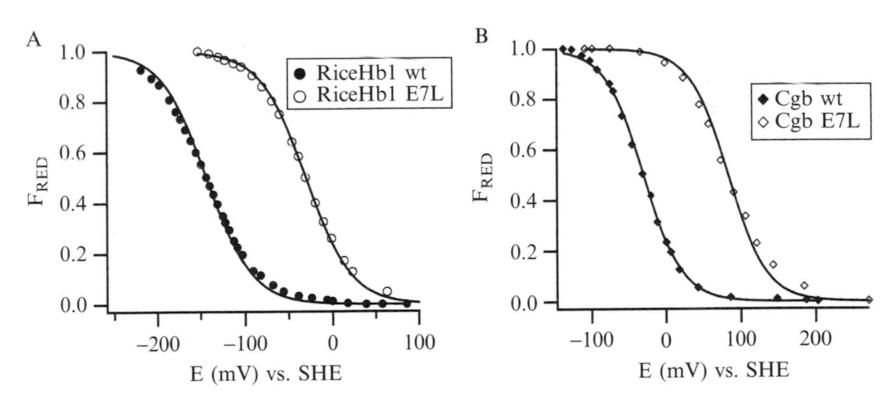

Figure 20.7 Potentiometric titration experiments. The reduction potential (E_{obs}) for RiceHb1/RiceHb1 E7L (A) and Cgb/Cgb E7L (B) is measured by sodium dithionite titration under anaerobic conditions and is plotted versus the fraction of reduced protein (F_{red}). The solid lines are fits to Eq. 20.4 with E_{mid} as the fitted parameter.

$$\left(\frac{1 + K_{L3}[L]}{1 + K_{L2}[L]}\right) = e^{-\left(\frac{nF(E_{mid,L} - E_{mid,L=0})}{RT}\right)} \quad (20.7)$$

When working with a ligand, like CO, which binds only in the ferrous oxidation state, Eq. (20.7) is simplified to Eq. (20.8)

$$E_{mid,L} = E_{mid} + \frac{RT}{nF}\ln(1 + K_{CO}[CO]) \quad (20.8)$$

This is a useful relationship between the change in midpoint reduction potential ($E_{mid,L}$ versus E_{mid}) in the presence of an exogenous ligand at some [CO], allowing for calculation of the equilibrium affinity constant for that ligand (K_{CO}).

7. Conclusions

Evaluation of hexacoordination in hxHbs can be approached in a hierarchical manner. In the ferrous oxidation state, a simple absorbance spectrum of the reduced, unliganded hxHbs can provide an estimate of K_{H2}. Flash photolysis can provide k'_{CO}, and stopped-flow experiments with CO will reveal the relative values of k_{H2}, k_{-H2}, and k'_{CO}. The stopped-flow experiments will also test the estimate of K_{H2}, as values < 10 should exhibit amplitude loss at higher [CO], and $K_{H2} > 10$ will yield ΔA_T at every [CO]. The asymptote from the mixing experiment will provide k_{-H2}, and the approach to that asymptote (along with k'_{CO}) will give k_{H2}. Investigations of K_{H3} are more difficult because of the lack of simple methods for determining k'_L in the ferric oxidation state. However, potentiometric titration that measures reduction midpoint potentials and site-directed mutagenesis can provide values of K_{H3}.

References

Altuve, A., Wang, L., Benson, D. R., and Rivera, M. (2004). Mammalian mitochondrial and microsomal cytochromes b5 exhibit divergent structural and biophysical characteristics. *Biochem. Biophys. Res. Commun.* **314,** 602–609.

Antonini, E., and Brunori, M. (1971). "Hemoglobin and myoglobin in their reactions with ligands." North-Holland Publishing Company, Amsterdam.

Brancaccio, A., Cutruzzolá, F., Allocatelli, C. T., Brunori, M., Smerdon, S. J., Wilkinson, A. J., Dou, Y., Keenan, D., Ikeda-Saito, M., Brantley, R. E., Jr., and

Olson, J. S. (1994). Structural factors governing azide and cyanide binding to mammalian metmyoglobins. *J. Biol. Chem.* **269,** 13843–13853.
Brunori, M., Giuffre, A., Nienhaus, K., Nienhaus, G. U., Scandurra, F. M., and Vallone, B. (2005). Neuroglobin, nitric oxide, and oxygen: Functional pathways and conformational changes. *Proc. Natl. Acad. Sci. USA* **848,** 3–8488.
Burmester, T., Ebner, B., Weich, B., and Hankeln, T. (2002). Cytoglobin: A novel globin type ubiquitously expressed in vertebrate tissues. *Mol. Biol. Evol.* **19,** 416–421.
Burmester, T., Weich, B., Reinhardt, S., and Hankeln, T. (2000). A vertebrate globin expressed in the brain. *Nature* **407,** 520–523.
Dewilde, S., Kiger, L., Burmester, T., Hankeln, T., Baudin-Creuza, V., Aerts, T., Marden, M. C., Caubergs, R., and Moens, L. (2001). Biochemical characterization and ligand binding properties of neuroglobin, a novel member of the globin family. *J. Biol. Chem.* **276,** 38949–38955.
Dou, Y., Admiraal, S. J., Ikeda-Saito, M., Krzywda, S., Wilkinson, A. J., Li, T., Olson, J. S., Prince, R. C., Pickering, I. J., and George, G. N. (1995). Alteration of axial coordination by protein engineering in myoglobin. *J. Biol. Chem.* **270,** 15993–16001.
Espenson, J. H. (1995). "Chemical kinetics and reaction mechanisms." McGraw Hill, New York.
Fago, A., Mathews, A. J., Dewilde, S., Moens, L., and Brittain, T. (2006). The reactions of neuroglobin with CO: Evidence for two forms of the ferrous protein. *J. Inorg. Biochem.* **100,** 1339–1343.
Gibson, Q. H., Olson, J. S., McKinnie, R. E., and Rohlfs, R. J. (1986). A kinetic description of ligand binding to sperm whale myoglobin. *J. Biol. Chem.* **261,** 10228–10239.
Halder, P., Trent, J. T., III., and Hargrove, M. S. (2006). The influence of the protein matrix on histidine ligation in ferric and ferrous hexacoordinate hemoglobins. *Proteins* **66,** 172–182.
Hargrove, M. (2005). Ligand binding with stopped-flow rapid mixing. *In* "Protein–ligand interactions, methods and applications" (H Press, ed.), pp. 323–342. Totowa, NJ.
Hargrove, M. S. (2000). A flash photolysis method to characterize hexacoordinate hemoglobin kinetics. *Biophys. J.* **79,** 2733–2738.
Hargrove, M. S., Brucker, E. A., Stec, B., Sarath, G., Arredondo-Peter, R., Klucas, R. V., Olson, J. S., Phillips, J., and George, N. (2000). Crystal structure of a nonsymbiotic plant hemoglobin. *Structure* **8,** 1005–1014.
Hvitved, A. N., Trent, J. T., Premer, S. A., III, and Hargrove, M. S. (2001). Ligand binding and hexacoordination in *Synechocystis* hemoglobin. *J. Biol. Chem.* **276,** 34714–34721.
Ioanitescu, A. I., Dewilde, S., Kiger, L., Marden, M. C., Moens, L., and Van Doorslaer, S. (2005). Characterization of nonsymbiotic tomato hemoglobin. *Biophys. J.* **89,** 2628–2639.
Jones, D. K., Patel, N., and Raven, E. L. (2002). Redox control in heme proteins: Electrostatic substitution in the active site of leghemoglobin. *Arch. Biochem. Biophys.* **400,** 111–117.
Kharitonov, V., Sundquist, A., and Sharma, V. (1994). Kinetics of nitric oxide autoxidation in aqueous solution. *J. Biol. Chem.* **269,** 5881–5883.
Kundu, S., and Hargrove, M. S. (2003). Distal heme pocket regulation of ligand binding and stability in soybean leghemoglobin. *Proteins* **50,** 239–248.
Kundu, S., Trent, J. T., III., and Hargrove, M. S. (2003a). Plants, humans and hemoglobins. *Trends in Plant Science* **8,** 387–393.
Kundu, S., Premer, S., Hoy, J., Trent, J., III., and Hargrove, M. (2003b). Direct measurement of equilibrium constants for high-affinity hemoglobins. *Biophys J.* **84,** 3931–3940.
Moore, G., and Pettigrew, G. (1990). "Cytochromes c: Evolutionary, structural and physicochemical aspects." Springer, Berlin.

Nienhaus, K., Kriegl, J. M., and Nienhaus, G. U. (2004). Structural dynamics in the active site of murine neuroglobin and its effects on ligand binding. *J. Biol. Chem.* **279,** 22944–22952.

Olson, J. S. (1981). Stopped-flow, rapid mixing measurements of ligand binding to hemoglobin and red cells. *Methods Enzymol.* **76,** 631–651.

Olson, J. S., Mathews, A. J., Rohlfs, R. J., Springer, B. A., Egeberg, K. D., Sligar, S. G., Tame, J., Renaud, J. P., and Nagai, K. (1988). The role of the distal histidine in myoglobin and haemoglobin. *Nature* **336,** 265–266.

Quillin, M. L., Arduini, R. M., Olson, J. S., and Phillips, G. N., Jr. (1993). High-resolution crystal structures of distal histidine mutants of sperm whale myoglobin. *J. Mol. Biol.* **234,** 140–155.

Sawicki, C., and Morris, R. (1981). Flash photolysis of hemoglobin. *Methods Enzymol.* **76,** 667–681.

Scott, N. L., Falzone, C. J., Vuletich, D. A., Zhao, J., Bryant, D. A., and Lecomte, J. T. J. (2002). Truncated hemoglobin from the cyanobacterium *Synechococcus* sp. PCC 7002: Evidence for hexacoordination and covalent adduct formation in the ferric recombinant protein. *Biochemistry* **41,** 6902–6910.

Smagghe, B. J., Sarath, G., Ross, E., Hilbert, J.-L., and Hargrove, M. S. (2006). Slow ligand binding kinetics dominate ferrous hexacoordinate hemoglobin reactivities and reveal differences between plants and other species. *Biochemistry* **45,** 561–570.

Trent, J. T., III., and Hargrove, M. S. (2002). A ubiquitously expressed human hexacoordinate hemoglobin. *J. Biol. Chem.* **277,** 19538–19545.

Trent, J. T., III., Hvitved, A. N., and Hargrove, M. S. (2001a). A model for ligand binding to hexacoordinate hemoglobins. *Biochemistry* **40,** 6155–6163.

Trent, J. T., III., Watts, R. A., and Hargrove, M. S. (2001b). Human neuroglobin, a hexacoordinate hemoglobin that reversibly binds oxygen. *J. Biol. Chem.* **276,** 30106–30110.

Van Dyke, B. R., Saltman, P., and Armstrong, A. (1996). Control of myoglobin electron–transfer rates by the distal (nonbound) histidine residue. *J. Am. Chem. Soc.* **118,** 3490–3492.

Van Doorslaer, S., Dewilde, S., Kiger, L., Nistor, S. V., Goovaerts, E., Marden, and Moens, L. (2003). Nitric oxide binding properties of neuroglobin: A characterization by EPR and flash photolysis. *J. Biol. Chem.* **278,** 4919–4925.

CHAPTER TWENTY-ONE

Purification of Class 1 Plant Hemoglobins and Examination of Their Functional Properties

Abir U. Igamberdiev *and* Robert D. Hill

Contents

Abstract

Class 1 hemoglobins are ubiquitous plant proteins induced under hypoxic conditions. They bind oxygen tightly and, as oxyhemoglobin, react with nitric oxide produced under hypoxic conditions. The reactions involved in NO production and scavenging help maintain the redox and energy status of the cell. This article describes the expression of class 1 barley (*Hordeum vulgare* L.) hemoglobin in *E. coli* cells and its purification to homogeneity. Methods for investigating the properties of purified hemoglobin and for measuring its nitric oxide scavenging activity are described. A method for isolation of a plant methemoglobin reductase is also presented.

Department of Plant Science, University of Manitoba, Winnipeg, Manitoba, Canada

Methods in Enzymology, Volume 436
ISSN 0076-6879, DOI: 10.1016/S0076-6879(08)36021-2

1. Introduction

Three classes of plant hemoglobins have been identified (Wittenberg *et al.*, 2002); class 1, class 2, and truncated. Class 1 hemoglobins are induced during hypoxia and are the most extensively studied of the three groups. The expression of a hemoglobin gene within 2 h of exposure to hypoxia was first demonstrated in barley (Taylor *et al.*, 1994). In germinating seeds, which can also be hypoxic, the gene is induced within 2 h of imbibition (Guy *et al.*, 2002). Barley class 1 hemoglobin is a homodimer with a monomeric molecular weight of 18.5 kDa (Duff *et al.*, 1997). A single cysteine-79 along with noncovalent interactions stabilize the dimer through an intramolecular disulfide bond (Bykova *et al.*, 2006). The O_2-dissociation constant of barley Hb is low (2.86 n*M*), indicating that it remains oxygenated at extremely low oxygen concentrations (Duff *et al.*, 1997) where cytochrome *c* oxidase is effectively nonfunctional (Cooper, 2002, 2003). The unique features of class 1 Hb result from the hexacoordination of the heme moiety during oxygen ligation, in comparison to the pentacoordination that occurs in leghemoglobins, erythrocytes, and muscle hemoglobins.

The tight binding of oxygen by class 1 hemoglobins excludes its function as an effective oxygen sensor, carrier, or store (Hill, 1998). Evidence (Igamberdiev *et al.*, 2004, 2006) points to oxyhemoglobin reacting with nitric oxide (NO), which is accumulated under hypoxic conditions (Dordas *et al.*, 2003), to convert NO to nitrate (Hargrove *et al.*, 2000). In the course of this reaction, the ferrous form of Hb is oxidized to the ferric form (methemoglobin). The ferrous form is regenerated by a methemoglobin reductase by the appropriate reducing agent (NADH or NADPH).

Here, we present a method of purification of recombinant barley class 1 hemoglobin from *E. coli* and discuss experimental approaches for investigation of its NO scavenging properties.

2. Purification of Barley Hemoglobin

The concentrations of class 1 hemoglobins in plant tissue are very low, which makes it impractical to purify large amounts of the proteins for study. In maize seedlings, Hb concentrations approach roughly 0.04% of total soluble protein, while in wheat, wild oat, and *Echinochloa crus-galli*, concentrations are 0.01% (Duff *et al.*, 1998). Hb amounts are the highest in the root (0.30%) and aleurone layer (0.12%) of barley seedlings, with very low concentrations (0.02%) in the coleoptile (Hebelstrup *et al.*, 2007).

2.1. Cloning barley hemoglobin in *E. coli*

E. coli strain DH-5α (Invitrogen, Canada) was used as a host for pUC19 plasmid (Invitrogen, Canada) into which Hb cDNA had been subcloned from Bluescript between the restriction sites *Sst*I and *Xba*I. The insert was removed from the Bluescript plasmid and then reinserted into pUC19 in such a way as to remove the 5′ region and have the coding sequence in the correct reading frame. The pUC19 ATG was used as the start codon.

The Bluescript plasmid containing the insert (approximately 100 μg) was digested for 6 h with *Sst*II (which cuts the insert at the 7th and 10th base pair inside the coding sequence) and then dephosphorylated with calf intestinal alkaline phosphatase. The 8-mer (single-stranded adapter, ATCGCCGG) was then phosphorylated and allowed to anneal with the 14-mer (single-stranded adapter, AGCTTAGCGGCCGC) to form a linker adapter containing an *Sst*II site at one end, a *Hind*III site at the other end, and a *Not*I site in the middle. The adapter was then ligated to the insert end of the *Sst*II-digested plasmid. The plasmid was then digested with *Sst*I, releasing the insert cDNA (with the linker adapter ligated to it to form a *Hind*III site at the end). The insert was separated from the plasmid by agarose gel electrophoresis and purified using GeneClean II Kit (Duff *et al.*, 1997).

Nonrecombinant pUC19 (approximately 100 μg) was double digested with *Sst*I and *Hind*III and then dephosphorylated. The insert was then ligated with the nonrecombinant pUC19, and the resulting ligation mix was used to transform DH5-α *E. coli* cells.

DH5-α cells were transformed according to the instructions for Invitrogen subcloning efficiency competent cells. Blue-white screening was unnecessary because all the colonies tested contained the recombinant plasmid (efficiency = 100%). Colonies were picked from the agar plates using a sterile loop and transferred to tubes containing 10 ml of sterile LB (with 150 μg/ml ampicillin). The cells were allowed to grow overnight. Cells containing the recombinant and nonrecombinant (pUC19) plasmids were grown for protein expression in sterile LB media containing 150 μg/ml ampicillin, 100 μg/ml δ-aminolevulinic acid, with or without 1 m*M* IPTG1, at 37° for 6 to 8 h. The bacterial cells were collected by centrifugation and frozen at −80° until used.

2.2. Extraction and purification of recombinant barley Hb

All procedures were performed at 4°. For all chromatographic separations, an FPLC system (GE Healthcare) can be used. All buffers should be degassed. Protein in the fractions was determined according to Bradford (1976) and using bovine serum albumin as a standard.

2.3. Purification protocol

1. Resuspend the bacterial cells (5 to 10 g, wet weight) in 40 ml of extraction buffer (50 m*M* Tris-HCl, pH 8.0, 100 m*M* NaCl, 10% sucrose (w/v), 1 m*M* dithiothreitol, 1 m*M* EDTA, 14 m*M* 2-mercaptoethanol, 1 m*M* phenylmethylsulfonyl fluoride, 10 μg/ml leupeptin, 10 μg/ml chymostatin, 10 μg/ml E-64) and disrupt by three passes through a chilled French pressure cell at approximately 20,000 psi.
2. Clarify the lysate by centrifugation at 27,000 g for 10 min. Dilute the supernatant with extraction buffer to 50 ml and fractionate with polyethylene glycol 8,000, collecting a fraction precipitating between 10 and 25% polyethylene glycol. Centrifuge at 15,000 g for 20 min. Redissolve the red-colored pellet in 30 ml of buffer A (50 m*M* Tris-HCl, pH 8.5, 1 m*M* dithiothreitol, 1 m*M* EDTA) and apply at a rate of 1 ml/min to a column (1.5 $\times$ 9 cm) of DEAE-Sephacel (GE Healthcare) preequilibrated with buffer A. After washing with 25 ml of buffer A, elute the protein with a 100-ml linear gradient of 0 to 250 m*M* KCl in buffer A.
3. The hemoglobin fraction elutes around 150 to 200 m*M* KCl, as evidenced by the red color in the fractions. Dilute these pooled fractions approximately 2.5- to 3-fold with buffer A. Add ammonium sulfate to a final concentration of 30% (saturated) and ensure that the ammonium sulfate is dissolved. Load the sample onto a prepacked Phenyl-Sepharose column HR 5/5 (1-ml volume) equilibrated with buffer A containing 30% (saturated) ammonium sulfate. Then elute the Hb at a flow rate of 1 ml/min with a 50-ml linear gradient of 30 to 0% (saturation) ammonium sulfate in buffer A.
4. Analyze the fractions eluting from the Phenyl-Sepharose column by their absorbance at A_{412}, pooling fractions with higher absorbance values. Concentrate to a final volume of $\approx$200 μl and buffer exchange into PBS (40 m*M* KH_2PO_4/K_2HPO_4, pH 7.0, 150 m*M* NaCl) using a Centricon 10 concentrator.
5. Apply the concentrated fraction to a prepacked Superose 12 HR 10/30 column (1 $\times$ 30 cm) using 50 m*M* Tris-HCl, pH 8.5, 150 m*M* NaCl as column buffer.
6. Elute at 0.2 ml/min collecting 0.5 ml fractions. Assay at A_{280} and A_{412}, calculating the A_{412} to A_{280} ratio to estimate hemoglobin purity.

The purified Hb can either be used immediately for analysis or stored at $-80°$ until needed. After purification, Hb in solution is present in the ferrous oxy form, but after storage at 4° it slowly turns to the ferric form with a half-life of approximately 3 days. At $-80°$ it can be stored for months without significant oxidation.

2.4. Molecular mass of recombinant Hb

Hb fractions after Superose 12 have a 412 to 280 nm absorbance ratio in the range of 3.0 to 3.2, which corresponds to 90 to 95% purity. A ratio of 3.4 corresponds to homogeneity.

The native molecular mass of the protein can be determined on a Superose 12 column using bovine serum albumin (66 kDa), ovalbumin (45 kDa), carbonic anhydrase (29 kDa), and cytochome *c* (12.3 kDa) as standards. SDS-PAGE electrophoresis can be performed using a BioRad mini-gel system with acrylamide concentrations of 15%.

The purification of barley Hb is shown in Table 21.1. Recombinant barley Hb was purified 16-fold with a yield of 5%.

2.5. Characterization of barley Hb

Digest, *in situ*, a Coomassie-stained Hb band from an SDS-PAGE gel with modified trypsin, as described in Shevchenko *et al.* (1996). Briefly, prior to digestion, reduce the Hb with 10 m*M* DTT in 100 m*M* NH_4HCO_3 buffer at 56° for 45 min and then incubate with 55 m*M* iodoacetamide at 25° for 30 min. Purify the tryptic peptide mixture on a reverse-phase POROS R2 (20- to 30- μm bead size, PerSeptive Biosystems, Framingham, CA) nano-column, eluting the fraction onto a MALDI probe with saturated matrix solution (2,5-dihydroxybenzoic acid in 50% [v/v] acetonitrile/5% formic acid). Although other similar machines may be used, we used a MALDI Qq-TOF mass spectrometer (Manitoba/Sciex prototype) for MS/MS analysis of the peptide. A Knexus automation software package (Proteometrics LLC, Canada) with a ProFound search engine can be used for peptide mass fingerprint analysis of MS spectra. Tandem MS spectra can

Table 21.1 Purification of recombinant barley Hb from *E. coli* transformed extracts

Fraction	Volume (ml)	Total protein (mg)	Hb* (mg)	Hb %	Purification	Yield
Total extract	55	765	46	6	1	100
PEG 10–25%	30	285	33.8	11.9	2	73.5
DEAE Sephacel	16	28.4	15.6	54.9	9	33.9
Phenyl Superose	9	11.1	9.3	83.8	14	20.2
Superose 12	0.5	2.4	2.3	95.8	16	5

* Amount of Hb determined by A_{412} measurements and a standard curve.

be analyzed using m/z (Proteometrics Ltd., New York, NY) software and a Sonar MS/MS (Proteometrics, Canada) search engine.

The recombinant Hb has an expected N-terminal sequence of Met-Ile-Thr-Pro-Ser-Leu-Ala-Ala-Ala-Glu, versus the native Hb N-terminal sequence of Met-Ser-Ala-Ala-Glu, and therefore has five additional amino acids. Native molecular mass as assessed by size-exclusion chromatography on Superose 12 was determined to be 40 kDa ($\pm$ 4 kDa), which suggests that barley Hb is a homodimer. SDS-PAGE yielded bands with molecular weights of 18.5 $\pm$ 0.5 kDa. From the mass-spectrometry data, the average molecular weight of the Hb monomer was determined to be 18,625 $\pm$ 2 Da. Translation of the recombinant protein sequence leads to a protein mass of 18,538, excluding the covalently bound heme.

The spectra of purified barley hemoglobin (Table 21.2, Figure 21.1) show characteristic maxima that differ for deoxy ferrous, oxy ferrous, and ferric forms. Addition of 0.1 m*M* NADH in the presence of 20 μM FAD to metHb (ferric) results in partial reduction accompanied by dioxygenation (see Figure 21.1, inset). A similar reduction can be observed in the presence of ascorbate.

3. Measurement of NO Scavenging Activity

Measurement of *in situ* levels of NO in plant tissue generally require chemiluminescence detection, fluorescent probes, or EPR spectroscopy, but the higher concentrations of NO employed when following *in vitro* scavenging can be measured effectively using an NO electrode (NOMK2, World Precision Instruments, USA) in 50 m*M* Tris-HCl buffer (pH 7.5). When FAD is present, addition of 1 mg ml^{-1} bovine Cu,Zn-superoxide dismutase (Sigma) is necessary to prevent formation of peroxynitrite via the interaction of NO and superoxide. This is especially important when NO is generated from sodium nitroprusside (SNP) by light.

NO can be delivered in several ways. The most common and least expensive donor is SNP (Sigma), generating NO under illumination at 200 μmol quanta m^{-2} s^{-1} with continuous stirring. Steady micromolar

Table 21.2 Spectral properties of recombinant barley hemoglobin (modified from Duff *et al.*, 1997)

Derivative	λ_{max}	$\varepsilon_m M$
Ferrous Hb	425, 529, 535, 555, 563	190, 15.0, 14.5, 20.5, 21.2
Ferrous oxyHb	412, 540, 576	149, 17.3, 17.3
Ferric Hb	411, 534, 565	141, 14.4, 11.5

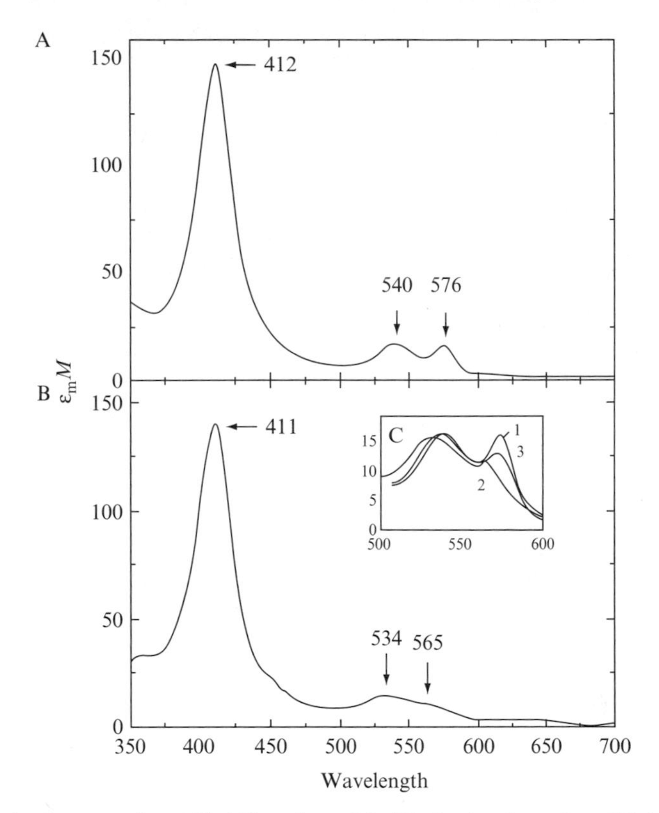

Figure 21.1 Spectra of oxyHb (A) and metHb (B). Reduction of metHb is shown on insert C (1: reduced oxyHb, 2: metHb, 3: partial reduction by NADH in the presence of $FADH_2$).

NO levels can be achieved after ≈5 min of illumination of a 1 m*M* SNP solution. This level is optimal for measuring NO scavenging activity. Measurement of NO scavenging can be initiated by the addition of 0.1 m*M* NADH or NADPH. Crude homogenates, particularly from green leaves, should not be used directly because of the potential presence of nonenzymic, low-molecular-weight NO scavengers. Gel filtration through a Sephadex G-25 column should be sufficient to solve this problem. A major shortcoming of SNP use is that, in addition to supplying NO, it releases iron and cyanide and can generate superoxide. Cyanide is a major inhibitor of many heme-containing enzymes, and iron may have metabolic effects (Kim *et al.*, 2006). The NO^+ moiety of SNP can bind directly to amine and thiol residues of proteins (Thomas *et al.*, 2002).

Other NO donors can be used to bypass this problem. Most of these donors, like the diazeniumdiolates termed NONOate-derivatives, generate NO independently of light and can be selected to deliver NO at predetermined rates. DEA/NO, sodium 2-(N, N-diethylamino)-diazenolate-2-oxide (Alexis Biochemicals) can be used at a concentration of 20 *μM*, delivering ≈1 *μM* NO at room temperature as the neutral radical NO. Preparation of the stock (X 100) solution of DEA/NO in an alkaline medium (10 m*M* NaOH) prevents early decomposition of the compound before it is added to the buffer (100 m*M* phosphate). Decomposition of DEA/NO is sensitive to pH and occurs in a few minutes at pH 7.2 to 7.6. Concentration of DEA/NO in stock solutions can be determined by measuring the absorbance at 250 nm ($\varepsilon = 8\ mM^{-1}\ cm^{-1}$) shortly before use (Thomas *et al.*, 2002).

Alternatively, NO can be delivered from compressed gas (Matheson Company Inc., USA) by bubbling NO through a buffer solution. This has disadvantages, however, because the gas, stored under compression, contains impurities, such as N_2O, NO_2, and N_2O_3 (Bonner, 1996), which tend to increase during storage, and additional purification may be necessary.

It is important to check whether NAD(P)H in the absence of the sample has any effect on NO levels. It is also desirable to follow NO scavenging using at least two different methods of NO supply (i.e., delivered by SNP, DEA/NO, or from compressed gas) to be certain that the effect is not due to the NO donor.

When working with the NO electrode, the initial NO concentrations around 1 *μM* are most desirable. Lower concentrations are possible, but the precision of the measurement may suffer. Higher concentrations may result in undesirable effects via nitrosylation of participating proteins and other side reactions.

4. Identification of a Methemoglobin Reductase

The reaction of NO with oxyhemoglobin results in the formation of methemoglobin. The protein in this form is inactive with respect to ligand binding and can potentially be reduced to an active form by a number of possible reductases in the plant cytoplasm. The following describes a procedure for isolating one such methemoglobin reductase from barley seedlings. This reductase has been identified as a cytosolic monodehydroascorbate (MDHA) reductase (Igamberdiev *et al.*, 2006). For purification, we used the method of Hossain and Asada (1985) with modifications. The original identification of this protein as a methemoglobin reductase involved measurement of NO-scavenging activity in the presence of barley hemoglobin and during its purification. It is also possible to directly follow the rate of MDHA reduction by monitoring absorbance at 340 nm using an extinction coefficient for NADH of $\varepsilon = 6.22\ mM^{-1}\ cm^{-1}$, in 50 m*M* Tris-HCl buffer, pH 8.0, containing 0.1 m*M* NADH, 2.5 m*M* ascorbate, and an amount of ascorbate oxidase (Sigma) to yield ≈3 μ*M* MDHA (Hossain *et al.*, 1984).

4.1. Purification protocol for MDHA reductase

1. Germinate 50 to 100 g barley (*Hordeum vulgare* L. cv Harrington) kernels in darkness on wet filter paper for 3 days. Isolate 40 g of root tissue from the seedlings and grind in liquid nitrogen.
2. Homogenize the powder in 50 m*M* Tris-HCl buffer, pH 8.0, containing 1 m*M* EDTA and 5 m*M* $MgCl_2$. After filtering through cheesecloth and centrifugation at 10,000 g, fractionate the supernatant using ammonium sulfate, collecting the fraction between 50 and 70% saturation by centrifugation. Redissolve the pellet in 10 m*M* Tris-HCl buffer, pH 8.0, and adjust the concentration of ammonium sulfate to 30% saturation.
3. Load the solution on a Phenyl-Superose column and elute using a decreasing gradient of ammonium sulfate from 30 to 0% saturation. Collect active fractions, pool, and concentrate in Centricon 10 tubes (Millipore, Nepean, Canada).
4. Load the combined fraction on a DEAE-cellulose column (1 × 5 cm) equilibrated with Tris-HCl, pH 8.0. A step gradient of NaCl is used to further purify Hb- and NADH-dependent NO-scavenging activity. The major part of activity is eluted between 35 and 50 m*M* NaCl.
5. Collect active fractions, concentrate, and load onto a Blue Sepharose CL–6B column (0.5 × 3 cm). Elute protein with 2 m*M* NADH in 10 m*M* Tris-HCl buffer (Hossain and Asada, 1985).

The single band from the Blue Sepharose column fraction has a molecular mass of ≈40 kDa. A single protein band from the SDS gel, analyzed by MALDI Qq-TOF single and tandem mass spectrometry, identified the protein as MDHA reductase (Igamberdiev *et al.*, 2006). MDHA reductase activity with the purified recombinant barley Hb (in metHb form) can be determined at 340 nm by the oxidation of NADH in the absence or presence of ascorbate (2 m*M*) and can be verified by increase of absorbance of Hb at 576 nm. Extinction coefficients of 17.3 and 9.2 mM^{-1} cm^{-1} for the reduced ferrous oxy form and oxidized ferric form of Hb, respectively, can be used (Duff *et al.*, 1997).

The K_m values of MDHA reductase with different substrates can be determined in Hanes coordinates (s/v, s) by varying concentrations of Hb, NADH, and ascorbate at constant concentrations of other substrates (5 μ*M* Hb, 0.1 m*M* NADH, 2 m*M* ascorbate).

Oxidation of NADH by the purified enzyme is slow in the presence of MetHb and NADH alone (Table 21.3), but the rate is much higher than the rate of met Hb alone with NADH (Fig. 21.2). Addition of ascorbate results

Table 21.3 NADH dehydrogenase activity of purified monodehydroascorbate (MDHA) reductase (μmol NADH oxidized per min min/per mg protein) with different substrates. NADH decline was monitored at 340 nm. Taken from Igamberdiev *et al.* (2006) with modifications

Substrate	Rate
Ascorbate free radical (MDHA)	185
MDHA + pHMB	22
MetHb	0.5
MetHb + ascorbate	9
MetHb + ascorbate + *p*HMB	0.5

Table modified from Igamberdiev *et al.* (2006). NADH decline was monitored at 340 nm. Concentrations applied: MDHA –, $\approx\sim 3$ μM;, *p*HMB, – 0.1 mM;, MetHb, – 3 μM;, ascorbate – 2.5 mM;, NADH – 0.1 mM. SE is $\leq$20% in all measurements.

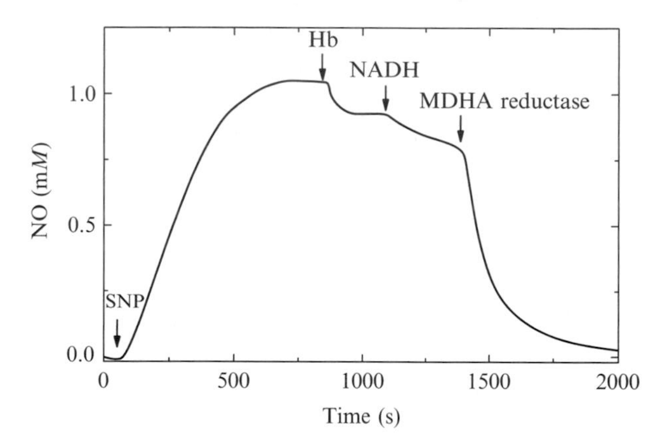

Figure 21.2 NO scavenging by barley hemoglobin. NO (concentration $\approx$1 μM) was generated by SNP. The addition of 50 μM barley Hb almost immediately decreased NO concentration stoichiometrically (by $\approx$50 μM), which resulted in conversion of Hb into metHb (ferric) form, not capable of further NO scavenging. The addition of NADH resulted in a slow sustained rate of NO scavenging caused by chemical reduction of metHb. Addition of MDHA reductase resulted in higher rate of NO scavenging due to the efficient reduction of metHb.

in close to a 20-fold increase in the reaction rate. The reaction is inhibited by SH-reagents (like para-hydroxymercurybenzoate, *p*HMB). Purified MDHA reductase exhibits a threefold higher rate with NADH than with NADPH, using either MDHA or metHb. NO scavenging in the presence of MDHA reductase, NADH, and Hb is significant even without externally added ascorbate (see Fig. 21.2).

Measuring NADH oxidation, the K_m value in the presence of ascorbate is 0.3 ± 0.1 μM for metHb. The K_m for ascorbate is 0.6 ± 0.3 mM and the

K_m for NADH is 4 ± 1 μM. Ascorbate interferes with NO amperometric measurements, precluding the possibility of measuring disappearance of NO with all reagents present (Igamberdiev *et al.*, 2006).

5. Conclusion

Purification of barley class 1 hemoglobin and investigation of its NO-scavenging activity have made it possible to understand some of the mechanism of the operation of Hb in plant cells. Our studies with barley Hb demonstrate that Hb alone is incapable of sustaining physiologically significant NAD(P)H-dependent NO-converting activity (Igamberdiev *et al.*, 2006). We have identified MDHA reductase as one candidate for a methemoglobin reductase. The reduction of methemoglobin in this reaction is mediated by ascorbate (Igamberdiev *et al.*, 2006). The total reaction of NO scavenging is described by the following equation:

$$2NO + 2O_2 + NAD(P)H \rightarrow 2NO_3^- + NAD(P)^+ + H^+ \qquad (21.1)$$

The reaction consists of several steps.

1. Oxygenated ferrous hemoglobin reacts with NO (NO dioxygenation), a reaction insensitive to SH reagents:

$$Hb(Fe^{2+})O_2 + NO \rightarrow Hb(Fe^{3+}) + NO_3^- \qquad (21.2)$$

2. Oxidized (Fe^{3+}) hemoglobin (ferric Hb or metHb) is reduced by MDHA reductase. This includes reduction at the active site of MDHA reductase, facilitated by ascorbate:

$$Hb(Fe^{3+}) + \text{Ascorbate} \rightarrow Hb(Fe^{2+}) + \text{MDHA} \qquad (21.3)$$

3. MDHA is immediately reduced by MDHA reductase (this step is sensitive to SH reagents):

$$2\text{MDHA} + NAD(P)H + H^+ \rightarrow 2\text{Ascorbate} + NAD(P)^+ \qquad (21.4)$$

4. Reduced (ferrous) hemoglobin is oxygenated immediately at very low oxygen concentrations and ready to oxygenate NO to nitrate:

$$Hb(Fe^{2+}) + O_2 \rightarrow Hb(Fe^{2+})O_2 \qquad (21.5)$$

5. Hypoxically induced nitrate reductase reduces nitrate to nitrite:

$$NO_3^- + NAD(P)H + H^+ \rightarrow NO_2^- + NAD(P)^+ + H_2O \quad (21.6)$$

Nitrite can be reduced to NO by nitrite: NO reductase. This enzyme is most likely associated with mitochondria, as evidenced by production of NO from nitrite under anaerobic conditions associated with complexes III and IV of the mitochondrial electron-transport chain (Gupta *et al.*, 2005; Planchet *et al.*, 2005; Stoimenova *et al.*, 2007). This reaction is linked to a limited rate of mitochondrial ATP synthesis (Stoimenova *et al.*, 2007). Other (probably minor) contributors are the side reaction of nitrate reductase with nitrite (Yamasaki and Sakihama, 2000) and a plasma membrane nitrite: NO reductase (Stöhr and Stremlau, 2006).

$$2NO_2^- + NAD(P)H + 3H^+ \rightarrow 2NO + NAD(P)^+ + 2H_2O \quad (21.7)$$

The complete sequence of these reactions (defined as the Hb/NO cycle) involves two NAD(P)H molecules oxidized per one NO and is linked to ATP synthesis (Stoimenova *et al.*, 2007). The pathway operates as an effective alternative to classic fermentation pathways under hypoxia (Igamberdiev and Hill, 2004), maintaining cell redox and energy status.

REFERENCES

Bonner, F. T. (1996). Nitric oxide gas. *Methods Enzymol.* **268,** 50–57.

Bradford, M. M. (1976). A rapid and sensitive method for the quantitation of microgram quantities of protein utilizing the principle of protein-dye binding. *Anal. Biochem.* **72,** 248–254.

Bykova, N. V., Igamberdiev, A. U., Ens, W., and Hill, R. D. (2006). Identification of an intermolecular disulfide bond in barley hemoglobin. *Biochem. Biophys. Res. Commun.* **347,** 301–309.

Cooper, C. E. (2002). Nitric oxide and cytochrome oxidase: Substrate, inhibitor or effector? *Trends Biochem. Sci.* **27,** 33–39.

Cooper, C. E. (2003). Competitive, reversible, physiological? Inhibition of mitochondrial cytochrome oxidase by nitric oxide. *IUBMB Life* **55,** 591–597.

Dordas, C., Hasinoff, B. B., Igamberdiev, A. U., Manac'h, N., Rivoal, J., and Hill, R. D. (2003). Expression of a stress-induced hemoglobin affects NO levels produced by alfalfa root cultures under hypoxic stress. *Plant J.* **35,** 763–770.

Duff, S. M. G., Wittenberg, J. B., and Hill, R. D. (1997). Expression, purification, and properties of recombinant barley (*Hordeum sp.*) hemoglobin: Optical spectra and reactions with gaseous ligands. *J. Biol. Chem.* **272,** 16746–16752.

Duff, S. M. G., Guy, P. A., Nie, X., Durnin, D. C., and Hill, R. D. (1998). Hemoglobin expression in germinating barley. *Seed Sci. Res.* **8,** 431–436.

Gupta, K. J., Stoimenova, M., and Kaiser, W. M. (2005). In higher plants, only root mitochondria, but not leaf mitochondria reduce nitrite to NO, in vitro and in situ. *J. Exp. Bot.* **56,** 2601–2609.

Guy, P. A., Sidaner, J.-P., Schroeder, S., Edney, M., MacGregor, A. W., and Hill, R. D. (2002). Embryo phytoglobin gene expression as a measure of germination in cereals. *J. Cereal Sci.* **36,** 147–156.

Hargrove, M. S., Brucker, E. A., Stec, B., Sarath, G., Arredondo-Peter, R., Klucas, R. V., Olson, J. S., and Phillips, G. N., Jr. (2000). Crystal structure of a nonsymbiotic plant hemoglobin. *Structure* **8,** 1005–1014.

Hebelstrup, K. H., Igamberdiev, A. U., and Hill, R. D. (2007). Metabolic effects of hemoglobin gene expression in plants. *Gene* **398**, 86–93.

Hill, R. D. (1998). What are hemoglobins doing in plants? *Can. J. Bot.* **76,** 707–712.

Hossain, M. A., and Asada, K. (1985). Monodehydroascorbate reductase from cucumber is a flavin adenine dinucleotide enzyme. *J. Biol. Chem.* **260,** 12920–12926.

Hossain, M. A., Nakano, Y., and Asada, K. (1984). Monodehydroascorbate reductase in spinach chloroplasts and its participation in regeneration of ascorbate for scavenging hydrogen peroxide. *Plant Cell Physiol.* **25,** 385–395.

Igamberdiev, A. U., and Hill, R. D. (2004). Nitrate, NO and haemoglobin in plant adaptation to hypoxia: An alternative to classic fermentation pathways. *J. Exp. Bot.* **55,** 2473–2482.

Igamberdiev, A. U., Bykova, N. V., and Hill, R. D. (2006). Scavenging of nitric oxide by barley hemoglobin is facilitated by a monodehydroascorbate reductase-mediated ascorbate reduction of methemoglobin. *Planta* **223,** 1033–1040.

Igamberdiev, A. U., Seregélyes, C., Manac'h, N., and Hill, R. D. (2004). NADH-dependent metabolism of nitric oxide in alfalfa root cultures expressing barley hemoglobin. *Planta* **219,** 95–102.

Kim, H. J., Tsoy, I., Park, M. K., Lee, Y. S., Lee, J. H., Seo, H. G., and Chang, K. C. (2006). Iron released by sodium nitroprusside contributes to heme oxygenase-1 induction via the cAMP-protein kinase A-mitogen-activated protein kinase pathway in RAW 264.7 cells. *Mol. Pharmacol.* **69,** 1633–1640.

Planchet, E., Gupta, K. J., Sonoda, M., and Kaiser, W. M. (2005). Nitric oxide emission from tobacco leaves and cell suspensions: Rate limiting factors and evidence for the involvement of mitochondrial electron transport. *Plant J.* **41,** 732–743.

Shevchenko, A., Jensen, O. N., Podtelejnikov, A. V., Sagliocco, F., Wilm, M., Vorm, O., Mortensen, A., Shevchenko, A., Boucherie, H., and Mann, M. (1996). Linking genome and proteome by mass spectrometry: Large-scale identification of yeast proteins from two-dimensional gels. *Proc. Natl. Acad. Sci. USA* **93,** 14440–14445.

Stöhr, C., and Stremlau, S. (2006). Formation and possible roles of nitric oxide in plant roots. *J. Exp. Bot.* **57,** 463–470.

Stoimenova, M., Igamberdiev, A. U., Gupta, K. J., and Hill, R. D. (2007). Nitrite-driven anaerobic ATP synthesis in barley and rice mitochondria. *Planta* **226**, 465–474.

Taylor, E. R., Nie, X. Z., MacGregor, A. W., and Hill, R. D. (1994). A cereal hemoglobin gene is expressed in seed and root tissues under anaerobic conditions. *Plant Molec. Biol.* **24,** 853–862.

Thomas, D. D., Miranda, K. M., Espey, M. G., Citrin, D., Jourd'heuil, D., Paolocci, N., Hewett, S. J., Colton, C. A., Grisham, M. B., Feelisch, M., and Wink, D. A. (2002). Guide for the use of nitric oxide (NO) donors as probes of the chemistry of NO and related redox species in biological systems. *Methods Enzymol.* **359,** 84–105.

Wittenberg, J. B., Bolognesi, M., Wittenberg, B. A., and Guertin, M. (2002). Truncated hemoglobins: A new family of hemoglobins widely distributed in bacteria, unicellular eukaryotes, and plants. *J. Biol. Chem.* **277,** 871–874.

Yamasaki, H, and Sakihama, Y. (2000). Simultaneous production of nitric oxide and peroxynitrite by plant nitrate reductase: *In vitro* evidence for the NR-dependent formation of active nitrogen species. *FEBS Lett.* **468,** 89–92.

CHAPTER TWENTY-TWO

USE OF *IN SILICO* (COMPUTER) METHODS TO PREDICT AND ANALYZE THE TERTIARY STRUCTURE OF PLANT HEMOGLOBINS

Sabarinathan Kuttalingam Gopalasubramaniam,* Verónica Garrocho-Villegas,* Genoveva Bustos Rivera,* Nina Pastor,† *and* Raúl Arredondo-Peter*

Contents

Abstract

Amino acid sequences for more than 60 plant hemoglobins (Hbs) are deposited in databases, but the tertiary structure of only 4 plant Hbs have been reported; thus, the gap between the reported sequences and structures of plant Hbs is large. Elucidating the structure of plant Hbs is essential to fully understanding the function of these proteins in plant cells. Determining the actual protein structure by experimental methods (i.e., by X-ray crystallography) requires considerable protein material and is expensive; thus, this type of work is limited to few laboratories around the world. *In silico* (computer) methods to predict the tertiary structure of proteins from amino acid sequences have been implemented and are helping reduce the sequence-structure gap. Thus, *in silico* methods are useful tools for predicting the tertiary structure of several plant

* Laboratorio de Biofísica y Biología Molecular, Facultad de Ciencias, Universidad Autónoma del Estado de Morelos, Cuernavaca, Morelos, Mexico
† Facultad de Ciencias, Universidad Autónoma del Estado de Morelos, Cuernavaca, Morelos, Mexico

Methods in Enzymology, Volume 436
ISSN 0076-6879, DOI: 10.1016/S0076-6879(08)36022-4

Hbs from amino acid sequences deposited in databases. In this chapter, we describe a method for predicting and analyzing the structure of a rice Hb2 from the template structure of native rice Hb1. This method is based on a comparative modeling method that uses programs from the SWISS-MODEL server.

1. Introduction

1.1. Overview

The amino acid sequence of a protein determines its tertiary/quaternary structure (Anfinsen *et al.*, 1961) and ultimately its function. One goal of structural biology is to elucidate the three-dimensional conformation of proteins, so protein function can be deduced from structural analysis. About 3 million protein sequences are deposited in the GenBank database (http://www.ncbi.nlm.nih.gov), a large reservoir of information for understanding protein function. However, the number of experimentally determined crystal structures is two orders of magnitude behind the number of protein sequences (Ginalski *et al.*, 2005). Although the experimental determination of a protein structure has become more efficient, the gap between the number of known protein sequences and structures is rapidly increasing, primarily as a result of numerous whole-genome sequencing programs. However, several methods for predicting the tertiary structure of proteins from amino acid sequences have been implemented and are helping reduce the sequence-structure gap, thereby allowing researchers to gather significant insights into protein folding and function.

Protein structure prediction began before the first protein structures were solved by experimental methods. Today it is possible to accurately predict the folding pattern of distantly related sequences: programs that predict secondary structures can distinguish helix, sheet, and turn residues with greater than 60% accuracy (Montgomerie *et al.*, 2006; Tsai, 2002). However, despite accuracy and wide availability (see the subsequent section), prediction methods are not expected to replace experimental determination of protein structures, but to complement and fill the gap between the number of sequences and structures and to guide experimental design for testable hypotheses. Structure prediction methods are often divided into three areas (Ginalski *et al.*, 2005): (1) methods based on sequence similarity, which assign a fold to a protein from a close homolog with a known structure; (2) threading methods, which thread the sequence of a query protein through a template structure; (3) hybrid methods, which combine sequence similarity and threading methods to identify homologs to a query protein in protein families; and (4) practical *ab initio* methods, which predict the native structure of a protein by simulating the biological process of protein folding.

The most accurate predictions of protein structure are obtained when a template structure with a high level of sequence similarity (>25%) to a target protein is identified in the Brookhaven Protein DataBase (PDB database) (Berman *et al.*, 2000; http://www.rcsb.org/pdb/) (i.e., by homology or comparative modeling (step 1 previously) (Petrey and Honig, 2005). Steps for protein structure prediction by comparative modeling can be summarized as follows (Kretsinger *et al.*, 2004; Petrey and Honig, 2005).

First is identification of a query (target) sequence with a known homologous (template) structure. Sequences that belong to the same family of proteins are assumed to be homologous and thus to fall into the same structural class. If the expected (E) value (probability of finding by chance a match with the query sequence) is less than 10^{-2} in a sequence alignment (or roughly 25% identity in a sequence alignment of proteins more than 200 residues in length), it is inferred that the two proteins are homologous.

Second is alignment of the target sequence to the template structure. The accuracy of alignment of homologous proteins is the key limitation for the quality and usefulness of models, as misalignment is the greatest source of error in comparative modeling. Thus, it is essential to perform optimal sequence alignment to maximize the identity or similarity at aligned positions. This is usually performed using reliable programs for sequence alignment (e.g., Clustal program; Thompson *et al.*, 1997), although manual verification by the user is always recommended. The ultimate choice of the template structure depends on the statistical significance of the alignment score (lowest E value) between the target and template sequences.

Third is building of an initial model for all or part of the target sequence based on the template structure. There are several programs that can be used to build an initial model. Table 22.1 summarizes selected servers and programs publicly available for prediction of protein structures.

Fourth is modeling of the side chains and loops in the target model that are different from the template structure.

Fifth is model refinement, which allows the predicted structure to adopt an (energetically) optimal conformation. Table 22.2 summarizes selected servers for protein structure analysis, including energy optimization and molecular dynamics.

Sixth is model evaluation. After structure prediction, fold recognition results are validated using independent observations, such as by modeling against several templates, using different modeling programs, or using validation programs from servers that test the reliability and quality of predicted structures (see Tables 22.1 and 22.2). Model evaluation methods give higher scores to models that are closer to native structures.

In this chapter, we describe a method for predicting the structure of plant Hbs based on comparative modeling. The interested reader should refer to reviews by Ginalski *et al.* (2005), Kretsinger *et al.* (2004), Petrey and Honig (2005), Tsai (2002), and Xiang (2006) for a detailed description of methods other than comparative modeling (steps 3 and 4, previously).

Table 22.1 Summary of selected servers for the prediction, evaluation, and visualization of protein structures

SERVER NAME	SERVICE	URL
	Servers for protein structure prediction and modeling	
3D-Jigsaw	Builds three-dimensional models for proteins based on homologs of known structure.	http://www.bmm.icnet.uk/servers/3djigsaw
BMERC	Predicts probable secondary structures and folding classes for a given amino acid sequence.	http://bmerc-www.bu.edu/psa/index.html
Geno3D	Predicts three-dimensional protein structure by satisfying spatial restraints (distances and dihedral).	http://geno3d-pbil.ibcp.fr/cgi-bin/geno3d_automat.pl?page=/GENO3D/geno3d_home.html
JPRED	Predicts protein secondary structure by consensus methods.	http://www.compbio.dundee.ac.uk/~www-jpred/
ModBase	Predicts three-dimensional protein models by comparative modeling.	http://modbase.compbio.ucsf.edu/modbase-cgi/index.cgi
MODELLER 4	Predicts protein three-dimensional structures by homology or comparative modeling.	http://salilab.org/modeller/
NNPredict	Predicts the secondary structure for each residue in an amino acid sequence.	http://alexander.compbio.ucsf.edu/~nomi/nnpredict.html
PROF	Predicts protein secondary structure by profile-based neural network method.	http://www.aber.ac.uk/~phiwww/prof/
PredictProtein	Retrieves similar sequences in the database and predicts the protein structure.	http://cubic.bioc.columbia.edu/predictprotein/
PROSPECT	Predicts protein structures based on threading-based methods, particularly for the recognition of the fold template whose sequence has insignificant homology to the target sequence.	http://compbio.ornl.gov/structure/prospect/

PSIpred	Predicts secondary structure by incorporating two feed-forward neural networks.	http://bioinf.cs.ucl.ac.uk/psipred/
SSpro	Predicts secondary structure based on an assemble of 100 1D-RNNs (one dimensional recurrent neural networks).	http://www.ics.uci.edu/%7Ebaldig/scratch/explanation.html#SSpro
SSThread	Predicts protein secondary structure by threading.	http://www.ddbj.nig.ac.jp/search/ssthread.html
Swiss-model	Fully automated protein structure homology-modeling server, accessible via the ExPASy web server.	http://swissmodel.expasy.org//SWISS-MODEL.html
	Servers for evaluating predicted protein models	
CastP	Calculates voids and pockets in the predicted protein structure.	http://cast.engr.uic.edu/cast/
MaxSub	Assesses the quality of predicted protein structures.	http://www.cs.bgu.ac.il/~dfischer/MaxSub/
PROCHECK	Checks stereochemical quality of predicted protein structure by analyzing its overall and residue by residue geometry.	http://www.biochem.ucl.ac.uk/~roman/procheck/procheck.html
PROSA II	Checks $C\alpha$ and $C\beta$ potentials and backbone of the protein structure.	http://www.lmcp.jussieu.fr/sincris-top/logiciel/prg-prosa.html
RAMPAGE	Visualizes and assesses the Ramachandran plot of a protein structure.	http://raven.bioc.cam.ac.uk/rampage.php
	Web servers for protein structure retrieval and visualization	
AISMIG	Generates and visualizes high resolution 3D images from PDB structure files.	http://www.dkfz-heidelberg.de/spec/aismig
MovieMaker	Generates short (~10 s) downloadable movies for protein motion and interactions.	http://wishart.biology.ualberta.ca/moviemaker/
PPG	Generates pictures and animations of protein structures from PDB files.	http://bioserv.rpbs.jussieu.fr/cgi-bin/PPG

(*continued*)

Table 22.1 *(continued)*

SERVER NAME	SERVICE	URL
VERIFY3D	Analyzes the quality of a proposed structure in PDB format.	http://www.doe-mbi.ucla.edu/Services/Verify_3D/
Swiss-PdbViewer	Analyzes several proteins at the same time. The proteins can be superimposed in order to deduce structural alignments and to compare their active sites or any other protein regions.	http://ca.expasy.org/spdbv/
VMD	Visualizes, displays, animates, and analyzes large biomolecular systems using 3-D graphics and built-in scripting.	http://www.ks.uiuc.edu/Research/vmd/
PyMOL	Real-time visualization and rapid generation of high-quality molecular graphics, images, and animations.	http://pymol.org/funding.html
Rasmol	Visualizes proteins, nucleic acids, and small molecules.	http://www.umass.edu/microbio/rasmol/

Table 22.2 Summary of selected servers for the analysis of protein structures

SERVER NAME	SERVICE	URL
	Web servers for protein analysis	
3D—PSSM	Predicts a protein function from a protein sequence.	http://www.sbg.bio.ic.ac.uk/3dpssm/html/ffrecog-simple.html
CHARMM	Performs macromolecular simulations, including energy minimization, molecular dynamics, and Monte Carlo simulations.	http://yuri.harvard.edu/
CATH	Performs hierarchical classification of protein domain structures, which clusters proteins at four major levels: Class(C), Architecture(A), Topology(T), and Homologous superfamily (H).	http://www.biochem.ucl.ac.uk/bsm/cath/
Diamond STING	Performs analysis of protein quality, structure, stability, and function.	http://trantor.bioc.columbia.edu/SMS
*i*MolTalk	Computes Ramachandran plot (Φ/Ψ angles), distance matrix of Cα atoms and interface between two chains of a protein structure.	http://i.moltalk.org/
SCRATCH	Computes solvent accessibility, disulfide bridges, stability effects of single amino acid mutations, disordered regions, domains, beta-residue and beta-strand pairings, and amino acid contact maps.	http://www.igb.uci.edu/servers/psss.html
SCR_FIND	Computes structurally conserved regions (SCRs) from superimposed protein structures and multiple sequence alignments.	http://schubert.bio.uniroma1.it/SCR_FIND/

(*continued*)

Table 22.2 (*continued*)

SERVER NAME	SERVICE	URL
Sride	Identifies stabilizing residues in a protein structure by using the parameters surrounding hydrophobicity, long-range order, stabilization center, and conservation score.	http://sride.enzim.hu/
	Web servers for protein docking	
Gold	Performs protein-ligand docking based on genetic algorithms; provides a ranked list of ligand positions from multiple binding sites.	http://www.ccdc.cam.ac.uk/products/life_sciences/gold/
Hex4.5	Performs protein-ligand docking.	http://www.csd.abdn.ac.uk/hex/
ICM-Dock	Performs ligand-protein docking, peptide-protein docking, protein-protein docking with step-by-step docking methods.	http://www.molsoft.com/docking.html
Relibase	Performs receptor-ligand interactions.	http://relibase.ebi.ac.uk/reli-cgi/rll?/reli-cgi/general_layout.pl+home
ZDOCK	Performs fast Fourier transform to search all possible binding modes for proteins, evaluating based on shape complementarity, desolvation energy, and electrostatics.	http://zlab.bu.edu/zdock/

1.2. Reported amino acid sequences and tertiary structures for plant hemoglobins

Early work and recent genome sequencing programs have generated nucleotide and amino acid sequences for more than 60 plant Hbs, but the tertiary structure of only 4 plant Hbs has been reported (3 [lupin and soybean leghemoglobins and rice nonsymbiotic Hb1] from X-ray analysis [PDB ID 1GDI, 1BIN, and 1D8U, respectively] and 1 [maize Hb] from prediction [*in silico*] analysis; Sáenz-Rivera *et al.*, 2004); thus, the gap between the reported sequence and structure of plant Hbs is very large. Elucidating the structure of plant Hbs is of interest because this information is essential to fully understanding the function of these proteins in plant cells. However, with the exception of leghemoglobins (Lbs), which are abundant proteins in the root nodules of nitrogen-fixing plants (Appleby, 1984; Appleby, 1992), plant Hbs are synthesized in very low concentrations in plant tissues (Ross *et al.*, 2001); thus, obtaining enough protein material for structural analysis is extremely difficult. An approach to obtain high quantities of protein has been to generate recombinant plant Hbs (Arredondo-Peter *et al.*, 1997a; Arredondo-Peter *et al.*, 1997b; Hargrove *et al.*, 1997), but this procedure is expensive and time consuming, and requires the availability of facilities for protein crystallization and analysis. Therefore, the use of *in silico* (computer) methods to predict protein structures is an alternative approach to understanding the structures of plant Hbs. Resultant models can be subjected to structural analysis, for example, to identify the position of specific amino acid residues and helices and loops, to measure atomic distances, and to predict the existence of hydrogen bonds and potential interaction with other cellular molecules (docking). There are many methods and programs for predicting the structure of a protein (see Tables 22.1 and 22.2); however, in this chapter, we describe a method based on comparative modeling that is routinely used in our laboratory to predict and analyze the structure of plant Hbs. This method is useful for predicting novel structures for plant Hbs because these proteins are conserved: sequence similarity between primitive (moss) and evolved (monocot and dicot) Hbs is about 35 to 50% (Arredondo-Peter *et al.*, 2000). Thus, the expected reliability of predicted plant Hb structures is high.

2. An *In Silico* Method for Predicting the Tertiary Structure of Rice Hb2

A family of *hb* genes exists in rice and consists of four (*hb1* to *4)* copies (Lira-Ruan *et al.*, 2002). The tertiary structure of rice Hb1 was elucidated by X-ray crystallography (PDB ID 1D8U) (Hargrove *et al.*, 2000), but the structure of Hb2, Hb3, and Hb4 is still not known. Sequence similarity between rice Hb1, Hb2, Hb3, and Hb4 is 93 to 97% (Lira-Ruan *et al.*, 2002); thus, it is possible to predict the structure of Hb2, Hb3, and Hb4 by comparative modeling using the structure of rice Hb1 as a template. We describe the use of the Web server SWISS-MODEL (Schwede *et al.*, 2003; http://swissmodel.expasy.org//SWISS-MODEL.html) to predict the structure of rice Hb2 using rice Hb1 as template. SWISS-MODEL predicts protein structures with the modeling program Promod II (Peitsch, 1996) and the energy minimization program Gromos96 (vanGunsteren *et al.*, 1996), and it is an easy-to-use fully automatic interface.

The first step in the process of protein structure prediction is the target and template selection:

1. The rice Hb2 (target) sequence to be modeled is obtained from the GenBank database with the accession number AAK72228.
2. The crystal structure of (template) rice Hb1 is obtained from the PDB database (http://www.rcsb.org/pdb/) with the ID number 1D8U.
3. If no template structure has been selected, a PSI-BLAST analysis of the target sequence is performed to identify the protein structure with highest similarity (lowest E value).

The second step is to access SWISS-MODEL. This server displays five menu options ("alignment interface," "project [optimize] mode," "oligomer modeling," "GPCR mode," and "first approach mode"). The first approach mode option allows the user to predict a preliminary protein structure by the following procedure:

1. Provide the user's e-mail address, full name, and job title.
2. Paste the sequence of the target protein (rice Hb2) in the FASTA format and provide the PDB ID number for the template (rice Hb1) structure (see the previous section). It is also possible to select a template structure using the program BLASTP2 and the database ExNRL-3D, which are available from the SWISS-MODEL server.
3. Select the output file as Swiss Pdb project or pdb files. A Swiss Pdb project file contains coordinates for both the predicted target sequence and template structure, and a pdb file contains coordinates for only the predicted target sequence. The user may also select the WhatCheck report option, which evaluates the preliminary model. A WhatCheck report provides the RMS Z- and structure Z-scores (and other structure evaluations) that quantitate the reliability of the model (Table 22.3). Submit the job request.
4. A first e-mail message is received at the user's account that contains the job title, date, target sequence, and output file.
5. A second e-mail message is received that contains a trace log file for the processing job. This file describes the processes performed by the server to model the structure of the target (rice Hb2) protein. The following are

Table 22.3 RMS (Root Mean Square) Z- and structure Z-scores[a] of predicted rice Hb2. RMS Z-score values close to (either above or below) 1.0 indicate reliable predicting values; positive structure Z-score values indicate that predicting values are better than average (Hooft *et al.*, 1996)

RMS Z-scores	
Bond lengths	0.625
Bond angles	0.844
Omega angle restraints	0.582
Side chain planarity	0.958
Improper dihedral distribution[b]	1.635
Inside/outside distribution	0.981
Structure Z-scores	
1st generation packing quality[c]	1.518
2nd generation packing quality[c]	0.172
Ramachandran plot appearance	0.511
chi-1/chi-2 rotamer normality[d]	1.473
Backbone conformation	0.345

[a] A Z-score corresponds to the number of standard deviations that the score deviates from the expected value.
[b] Improper dihedral distribution is a measure of the chirality/planarity of the structure at a specific atom.
[c] 1st and 2nd generation packing quality compare the packing environment of the residues with the average packing environment for all residues of the same type in good PDB files.
[d] chi-1/chi-2 rotamer normality expresses how well the chi-1/chi-2 angles of all residues are corresponding to the populated areas in the database within expected ranges for well refined structures.

the programs and program activities at the SWISS-MODEL server that predict a preliminary model.

- AlignMaster: aligns target and template (rice Hb2 and Hb1, respectively) sequences and provides the sequence identity of target and template sequences. This program creates a batch file for the Promod II program.
- Promod II: models the target sequence by aligning and refining the raw sequence, assessing backbone positions, adding missing side groups, building and refining loops, optimizing the position of side groups, and creating a preliminary model.
- Gromos96: minimizes the energy of the preliminary model throughout two processes. One process uses the steepest descent method with 200 cycles and 25/C-factor constraints, and the other uses the conjugate gradient method with 300 cycles and 2500/C-factor constraints (vanGunsteren *et al.*, 1996). Energy minimization of the preliminary model by Gromos96 program generates the second version of the predicted (target) protein (rice Hb2) structure.

6. A third e-mail message is received that contains coordinates for the second version of the predicted (target) protein (rice Hb2) structure. This structure can be visualized and analyzed with options from the SwissPDBViewer program (see the subsequent section).
7. A fourth e-mail message is received that contains a WhatCheck Report file (if the user chose this option).

Because the previously mentioned processes do not incorporate prosthetic groups into the predicted (rice Hb2) structure, heme should be incorporated into rice Hb2 by pasting the heme coordinates from the template rice Hb1 into the predicted structure of rice Hb2. Program SwissPDBViewer accommodates the heme into the rice Hb2 heme pocket at an average of 2 Å from surrounding atoms.

The third step is to optimize the structure (of predicted rice Hb2) conformation:

1. Upload the second version of the predicted structure to SWISS-MODEL, select the "project optimize mode" option from the menu, and submit the job request. Note that proteins with low sequence identity with the template (e.g., around 50%) and poor Z-scores require optimization before proceeding to the automated optimization. In this case, the second version of the predicted structure has to be analyzed by SwissPDBViewer, optimized to improve the model, and converted into a format readable by the project optimize mode.
2. An e-mail message is received that contains coordinates for the optimized model.
3. The final version of the predicted (target) protein (rice Hb2) structure is visualized and analyzed with options from SwissPDBViewer or any other programs for visualization and analysis of protein structures (see Table 22.2).

Despite the availability of evaluation programs for predicted structures (see Table 22.2), we tested the reliability of the method described previously by predicting the structure of soybean Lb*a* (whose crystal structure is deposited in the PDB database with the ID 1BIN) from the template crystal structure of lupin Lb (PDB ID 1GDI) and rice Hb1 (PDB ID 1D8U) (which are 57 and 43% similar to soybean Lb*a*, respectively), and by comparing the structure of predicted and native soybean Lb*a*. Figure 22.1 shows that the folding of predicted and native soybean Lb*a* is highly similar. However, a major and minor difference was detected at the helix F and CD loop folding, respectively, when the structure of the soybean Lb*a* was predicted from the lupin Lb template (Fig. 22.1A); also, major and minor differences were detected at the CD loop and the helix F and GH loop, respectively, when the structure of the soybean Lb*a* was predicted from the rice Hb1 template (Fig. 22.1B). Folding differences between predicted and

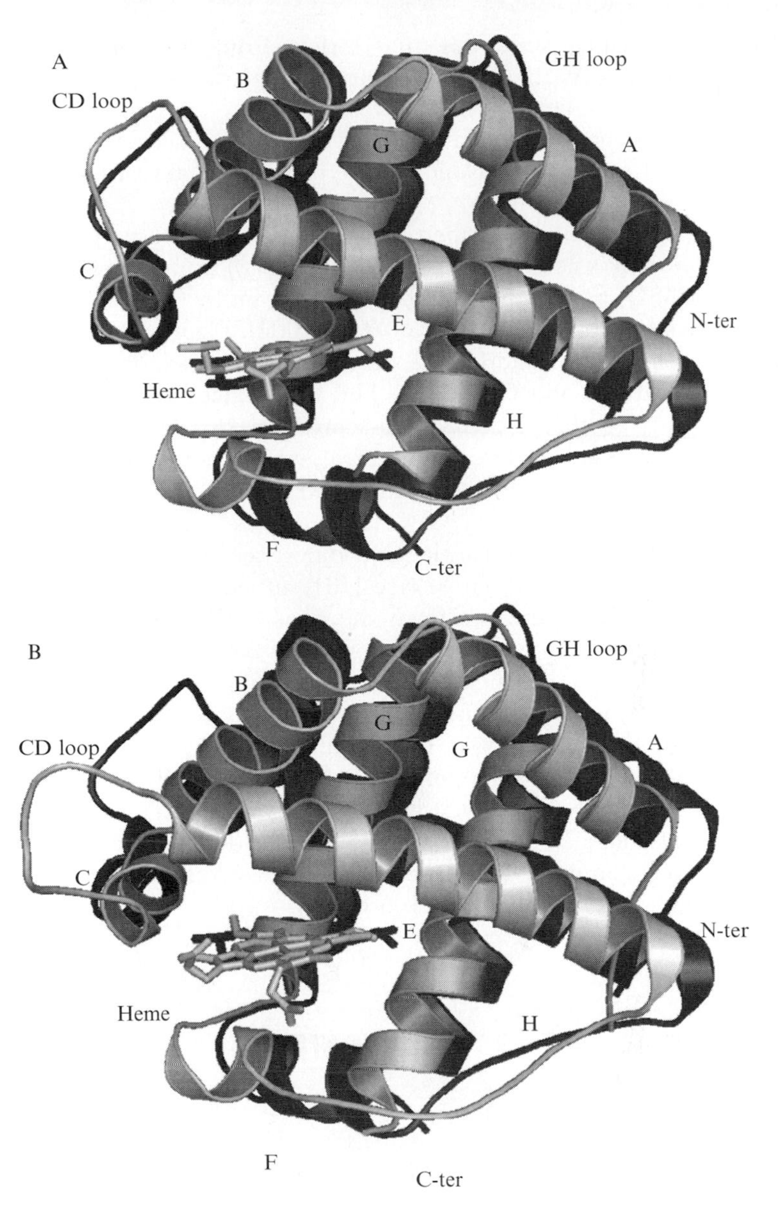

Figure 22.1 Effect of template selection on the prediction of the soybean Lb*a* structure. Soybean Lb*a* structure was predicted from the template lupin Lb (A) and rice Hb1 (B) crystal structures. Predicted and native soybean Lb*a* structures are shown in gray and black, respectively. Helices are indicated with letters A to H.

native soybean Lb*a* structures may result from structural differences between native soybean Lb*a* and lupin Lb and rice Hb1, specifically at the CD loop and helix F. However, with the exception of specific protein regions, predicted folding of soybean Lb*a* is acceptable, and thus the method described is reliable for predicting the tertiary structure of plant Hbs.

We used the method described above to predict the tertiary structure of rice Hb2 (GenBank accession number AAK72228) from template rice Hb1 (PDB ID 1D8U). Rice Hb1 and Hb2 are 97% similar (Fig. 22.2A), and are thus highly homologous (Lira-Ruan *et al.*, 2002). Figure 22.2B shows that the folding of rice Hb1 and Hb2 is almost identical, including the unstable prehelix A and CD and GH loops. The WhatCheck report of predicted rice Hb2 showed that the RMS Z-scores are close to 1 and that the structure Z-scores are positive (Table 22.3), which indicates that the quality of the predicted rice Hb2 is acceptable and that it conforms to common refinement constrains.

In order to identify similarities and differences between the structure of predicted rice Hb2 and native rice Hb1, an analysis of protein regions was performed using routines from the SwissPDBViewer program. For example, we measured distances of amino acid residues to Fe at the heme prosthetic group by using the "distances" option from the toolbar. Results showed that distance and orientation of amino acids that either interact with Fe or affect binding of ligands to Fe are highly similar in rice Hb1 and Hb2 (Fig. 22.3). Several additional analyses of the predicted structure can be performed by using routines available from the SwissPDBViewer and other programs (see Table 22.2); these analyses are not described in this chapter; however, users can easily familiarize themselves with program options by going through the help text of each program.

3. Image Editing to Generate High-Quality Figures

After structure prediction and analysis, high-quality figures should be generated for publication and presentation purposes. Several editing programs are publicly available (Table 22.2). Here we describe the use of Pymol (http://www.pymol.org/funding.html), which was used to generate the figures presented in this contribution (this program also allows users to perform structure analyses); specifically, we describe the editing procedure to generate high-quality figures of predicted structures, including helices and loops and the heme pocket of the predicted rice Hb2 structure.

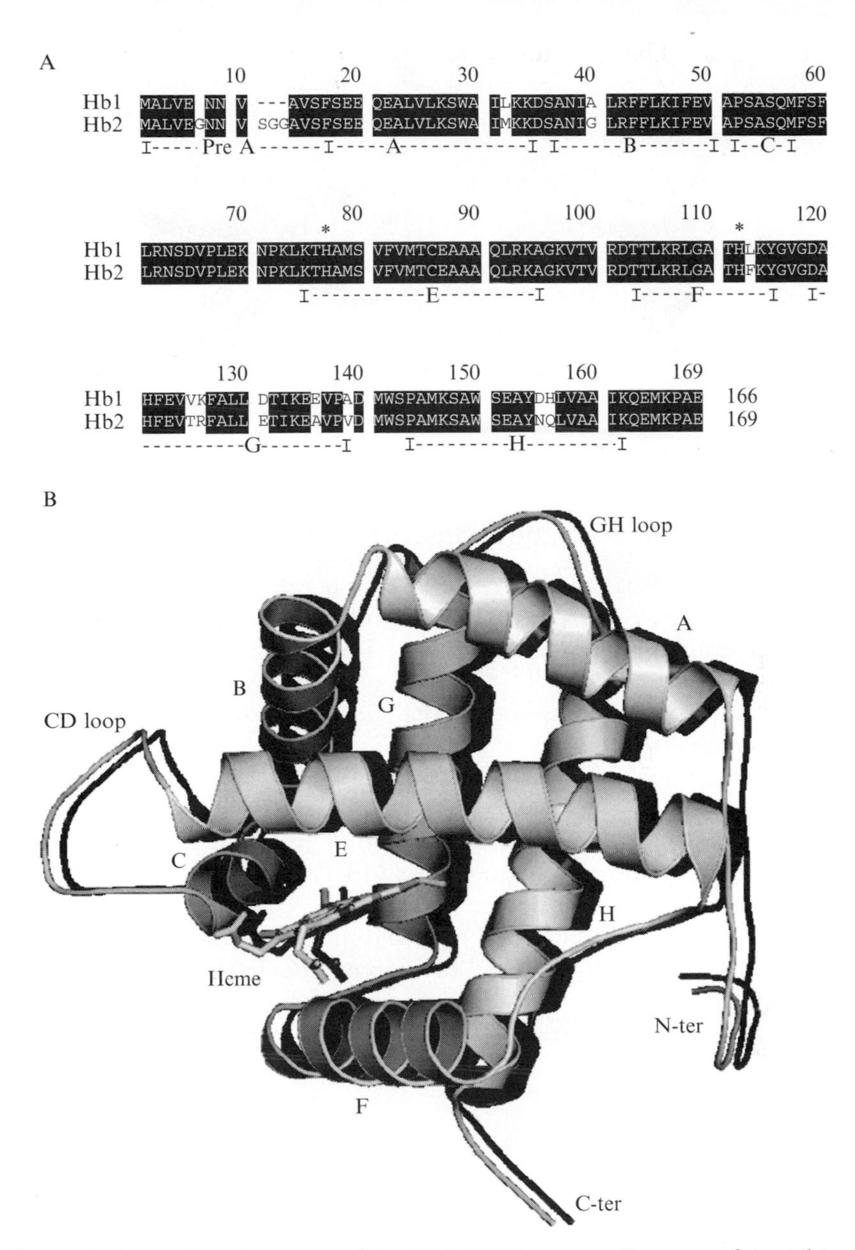

Figure 22.2 Predicted structure of rice Hb2. (A) Sequence alignment of rice Hb1 and Hb2; residues marked with black background are conserved amino acids; asterisks show the position of distal and proximal His (H77 and H112, respectively). (B) Overlapping of the structure of predicted rice Hb2 (gray) and native rice Hb1 (black); helices are indicated with letters A to H.

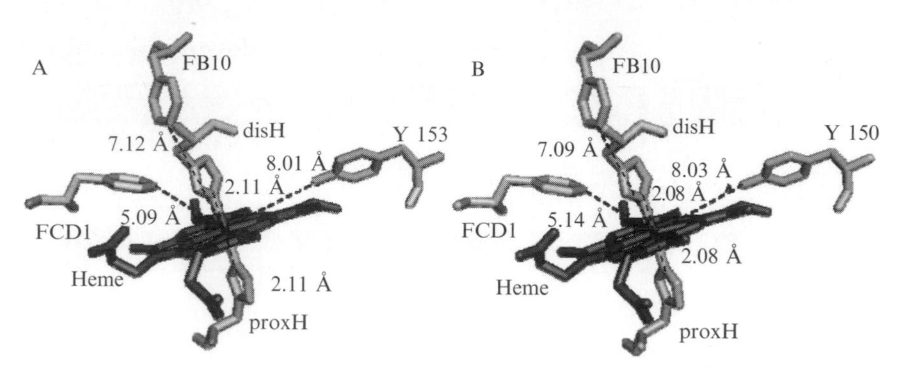

Figure 22.3 Distance of amino acid residues (gray) that either interact with Fe at the heme prosthetic group (black) or affect binding of ligands to Fe in predicted rice Hb2 (A) and native rice Hb1 (B).

3.1. Editing of helices and loops

1. Open the file that contains coordinates for the predicted structure by using the "open" option in the PyMol Tcl/Tk GUI window. The protein structure is displayed (in the wire-frame representation) in the PyMol Viewer.
2. Select the "cartoon" option from the Show menu to display the protein structure in the ribbons and sheets representation, and select the "organic" option to display the heme prosthetic group into the (edited) protein structure.
3. Select the "coloring by structure" option from the Color menu to color ribbons and loops; heme can also be selected and colored from the color menu.
4. Click on the protein structure; hold and move the mouse in any direction to position the structure to the desired orientation.
5. Select the "save image" option from the File menu to save the edited structure as an image file.

3.2. Editing of the amino acid residues at the heme pocket

To display only the amino acid residues that form the heme pocket (or any other protein region of interest), select the full protein structure and then choose "hide everything" from the Hide menu. Next, the amino acid sequence of the predicted structure is displayed by using the "sequence" option from the display menu, and the amino acids that form both the heme pocket and the heme group are selected from the amino acid sequence and displayed using the "show" option. The heme pocket (containing the heme group) can be zoomed in or out by clicking and holding the right button of the mouse and by displacing the mouse upward or downward until the desired size of the heme pocket is obtained.

The highest figure quality for the predicted structure is obtained by selecting the "maximum quality" option from the Display menu. Finally, the image is saved as described previously. Additional editing, such as the labeling of figures, is performed using commercial programs.

ACKNOWLEDGMENTS

The authors wish to express their gratitude to Fernando Violante and Kalyan C. Kondapalli for testing the method described in this contribution. Work in Raúl Arredondo-Peter's laboratory is funded by Mexico's Consejo Nacional de Ciencia y Tecnología (CONACYT Project No. 42873-Q).

REFERENCES

Anfinsen, C. B., Haber, E., Sela, M., and White, F. H. (1961). The kinetics of formation of native ribonuclease during oxidation of the reduced polypeptide chain. *Proc. Natl. Acad. Sci. USA* **47,** 1309–1314.

Appleby, C. A. (1984). Leghemoglobin and *Rhizobium* respiration. *Annu. Rev. Plant Physiol.* **35,** 443–478.

Appleby, C. A. (1992). The origin and functions of haemoglobin in plants. *Sci. Progress* **76,** 365–398.

Arredondo-Peter, R., Moran, J. F., Sarath, G., Luan, P., and Klucas, R. V. (1997a). Molecular cloning of the cowpea (*Vigna unguiculata*) leghemoglobin II gene and expression of its cDNA in *Escherichia coli*: Purification and characterization of the recombinant protein. *Plant Physiol.* **114,** 493–500.

Arredondo-Peter, R., Hargrove, M. S., Sarath, G., Moran, J. F., Lohrman, J., Olson, J. S., and Klucas, R. V. (1997b). Rice hemoglobins: Gene cloning, analysis and oxygen-binding kinetics of a recombinant protein synthesized in *Escherichia coli. Plant Physiol.* **115,** 1259–1266.

Arredondo-Peter, R., Ramírez, M., Sarath, G., and Klucas, R. V. (2000). Sequence analysis of an ancient hemoglobin cDNA isolated from the moss *Physcomitrella patens* (Accession No. AF218049). *Plant Physiol.* **122,** 1457.

Berman, H. M., Westbrook, J., Feng, Z., Gilliland, G., Bhat, T. N., Weissig, H., Shindyalov, I. N., and Bourne, P. E. (2000). The protein data bank. *Nucl. Acids Res.* **28,** 235–242.

Ginalski, K., Grishin, N. V., Godzik, A., and Rychlewski, L. (2005). Practical lessons from protein structure. *Nucl. Acids Res.* **33,** 1874–1891.

Hargrove, M. S., Brucker, E. A., Stec, B., Sarath, G., Arredondo-Peter, R., Klucas, R. V., Olson, J. S., and Philips, G. N., Jr. (2000). Crystal structure of a non-symbiotic hemoglobin. *Structure* **8,** 1005–1014.

Hargrove, M. S., Barry, J. K., Brucker, E. A., Berry, M. B., Phillips, G. N., Olson, J. S., Arredondo-Peter, R., Dean, J. M., Klucas, R. V., and Sarath, G. (1997). Characterization of recombinant soybean leghemoglobin *a* and apolar distal histidine mutants. *J. Mol. Biol.* **267,** 1032–1042.

Hooft, R. W. W., Vriend, G., Sander, C., and Abola, E. E. (1996). Errors in protein structures. *Nature* **381,** 272.

Kretsinger, R. H., Ison, R. E., and Hovmöller, S. (2004). Prediction of protein structure. *Methods. Enzymol.* **383,** 1–27.

Lira-Ruan, V., Ross, E., Sarath, G., Klucas, R. V., and Arredondo-Peter, R. (2002). Mapping and analysis of a hemoglobin gene family from rice (*Oryza sativa*). *Plant Physiol. Biochem.* **40,** 199–202.

Montgomerie, S., Sundararaj, S., Gallin, W. J., and Wishart, D. S. (2006). Improving the accuracy of protein secondary structure prediction using structural alignment. *BMC Bioinformatics* **7,** 301.

Peitsch, M. C. (1996). ProMod and Swiss-Model: Internet-based tools for automated comparative protein modeling. *Biochem. Soc. Trans.* **24,** 274–279.

Petrey, D., and Honig, B. (2005). Protein structure prediction: Inroads to biology. *Mol. Cell* **20,** 811–819.

Ross, E. J. H., Shearman, L., Mathiesen, M., Zhou, J., Arredondo-Peter, R., Sarath, G., and Klucas, R. V. (2001). Non-symbiotic hemoglobins are synthesized during germination and in differentiating cell types. *Protoplasma* **218,** 125–133.

Sáenz-Rivera, J., Sarath, G., and Arredondo-Peter, R. (2004). Modeling the tertiary structure of a maize (*Zea mays* ssp. *mays*) non-symbiotic hemoglobin. *Plant Physiol. Biochem.* **42,** 891–897.

Schwede, T., Kopp, J., Guex, N., and Peitsch, M. C. (2003). SWISS-MODEL: An automated protein homology-modeling server. *Nucl. Acids Res.* **31,** 3381–3385.

Thompson, J. D., Gibson, T. J., Plewniak, F., Jeanmougin, F., and Higgins, D. G. (1997). The clustal X windows interface: Flexible strategies for multiple sequence alignment aided by quality analysis tools. *Nucl. Acids Res.* **24,** 4876–4882.

Tsai, C. S. (2002). "An introduction to computational biochemistry." Wiley-Liss Inc., New York.

vanGunsteren, W. F., Billeter, S. R., Eising, A., Hünenberger, P. H., Krüger, P., Mark, A. E., Scott, W. R. P., and Tironi, I. G. (1996). Biomolecular simulations: The GROMOS96 manual and user guide Zurich: VdF Hochschulverlag ETHZ, Zurich.

Xiang, Z. (2006). Advances in homology protein structure modeling. *Curr. Prot. Pept. Sci.* **7,** 217–227.

CHAPTER TWENTY-THREE

A Self-Induction Method to Produce High Quantities of Recombinant Functional Flavo-Leghemoglobin Reductase

Estibaliz Urarte,* Iñigo Auzmendi,* Selene Rol,* Idoia Ariz,* Pedro Aparicio-Tejo,* Raúl Arredondo-Peter,† *and* Jose F. Moran*

Contents

Abstract

Ferric leghemoglobin reductase (FLbR) is able to reduce ferric leghemoglobin (Lb^{3+}) to ferrous (Lb^{2+}) form. This reaction makes Lb functional in performing its role since only reduced hemoglobins bind O_2. FLbR contains FAD as

* Instituto de Agrobiotecnologia, Universidad Pública de Navarra-CSIC-Gobierno de Navarra, Pamplona, Navarre, Spain
† Laboratorio de Biofísica y Biología Molecular, Facultad de Ciencias, Universidad Autónoma del Estado de Morelos, Cuernavaca, Morelos, México

Methods in Enzymology, Volume 436
ISSN 0076-6879, DOI: 10.1016/S0076-6879(08)36023-6

prosthetic group to perform its activity. FLbR-1 and FLbR-2 were isolated from soybean root nodules and it has been postulated that they reduce Lb^{3+}. The existence of Lb^{2+} is essential for the nitrogen fixation process that occurs in legume nodules; thus, the isolation of FLbR for the study of this enzyme in the nodule physiology is of interest. However, previous methods for the production of recombinant FLbR are inefficient as yields are too low.

We describe the production of a recombinant FLbR-2 from *Escherichia coli* BL21(DE3) by using an overexpression method based on the self-induction of the recombinant *E. coli*. This expression system is four times more efficient than the previous overexpression method. The quality of recombinant FLbR-2 (based on spectroscopy, SDS-PAGE, IEF, and native PAGE) is comparable to that of the previous expression system. Also, FLbR-2 is purified near to homogeneity in only few steps (in a time scale, the full process takes 3 days). The purification method involves affinity chromatography using a Ni-nitrilotriacetic acid column. Resulting rFLbR-2 showed an intense yellow color, and spectral characterization of rFLbR-2 indicated that rFLbR-2 contains flavin. Pure rFLbR-2 was incubated with soybean Lba and NADH, and time drive rates showed that rFLbR-2 efficiently reduces Lb^{3+}.

1. Introduction

Ferric leghemoglobin reductase (FLbR) is an enzyme that catalyzes the reduction of ferric leghemoglobin (Lb) to its functional ferrous form using pyridine nucleotides as reductants. FLbR-1 and FLbR-2 proteins were isolated from soybean nodules, characterized, and showed to be homodimers of near 110 kD, which contain FAD as a prosthetic group and a redox-active disulfide group per subunit (Ji *et al.*, 1991; Moran *et al.*, 2002).

Several families of hemoglobins (Hbs) with differentiated functions are known to coexist in plants (Ross *et al.*, 2002). Among them, leghemoglobins (Lbs) were initially found in legume nodules and later in nonlegume plants. The presence of abundant functional Lb is essential for the N_2 fixation process, since it controls the free O_2-concentration in the nodule and thus avoids oxidative damage to the O_2-labile nitrogenase (for a review, see Appleby, 1984, 1992; Arredondo-Peter *et al.*, 1998). Nonsymbiotic hemoglobins (nsHbs) and truncated hemoglobins (trHbs) were discovered at a later stage in nonsymbiotic tissues of legume and nonlegume plants; all of them show reversible binding of O_2 (Arredondo-Peter *et al.*, 1998; Ross *et al.*, 2002).

To bind O_2, Hbs must be in the ferrous (Hb^{2+}) form. Mechanisms for maintaining Hbs in the reduced form are essential for their function *in vivo*. However, Lbs tend to spontaneously autoxidize to Lb^{3+}, which is favored during stress and senescence conditions in the plant nodule (Jun *et al.*, 1994;

Puppo *et al.*, 1981). This tendency is analogous to that of other Hbs (Misra and Fridovich, 1972).

Studies on the enzymatic mechanisms for Lb reduction were initially carried out with lupine (Kretovich *et al.*, 1982) and soybean (Saari and Klucas, 1984) nodule Lbs. Young, actively N_2-fixing nodules contain only Lb^{2+} or $Lb^{2+}O_2$ (Appleby, 1984; King *et al.*, 1988; Monroe *et al.*, 1989). However, Lb^{3+} is present in senescent sweet clover (*Melilotus officinalis* [L.] Lam.) and soybean nodules (Lee *et al.*, 1995). Using diffuse reflectance and direct transmission spectroscopy, Lee and Klucas (1984) showed that Lb^{3+} generated in soybean (*Glycine max* [L.] Merr.) nodule slices treated with hydroxylamine is rapidly reduced to Lb^{2+}. These observations indicated that mechanisms exist in the nodules to maintain Lb in the functional, reduced state (Becana and Klucas, 1990).

In soybean nodules, three isoenzymes with FLbR activity (Ji *et al.*, 1991) and at least two gene copies (Ji *et al.*, 1994a) have been detected. Soybean *flbr-1* and *flbr-2* genes have been isolated, and the deduced protein sequence showed high homology with dihydrolipoamide dehydrogenase (DLDH) and, to a lesser extent, with glutathione reductase, mercuric reductase, and trypanothione reductase from various organisms (Ji *et al.*, 1994a; Moran *et al.*, 2002; Pullikuth and Gill, 1997). All these enzymes belong to the NAD:disulfide oxidoreductase family and are homodimers that contain FAD and a pair of redox-active Cys residues involved in the electron transfer from NAD(P)H and flavin to the substrates (Williams, 1991). Also in the legume lupine, at least two different kinds of flavoproteins were isolated from nodules, and both groups showed Lb reductase activity (Topunov *et al.*, 2002)

It has also been shown that another flavoprotein from plants, the monodehydroascorbate reductase (MDA-R), is able to mediate in the reduction of nsHbs using ascorbate as a reductant (Igamberdiev *et al.*, 2006). It has been proposed that in nsHbs and trHbs functions regulating NO homeostasis, and thus maintaining the mechanism active, ferric nsHb or trHb must be reduced back to the ferrous form (Dordas *et al.*, 2003; Perazzolli *et al.*, 2004).

The cDNAs coding for FLbRs have been cloned and expressed in recombinant *E. coli* (Ji *et al.*, 1994b; Luan *et al.*, 2000; Moran *et al.*, 2002) using commercial expression vectors, but the yield was low. The main limitation in using the systems in the previously mentioned reports was that most of the recombinant FLbR formed inclusion bodies (Ji *et al.*, 1994b; Moran *et al.*, 2002). Furthermore, long protocols were needed to purify the recombinant FLbR (Ji *et al.*, 1991; Luan *et al.*, 2000). In this work, we describe a new expression and purification method that can produce higher amounts of pure FLbR that is active in Lb reduction. This method could also be used to overexpress other Hbs different from rFLbR2.

2. A Self-Induction System for Overexpression of FLbR2

A recombinant *E. coli* BL21(DE) cell line containing pET28a(+):: FLbR2 vector is able to overexpress the FLbR2 cDNA with $(Hys)_6$ tagged to the N-terminus based on IPTG induction (Moran *et al.*, 2002). In this work, the expression system for rFLbR2 is improved using a new overexpression method (Studier, 2005) that is able to produce high amounts of prosthetic groups containing enzymes. This expression system is called self-inducible, as it does not need IPTG for induction. The expression system is compatible with the typical expression vector, which transcribes the recombinant genes under the control of the T7 polymerase as pET series vector, which uses a *lac*-inducible promotor. It has often been reported that proteins containing prosthetic groups are not easy to synthesize in recombinant expression systems due to the impairment in the synthesis of the protein backbone and the prosthetic group (Arredondo-Peter *et al.*, 1997; Sikorski *et al.*, 1995).

Thus, when using inducible systems based on the induction with IPTG, some authors had employed denaturing buffers to isolate the apo-proteins, and at a later stage, prosthetic group was added (Hargrove *et al.*, 1997). In this method, the overexpression system is based on the self-induction of the recombinant *E. coli*, which improves the overexpression and purification of enzymes containing prosthetic groups.

2.1. Medium growth and incubating conditions of the recombinant cells

Transformed *E. coli* BL21(DE3) cells, containing the pET28a(+)::FLbR-2 construct (Moran *et al.*, 2002) are grown in two 2-l flasks, in a ZYP-5052 media essentially as described by Studier (2005) but with some modifications. Briefly, to obtain ZYP-5052, the following reagents were mixed in the following sequence: 465 ml of ZY reagent, 0.5 ml of 1 M $MgSO_4$, 50 μl of 1000x metals mixture, 10 ml of 50x5052, 25 ml of 20xNPS, and 0.5 ml of kanamycin (100 mg/ml). ZY reagent contained 5 g peptone and 2.5 g yeast extract in 465 ml of deionized distilled water. 50×5052 reagent contained 19.9 ml glycerol ($\approx$25 g), 2.5 g glucose, 10 g a-lactose, and 36.5 ml deionized water. 1000x metals mixture was obtained as described by Studier (2005) except for the addition of Na_2SeO_3, which was omitted. 20xNPS reagent contained 6.6 g $(NH_4)_2SO_4$, 13.6 g KH_2PO_4, 14.2 g Na_2HPO_4, and 90 ml of deionized distilled water. It must be added at the end to avoid precipitate formation. Solutions were autoclaved for 15 min and stored at room temperature. In this method, kanamycin is used at significantly higher concentrations (100 μg/ml) than is normally the case

(25 to 40 μg/ml), and 0.5 ml were included in each flask after mixing all the reagents (Studier, 2005). Finally, inoculation of the recombinant cell line was made by transferring of 50 μl from a glycerol stock into each of the 2-l flasks.

Cultures are incubated at 37° for 20 h with shaking at 200 rpm. Cells are harvested by centrifugation at 4000 × *g* and the cell paste is frozen in liquid N_2 and stored at −80° until used.

2.2. Harvest and breaking of the *E. coli* cells

The cell paste (1.4 g) is thawed on ice and resuspended in 10 ml of the equilibration buffer containing 20 mM sodium phosphate (pH 7.5) and 500 mM sodium chloride; 10 μL DNAse (25 $\mu g\ ml^{-1}$) and 40 mg lysozyme (2 $\mu g\ ml^{-1}$) are added. The suspension is incubated at 4° with gentle agitation for 15 min, and afterward sonicated on ice with three 30-s pulses at 70% power (Soniprep 150 sonicator, Sanyo). After sonication, the lysed cell suspension is cleared by centrifugation at 48,000 × *g* at 4° for 20 min.

2.3. Purification of recombinant FLbR2

The resulting supernatant is chromatographed on a 5-ml Ni-nitrilotriacetic acid resin-based column (His-Trap Chelating, GE Healthcare, Uppsala, Sweden), which selectively binds the poly-His tag-fused rFLbR-2. The column is preequilibrated with 5 column volumes of 20 mM sodium phosphate buffer (pH 7.5) containing 500 mM sodium chloride. Chromatography is performed at a 5 ml/min flow rate. Recombinant FLbR-2 is eluted with the equilibrating buffer that contains 500 mM imidazole. A linear gradient (100 ml) from 50 to 500 mM imidazole is applied and salt concentration is detected by conductance (mS). Total protein is detected at a 280-nm wavelength. The collected yellowish fractions are sterilized through a 0.20 μm filter syringe to avoid any bacterial contamination. Subsequently, the fractions are dialyzed overnight against 50 mM Tris-HCl (pH 7.5), containing 1 mg thrombin (Sigma-Aldrich, St. Louis) at 4° in order to remove the $(His)_6$ tag. Protein purification is assessed by gel electrophoresis and, accordingly, fractions are pooled.

As a second purification step, pooled fractions are loaded onto a 5 ml Hi-Trap QFF Sepharose column (GE Healthcare, Uppsala, Sweden), equilibrated with 50 mM Tris-HCl (pH 7.5). Chromatography is performed at a 5 ml/min flow rate. A 20 column-volumes (100 ml) linear sodium chloride gradient (0 to 500 mM) is applied to the column. Recombinant FLbR-2 elutes at ≈350 mM sodium chloride (see subsequent section). Collected fractions are pooled and dialyzed against 50 mM Tris-HCl and stored at −80°. Protein concentration is determined by a dye-binding assay (Bio-Rad Laboratories, Hercules, CA) using BSA as a standard (Fig. 23.1).

Purified rFLbR-2 is electrophoresed in SDS polyacrylamide gels (10% PAGE, w/v) (Laemmli, 1970) in order to monitor the protein purification. Either 50 μg or 4 μg of protein are loaded into the gels for crude extracts or purified fractions of rFLbR, respectively. Assessment of the purification can be also performed by Western blot analysis using a monoclonal anti-$(Hys)_6$ antibody from Sigma (Sigma-Aldrich, St. Louis, MO) at a 1:2,000 dilution (v/v). As secondary antibody anti-mouse IgG alkaline phosphatase conjugate (Sigma-Aldrich) is used at a 1:20,000 dilution (v/v). Cross-reacting protein bands are visualized using nitro blue tetrazolium and 5-bromo-4-chloro-3-indolyl phosphate (BCIP/NBT, Sigma-Aldrich) (Fig. 23.2).

3. Spectroscopic Methods

3.1. UV-Visible spectra

UV-Vis spectra of the rFLbR in 50 mM phosphate buffer, pH 7.0, containing 2 m*M* EDTA (Fig 23.3) were carried out in a double-beam spectrophotometer (Uvikon XS, Bio-Tek Instruments, Saint Quentin Yvelines, France). The oxidized form of FLbR-2 is generated by the addition of excess of 5 mM NAD^+ to the medium under aerobic conditions, and the reduced form is obtained by the addition of 5 mM NADH. Molar absorptivity is calculated after weighing the dry fraction of the rFLbR-2 to establish the concentration of the enzyme in solution.

3.2. Assays for the catalysis of Lb^{3+} reduction

Soybean Lba reduction is assayed following the decreases of A574 (Ji *et al.*, 1994b; Saari and Klucas, 1984). A dual-beam spectrophotometer is used (Uvikon XS, Bio-Tek Instruments). All measurements are performed in a 1-ml cuvette (1.0 cm, path length) with a final assay buffer containing 50 mM potassium phosphate (pH 6.5), 2 mM EDTA, 50 μM Lba^{3+}, and 10 mM NADH. Pure rFLbR (1.5 to 2 μg) is added as indicated (Fig. 23.4 and Fig. 23.5), and appropriate controls lacking either rFLbR2 enzyme, reductant (NADH), or both are also included (see Figs. 23.4 and 23.5). The Lb reduction assay has been sometimes performed at pH 7.0 (Ji *et al.*, 1991), but pH 6.5 seems to improve the differences in reaction rates (Luan, 2000; Moran *et al.*, 2002). The lower pH is more comparable to the physiological pH of the legume root nodule (Pladys *et al.*, 1988).

4. Results

4.1. Representative purification and protein production yield

Figure 23.1 shows the typical chromatogram obtained for recombinant protein purification on Ni-NTA resin. First, the affinity purification step purifies the protein almost to homogeneity. The use of a second chromatography step after digestion with thrombin is essential to remove thrombin and $(His)_6$-tag from the digestion mix. The overnight purification dialysis saves time because it allows for thrombin digestion and buffer change simultaneously. Using this method, we have been able to obtain quantities of 3 to 4 g of cell paste from a 2-l flask using 500 ml of the ZYP-5052 broth. Purification procedure started from 1.4 g of cell paste and allowed to obtain approximately 4 mg of pure FLbR-2 in two purification steps. The protein can be also lyophilized or frozen in 10% glycerol at −80°, maintaining its activity and structural integrity (not shown). Previous system reported yields of 1.9 mg of rFLbR from 3 g of cell paste (Moran *et al.*, 2002).

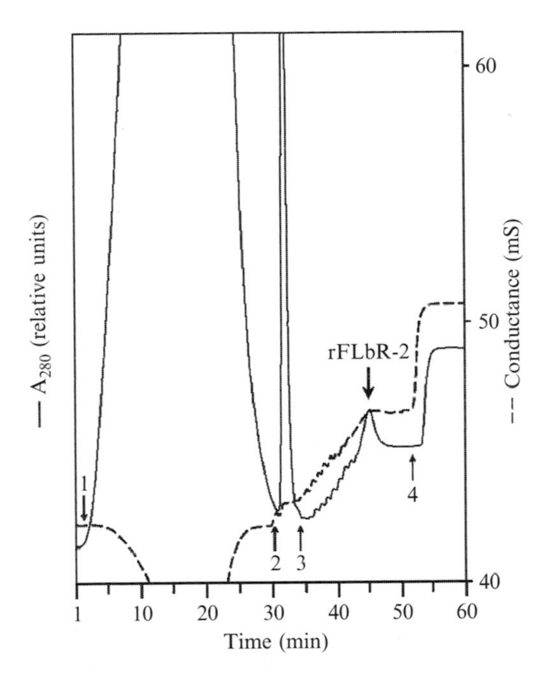

Figure 23.1 Chromatogram of the purification of the rFLbR-2. Arrow 1, cell lysate injection (10 ml); arrow 2, 50 mM imidazole application; arrow 3, gradient start (50 to 500 mM imidazole); arrow 4, column washed with 500 mM imidazole.

4.2. Purification of rFLbR-2: Purification assessment

The purification steps of FLbR-2 are shown in Fig. 23.2. Coomassie-stained gel shows that rFLbR-2 is the main overexpressed fraction in the bacterial extract solution (see Fig. 23.2A and 23.2B, lane 2). However, contrary to what is previously seen for rFLbR-2, the amount of recombinant protein in the pellet fraction of the bacterial extract is low. Pure protein is efficiently purified as seen in Western blot and SDS-PAGE (Fig. 23.2A, lanes 3 and 4). The Western blot confirmed that the use of thrombin during overnight dialysis effectively cut the $(Hys)_6$ tag off the rFLbR-2, and thus is not detected by the monoclonal antibody (see Fig. 23.2B, lane 4). Several replica purifications originated equivalent results, indicating high reproducibility. The quality of recombinant FLbR-2 was comparable to the previous expression system for rFLbr-2 (Moran *et al.*, 2002) and purity was near homogeneity, but higher quantities of FLbR-2 are achieved.

4.3. UV-Vis spectra of the flavoenzyme

This overexpression system is clearly an improved method for producing the flavin prosthetic group of the enzyme and allows for following of the chromatography visually (Fig. 23.3A). Pure rFLbR-2 showed an intense yellow color (see Fig. 23.3A). Spectral characterization of rFLbR-2

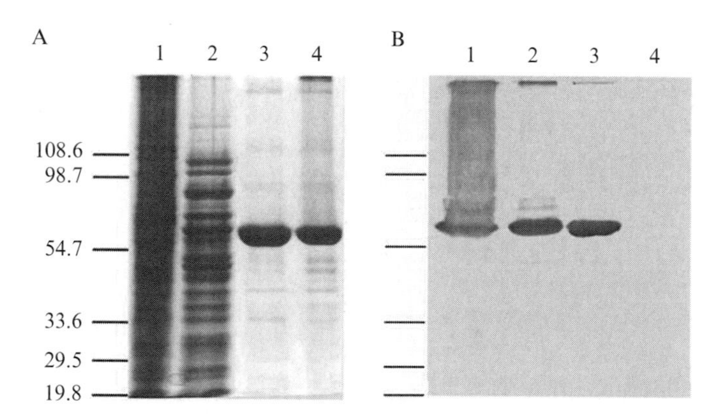

Figure 23.2 Purification of rFLbR-2: (A) Samples were electrophoresed in a 12.5% SDS-PAGE gel and stained with Coomassie blue. (B) Western-blot of the gel using a monoclonal anti-$(Hys)_6$ antibody. Lane 1, resuspension in 2% SDS of proteins from the pellet of the lysate of *E. coli* BL21(DE3) clone containing pETs28a(+)::FLbR-2 plasmid; lane 2, soluble fraction of the above lysate: lane 3, purified rFLbR-2 by Ni-NTA resin; lane 4, rFLbR-2 digested with thrombin and purified by Q-resin as described in the text. Molecular weights in KDa are indicated. Lane 4, rFLbR is not immunoblot detected since $(Hys)_6$ tag has been thrombin remove from the protein. (See color insert.)

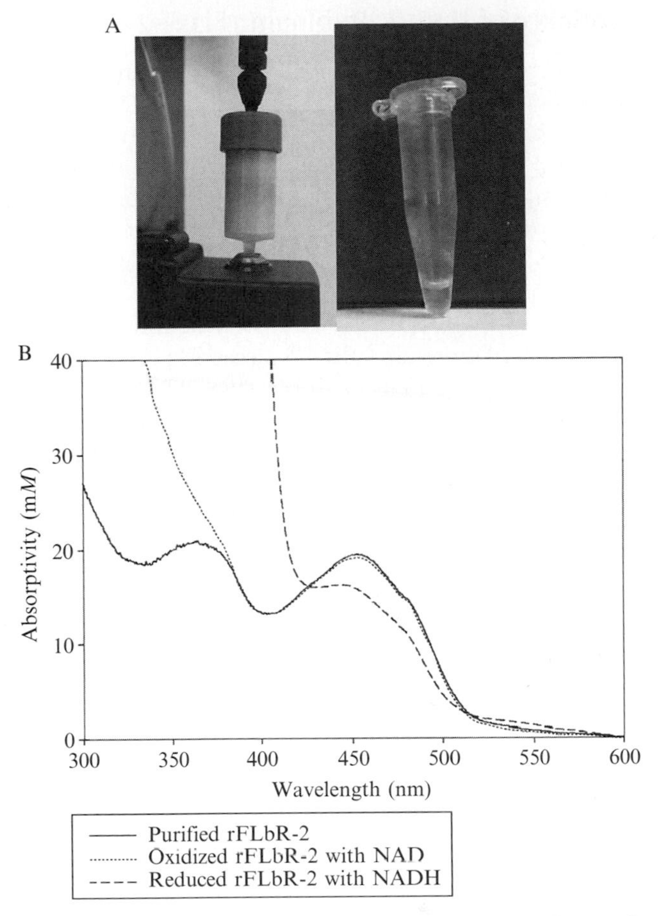

Figure 23.3 (A) Illustration of the yellowish color obtained for the purification of the rFLbR, both during the Ni-NTA chromatography and at one of the collected fractions thereafter. (B) UV-Vis spectra of rFLbR-2. (See color insert.)

confirmed the presence of flavin (Fig. 23.3B). Although native FLbR and recombinant FLbR-1 have been purified in a partially reduced form, rFLbR-2 is purified in oxidized form under this expression system used in this work. The expressed rFLbR-2 can be reduced with NADH (see Fig. 23.3). UV-visible spectra of rFLbR-2 revealed high similarities to those of pig and yeast DLDH and to those of oxidized soybean FLbR-1 (Ji *et al.*, 1994b). The rFLbR-2 shows UV absorption peak at 278 nm and a visible absorption peak at 460 nm, with two shoulders at 435 and 485 nm, consistent with the presence of FAD as a prosthetic group. The molar absorption coefficient was 19 $mM^{-1}cm^{-1}$ at 460 nm (Fig. 23.3).

4.4. Functionality of the recombinant rFLbR-2

Leghemoglobin (Lba^{3+}) can be purified from soybean root nodules (Jun *et al.*, 1994). Purified Lba^{3+} needs to be fully oxidized by using an excess of potassium ferricyanide and finally this oxidizer is removed from solution (Jun *et al.*, 1994). Lb can be stored frozen in this condition. The used reduction assay originates Lb^{2+} from Lb^{3+}, which becomes oxygenated during the reduction process. Thus, the measures were made by following absorbance increase at 574 nm, corresponding to the α peak of the spectrum of the Lb^{2+}-O_2.

Time drive measurements of the Lba^{3+} reduction were taken in both Figs. 23.4 and 23.5. In Figure 23.4 (line A), reaction rate of Lba reduction is compared to controls lacking either FLbR (Fig. 23.4, line B) or reductant (Fig. 23.4, line C). Results clearly indicated that in absence of rFLbR and NADH no reaction occurs. NADH (10 mM) is able to slowly reduce Lbs, but the reaction rate increases more than three times when the rFLbR-2 enzyme is also included into the reaction mix. The reduction of Lb^{3+} is also evident when either rFLbR-2 or rFLbR-2 + NADH were added at 1 min after the start of the reaction (see Fig. 23.5). Again reaction rate (reduction of Lb^{3+} to Lb^{2+}) was faster in the presence of the rFLbR-2 plus NADH. Addition of

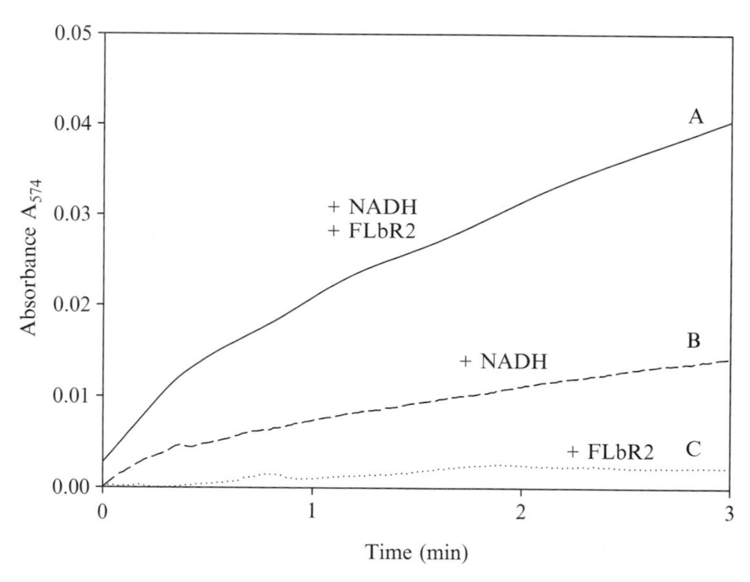

Figure 23.4 Assay for the enzymatic activity of the purified rFLbR-2. Reaction mixtures are described in the text. A, reaction rate for the reaction mix including NADH (10 mM) and FLbR-2. B, reaction rate for the reaction mix including NADH, without rFLbR-2. C, reaction rate for the reaction in the presence of FLbR-2 without NADH.

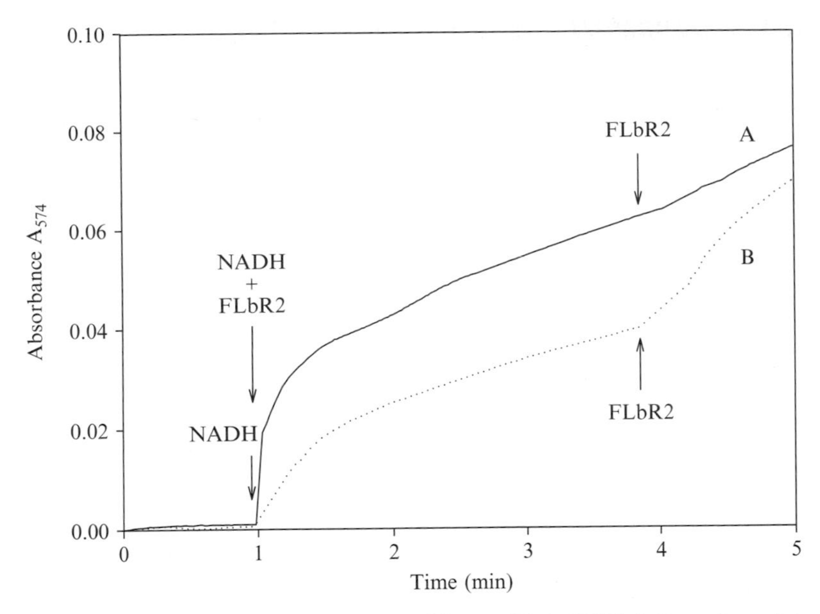

Figure 23.5 Assay for enzymatic activity of the purified rFLbR-2. Reaction mixtures are described in the text. A, at 1 min, we added NADH 10 mM and FLbR2. B, only NADH was added. At 4 min, FLbR2 was added in both reactions.

rFLbR-2 beyond 4 μg originated no change in the reduction of Lb3+, indicating that the catalyst concentration was saturated. However, the reaction rate for the sample containing only the reductant NADH increased with the addition of rFLbR, indicating that this factor was limiting.

5. Conclusions

A new system to overexpress a recombinant FLbR has been employed. This overexpression system makes it possible to obtain a much higher yield than previous overexpression of FLbR-2 (Ji *et al.*, 1991, 1994b; Moran *et al.*, 2002). This method is simpler: it does not require IPTG for induction, the expression system itself produces the induction of the protein, and it is four times more productive than the previous overexpression systems. Furthermore, assays to check for functionality of the hemoglobin reductase activity are also shown in this chapter. Protein originated by this system is functional and might be used for reduction of diverse Hbs.

ACKNOWLEDGMENTS

This work was financed by a grant from the Department of Education, Government of Navarre, Spain, and by CONACYT (project no. 42873-Q), México. E. Urarte and I. Auzmendi thank the Department of Industry of the Government of Navarre for fellowships within the Technics Formation Program. I. Ariz is the recipient of a Ph.D. fellowship from the Public University of Navarre, Spain. J. Moran is contracted within the Ramón y Cajal program, MEC, Spain.

REFERENCES

Appleby, C. A. (1984). Leghemoglobin and *Rhizobium* respiration. *Annu. Rev. Plant Physiol.* **35,** 443–478.

Appleby, C. A. (1992). The origin and functions of haemoglobin in plants. *Sci. Prog.* **76,** 365–398.

Arredondo-Peter, R., Hargrove, M., Moran, J. F., Sarath, G., and Klucas, R. V. (1998). Plant hemoglobins. *Plant Physiol.* **118,** 1121–1125.

Arredondo-Peter, R., Moran, J. F., Sarath, G., Luan, P., and Klucas, R. V. (1997). Molecular cloning of the cowpea leghemoglobin II gene and expression of its cDNA in *Escherichia coli*: Purification and characterization of the recombinant protein. *Plant Physiol.* **114,** 493–500.

Becana, M., and Klucas, R. V. (1990). Enzymatic and non-enzymatic mechanisms for ferric leghemoglobin reduction in legume root nodules. *Proc. Natl. Acad. Sci. USA* **87,** 7295–7299.

Dordas, C., Rivoa, J., and Hill, R. D. (2003). Plant haemoglobins, nitric oxide and hypoxic stress. *Ann. Bot.* **91,** 173–178.

Hargrove, M. S., Barry, J. K., Brucker, E. A., Berry, M. B., Phillips, G. N., Jr., Olson, J. S., Arredondo-Peter, R., Dean, J. M., Klucas, R. V., and Sarath, G. (1997). Characterization of recombinant soybean leghemoglobin-a and apolar distal histidine mutants. *J. Mol. Biol.* **266,** 1032–1042.

Igamberdiev, U., Bykova, N. V., and Hill, R. D. (2006). Nitric oxide scavenging by barley haemoglobin is facilitated by a monodehydroascorbate reductase-mediated ascorbate reduction of methemoglobin. *Planta* **223,** 1033–1040.

Ji, L., Becana, M., Sarath, G., and Klucas, R. V. (1994a). Cloning and sequence analysis of a cDNA encoding ferric leghemoglobin reductase from soybean nodules. *Plant Physiol.* **104,** 453–459.

Ji, L., Becana, M., Sarath, G., Shearman, L., and Klucas, R. V. (1994b). Overproduction in *Escherichia coli* and characterization of a soybean ferric leghemoglobin reductase. *Plant Physiol.* **106,** 203–209.

Ji, L., Wood, S., Becana, M., and Klucas, R. V. (1991). Purification and characterization of soybean root nodule ferric leghemoglobin reductase. *Plant Physiol.* **96,** 32–37.

Jun, H. K., Sarath, G., Moran, J. F., Becana, M., Klucas, R. V., and Wagner, F. (1994). Characteristics of modified Leghemoglobins isolated from soybean (*Glycine max* Merr.) root nodules. *Plant Physiol.* **104,** 1231–1236.

King, B. J., Hunt, S., Weagle, G. E., Walsh, K. B., Pottier, R. H., Canvin, D. T., and Layzell, D. B. (1988). Regulation of O_2 concentration in soybean nodules observed by *in situ* spectroscopic measurement of leghemoglobin oxygenation. *Plant Physiol.* **87,** 296–299.

Kretovich, W. L., Melik-Sarkissyan, S. S., Bashirova, N. F., and Topunov, A. F. (1982). Enzymatic reduction of leghemoglobin in lupin nodules. *J. Appl. Biochem.* **4,** 209–217.

Laemmli, U. K. (1970). Cleavage of structural proteins during the assembly of the head of bacteriophage T4. *Nature* **227,** 680–685.

Lee, K. K., and Klucas, R. V. (1984). Reduction of ferric leghemoglobin in soybean root nodules. *Plant Physiol.* **74,** 984–988.

Lee, K. K., Shearman, L. L., Erickson, B. K., and Klucas, R. V. (1995). Ferric leghemoglobin in plant-attached leguminous nodules. *Plant Physiol.* **109,** 261–267.

Luan, P., Aréchaga-Ocampo, E., Sarath, G., Arredondo-Peter, R., and Klucas, R. V. (2000). Analysis of a ferric leghemoglobin reductase from cowpea (*Vigna unguiculata*) root nodules. *Plant Sci.* **154,** 161–170.

Misra, H. P., and Fridovich, I. (1972). The generation of superoxide radical during the autoxidation of hemoglobin. *J. Biol. Chem.* **247,** 6960–6962.

Monroe, J. D., Owens, T. G., and LaRue, T. A. (1989). Measurement of the fractional oxygenation of leghemoglobin in intact detached pea nodules by reflectance spectroscopy. *Plant Physiol.* **91,** 598–602.

Moran, J. F., Sun, Z., Sarath, G., Arredondo-Peter, R., James, E. K., Becana, M., and Klucas, R. V. (2002). Molecular cloning, functional characterization, and subcellular localization of soybean nodule dihydrolipoamide reductase. *Plant Physiol.* **128,** 300–313.

Perazzolli, M., Dominici, P., Romero-Puertas, M. C., Zago, E., Zeier, J., Sonoda, M., Lamb, C., and Delledonne, M. (2004). *Arabidopsis* nonsymbiotic hemoglobin AHb1 modulates nitric oxide bioactivity. *Plant Cell.* **16,** 2785–2794.

Pladys, D., Barthe, P., and Rigaud, J. (1988). Changes in intracellular pH in French-bean nodules induced by senescence and nitrate treatment. *Plant Sci.* **56,** 99–106.

Pullikuth, A. K., and Gill, S. S. (1997). Primary structure of an invertebrate dihydrolipoamide dehydrogenase with phylogenetic relationship to vertebrate and bacterial disulfide oxidoreductases. *Gene* **200,** 163–172.

Puppo, A., Rigaud, J., and Job, D. (1981). Role of superoxide anion in leghemoglobin autoxidation. *Plant Sci. Lett.* **22,** 353–360.

Ross, E. J. H., Lira-Ruan, V., Arredondo-Peter, R., Klucas, R. V., and Sarath, G. (2002). Recent insights into plant hemoglobins. *Rev. Plant Biochem. Biotechnol.* **1,** 173–189.

Saari, L. L., and Klucas, R. V. (1984). Ferric leghemoglobin reductase from soybean root nodules. *Arch. Biochem. Biophys.* **231,** 102–113.

Sikorski, M. M., Topunov, A. F., Strozycki, P. M., Vorgias, C. E., Wilson, K. E., and Legocki, A. B. (1995). Cloning and expression of plant leghemoglobin cDNA of *Lupinus luteus* in *Escherichia coli* and purification of the recombinant protein. *Plant Sci.* **108,** 109–117.

Studier, F. W. (2005). Protein production by auto-induction in high-density shaking cultures. *Protein Expr. Pur.* **41,** 207–234.

Topunov, A. F., Shleev, S., Petrova, N., Rozov, F., Zhabaevva, M., and Bach, A. N. (2002). A self-induction method to produce high quantities of recombinant functional ferric-leghemoglobin reductase. 5th European Nitrogen Fixation Congress. Norwich, UK.

Williams, C. H., Jr. (1991). Lipoamide dehydrogenase, glutathione reductase, thioredoxin reductase and mercuric ion reductase: Family of flavoenzyme transhydrogenases. *In* "Chemistry and biochemistry of flavoenzymes," Vol. 3 (F. Muller, ed.), 121–212. CRC Press, Boca Raton, FL.

CHAPTER TWENTY-FOUR

Spectroscopic and Crystallographic Characterization of bis-Histidyl Adducts in Tetrameric Hemoglobins

Alessandro Vergara,*,† Luigi Vitagliano,† Cinzia Verde,‡
Guido di Prisco,‡ *and* Lelio Mazzarella*,†

Contents

Abstract

Hemoglobins (Hbs) are important proteins devoted to oxygen transport. Hbs carry out their function by keeping the iron atom, which binds the oxygen molecule, in its reduced Fe(II) state. Nonetheless, it is well known that Hbs frequently undergo, even under physiological conditions, spontaneous oxidation. Although these processes have been widely investigated, their role and impact in different biological contexts are still highly debated. In vertebrate Hbs, assembled in $\alpha_2\beta_2$ tetramers, it has traditionally been assumed that

* Department of Chemistry and Consorzio Bioteknet, University of Naples Federico II, Naples, Italy
† Institute of Biostructures and Bioimaging, CNR, Naples, Italy
‡ Institute of Protein Biochemistry, CNR, Naples, Italy

Methods in Enzymology, Volume 436
ISSN 0076-6879, DOI: 10.1016/S0076-6879(08)36024-8

oxidized forms endowed with nativelike structures are either aquo-met or hydroxy-met states, depending on the pH of the medium. This view has been questioned by several independent investigations. In the past, indirect evidence of the existence of alternative nativelike oxidized forms was obtained from spectroscopic analyses. Indeed, it was suggested that, in tetrameric Hbs, bis-histidyl hemichrome states could be compatible with folded structures. Recent studies performed by complementing spectroscopic and crystallographic methodologies have provided a detailed picture of hemichrome structure and formation in these proteins. Here we review the methodological approaches adopted to achieve these results, the main structural features of these states, and the current hypotheses on their possible functional implications.

1. Introduction

The autoxidation process in Hbs isolated from different organisms shows similarities and differences. The product of heme oxidation is dependent on many factors (e.g., globin matrix, pH, temperature, oxygen pressure, denaturant agents). In the Hb superfamily, the most common ferric form, as a product of autoxidation, is the exogenous hexacoordinated (6C) aquo-met in equilibrium with the hydroxy-met form, whose pKa depends on the globin matrix. However, a variety of coordination states have been detected and characterized. In fact, in addition to non-aquo exogenous coordination (e.g., F^-, CN^-, N_3^-), both pentacoordination (5C) and endogenous hexacoordination (hemichromes) states have been observed. Within the large Hb superfamily, the occurrence and relevance of these products is variable.

The heme 5C coordination is relatively rare in Hb. This state has been initially reported for monomeric (Ilari *et al.*, 2002) and dimeric (Boffi *et al.*, 1994) Hbs, and more recently for Hbs with higher structural complexity, such as tetrameric Hbs from polar fish (Giordano *et al.*, 2007; Vergara *et al.*, 2007) and giant Hbs (Marmo Moreira *et al.*, 2006). The occurrence of endogenous hexacoordination is more widespread. The specific endogenous coordination under native conditions depends on the distal residue, most frequently His (bis-His adduct, or His-hemichrome). More rarely, in Hbs with lower structural complexity (Milani *et al.*, 2005) or in mutants of human Hb (HbA) (Nagai *et al.*, 2000), Tyr has been found to act as the sixth ligand at the iron site. Because in tetrameric wild-type Hbs the proximal and distal residues are both His, in this chapter the only hemichrome states described are those formed by His residues.

In initial analyses of hemichromes in tetrameric Hbs, the states were strictly associated with denaturation. More recent investigations have indicated that a hemichrome state may be considered a substate of the nativelike

Hb population. These states can be distinguished from those associated with denatured conditions by reversibility and the absence of precipitation and heme release (Rifkind *et al.*, 1994). A detailed description of the hemichrome formation and denaturation, based on HbA experiments, has been previously reported (Rifkind *et al.*, 1994).

More recently, two new areas have emerged in this field: (1) new hypotheses on the physiological role of hemichrome and (2) the elucidation of Hb three-dimensional structures containing a bis-His adduct.

The ferric forms of Hb are physiologically inert to further oxygenation, but several subsequent side reactions in the Hb autoxidation may interfere with or merge into other biochemical pathways, including the formation of a hemichrome. The relevance of this species spans from biomedical to physiological aspects. For example, autoxidation is a serious problem that limits the storage time of acellular Hb-based blood substitutes (Ray *et al.*, 2002); also, hemichrome detection has been suggested as a valuable tool for tumor diagnosis (Croci *et al.*, 2001). The reaction of acetylphenylhydrazine (APH) with erythrocytes leads to hemichrome formation in healthy people and not in breast-cancer patients (Croci *et al.*, 2001). Some of these hypotheses were reviewed in Rifkind *et al.* (1994). The bis-His complex can be involved in ligand binding (de Sanctis *et al.*, 2004a; Pesce *et al.*, 2004), in the *in vivo* reduction of met-Hb, in Heinz body formation (Rifkind *et al.*, 1994), and in nitric oxide scavenging.

It has also been suggested that hemichrome can be involved in Hb protection from peroxidation attack (Feng *et al.*, 2005). The hemichrome species of isolated human α subunits complexed with the α-helix stabilizing protein (AHSP) does not exhibit peroxidase activity (Feng *et al.*, 2005).

2. Antarctic Fish Hemoglobins

Mammalian Hbs show a low level of hemichrome in physiological conditions at room temperature. In contrast, Antarctic fish Hbs are easily oxidized to hemichrome (Riccio *et al.*, 2002; Vergara *et al.*, 2007; Vitagliano *et al.*, 2004). Compared to temperate and tropical fish, most species of the dominant Antarctic suborder Notothenioidei have evolved reduced erythrocyte numbers and Hb concentration and multiplicity (Everson and Ralph, 1968; Wells *et al.*, 1980). The family Channichthyidae (the notothenioid crown group) is devoid of Hb (Ruud, 1954).

Although a report in the literature (Sidell and O'Brien, 2006) challenges the ensuing hypothesis, we believe that the adaptive reduction in Hb content/multiplicity and erythrocyte numbers in the blood of Antarctic notothenioids counterbalances the increase in blood viscosity produced by subzero seawater temperature (Wells *et al.*, 1990) with potentially negative physiological effects (i.e., higher demand of energy needed for circulation). The oxygen affinity of the Hbs of many Antarctic species (which controls binding of oxygen at the exchange surface and release to the tissues) is quite low (di Prisco *et al.*, 1988; Verde *et al.*, 2006), as indicated by the values of p_{50} (the oxygen partial pressure required to achieve half saturation). This feature is probably linked to the high oxygen concentration in the cold Antarctic waters. The decreased oxygen affinity of Hb at the lowest pH values of the physiological range is known as the alkaline Bohr effect, reviewed in Riggs (1988). In many teleost fish, when the pH is lowered, the oxygen affinity of Hb decreases to such an extent that it cannot be fully saturated even at very high oxygen pressure. At low pH, cooperativity is totally lost and the oxygen capacity of blood undergoes reduction of 50% or more of the value measured at alkaline pH. This feature is known as the Root effect, reviewed in Brittain (2005). Most Antarctic fish Hbs display a strong Root effect (Verde *et al.*, 2007). Current work in our laboratory, based on a combined electron paramagnetic resonance (EPR)–crystallography approach, suggests a correlation between Root effect and hemichrome stability in Antarctic fish Hbs.

To date, crystallographic structures of cold-adapted Hbs in canonical R and T states (Camardella *et al.*, 1992; Ito *et al.*, 1995; Mazzarella *et al.*, 1999, 2006a, 2006b) have highlighted the classical conformation and high similarity with their mesophilic homologues. Cold adaptation of the oxygen-transport system in Antarctic fish seems to be based on evolutionary changes involving levels of biological organization higher than the structure of Hb. These include changes in the rate of Hb synthesis or in regulation by allosteric effectors, which affect the amount of oxygen transported in blood. These factors are currently believed to be more important for short-term response to environmental challenges than was previously believed.

3. Overview of the Combined Spectroscopic/Crystallographic Approach to the Characterization of Hemichrome in Tetrameric Hbs

This chapter will focus on the methods to detect His-hemichrome in tetrameric Hbs by the combined spectroscopic/crystallographic approach, and provides data and information useful for its optimal use.

In monomeric or homo-oligomeric Hbs, spectroscopic analysis clearly reveals heme states. However, in tetrameric $\alpha_2\beta_2$ heterodimers, the distribution of ferric-heme states among molecules and between α and β subunits cannot be unambiguously assigned by spectroscopic analysis (unless perturbative hybrid synthesis is conducted); thus, the three-dimensional structure is required.

The history of the three-dimensional structure of hemichromes in members of the Hb superfamily is quite recent; the first crystal structure of a monomeric Hb dates to 1995, by Mitchell *et al.* (1995a). Only in 2002 was a structure of a tetrameric Hb in a hemichrome state reported (Riccio *et al.*, 2002). Crystallographic studies on tetrameric Hbs (Riccio *et al.*, 2002; Robinson *et al.*, 2003; Vergara *et al.*, 2007; Vitagliano *et al.*, 2004) have provided a significant contribution (1) by showing that these Hbs tend to form partial hemichrome states (in α or β) and (2) by providing a clear assignment of the binding states of the individual chains within the tetramer. The correspondence of the protein features detected in solution and in the crystal state has been supported by Raman experiments performed in both states (unpublished results).

The combination of high-resolution crystal structure and spectroscopic techniques finely tuned by heme structural modifications (EPR and Raman spectroscopy) provides new tools for identifying stereochemical parameters of hemichrome coordination. However, details of the quaternary structure and its compatibility with endogenous coordination are as yet a unique field of three-dimensional structural characterization. Methodological details of this investigation are reported in the following sections.

4. Hemichrome Formation and Detection

4.1. Hemichrome occurrence in Hbs

The occurrence of hemichrome is frequent in nonvertebrate (Weber and Vinogradov, 2001), monomeric, and dimeric (Milani *et al.*, 2005) Hbs. Also, mammalian (human and mouse) monomeric neuroglobin and dimeric human cytoglobin (de Sanctis *et al.*, 2004a), and the complex human α-subunit AHSP (Feng *et al.*, 2005) have been described as forming bis-His adducts.

In mammals, tetrameric Hbs forms hemichromes in a number of chemical conditions, classified by Rifkind *et al.* (1994) as nondisruptive and disruptive of the globin structure. Irreversible hemichrome formation is facilitated by subunit dissociation, Hb instability (e.g., in several abnormal HbA mutants reviewed in Rifkind *et al.* [1994] and Jeong *et al.* [1999]), addition of denaturants (e.g., dodecyl trimethylammonium bromide [Ajloo *et al.*, 2002], APH [Croci *et al.*, 2001]), and dehydration (Rifkind *et al.*, 1994). Reversible hemichrome of HbA is obtained at low temperature (Rifkind *et al.*, 1994; Svistunenko *et al.*, 2000), at high oxygen pressure (Rifkind *et al.*, 1994), or by oxygenation of NO adducts (Arnold *et al.*, 1999). There is a strong, nonsystematic pH dependence of hemichrome stability in monomeric Hb (Couture *et al.*, 1999), dimeric Hb (Boffi *et al.*, 1994), human HbA (Svistunenko *et al.*, 2000), and Antarctic fish Hbs (unpublished results).

Promising perspectives are linked to the ability to trap hemichromes of HbA and myoglobin (Mb) via thermal generation (15 min at 75°) of trialose glass-encapsuled samples (Ray *et al.*, 2002). It is interesting that Mb thin films provide hemichromes with efficient electron-transfer properties (Feng and Tachikawa, 2001). As an alternative encapsulation mechanism, hemichrome formation in the α subunits of horse aquo-met-Hb crystals is reported following a drop in pH to 5.4 (Robinson *et al.*, 2003; see the section "Crystallization" herein).

4.2. The exceptional behavior of Antarctic fish Hbs

Within the realm of tetrameric Hbs, Antarctic fish Hbs are exceptions. Under native conditions, they quantitatively form hemichrome in the β subunits (Riccio *et al.*, 2002; Vitagliano *et al.*, 2004). However, the bis-His coordination in the ferrous state, called hemochrome (Rifkind *et al.*, 1994), has never been observed. Thus, under reduction, the hemichrome species is reversibly converted to the classical pentacoordinated deoxy form (Vitagliano *et al.*, 2004). In contrast, other monomeric or dimeric Hbs display bis-His coordination at the iron atom in both the ferric state and the ferrous state (Couture *et al.*, 1999; Das *et al.*, 1999; Sawai *et al.*, 2005).

As yet, there is no a clear understanding of the molecular forces that prevent hemochrome formation in the Antarctic fish Hbs. Approaches based on quantum mechanics that take into account the flexibility of the protein matrix have been developed (Amadei *et al.*, 2005; Bikiel *et al.*, 2006) for heme-containing models. These methodologies could be applied to investigate bis-His coordination.

There are two protocols for obtaining the hemichrome in *Trematomus bernacchii* Hb (HbTb), both at neutral pH and at room temperature. One is based on exposure of the CO form to air for a few days, and the second on the ready oxidation by potassium ferricyanide of the CO form. The spectral properties and the crystal structures of these two forms of HbTb are the same (Vitagliano *et al.*, 2004). However, extensive analysis of the major Hb of the Antarctic fish *Trematomus newnesi* (Hb1Tn) at different times of air exposure reveals that the β-heme coordination and geometry undergo extensive changes along air exposure from 3 h to a week or months (Riccio *et al.*, 2002).

When Antarctic fish Hbs are oxidized by ferricyanide, a large variety of ferric species (observed by EPR) are formed (Vergara *et al.*, 2007), namely aquo-met, hydroxy-met, and two distinct hemichromes (see the section "Atomic resolution structure of HbTb in a partial hemichrome state" herein). As exceptions, the cathodic component of *T. newnesi* (HbCTn)

(Vergara *et al.*, 2007) and Arctic Hb from *Liparis tunicatus* (unpublished) revealed a 5C ferric form even at physiological pH, possibly related to the substitution Val E11(β67) → Ile. Preliminary spectroscopic studies on this Hb reveal an unligated ferric state (unpublished). The prominent high-spin signal corresponds to that of an aquo-met form. Probably, the replacement of Val E11(67β) with bulkier Ile in *Liparis tunicatus* Hb generates unfavorable interactions in the heme pocket, and thus negatively affects the bis-His adduct formation in the β chains (Vergara *et al.*, 2007).

4.3. Spectroscopic markers

An optical spectrum of a hemichrome complex shows a peak at 535 and a shoulder at 565 nm, not superimposing the peaks of the aquo-met complex at around 500 and 631 nm. It is worth mentioning that hydroxy-met displays peaks at around 541 and 576 nm (Feis *et al.*, 1994).

Few Raman spectroscopy analyses are available for tetrameric Hbs in hemichrome state (Nagai *et al.*, 2000; Verde *et al.*, 2004), but there is abundant literature on cytochromes (Smulevich *et al.*, 2005), truncated Hbs (Egawa and Yeh, 2004) and dimeric Hbs (Sawai *et al.*, 2005). The Raman markers of hemichrome are the [ν_3] 1508, [ν_2] 1578, [ν_{37}] 1604, and [ν_{10}] 1638 cm^{-1} bands, which differ from those of aquo-met ([ν_3] 1478, [ν_{38}]1512, [ν_2] 1560, and [ν_{37}] 1578 cm^{-1}) (Verde *et al.*, 2004).

Raman microscopy has been fulfilling a role in characterizing single protein crystals; it is a valuable tool for secondary-structure analysis and for following binding processes (Carey, 2006). In combination with X-ray diffraction and spectroscopic analysis, Raman microscopy is also valuable for bridging results from the solution to the crystal phase (Smulevich *et al.*, 1990), which solves possible conflicts (Smulevich, 1998) between spectroscopic and crystallographic evidence. Studies on peroxidase are an example of how the combination of spectroscopic and crystallographic characterizations may strengthen conclusions (Marzocchi and Smulevich, 2003; Smulevich, 1998).

As supporting evidence, we report the comparison between the solution and single-crystal Raman spectra of the ferric form of Hb1Tn (Fig. 24.1). The similarity of the two spectra extends to every band of the spectrum, except for the scattering assigned to the vinyl groups, which gives rise to two bands ($\nu_{C=C}$) in the solution state at 1620 and 1629 cm^{-1}, respectively, and to only one in the crystal phase at 1620 cm^{-1} (unpublished). Extensive correlation between $\nu_{C=C}$ and orientation of the vinyl group in respect to the heme plane is available (Marzocchi and Smulevich, 2003).

EPR spectroscopy is a long-standing approach to the analysis of hemichromes (Blumberg and Peisach, 1972). The low-spin (S = 1/2) ferric hemichrome is EPR active and exhibits signals that differ from those of

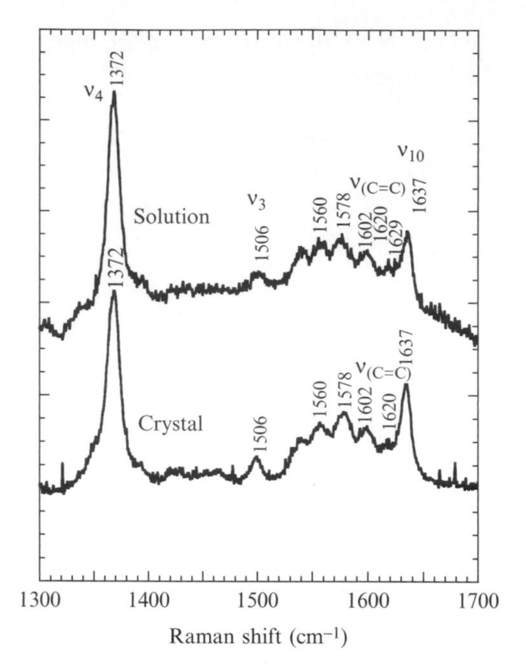

Figure 24.1 Raman spectra (excitation line 488 nm, laser power at the sample 3 mW) of ferric Hb1Tn at pH 7.6. in a single crystal and in solution (unpublished).

high-spin 5C and aquo-met forms and from those of low-spin hydroxy-met complexes. The theory of the EPR band shape for low-spin ferric forms has been reported (Palmer, 1985). Depending on the charge state of His side chains and the geometry of bis-His coordination, the g anisotropy of hemichrome signals may vary (Walker, 2004; Walker *et al.*, 1984). According to the Blumberg and Peisach (1972) classification of the g values in different groups, hemichromes belong to class B, with g_{max} from 3.0 to 3.4, and g_{mid} from 2.1 to 2.3. Pulsed techniques have been applied to investigate hemichromes from members of the Hb superfamily (Ioanitescu *et al.*, 2005), providing another valuable tool to obtain stereochemical parameters of bis-His coordination.

EPR data have been compared to the structural data of many model compounds (Walker, 2004; Walker *et al.*, 1984). However, the increasing number of crystal structures of Hbs in hemichrome state has allowed for correlating g anisotropy and heme stereochemistry (see the section "Atomic resolution structure of HbTb in a partial hemichrome state" herein). Reversible hemichrome formation (Vitagliano *et al.*, 2004) and reduction (Weiland *et al.*, 2004) have also been investigated. It is interesting that in monomeric Hbs the reduction kinetics of hemichromes is faster than that of pentacoordinated unligated states.

In HbA, the α chains are oxidized faster than the β chains (Tsuruga *et al.*, 1998) and the hemichrome formation passes through the aquo-met form (Rachmilewitz *et al.*, 1971).

Hemichrome formation of Antarctic fish HbTb has been studied by optical spectroscopy (Vitagliano *et al.*, 2004), as shown in Fig. 24.2. In the starting CO form, the Soret band occurs at 418 nm and the Q bands at 538 and 567 nm. Upon air exposure, the spectrum begins to change. First, the Soret band broadens and blue-shifts to 414 nm, and the Q band red-shifts to 575 nm, indicating formation of an oxy form. Subsequently, a weak shoulder at 630 nm (corresponding to the charge transfer band of a hexacoordinated high-spin aquo-met form) becomes evident; the Soret band downshifts further to 407 nm, and the Q bands broaden, shifting to 530 and 565 nm.

Preliminary results of hemichrome formation in Hb1Tn have also been obtained by Raman spectroscopy in solution and in the crystal phase. The ν_4 band, very sensitive to the oxidation state, reveals an intermediate species, intermediate between the initial ferrous (1353 cm^{-1}) and the final aquo-met/hemichrome form (1372 cm^{-1}). Hemichrome formation, studied by X-ray crystallography, identifies this intermediate as a ferric unligated 5C species (unpublished).

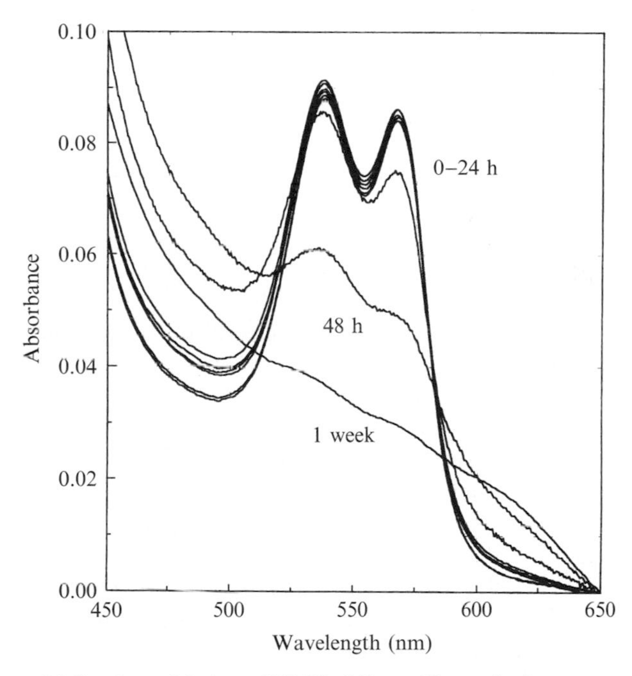

Figure 24.2 Autoxidation of HbTb, followed by optical spectroscopy.

5. Structural Characterization of Hemichrome in Tetrameric Hbs

The first crystallographic characterization of tetrameric Hb in hemichrome state was reported for Hb1Tn (Riccio *et al.*, 2002). Subsequent studies described hemichrome structures in horse Hb (Robinson *et al.*, 2003) and in HbTb (Vergara *et al.*, 2007; Vitagliano *et al.*, 2004). The hemichromes of Hb1Tn, HbTb, and horse Hb were obtained and crystallized under remarkably different conditions.

5.1. Crystallization

Crystallization of Hb1Tn hemichrome was undertaken on the basis of UV-visible spectra in solution (Riccio *et al.*, 2001), which suggested the presence of a significantly populated hemichrome species along the oxidation pathway of the protein. Because at that time there was no evidence of the folded state of these species, initial crystallization attempts were performed on air-exposed samples. Despite the presence of a multitude of different forms, as suggested by the spectra, well-shaped crystals were obtained using monomethyl polyethylene glycol 5,000 as a precipitant in 50 m*M* Tris-HCl, pH 7.6. The crystals displayed a diffraction pattern characteristic of well-ordered crystals (Riccio *et al.*, 2001). The indexing of the diffraction spot surprisingly showed that the crystals were isomorphous to those of the carbonmonoxy form (Mazzarella *et al.*, 1999), grown under completely different conditions (2.0 ammonium sulfate, 50 m*M* Tris-HCl, pH 8.0). Even more surprising, the electron density maps showed the presence of a bis-histidyl adduct at the β iron heme. An exogenous ligand, initially identified as a CO molecule (successively assigned as a mixture of CO/O_2/aquo-met forms present in the crystal state), was present at the α iron (Riccio *et al.*, 2002). These data provided the first direct evidence that a partial hemichrome state, associated to the β subunits, was compatible with a well-defined three-dimensional structure of the protein. Therefore, a similar crystal packing, obtained under completely different conditions, can allocate two different Hb1Tn structures, namely the carbomonoxy and the partial hemichrome, that exhibit significant structural differences at quaternary-structure level (see the subsequent section for details). These results prompted further investigations, both in solution and in the crystal state, on other Antarctic fish Hbs. Crystals suitable for X-ray diffraction analyses were obtained with HbTb (Vitagliano *et al.*, 2004), closely related to Hb1Tn. In addition to crystallization of the air-exposed protein, a second strategy was adopted in this case. Crystallization was also attempted on fully oxidized HbTb obtained with sodium ferricyanide (see the previous section).

Diffracting crystals were obtained using either air-exposed or ferricyanide-treated HbTb. As for Hb1Tn, a bis-histidyl adduct was formed only at the β iron. In contrast, the α iron was in aquo/hydroxy-met state.

The characterization of the horse Hb hemichrome state was accomplished under totally different conditions (Robinson *et al.*, 2003). These studies were based on the observation (Perutz, 1954) that crystalline horse met-Hb undergoes a large lattice transition as pH is decreased from 7.1 to 5.4. Analysis of the crystals grown at neutral pH showed that both α and β iron adopted a standard aquo-met state. However, the decrease of pH to 5.4 produced a transition of the α iron to a hemichrome state. No variation was detected for the β iron, which maintained its aquo-met state. Like Antarctic fish Hb, horse Hb is able to form a partial hemichrome with a nativelike structure. Notably, in these two proteins, the chain that forms the bis-histidyl complex is different. It is not known whether this puzzling behavior is dictated by intrinsic differences of the individual chains, by differences in the tetramer flexibility, or by different conditions of hemichrome formation. It is worth mentioning that, in both cases, only partial hemichrome states were observed. This may suggest that tetrameric Hbs retain their native structure only in states with a single-chain type in the hemichrome state. Although data available are still limited, it can be proposed that the simultaneous formation of bis-histidyl adducts in all chains of the tetramer has strong disturbing effects on the structure of these Hbs.

5.2. The available three-dimensional structural models

Comparative analysis of the structural alterations induced by hemichrome formation in horse and Antarctic fish Hbs reveals differences and analogies. In all cases, a scissoring motion of the EF fragment of the chain that forms the bis-histidyl complexes is observed (Fig. 24.3). This is an expected rearrangement of the globin structure required for the coordination of distal His to the iron atom. It is worth mentioning that this distortion of the tertiary structure is costly from an energetic point of view. Indeed, if the distance between the C^{α} atoms of distal and proximal His is taken to estimate the entity of the structural variation (Fig. 24.4), it appears that the structural alteration associated with hemichrome formation is much larger than that exhibited by liganded/unliganded tetrameric Hbs in different packing contexts. It is interesting that in horse Hb, HbTb, and Hb1Tn, the modifications of the EF fragment are associated with a similar sliding of the heme that becomes more exposed to the solvent (Fig. 24.5). On the basis of this observation, it has been proposed that the hemichrome formation facilitates the involvement of the heme in electron-transfer processes (Robinson *et al.*, 2003), which suggests possible functional roles of the native-like hemichrome structures.

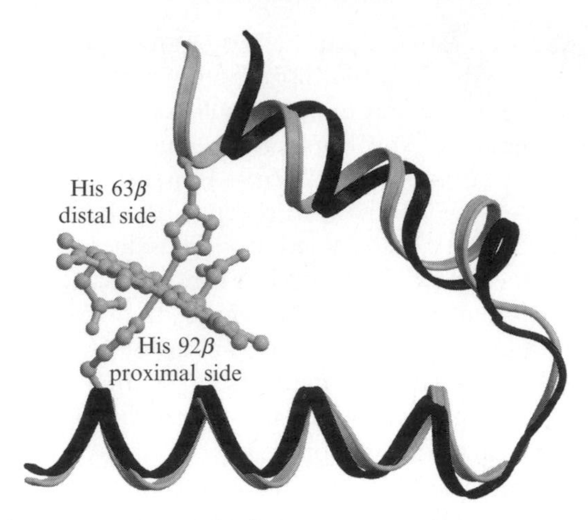

Figure 24.3 Scissoring motion of the β EF region occurring upon hemichrome formation. The local structure of the partial hemichrome state of HbTb (gray) is compared to that of HbTbCO (black). Residues belonging to helices F of the two structures have been superimposed. The drawing has been generated using the programs MOLSCRIPT (Kraulis, 1991) and RASTER3D (Merritt and Bacon, 1997). Proximal/distal His and the heme of HbTbCO are not shown for the sake of clarity.

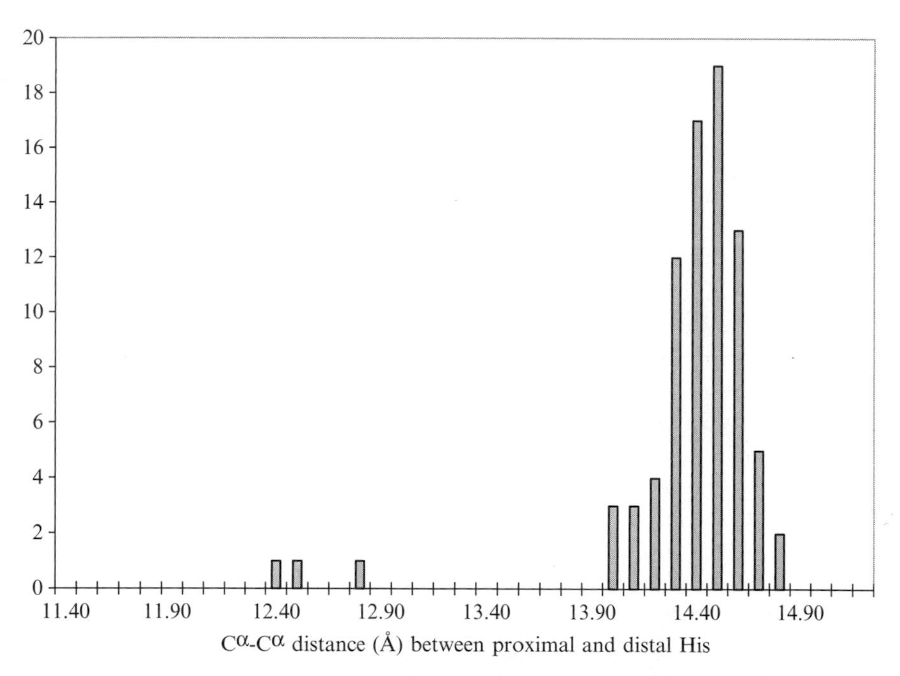

Figure 24.4 Distribution of the C^{α}-C^{α} distances between proximal and distal His in tetrameric Hbs reported in the Protein Data Bank (PDB). The three lowest C^{α}-C^{α} distances values correspond to chains forming bis-His adducts.

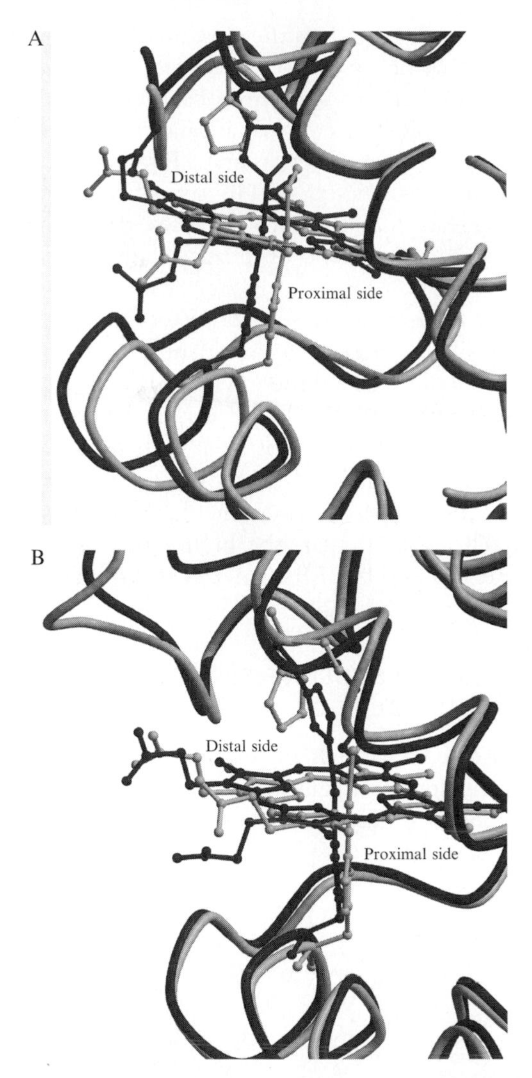

Figure 24.5 Shift of the heme upon hemichrome formation in HbTb (A) and horse Hb (B). For HbTb, the bis-His complex (black) is compared to HbTbCO (gray). All residues of the β chain have been superimposed. For horse Hb, the bis-His complex (black) has been compared to the aquo-met state obtained at pH 7.1; all residues of the α chain have been superimposed.

A strikingly different behavior is displayed by the quaternary-structure variations of mammalian and Antarctic fish Hbs upon hemichrome formation. Although the changes observed in horse Hb are limited to the tertiary structure, the overall structure of Antarctic fish Hbs exhibits significant variations. Indeed, a rotation of about 4 to 5° of the $\alpha_1\beta_1$ dimer relative to the $\alpha_2\beta_2$ dimer is observed when the structure of the partial hemichrome state is compared to that of the carbomonoxy form. It is interesting that the modification of the quaternary structure observed upon hemichrome formation falls in the physiological conformational transition between liganded (R) and unliganded (T) states of tetrameric Hbs (Fig. 24.6). This similarity of structural variations associated to Hb oxidations and those linked to oxygen release suggests that this protein uses the same conformational switch in different scenarios and is possibly involved in the T $\rightarrow$ R transition.

In principle, the differences observed between horse Hb and Hb1Tn/HbTb could be ascribed either to a higher intrinsic flexibility of Antarctic fish Hbs or to the peculiar crystalline environment in which the horse Hb hemichrome was formed. It should be noted, however, that studies in solution show that Antarctic fish Hbs clearly show a natural tendency to form bis-His (Vitagliano *et al.*, 2004). Hypotheses on the regions responsible for this distinctive behavior of Antarctic fish Hbs have been proposed (Riccio *et al.*, 2002). In particular, the higher flexibility of these Hbs has been related to the CD and the EF corner of the β chain.

5.3. Atomic resolution structure of HbTb in partial hemichrome state

Initial data collections on crystals of tetrameric Hbs in hemichrome state were collected using standard rotating-anode X-ray generators. Synchrotron radiation has been successfully used to increase the resolution of the

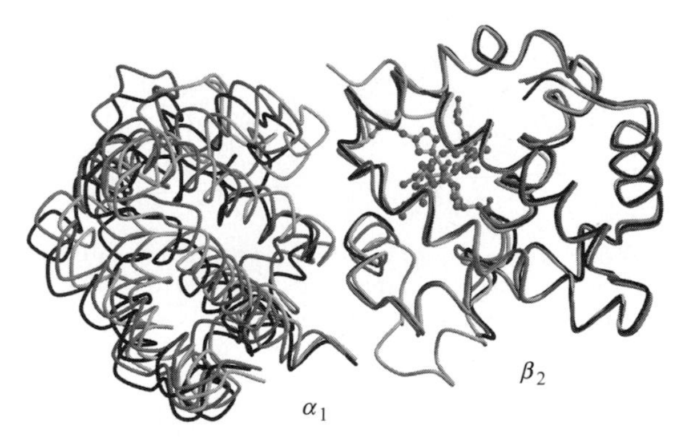

Figure 24.6 Quaternary-structure variations induced by hemichrome formation in Antarctic fish Hbs, in the framework of the R-T transition. The β_1 chain of HbTbCO (light gray), HbTbdeoxy (black), and HbTb in partial hemichrome state (dark gray) have been superimposed. The differences in the orientation of the α chains indicate that the quaternary structure of the partial hemichrome state is intermediate between the R and T states.

data and to provide more accurate structures. Although structure analyses carried out using synchrotron data have mostly confirmed previous results, this study provided a detailed picture of the iron coordination (Fig. 24.7). This information has been used to assign plausible structures to the two HbTb hemichromes detected in solution by EPR. A recent analysis, based on the 11 crystal structures of Hb in hemichrome states, led to conclusions that the two most relevant stereochemical parameters in tuning the g anisotropy are the tilt angle between the heme plane and the imidazole plane of distal His, θ_d, and the dihedral angle between the two imidazole planes, ω (Vergara *et al.*, 2007). Indeed, following procedures previously adopted for cytochromes and using available literature data, correlation between EPR signals and iron coordination geometry of hemichrome adducts in Hb structures has been established. In general, two different classes, corresponding to either ideal or distorted iron coordination states, have been identified. The analysis of the high-resolution HbTb hemichrome structure clearly indicates that the adduct falls among distorted hemichromes. As a result, it has been inferred that the other HbTb hemichrome is endowed with the ideal geometry of iron coordination.

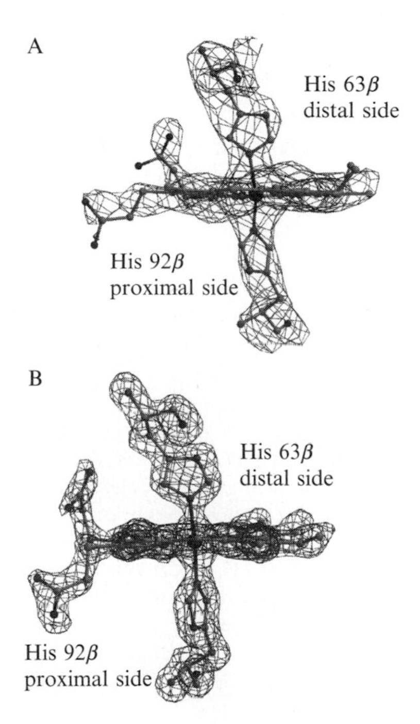

Figure 24.7 Electron density of HbTb hemichrome resulting from data collected using conventional X-ray generators (A, contoured at 3.0 σ) and synchrotron sources (B, contoured at 2.5 σ).

6. Conclusions

The past decade provided 11 independent crystal structures of Hbs in bis-His coordination (de Sanctis *et al.*, 2004b, 2005; Feng *et al.*, 2005; Hargrove *et al.*, 2000; Hoy *et al.*, 2004; Mitchell *et al.*, 1995b; Pesce, 2003; Riccio *et al.*, 2002; Robinson *et al.*, 2003; Vallone *et al.*, 2004; Vitagliano *et al.*, 2004). The stereochemical details of high-resolution crystal structure highlighted several correlations between spectroscopic properties in solution, theoretical predictions, and experimental structures. The most recent results stressed the Raman and EPR efficiency in predicting structural details. Raman analysis seems to provide a valuable bridge for solving possible discrepancies between solution- and crystal-derived information.

The occurrence of hemichrome/hemochrome in members of the Hb superfamily is not uniform. Together with different stability conditions, these findings suggest that the functional role of the bis-His coordination is not singular, and that it is possibly a tool for modulating ligand-binding or redox properties. In Antarctic fish Hbs, it is still disputed whether hemichromes have a biological function or are merely an evolutionary memory.

Acknowledgments

This study is financially supported by the Italian National Programme for Antarctic Research (PNRA), part of the Evolution and Biodiversity in the Antarctic (EBA) program endorsed by the Scientific Committee on Antarctic Research (SCAR).

References

Ajloo, D., Moosavi-Movahedi, A. A., Hakimelahi, G. H., Saboury, A. A., and Gharibi, H. (2002). The effect of dodecyl trimethylammonium bromide on the formation of methemoglobins and hemichrome. *Coll. Surf. B: Biointerfaces* **26,** 185–196.

Amadei, A., Marinelli, F., D'Abramo, M., D'Alessandro, M., Anselmi, M., Di Nola, A., and Aschi, M. (2005). Theoretical modeling of vibroelectronic quantum states in complex molecular systems: Solvated carbon monoxide, a test case. *J. Chem. Phys.* **122,** 124506/1–124506/10.

Arnold, E. V., Bohle, D. S., and Jordan, P. A. (1999). Reversible and irreversible hemichrome generation by the oxygenation of nitrosylmyoglobin. *Biochemistry* **38,** 4750–4756.

Bikiel, D. E., Boechi, L., Capece, L., Crespo, A., De Biase, P. M., Di Lella, S., GonzalezLebrero, M. C., Marti, M. A., Nadra, A. D., Perissinotti, L. L., Scherlis, D. A., and Estrin, D. A. (2006). Modeling heme proteins using atomistic simulations. *Phys. Chem. Chem. Phys.* **8,** 5611–5628.

Blumberg, W. E., and Peisach, J. (1972). Low-spin ferric forms of hemoglobin and other heme proteins. *Wenner-Gren Center International Symposium Series* **18,** 219–225.

Boffi, A., Takahashi, S., Spagnuolo, C., Rousseau, D. L., and Chiancone, E. (1994). Structural characterization of oxidized dimeric *Scapharca inaequivalvis* hemoglobin by resonance Raman spectroscopy. *J. Biol. Chem.* **269,** 20437–20440.

Brittain, T. (2005). The Root effect in hemoglobins. *J. Inorg. Biochem.* **99,** 120–129.

Camardella, L., Caruso, C., D'Avino, R., di Prisco, G., Rutigliano, B., Tamburrini, M., Fermi, G., and Perutz, M. F. (1992). Hemoglobin of the Antarctic fish *Pagothenia bernacchii*: Amino acid sequence, oxygen equilibria and crystal structure of its carbonmonoxy derivative. *J. Mol. Biol.* **224,** 449–460.

Carey, P. R. (2006). Raman crystallography and other biochemical applications of Raman microscopy. *Annu. Rev. Phys. Chem.* **57,** 527–554.

Couture, M., Das, T. K., Lee, H. C., Peisach, J., Rousseau, D. L., Wittenberg, B. A., Wittenberg, J. B., and Guertin, M. (1999). *Chlamydomonas* chloroplast ferrous hemoglobin: Heme pocket structure and reactions with ligands. *J. Biol. Chem.* **274,** 6898–6910.

Croci, S., Pedrazzi, G., Passeri, G., Piccolo, P., and Ortalli, I. (2001). Acetylphenylhydrazine induced haemoglobin oxidation in erythrocytes studied by Mossbauer spectroscopy. *Biochim. Biophys. Acta* **1568,** 99–104.

Das, T. K., Lee, H. C., Duff, S. M. G., Hill, R. D., Peisach, J., Rousseau, D. L., Wittenberg, B. A., and Wittenberg, J. B. (1999). The heme environment in barley hemoglobin. *J. Biol. Chem.* **274,** 4207–4212.

de Sanctis, D., Pesce, A., Nardini, M., Bolognesi, M., Bocedi, A., and Ascenzi, P. (2004a). Structure-function relationships in the growing hexa-coordinate hemoglobin sub-family. *IUBMB Life* **56,** 643–651.

de Sanctis, D., Dewilde, S., Pesce, A., Moens, L., Ascenzi, P., Hankeln, T., Burmester, T., and Bolognesi, M. (2004b). Crystal structure of cytoglobin: The fourth globin type discovered in man displays heme hexa-coordination. *J. Mol. Biol.* **336,** 917–927.

de Sanctis, D., Dewilde, S., Vonrhein, C., Pesce, A., Moens, L., Ascenzi, P., Hankeln, T., Burmester, T., Ponassi, M., Nardini, M., and Bolognesi, M. (2005). Bishistidyl heme hexacoordination, a key structural property in *Drosophila melanogaster* hemoglobin. *J. Biol. Chem.* **280,** 27222–27229.

di Prisco, G., Giardina, B., D'Avino, R., Condo, S. G., Bellelli, A., and Brunori, M. (1988). Antarctic fish hemoglobin: An outline of the molecular structure and oxygen binding properties: II. Oxygen binding properties. *Comp. Biochem. Physiol. B* **90,** 585–591.

Egawa, T., and Yeh, S. (2004). Structural and functional properties of hemoglobins from unicellular organism as revealed by resonance Raman spectroscopy. *J. Inorg. Biochem.* **99,** 72–76.

Everson, I., and Ralph, R. (1968). Blood analyses of some Antarctic fish. *Br. Antarctic Surv. Bull.* **15,** 59–62.

Feis, A., Marzocchi, M. P., Paoli, M., and Smulevich, G. (1994). Spin state and axial ligand bonding in the hydroxide complexes of metmyoglobin, methemoglobin, and horseradish peroxidase at room and low temperatures. *Biochemistry* **33,** 4577–4583.

Feng, L., Zhou, S., Gu, L., Gell, D., Mackay, J., Weiss, M., Gow, A., and Shi, Y. (2005). Structure of oxidized a-haemoglobin bound to AHSP reveals a protective mechanism for haem. *Nature* **435,** 697–701.

Feng, M., and Tachikawa, H. (2001). Raman spectroscopic and electrochemical characterization of myoglobin thin film: Implication of the role of Histidine 64 for fast heterogeneous electron transfer. *J. Am. Chem. Soc.* **123,** 3013–3020.

Giordano, S., Vergara, A., Lee, H. C., Peisach, J., Balestrieri, M., Mazzarella, L., Parisi, E., Balestrieri, M., di Prisco, G., and Verde, C. Hemoglobin structure/function and globin-gene evolution in the Arctic fish. *Liparis tunicatus. Gene,* in press.

Hargrove, M. S., Brucker, E. A., Stec, B., Sarath, G., Arredondo-Peter, R., Klucas, R. V., Olson, J. S., and Phillips, G. N. (2000). Crystal structure of a nonsymbiotic plant hemoglobin. *Structure* **8,** 1005–1014.

Hoy, J. A., Kundu, S., Trent, J. T., III, Ramaswamy, S., and Hargrove, M. S. (2004). The crystal structure of *Synechocystis* hemoglobin with a covalent heme linkage. *J. Biol. Chem.* **279,** 16535–16542.

Ilari, A., Bonamore, A., Farina, A., Johnson, K., and Boffi, A. (2002). The X-ray structure of ferric *Escherichia coli* flavohemoglobin reveals an unexpected geometry of the distal heme pocket. *J. Biol. Chem.* **26,** 23725–23732.

Ioanitescu, A. I., Dewilde, S., Kiger, L., Marden, C. M., Moens, L., and Van Doorslaer, S. (2005). Characterization of nonsymbiotic tomato hemoglobin. *Biophys. J.* **89,** 2628–2639.

Ito, N., Komiyama, N. H., and Fermi, G. (1995). Structure of deoxyhemoglobin of the Antarctic fish *Pagothenia bernacchii* with an analysis of the structural basis of the root effect by comparison of the liganded and unliganded hemoglobin structures. *J. Mol. Biol.* **250,** 648–658.

Jeong, S. T., Ho, N. T., Hendrich, M. P., and Ho, C. (1999). Recombinant hemoglobin (a29Leucine → Phenylalanine, a96Valine → Tryptophan, b108Asparagine → Lysine) exhibits low oxygen affinity and high cooperativity combined with resistance to autoxidation. *Biochemistry* **38,** 13433–13442.

Kraulis, P. J. (1991). MOLSCRIPT: A program to produce both detailed and schematic plots of protein structures. *J. Appl. Crystallogr.* **24,** 946–950.

Marmo Moreira, L., Lima Poli, A., Costa-Filho, A. J., and Imasato, H. (2006). Pentacoordinate and hexacoordinate ferric hemes in acid medium: EPR, UV-Vis and CD studies of the giant extracellular hemoglobin of *Glossoscolex paulistus. Biophys. Chem.* **124,** 62–72.

Marzocchi, M. P., and Smulevich, G. (2003). Relationship between heme vinyl conformation and the protein matrix in peroxidases. *J. Raman Spectr.* **34,** 725–736.

Mazzarella, L., Vergara, A., Vitagliano, L., Merlino, A., Bonomi, G., Scala, S., Verde, C., and di Prisco, G. (2006a). High resolution crystal structure of deoxy hemoglobin from *Trematomus bernacchii* at different pH values: The role of histidine residues in modulating the strength of the Root effect. *Proteins* **65,** 490–498.

Mazzarella, L., Bonomi, G., Lubrano, M. C., Merlino, A., Vergara, A., Vitagliano, L., Verde, C., and di Prisco, G. (2006b). Minimal structural requirement of Root effect: Crystal structure of the cathodic hemoglobin isolated from *Trematomus newnesi. Proteins* **62,** 316–321.

Mazzarella, L., D'Avino, R., di Prisco, G., Savino, C., Vitagliano, L., Moody, P. C. E., and Zagari, A. (1999). Crystal structure of *Trematomus newnesi* hemoglobin re-opens the Root effect question. *J. Mol. Biol.* **287,** 897–906.

Merritt, E. A., and Bacon, D. J. (1997). Raster3D: Photorealistic molecular graphics. *Methods Enzymol.* **277,** 505–524.

Milani, M., Pesce, A., Nardini, M., Ouellet, H., Ouellet, Y., Dewilde, S., Bocedi, A., Ascenzi, P., Guertin, M., Moens, L., Friedman, J. M., Wittenberg, J. B., and Bolognesi, M. (2005). Structural bases for heme binding and diatomic ligand recognition in truncated hemoglobins. *J. Inorg. Biochem.* **99,** 97–109.

Mitchell, D. T., Kitto, G. B., and Hackert, M. L. (1995a). Structural analysis of monomeric hemichrome and dimeric cyanomet hemoglobins from *Caudina arenicola. J. Mol. Biol.* **251,** 421–431.

Mitchell, D. T., Ernst, S. R., Wu, W.-X., and Hackert, M. L. (1995b). Three-dimensional structure of a hemichrome hemoglobin from *Caudina arenicola. Acta Crystallogr. D* **51,** 647–653.

Nagai, M., Aki, M., Li, R., Jin, Y., Sakai, H., Nagatomo, S., and Kitagawa, T. (2000). Heme structure of hemoglobin M Iwate [R87(F8)HisfTyr]: A UV and visible resonance Raman study. *Biochemistry* **39,** 13083–13105.

Palmer, G. (1985). The electron paramagnetic resonance of metalloproteins. *Biochem. Soc. Trans.* **13,** 548–560.

Perutz, M. F. (1957). The structure of hemoglobin III. Direct determination of the molecular transform. *Proc. Roy. Soc. London Ser. A* **225,** 264–286.

Pesce, A., Dewilde, S., Nardini, M., Moens, L., Ascenzi, P., Hankeln, T., Burmester, T., and Bolognesi, T. (2003). Human brain neuroglobin three-dimensional structure. *Structure* **11,** 1087.

Pesce, A., De Sanctis, D., Nardini, M., Dewilde, S., Moens, L., Hankeln, T., Burmester, T., Ascenzi, P., and Bolognesi, M. (2004). Reversible hexa- to penta-coordination of the heme Fe atom modulates ligand binding properties of neuroglobin and cytoglobin. *IUBMB Life* **56,** 657–664.

Rachmilewitz, E. A., Peisach, J., and Blumberg, W. E. (1971). Stability of oxyhemoglobin A and its constituent chains and their derivatives. *J. Biol. Chem.* **246,** 3356–3366.

Ray, A., Friedman, B. A., and Friedman, J. M. (2002). Trehalose glass-facilitated thermal reduction of metmyoglobin and methemoglobin. *J. Am. Chem. Soc.* **124,** 7270–7271.

Riccio, A., Vitagliano, L., di Prisco, G., Zagari, A., and Mazzarella, L. (2001). Liganded and unliganded forms of Antarctic fish haemoglobins in polyethylene glycol: Crystallization of an R-state haemichrome intermediate. *Acta Crystallogr. D* **57,** 1144–1146.

Riccio, A., Vitagliano, L., di Prisco, G., Zagari, A., and Mazzarella, L. (2002). The crystal structure of a tetrameric hemoglobin in a partial hemichrome state. *Proc. Natl. Acad. Sci. USA* **99,** 9801–9806.

Rifkind, J. M., Abugo, O., Levy, A., and Heim, J. M. (1994). Detection, formation, and relevance of hemichrome and hemochrome. *Methods Enzymol.* **231,** 449–480.

Riggs, A. (1988). The Bohr effect. *Annu. Rev. Physiol.* **50,** 181–204.

Robinson, V. L., Smith, B. B., and Arnone, A. (2003). A pH-dependent aquomet-to-hemichrome transition in crystalline horse methemoglobin. *Biochemistry* **42,** 10113–10125.

Ruud, J. T. (1954). Vertebrates without erythrocytes and blood pigment. *Nature* **173,** 848–850.

Sawai, H., Makino, M., Mizutani, Y., Ohta, T., Hiroshi, S., Uno, T., Kawada, N., Yoshizato, K., Kitagawa, T., and Shiro, Y. (2005). Structural characterization of the proximal and distal histidine environment of cytoglobin and neuroglobin. *Biochemistry* **44,** 13257–13265.

Sidell, B. D., and O'Brien, K. M. (2006). When bad thing happen to good fish: The loss of hemoglobin and myoglobin expression in Antarctic icefishes. *J. Exp. Biol.* **209,** 1791–1802.

Smulevich, G. (1998). Understanding heme cavity structure of peroxidases: Comparison of electronic absorption and resonance Raman spectra with crystallographic results. *Biospectr.* **4,** S3–S17.

Smulevich, G., Feis, A., and Howes, B. D. (2005). Fifteen years of Raman spectroscopy of engineered heme containing peroxidases: What have we learned? *Acc. Chem. Res.* **38,** 433–440.

Smulevich, G., Wang, Y., Mauro, J. M., Wang, J., Fishel, L. A., Kraut, J., and Spiro, T. G. (1990). Single-crystal resonance Raman spectroscopy of site-directed mutants of cytochrome c peroxidase. *Biochemistry* **29,** 7174–7180.

Svistunenko, D. A., Sharpe, M. A., Nicholls, P., Blenkinsop, C., Davies, N. A., Dunne, J., Wilson, M. T., and Cooper, C. E. (2000). The pH dependence of naturally occurring low-spin forms of methaemoglobin and metmyoglobin: An EPR study. *Biochem. J.* **351,** 595–605.

Tsuruga, M., Matsuoka, A., Hachimori, Y., Sugawara, Y., and Shikama, K. (1998). The molecular mechanism of autoxidation for human oxyhemoglobin. *J. Biol. Chem.* **273,** 8607–8615.

Vallone, B., Nienhaus, K., Matthes, K., Brunori, M., and Nienhaus, G. (2004). The structure of murine neuroglobin: Novel pathways for ligand migration and binding. *Proteins* **56,** 85–92.

Verde, C., Howes, B. D., de Rosa, M. C., Raiola, L., Smulevich, G., Williams, R., Giardina, B., Parisi, E., and di Prisco, G. (2004). Structure and function of the Gondwanian hemoglobin of *Pseudaphritis urvillii*, a primitive notothenioid fish of temperate latitudes. *Prot. Sci.* **13,** 2766–2781.

Verde, C., Parisi, E., and di Prisco, G. (2006). The evolution of thermal adaptation in polar fish. *Gene* **385,** 137–145.

Verde, C., Vergara, A., Parisi, E., Giordano, D., Mazzarella, L., and di Prisco, G. (2007). The Root effect: A structural and evolutionary perspective. *Antarctic Sci.* **19,** in press.

Vergara, A., Frazese, M., Merlino, A., Vitagliano, L., Verde, C., di Prisco, G., Lee, H. C., Peisach, J., and Mazzarella, L. (2007). Structural characterization of ferric hemoglobins from three Antarctic fish species of the suborder Notothenioidei. *Biophys. J.,* **93,** 2822–2829.

Vitagliano, L., Bonomi, G., Riccio, A., di Prisco, G., Smulevich, G., and Mazzarella, L. (2004). The oxidation process of Antarctic fish hemoglobins. *Eur. J. Biochem.* **271,** 1651–1659.

Walker, F. A. (2004). Models of the bis-Histidine-ligated electron-transferring cytochromes. Comparative geometric and electronic structure of low-spin ferro- and ferrihemes. *Chem. Rev.* **104,** 589–615.

Walker, F. A., Reis, D., and Balke, V. L. (1984). Models of the cytochromes b. 5. EPR studies of low-spin iron(III) tetraphenylporphyrins. *J. Am. Chem. Soc.* **106,** 6888–6898.

Weber, R. E., and Vinogradov, S. N. (2001). Nonvertebrate hemoglobins: Functions and molecular adaptations. *Physiol. Rev.* **81,** 569–628.

Weiland, T. R., Kundu, S., Trent, J. T., III, Hoy, J. A., and Hargrove, M. S. (2004). Bis-histidyl hexacoordination in hemoglobins facilitates heme reduction kinetics. *J. Am. Chem. Soc.* **126,** 11930–11935.

Wells, R. M. G., Macdonald, J. A., and di Prisco, G. (1990). Thin-blooded Antarctic fishes: A rheological comparison of the hemoglobin-free icefishes *Chionodraco kathleenae* and *Cryodraco antarcticus* with a red-blooded notothenioid, *Pagothenia bernacchii*. *J. Fish Biol.* **36,** 595–609.

Wells, R. M. G., Ashby, M. D., Duncan, S. J., and MacDonald, J. A. (1980). Comparative studies of the erythrocytes and haemoglobins in notothenioid fishes from Antarctica. *J. Fish. Biol.* **17,** 517–527.

CHAPTER TWENTY-FIVE

Dinitrosyl Iron Complexes Bind with Hemoglobin as Markers of Oxidative Stress

Konstantin B. Shumaev,* Olga V. Kosmachevskaya,* Alexander A. Timoshin,† Anatoly F. Vanin,‡ *and* Alexey F. Topunov*

Contents

* A. N. Bach Institute of Biochemistry, Russian Academy of Sciences, Moscow, Russia
† Russian Cardiology Scientific Research Complex, Moscow, Russia
‡ N. N. Semenov Institute of Chemical Physics, Russian Academy of Sciences, Moscow, Russia

Methods in Enzymology, Volume 436
ISSN 0076-6879, DOI: 10.1016/S0076-6879(08)36025-X

Abstract

Prooxidant and antioxidant properties of nitric oxide (NO) during oxidative stress are mostly dependent on its interaction with reactive oxygen species, Fe ions, and hemoproteins. One form of NO storage and transportation in cells and tissues is dinitrosyl iron complexes (DNIC), which can bind with both low-molecular-weight thiols and proteins, including hemoglobin. It was shown that dinitrosyl iron complexes bound with hemoglobin (Hb-DNIC) were formed in rabbit erythrocytes after bringing low-molecular-weight DNIC with thiosulfate into blood. It was ascertained that Hb-DNIC intercepted free radicals reacting with hemoglobin SH-groups and prevented oxidative modification of this protein caused by hydrogen peroxide. Destruction of Hb-DNIC can take place in the presence of both hydrogen peroxide and *tert*-butyl hydroperoxide. Hb-DNIC can also be destroyed at the enzymatic generation of superoxide-anion radical in the xanthine–xanthine oxidase system. If aeration in this system was absent, formation of the nitrosyl R-form of hemoglobin could be seen during the process of Hb-DNIC destruction. Study of Hb-DNIC interaction with reactive oxygen metabolites is important for understanding NO and Hb roles in pathological processes that could result from oxidative stress.

1. Introduction

NO takes part in normal and pathological processes. It works as a signal molecule and shows prooxidant and antioxidant properties (Droge, 2003; Joshi, 1999; Ma *et al.*, 1999; O'Donnell and Freeman, 2001). The character of NO's influence on free radical oxidation reactions greatly depends on its interaction with superoxide radical, lipid-derived radicals, Fe ions, and hemoproteins (Gorbunov *et al.*, 1995; Kagan *et al.*, 2001; Rubbo *et al.*, 1994). It is known now that not only reactive oxygen (ROS) and reactive nitrogen (RNS) species play extremely important role in oxidative stress processes but also iron complexes, both low-molecular-weight and protein-bound forms (Jeney *et al.*, 2002; Kruszewski, 2004). One of the forms of NO storage and transportation in cells and tissues is the dinitrosyl iron complex (DNIC) (Muller *et al.*, 2002). The first evidence of the existence of such compounds was obtained by Vanin and coworkers (Vanin and Chetverikov, 1968; Vanin *et al.*, 1967). The general formula of thiol containing DNIC is $([RS^-]_2Fe^+[NO^+]_2)^+$. Hemoglobin can react with ROS and RNS and affects free radical oxidation processes (Reeder and Wilson, 2005). During these processes, oxidative modification of hemoglobin takes place. As a result, heme degrades and lets out Fe ions, which catalyze the formation of free radicals in Fenton and Haber-Weiss reactions (Kagan *et al.*, 2001; Nagababu and Rifkind, 2000). After Fe is released from heme, it can be included in thiol-containing DNIC (Jucket *et al.*, 1998). Hemoglobin is a unique protein that can bind -NO in three different ways. NO can react with heme and with cysteine residues (cys-β93) to form nitrosylhemoglobin ($HbFe^{II}NO$) and S-nitrosohemoglobin, respectively (Luchsinger, 2003). The thiol group of cysteine β93 can also form DNIC (Gow *et al.*, 1999; Vanin *et al.*, 1998). It has been shown that thiol-containing DNIC can be destroyed by superoxide, oxoferrylmyoglobin, and other strong oxidants (Shumaev *et al.*, 2004; Shumaev *et al.*, 2006), which may explain the low level of these complexes in biological systems. In contrast, the functioning of superoxide dismutase (SOD) leads to increased nitrosylhemoglobin and S-nitrosohemoglobin yields during the interaction of oxyhemoglobin with NO (Gow *et al.*, 1999). It is well known that under oxidative stress, including ischemia and tissue reperfusion injury, ROS and NO production are intensive. In these conditions, Hb-bound DNIC can be a factor affecting pathological processes during oxidative stress. This article describes our studies on methods for investigating the interaction of DNIC and hemoglobin under oxidative stress and its influence on the Hb state in these conditions.

1.1. Materials

In the experiments, we used 5-diethoxyphosphoryl-5-methyl-1-pyrroline-N-oxide (DEPMPO) from Oxis (USA), 4-hydroxy-2,2,6,6-tetramethylpiperidine-1-oxyl (TEMPOL) from Merck (Germany), sodium ascorbate, sodium thiosulfate, reduced L-glutathione, hydrogen peroxide, bovine methemoglobin, PAPA/NONOate, diethylenetriaminepentaacetic acid (DTPA), xanthine, xanthine oxidase, SOD, horseradish peroxidase type VI-A (HRP), L-tyrosine and sodium 1,2-dihydroxybenzene-3,5-disulfonate (TIRON) from Sigma (USA), tert-butyl hydroperoxide from Aldrich (USA), ferrous sulfate from Fluka (Switzerland), Angeli's salt ($Na_2N_2O_3$) from Cauman Chemical (USA), and reagents for electrophoresis from Panreac (Spain). All chemicals were of the highest purity.

1.2. Synthesis of DNIC

Paramagnetic DNIC with phosphate was obtained by treating the 5 m*M* ferrous sulfate ($FeSO_4$) in 100 m*M* K-Na phosphate buffer (pH 6.8) with gaseous NO in a Thunberg tube containing 100 mL gas phase. The 1 mL of $FeSO_4$ solution in distilled water (pH 5.5) and 4.5 mL of phosphate buffer were placed in the bottom and upper parts of the Thunberg tube, respectively. The apparatus was evacuated, and NO gas was added at 100 mm Hg pressure. These solutions were mixed and shaken (5 min), and then NO evacuated from the Thunberg tube. A similar approach was used for synthesis of DNIC with thiosulfate (Sanina *et al.*, 2005). Before the experiments, the crystals of tetranitrosyl iron complex with thiosulfate were dissolved in the physiological solution containing a 20-fold molar excess of thiosulfate, which resulted in the transformation of tetranitrosyl iron complex to DNIC with thiosulfate. Hb-DNIC were obtained by adding 950 μL of 1 mM Hb solution in 0.1 M K-Na phosphate buffer (pH 7.4) to 300 μL of 5 mM phosphate DNIC. After 5 min incubation, Hb-DNIC at $\approx$1 mM concentration appeared ($\approx$1.4 molecules of DNIC per 1 tetrameric Hb). Solutions of phosphate DNIC and Hb-DNIC were divided into aliquots and were stored at liquid nitrogen temperature. The concentration of synthesized DNIC was measured using the EPR method by the intensity of the EPR signal from the complex and by performing double integration. The stable nitroxyl radical TEMPOL was used as a standard paramagnetic sample.

1.3. EPR assay

DNIC in whole blood and the model systems were detected by EPR spectroscopy. EPR spectra were recorded at room temperature (25°) or at 77°K (in a quartz finger-type Dewar flask filled with liquid nitrogen) using an X-band EPR spectrometer E-109E (Varian, USA). For EPR measurements involving superoxide generation, the samples (80 μL) were injected into gas-permeable PTFE 22 capillary tubes (Zeus Industrial Products Inc., USA). These samples were placed into the quartz tube in the resonator of the spectrometer, and measurements were taken at constant air passage through the tube. For other experiments, the samples (80 μL) were injected into glass capillary tubes. Instrument settings were as follows: modulation frequency, 100 kHz; time constant, 0.032; microwave power, 10 mW; microwave frequency, 9.15 GHz; and modulation amplitudes, 0.2 or 0.1 mT for DNIC and DEPMPO spin adducts, respectively. For TIRON semiquinone EPR signal registration, the settings were as follows: modulation frequency, 100 kHz; time constant, 0.032; microwave power, 5 mW; microwave frequency, 9.14 GHz; and modulation amplitude, 0.05 mT.

1.4. Experiments on animals

Adult female rabbits weighing 3 to 4 kg were used. At the beginning of each experiment, a catheter was implanted in the marginal ear vein under ketamine narcosis (100 mg/kg), and the initial blood sample was taken. In all experiments, intravenous introduction of DNIC with thiosulfate in 1 mL of physiological solution was performed during 1 to 2 min. The dose of bolus introduction of the preparation was 5 μmol of DNIC per kg of body weight. For 5 min after DNIC introduction, the samples of venous blood were taken through the catheter. To study DNIC in plasma and packed red blood cells separately, part of the blood sample was centrifuged after preparation and heparin (20 ME/mL) addition; plasma was then separated from erythrocytes, which were then frozen and stored in liquid nitrogen to be thawed directly before EPR spectra recording.

2. Formation of Hb-DNIC *In Vivo*

2.1. Overview

After 5 min of introducing low-molecular-weight DNIC to thiosulfate ligands in blood, an EPR signal of protein DNIC was observed (Fig. 25.1A). After blood fractionation, it was shown that ≈10% of protein DNIC is localized in erythrocytes (Fig. 25.1B, spectrum 1). It is interesting that this spectrum is identical to that of Hb-DNIC obtained in model conditions (see Fig. 25.1B, spectrum 2). The rest of the DNIC was bound with serum proteins, mostly albumin (Fig. 25.1C). The data obtained in experiments with animals show that protein DNICs are more stable than low-molecular-weight complexes and that most DNIC in cells and tissues is of a protein form (Muller *et al.*, 2002).

2.2. Spin adducts of DEPMPO with free radicals

1. Studies of the generation of free radicals during the interaction of hemoglobin and *tert*-butyl hydroperoxide were made with the spin-trap DEPMPO in medium containing 150 m*M* K-Na phosphate buffer (pH 7.4), 40 m*M* DEPMPO, 1.6 m*M* DTPA, 235 μM hemoglobin, and *tert*-butyl hydroperoxide at 0.4 and 0.8 m*M* concentrations. In experiments with Hb-DNIC, the concentration was 340 μM.

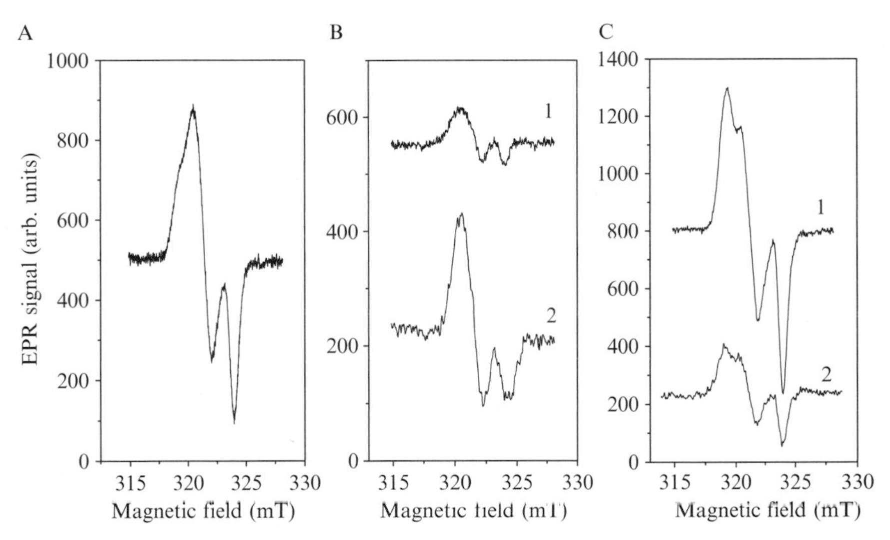

Figure 25.1 EPR spectra of protein-bound DNIC in blood and model systems. (A) EPR signal of rabbit whole blood obtained 5 min after intravenous introduction of DNIC with thiosulfate. (B) EPR signal of packed red blood cells (spectrum 1) and Hb-DNIC in the model system (spectrum 2). (C) EPR signal of DNIC bound with plasma proteins (spectrum 1) and albumin-bound DNIC in the model system (spectrum 2).

2. The standard EPR spectrum of the DEPMPO adduct with thiyl radical was registered in a mixture containing 0.5 m*M* hydrogen peroxide, 0.8 m*M* reduced glutathione, and horseradish peroxidase (0.2 mg/mL) (Fig. 25.2B, spectrum 1). Using a similar system but containing 2 m*M* tyrosine instead of glutathione, a DEPMPO adduct with the tyrosine radical was obtained (Fig. 2B, spectrum 2). Spectra of DEPMPO adducts with alkoxyl and alkylperoxyl radicals were registered when 1.4 m*M* *tert*-butyl hydroperoxide interacted with 0.45 hemin (Fig. 25.2B, spectrum 3). In all cases, reaction mixtures contained 150 m*M* K-Na phosphate buffer (pH 7.4), 40 m*M* DEPMPO, and 1.6 m*M* DTPA. EPR spectra were recorded 3 min after mixing of all components.
3. Standard spectra of spin DEPMPO adducts were used to identify different free radical intermediates of Hb's reaction with *tert*-butyl hydroperoxide (Fig. 25.2A, spectra 1 and 2).

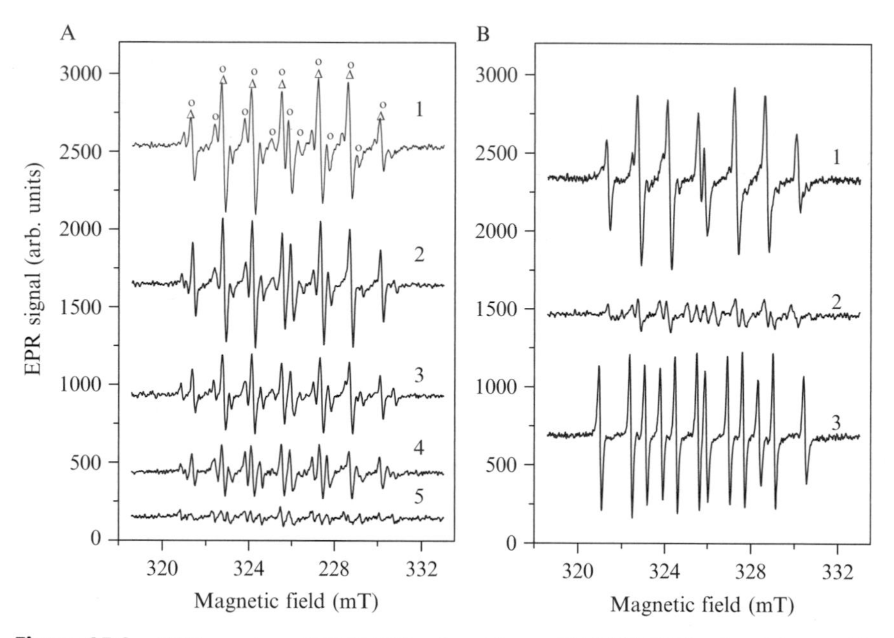

Figure 25.2 EPR spectra of DEPMPO spin adducts. (A) DEPMPO spin adducts with free radicals formed after the addition of 235 μM metHb and 0.8 m*M* (2) or 0.4 m*M* (4) *tert*-butyl hydroperoxide, and the addition of 340 μM Hb-DNIC and 0.8 m*M* (3) or 0.4 m*M* (5) *tert*-butyl hydroperoxide to the reaction mixture. The reaction mixture contained 0.15 *M* phosphate buffer (pH 7.4), 40 m*M* DEPMPO, and 1.6 m*M* DTPA. (B) DEPMPO spin adducts with thiyl (1) and tyrosine radicals (3), and free radical derivatives of *tert*-butyl hydroperoxide (2) obtained in various model systems. Spectrum (A1) shows the sum of EPR spectra of the spin adducts calculated according to the following equation: (A1) – 0.75$^{\times}$ (B1) + (B2) + 0.2$^{\times}$ (B3). Symbols (Δ) and (O) show the input of (B1) and (B3), spectra respectively.

3. Influence of Hb-DNIC on Free Radical Generation During Hb's Interaction with *tert*-Butyl Hydroperoxide

DNIC can be bound with thiol (-SH) protein groups. These groups are among the first to be oxidatively modified when ROS and RNS are generated. One of the main processes resulting in damage of biological structures during oxidative stress is the interaction of hemoglobin and myoglobin with hydrogen peroxide and lipid hydroperoxides (Gorbunov *et al.*, 1995; Reeder and Wilson, 2005). Reaction of hemoglobin with organic hydroperoxides leads to formation of such active intermediates as oxoferrylhemoglobin (Hb-heme-$Fe^{IV} = O$), alkoxyl ($RO^{\bullet}$), and alkylperoxyl ($ROO^{\bullet}$) (Kagan *et al.*, 2001; Zee, 1997):

$$\text{Hb-Fe}^{III} + \text{ROOH} \rightarrow \text{Hb-Fe}^{IV}{=}\text{O} + \text{RO}^{\bullet} + \text{H}^{+} \qquad (25.1)$$

$$\text{Hb-Fe}^{III} + \text{ROOH} \rightarrow \text{Hb}^{\bullet}\text{-Fe}^{IV} = \text{O} + \text{ROH} \qquad (25.2)$$

$$\text{Hb}^{\bullet}\text{-Fe}^{IV} = \text{O} + \text{ROOH} \rightarrow \text{Hb-Fe}^{IV}{=}\text{O} + \text{ROO}^{\bullet} + \text{H}^{+} \qquad (25.3)$$

$$\text{Hb-Fe}^{IV} = \text{O} + \text{ROOH} \rightarrow \text{Hb-Fe}^{III}\text{-OH} + \text{ROO}^{\bullet} \qquad (25.4)$$

After analysis of the EPR spectra of spin DEPMPO adducts, it can be stated that that during Hb's reaction with *tert*-butyl hydroperoxide, alkylperoxyl radicals and free amino acid radicals (including tyrosine and thiyl radicals) are generated (Figs. 25.2A and 25.2B). The latter probably form as result of cysteine-β93 residue oxidation after action of alkoxyl and alkylperoxyl radicals. But during interaction of *tert*-butyl hydroperoxide with Hb-DNIC, the yield of free radical intermediates decreases (Fig. 25.2A, spectra 2 to 5). Maximal decrease of thiyl radical level was observed when concentrations of *tert*-butyl hydroperoxide and Hb-DNIC were close to equimolar Hb (Fig. 25.2A, spectra 4 and 5). These data show that DNICs intercept free radical intermediates that react with the SH-group of cysteine-β93.

4. Study of Oxidative Hb Modification by SDS Electrophoresis

Electrophoresis of samples was made in 15% gel with SDS on 15 × 15 cm plates with 1-mm gel thickness according to the method of Laemmli (1970), with modifications. The reaction mixture (45 μL) contained hemoglobin (Fig. 25.3A), Hb-DNIC (Fig. 25.3B), or Hb + destroyed phosphate

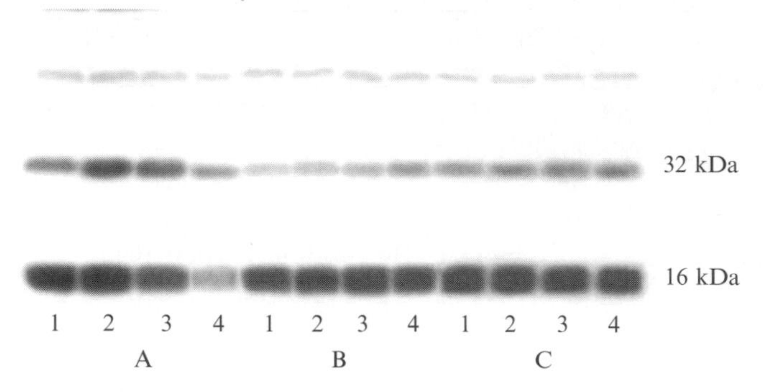

Figure 25.3 Oxidative modification of hemoglobin by hydrogen peroxide detected by SDS-PAGE. Protein samples contained the following: (A) hemoglobin; (B) Hb-DNIC; C, destroyed Hb-DNIC. (1) Control without hydrogen peroxide; (2) 400 μM; (3) 800 μM; and (4) 1600 μM of hydrogen peroxide.

DNIC (Fig. 25.3C); 0.2 *M* K-Na phosphate buffer (pH 7.4); and 5 m*M* DTPA. Hydrogen peroxide at different final concentrations was added to the mixture for a final volume of 45 μL. The final concentration of Hb in all cases was 560 μM (780 μM for Hb-DNIC). Mixture was incubated at room temperature for 5 min. After incubation, 45 μL of sample buffer was added and a standard electrophoretic procedure was carried out. Conditions of electrophoresis were as follows: in concentrating gel, I = 30 mA, U = 300 V; in dividing gel, I = 60 mA, U = 360 V. Gels were stained with Coomassie Brilliant Blue R-250.

Destroyed phosphate DNICs were obtained by 5-h incubation at room temperature. Their concentration was controlled by EPR spectroscopy.

5. Influence of Hb-DNIC on Oxidative Modification of Hb Under the Action of Hydrogen Peroxide

Using SDS-electrophoresis, it was shown that as a result of methemoglobin treatment with hydrogen peroxide of various concentrations, oxidative modification of the protein took place (Fig. 25.3). Hydrogen peroxide in 400 and 800 μM concentrations caused the dimerization of Hb subunits (Fig. 25.3A, lines 2 and 3). It is known that formation of such dimers during Hb reaction with ROS and RNS can be the result of the appearance of disulfide bonds between cysteine-β93 residues and dimerization of free tyrosine radicals (Romero *et al.*, 2003). Under the action of 1600 μM hydrogen peroxide, intensity of both Hb subunit monomers and bound dimers bands was decreased (Fig. 25.3A, line 4), which could be a result of oxidative protein degradation. At the same time, DNIC effectively inhibited oxidative modification of Hb bound to it (Fig. 25.3B).

Indeed, if hydrogen peroxide concentration is equimolar or lower (400 and 800 μM) than Hb-DNIC concentration (780 μM), Hb subunit dimers are almost absent (Fig. 25.3B, lines 2 and 3). With high hydrogen peroxide concentration (1600 μM), subunit dimers appeared, but Hb-DNIC prevented protein degradation (Fig. 25.3B, line 4). In the presence of destroyed Hb-DNIC (5-h incubation of Hb-DNIC at room temperature), oxidative modification of Hb is not inhibited (Fig. 25.3C).

These results show that Hb-DNICs function as specific antioxidants to protect bound proteins from oxidative modification.

5.1. Study of Hb-DNIC destruction by hydroperoxides

1. Changes of Hb-DNIC concentration in the presence of *tert*-butyl hydroperoxide and hydrogen peroxide were estimated by the intensity of the EPR signal (Fig. 25.4). The reaction mixture included 150 m*M* K-Na phosphate buffer (pH 7.4), 1.6 m*M* DTPA, 410 μM Hb-DNIC (Hb concentration of 285 μM), and hydroperoxides studied in the range of 0.2 to 3.0 m*M* concentration. Time between component mixing and EPR spectra registration was 2 min. In all experiments, a DTPA-chelating

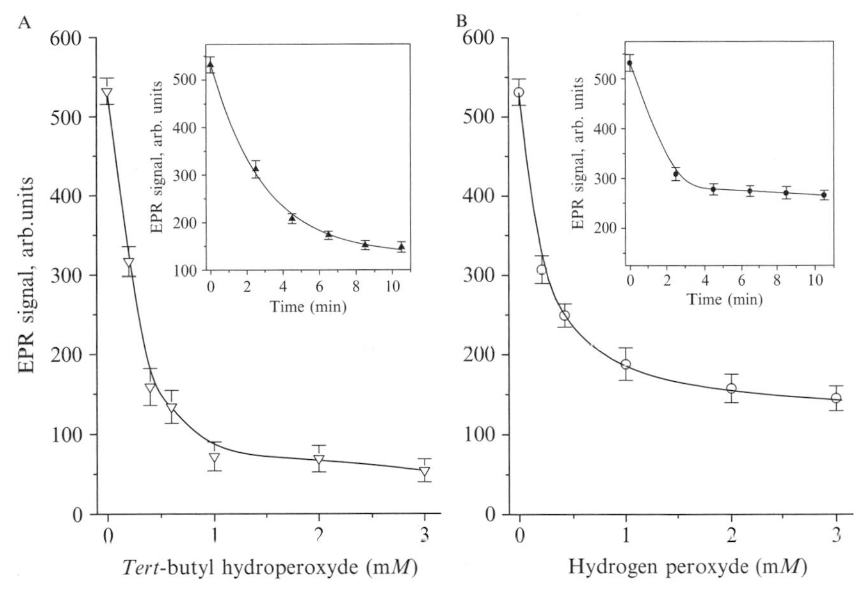

Figure 25.4 Destruction of Hb-DNIC at the presence of *tert*-butyl hydroperoxide (A) and hydrogen peroxide (B) of different concentrations. Inserts show kinetics of Hb-DNIC destruction if concentration of hydroperoxide was 200 μM. Initial Hb-DNIC concentration in reaction mixture was 410 μM.

agent forming redox inactive complexes with Fe ions was added to the medium to prevent generation of radicals in Fenton-type reactions.

2. Low-temperature EPR-spectroscopy was used to study Hb protein radicals (Fig. 25.5). After 1 min of mixing methemoglobin or Hb-DNIC with 1 m*M* hydrogen peroxide, 300-mL aliquots were frozen in liquid nitrogen.

6. Influence of *tert*-Butyl Hydroperoxide and Hydrogen Peroxide on Hb-DNIC

Interaction of hydroperoxides with Hb-DNIC is accompanied by their destruction, which correlated with the protective properties of these complexes. In Fig. 25.4, kinetic curves of Hb-DNIC destruction under

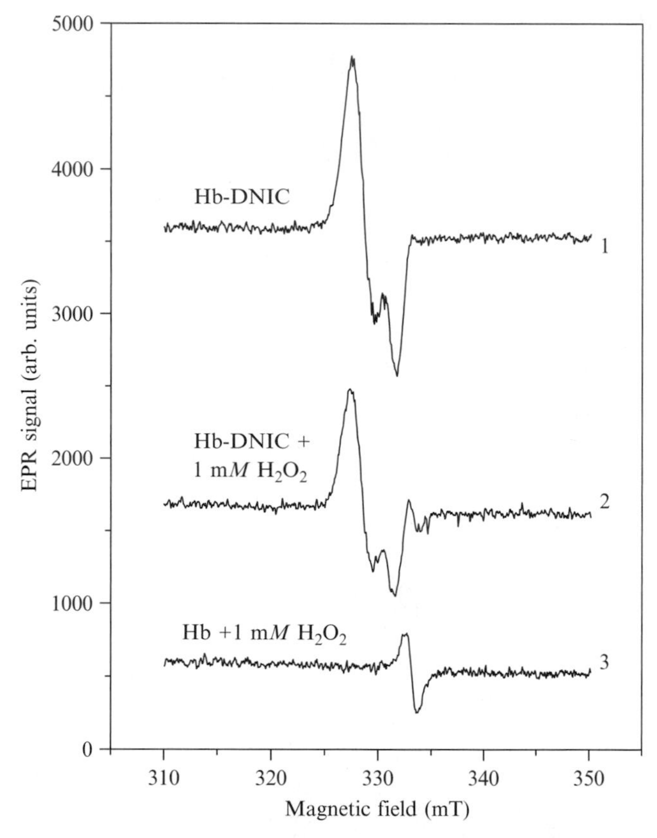

Figure 25.5 EPR spectra of Hb-DNIC (1) and mixtures of Hb-DNIC (2) and methemoglobin (3) with hydrogen peroxide recorded at liquid nitrogen temperature. Concentrations of hemoglobin and hydrogen peroxide in reaction mixture were 0.3 and 1 m*M*, respectively. The time of sample incubation before freezing was 1 min.

action of *tert*-butyl hydroperoxide and hydrogen peroxide are shown. The *tert*-butyl hydroperoxide was more active than hydrogen peroxide, and its kinetic curve was close to an exponential law (Figs. 25.4A and 25.4B). Destruction of Hb-DNIC in the presence of *tert*-butyl hydroperoxide may result from the strong oxidants that appeared in 25.1–25.4 taking place on these complexes.

It is known that NO interacts with alkoxyl and alkylperoxyl radicals in such reactions (Chamulitrat, 2001; Rubbo *et al.*, 1994; Schafer *et al.*, 2002):

$$ROO^{\bullet} + NO^{\bullet} \rightarrow ROONO \rightarrow [RO^{\bullet} + NO_2^{\bullet}] \rightarrow RONO_2 \quad (25.5)$$

$$RO^{\bullet} + NO^{\bullet} \rightarrow RNO_2 \quad (25.6)$$

Rate constants of these diffusion-controlled reactions are very high ($10^9 - 10^{11}\ M^{-1}c^{-1}$) (O'Donnell and Freeman, 2001; Schafer *et al.*, 2002), 10 times higher than the constant of the NO reaction with oxoferrylhemoglobin (Herold and Rehmann, 2003). In this case, degradation of Hb-DNIC by *tert*-butyl hydroperoxide is probably caused by alkoxyl and alkylperoxyl radicals formed from it.

During the interaction of hemoproteins with hydrogen peroxide, one can observe the appearance of a hydroxyl radical, but many scientists have reported the absence of such a process (Giulivi and Cadenas, 1998; Gunter *et al.*, 1998; Winterbourn, 1985). At the same time, in the reaction of hemoglobin with hydrogen peroxide, oxoferryl and probably perferryl forms of heme appeared, which can oxidize amino acid residues of this protein. During this reaction, protein radicals that could be the cause of destruction of Hb-DNIC and oxidative modification of "free" Hb (not bound with DNIC) were formed (see Fig. 25.3). The kinetics of Hb-DNIC destruction in the presence of hydrogen peroxide shows that the time for generation of these radical intermediates is relatively short (not more than 1 to 2 min) (see Fig. 25.4B, insert). The EPR signal of a protein radical forming during interaction of Hb with hydrogen peroxide was detected at the temperature of liquid nitrogen (see Fig. 25.5, spectrum 3). However, it is difficult to assess the influence of Hb-DNIC on properties of free amino acid radicals of hemoglobin because of the superposition of EPR signals of these radicals and DNIC. The protein radical found is most probably an Hb tryptophan or tyrosine residue. It was ascertained that phenoxyl radical Tyr-103 of human hemoglobin can oxidize Cys-110 to form a thiyl radical (Michael *et al.*, 2004). Thus, it is possible that Hb-DNIC also reacts with the tyrosine radical. Participation of $O_2^{-\bullet}$ (forming in the heme pocket at the reaction of ferryl hemoglobin with hydrogen peroxide) in destroying DNIC is also possible (Nagababu and Rifkind, 2000).

6.1. Interaction between superoxide and Hb-DNIC

1. Superoxide was generated by the one-electron reduction of oxygen catalyzed by xanthine oxidase. The reaction mixture contained 150 m*M* K-Na phosphate buffer (pH 7.4), 4 m*M* xanthine, xanthine oxidase (0.6 U/mL), and Hb-DNIC (400 or 780 μM). $O_2^{-\bullet}$ generation was controlled with the spin-trap TIRON. In this case, TIRON was included in the reaction mixture (10 m*M*) rather than Hb-DNIC. During the reaction of TIRON with superoxide, free radical (TIRON semiquinone) appeared, allowing for measurement of the kinetics of $O_2^{-\bullet}$ generation (Ruuge *et al.*, 2002).
2. To decrease $O_2^{-\bullet}$ concentration, SOD (500 U/mL) or Na ascorbate (4.5 m*M*) was added to reaction mixture. Catalase (600 U/mL) was added to study the influence of hydrogen peroxide on Hb-DNIC destruction.
3. Components of the reaction mixture were mixed for 1 min. A sample (80 μL) of mixture was placed in gas-permeable capillary tubes (for experiments with permanent aeration) or in glass capillary tubes (for hypoxia modeling experiments).
4. Nitrosylhemoglobin was obtained by addition of Angeli's salt or PAPA/NONOate (final concentration 10 m*M*) to 0.5 m*M* Hb solution in 150 m*M* K-Na phosphate buffer (pH 7.4). The maximum level of nitrosylhemoglobin was observed after 8 min incubation.

7. Superoxide-Dependent Destruction of Hb-DNIC

Superoxide ($O_2^{-\bullet}$) is a predecessor of the majority of other ROS and probably the reason for NO elimination in biological systems. In these systems, generation of $O_2^{-\bullet}$ is the result of the one-electron reduction of oxygen by the mitochondrial respiratory chain, NADPH-oxidase of macrophages, and other enzymatic systems (Droge, 2003). In our experiments for $O_2^{-\bullet}$ generation, the xanthine–xanthine oxidase enzymatic system was used. Figure 25.6A shows kinetic curves of Hb-DNIC disintegration in this system under permanent aeration. $O_2^{-\bullet}$ generation took place throughout the experiment time ($\approx$14 min). It was confirmed by the formation of TIRON semiquinone under the same conditions in which we saw Hb-DNIC destruction (Fig. 25.6B, curves 1 and 2). The strong evidence that Hb-DNIC destruction is the result of $O_2^{-\bullet}$ action is inhibition of this process by SOD (see Fig. 25.6A, curve 3). Superoxide-dependent destruction of DNIC is effectively inhibited by an antioxidant such as ascorbate (Fig. 25.6A, curve 4). It is interesting that catalase also protected Hb-DNIC

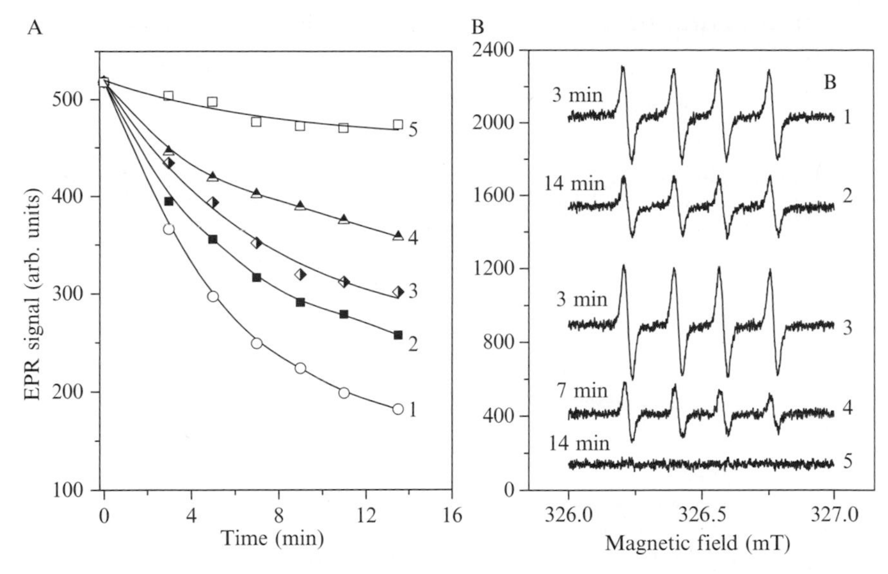

Figure 25.6 A: kinetics of Hb-DNIC destruction. (1) Permanent aeration in the presence of xanthine (4 m*M*) and xanthine oxidase (0.6 U/mL); (2) the same as (1) + catalase (600 U/mL); (3) the same as (1) + SOD (500 U/mL); (4) the same as (1) + Na ascorbate (4.5 m*M*); and (5) incubation of Hb-DNIC in 150 m*M* K-Na phosphate buffer (pH 7.4) without additions. B: TIRON semiquinone formation during enzymatic generation of superoxide at permanent aeration (spectra 1 and 2) and without aeration (spectra 3 to 5). In the reaction mixture (similar to A1), Hb-DNIC are changed toTIRON.

under the same conditions (Fig. 25.6A, curve 2). Thus, the formation of hydrogen peroxide during enzymatic and nonenzymatic SOD probably also takes part in Hb-DNIC destruction. This fact is consistent with results described previously on the destruction of Hb-DNIC in the presence of hydrogen peroxide. Note that the reaction of Hb-DNIC with $O_2^{-\bullet}$ *in vivo* can regulate physiological functions of NO, decreasing concentration of this storage form.

In the absence of aeration in xanthine- and xanthine oxidase–containing medium, we also could see destruction of Hb-DNIC, but the rate of this process was slower than under aeration (Fig. 25.7). During decreasing Hb-DNIC EPR signal intensity, the new EPR signal appeared in the reaction medium (Fig. 25.7D). This signal is typical for the R-form of hemoglobin with nitrosylated Fe of heme (Gow *et al.*, 1999). It is possible that this change of NO binding is NO regeneration from products of reaction of Hb-DNIC with $O_2^{-\bullet}$. These products could be nitrite or nitrate, which are reducing in hypoxic conditions (Li *et al.*, 2001; Millar *et al.*, 1998). The appearance of hypoxic conditions in the system studied could be detected by the stopping of the formation of the free radical TIRON (Fig. 25.6B,

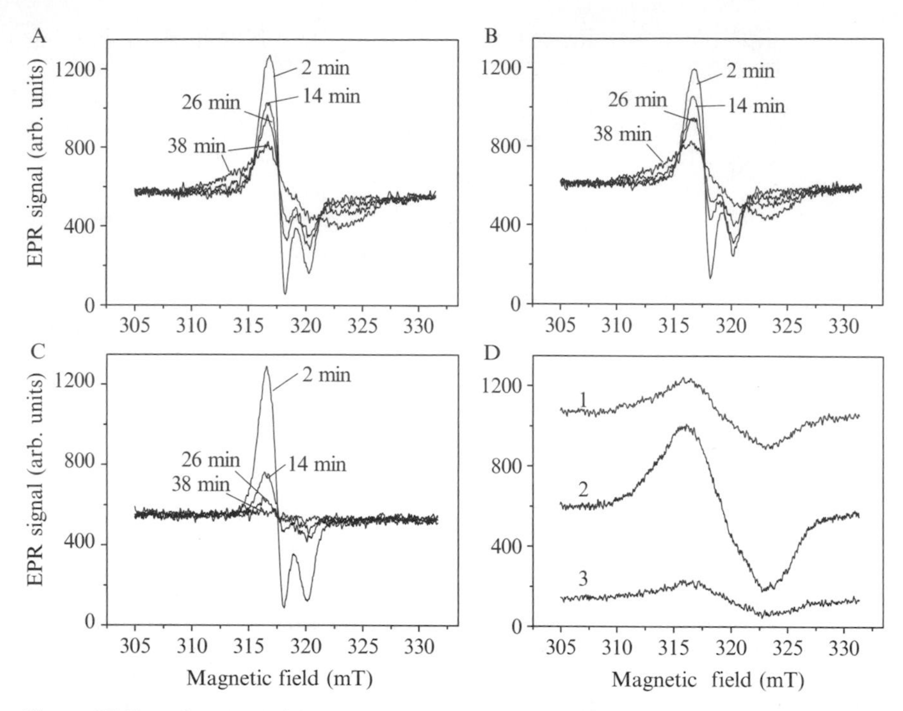

Figure 25.7 Hb-DNIC destruction and nitrosylhemoglobin formation in xanthine–xanthin oxidase model system. (A): System without aeration. The reaction mixture contained 150 m*M* K-Na phosphate buffer (pH 7.4), 4 m*M* xanthine, xanthin oxidase (0.6 U/mL), and 780 μM Hb-DNIC. (B): System (A) + 1.6 m*M* DTPA. (C): System (A) with aeration. (D): EPR spectra of nitrosylhemoglobin obtained 7 min after addition of 10 m*M* Angeli's salt (D2) or 10 m*M* PAPA/NONOate (D3) to 0.5 m*M* methemoglobin solution in 150 m*M* K-Na phosphate buffer (pH 7.4). (D1) shows EPR spectrum of nitrosylhemoglobin registered after 38-min incubation of (B) mixture.

spectra 3 to 5). The concentration of oxygen, which was the substrate for xanthine oxidase, decreased to a level insufficient for $O_2^{-\bullet}$ generation. In this case, NO_2^- and NO_3^- began to work as electron acceptors in the xanthine oxidase reaction. It is necessary to note that addition of DTPA, sharply reducing redox activity of Fe ions, did not really influence nitrosylhemoglobin formation (Fig. 25.7B). Thus, in our experiments, NO reduction was not connected to the action of free iron. Hb nitrosylation as a result of Hb-DNIC destruction in hypoxic conditions in the presence of xanthine and xanthine oxidase was 40% of the maximum nitrosylhemoglobin level. This maximum could be reached by addition of Angeli's salt (donor of nitroxyl anion NO^-) to Hb (Fig 25.7D, spectrum 2).

Hb-DNIC could be NO donors. During the interaction of superoxide with one nitrosyl ligand, the second would be released to the medium as NO. It is known that superoxide can reduce methemoglobin (Winterbourn, 1985) and the forming of oxygenated Hb interacts with free NO in the following reaction:

$$Hb\text{-}Fe^{II}\text{-}O_2 + NO \rightarrow Hb\text{-}Fe^{III} + NO_3 \qquad (25.7)$$

Methemoglobin and nitrate are also products of nitrosylhemoglobin oxidation by molecular oxygen (Gow *et al.*, 1999; Herold and Rock, 2005). Thus, if oxygen concentration in our reaction mixture is enough for these reactions, there is no accumulation of nitrosylated hemoglobin in the mixture. Nitrosylhemoproteins were detected in cardiac ischemia muscle (Tiravanti *et al.*, 2004). Thus, it is possible that nitrosylhemoglobin forms *in vivo* under hypoxic conditions in ischemically degenerated tissues. In this case, NO is a product not only of NO synthases but also of enzymatic reduction of nitrate and nitrite and a result of release from DNIC.

8. Conclusion

Hb-DNICs react with many active metabolites that take part in processes that are the basis of oxidative stress. One of the important functions of Hb-DNIC is its protection of oxidative modification of hemoglobin in the unbalanced pro- and antioxidant reactions in blood. Stable concentration of different DNIC forms in biological systems is usually lower than can be detected with EPR, which is the only method for their detection. However, the level of Hb-DNIC can be increased by the addition of exogenic NO donors to an animal organism. In this case, we propose the use of Hb-DNIC as a marker of ROS formation in model systems *in vivo* and *in vitro*. The possibility of simultaneous registration of NO both included in Hb-DNIC and bound with heme of nitrosohemoglobin can be used to study transformations of NO metabolites. It is interesting that modified (glycosylated) hemoglobin is one of the main molecular markers in studies of diabetes. In addition, glycosylation of Hb influences its interaction with NO (James *et al.*, 2004). It is known that the actions of different NO donors on pathological changes in cells and tissues during oxidative stress are various. Thus, a study of the mechanisms of interaction of ROS with Hb-DNIC is important to the investigation of the role of these physiological complexes of NO in the processes of free radical oxidation.

Acknowledgments

The work was carried out with the support of the Russian Foundation for Basic Research (grants 05–04–49751 and 06–04–81054Bel_a).

REFERENCES

Chamulitrat, W. (2001). EPR studies of nitric oxide interactions of alkoxyl and peroxyl radicals in *in vitro* and *ex vivo* model systems. *Antioxid. Redox Signal.* **3,** 177–187.

Droge, W. (2003). Free radicals in the physiological control of cell function. *Physiol. Rev.* **82,** 47–95.

Giulivi, C., and Cadenas, E. (1998). Heme protein radicals: Formation, fate, and biological consequences. *Free Radic. Biol. Med.* **24,** 269–279.

Gorbunov, N. V., Osipov, A. N., Day, B. W., Zayas-Rivera, B., Kagan, V. E., and Elsayed, N. M. (1995). Reduction of ferrylmyoglobin and ferrylhemoglobin by nitric oxide: A protective mechanism against ferrylhemoprotein-induced oxidations. *Biochem.* **34,** 6689–6699.

Gow, A. J., Luchsinger, B. P., Pawloski, J. R., Singel, D. J., and Stamler, J. S. (1999). The oxyhemoglobin reaction of nitric oxide. *Biochem.* **96,** 9027–9032.

Gunter, M. R., Tschirret-Gutt, R. A., Witkowska, H. E., Fann, Y. C., Barr, D. P., Oritz de Montellano, P. R., and Mason, R. P. (1998). Site-specific spin trapping of tyrosine radicals in the oxidation of metmyoglobin by hydrogen peroxide. *Biochem. J.* **330,** 1293–1299.

Herold, S., and Rehmann, F.-J. K. (2003). Kinetics of the reactions of nitrogen monoxide and nitrite with ferryl hemoglobin. *Free Radic. Biol. Med.* **34,** 531–545.

Herold, S., and Rock, G. (2005). Mechanistic studies of the oxygen-mediated oxidation of nitrosylhemoglobin. *Biochem.* **44,** 6223–6231.

James, P. E., Lang, D., Tufnell-Barret, T., Milsom, A. B., and Frenneaux, M. P. (2004). Vasorelaxation by red blood cells and impairment in diabetes: Reduced nitric oxide and oxygen delivery by glycated hemoglobin. *Circ. Res.* **94,** 976–983.

Jeney, V., Balla, J., Yachie, A., Varga, Z., Vercellotti, G. M., Eaton, J. W., and Balla, G. (2002). Pro-oxidant and cytotoxic effects of circulating heme. *Blood* **100,** 879–887.

Joshi, M. S., Ponthier, J. L., and Lancaster, J. R., Jr. (1999). Cellular antioxidant and pro-oxidant action of nitric oxide. *Free Radic. Biol. Med.* **27,** 1357–1366.

Jucket, M., Zheng, Y., Yuan, H., Pastor, T., Antholine, W., Weber, M., and Vercellotti, G. (1998). Heme and the endothelium: Effects of nitric oxide on catalytic iron and heme degradation by heme oxygenase. *J. Biol. Chem.* **273,** 23388–23397.

Kagan, V. E., Kozlov, A. V., Tyurina, Y. Y., Shvedova, A. A., and Yalowich, J. C. (2001). Antioxidant mechanisms of nitric oxide against iron-catalyzed oxidative stress in cells. *Antioxid. Redox Signal.* **3,** 189–202.

Kruszewski, M. (2004). The role of labile iron pool in cardiovascular diseases. *Acta Biochim. Pol.* **51,** 471–480.

Laemmli, U. K. (1970). Cleavage of structural proteins during the assembly of the head of bacteriophage T4. *Nature* **227,** 680–685.

Li, H., Samouilov, A., Liu, X., and Zweier, J. L. (2001). Characterization of the magnitude and kinetics of xanthine oxidase catalyzed nitrite reduction: Evaluation of its role in nitric oxide generation in anoxic tissues. *J. Biol. Chem.* **276,** 24482–24489.

Luchsinger, B. P., Rich, E. N., Gow, A. J., Williams, E. M., Stamler, J. S., and Singel, D. J. (2003). Routes to S-nitroso-hemoglobin formation with heme redox and preferential reactivity in the β-subunits. *Proc. Natl. Acad. Sci. USA* **100,** 461–466.

Ma, X. L., Gao, F., Liu, G.-L., Lopez, B. L., Christopher, T. A., Fukuto, J. M., Wink, D. A., and Feelisch, M. (1999). Opposite effects of nitric oxide and nitroxyl on postischemic myocardial injury. *Proc. Natl. Acad. Sci. USA* **96,** 14617–14622.

Michael, G. R. (2004). Probing the free radicals formed in the metmyoglobin-hydrogen peroxide reaction. *Free Radic. Biol. Med.* **36,** 1345–1354.

Millar, T. M., Stevens, C. R., Benjamin, N., Eisental, R., Harrison, R., and Blake, D. R. (1998). Xanthine oxidoreductase catalyses the reduction of nitrates and nitrite to nitric oxide under hypoxic conditions. *FEBS Lett.* **427,** 225–228.

Muller, B., Kleschyov, A. L., Alencar, J. L., Vanin, A. F., and Stoclet, J. C. (2002). Nitric oxide transport and storage in the cardiovascular system. *Ann. NY Acad. Sci.* **962,** 131–139.

Nagababu, E., and Rifkind, J. M. (2000). Reaction of hydrogen peroxide with ferrylhemoglobin: Superoxide production and heme degradation. *Biochemistry* **39,** 12503–12511.

O'Donnell, V.B, and Freeman, B. A. (2001). Interaction between nitric oxide and lipid oxidation pathways. *Circ. Res.* **88,** 12–21.

Reeder, B. J., and Wilson, M. T. (2005). Hemoglobin and myoglobin associated oxidative stress: From molecular mechanisms to disease states. *Curr. Med. Chem.* **12,** 2741–2751.

Romero, N., Radi, R., Linares, E., Augusto, O., Detweiler, C. D., Mason, R. P., and Denicola, A. (2003). Reaction of human hemoglobin with peroxynitrite: Isomerization to nitrate and secondary formation of protein radicals. *J. Biol. Chem.* **278,** 44049–44057.

Rubbo, H., Radi, R., Trujillo, M., Telleri, R., Kalyanaraman, B., Barnes, S., Kirk, M., and Freeman, B. A. (1994). Nitric oxide regulation of superoxide and peroxinitrite-dependent lipid oxidation: Formation of novel nitrogen-containing oxidized lipid derivatives. *J. Biol. Chem.* **269,** 26066–26075.

Ruuge, E. K., Zabbarova, I. V., Korkina, O. V., Khatkevich, A. N., Lakomkin, V. L., and Timoshin, A. A. (2002). Oxidative stress and myocardial injury: Spin-trapping and low-temperature EPR study. *Curr. Top. Biophysics* **26,** 145–155.

Sanina, N. A., Aldoshin, S. M., Rudneva, T. N., Golovina, N. I., Shilov, G. V., Shul'ga Yu, M., Martynenko, V. M., and Ovanesyan, N. S. (2005). Synthesis, structure and solid-phase transformations of Fe nitrosyl complex $Na_2[Fe_2(S_2O_3)_2(NO)_4]^{\bullet}\ 4H_2O$. *Russ. J. Coord. Chem.* **31,** 323–328.

Schafer, F.Q, Wang, H. P., Kelley, E. E., Gueno, K. L., Martin, S.M, and Buetter, G. R. (2002). Comparing β-carotene, vitamin E and nitric oxide as membrane antioxidants. *J. Biol. Chem.* **383,** 671–681.

Shumaev, K. B., Petrova, N. E., Zabbarova, I. V., Vanin, A. F., Topunov, A. F., Lankin, V. Z., and Ruuge, E. K. (2004). Interaction of oxoferrylmyoglobin and dinitrosyl-iron complexes. *Biochem. (Moscow)* **69,** 569–574.

Shumaev, K. B., Gubkin, A. A., Gubkina, S. A., Gudkov, L. L., Sviryaeva, I. V., Timoshin, A. A., Topunov, A. F., Vanin, A. F., and Ruuge, E. K. (2006). Interaction between dinitrosyl-iron complexes and intermediates of oxidative stress. *Biophysics* **51,** 423–428.

Tiravanti, E., Samouilov, A., and Zweier, J. L. (2004). Nitrosyl-heme complexes are formed in the ischemic heart: Evidence of nitrite-derived nitric oxide formation, storage, and signaling in post-ischemic tissues. *J. Biol. Chem.* **279,** 11065–11073.

Vanin, A. F., and Chetverikov, A. G. (1968). Paramagnetic nitrosyl complexes of heme and non-heme. *Biofizika (Rus.)* **13,** 608–616.

Vanin, A. F., Blumenfeld, L. A., and Chetverikov, A. G. (1967). Investigation of non-heme iron complexes in cells and tissues by the EPR method. *Biofizika (Rus.)* **12,** 829–841.

Vanin, A. F., Serezhenkov, V. A., Mikoyan, V. D., and Genkin, M. V. (1998). The 2,03 signal as an indicator of dinitrosyl-iron complexes with thiol-containing ligands. *Nitric Oxide* **2,** 224–234.

Winterbourn, C. C. (1985). Free radical production and oxidative reaction of hemoglobin. *Environ. Health Perspect.* **64,** 321–330.

Zee, J. (1997). Formation of peroxide- and globin-derived radicals from the reaction of methemoglobin and metmyoglobin with *t*-butyl hydroperoxide: An ESR spin-trapping investigation. *Biochem. J.* **322,** 633–639.

CHAPTER TWENTY-SIX

Linked Analysis of Large Cooperative, Allosteric Systems: The Case of the Giant HBL Hemoglobins

Nadja Hellmann,* Roy E. Weber,† *and* Heinz Decker*

Contents

Abstract

Homotropic and heterotropic allosteric interactions are important mechanisms that regulate protein function. These mechanisms depend on the ability of oligomeric protein complexes to adopt different conformations and to transmit conformation-linked signals from one subunit of the complex to the neighboring ones. An important step in understanding the regulation of protein function is to identify and characterize the conformations available to the protein complex. This task becomes increasingly challenging with increasing numbers of interacting binding sites. However, a large number of interacting binding sites allows for high homotropic interactions (cooperativity) and thus represents the most interesting case. Examples of very large, cooperative protein complexes are the

* Institute for Molecular Biophysics, Johannes Gutenberg University, Mainz, Germany
† Zoophysiology, Institute of Biological Sciences, University of Aarhus, Denmark

Methods in Enzymology, Volume 436
ISSN 0076-6879, DOI: 10.1016/S0076-6879(08)36026-1

giant hexagonal bilayer hemoglobins of annelid worms that contain 144 oxygen-binding sites. Moreover, these proteins show strict hierarchy in structure. In order to understand the interaction of various ligands such as oxygen, CO, or nitric oxide (NO), the principle binding behavior of these protein complexes has to be understood. For the hemoglobins of two species, the hierarchical structure is shown to have functional implications. By employing simultaneous analysis of several oxygen-binding curves, it could be shown that the nested MWC model provides a good description of the functional data. A strategy for the experimental setup and data analysis is suggested that allows for a reduction in the number of free parameters. Possible advantages of a hierarchical cooperative model compared to a linear extension of the MWC model are discussed.

1. Introduction

Hexagonal bilayer (HBL) hemoglobins (Hbs) are large hetero-oligomeric complexes that occur freely dissolved in the hemolymph of annelid (e.g., leeches, earthworms, polychaetes) and vestimentiferan worms (Weber and Vinogradov, 2001). These oligomers have molecular weights of approximately 3.6 MDa and are formed by a hierarchical assembly of 144 globin chains and a number of linker chains (Lamy *et al.*, 1996; Vinogradov, 1985). The HBL architecture is characterized by two superimposed rings that each contains six dodecamers and surround a central cavity; each dodecamer comprises three tetramers (12 oxygen-binding heme groups [de Haas *et al.*, 1996b; Fushitani and Riggs, 1988; Kapp *et al.*, 1990; Lamy *et al.*, 1996; Weber and Vinogradov, 2001]) (Fig. 26.1). The hierarchical arrangement can thus be written as $2 \times [6 \times (3 \times 4)]$. This quaternary structure, which is based on EM-reconstructions of different HBLs (de Haas *et al.*, 1996a,b,c; Zal *et al.*, 1996b), is supported by the solution of recent X-ray structure of the HBL Hb from the earthworm *Lumbricus terrestris* (Royer *et al.*, 2000, 2006). The 144 heme-bearing globin chains within this oligomer bind oxygen cooperatively, a function that which is allosterically regulated by protons and divalent cations such as Ca^{2+} and Mg^{2+} (Ochiai and Weber, 2002; Weber, 1981).

An obvious question is whether the complex hierarchical structure of the HBL Hbs has functional implications. An extensive part of the available data on HBL Hbs pertains to native HBL molecules from the earthworm *L. terrestris*. Understanding the functional implications of the hierarchical structure requires knowledge about the oxygen-binding properties of both the native 144-mer structures and of its dissociation products. The size of HBL dissociation products induced by chemical treatment depends on the species. In the presence of urea, leech *Macrobdella* Hb dissociates predominantly into tetrameric subunits (Weber *et al.*, 1995), whereas earthworm *Lumbricus* Hb dissociates predominantly into dodecameric structures that dissociate further into tetramers and monomers in the presence of guanidinium hydrochloride (Krebs *et al.*, 1996). Again, whereas dissociation experiments indicate that earthworm and polychaete (*Alvinella*) dodecamers consist of disulfide-bonded trimers and monomers, having an $[a + b + c]_3[d]_3$ stoichiometry, vestimentiferan (*Riftia*) and leech Hbs consist of disulfide bonded dimers and monomers (Lamy *et al.*, 1996; Zal *et al.*, 1996a).

Although the dissociation products may not occur *in vivo*, a study of their molecular and functional properties yields important information about the minimal oligomeric structure required to exhibit the functional properties of the native oligomer, or alternatively, to what extent the functional properties manifested in the HBL structures are expressed in the constituent subassemblies and subunits. However, one needs to distinguish between those intersubstructure interactions that maintain the oligomeric structure and are independent of oxygen and effector binding and those interactions that transmit the information about the conformational state of the neighboring substructures, giving rise to the cooperative effects. Because for HBL Hbs no dissociation/association was observed upon changes in oxygen partial pressure, one can assume that only a small fraction of all interactions are responding to conformational changes. Thus, to probe the influence of molecular hierarchy on the functional properties of HBLs, one has to analyze the oxygen-binding curves of the native 144-mer on the basis of appropriate models for cooperativity.

Past analyses of oxygen-binding data of HBL hemoglobins have been almost entirely restricted to the application of the MWC model. However, if oxygen-binding curves under different conditions (e.g., pH) are compared, it becomes evident the MWC model provides an inadequate description of the oxygen-binding behavior of these respiratory proteins. Indications for this mismatch are, for example, the dependence of the allosteric size on pH (Fushitani and Riggs, 1991; Fushitani *et al.*, 1986) with values between 5 and 12 coupled subunits, and variations in the values for the affinities of the two conformations (T and R) upon changes in effector concentration (Fushitani *et al.*, 1986; Igarashi *et al.*, 1991). In contrast, binding curves obtained under a single set of conditions might well be described by the MWC model.

Thus, there is a need for strategies to analyze a number of binding curves simultaneously using possibly complicated models based on a larger number of parameters. Here, we present such a strategy as applicable to, for example, the HBL Hb of *L. terrestris*. The models used are exclusively extensions of the MWC model, since for complexes such as HBL molecules containing a large number of binding sites, other approaches would include too many degrees of freedom.

2. Materials

2.1. Hb preparation and oxygen-binding curves

The Hb was prepared according to Vinogradov *et al.* (1991). Oxygen-binding curves were determined using a modified diffusion-chamber technique (Weber, 1981; Weber *et al.*, 1987) in the presence of Tris/bisTris buffers at 20° at pH values near 7.0 and 7.8, and in the presence of either 1.6 mM $CaCl_2$ and 0.15 mM $MgCl_2$, or of 44 mM $CaCl_2$ in the absence of magnesium.

3. Theory

3.1. Fitting function

For each model applied to the binding data, the corresponding binding polynomial was constructed. The saturation degree $\bar{x}$ with respect to oxygen partial pressure x was derived from the binding polynomial P as follows (Wyman and Gill, 1990):

$$\bar{x} = \frac{\partial \ln P}{n \partial \ln x} \tag{26.1}$$

Furthermore, slight experimental uncertainties in the 0 and 100% saturation values were allowed. Thus, the function f fitted to the data was

$$f = (A - Ao)\bar{x} + Ao \tag{26.2}$$

The fitting routine was based on a nonlinear regression analysis (Levenberg-Marquardt routine) incorporated into the program Sigma Plot 2001 (SPSS, Illinois, USA). The iterations stopped when the change in the sum of the square of residuals was less than 0.0001.

3.2. Models

Two different strategies were followed: the combined analysis of two different oligomeric states of leech *Macrobdella decora* Hb (Hellmann *et al.*, 2003b) and the combined analysis of one oligomeric state at four different linked sets of effector concentration of earthworm *L. terrestris* Hb. All models are variants or extensions of the MWC model.

3.2.1. MWC model

The MWC model for an allosteric unit of size q is given by the following binding polynomial (Monod *et al.*, 1965):

$$P_{mwc,q} = (1 + K_r x)^q + l_o(1 + K_t x)^q = P_r^q + l_o P_t^q \qquad (26.3)$$

Thus, the molecule can adopt two different conformations t and r that are characterized by binding constants K_t and K_r. The equilibrium constant describing the ratio between the two conformations in the unligated state is denoted $l_o = [t_o]/[r_o]$.

In order to avoid confusion, we point out here the difference between the Hill coefficient (n_H) and q, which is the size of the allosteric unit in a concerted model for cooperativity based on the MWC model. In such a model, it is assumed that under equilibrium conditions, a number of q subunits always adopt the same conformation. Thus, the parameter q of the MWC model describes exactly how many subunits are coupled. All subunits in this allosteric unit have the same affinity for ligands. In contrast, the Hill coefficient is a measure of cooperativity and depends on both the size of the allosteric unit and all relevant equilibrium constants (binding constants and allosteric equilibrium constants). The Hill coefficient can never exceed the number of interacting sites, which is the size of the allosteric unit in a concerted model ($n_H < q$). Thus, the Hill coefficient represents a lower limit for the size of the allosteric unit.

3.2.2. 3-state model

The 3-state model, the simplest extension of the MWC model, assumes three different conformations (r, t, s) that are in an equilibrium described by the corresponding allosteric equilibrium constants. The binding polynomial is given by

$$P_{3state} = (1 + K_R x)^q + l_T(1 + K_T x)^q + l_S(1 + K_S x)^q \qquad (26.4)$$

Here, the allosteric equilibrium constants are defined as $l_t = [t_o]/[r_o]$ and $l_s = [s_o]/[r_o]$.

3.2.3. Nested MWC model

For analysis of the whole 144-meric molecule, the nested MWC model (Decker *et al.*, 1986; Robert *et al.*, 1987) was applied in the following form:

$$\begin{aligned} P_{nest} &= P_R^s + \Lambda P_T^s = (P_{rR}^w + l_R P_{tR}^w)^s + \Lambda(P_{rT}^w + l_T P_{tT}^w)^s \\ \Lambda &= L\frac{(1 + l_R)^s}{(1 + l_T)^s} \\ P_{\alpha\beta} &= 1 + K_{\alpha\beta} x \\ \alpha\beta &= rR, tR, rT, tT \end{aligned} \qquad (26.5)$$

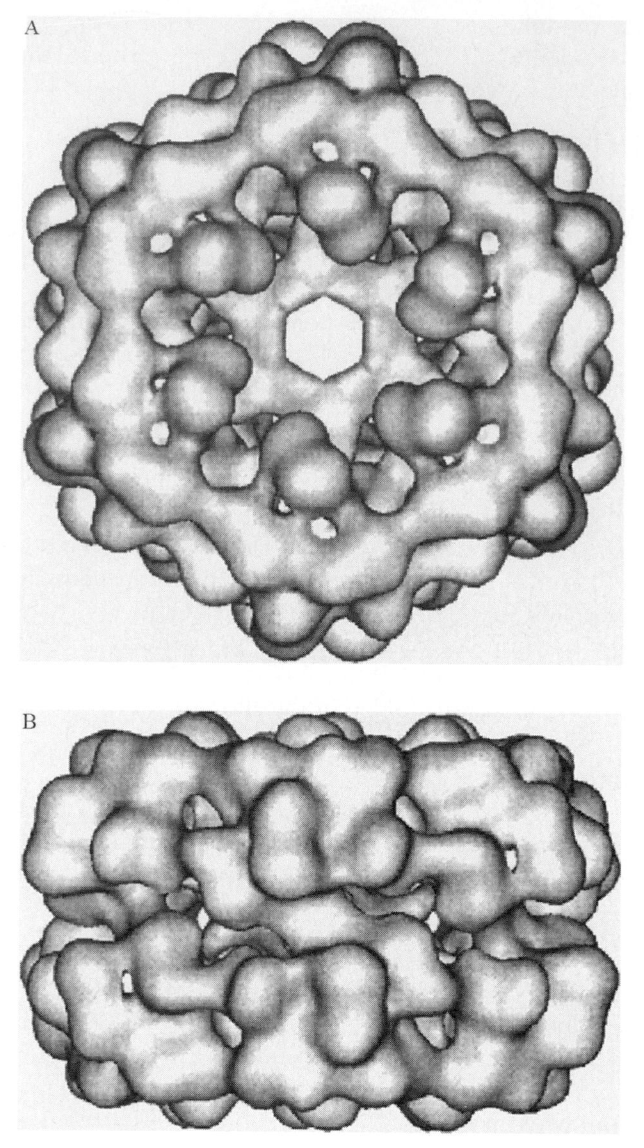

Figure 26.1 Continued

In this model, a number of w subunits form an allosteric unit according to the MWC model and can adopt two basic conformations r and t. However, when these allosteric units assemble to a larger structure, consisting of s copies of the w-sized allosteric units, additional constraints are imposed on the conformations. The $w \times s$ structure can in turn be considered a (large) allosteric unit with size $q = w \times s$, which can adopt two

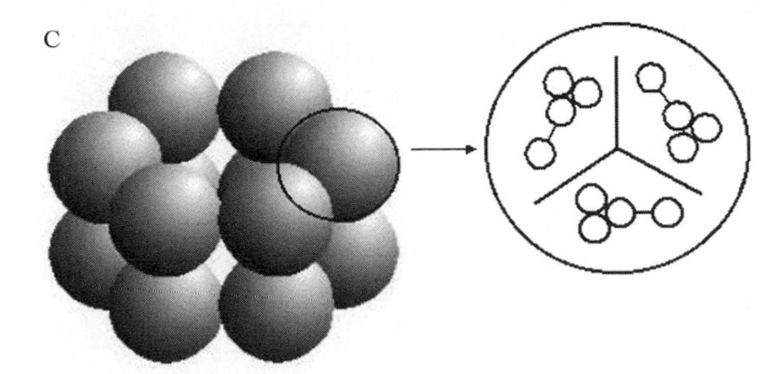

Figure 26.1 HBL hemoglobin assembly. (A) and (B): EM reconstructions, taken from Weber and Vinogradov (2001). (C): Schematic drawing of the 12 × 12-mer and the 3 tetramers within a 12-mer.

conformations R and T. When the $w \times s$-mer is in the R state, the nested w-meric allosteric units can adopt the two conformations rR and tR. When the $w \times s$-mer is in the T state, two other conformations, rT and tT, are available. Thus, the conformation of the $w \times s$-mer defines the conformation of the nested w-mers in a hierarchical manner. The equilibrium between unliganded R_o and T_o states is given by $L = [T_o]/[R_o]$. The allosteric equilibrium constants l_R and l_T correspond to $l_R = [tR_o]/[rR_o]$ and $l_T = [tT_o]/[rT_o]$. The hierarchical cooperative model for three cooperatively linked tetramers ($s = 3$, $w = 4$) is shown in Fig. 26.2. It should be kept in mind that the largest allosteric unit is not necessarily the whole molecule. The largest allosteric unit might be smaller if further oligomerization of these allosteric units to the final structure does not induce functional coupling.

In the case of HBL extracellular annelid Hb, several possibilities for such hierarchical structures exist. We applied two different models. First, the tetramer is described as the smallest allosteric unit. At the next level, the 3 tetramers interact and form a larger allosteric unit: the tetrameric allosteric unit is nested into the 12-meric allosteric unit (see Fig. 26.2). This model assumes the absence of heterogeneity in the individual subunits and the absence of interactions between the 12 dodecamers in the 144-meric molecule. Second, a model where the smallest allosteric unit is the 12-mer was applied. It is assumed that the allosteric units nest into the 6 × 12-meric half-molecule and that there are no further interactions between the half molecules.

3.3. Influence of allosteric effectors

In any purely concerted model for cooperativity, the influence of effectors on the conformational distribution can be described as a shift in the allosteric equilibrium constants. Each allosteric equilibrium constant L_{AB} describes

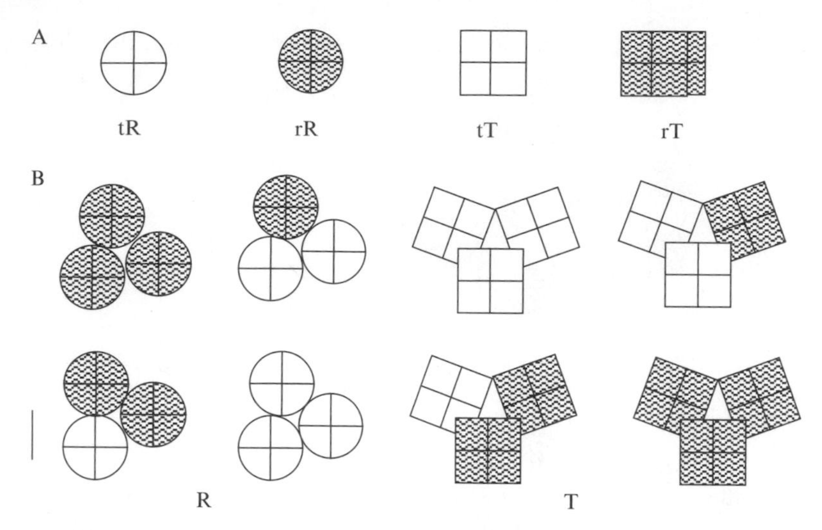

Figure 26.2 Scheme for the 3 × 4 nested MWC model. (A): Conformations available for the small allosteric unit (tetramer). (B) Composition of tetramer conformations restrained by the conformation of the 3 × 4 tetramer. A number of s = 3 copies of the allosteric units (tetramers) assemble to a larger allosteric unit the s × w-mer (here the 3 × 4-mer). The two s × w substructures in the R and T states are in an allosteric equilibrium defined by the equilibrium constant $L = [R_o]/[T_o]$. Within these two conformations, the nested allosteric units with w = 4 subunits can adopt two conformations, r and t. Thus, the allosteric units can adopt four conformations for T and R, respectively.

the equilibrium between two conformations A and B in the absence of ligand:

$$L_{AB} = \frac{[A]_o}{[B]_o} \tag{26.6}$$

An allosteric effector has, by definition, different binding affinities for these two conformations, and therefore the conformational distribution is shifted in the presence of this effector (y). Let L^o_{AB} be the allosteric equilibrium constant in the absence of this effector. Then the allosteric equilibrium constant $L_{AB}(y)$ in the presence of the effector y is given by

$$L_{AB}(y) = L^o_{AB}\frac{Q_A}{Q_B} = L^o_{AB}F_{AB} \tag{26.7}$$

and

$$\frac{Q_A}{Q_B} = F_{AB}$$

Here, Q_A and Q_B are the binding polynomials of the effector for the two conformations. These polynomials contain the concentration of the effector and the binding constants of the binding sites of the allosteric unit to which the allosteric equilibrium constant refers. In the simplest case, the allosteric unit offers m identical binding sites. Then, the binding polynomial is written as

$$Q_{\alpha\beta} = (1 + z_{\alpha\beta}\gamma)^{m} \quad \alpha\beta = A, B \tag{26.8}$$

Here, the binding constants of the effector for the two conformations are z_A and z_B, respectively.

For the present purpose, the effector-binding polynomials do not need to be specified. The important thing is that these polynomials are a constant for a given free effector concentration.

In order to describe the effect of more than one effector that binds to independent binding sites, a pair of binding polynomials Q_A and Q_B has to be introduced for each effector:

$$L_{AB}(\gamma_1, \gamma_2) = L^{o}_{AB} \frac{Q_A^{(\gamma_1)}}{Q_B^{(\gamma_1)}} \frac{Q_A^{(\gamma_2)}}{Q_B^{(\gamma_2)}} \tag{26.9}$$

Again, one may express the change in the allosteric equilibrium constant in terms of the effector-dependent factors $F^{(y1)}$ and $F^{(y2)}$:

$$L_{AB}(\gamma_1, \gamma_2) = L^{o}_{AB} F^{(\gamma 1)} F^{(\gamma 2)} \tag{26.10}$$

In some cases (e.g., with protons as effector), the allosteric equilibrium constant in the absence of effector (L^{o}_{AB}) is not very useful. Then, one needs the relationship between the allosteric equilibrium constant of an arbitrary effector concentration (y) relative to a reference state (y_{ref}). This can be obtained by applying Eq. (26.7):

$$L_{AB}(\gamma_{ref}) = L^{o}_{AB} \frac{Q_A(\gamma_{ref})}{Q_B(\gamma_{ref})} \quad L_{AB}(\gamma) = L^{o}_{AB} \frac{Q_A(\gamma)}{Q_B(\gamma)},$$

which yields the following equation:

$$\begin{aligned} L_{AB}(\gamma) &= L^{ref}_{AB} F^{(\Delta\gamma)} \\ F^{(\Delta\gamma)} &= \frac{Q_A^{(\gamma)}}{Q_B^{(\gamma)}} \frac{Q_B^{(\gamma_{ref})}}{Q_A^{(\gamma_{ref})}} \end{aligned} \tag{26.11}$$

3.4. Constraints on the allosteric equilibrium constants

A factor that describes the changes in allosteric equilibrium constants resulting from changes in effector concentration can be employed to reduce the number of free parameters describing a particular model. The strategy is

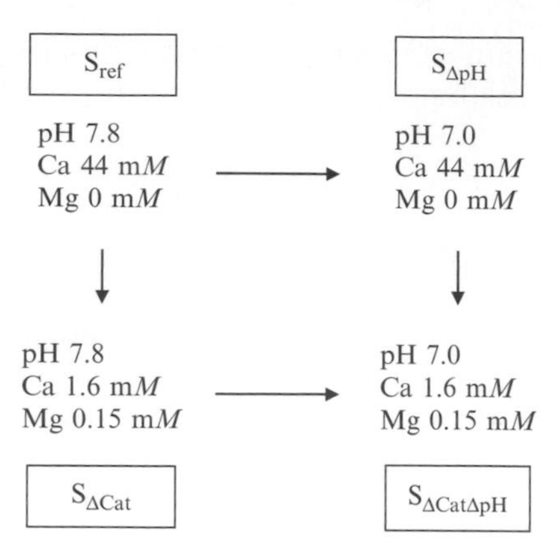

Figure 26.3 Set of effector concentrations chosen to enable a linked analysis. Starting from the reference state, the different experimental conditions can be grouped in a circle, so that from one state to the next only one type of condition (either pH or divalent ion levels) is changed.

demonstrated exemplarily for the conditions employed in the study of the HBL Hb of *L. terrestris* as depicted in Fig. 26.3, for the case of the MWC model. Each pair of conditions has either the same pH or the same Ca^{2+}/Mg^{2+} concentrations. Denote, for example, the allosteric equilibrium constant L at pH 7.8 in the presence of 44 m*M* Ca^{2+} as reference state (L^{ref}). When the same experiment is performed in the presence of 1.6 m*M* Ca^{2+} and 0.15 m*M* Mg^{2+}, the allosteric equilibrium constant is changed by a factor $F^{cat} = F^{\Delta Ca}F^{\Delta Mg}$. Similarly, if the allosteric equilibrium constants obtained at pH 7.0 in the presence of 44 m*M* Ca^{2+} and of 1.6 m*M* Ca^{2+} and 0.15 m*M* Mg^{2+} are compared, they differ by the same factor F^{Cat}. Thus, if the allosteric equilibrium constants for the four data sets are expressed relative to the reference state, the following relations are obtained for L:

$$\begin{aligned} L_{S\Delta pH\Delta Cat} &= L^{ref} F^{cat} F^{\Delta pH} \\ L_{S\Delta pH} &= L^{ref} F^{\Delta pH} \\ L_{S\Delta Cat} &= L^{ref} F^{cat} \\ L_{s,ref} &= L^{ref} \end{aligned} \tag{26.12}$$

The indices refer to labels in Fig. 26.3.

3.5. Statistics

In order to evaluate the applicability of the different models, the probability Q that the observed squared difference between data and the specific model, as measured by χ^2, occurs by chancc was calculated using the complement of the incomplete Gamma function (Press *et al.*, 1989):

$$\chi^2 = \sum_{1}^{n} \frac{(y_i - f_i)^2}{\sigma_i^2}$$
$$P(a,b) = \frac{\Gamma(a,b)}{\Gamma(a)} = \frac{1}{\Gamma(a)} \int_b^\infty e^{-t} t^{a-1}\, dt \qquad (26.13)$$

where a = (N − 2)/2 and b = $\chi^2/2$. Here, N is the degree of freedom = number of data points (n) minus number of parameters. This approach is based on the assumption that the errors show a normal distribution. Then, a value of 0.1 can be considered an indication that the goodness-of-fit is acceptable. If it is greater than about 0.001, the fit may be acceptable if the errors are nonnormal or have been moderately underestimated (Press *et al.*, 1989).

For the calculation of χ^2, an error (σ_i) of 5% in the saturation values was assumed for saturation values greater than 0.2 and an absolute error of 0.01 for values less than 0.2.

4. HBL Hb from *L. terrestris*

For *L. terrestris* HBL Hb, four binding curves obtained at different experimental conditions with respect to effector concentration were analyzed. In order to reduce the number of free parameters, the experimental conditions were chosen such that the four allosteric equilibrium constants for each conformational equilibrium can be described by one equilibrium constant plus two factors ([Eq. 26.12], Fig. 26.3).

Because individual analysis of the binding curves in terms of the MWC model gave a high correlation between the values for L and K_r for each single data set, the four data sets were analyzed simultaneously. In this global approach, the oxygen-binding constants were assumed to be identical for all four data sets, and the allosteric equilibrium constants were allowed to be specific for each data set, reflecting the different conditions with respect to pH and Ca^{2+} and Mg^{2+} concentration. For comparison, the MWC model was also fitted to the data in a global approach where the additional constraints [Eq. (26.12)] are taken into account directly. When no constraints were employed *a priori*, an acceptable agreement was found for the MWC model (Table 26.1). However, when the values for L were tested while applying the constraints, they did not pass. This is also reflected by the significant deviation between data and fit, if the constraints are implemented in the fitting routine (see residuals in Fig. 26.4).

Table 26.1 Comparison of different cooperative models

Model	r^2	N	Q	χ^2
MWC★	0.99899	65	0.95	47.5
MWC	0.9866	66	10^{-14}	191
3state	0.99948	62	0.99948	30.4
Nested MWC, 3x4	0.99979	59	0.9997	27
Nested MWC, 6x12	0.9987	59	10^{-9}	2.4

The following parameters are listed: the linear regression coefficient r^2, describing the agreement between data and fit; the degrees of freedom N = number of data points – number of free parameters; constraints are included as given in Eq.10 except for MWC★; the probability Q that the observed value of χ^2 occurs by chance.

Therefore, as the simplest extension of the MWC model, a 3-state model was tested again including the constraints on the allosteric equilibrium constants. Here, a reasonable agreement could only be achieved with $q = 8$. This would indicate that the tetramers are coupled in pairs. Because the oligomer is organized as a 12 × 12-mer, this model would imply that the 12-meric structure does not play a role in the functional hierarchy.

Alternatively, the nested MWC model was applied to the data. The structure suggests a coupling of three tetramers, which gave very good agreement with the data (see Fig. 26.4B). In contrast, attempts to fit a nested MWC model based on larger allosteric units, such as a 6 × 12mer, failed: the χ^2 value did not yield a high probability of this model (see Table 26.1). Thus, the most likely candidate for this set of oxygen-binding curves for *L. terrestris* HBL Hb is a 3 × 4 nested MWC model. The parameters for this model are given in Table 26.2.

The binding constants K_{tR} and K_{tT} are quite well defined. In contrast, for the parameters K_{rR} and K_{tR}, a rather broad range, between 2 and 10 $Torr^{-1}$, is possible. In order to determine whether the exact values of these two parameters have a strong impact on the values of the remaining

Table 26.2 Fit-parameters for the 3x4 Nested MWC model

Reference state parameters		Effect of pH		Effect of Ca^{2+}, Mg^{2+}	
$\log l_{R,ref}$	2.0 ± 0.6	$C_{\Delta ph,R}$	2.0 ± 0.4	$C_{cat,R}$	1.7 ± 0.3
$\log l_{T,ref}$	5.1 ± 3.3	$C_{\Delta ph,T}$	1.4 ± 0.2	$C_{cat,T}$	1.5 ± 0.3
$\log l_{ref}$	-4 ± 9	$C_{\Delta ph,\Lambda}$	0.8 ± 0.6	$C_{cat,\Lambda}$	1.8 ± 1.8

$C_{\Delta pH}$ describes the shift in the allosteric equilibrium constants due to a pH shift, C_{cat} the change due to the change in Ca^{2+} and Mg^{2+} concentrations. Since the fitting routine worked more stably in this variant, we used the parameter C instead of the parameter F as defined in Eq. 26.6. C_w is defined as $1+(\log F_w/\log L_w^{o})$ with the index "w" denoting the respective allosteric equilibrium constant (R,T,Λ). The deviations from 0 and 100% saturation were less than 3%. The oxygen binding constants were $K_{rT}=7 \pm 14$; $K_{tT}=0.051 \pm 0.002$; $K_{rR}=4.9 \pm 1.1$; $K_{tR}=0.42 \pm 0.03$.

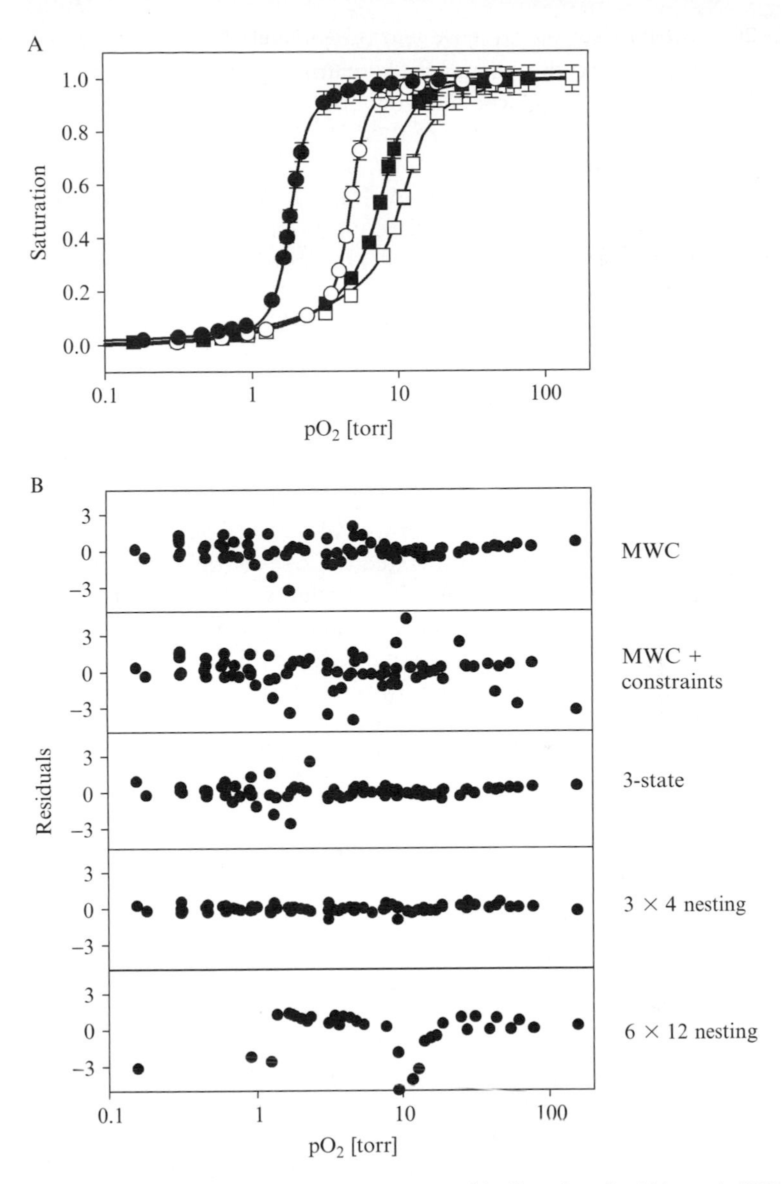

Figure 26.4 Oxygen-binding data. (A): Oxygen-binding data for 144-meric HBL Hb from *L. terrestris* at four different effector concentrations: pH 7.8, 44 mM Ca^{2+} (solid circle), pH 7.8, 1.6 mM Ca^{2+}, 0.15 mM Mg^{2+} (open circle), pH 7.0, 44 mM Ca^{2+} (solid square), and pH 7.0, 1.6 mM Ca^{2+}, 0.15 mM (open square). The lines correspond to the fit for the 3 × 4 nested MWC model with the parameters as given in Table 26.2. (B): Normalized sum of residuals (chi^2) for the different models shown in Table 26.1.

parameters, the following strategy was carried out. Different pairs of values for K_{rT} and K_{rR} were chosen and the remaining parameters were optimized in the fitting procedure. Those fits where the χ^2 was similar to the one given in Table 26.2 were selected. Then, the values of the remaining parameters were averaged between these sets and the standard error calculated. The results are shown in Table 26.3. As can be seen by comparison with Table 26.2, the values of the remaining parameters are reasonably well defined, despite the uncertainties in the values for K_{rR} and K_{rT}.

5. HBL Hb from *M. decora*

The analysis of the oxygen-binding behavior of leech hemoglobin is described in more detail elsewhere (Hellmann *et al.*, 2003b). Only a summary of the results is given here. For the 144-meric oligomer, the application of the MWC model again indicated the size of the allosteric unit (q) to be about 8 (Hellmann *et al.*, 2003b). If the value was held constant at $q = 12$, a slightly smaller agreement between fit and data was observed. Alternatively, the nested MWC model was applied with different combinations of the sizes of the two allosteric units. Inspired by the structural information, the following combinations were tried: 3×4, 6×12, and 12×12. The agreement between fit and data was similar in all cases, the variant 6×12 having the lowest value for χ^2. The tetrameric dissociation product could best be described by assuming a MWC model with a trimeric allosteric unit and a conformation-independent monomer.

In two arthropod hemocyanins (Decker and Sterner, 1990) it was found that isolated oligomers corresponding to the size of the smaller allosteric unit can be described by a MWC model, and the binding properties of these two conformations were very similar to two of the conformational states available for the fully assembled, hierarchical structure. In the case of the tetramer and the 144-mer of leech, such a comparison is difficult because of the large errors in the parameters determined for the tetramer. Thus, the tetramer was reanalyzed, with values fixed to those binding constants found in the native 144-mer. On the basis of these results, the following scenario is suggested: the isolated tetramer behaves similarly to the small allosteric unit in the T state and undergoes conformational transitions between the tT state and the rT state. The assembly to the 144-form enlarges the conformational space of the tetramer, allowing it to adopt two more conformations rR and tR. Furthermore, the assembly seems to enlarge the allosteric unit so that all subunits of the tetramer are part of the allosteric unit; additionally, three of the tetramers are coupled, yielding an allosteric unit composed of 12 binding sites.

Table 26.3 Parameters for the 3x4 Nested MWC model with fixed values for K_{rT} and K_{rR}

	K_{rT} [$Torr^{-1}$]	K_{rR} [$Torr^{-1}$]	l_{Rref}	l_{Tref}	l_{ref}	$C_{\Delta ph,R}$	$C_{\Delta ph,T}$	$C_{\Delta ph,\Lambda}$	$C_{cat,R}$	$C_{cat,T}$	$C_{cat,\Lambda}$
Average	0.0442	0.310	1.50	4.80	-3.69	2.14	1.35	0.58	1.92	1.33	1.16
Std.error	0.0001	0.001	0.28	0.42	0.63	0.15	0.02	0.15	0.12	0.02	0.03

For pairwise fixed values of the parameters K_{rT} and K_{rR} the remaining parameters were optimized. The following pairs of values for k_{rT} and k_{rR} were used {5;4},{5;3},{10;5},{2.5;2},{10;7} $Torr^{-1}$. The remaining optimized parameters were averaged between these sets.

6. What Is the Advantage of Hierarchical Function?

For two examples of giant HBL Hbs, the detailed analysis of oxygen-binding curves indicates that the structural hierarchy is coupled to a functional hierarchy, which in these cases fits the nested MWC model. Although a linear extension such as the 3-state MWC model cannot be excluded on the basis of the binding data themselves, it seems unlikely in terms of the size of the allosteric unit. It is interesting that in both *L. terrestris* and *M. decora* the 3-state model or the MWC model seem to imply the existence of an 8-meric allosteric unit. One should not discard this interpretation entirely.

In both cases, the nested MWC model provides a better description of the data than the original MWC model. For the Hb of *L. terrestris*, the 3 × 4 nested MWC model seems to be the only variant suitable to describe the data. Here, binding curves at four different conditions were analyzed. Notably, the natural dissociation product of this HBL hemoglobin is the 12-mer, which supports the idea that the 12-mer forms a structural and functional entity. Indeed, the 12-meric subassemblies, prepared by mild dissociation at neutral pH in 4 *M* urea, show similar oxygen-binding properties as the native 144-meric molecule: the p_{50} is similar and the same Bohr-effect and Mg^{2+} and Ca^{2+} sensitivities were found. The cooperativity was slightly reduced in the dissociation product ($n_{50} = 3$ to 4 for 12-mers, ≈5 for the 144-mer, [Krebs *et al.*, 1996]). However, the HBL Hb from *M. decora* does not form stable dodecamers, and removal of Ca^{2+}/Mg^{2+} causes dissociation to tetramers. Nevertheless, in terms of functional hierarchy, a nested MWC model based on at least 3 × 4 structures has to be assumed for the intact 144-meric assembly, and even larger allosteric units cannot be excluded. Thus, the dissociation products observed in *M. decora* do not correspond to the functional hierarchy.

Although the final answer cannot be given as to whether the complicated HBL structures evolved to serve an hierarchical function, the question arises whether specific advantages or properties distinguish the hierarchical cooperative model from the linear extension of the MWC model.

A closer inspection of the changes in the allosteric equilibrium constants at different effector concentrations in *L. terrestris* reveals an interesting pattern: both a decrease in pH and a combined decrease in Ca^{2+} and increase in Mg^{2+} concentration seem mainly to affect the allosteric equilibrium constants of the smaller allosteric unit. This is depicted in Fig. 26.5, based on the values from Table 26.3.

The same behavior has also been observed for some arthropod hemocyanins. A very clear, albeit physiologically irrelevant, example is the effect of TRIS on the oxygen-binding properties of tarantula 4 × 6-mer

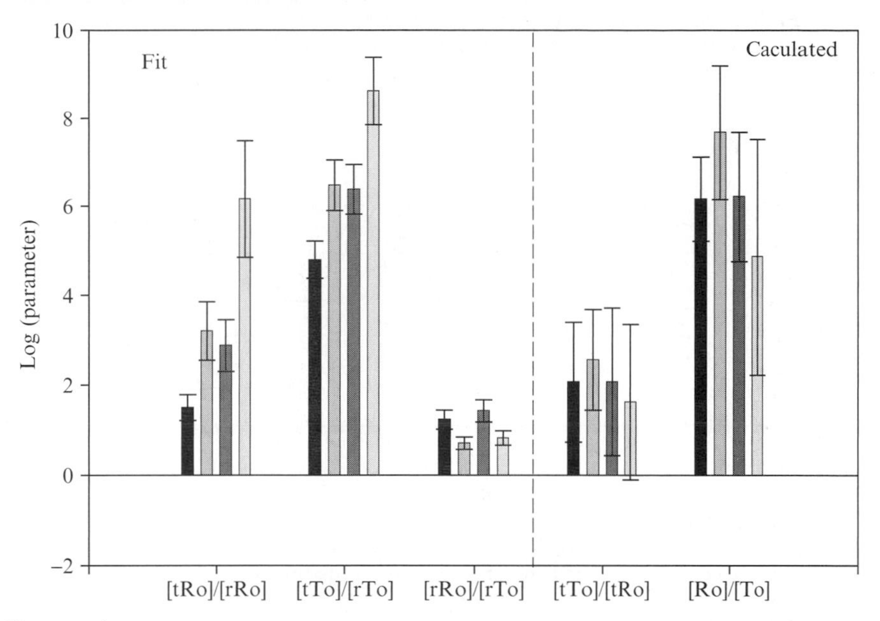

Figure 26.5 Comparison of conformational equilibria for the four different conditions. Logarithmic value of the allosteric equilibrium constants as obtained from analysis of the four binding curves are depicted: $l_r = [tR_o]/[rR_o]$, $l_t = [tT_o]/[rT_o]$ and $L^{-\{1/3\}} = [rR_o]/[rT_o]$. From these values, the equilibria between $[tT_o]$ and $[tR_o]$ and between $[T_o]$ and $[R_o]$ can be calculated. The change of pH or cation concentration mainly influences the allosteric equilibria between the small allosteric units (l_r, l_t).

hemocyanin (Sterner *et al.*, 1994). Here, only one of the three allosteric equilibrium constants of the nested MWC model is shifted by the effector. For protons as effectors, typically all three allosteric equilibrium constants are shifted, but L usually is much less than the two allosteric equilibrium constants than the small allosteric units (Dainese *et al.*, 1998; Decker and Sterner, 1990; Molon *et al.*, 2000). The effect of hydration/osmotic pressure also seems to operate mainly at the level of the small allosteric unit (Hellmann *et al.*, 2003a). In the case of urate binding to hemocyanin from *H. vulgaris*, no strong preference for modulation of one of the hierarchical levels has been observed. Rather, preferential binding to conformation rR was observed (Menze *et al.*, 2005). Similar results were found for the dye neutral red (Sterner and Decker, 1990) and the binding of urate and L-lactate to hemocyanin from *H. americanus* (unpublished results).

Another possible advantage pertains to the way conformational distributions can be manipulated. Calculation of the conformational distribution of HBL from *L. terrestris* at pH 7, 44 m*M* Ca^{2+}, reveals a peculiar pattern: the conformation with the highest affinity (rT) has a maximum at medium ligand concentration (Fig. 26.6). The oligomeric structure responsible for this maximum is depicted (Fig. 26.6). Thus, one can state that the nested

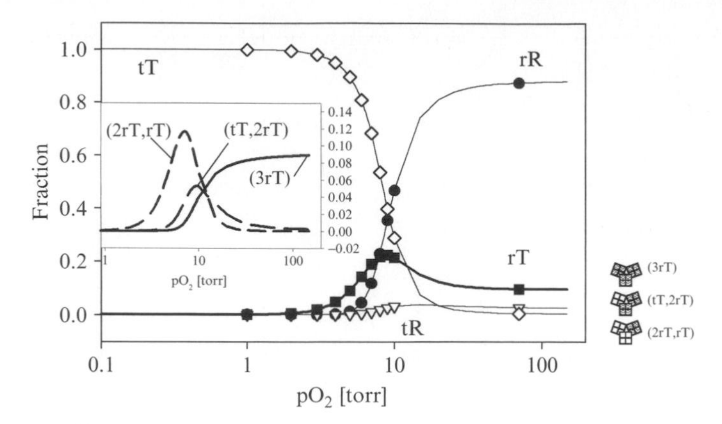

Figure 26.6 Conformational distribution at pH 7.0, 44 mM Ca^{2+}. Fractions of molecules that are in conformation tT, tR, rR, and rT. Inset: fraction of 12-meric structures, which contain 1 tetramer of conformation rT and 2 tetramers with conformation tT (2tT, rT), 2 tetramers of conformation rT, and 1 tetramer with conformation tT (tT, 2rT), or 3 tetramers with conformation rT (3rT). These are the species from which the rT component adds up to yield the conformation rT (thick line in the main panel).

MWC model is able to accommodate such a conformational distribution, whereas a linear extension of the MWC model is not. The latter statement can be proved mathematically and is given in the appendix.

The advantage of having such a conformational distribution is not clear yet. One might speculate that under certain physiological conditions, particular conformations with specific properties are advantageous. These properties could be the affinity for binding certain effectors and/or specific kinetic properties. Because very little is known about the kinetic properties of the various conformational states of HBLs, the understanding of the role of conformational distributions has to await further experiments.

7. Concluding Remarks

A general problem in the analysis of cooperative binding data is that often the values of groups of parameters exhibit a high dependence on one another. The situation can be improved by analyzing several data sets simultaneously. In the simultaneous approach, the basic idea of concerted models, that the oxygen-binding constants for the different conformations do not depend on effector concentration, is included *a priori*. Furthermore, as here demonstrated, the dependence of the allosteric equilibrium constants on the effector concentrations can be included explicitly in a coupled manner.

Other strategies for verifying such a model and the properties of the conformations therein are available in the literature. A promising approach is to study trapped conformations, for example using hydrogels, which allow for verification of the ligand- and effector-binding properties as deduced by the analysis of ligand-binding data (Bettati and Mozzarelli, 1997). Another approach is to combine direct effector-binding studies with the shift in ligand-binding curves induced by these effectors. Not all effectors are amenable to such an approach, but examples can be found in the literature (Menze *et al.*, 2005).

In this case, the binding polynomials for the effector as used in Eq. (26.7) have to be specified. An example for the simplest case is given in Eq. (26.8). The values for the effector binding constants (z_A and z_B) can be derived from the shifts in the allosteric equilibrium constants and can be compared to those obtained by direct binding experiments to oxygenated and deoxygenated forms. This type of analysis is not based on the assumption of symmetrical binding curves, as is the case in the most common approach based on the shift in P_{50} (Wyman and Gill, 1990).

Thus, with the aid of additional experiments addressing the function and a strategic choice of experimental conditions, it is possible to gain information about the conformational states even in large cooperative systems. Knowledge of the number, properties, and coupling of conformations of cooperative proteins is required to understand how these complex machines work, and why nature created such complicated structures. The question whether any of these conformations play an important role in the regulation of NO concentrations remains unanswered. NO does not bind cooperatively (Gow *et al.*, 2005), indicating that its binding affinity does not differ significantly for the different conformations. Nevertheless, surrounding conditions as pH or cation concentration might have a regulatory impact via their allosteric properties.

Appendix A

We want to show here that the situation as shown in Fig. 26.6 where the conformation with the highest affinity is mostly populated at intermediate ligand concentrations and decreases if the ligand concentration is increased further, cannot occur if a linear extension of the MWC model is applied.

The general form of a linear extension of the MWC model, describing m conformations instead of two as assumed in the original form (Monod *et al.*, 1965), is given by the following binding polynomial

$$P(x) = \sum_{i=1}^{m} L_i P_i(x) \quad P_i(x) = (1 + K_i x)^q \qquad (26.A1)$$

Here, K_i is the binding constant of the ligand to the i-th conformation, x is the ligand concentration, and q is the size of the allosteric unit. The fraction of molecules in the i-th conformation is given by

$$\alpha_i(x) = \frac{L_i P_i(x)}{P(x)} \tag{26.A2}$$

We want to prove the following statement: if for two arbitrary conformations i and j with $K_j > K_i$, a ligand concentration x_o can be found, where the following relationship holds true,

$$\alpha_j(x_o) > \alpha_i(x_o) \tag{26.A3}$$

then the following relationship is true for all conditions where $x > x_o$:

$$\alpha_j(x) > \alpha_i(x) \tag{26.A4}$$

Proof

Equation (26.A4) is equivalent to

$$\frac{L_j(1 + K_j x)^q}{P(x)} > \frac{L_i(1 + K_i x)^q}{P(x)}. \tag{26.A4}$$

Thus, it is sufficient to show that $L_j(1 + K_j x)^q < L_i(1 + K_i x)^q$ because $P(x) > 0$.
We do this by showing that the assumption $L_j(1 + K_j x)^q < L_i(1 + K_i x)^q$ leads to contradictions with the prerequisite $K_j > K_i$.

From [Eq. (26.A3)] it follows that $L_j(1 + K_j x_o)^q > L_i(1 + K_i x_o)^q$, which yields the inequality

$$Z := \left(\frac{L_j}{L_i}\right)^{1/q} > \frac{1 + K_i x_o}{1 + K_j x_o} \tag{26.A5}$$

Now assume that $L_j(1 + K_j x)^q < L_i(1 + K_i x)^q$ for any $x > x_o$. Then it follows that

$$\frac{1 + K_i x}{1 + K_j x} > \left(\frac{L_j}{L_i}\right)^{1/q} = Z$$

Combination with [Eq. (26.A5)] yields

$$\frac{1 + K_i x}{1 + K_j x} > \left(\frac{L_j}{L_i}\right)^{1/q} = Z > \frac{1 + K_i x_o}{1 + K_j x_o},$$

and therefore

$$\frac{1 + K_i x}{1 + K_j x} > \frac{1 + K_i x_o}{1 + K_j x_o}$$

or

$$(1 + K_i x)(1 + K_j x_o) > (1 + K_j x)(1 + K_i x_o).$$

Factoring out leads to the following expression:

$$K_i x + K_j x_o > K_j x + K_i x_o$$

or

$$(K_i - K_j)x > (K_i - K_j)x_o$$

Because x > xo >0, this can only be true if $K_i > K_j$. This contradicts the prerequisite $K_j > K_i$.

Stated in words: In a linear extension of the MWC model, the situation where the conformation with the highest affinity (K_j) is mainly populated at intermediate ligand concentrations cannot arise. For this to occur, at higher ligand concentrations a conformation with a lower affinity has to replace conformation j. As shown previously, this is not possible.

REFERENCES

Bettati, S., and Mozzarelli, A. (1997). T state hemoglobin binds oxygen noncooperatively with allosteric effects of protons, inositol hexaphosphate, and chloride. *J. Biol. Chem.* **272,** 32050–32055.

Dainese, E., Di Muro, P., Beltramini, M., Salvato, B., and Decker, H. (1998). Subunits composition and allosteric control in *Carcinus aestuarii* hemocyanin. *Eur. J. Biochem.* **256,** 350–358.

de Haas, F., Zal, F., Lallier, F. H., Toulmond, A., and Lamy, J. N. (1996a). Three-dimensional reconstruction of the hexagonal bilayer hemoglobin of the hydrothermal vent tube worm *Riftia pachyptila* by cryoelectron microscopy. *Proteins* **26,** 241–256.

de Haas, F., Biosset, N., Taveau, J. C., Lambert, O., Vinogradov, S. N., and Lamy, J. N. (1996b). Three-dimensional reconstruction of *Macrobdella decora* (leech) hemoglobin by cryoelectron microscopy. *Biophys. J.* **70,** 1973–1984.

de Haas, F., Zal, F., You, V., Lallier, F., Toulmond, A., and Lamy, J. N. (1996c). Three-dimensional reconstruction by cryoelectron microscopy of the giant hemoglobin of the polychaete worm *Alvinella pompejana*. *J. Mol. Biol.* **264,** 111–120.

Decker, H., and Sterner, R. (1990). Nested allostery of arthropodan hemocyanin (*Eurypelma californicum* and *Homarus americanus*). The role of protons. *J. Mol. Biol.* **211,** 281–293.

Decker, H., Robert, C. H., and Gill, S. J. (1986). Nesting: An extension of the allosteric model and its application to tarantula hemocyanin. *Invert. Oxygen Carriers* **38,** 383–388.

Fushitani, K., and Riggs, A. F. (1988). Non-heme protein in the giant extracellular hemoglobin of the earthworm *Lumbricus terrestris*. *Proc. Natl. Acad. Sci. USA* **85,** 9461–9463.

Fushitani, K., and Riggs, A. F. (1991). The extracellular hemoglobin of the earthworm, *Lumbricus terrestris*: Oxygenation properties of isolated chains, trimer, and a reassociated product. *J. Biol. Chem.* **266,** 10275–10281.

Fushitani, K., Imai, K., and Riggs, A. F. (1986). Oxygenation properties of hemoglobin from the earthworm, *Lumbricus terrestris*: Effects of pH, salts, and temperature. *J. Biol. Chem.* **261,** 8414–8423.

Gow, A. J., Payson, A. P., and Bonaventura, J. (2005). Invertebrate hemoglobins and nitric oxide: How heme pocket structure controls reactivity. *J. Inorg. Biochem.* **99,** 903–911.

Hellmann, N., Raithel, K., and Decker, H. (2003a). A potential role for water in the modulation of oxygen-binding by tarantula hemocyanin. *Comp. Biochem. Physiol. A* **136,** 725–734.

Hellmann, N., Weber, R. E., and Decker, H. (2003b). Nested allosteric interactions in extracellular hemoglobin of the leech *Macrobdella decora*. *J. Biol. Chem.* **278,** 44355–44360.

Igarashi, Y., Kimura, K., and Kajita, A. (1991). Analysis of oxygen equilibria of the giant hemoglobin from the earthworm *Eisenia foetida* using the Adair model. *J. Biochem. (Tokyo)* **109,** 256–261.

Kapp, O. H., Qabar, A. N., Bonner, M. C., Stern, M. S., Walz, D. A., Schmuck, M., Pilz, I., Wall, J. S., and Vinogradov, S. N. (1990). Quaternary structure of the giant extracellular hemoglobin of the leech *Macrobdella decora*. *J. Mol. Biol.* **213,** 141–158.

Krebs, A., Kuchumov, A. R., Sharma, P. K., Braswell, E. H., Zipper, P., Weber, R. E., Chottard, G., and Vinogradov, S. N. (1996). Molecular shape, dissociation, and oxygen binding of the dodecamer subunit of *Lumbricus terrestris* hemoglobin. *J. Biol. Chem.* **271,** 18695–18704.

Lamy, J. N., Green, B. N., Toulmond, A., Wall, J. S., Weber, R. E., and Vinogradov, S. N. (1996). Giant hexagonal bilayer hemoglobins. *Chem. Rev.* **96,** 3113–3124.

Menze, M. A., Hellmann, N., Decker, H., and Grieshaber, M. K. (2005). Allosteric models for multimeric proteins: Oxygen-linked effector binding in hemocyanin. *Biochemistry* **44,** 10328–10338.

Molon, A., Di Muro, P., Bubacco, L., Vasilyev, V., Salvato, B., Beltramini, M., Conze, W., Hellmann, N., and Decker, H. (2000). Molecular heterogeneity of the hemocyanin isolated from the king crab *Paralithodes camtschaticae*. *Eur. J. Biochem.* **267,** 7046–7057.

Monod, J., Wyman, A., and Changeux, J. P. (1965). On the nature of allosteric transitions: A plausible model. *J. Mol. Biol.* **12,** 88–118.

Ochiai, T., and Weber, R. E. (2002). Effects of magnesium and calcium on the oxygenation reaction of erythrocruorin from the marine polychaete *Arenicola marina* and the terrestrial oligochaete *Lumbricus terrestris*. *Zoolog. Sci.* **19,** 995–1000.

Press, W. H., Flannery, S. A., Teukalsky, S. A., and Vetterling, W. T. (1989). Numerical recipes (Fortran Version), pp. 160–163. Cambridge University Press, Cambridge.

Robert, C. H., Decker, H., Richey, B., Gill, S. J., and Wyman, J. (1987). Nesting: Hierarchies of allosteric interactions. *Proc. Natl. Acad. Sci. USA* **84,** 1891–1895.

Royer, W. E., Jr., Strand, K., van Heel, M., and Hendrickson, W. A. (2000). *Proc. Natl. Acad. Sci. USA* **97,** 7107–7111.

Royer, W. E., Jr., Sharma, H., Strand, K., Knapp, J. E., and Bhyravbhatla, B. (2006). Lumbricus erythrocruorin at 3.5 Å resolution: Architecture of a megadalton respiratory complex. *Structure* **14,** 1167–1177.

Sterner, R., and Decker, H. (1990). Conformational transition of Carcinus maenas monitored with an organic dye (neutral red). *In* "Invertebrate Dioxygen Carriers" (G. Preaux and R. Lontie, eds.), pp. 193–196. Leuven University Press, Leuven, the Netherlands.

Sterner, R., Bardehle, K., Paul, R., and Decker, H. (1994). Tris: An allosteric effector of tarantula haemocyanin. *FEBS Lett.* **339,** 37–39.

Vinogradov, S. N. (1985). The structure of invertebrate extracellular hemoglobins (erythrocruorins and chlorocruorins). *Comp. Biochem. Physiol. B* **82,** 1–15.

Vinogradov, S. N., Sharma, P. K., Qabar, A. N., Wall, J. S., Westrick, J. A., Simmons, J. H., and Gill, S. J. (1991). A dodecamer of globin chains is the principal functional subunit of the extracellular hemoglobin of *Lumbricus terrestris. J. Biol. Chem.* **266,** 13091–13096.

Weber, R. E. (1981). Cationic control of O_2 affinity in lugwonm erythrocruorin. *Nature* **292,** 386–387.

Weber, R. E., and Vinogradov, S. N. (2001). Nonvertebrate hemoglobins: Functions and molecular adaptations. *Physiol. Rev.* **81,** 569–628.

Weber, R. E., Jensen, F. B., and Cox, R. P. (1987). *J. Comp. Physiol. B* **157,** 145–152.

Weber, R. E., Malte, H., Braswell, E. H., Oliver, R. W., Green, B. N., Sharma, P. K., Kuchumov, A., and Vinogradov, S. N. (1995). Mass spectrometric composition, molecular mass and oxygen binding of Macrobdella decora hemoglobin and its tetramer and monomer subunits. *J. Mol. Biol.* **251,** 703–720.

Wyman, J., and Gill, S. J. (1990). "Binding and linkage." University Science Books. Mill Valley, CA.

Zal, F., Lallier, F. H., Green, B. N., Vinogradov, S. N., and Toulmond, A. (1996a). The multi-hemoglobin system of the hydrothermal vent tube worm *Riftia pachyptila.* II. Complete polypeptide chain composition investigated by maximum entropy analysis of mass spectra. *J. Biol. Chem.* **271,** 8875–8881.

Zal, F., Lallier, F. H., Wall, J. S., Vinogradov, S. N., and Toulmond, A. (1996b). The multi-hemoglobin system of the hydrothermal vent tube worm *Riftia pachyptila.* I. Reexamination of the number and masses of its constituents. *J. Biol. Chem.* **271,** 8869–8874.

CHAPTER TWENTY-SEVEN

Mass Mapping of Large Globin Complexes by Scanning Transmission Electron Microscopy

Joseph S. Wall,* Martha N. Simon,* Beth Y. Lin,* *and* Serge N. Vinogradov†

Contents

Abstract

Scanning transmission electron microscopy (STEM) of unstained, freeze-dried biological macromolecules in the dark-field mode provides an image based on the number of electrons elastically scattered by the constituent atoms of the macromolecule. The image of each isolated particle provides information about the projected structure of the latter, and its integrated intensity is directly related to the mass of the selected particle. Particle images can be sorted by shape, providing independent histograms of mass to study assembly/disassembly

* Biology Department, Brookhaven National Laboratory, Upton, New York
† Department of Biochemistry and Molecular Biology, School of Medicine, Wayne State University, Detroit, Michigan

Methods in Enzymology, Volume 436
ISSN 0076-6879, DOI: 10.1016/S0076-6879(08)36027-3

intermediates. STEM is optimized for low-dose imaging and is suitable for accurate measurement of particle masses over the range from about 30 kDa to 1,000 MDa. This article describes the details of the method developed at the Brookhaven National Laboratory STEM facility and illustrates its application to the mass mapping of large globin complexes.

1. Introduction

The concept of scanning transmission electron microscopy (STEM) was described by Von Ardenne in the 1930s at about the same time that Ruska developed the transmission electron microscope (TEM). However, the experimental realization of STEM did not start until the availability of electronics and adequate electron sources in the 1960s. Crewe *et al.* (1970) at the University of Chicago developed the cold field emission electron source and built a STEM able to visualize single heavy atoms on thin carbon substrates. The first application of this method to the imaging of biological molecules was demonstrated soon thereafter (Wall, 1971), and this work was continued at Brookhaven National Laboratory (BNL) with the construction of STEM1 as a user facility in 1977. Since then, the method has been widely used to solve a number of structural problems in molecular biology (Hainfeld and Wall, 1988; Thomas *et al.*, 1994; Wall and Hainfeld, 1986; Wall and Simon, 2001). The BNL STEM facility operates as a National Institutes of Health (NIH) Biotechnology Resource, and its use is available without charge to anyone with an appropriate peer-reviewed research project.

2. Description of STEM Mass Mapping

Samples of unstained biological material (see the section herein "STEM specimen preparation") are deposited on a 2- to 3-nm-thick carbon film, washed with volatile buffer, fast frozen, and freeze-dried overnight in an ultrahigh vacuum. Grids are then transferred under vacuum to the STEM operating at 40 keV (STEM1, BNL) and examined under vacuum at $-160°$. STEM produces an image one point at a time by probing the sample with a focused (0.25-nm diameter) electron beam and by counting the emerging electrons on an array of detectors that measure how much they are deflected (scattered) by the mass in the specimen (Fig. 27.1). Detectors measuring large-angle scattering give a signal proportional to the mass of atoms in the path of the beam. Therefore, the STEM (dark-field) image is essentially a map of the mass in each small element (pixel) of the image. Because the image is in focus and there is no phase contrast, STEM images of biological specimens are easily interpreted. Any

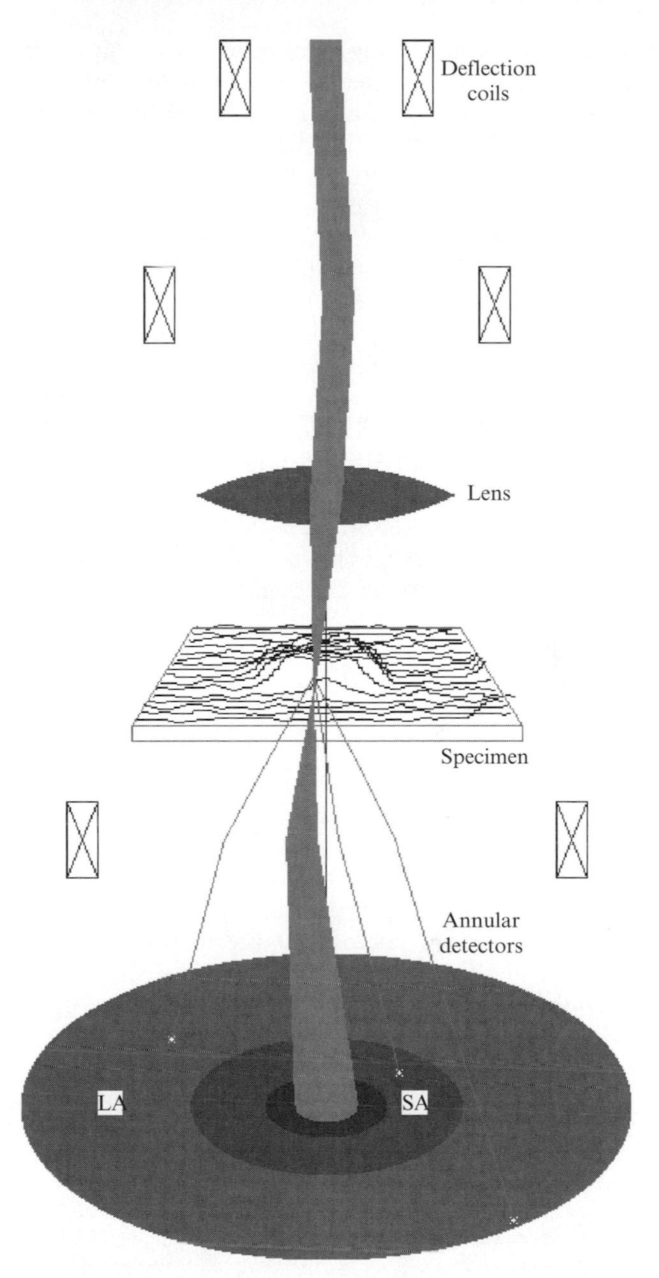

Figure 27.1 Schematic representation of STEM operation. A 40-keV electron beam (green) is focused by a high-quality electron lens (focal length 1 mm, spherical aberration constant 0.6 mm) onto a specimen mounted on a thin substrate. Scattered electrons (red) are generated when an incident electron interacts with an atom in the specimen and are removed from the beam, striking one of the annular detectors (darker red), where they create a flash of light recorded by a photomultiplier tube. STEM forms an

residual material is also immediately evident, so one can judge the specimen quality directly. Mass measurement is performed offline using custom software, PCMass, which runs on a PC using Windows 95 or higher. This program is available at the STEM FTP site (ftp.stem.bnl.gov). Mass measurements are made by selecting suitable particles with an adjustable circular or rectangular cursor (Fig. 27.2). The program sums intensity minus background for each pixel inside the boundary and multiplies the net signal by the STEM calibration constant. The background is subtracted on the assumption that the substrate under the particle has the same mass per unit area as that observed in "clean" areas between particles. Once suitable parameters are selected, the program can be run in an automated mode to make particle selection faster and more reproducible. The STEM calibration constant is checked using tobacco mosaic virus (TMV) included in every specimen. This virus is an easily distinguishable rod, 300 nm long and 18 nm in diameter, which is expected to have the mass per unit length of 131 kD/nm. Deviations from this value indicate impurities in the specimen or buffers. Absolute accuracy ranges from +10% for 100 kDa particles to 4% for 1 MDa and 2% for particles greater than 5 MDa.

3. Advantages of STEM

Because STEM images isolated and unstained biological molecules after freeze-drying, STEM has a unique advantage over other methods of mass determination, in that a large number of particles can be sorted by shape and measured to obtain excellent statistics. Furthermore, STEM requires only a small amount of material, typically 10 to 50 μl of 50 to 300 μg/ml. The STEM image intensity is directly proportional to the local mass per unit area in the corresponding region of the specimen. Given absolute scattering cross-sections or a calibration specimen of known mass,

---

image one point at a time by positioning the beam with a pair of deflection coils above the specimen (balanced to make the beam cross the electron optical axis at the center of the objective lens) and by recording the detected signals digitally. A third deflection coil (un-scan) below the specimen recenters the deflected beam on the detector array. Each measurement takes 30 μs, after which the beam position is incremented across the specimen in a raster pattern, requiring 8.5 s for a 512 $\times$ 512 pixel image. The large-angle (LA) detector counts electrons scattered 40 to 200 mR; the small angle (SA), 15 to 40 mR; and the bright field (dark red), 0 to 15 mR. The LA is preferred for mass measurements because it is relatively immune to phase contrast (i.e., the signal depends only on the number of atoms in the beam path, not on their relative spatial arrangement). The other detector signals are used primarily to measure beam current for absolute calibration of detector signals. Image magnification is controlled precisely by attenuation of currents in the deflection coils. (See color insert.)

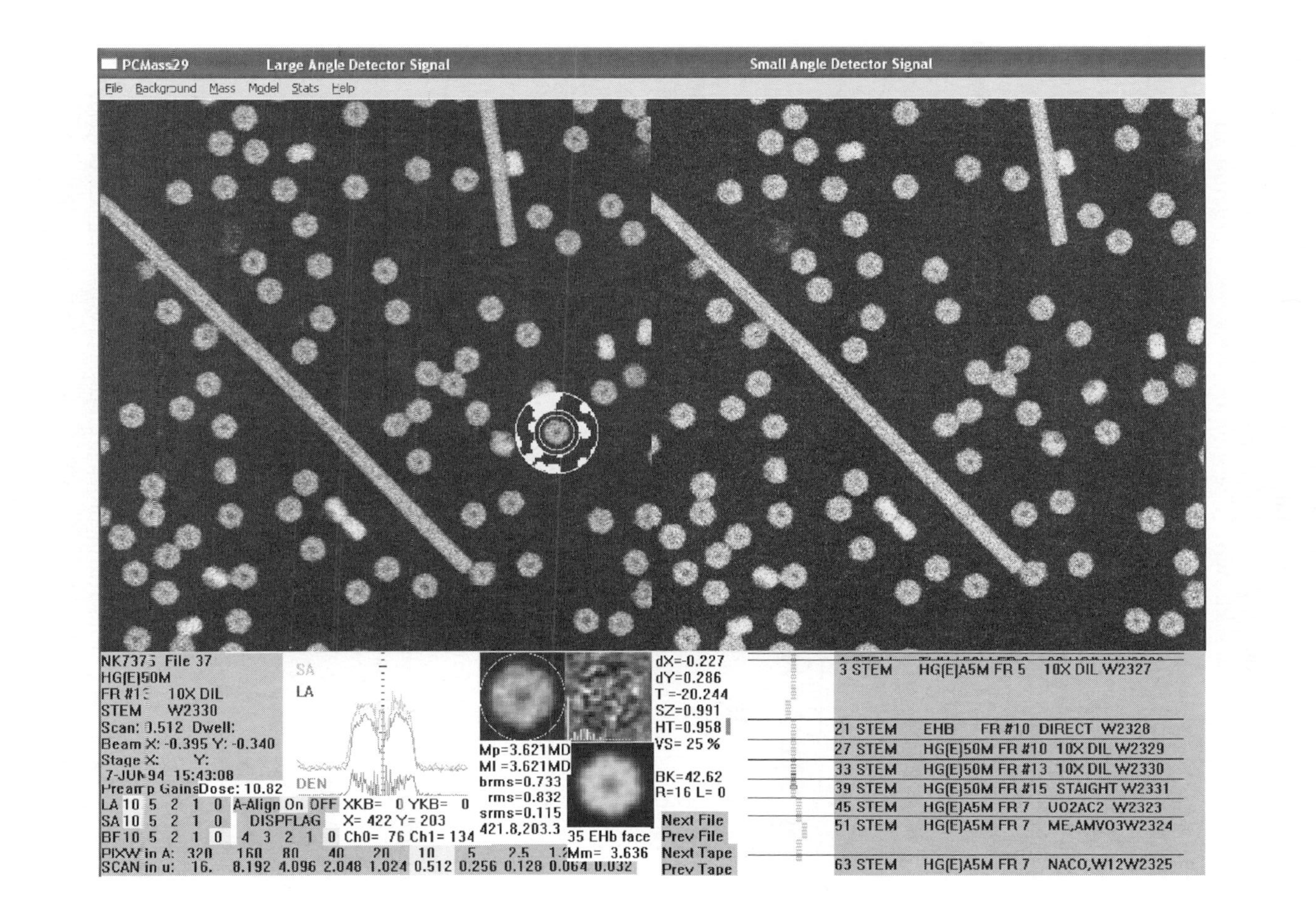
PCMass29
Large Angle Detector Signal
Small Angle Detector Signal
File Background Mass Model Stats Help
NK7375 File 37
HG(E)50M
10X DIL
STEM W2330
Dwell:
Beam X: -0.395 Y: -0.340
Stage X: Y:
15:43:08
Dose: 10.82
LA 10 5 2 1 0
SA 10 5 2 1 0
BF10 5 2 1 0
A-Align On OFF
DISPFLAG
4 3 2 1 0
XKB= 0 YKB= 0
X= 422 Y= 203
Ch0= 76 Ch1= 134
SA
LA
DEN
Mp=3.621MD
Ml =3.621MD
brms=0.733
rms=0.832
srms=0.115
421.8,203.3
35 EHb face
dX=-0.227
dY=0.286
T =-20.244
SZ=0.991
HT=0.958
VS= 25 %
BK=42.62
R=16 L= 0
Next File
Prev File
Next Tape
Prev Tape
3 STEM HG(E)A5M FR 5 10X DIL W2327
21 STEM EHB FR #10 DIRECT W2328
27 STEM HG(E)50M FR #10 10X DIL W2329
33 STEM HG(E)50M FR #13 10X DIL W2330
39 STEM HG(E)50M FR #15 STAIGHT W2331
45 STEM HG(E)A5M FR 7 UO2AC2 W2323
51 STEM HG(E)A5M FR 7 ME,AMVO3W2324
63 STEM HG(E)A5M FR 7 NACO,W12W2325

intensity in the STEM can be integrated over an isolated particle, a known length of filament, or any arbitrary area and converted to a molecular weight. This provides a direct link to biochemistry if total or subunit molecular weights are known. The homogeneity or variation in mass from one object to another may be interesting in itself or can provide a further check on the quality of the specimen. Internal details are faithfully rendered in the projected mass distribution to a resolution of 2 to 4 nm. For example, the radial mass (density) profile of control particles, such as TMV (included in most STEM specimens), provides a sensitive measure of the quality of specimen preparation and of the extent of radiation damage.

In cases where freeze-dried specimens give good mass measurements but limited visualization of internal detail, parallel grids prepared with negative stains such as uranyl acetate or methylamine vanadate can give additional information. The freedom from phase contrast fringes in STEM permits easier interpretation of such images than does conventional EM.

4. STEM Specimen Preparation

Freeze-dried specimens for mass measurements in the STEM are prepared by the wet film method. Titanium grids (2.3 mm) are coated with a thick holey carbon film. The open areas are 5 to 10 μm in diameter, which will support the thin (2 to 3 nm thick) carbon film substrate. The thin carbon film is prepared by ultrahigh vacuum evaporation onto a freshly

Figure 27.2 Display screen of PCMass program used for STEM mass measurement. The specimen is earthworm hemoglobin with TMV controls (rods). The left image is the signal from the large-angle annular detector and the right is from the small-angle detector. The width of the image is 0.512 μm, as indicated in the green area below the left image listing specimen and microscope parameters for the currently displayed image. The green area below the right image lists parameters of the 64 images in the same group as the current image. The black horizontal bars show specimen changes, and the narrow green bars indicate relative magnification (changed to purple when measurements have been made). A set of three circles (positioned by the mouse) marks the particle currently being measured. Areas in white mask are surrounding particles and "dirt" on the background. The selected particle is shown magnified immediately below the left image with its projected mass profile (LA and SA) to the left (the red curve is computed radial density, which is only meaningful for particles with spherical or cylindrical symmetry). Below and to the right is a mathematical model (Model #35 of 80 available) that the program fit to the particle image shown. To the right of the selected particle and above the model is the difference image, particle minus model (useful in highlighting defects in the particle which might affect mass measurement). To the right are listed various fitting parameters. An insert in the difference image shows the rotational power spectrum. Immediately below the zoomed image is the measured mass, computed using the background for the entire picture (Mp) or the local background (Ml, computed in unmasked areas between outermost and next circles). Below the mass measurements are additional fitting parameters indicating the quality of the fit between selected particle and model. Pressing the left mouse button or the "Enter" key records all parameters for the selected particle in a database for statistical or further image analysis. (See color insert.)

cleaved crystal of rock salt. The thin carbon film is then floated on a dish of clean water located in a laminar flow clean hood. Grids covered with holey film, assembled in rings and caps for handling in the microscope, are placed face down on the floating thin carbon film. The grids are picked up from above one at a time such that the thin carbon film retains a droplet of water. This water is exchanged by washing and wicking, either with water or with injection buffer for the specimen. TMV solution (2 μl), at a concentration of 100 μg/ml, is injected into the drop and allowed to stand for 1 min. The TMV is both a qualitative and a quantitative internal control for all specimens.

After four washings, 2 μl of the specimen (its concentration is determined primarily by its size) is injected into the drop and allowed to stand for 1 min. Ten more washings are performed. Half of these can be with injection buffer, but the final few must be with a volatile buffer such as 20 m*M* ammonium acetate or water. After the final wash, the grid is pinched between two pieces of filter paper (leaving a retained layer of solution less than 1 μm thick) and immediately plunged into liquid nitrogen slush. Six grids are transferred under liquid nitrogen to an ion-pumped freeze-dryer with an efficient cold trap, freeze-dried overnight by gradually warming to $-80°$, and transferred under vacuum to the STEM.

5. Limitations of STEM

5.1. Molecular mass

STEM can provide valuable information on individual particles in the molecular weight range from 30 kDa to 1,000 MDa. In general, the larger the particle, the more accurate is the mass measurement. Success in looking at small proteins (less than 100 kDa) will depend on their conformation. If parts of them are very extended, it may be difficult to define the measuring area. Such particles may suffer more mass loss and the mass measurements will be low. Proteins of less than 30 kDa molecular weight are likely to be too small to yield good mass measurements unless they are unusually dense and compact. tRNA can be seen in the STEM, but it is very compact, and nucleic acids are more radiation resistant than proteins. Double-stranded DNA is also visible in the STEM for the same reasons. At the other extreme, while viruses (and subviral particles of large viruses) such as herpes simplex have given excellent results, STEM is probably not the best choice for looking at very large structures such as bacteria, cells, or organelles. These are essentially opaque (i.e., saturated) in the STEM, and therefore yield little information. For filaments or two-dimensional arrays, the actual size is not a problem. It is only the mass density that is of concern, so they are fine if they are spread out enough and not too dense.

5.2. Purity

The physical purity of the specimens is very important for STEM preparations. Essentially, everything is visible in the STEM, and impurities may not wash off the grid. Most biologists have some assay for the biological activity of their specimens. However, many preparations contain biologically inert but physically visible material that may make it difficult to obtain good STEM data. Particulates from columns or gels are a problem. Additives for activity such as bovine serum albumin (BSA), polyethylene glycol (PEG), or trypsin are also a problem. Precipitations can concentrate the specimen but also enrich contaminants. If a multisubunit enzyme has 90% activity, but the remaining 10% has fallen apart, that can be visible in the STEM. Frequently, a final purification over an appropriate sizing column (e.g., BioGel A5M) where the intact specimen comes off in the void volume gives adequate purity. At times, the particles of interest can be distinguished from contaminants. However, it is clearly preferable to eliminate contaminants when possible.

5.3. Concentration

The STEM mass mapping ability depends on a clean background between isolated particles. The necessary concentration of a given specimen for a good distribution of particles on a grid is somewhat empirical. However, for data analysis it is very important. If the particles are too close together, they cannot be measured individually; if they are too sparse, it is very difficult to obtain both particles and the control TMV in an image. Also, searching for rare events in the STEM is slow. As a guide, large particles such as viruses need to be applied to the grid at a relatively high concentration such as 200 to 300 μg/ml. For smaller proteins, a concentration of 50 μg/ml might be adequate. It is not known how well any given specimen will adhere to the carbon grid. For a new specimen, a series of dilutions is usually made. A highly concentrated specimen can always be diluted, but there is little that can be done with a specimen that is too dilute.

5.4. Buffers and salts

The ideal specimen for the STEM is one that can be applied to the grid in water or 20 m*M* ammonium acetate, which volatilizes during freeze-drying. However, many biological specimens have additional requirements. Most salts and buffers can be tolerated in the injection buffer for the specimen. The assumption is that if the specimen is absorbed to the carbon grid in a few washes of its own buffer, it will be able to tolerate some final washes with a volatile solution such as water or ammonium acetate. For example, low levels (1 to 2 m*M*) of Mg^{2+}, Ca^{2+}, ATP (or nonhydrolyzable analogues) are often needed for biological activity and usually will wash away. Low concentrations of glycerol or sucrose also usually wash off, as will DTT or mercaptoethanol. NaCl at levels as high as 100 m*M* usually washes off with ammonium acetate, but KCl at more than a few m*M* occasionally causes problems. Phosphate buffers should be avoided if possible since they often leave a background high in "spots." Tris buffers often, but not always, have a bad background, whereas Hepes, Mops, and Pipes are usually better. For a new specimen in an unusual buffer, a grid of TMV by itself washed with the buffer is made as a control. Some specimens, such as membrane proteins, must be solubilized in a detergent, and these are always difficult to work with, though some successful results have been obtained.

5.5. Stability of sample and denaturation

Some specimens that have met all the previously mentioned criteria simply do not work out. If the sample is unstable at low concentration and contains significant monomer, mass measurements may be compromised. Harm can come to the specimen during its attachment to the carbon film (a part can attach firmly while the rest of the molecule flails around), during the freeze-drying process, and during the data-taking process (from radiation damage in the electron beam). An occasional specimen will not attach to the carbon film. Double-stranded DNA, for example, does not attach reliably even at high concentrations. However, with polylysine pretreatment of the carbon grid, DNA will bind at concentrations of less than 1 μg/ml. Polylysine pretreatment will also help some other types of specimens. All proteins denature to some extent at the air–water interface of the drop on the grid. Usually, washing and wicking removes this denatured film. However, some proteins continue to denature at the interface until there is nothing left on the grid. Sometimes washing with low concentrations of organic solvents such as ethanol or acetonitrile will help. Some specimens that fall apart on the carbon film benefit from a brief fixation with glutaraldehyde just before injection. If this does help, the increase in mass is usually less than 10%. Occasionally specimens will be all right on some parts of the grid but not others.

5.6. Substrate

The thinnest- and cleanest-possible substrates as well as freeze-drying are critical for accurate mass measurements. The BNL STEM group has invested considerable effort in developing this methodology. Also, dark-field imaging is of limited usefulness for specimens on a thick background, such as sectioned or ice-embedded material.

6. STEM Operation

The STEM is operated at 40 keV with a probe focused to 0.25 nm. The sample is maintained at −160° to eliminate contamination and to reduce mass loss. The scan raster is 512 × 512 points with a dwell time of 30 μs per pixel (8.5 s per scan). For most mass measurements, the pixels are separated by 1 nm giving a scan width of 0.512 μm. These scanning conditions usually provide the best compromise between resolution of the specimen and radiation damage.

Electrons elastically scattered through large angles (40 to 200 mR), through small angles (15 to 40 mR), or unscattered (0 to 12 mR) are collected on separate detectors consisting of scintillators and photomultipliers. Both large and small angle signals (normalized by the total beam current) are recorded digitally. The dose is kept below 1,000 electrons/nm^2 to ensure that mass loss from radiation damage is less than 2.5% at the −150° specimen temperature employed.

Areas of the grid to be used for mass measurements are selected for clean background and adequate numbers of both TMV and specimen particles. The background is computed in clear areas in each image, and the intensity minus background is summed over each particle and multiplied by a calibration factor for a mass value.

The STEM calibration factor is checked in each image using TMV, which is included as an internal control. The TMV also serves as a qualitative control for the whole process of specimen preparation. If the TMV looks good and has the expected radial density profile and mass per unit length, the specimen has a chance of being good, but if the TMV has defects, some step in the specimen preparation has gone wrong. Different kinds of mass measurements can be obtained from dark-field STEM data. The total mass of the individual particles can be used to determine their oligomeric state. For filamentous specimens, the mass per length can be used to determine the spacing of repeating units. Similarly, the mass per area can be a useful measurement on two-dimensional arrays.

7. Interactive Mass Mapping with PCMass

The PCMass program is written in C and runs in Windows 95 or higher for visualization and analysis of BNL STEM images. Any PC with greater than 16 MB memory and a 1024 × 768 high-color or true-color display should be able to run the program. A 1280 × 1024 display provides better display of mass statistics. A processor speed of >200 MHz gives minimal delays for most operations. A good Internet connection should allow one to receive images by Internet (ftp://ftp.stem.bnl.gov). STEM images are 528,384 bytes, and one "tape" or folder containing 64 images occupies 32 MB.

STEM images consist of a 4096-byte header followed by 512 × 512 pixels of image data consisting of two channels (8-bit each) interleaved. PCMass reads a STEM image, displaying both image channels side by side with the header information below the left image. The program also recognizes the magnification and sets calibration automatically. Any image or portion of the display screen can be copied to the clipboard and pasted into a document or image, as will be described subsequently. The current version of PCMass does not print directly and has limited image-filtering and annotation capabilities, so a second standard image package (e.g., PhotoShop, Canvas, Paint) is recommended for publication-quality output. Similarly, graphs and histograms are intended to give a rapid view of the data in hand, and publication quality can be obtained by importing results files into SigmaPlot or similar software.

STEM images can be displayed, annotated, and printed with PhotoShop using the Open As and RAW import mode from the File menu with the previously mentioned parameters entered into the specified fields (512, 512, 2 channels, interleaved, 4096 header). Some care is required to determine which image is the large-angle detector signal (the channel most reliable for mass measurements) when using PhotoShop. PhotoShop also permits saving a flattened image (use the Flatten option in the Layers Menu) as a JPG or TIF file, by using the Save As option for distribution to other programs or users. This allows image compression to save space.

The key to reliable STEM mass measurement is accurate background determination. With even the thinnest carbon film substrate, the particle is usually less than half the mass (particle + background) within the measuring area. Two issues arise in background determination: (1) deciding which pixels are far enough away from an object to be considered background and (2) deciding whether the background measured away from the particle is valid underneath the particle. The second issue becomes critical if the sample contains detergent, salt, sugar, denatured protein, or any material other than the specimen of interest. PCMass contains a background program, originally written by J. Hainfeld with improvements by T. Baker, which masks out particles efficiently as long as the specimen is not too crowded or contaminated. Several additional means, which will be discussed subsequently, are provided to test the reliability of the background determination.

Actual mass measurement consists of selecting suitable particles with an adjustable circle or rectangle. A mouse positions the circle or rectangle and keystrokes adjust size and rectangle length. The program sums intensity minus background for each pixel inside the boundary, then multiplies by the STEM calibration constant. In PCMass, the displayed mass value is updated as parameters are changed. A mouse click or "=" keystroke mark the particle on the display and save the measurement parameters. Showing the mass value places the burden on the user to be objective in particle selection, instead of excluding particles simply because they spoil the standard deviation (SD). This is a deliberate decision, as focusing special attention on deviant particles is important to improving future preparations and understanding the specimen as well as its limits of interpretation.

TMV is included in nearly all STEM samples as an internal control. In the past, we used a TMV calibration for each image. However, extensive measurements have demonstrated that the STEM mass calibration is very stable and fluctuations in TMV mass per unit length (M/L) are primarily due to variation in specimen quality. Therefore PCMass uses a standard STEM calibration value (115 Dalton/intensity unit with 1-nm pixels) instead of a TMV calibration. If extensive measurements suggest that a different TMV value should be used, we advise scaling with great caution and reporting the scaling method in any publication. It is always a good idea to check all unobstructed TMV segments for M/L and shape and to examine the squareness of ends (most sensitive to damage). Many problems that could compromise data quality can be sorted out using the known TMV structure.

PCMass provides a number of models as an aid in selecting particles of certain types. It automatically adjusts the size, amplitude, and orientation of the selected model, displaying image, model, and difference image in separate windows. Rotational power spectrum, radial mass profile, and density profile are also displayed. The difference image can be especially useful in identifying damaged particles or salt. Separate mass statistics are maintained for measurements made relative to each of the 80 available models. Extensive capabilities are provided for viewing mass measurements and histograms of single images and groups of images.

8. Performance Checks

PCMass provides histograms for each of up to five models on each image, as well as global histograms over a range of images. This gives a rapid overview of trends and highlights suspect particles or images. Mass loss is typically 2.5% at a dose of 10 el/A^2 normally used on a 0.512-μm micron scan. A 0.256-μm scan at the same beam current would deliver 40 el/A^2 and cause 10% mass loss, so it is usually best not to combine data recorded with different scan sizes without careful consideration. If the specimen contains detergent or salt, the background may fluctuate from image to image and the TMV M/L may fluctuate in the same or the opposite direction depending on the affinity of the contaminant for carbon film or protein. The TMV M/L should be 13.1 kDa/A with an SD of 2% or less under ideal conditions.

If the sample particles are known to have a rigid shape similar to that of the selected model but many are failing the RMS test, the particles may be falling apart. That would also tend to show up as a skew in the overall histogram. For reference in judging this, a Gaussian with the same mean, SD, and integrated area is displayed in the main histogram window. With a large number of measurements, it is tempting to report the standard error of the mean, rather than the standard deviation. That assumes the errors are purely random, and we consider the SD to be a more realistic measure of overall accuracy.

9. Applications

STEM mass mapping has been used extensively by one of us (SNV) to determine the mass of annelid hexagonal bilayer hemoglobins (Lamy *et al.*, 1996). Freeze-dried *Lumbricus terrestris* blood, which is essentially a pure solution of its 3.56 mDa hexagonal bilayer hemoglobin (Lamy *et al.*, 1996), is shown in Figure 27.3 as an example of a good STEM specimen. STEM mass measurements can also be used to determine the mass of its globin dodecamer subunit (Martin *et al.*, 1996) and to characterize the uniformity and mass distribution of *in vitro* reassembled structures (Kuchumov *et al.*, 1999). The stoichiometry of the *Lumbricus* hemoglobin is now known from its crystal structure (Royer *et al.*, 2006); thus, the mass of the native complex can be calculated with high precision from the electrospray ionization mass-spectrometric determination of the masses of its constituent subunits (Martin *et al.*, 1996): the 4 $\times$ 36 globin chains a to d and 36 nonglobin linker chains L1 to L4. Because some of the subunits have isoforms differing in molecular mass and because of posttranslational glycosylation, *Lumbricus* hemoglobin exhibits complicated combinatorics (Hanin and Vinogradov, 2000); thus, there is a distribution of masses ranging from 3,418 to 3,725 kDa, with an expected mass of 3,517 $+$ 16 kDa (Hanin *et al.*, 2003). Nevertheless, because of the stability of the hemoglobin and the ease of obtaining a sample of *Lumbricus* blood, it provides a useful alternative to TMV as an internal standard in STEM mass mapping of proteins and is used as such at the BNL STEM facility. It is worth pointing out that of all the methods used to determine the mass of annelid extracellular hemoglobins (e.g., sedimentation equilibrium, light scattering, multiple-angle laser light scattering, gel filtration, small-angle X-ray scattering), STEM mass mapping was the most accurate: its results were closest to the masses calculated using the *Lumbricus* hemoglobin stoichiometry and the subunit masses determined by electrospray ionization mass spectrometry (Hanin *et al.*, 2003).

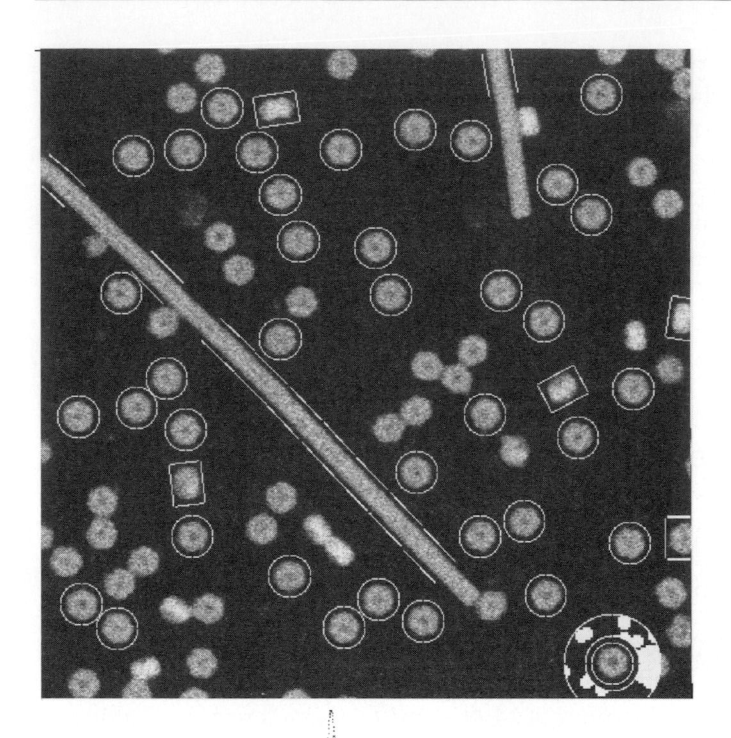

3769.554 +/- 94.312 N= 37

2500.000 5000.000 7500.000

Summary of 302 measurements on File C:\MSC\PCMass\MassMeas\nk737537.smm

Dose= 10.82	Model 21 TMV Rod	Model 35 EHb face	Model 49 EHb side
Manual all	———— -	———— -	———— -
Manual pass b	———— -	———— -	———— -
Manual pass	———— -	———— -	———— -
AutoSel all	12.343M +/-18.8% 24	3.783M +/- 7.4% 57	3.704M +/- 2.3% 53
AutoSel pass b	12.993M +/- 5.2% 10	3.770M +/- 2.5% 42	———— -
AutoSel pass	———— -	3.743M +/- 2.9% 19	———— -
Align all	13.089M +/-12.0% 24	3.777M +/- 6.3% 56	3.754M +/- 4.9% 54
Align pass b	13.424M +/- 6.3% 10	3.774M +/- 2.5% 42	3.682M +/- 3.7% 9
Align pass	13.257M +/- 2.2% 8	3.770M +/- 2.5% 37	3.818M +/- 3.0% 3

Use Arrow Keys to select, <Enter> to display, <Shift><Enter>with mass, <SpaceBar><Enter>- failing pai

Figure 27.3 Summary screen of PCMass program (reformatted for clarity). The large-angle dark-field detector signal is marked with circles indicating face-on hemoglobin particles (Model #35), rectangles indicating edge-on particles (Model #49), and parallel lines indicating regions of TMV (Model #21) used for measurement of mass per unit length. The mean, SD, and number of accepted particles are tabulated in the "Align pass" row generated by automated mass measurement of the image shown. Manual measurements would be tabulated in the first three rows, depending on how many of the selection criteria they passed. Below each tabulation is a miniature histogram for the selected category. A larger histogram of the measurements highlighted in yellow is shown in the middle of the figure together with a Gaussian curve having the same mean, standard deviation, and integrated area. (See color insert.)

STEM measures both mass and length, so it is also suitable for studying filaments, giving mass per unit length. Furthermore, the distance of every pixel from the center line of a filament or center of a particle is accurately determined, giving a radial mass profile. If the object has spherical or cylindrical symmetry, the mass profile can be transformed to give a radial density profile. STEM has proven particularly useful in studies of self-assembly of paired helical filaments found in Alzheimer's disease (Ksiesak-Reding and Wall, 2005) and prions (Baxa *et al.*, 2003).

REFERENCES

Baxa, U., Taylor, K. L., Wall, J. S., Simon, M. N., Cheng, N., Wickner, R. B., and Steven, A. C. (2003). Architecture of Ure2p prion filaments: The N-terminal domains form a central core fiber. *J. Biol. Chem.* **278,** 43717–43727.

Crewe, A. V., Wall, J. S., and Langmore, J. (1970). Visibility of single atoms. *Science* **168,** 1338–1343.

Hainfeld, J. F., and Wall, J. S. (1988). High resolution electron microscopy for structure and mapping. *Basic Life Sci.* **46,** 131–147.

Hanin, L. G., and Vinogradov, S. N. (2000). Combinatorics of giant hexagonal bilayer hemoglobins. *Math. Biosci.* **163,** 59–73.

Hanin, L. G., Zal, F., Green, B. N., and Vinogradov, S. N. (2003). Mass distribution of a macromolecular assembly based on electrospray ionization molecular masses of its constituent subunits. *J. Biosci.* **28,** 101–112.

Ksiesak-Reding, H., and Wall, J. S. (2005). Characterization of paired helical filaments by scanning transmission electron microscopy. *Microsc. Res. Tech.* **67,** 126–140.

Kuchumov, A. R., Taveau, J.-C., Lamy, J. N., Wall, J. S., Weber, R. E., and Vinogradov, S. N. (1999). The role of the linker subunits in the reassembly of the 3.6MDa hexagonal bilayer hemoglobin from *Lumbricus terrestris*. *J. Mol. Biol.* **289,** 1361–1374.

Lamy, J. N., Green, B. N., Toulmond, A., Wall, J. S., Weber, R. E., and Vinogradov, S. N. (1996). Giant hexagonal bilayer hemoglobins. *Chem. Rev.* **96,** 3113–3124.

Martin, P. D., Kuchumov, A. R., Green, B. N., Oliver, R. W., Braswell, E. H., Wall, J. S., and Vinogradov, S. N. (1996). Mass spectrometric composition and molecular mass of *Lumbricus terrestris* hemoglobin: A refined model of its quaternary structure. *J. Mol. Biol.* **255,** 154–169.

Royer, W. E., Jr., Sharma, H., Strand, K., Knapp, J. E., and Bhyravbhatla, B. (2006). Lumbricus erythrocruorin at 3.5 A resolution: Architecture of a megadalton respiratory complex. *Structure* **14,** 1167–1177.

Thomas, D., Schultz, P., Steven, A. C., and Wall, J. S. (1994). Mass analysis of biological macromolecular complexes by STEM. *Biol. Cell.* **80,** 181–192.

Wall, J. S. (1971). A high resolution scanning microscope for the study of biological molecules. Ph. D. thesis, University of Chicago.

Wall, J. S., and Hainfeld, J. F. (1986). Mass mapping with the scanning transmission electron microscope. *Annu. Rev. Biophys. Chem.* **15,** 355–376.

Wall, J.S, and Simon, M. N. (2001). Scanning transmission electron microscopy of DNA-protein complexes. *Methods Mol. Biol.* **148,** 589–601.

CHAPTER TWENTY-EIGHT

Mini-Hemoglobins from Nemertean Worms

Thomas L. Vandergon* and Austen F. Riggs†

Contents

Abstract

Hemoglobins (Hbs) found in members of the phylum Nemertea are smaller than any other known Hb molecules. These mini-Hbs have been of great interest because of their unique three-dimensional structure and their stable ligand-binding properties. Also of interest is the expression of mini-Hb in neural tissue, body wall muscle tissue, and red blood cells. This chapter outlines methods that may be used to isolate and purify functional mini-Hbs from all three tissue types in nemertean worms.

1. Background

Animals in the phylum Nemertea are related to annelids, mollusks, and other protostomous coelomates on the basis of both molecular and anatomical data (Sundberg *et al.*, 2001; Thollesson and Norenburg, 2003; Turbeville *et al.*, 1992). They are all soft-bodied, vermiform organisms that live primarily in marine and estuarine habitats, although both freshwater and

* Natural Science Division, Pepperdine University, Malibu, California
† Section of Neurobiology, University of Texas at Austin, Austin, Texas

Methods in Enzymology, Volume 436
ISSN 0076-6879, DOI: 10.1016/S0076-6879(08)36028-5

terrestrial representatives also occur. Many nemerteans express intracellular hemoglobin (Hb) in tissues that include neural tissue, body wall muscles, and circulating red blood cells (coelomocytes). Neural tissue Hb is most abundant in the brain ganglion and main nerve cords that run laterally or ventrolaterally along the body. Body wall muscle Hb is expressed along the length of the body. The level of expression appears to vary with tissue location and metabolic demand in each animal. Hb is expressed in red blood cells (RBC) of the circulatory system (coelomic channels), with major vessels running laterally along the body, medial to nerve cords, and forming small lacunae at the anterior and posterior ends of the body. The neural and body wall Hbs of *Cerebratulus lacteus* have been extensively studied in their native (Vandergon *et al.*, 1998) and recombinant (Marti *et al.*, 2006; Pesce *et al.*, 2001, 2002, 2004) forms. The intracellular Hb in the neural tissues and body wall muscle cells of *C. lacteus* are the smallest native Hbs identified to date in any organism and have been dubbed "mini-Hbs" (Vandergon *et al.*, 1998). Both tissue globins are 109 residues in length, and they both lack a terminal methionine in the mature protein.

2. Isolation and Purification of *C. lacteus* Hb

2.1. Anesthesia

Animals should be relaxed in seawater (SW) containing an anesthetic such as tricane methane sulfonate (MS-222) (Ross and Ross, 1999). Dissolve MS-222 to a concentration of 0.1 to 1% in 0.45 μm filtered SW or 0.45 μm filtered artificial seawater (FASW) isotonic to the ambient water made up as shown in Table 28.1. The higher concentrations of MS-222 may speed up anesthesia. Powdered concentrates for FASW may also be purchased at aquarium supply stores.

Table 28.1 Composition of artificial seawater

Component	[C] m*M*	Molecular Formula	Formula Weight	g/L
NaCl	423.00	NaCl	58.44	24.72
KCl	9.00	KCl	74.55	0.67
$CaCl_2$	9.27	$CaCl_2{\bullet}2H_2O$	147.02	1.36
$MgCl_2$	22.94	$MgCl_2{\bullet}6H_2O$	203.31	4.66
$MgSO_4$	25.50	$MgSO_4{\bullet}7H_2O$	246.48	6.29
$NaHCO_3$	2.14	$NaHCO_3$	84.01	0.18

Notes: For artificial seawater, add $NaHCO_3$ last, and then filter; from Cavanaugh (1975) and Lyman and Fleming (1940).

Total unresponsiveness may take 1 to several hours depending upon the size of the animal and the concentration of MS-222. Alternative anesthetics that have been used successfully on nemerteans include 5% chloral hydrate in FASW (Russell, 1963) and 7.5% $MgCl_2$ in FASW (Gibson, 2001).

2.2. Tissue Hb isolation

In many species of nemerteans, including *C. lacteus*, the bright pink or red ganglion is visible through the head. To isolate the neural hemoglobin, first anesthetize the animals until unresponsive as described previously, and then place the animal on a glass slide with a few drops of FASW containing the anesthetic. For large *C. lacteus*, the anterior 10 to 15 cm of the animal should be severed from the rest of the body and placed on the glass slide. While viewing under a dissecting microscope, use a sharp scalpel or razor blade to cut just behind the cerebral ganglion. Continue to remove excess tissue from the anterior, lateral, and then dorsal and ventral sides of the head surrounding the ganglion until the four-lobed reddish ganglion is isolated. Careful dissection of excess tissue surrounding the neural ganglion is necessary to prevent introduction of body wall or RBC hemoglobin into the neural hemoglobin preparation. Each lateral nerve cord also may be dissected from the body by first making a series of crosscuts along the anterior part of the body behind the brain ganglion to create several 2-cm-long pieces. (Longer segments tend to bend and are more difficult to dissect.) Take each 2-cm body segment and make a longitudinal cut just medial to the nerve cords on each side, and lay the cord-containing segments on the slide with the cut side down. Continue to dissect away tissue surrounding the cord on each side until all excess tissue is removed.

To isolate body wall muscle hemoglobin, begin with anesthetized animals and carefully dissect segments of body wall muscle posterior to the region of the rhynchocoel, where the expression level typically is higher. The rhynchocoel is the sac bearing the proboscis (feeding apparatus) and typically occupies the anterior third of the body. Care should be exercised to avoid portions of the lateral blood channels and the lateral (or middorsal) nerve cords. This may not be possible for small specimens, but it can be accomplished with larger specimens by dissecting the dorsolateral body regions while avoiding the ventrolateral nerve cords, the coelomic channels, and the middorsal nerve.

Transfer the isolated ganglion/nerve segments or body wall segments to clean tubes containing a small amount of the lysis buffer, 0.05 *M* Tris, 0.1 *M* NaCl, pH 7.6. Tissue hemoglobin is more difficult to extract than is RBC hemoglobin and usually requires homogenizing the tissues. Homogenization of fresh tissue in the lysis buffer may be performed at 0° with a Dounce or Potter-Elvehjem tissue grinder. Care must be exercised to prevent excess heating of the tissue sample during extraction to prevent degradation of the Hb. Alternatively, tissues may be ground in liquid nitrogen to a fine powder, with subsequent resuspension of the powder in lysis buffer at 0°. Crude Hb extracts from ground tissue should be centrifuged at 12,000 to 15,000 g-force for 10 min at 0° to pellet the cell debris. Transfer the clear supernatant to clean tubes for use or long-term storage at −80°.

Native nemertean Hbs are reasonably stable in our experience, and the level of autoxidation is low so that protease inhibitors (e.g., phenyl methyl sulfonyl fluoride [PMSF]) and reducing agents (e.g., dithiothreitol) are unnecessary. If autoxidation does become a problem in a preparation, CO can be added to the lysis buffer to convert all the HbO_2 to the more stable HbCO form.

2.3. Tissue Hb purification

Cleared Hb extracts may be purified by size-exclusion chromatography in a column (0.7- to 1.0-cm diameter by 50 to 120 cm) of Sephadex G-75–120 (Sigma-Aldrich, St. Louis, MO) equilibrated with 0.05 *M* Tris, 0.1 *M* NaCl, 10 mM EDTA, pH 7.6 (Sephadex G50–150 may be used, although we find better separation with a G75–120 matrix). The sample loading volume should not exceed 1% of the column volume. Chromatographic separation is accomplished at 4 to 12° at a flow rate of about 0.2 ml/min (varies with column height and packing). The Hb-containing fraction(s) may be identified by absorbance at both 280 and 415 nm.

An alternative method for purification of the crude Hb extracts is by native gel electrophoresis. Small samples (12 to 24 μl) of crude extract may be mixed with 3 to 6 μl of a nondenaturing loading buffer (10% glycerol in 50 m*M* Tris/0.192 *M* glycine buffer, pH 7.6, with 0.5% bromophenol blue dye) and electrophoresed on a 14% acrylamide gel in Tris/glycine running buffer for 90-120 V-h. The running buffer is 0.025 *M* Tris, 0.192 *M* glycine, pH 8.3. Premade Tris/glycine gels without stacking gels may be purchased from a variety of manufacturers or may be made using standard protocols (Garfin, 1990). Human Hb and horse or sperm whale myoglobin may be used as markers on these gels, but the samples and markers do not necessarily migrate relative to their molar masses and should be spaced away from sample lanes. Both neural and body wall muscle Hb from *C. lacteus* yield two colored bands on native gels. These bands correspond to the monomeric and dimeric Hb forms. Both bands have the absorbances at 280 and 415 nm, characteristic of Hb. To elute Hb from the gel, excise the bands of interest from the gel matrix with a scalpel or spatula, and place each in a 1.5-ml microfuge tube containing a small volume (100 to 200 μl) of lysis buffer (0.05 *M* Tris, 0.1 *M* NaCl, pH 7.6). Break up the acrylamide gel into small pieces with a pipette tip or sterile inoculating loop to create a slurry, and incubate at 4 to 12° for 12 to 24 h. Carefully remove the fluid from the gel pieces and transfer to a clean tube with a fine-tipped pipette.

Product from equivalent bands in separate lanes may be combined in a single tube. The eluted fractions should be centrifuged at 12,000 to 15,000 g for 10 min at 0° to pellet any remaining acrylamide and the cleared supernatant transferred to a clean tube for use or storage.

Hb samples purified via size-exclusion chromatography or native gel electrophoresis are dilute but may be concentrated using Centricon YM-10 centrifugal filter units (Millipore Corp.) following the manufacturer's recommended protocols.

3. Other Nemertean Hbs

Although *C. lacteus* has only a small number of coelomic red blood cells relative to its body size, other species of nemerteans, particularly *Lineus* spp. and *Amphiporus* spp. produce RBCs in sufficient quantities for potential isolation.

3.1. Red Blood Cells

RBCs are found mainly in the lateral blood channels and can be difficult to obtain in quantity because the channels are very small in diameter and compress when the animals are dissected, particularly if animals are not anesthetized prior to cutting. To collect red blood cells, first anesthetize animals until unresponsive (as described previously) and place them in a shallow dish containing FASW with anesthetic. While viewing the specimen under a dissecting microscope, use a sharp scalpel or razor blade to crosscut one or more times in the middle body region and near anterior and posterior lacunae. Gentle squeezing of the body segments using a bent glass pipette may help push red blood cells out of the channels, where they will fall to the bottom of the dish. RBCs that settle to the bottom of the dish may then be taken up in a pipette and transferred to a small centrifuge tube. When full of liquid, centrifuge the tube at about 3,000 g force for 2 to 5 min at 4 to 12° to pellet the RBCs. Carefully remove the supernatant and discard. This procedure may be repeated several times in the same tube to concentrate the RBCs. RBCs may remain intact and functional for 1 to 2 days at 4° in FASW.

To lyse the RBCs, resuspend the pelleted cells in 5 to 10 times the pellet volume of lysis buffer (0.05 *M* Tris, 0.1 *M* NaCl, pH 7.6) at 0° by repeated gentle vortexing. Care should be exercised to prevent overheating of sample and protein denaturation by too-vigorous shaking. The cell suspension may be freeze-thawed to assist in lysis of RBCs. Centrifuge the lysed cell solution at 15,000 RPM for 10 min at 4° to pellet the cell debris and assess the degree of cell lysis. The pellet should be clear or a pale pink color. A dark red pellet will indicate insufficient lysis of cells. The clear reddish-orange supernatant may be transferred to a clean tube for storage or use. Lysed material should be kept on ice or frozen to prevent degradation. Gently equilibrate the lysed solution with CO to form HbCO as an aid to prevent oxidation and degradation.

3.2. Collecting Nemerteans

C. lacteus along with the much smaller species *Lineus* spp. and *Amphiporus* spp. have been available for purchase in limited supply from the Marine Biological Laboratories at Woods Hole in past years, although recently are not always available. In nature, nemerteans are common in both mussel and oyster beds and in surface estuarine sediment, and they can be found crawling under rocks or debris in shallow marine or estuarine waters. Several references have been published to aid in identification of nemerteans (Brunberg, 1964; Coe, 1905; Corrêa, 1961; Gibson, 1994, 1995, 2001; Gibson and Moore, 1976). To collect smaller nemerteans, remove clumps of mussels or other encrusting plant and animal material from hard surfaces, or gather piles of shell debris (subtidal) or other sediment down to about 20-cm depth and place this material in a shallow container, just covered with water collected at the same location. Allow the container to become anoxic by not stirring or aerating the water. Each day for the next week, examine the surface of the sediments, and the container walls especially along the water surface. Nemerteans will typically abandon the decaying material and may be observed gliding on the edges of the container, usually at or near the surface. They also may be observed crawling on the surface of the sediments or on the water surface itself. These animals can easily be collected with a plastic pipette and transferred into shallow dishes containing clean aerated water collected with the specimens or in FASW isotonic to the ambient water. A shovel and sieve may be used to collect larger nemerteans, but the animals tend to break apart with this procedure, so obtaining whole, unfragmented individuals is difficult. Isolated animals may be maintained at 4 to 12° in the laboratory for several days if fresh medium (FASW) is provided.

REFERENCES

Brunberg, L. (1964). On the nemertean fauna of Danish waters. *Ophelia* **1,** 77–111.

Cavanaugh, G. M. (1975). "Formulae and methods VI." The Marine Biological Laboratory, Woods Hole, MA.

Coe, W. R. (1905). Nemerteans of the west and northwest coasts of America. *Bull. Mus. Comp. Zool. Harvard* **47,** 1–318.

Corrêa, D. D. (1961). Nemerteans from Florida and Virgin Islands. *Bull. Mar. Sci. Gulf Carib.* **11,** 1–44.

Garfin, D. E. (1990). One-dimensional gel electrophoresis. *Methods Enzymol.* **182,** 425–441.

Gibson, R. (1994). Nemerteans. Synopses of the British Fauna, n.s., 24 2d ed. Shrewsbury, UK, Field Studies Council.

Gibson, R. (1995). Nemertean genera and species of the world: An annotated checklist of original names and description citations, synonyms, current taxonomic status, habitats and recorded zoogeographic distribution. *J. Nat. Hist.* **29,** 271–562.

Gibson, R. (2001). Phylum Nemertea (Nemertinea, Nemertini, Rhynchocoela). *In* "Keys to the marine invertebrates of the Woods Hole region" (Smith, R. I., ed.), at http://www.mbl.edu/BiologicalBulletin/KEYS/INVERTS/Gibson/index.html.

Gibson, R., and Moore, J. (1976). Freshwater nemerteans. *Zool. J. Linn. Soc.* **58,** 177–218.

Lyman, J., and Fleming, R. H. (1940). Composition of seawater. *J. Mar. Res.* **3,** 134–146.

Marti, M. A., Bikiel, D. E., Crespo, A., Nardini, M., Bolognesi, M., and Estrin, D. A. (2006). Two distinct heme distal site states define *Cerebratulus lacteus* mini-hemoglobin oxygen affinity. *Proteins* **62,** 641–648.

Pesce, A., Nardini, M., Dewilde, S., Ascenzi, P., Riggs, A. F., Yamauchi, K., Geuens, E., Moens, L., and Bolognesi, M. (2001). Crystallization and preliminary X-ray analysis of neural haemoglobin from the nemertean worm *Cerebratulus lacteus. Acta Crystallogr.* **57,** 1897–1899.

Pesce, A., Nardini, M., Dewilde, S., Geuens, E., Yamauchi, K., Ascenzi, P., Riggs, A. F., Moens, L., and Bolognesi, M. (2002). The 109-residue nerve tissue minihemoglobin from *Cerebratulus lacteus* highlights striking structural plasticity of the alpha-helical globin fold. *Structure* **10,** 725–735.

Pesce, A., Nardini, M., Ascenzi, P., Geuens, E., Dewilde, S., Moens, L., Bolognesi, M., Riggs, A. F., Hale, A., Deng, P., Nienhaus, G. U., Olson, J. S., Nienhuis, K. (2004). Thr-E11 regulates O_2 affinity in *Cerebratulus lacteus* mini-hemoglobin. *J. Biol. Chem.* **279,** 33662–33672. Erratum in: *J. Biol. Chem.* **279**, 41258.

Ross, L. G., and Ross, B. (1999). "Anaesthetic and sedative techniques for aquatic animals" 2d ed. Blackwell Science, UK.

Russell, H. D. (1963). Notes on methods for the narcotization, killing, fixation, and preservation of marine organisms. Systematics-Ecology Program. Marine Biological Laboratory, Woods Hole, MA. Available at http://www.mbl.edu/BiologicalBulletin/CLASSICS/Russell/Russell-TitlePage.html.

Sundberg, P., Turbeville, J. M., and Lindh, S. (2001). Phylogenetic relationships among higher nemertean (Nemertea) taxa inferred from 18S rDNA sequences. *Mol. Phylogen. Evol.* **20,** 327–334.

Thollesson, M., and Norenburg, J. L. (2003). Ribbon worm relationships: A phylogeny of the phylum Nemertea. *Proc. Roy. Soc. B.* **270,** 407–415.

Turbeville, J. M., Field, K. G., and Raff, R. A. (1992). Phylogenetic position of phylum Nemertini, inferred from 18s rRNA sequences: Molecular data as a test of morphological character homology. *Mol. Biol. Evol.* **9,** 235–249.

Vandergon, T. L., Riggs, C. K., Gorr, T. A., Colacino, J. M., and Riggs, A. F. (1998). The mini-hemoglobins in neural and body wall tissue of the nemertean worm, *Cerebratulus lacteus. J. Biol. Chem.* **273,** 16998–17011.

CHAPTER TWENTY-NINE

Comparative and Evolutionary Genomics of Globin Genes in Fish

Enrico Negrisolo,* Luca Bargelloni,* Tomaso Patarnello,* Catherine Ozouf-Costaz,† Eva Pisano,‡ Guido di Prisco,§ *and* Cinzia Verde§

Contents

* Department of Public Health, Comparative Pathology, and Veterinary Hygiene, University of Padova, Legnaro, Italy
† Department Systematics and Evolution, Paris, France
‡ Department of Biology, University of Genoa, Genoa, Italy
§ Institute of Protein Biochemistry, CNR, Naples, Italy

Methods in Enzymology, Volume 436
ISSN 0076-6879, DOI: 10.1016/S0076-6879(08)36029-7

Abstract

Sequencing genomes of model organisms is a great challenge for biological sciences. In the past decade, scientists have developed a large number of methods to align and compare sequenced genomes. The analysis of a given sequence provides much information on the genome structure but to a lesser extent on the function. Comparative genomics are a useful tool for functional and evolutionary annotation of genomes. In principle, comparison of genomic sequences may allow for identification of the evolutionary selection (negative or positive) that the functional sequences have been subjected to over time. Positively selected genome regions are the most important ones for evolution, because most changes are adaptive and often induce biological differences in organisms. The draft genomes of five fish species have recently become available. We herewith review and discuss some new insights into comparative genomics in fish globin genes. Special attention will be given to a complementary methodological approach to comparative genomics, fluorescence *in situ* hybridization (FISH). Internet resources for analyzing sequence alignments and annotations and new bioinformatic tools to address critical problems are thoroughly discussed.

1. Introduction

The very fast development of genomic technologies has brought biological sciences to the threshold of a revolution. The availability of genome sequences greatly enhances our understanding of the structure, content, and evolution of genomes. Over the past few years, many organisms (about 1,000 species) have had their genomes completely sequenced; in many others, sequencing is in progress (for a list, see genomes@ncbi.nlm.nih.gov). Clearly, the success of the publicly and privately financed genome projects is largely due to the development of high-throughput, low-cost DNA sequencing methodologies and to bioinformatic tools available for genome annotations. Unfortunately, the billions of sequenced DNA bases do not tell us what the ensemble of genes does or how cells form organisms. However, where the mere sequence fails, other features of the genome offer help in the evolutionary analysis and functional inference. For example, the gene positions and changes in gene order may indicate evolutionary splits.

Comparative genomics is a relatively young discipline that aims to identify structural and functional genomic elements conserved across different species (Nobrega and Pennacchio, 2004). The first sequenced genomes were so phylogenetically distant that they did not allow for the study of function and evolution or, above all, for comparison among noncoding sequences. In the recent past, the comparison between phylogenetically related genomes has provided additional insights into gene function, structure, and evolution. The comparison among sequenced genomes not only led to the discovery of a new domain of life, the Archaea, but also revealed that chimpanzees, not gorillas as previously believed, are our closest relative. With the availability of additional vertebrate genomes, it will also be possible to explore intermediate nodes such as the last common ancestor of amniotes, sarcopterygians, and actinopterygians.

The study of the chromosomal organization and regulation of globin genes in teleost fish provides a nice example of how such comparative methods can be applied. The goal of comparative genomics is to provide information on function and particularly on evolution. The functional sequences experienced evolutionary selection that left a mark (signature) on the aligned DNA sequences. The search for these signatures when comparing genomic sequences may help identify functional DNA.

However, the success of future genomics research will depend not only on the new available technologies but also on the expertise of researchers and the integration and synthesis of knowledge on the genome, physiology, and biochemistry of model organisms.

In this chapter, we review and discuss some of the new insights about comparative genomics in fish globin genes. Bioinformatic tools to delve more deeply into such issues and to use comparative genomics are summarized.

2. Comparative Genomics of Fish Globin Genes

2.1. Available data sets

In this section, we will focus on comparison of genomes at the sequence level. Molecular cytogenetic approaches will be reviewed in a subsequent section (see the "Cytogenetic methods for comparative genomics: FISH").

Technological progress in DNA cloning and sequencing has been accompanied by a large data set of genomic sequence information. In the past few years, draft genome sequences have become available for five fish species: two puffer fish (*Fugu rubripes* and *Tetraodon nigroviridis*) (Aparicio *et al.*, 2002; Jaillon, *et al.*, 2004), zebra fish (*Danio rerio*), medaka (*Oryzias latipes*), and threespine stickleback (*Gasterosteus aculeatus*). Genome sequences of these species can be accessed through dedicated Web sites (Table 29.1). Additional species are currently being sequenced, and low- or medium-coverage genome sequences are expected in the near future. For instance, sequencing (5X coverage) the genomes of the Nile tilapia (*Oreochromis niloticus*) and three haplochromine cichlids (2X coverage) will begin this year (Cichlid Genome Consortium, see Table 29.1). Low-coverage (1.5X) sequencing of the genome of the European sea bass (*Dicentrarchus labrax*) is under way at the Max Planck Institute for Molecular Genetics in Berlin (Marine Genomics in Europe; R. Reinhardt,

Table 29. 1 List of programs/servers and their web sites

Program/server	web site
Adaptive evolution server	http://www.datamonkey.org/
ADAPTSITE	http://www.bio.psu.edu/People/Faculty/Nei/Lab/
ANCESCON	ftp://iole.swmed.edu/pub/ANCESCON/
ANCESTOR	http://www.bio.psu.edu/People/Faculty/Nei/Lab/ancestor2.htm
BayesPhylogenies	http://www.evolution.rdg.ac.uk/BayesPhy.html
Cichlid Genome Consortium	http://hcgs.unh.edu/cichlid/
ClustalW	http://pbil.ibcp.fr/htm/index.php
ClustalX	http://bips.u-strasbg.fr/fr/Documentation/ClustalX/
Codon Align 2.0	http://homepage.mac.com/barryghall/CodonAlign.html
CONREAL	http://conreal.niob.knaw.nl/
CONSEL	http://www.is.titech.ac.jp/~shimo/prog/consel/
ConSurf	http://consurf.tau.ac.il/
covARES	available upon request to the authors
DAMBE	http://dambe.bio.uottawa.ca/software.asp
Danio rerio Sequencing Project	http://www.sanger.ac.uk/Projects/D_rerio/
Dcode.org	http://www.dcode.org/
dde	http://www.bio.psu.edu/People/Faculty/Nei/Lab/
DIVERGE2	http://xgu.gdcb.iastate.edu/
ECR Browser	http://ecrbrowser.dcode.org/
Ensemble Genome Browser	http://www.ensembl.org/index.html
GenBank	http://www.ncbi.nlm.nih.gov/
Geneious	http://www.geneious.com/
HyPhy	http://www.hyphy.org/
MEGA	http://www.megasoftware.net/
Mesquite	http://mesquiteproject.org./mesquite/mesquite.html
ModelTest 3.7	http://darwin.uvigo.es/
MrBayes 3.1	http://mrbayes.csit.fsu.edu/
MrModeltest 2.2	http://www.abc.se/~nylander/
MUSCLE	http://www.drive5.com/muscle/
PAML	http://abacus.gene.ucl.ac.uk/software/paml.html
PAUP* 4.10	http://paup.csit.fsu.edu/

Table 29. 1 (*continued*)

Program/server	web site
phylemon	http://phylemon.bioinfo.cipf.es/
PHYLIP	http://evolution.gs.washington.edu/phylip.html
Phylogeny programs	http://evolution.genetics.washington.edu/phylip/software.html
PHYML	http://atgc.lirmm.fr/phyml/
PLATCOM	http://platcom.informatics.indiana.edu/platcom/
ProtTest	http://darwin.uvigo.es/
Rate shift analysis server	http://www.daimi.au.dk/~compbio/rateshift/
SplitTester	http://www.public.iastate.edu/~voytas/SplitTester/
Stickleback (*Gasterosteus aculeatus*)	http://www.ensembl.org/Gasterosteus_aculeatus
SynBrowse	http://www.gmod.org/home
T-COFFEE	http://www.ch.embnet.org/software/TCoffee.html
TOUCAN 2	http://homes.esat.kuleuven.be/~saerts/software/toucan.php
TreeDyn	(http://www.treedyn.org/)
TreeView	http://taxonomy.zoology.gla.ac.uk/rod/treeview.html
TREEFINDER	http://www.treefinder.de/
UT Genome Browser (Medaka)	http://medaka.utgenome.org/
VerAlign	http://zeus.cs.vu.nl/programs/veralignwww/
VISTA	http://genome.lbl.gov/vista/index.shtml

personal communication). Other fish genome initiatives involve the Atlantic cod (*Gadus morhua*), several salmonids [*Oncorhynchus mykiss* (rainbow trout), *O. tshawytscha* (chinook salmon), *O. kisutch* (coho salmon), *Salmo salar* (Atlantic salmon), *S. trutta* (brown trout), and *Salvelinus alpinus* (Arctic char)], as well as carp, catfish, and flatfish species.

Information on genomic organization of α- and β-globin genes in teleost fish can be obtained by searching the available genome-sequence databases (e.g., Ensemble Genome Browser, Table 29.1). Several search strategies can be adopted, including sequence similarity searches (e.g., BLAST, BLAT, SSAHA) (Altschul *et al.*, 1990; Kent, 2002; Ning *et al.*, 2001), simple string search, or more sophisticated data-mining tools (e.g., BioMart) (Durinck *et al.*, 2005). Further bioinformatic approaches are briefly outlined in the following sections.

2.2. Bioinformatic tools for comparative genomics

Genome browsers are essential tools for molecular and genetic research. Several bioinformatic tools for comparative analyses are mainly available on the Web. They are usually hosted on very powerful servers, because they are computationally very demanding. However, researchers can also install them locally, provided that suitable computers are available. These packages allow for many kinds of genomic investigations. Analyses include full annotation of genomic sequences, prediction of gene boundaries, identification of promoters and regulative elements, fitting of expressed sequence tags (ESTs) to genome sequences, and identification of syntenic regions among genomes.

Although there are many similarities among different browsers, some information can be drawn from a given browser but not from others. Packages including many tools for comparative genomic studies are CONREAL (Berezikov *et al.*, 2005), Dcode.org (Loots and Ovcharenko, 2005), ECR Browser (Ovcharenko *et al.*, 2004), PLATCOM (Choi *et al.*, 2005), SynBrowse (Pan *et al.*, 2005), TOUCAN 2 (Aerts *et al.*, 2005), and VISTA (Frazer *et al.*, 2004). All these programs are accessible through dedicated Web sites (see Table 29.1).

The alignment of genomic sequences obtained from different organisms is an excellent example of comparative genomics analysis. It can be fruitfully applied to the genomic regions encoding globin genes. Multiple alignments can be performed by means of several programs.

Many browsers are specific for one major purpose. Among the newly developed tools, we mention AVID (Bray *et al.*, 2003), CHAOS (Brudno *et al.*, 2003a), LAGAN and Multi-LAGAN (Brudno *et al.*, 2003b), and MULAN (Ovcharenko *et al.*, 2005).

2.3. Identifying orthologous and paralogous genes

A main issue in comparative genomics is to identify orthologous genes and distinguish them from paralogous ones. Two sequences are orthologous if they correspond to the same (in evolutionary terms) gene in different species. This implies that they derived from a speciation process in which the gene was present in the last common ancestor and was transmitted to descendant species. Two sequences are paralogous if they are the product of a gene-duplication event (Fig. 29.1). Duplication may have occurred in the past (thus, we can find paralogous genes in several species), or it can be a very recent event, restricted to a given species. Considering the above definitions, it is essential to be able to discriminate among orthologous and paralogous genes. This is in fact a fundamental prerequisite to assess the level of gene-order conservation among different genomes. Even if there are specific programs to identify the syntenic portion of genomes

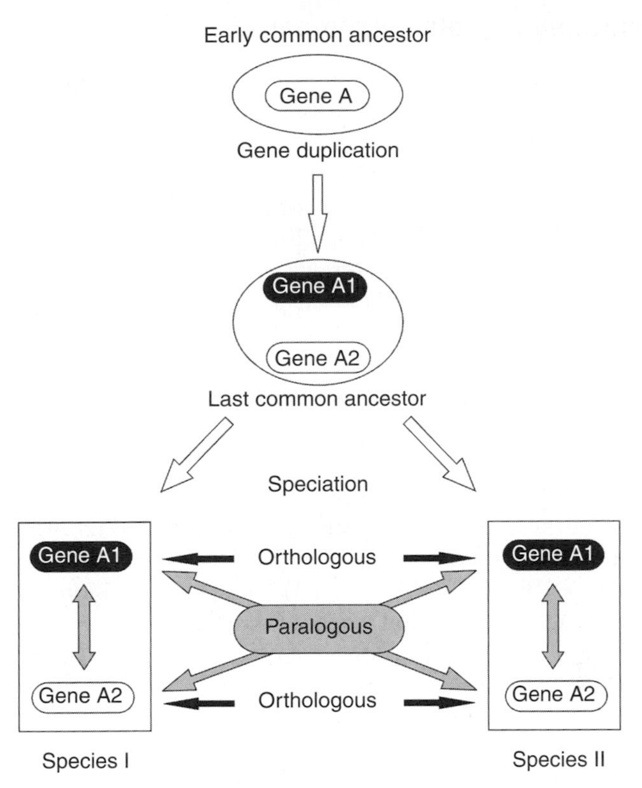

Figure 29.1 Schematic representation of the evolutionary process leading to orthologous and paralogous genes in sister species.

(e.g., SynBrowse), the best way to establish the relationship of orthology/paralogy among a set of genes is to perform a phylogenetic analysis. This requires a series of steps that are detailed in the following sections. To perform this analysis, it is necessary (1) to create a data set, (2) to align the sequences of interest, (3) to establish the phylogenetic relationships among the aligned sequences, (4) to test the reliability of the tree generated by the phylogenetic reconstruction, and (5) to evaluate the consistency of possible alternative evolutionary hypotheses.

2.4. Data set production

Globin genes provide an excellent case for practical exemplification of the analytical tools described previously. They have been sequenced in several fish species and are suitable material for our purpose. Available sequences can be downloaded from the Genbank Web site (see Table 29.1) by using the BLAST package (Altschul *et al.*, 1990; McGinnis and Madden, 2004).

2.5. Performing multiple alignment

Multiple alignment of nucleic and/or amino acid sequences can be performed with a vast number of programs. ClustalW (Thompson *et al.*, 1994) is the most widespread among many available algorithms. It can be accessed through several Web sites or downloaded locally in a personal computer. A user-friendly online version is available at the Pôle BioInformatique Lyonnais Gerland (see Table 29.1). ClustalW must be searched in the NPS@ section. A graphical interface of ClustalW, named ClustalX, can be used on a personal computer (Thompson *et al.*, 1997) (see Table 29.1). ClustalW algorithm is implemented also in several other programs dealing with evolutionary analyses (e.g., MEGA).

An interesting strategy when aligning nucleic acid sequences is to use the multiple alignment of the corresponding polypeptides as backbone. This approach does not permit gaps on the second and third positions in codons. Thus, gaps are inserted only between two adjacent codons. This is a very useful feature in the alignment, because it is biologically significant, provides reliable phylogenetic reconstructions, and allows direct analyses on codon sequences. DAMBE (Xia and Xie, 2001) and Codon Align (Hall, 2007) are also very useful to perform this two-step alignment.

When divergence among protein sequences is high (identity <30%), T-COFFEE is a very useful program for multiple alignments (Notredame *et al.*, 2000) (see Table 29.1). The MUSCLE package is a good alternative (Edgar, 2004). It is also possible to compare the accuracy of multiple alignments generated by different algorithms by means of the VerAlign program (Simossis *et al.*, 2005) (Table 29.1).

2.6. Phylogenetic analysis

Rigorous phylogenetic analyses are fundamental to properly infer the evolutionary history of globin genes/proteins. Many algorithms, methodologies, and philosophies have been developed to deal with this topic. An overview of these methods is beyond the goal of this contribution; the reader is referred to Hall (2007), Felsenstein (2004), Li (1997), Nei and Kumar (2000), Nielsen (2005), Page and Holmes (1998), Salemi and Vandamme (2003), and Swofford *et al.* (1996). The phylogenetic methods may be roughly grouped as distance-based and character-based. All approaches that convert differences among sequences into distances calculated according to a specific evolutionary model are included in the first group. The resulting matrix is successively used to build a distance-based tree. Character-based methods consider the variation of single characters, namely a character corresponding to each position of a multiple alignment. Within this framework, each nucleotide/amino acid residue at each specific position represents the character state. Character-based methods may explicitly use a molecular evolutionary model (this holds true for maximum likelihood [ML] and Bayesian inference approaches) or implicitly assume that all nucleotide/amino acid residue changes occur at the same rate. In the latter case, the best phylogenetic tree is that minimizing the amount of changes in the analyzed data set. The maximum-parsimony method uses this latter approach.

When we choose an approach explicitly using evolutionary models, it is necessary to identify the molecular model best fitting the analyzed data set before beginning the phylogenetic analysis (Kelchner and Thomas, 2007; Sullivan and Joyce, 2005). The ProtTest and ModelTest programs have been developed to perform this analysis (Abascal *et al.*, 2005; Posada and Crandall, 1998). ProtTest is devoted to the selection of models dealing with protein data sets. It is available as a stand-alone Java application or through its Web server. ModelTest works with nucleic acid sequences. It must be used coupled with the phylogenetic program PAUP* 4.10 (Swofford, 2002) and allows for selection among 56 evolutionary models. The Akaike information criterion and Bayesian approaches should be preferred over likelihood ratio tests in the selection of best-fitting models (Posada and Buckley, 2004).

The most popular algorithm working with distances is the neighbor-joining method implemented in several programs. PHYLIP (Felsenstein, 2005), MEGA (Kumar *et al.*, 2004), and PAUP* (Swofford, 2002) are the most commonly used (Table 29.1), also for parsimony analyses. The ML method is flexible and statistically robust (Felsenstein, 2004; Sullivan, 2005). Moreover, recent advances in computer programming allow one to bypass the overlong computational time that previously prevented application of the ML approach to large data sets. ML phylogenetic analysis can be performed using the PHYLIP, PAUP* 4.10, TREEFINDER (Jobb *et al.*, 2004) or PHYML (Guindon and Gascuel, 2003) packages. These programs are rather flexible in applying the best-fitting evolutionary model. TREEFINDER and PHYML are much faster than the first two packages in identifying the ML tree topology. This useful property is particularly valid for PHYML, based on a heuristic algorithm, which is why it is so fast. However, a further step is necessary whenever the goal is not only to determine the tree topology but also to calculate the branch lengths precisely. The tree determined with PHYML needs to be loaded into PAUP* in order to perform a more refined estimation of branch length (Sullivan *et al.*, 2005).

Irrespective of the tree-building method, the final topology ought to be further checked to assess its robustness. The nonparametric bootstrap test is the most widespread method for this purpose (Felsenstein, 1985). This test is included in all programs performing phylogenetic analysis. The needed number of replicates is usually 1,000. Nodes supported by more than 50% replicates are considered suitably supported. In particular, a node with support >70% is well supported; strong support is >85%. It should be noted that the bootstrap test is a measure of robustness but not necessarily a measure of truth (see Nei and Kumar, 2000).

Bayesian inference is another promising phylogenetic method. It allows use of sophisticated evolutionary models, thus adding realism to the analysis (Holder and Lewis, 2003; Huelsenbeck *et al.*, 2001, 2002; Ronquist, 2004). In Bayesian inference, the best-fitting evolutionary model can be selected through the MrModeltest (Nylander, 2004). Subsequent phylogenetic reconstruction can be performed with the programs MrBayes (Ronquist and Huelsenbeck, 2003) and BayesPhylogenies (Pagel and Meade, 2004). It should be noted that, in Bayesian analysis, support to the nodes of tree topology is provided as posterior probabilities.

Tree topology can be visualized using the packages TreeView (Page, 1996), Njplot (Perrière and Gouy, 1996), Geneious, or TreeDyn (Chevenet *et al.*, 2006) (Table 29.1). The latter program also allows for a very accurate annotation of the tree.

2.7. Evaluating alternative phylogenetic hypotheses

Once a tree topology has been generated, it may be important to verify whether the data set also supports alternative phylogenetic hypotheses (i.e., different tree topologies). Several tests have been created for this purpose. An extensively used one is the approximately unbiased unbiased (AU) test (Shimodaira, 2002). This test can be performed using the CONSEL program (Shimodaira and Hasegawa, 2001), which must be coupled with PAUP* or PAML (Yang, 2007).

3. Evolution of Globins and Identification of Functionally Relevant Amino Acid Residues

3.1. Type of selection acting on codons/amino acid residues

Once the phylogenetic relationships among a group of globin sequences have been identified, it is of interest to push the study of their evolution further. Detecting the selection type that may have acted on genes/polypeptides is a further step in the evolutionary analysis. Selection working on protein-encoding genes can be investigated in terms of the nonsynonymous/synonymous substitution rate occurring at the codon level. Nonsynonymous substitutions imply the change of coded residues, whereas synonymous ones do not produce any residue modification. These selection types can shape the evolution of protein-encoding genes, namely purifying or conservative, neutral, and positive or diversifying selections. The nonsynonymous (dN) to synonymous substitution (dS) rate ratio ω (=dN/dS) is used to discriminate among the three selection types. Purifying selection prevents the change of an amino acid residue at a given position in a multiple alignment, thus favoring an excess of synonymous versus non-synonymous substitutions. As a consequence, ω will have values <1; the lower the value, the higher the purifying selection acting on the residue. Neutral selection is essentially driven by mutation and genetic drift and implies an equal number of synonymous versus nonsynonymous substitutions; the ω values will be approximately 1. Positive selection implies an excess of nonsynonymous versus synonymous substitutions with change of residue at a given position. This implies that ω will be >1; the higher the values, the stronger the positive selection acting on the proteins.

Several methods and programs have been devised to detect different types of selection acting on coding sequences and their protein products. A brief overview is provided subsequently. For full description of each algorithm, please consult the references.

The PAML package implements a combination of Bayesian and likelihood methods (Yang, 2007). This program allows for the estimation of ω in light of several evolutionary models (Yang and Bielawski, 2000; Yang *et al.*, 2000). Before using PAML, the reader should refer to the references cited previously as well as to Anisimova *et al.* (2002), which provide useful guidelines for performing the analyses.

The type of selection may also be studied using the recently developed likelihood-based approaches (Kosakovsky Pond and Frost, 2005; Kosakovsky Pond and Muse, 2005). These algorithms have partially been implemented in the stand-alone package HyPhy (Kosakovsky Pond *et al.*, 2005) or can be accessed through a Web server (adaptive evolution server; see Table 29.1).

A third computer package developed to investigate types of selection is ADAPTSITE, which is based on the maximum-parsimony method (Suzuki and Gojobori, 1999; Suzuki and Nei, 2002; Suzuki *et al.*, 2001). Another approach is to investigate forces shaping codon evolution by directly studying amino acid sequences, provided that information on the protein structure is available (Suzuki, 2004).

All the previously mentioned methods have strengths and weaknesses (see Nunney and Schuenzel, 2006). Therefore, in each case, the reader ought to critically consider the results obtained with a single approach.

3.2. Detecting functionally diverging amino acid residues

One of the major goals in studying functional genomics is to identify functional divergence among member genes belonging to a gene family (Gu, 2001). This alternative method allows for study of the globin evolutionary history at the amino acid level. Following gene duplication, one copy maintains the original function and the other copy is free to change the amino acid structure by functional divergence (Li, 1983). Let us consider a multiple alignment of globin sequences. These may all be orthologous if the goal is to study the same polypeptide in different species or paralogous if the goal is to investigate the evolution of the protein family. Let us divide the data set in two monophyletic groups A and B. Each position of the alignment can be classified as type 0, I, II, and U (Gu, 2001). In the type 0 position, the same residue is present through the whole data set. In the type I position, residues are very much conserved in group A, whereas they are highly variable in group B or *vice versa*, implying that such residues have experienced altered functional constraints. In the type II position, groups A and B each exhibit two distinct kinds of residues. Within each group, residues are very much conserved, whereas they drastically differ between groups. This means that they display very different biochemical properties, implying that they may be responsible for functional differentiation. Finally, the type-U (unclassified) position contains many residues that are not distributed clearly between A and B. Residues in type I and II positions are good candidates as responsible for functional divergence. Type I functional divergence results in altered functional constraints between protein groups. Type II functional divergence results in no altered functional constraints within each group but radical change in the residue properties between A and B (Gu, 2001). Several methods based on different theoretical assumptions have been devised to identify these residues (Abhiman *et al.*, 2006; Gu, 1999, 2001, 2006; Kalinina *et al.*, 2004; Knudesen and Miyamoto, 2001; Knudsen *et al.*, 2003; Mayrose *et al.*, 2004; Nam *et al.*, 2005; Soyer and Goldstein, 2004). Packages allowing search for functionally diverging residues include DIVERGE2 (Gu, 2006), SplitTester (Gao *et al.*, 2005), dde (Nam *et al.*, 2005), and covARES (Blouin *et al.*, 2003), the servers Rate shift analysis (Knudsen and Miyamoto, 2001; Knudsen *et al.*, 2003), and ConSurf (Landau *et al.*, 2005). All these packages also use information from protein structure (possibly crystallographic) to identify functionally diverging residues.

As a final remark, readers should keep in mind that the statistical power of all programs listed herein depends on the size of the analyzed data set. The larger the data set, the more accurate is the estimation of parameters. Thus, if a limited number of sequences is available, the results should be treated with caution.

3.3. Inferring ancestral globin sequences

For some purposes, it is useful to infer the ancestral globin sequence, to overexpress it, and to check its behavior. ANCESTOR (Zhang and Nei, 1997) is a suitable computer package, available through its Web site (see Table 29.1). The PAML package also allows for performance of such analysis (Yang, 2007). Mesquite (Maddison and Maddison, 2006) is a third package based on parsimony methods suitable to reconstruct ancestral sequences (Table 29.1). A Mesquite analysis on fish globin was recently published (Near *et al.*, 2006). ANCESCON (Cai *et al.*, 2004) is another package dealing with a distance-based approach. Approaches suitable for reconstructing ancestral sequences are continuously developed (see Hall, 2006) and, generally speaking, this is a fast-developing field in molecular evolution. For a general assessment of merits and limits of currently available methods, see Williams *et al.* (2006).

3.4. General remarks

Readers should periodically refer to the Phylogeny Programs server, constantly updated and maintained by J. Felsenstein, University of Washington (see Table 29.1), where programs dealing with molecular evolution and phylogenetics can be found.

The Phylemon Web server also contains a suite of tools for molecular evolution, phylogenetics, and phylogenomics that includes several of the programs listed herein. It can be accessed through its Web site (see Table 29.1).

4. An Example of Comparative Genomics in Fish Globins

An application of the methods described previously is outlined in this section. We applied a comparative approach to study the ($\alpha + \beta$)-globin genomic clusters of medaka (*O. latipes*) and threespine stickleback (*G. aculeatus*). The genomic organization of α- and β-globin clusters had been previously studied in medaka (Maruyama *et al.*, 2004a, b). Protein sequences of the medaka globins were downloaded from GenBank according to the accession numbers provided by Maruyama *et al.* (2004a). Using the bioinformatic tools available at the ENSEMBL server (Table 29.1), we created a map of these sequences in the genomic portions of chromosomes 8 and 19 (Maruyama *et al.*, 2004a) (Fig. 29.2). The map also includes the flanking genes of DNA-glycosylase, UPF0171, and ankyrin, which allows for a finer delimitation of the ($\alpha + \beta$)-globin major cluster located on chromosome 8 (see Fig. 29.2). The sequences of medaka globins were subsequently used to identify their homologous counterparts in the genome of threespine stickleback (ENSEMBL server; Table 29.1). By iterative BLAST searches, we found that the α- and β-globin genes in the threespine stickleback are also organized in a single cluster (see Fig. 29.2), whose boundaries are defined by the same flanking regions as in medaka chromosome 8 (Fig. 29.2). The globin cluster in *G. aculeatus* appears to be located within a broader genomic portion, currently named group XI

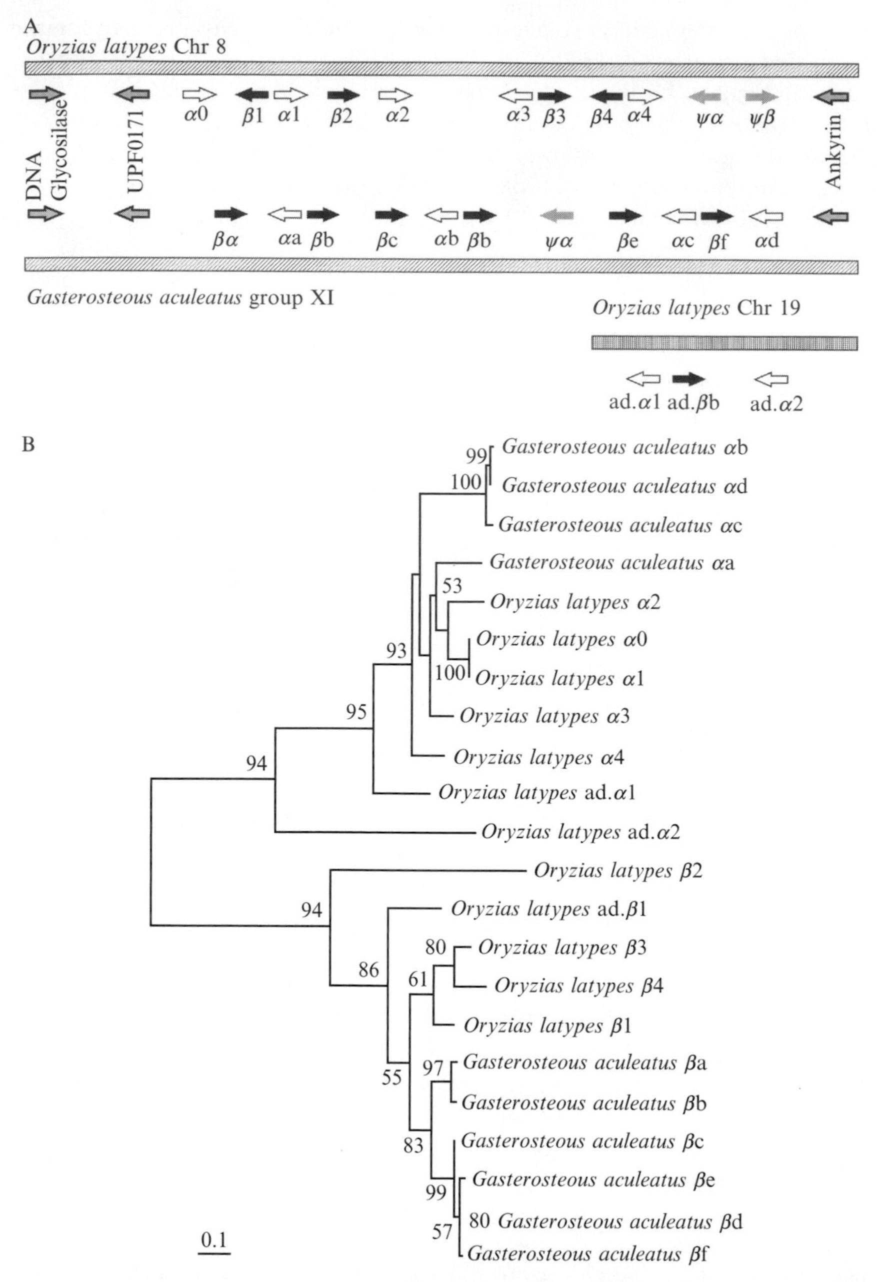

Figure 29.2 (A) Genomic organization and (B) phylogenetic relationships in α and β globin genes of *O. latipes* and *G. aculeatus*. The nomenclature of *O. latipes* α and β globins is identical to that published by Maruyama *et al.* (2004a). The *G. aculeatus* globin sequences have the following entries in ENSEMBL server: *G. aculeatus* αa (ENSGACP00000018378), *G. aculeatus* αb (ENSGACP00000018341), *G. aculeatus* αc

according to the nomenclature provided at the ENSEMBL server for the *G. aculeatus* genome (release 28/02/2007).

The maps of Figure 29.2 show that the genomic organizations of (α + β)-globin genes of *O. latipes* and *G. aculeatus* differ from each other. In fact, in medaka they are split in two distinct groups, whereas in the threespine stickleback they are all grouped in a single cluster.

Phylogenetic analysis was performed to assess the relationships of orthology and paralogy among α and β-globins of the two species. Multiple alignment of sequences was performed by ClustalW (Thompson *et al.*, 1994). The obtained data set was processed by ProtTest (Abascal *et al.*, 2005) to identify the evolutionary model best fitted to analyze these globins. The best-fitting model was WAG + G. This model implies the use of the empirical substitution matrix WAG (Whelan and Goldman, 2001) and modeling of site heterogeneity by the statistical gamma distribution (Yang, 1993).

ML phylogenetic analysis was performed by the PHYML program (Guindon and Gascuel, 2003) by applying the evolutionary model described previously. The robustness of the tree topology was assessed through non-parametric bootstrap (1000 replicates) (Felsenstein, 1985). The tree is shown in Figure 29.2. The most relevant phylogenetic outputs for α and β globins are discussed separately.

A clear relationship of orthology among α globins of *O. latipes* and *G. aculeatus* is not available. We can identify two evolutionary patterns. The first one is characterized by events of α-globin dupli-multiplications restricted to one species (e.g., *G. aculeatus* globins αb, αc, and αd). The second pattern can be observed in the close relationship between *G. aculeatus* globin αa and the cluster of α0, α1, and α2 globins of *O. latipes*. In this case we have a single gene in one species closely related to a group of genes in the other species. This is a nice example of pro-orthology (Mindel and Meyer, 2001).

In the β-globin group, the relevant pattern is that observed between the *O. latipes* β1, β3, and β4 cluster and the set including six globins of *G. aculeatus* (βa to βf). The phylogenetic relationship between these two groups of sequences is named co-orthology. The term *co-orthologous* is used

(ENSGACP00000018321), *G. aculeatus* αd (ENSGACP00000018313), and *G. aculeatus* βa (ENSGACP00000018389), *G. aculeatus* βb (ENSGACP00000018354), *G. aculeatus* βc (ENSGACP00000018346), *G. aculeatus* βf (ENSGACP00000018339), *G. aculeatus* βe (ENSGACP00000018339), *G. aculeatus* βf (ENSGACP00000018318). (A). White arrow, α globin gene, black arrow, β globin gene, gray arrow, α and β globin pseudogene, black-bordered gray arrow, flanking genes. Orientation of gene is indicated by the arrow point. Genes are not drawn to scale. (B). ML tree (−ln= −2535.80921) created with PHYML (Guindon and Gascuel, 2003). Bar represents 0.1 substitution per site. Bootstrap values are in percentages.

to define paralogous sequences arising from duplications of orthologous sequences following a given speciation (Sonnhammer and Koonin, 2002). It is worth mentioning that there are no adult α and β globins in the *G. aculeatus* cluster.

Four major α- and β-globin groups are currently known in teleost fishes (Maruyama *et al.*, 2004a): embryonic Hb (group I), notothenioid major adult Hb (group II), anodic adult Hb (group III), and cathodic adult Hb (group IV). Groups I and IV are still present in *O. latipes*, whereas Groups II and III were lost during the evolution of this species. Group I and IV are located, respectively, on chromosomes 8 and 19 (see Fig. 29.2; Maruyama *et al.*, 2004a). Comparison of genomic globin clusters and phylogenetic analysis suggest very different evolutionary histories of α and β globins in medaka and threespine stickleback. In fact, while in *O. latipes* there are two kinds of α and β globins with distinct locations of adult and embryonic forms, only sequences belonging to group I can be found in *G. aculeatus*. The threespine stickleback α and β globins are placed within a single genomic cluster (Fig. 29.2) that corresponds to the embryonic cluster of medaka. This is supported by two independent lines of evidence, namely phylogenetic analysis outputs and flanking genes defining the cluster boundaries in both species (Fig. 29.2).

5. Cytogenetic Methods for Comparative Genomics: Fluorescence *In Situ* Hybridization

As shown in the previous sections, high-throughput sequencing technologies in genomic analysis, large-scale genetic mapping, and the exponential growth in the number of powerful bioinformatic algorithms are the tools that help the most in understanding how the genomes are shaped and evolve, thus providing the base for the development of comparative and evolutionary genomics in fish, as well as in other organisms.

In this section we consider a very different but complementary methodological approach to comparative genomics, namely cytogenomics. The term *cytogenomics* was coined to define a new methodology that links high-tech molecular biology and so-called classical cytogenetics, for a better understanding of the genome structure and function. The main bulk of such new cytogenetics *sensu* Speicher and Carter (2005) is fluorescence *in situ* hybridization (FISH).

FISH and its derivative technologies are in constant progress, especially because of advances in medical research (reviewed in Speicher and Carter, 2005). As the most powerful tool currently available in cytogenectics, FISH is largely utilized in mammals and recently also in nonmammalian vertebrates. The use of FISH on fish, especially DNA-FISH (see subsequently) has rapidly increased from the first applications in salmonids in the 1990s (Pendas *et al.*, 1993, 1994) to date (Pisano *et al.*, 2007). With FISH, we are tackling a completely different perspective with respect to the other genetic and molecular methods, because it allows not only for detection of nucleic acid sequences but also for visualization of them in their context of morphologically preserved tissues, cells, cell nuclei, or chromosomes. The method is based on the ability of a known nucleotide sequence (probe) to hybridize the complementary nucleic-acid sequence (target) preserved *in situ*. The probe is directly or indirectly labeled with a fluorochrome, so that the product of hybridization can be revealed and visualized as a fluorescent signal. According to size and abundance of the target sequence, the hybridization signal can be visualized with an epifluorescent microscope, either optically or indirectly with a video camera coupled with a suitable digital imaging system. Both DNA and RNA sequences can be visualized; in the first case, we are dealing with DNA-FISH and in the second with RNA-FISH.

Designed to recognize and highlight targets on the chromosomes, DNA-FISH can be applied to obtain a large array of structural genomic information, allowing one to visualize chromatin structures, from entire chromosomes and chromosome arms to chromosome bands, and down to the level of individual genes (e.g., Raap and Tanke, 2006; Trask, 1991; van der Ploeg, 2000). Although its potential to fish-genome mapping was acknowledged several years ago (Phillips, 2001), DNA-FISH is not routinely used as yet, largely because of the high cost of the equipment. Its current application mostly deals with fish comparative karyotyping through mapping of highly repetitive DNA regions, such as the telomeric ones and clusters of ribosomal genes onto the chromosomes (reviewed in Pisano *et al.*, 2007). A very recent acquisition has been the chromosomal location of protein-coding genes through DNA-FISH in some fish species, including the mapping of globin genes in puffer fish (Gillemans *et al.*, 2003) and in four species of the Antarctic family Nototheniidae (Pisano *et al.*, 2003). In contrast, RNA-FISH was developed for studies of gene expression and organization of RNAs during transcription, processing, and transport in cells (Dirks *et al.*, 2001). Although it has provided important insights into globin-gene expression and regulation in mammals (Du *et al.*, 2003), to our knowledge, this methodology is not widely applied to similar studies on fish globins. Thus, we focus herewith on DNA-FISH.

5.1. Mapping of globin genes by FISH on fish

The molecular organization of α- and β-globin genes has been studied in several teleost species and can easily be obtained by searching the available genome-sequence databases (see previous sections for search strategies and bioinformatic approaches). The studied species include Atlantic salmon

(Wagner *et al.*, 1994), rainbow trout (Yoshizaki *et al.*, 1996), zebra fish (Chan *et al.*, 1997), carp (Miyata and Aoki, 1997), yellowtail (Okamoto *et al.*, 2001), puffer fish (Gillemans *et al.*, 2003), Antarctic perciform species (Cocca *et al.*, 2000; Lau *et al.*, 2001), and medaka (Maruyama *et al.*, 2004b). However, we are still far from having a clear picture of the chromosomal organization of globin loci in fish. For instance, in zebra fish the location of the α- and β-globin genes (both embryonic and adult) on the same chromosome has been predicted from the analysis of genomic clones (Chan *et al.*, 1997). On the other hand, in medaka, linkage analysis indicated that the embryonic and adult globin-gene clusters are located on two different chromosomes (Maruyama *et al.*, 2004b).

The "direct" *in situ* approach to globin-gene mapping, so far applied to a minority of fish species, also provided intriguing results. Pisano *et al.* (2003) demonstrated by FISH that an embryonic/juvenile cluster of α- and β-globin genes maps on a single chromosome pair in four notothenioid species. The arrangement of the α- and β-globin clusters on the same chromosome confirmed the linked configuration of the genomic globin clones described for many teleosts and was also coherent with indirect evidence in zebra fish (Chan *et al.*, 1997). However, Gillemans *et al.* (2003), using FISH to map an α/β-globin locus and an α globin locus onto the chromosomes of *F. rubripes*, showed that the two loci are located onto separate chromosomes. The parsing of α- and β-globin genes of puffer fish into asymmetrical gene sets located on two chromosomes suggested that the α- and β-globin genes, linked in strict configuration in most of the fish species, could have independently split in different chromosomes in distinct fish lineages groups during evolution (Pisano *et al.*, 2003).

This fragmentary picture of genomic and chromosomal organization of globin genes highlights the need for more comparative studies on both closely and distantly related species through various complementary approaches, including FISH. More detailed information on globin genes in fish will provide, in addition, important contributions to large-scale comparative genomic studies between fish and other vertebrates (e.g., Naruse *et al.*, 2004; Postlethwait *et al.*, 2000).

5.2. The FISH procedure

It is impossible to provide a standardized and detailed FISH protocol suitable for any fish species. To perform DNA-FISH onto metaphase chromosomes, it is necessary (a) to prepare high-quality metaphase chromosomes, (b) to choose and label an appropriate probe sequence, (c) to denature probe and chromosomes, (d) to allow hybridization between these, (e) to detect the hybridization signal, and (f) to counterstain the chromosomes and to analyze the FISH signals.

In providing some details for each step, we refer to the protocol described in Pisano *et al.* (2003) for mapping the chromosomal locus of a cluster of embryonic-juvenile α/β-globin genes in *Notothenia coriiceps* (Fig. 29.3), then successfully applied in three additional species of the family Nototheniidae.

5.2.1. Preparation of high-quality metaphase chromosomes

Chromosomes can be obtained from fish by two alternative methods: (1) direct preparation from naturally dividing cells (e.g., gill, gut, hematopoietic tissues) after *in vivo* treatment of the fish specimens or (2) from cultured cells (either fibroblasts or lymphocytes). Direct preparation is still the most common technique to obtain fish metaphase plates (Blaxall, 1975; Denton, 1973; Klinkhardt, 1991; Ozouf-Costaz *et al.*, 2008), but the results are not always adequate in terms of chromosome morphological quality and mitotic index (frequency of metaphase cells in the preparation), depending on the fish species (freshwater, marine, warm or cold water) and even on individuals. These difficulties may be overcome by using cultured cells.

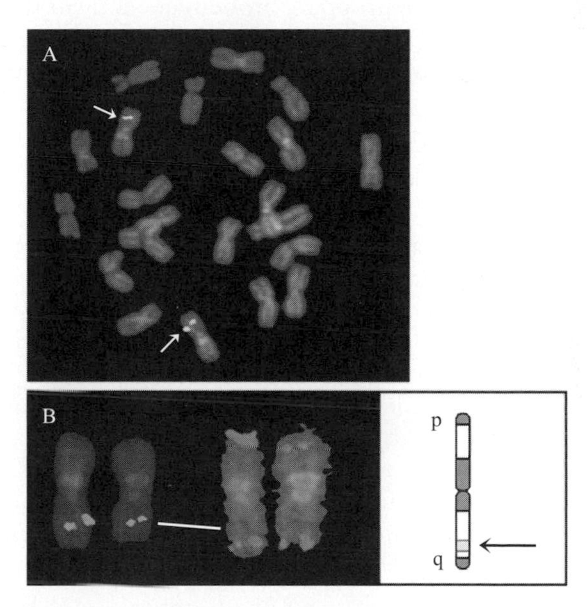

Figure 29.3 Cytogenetic mapping of globin genes in the Antarctic black rockcod *Notothenia coriiceps* (order Perciformes). The position of the globin-gene cluster onto a pair of homologous chromosomes is indicated by arrows in (A) and more precisely assigned to a subtelomeric region of the p arm of a metacentric element (B) as a result of sequential FISH of a embryonic/juvenile α/β-globin-gene probe (green signals) and a telomeric probe (red signals). DAPI counterstain (from Pisano *et al.*, 2003, with permission). (See color insert.)

The chromosomes of *N. coriiceps* (see Fig. 29.3) were from head-kidney cells, following *in vivo* treatment with colchicine and spreading on to superfrost slides, according to modifications of the conventional air-drying method (Ozouf-Costaz *et al.*, 2008).

5.2.2. Probe preparation and labeling

In general, the globin-gene sequence selected as probe should not contain many repeats of intronic regions to avoid unspecific hybridization to other chromosome sequences.

Protein-coding genes present in clusters of multiple copies, such as globins, can be mapped using λ clones. As an alternative to plasmids, bacterial artificial chromosomes (BACs) containing the globin gene sequence of interest can be used as probes. Clones from large insert libraries such as BACs are usually required to localize single copy genes in chromosomes, because BACs clones enhance the FISH signal and its visualization. Compared to the plasmid-FISH procedure, the protocol for BAC-FISH includes an additional step to saturate the repetitive elements generally present in the large inserts of these vectors, namely the use of suitable proportions of DNA-specific competitor from the species under investigation and carrier from a different group. The most appropriate probe sequence can then be labeled directly (Wiegant *et al.*, 1991) or preferably indirectly (Kessler, 1990) with fluorescently modified nucleotides. Currently, the most used labeling method is nick-translation (Rigby *et al.*, 1977), suitable for the DNA probes larger than 1 kb and subcloned in plasmids.

5.2.3. Denaturation

A prerequisite for hybridization of a double-stranded DNA probe with the complementary double-stranded chromosomal DNA sequence, denaturation is a crucial step. Fish chromosomes are small and generally prone to denaturation. The smaller they are, the shorter this treatment should be. We suggest denaturation of probe and chromosomes separately. The probe is denatured at 75° for 10 min, then ice chilled to prevent reannealing. For chromosomes, suitable denaturation conditions must be selected for each species. For freshly thawed preparations from Antarctic species, we denature the chromosomes for 20 to 90 sec at 68 to 72°. If chromosomes turn out to be too fragile, the slides are aged for two days at room temperature.

5.2.4. Hybridization

Hybridization is performed at 37° in a sealed moist chamber. A volume of 10 to 25 μl of the hybridization mixture is dropped onto each slide and covered with a coverslip. Small probes such as globin-gene sequences should be incubated for 48 to 72 h, in order to optimize the opportunity for the probe to find its target onto the chromosomes. Several variants to the basic one-color FISH can theoretically be performed (double, multicolor FISH), according to number of available probes and to the digital imaging microscopy equipment.

5.2.5. Detection, counterstain, and microscopic analysis

The biotinylated globin probe used in mapping *N. coriiceps* genes (see Fig. 29.3) was detected with streptavidin-FITC (greenish fluorescent signal). Amplifications of the signals by successive reactions with antibodies, as used in conventional immunocytochemical approaches, should be avoided, since they often produce an important fluorescent background. A high-resolution camera and adequate software allowing one to electronically amplify the fluorescent signals are recommended. Postdetection washing can be performed at room temperature. Postdetection rinsing ought to be brief to avoid the risk of losing the FISH signals.

The most commonly used fluorescent counter-stain is 4,4′,6-diamidino-2-phenylindole (DAPI), mixed with an antifade to avoid rapid fluorescence bleaching, used at higher concentration than that needed for mammalian chromosomes, and stored at 4° in the dark. Before microscopic analysis, the mounted FISH preparation can be stored at −20° in the dark. Analyses require a fluorescence microscope with a 100 W HbO lamp, equipped with a high-speed, high-resolution digital camera. The camera must resolve fine details (1392 [h] × 1040 [v] imaging array and 6.45 μm square pixels), allowing one to quantify bright and dim signals in the same image (12-bit digitalization). Several software packages are commercially available, such as Genus (Applied Imaging).

5.3. General remarks

FISH methodologies are in constant progress. However, the success of a DNA-FISH mapping design often goes beyond strictly following a given procedure. When working on low-repeat target sequences, such as globin and single-copy genes, mapping can be difficult due to (besides accurate chromosome identification) target size and molecular features, preparation of suitable probes, and resolution limits of the microscope. Thus, mapping of protein-coding genes has been experimented on so far in few fish species (Phillips, 2007). However, genome-mapping projects currently involve a large number of anonymous fragment length polymorphisms (AFLPs) and microsatellite markers to produce detailed maps. The large clones generated by these projects can be readily localized on fish chromosomes using appropriate FISH protocols. BAC clones were used on chromosomes, for controlling contiguous assemblages and orientation in the puffer fish genome (Jaillon *et al.*, 2004). BAC libraries are currently available not only for model species (e.g., puffer fish, zebra fish) but also increasingly for other fish (e.g., Phillips, 2007; Thorsen *et al.*, 2005; Watanabe *et al.*, 2003).

In the near future, new derivatives of FISH and FISH in combination with other approaches will be applied. COBRA-FISH (combined binary ratio-fluorescence *in situ* hybridization) is a multicolor method currently limited to human chromosomes that enables the color-based recognition of chromosome arms and allows gene integration site mapping in the context of chromosome arm painting (Raap and Tanke, 2006). Immuno-FISH, linking FISH and immunocytochemical methods, provides simultaneous detection of DNA, RNA, proteins, or other cellular factors (Leger *et al.*, 1994); RNA TRAP (tagging and recovery of associated proteins) is based on tagging and analyzing chromatin *in situ* in a transcriptionally active gene (Carter *et al.*, 2002).

REFERENCES

Abascal, F., Zardoya, R., and Posada, D. (2005). ProtTest: Selection of best-fit models of protein evolution. *Bioinformatics* **21,** 2104–2105.

Abhiman, S., Daub, C. O., and Sonnhammer, E. L. L. (2006). Prediction of function divergence in protein families using the substitution rate variation parameter alpha. *Mol. Biol. Evol.* **23,** 1406–1413.

Aerts, S., Van Loo, P., Thijs, G., Mayer, H., de Martin, R., Moreau, Y., and De Moor, B. (2005). TOUCAN 2: The all-inclusive open source workbench for regulatory sequence analysis. *Nucleic Acids Res.* **33,** W393–W396.

Altschul, S. F., Gish, W., Miller, W., Myers, E. W., and Lipman, D. J. (1990). Basic local alignment search tool. *J. Mol. Biol.* **215,** 403–410.

Anisimova, M., Bielawski, J. P., and Yang, Z. (2002). Accuracy and power of Bayes prediction of amino acid sites under positive selection. *Mol. Biol. Evol.* **19,** 950–958.

Aparicio, S., Chapman, J., Stupka, E., Putnam, N., Chia, J. M., Dehal, P., Christoffels, A., Rash, S., Hoon, S., Smit, A., *et al.* (2002). Whole-genome shotgun assembly and analysis of the genome of *Fugu rubripes*. *Science* **297,** 1301–1310.

Berezikov, E., Guryev, V., and Cuppen, E. (2005). CONREAL web server: Identification and visualization of conserved transcription factor binding sites. *Nucleic Acids Res.* **33,** W447–W450.

Blaxhall, P. C. (1975). Fish chromosome techniques: A review of selected literature. *J. Fish Biol.* **7,** 315–320.

Blouin, C., Boucher, Y., and Roger, A. J. (2003). Inferring functional constraints and divergence in protein families using 3D mapping of phylogenetic information. *Nucleic Acids Res.* **31,** 790–797.

Bray, N., Dubchak, I., and Pachter, L. (2003). AVID: A global alignment program. *Genome Res.* **13,** 97–102.

Brudno, M., Chapman, M., Göttgens, B., Batzoglou, S., and Morgenstern, B. (2003a). Fast and sensitive multiple alignment of large genomic sequences. *BMC Bioinformatics* **4,** 66.

Brudno, M., Do, C. B., Cooper, G. M., Kim, M. F., Davydov, E., NISC Comparative Sequencing ProgramGreen, E. D., Sidow, A., and Batzoglou, S. (2003b). LAGAN and Multi-LAGAN: Efficient tools for large-scale multiple alignment of genomic DNA. *Genome Res.* **13,** 721–731.

Cai, W., Pei, J., and Grishin, N. V. (2004). Reconstruction of ancestral protein sequences and its applications. *BMC Evol. Biol.* **4,** 33.
Carter, D., Chakalova, L., Osborne, C. S., Dai, Y. F., and Fraser, P. (2002). Long-range chromatin regulatory interactions *in vivo*. *Nature Genet.* **32,** 623–626.
Chan, F. Y., Robinson, J., Brownlie, A., Shivadasani, R. A., Donovan, A., Brugnara, C., Kim, J., Lau, B. C., Witkowska, H. E., and Zon, L. I. (1997). Characterization of adult α- and β-globin genes in the zebrafish. *Blood* **89,** 688–700.
Chevenet, F., Brun, C., Banuls, A. L., Jacq, B., and Christen, R. (2006). TreeDyn: Towards dynamic graphics and annotations for analyses of trees. *BMC Bioinformatics* **7,** 439.
Choi, K., Ma, Y., Choi, J.-H., and Kim, S. (2005). PLATCOM: A platform for computational comparative genomics. *Bioinformatics* **21,** 2514–2516.
Cocca, E., Detrich, H. W., III, Parker, S. W., and di Prisco, G. (2000). A cluster of four globin genes from the Antarctic fish *Notothenia coriiceps*. *J. Fish. Biol.* **57,** 33–50.
Denton, E. T. (1973). "Fish Chromosome Methodology." Charles Thomas Publisher, Springfield, IL.
Dirks, R. W., Molenaar, C., and Tanke, H. J. (2001). Methods for visualizing RNA processing and transport pathways in living cells,. *Histochem. Cell Biol.* **115,** 3–11.
Du, M.-J., Liu, D.-P., and Liang, C.-C. (2003). The impact of FISH on globin gene regulation research. *Exp. Cell Res.* **291,** 267–274.
Durinck, S., Moreau, Y., Kasprzyk, A., Davis, S., De Moor, B., Brazma, A., and Huber, W. (2005). BioMart and Bioconductor: A powerful link between biological databases and microarray data analysis. *Bioinformatics* **21,** 3439–3440.
Edgar, R. (2004). MUSCLE: Multiple sequence alignment with high score accuracy and high throughput. *Nucleic Acids Res.* **32,** 1792–1797.
Felsenstein, J. (1985). Confidence limits on phylogenies: An approach using bootstrap. *Evolution* **39,** 783–791.
Felsenstein, J. (2004). "Inferring phylogenies." Sinauer, Sunderland, MA.
Felsenstein, J. (2005). PHYLIP (Phylogeny Inference Package) version 3.6. Distributed by the author. Department of Genome Sciences, University of Washington.
Frazer, K. A., Pachter, L., Poliakov, A., Rubin, E. M., and Dubchak, I. (2004). VISTA: Computational tools for comparative genomics. *Nucleic Acids Res.* **32,** W273–W279.
Gao, X., Vander Velden, K.A, Voytas, D. F., and Gu, X. (2005). SplitTester: Software to identify domains responsible for functional divergence in protein family. *BMC Bioinformatics* **6,** 137.
Gillemans, N., McMorrow, T., Tewari, R., Wai, A. W. K., Burgtorf, C., Drabek, D., Ventress, N., Langeveld, A., Higgs, D., Tan-Un, K., Grosveld, F., and Philipsen, S. (2003). Functional and comparative analysis of globin loci in pufferfish and humans. *Blood* **101,** 2842–2849.
Gu, X. (1999). Statistical methods for testing functional divergence after gene duplication. *Mol. Biol. Evol.* **16,** 1664–1674.
Gu, X. (2001). Maximum likelihood approach for gene family evolution under functional divergence. *Mol. Biol. Evol.* **18,** 453–464.
Gu, X. (2006). A simple statistical method for estimating type-II (cluster-specific) functional divergence of protein sequences. *Mol. Biol. Evol.* **23,** 1937–1945.
Guindon, S., and Gascuel, O. (2003). A simple, fast, and accurate algorithm to estimate large phylogenies by maximum likelihood. *Sys. Biol.* **52,** 696–704.
Hall, B. G. (2006). Simple and accurate estimation of ancestral protein sequences. *Proc. Natl. Acad. Sci. USA* **103,** 5431–5436.
Hall, B. G. (2007). "Phylogenetic trees made easy: A how-to manual," 3rd ed. Sinauer, Sunderland, MA.
Holder, M., and Lewis, P. O. (2003). Phylogeny estimation: Traditional and Bayesian approaches. *Nature Rev. Gen.* **4,** 275–284.

Huelsenbeck, J. P., Ronquist, F., Nielsen, R., and Bollback, J. P. (2001). Bayesian inference of phylogeny and its impact on evolutionary biology. *Science* **294,** 2310–2314.

Huelsenbeck, J. P. H., Larget, B., Miller, R. E., and Ronquist, F. (2002). Potential applications and pitfalls of Bayesian inference of phylogeny. *Syst. Biol.* **51,** 673–688.

Jaillon, O., Aury, J. M., Brunet, F., Petit, J. L., Stange-Thomann, N., Mauceli, E., Bouneau, L., Fischer, C., Ozouf-Costaz, C., Bernot, A., Nicaud, S., Jaffe, D., *et al.* (2004). Genome duplication in the teleost fish *Tetraodon nigroviridis* reveals the early vertebrate proto-karyotype. *Nature* **431,** 946–957.

Jobb, G., von Haeseler, A., and Strimmer, K. (2004). TREEFINDER: A powerful graphical analysis environment for molecular phylogenetics. *BMC Evol. Biol.* **4,** 18.

Kalinina, O. V., Mironov, A. A., Gelfand, M. S., and Rakhmaninova, A. B. (2004). Automated selection of positions determining functional specificity of proteins by comparative analysis of orthologous groups in protein families. *Protein Sci.* **13,** 443–456.

Kelchner, S. A., and Thomas, M. A. (2007). Model use in phylogenetics: Nine key questions. *Trends Ecol. Evol.* **22,** 87–94.

Kent, W. J. (2002). BLAT: The BLAST-like alignment tool. *Genome Res.* **12,** 656–664.

Kessler, C. (1990). The digoxigenin system: Principle and application of the novel non-radioactive DNA labelling and detection system. *Biotechnology Int.* **1990,** 183–194.

Klinkhardt, M. B. (1991). A brief comparison of methods for preparing fish chromosomes: An overview. *Cytobios* **67,** 193–208.

Knudsen, B., and Miyamoto, M. M. (2001). A likelihood ratio test for evolutionary rate shifts and functional divergence among proteins. *Proc. Natl. Acad. Sci. USA* **98,** 14512–14517.

Knudsen, B., Miyamoto, M. M., Laipis, P. J., and Silverman, D. N. (2003). Using evolutionary rates to investigate protein functional divergence and conservation: A case study of the carbonic anhydrases. *Genetics* **164,** 1261–1269.

Kosakovsky Pond, S. L., and Frost, S. D. W. (2005). A genetic algorithm approach to detecting lineage-specific variation in selection pressure. *Mol. Biol. Evol.* **22,** 478–485.

Kosakovsky Pond, S. L., and Muse, S. V. (2005). Not so different after all: A comparison of methods for detecting amino-acid sites under selection. *Mol. Biol. Evol.* **22,** 1208–1222.

Kosakovsky Pond, S. L., Frost, S. D. W., and Muse, S. V. (2005). HyPhy: Hypothesis testing using phylogenies. *Bioinformatics* **21,** 676–679.

Kumar, S., Tamura, K., and Nei, M. (2004). MEGA3: Integrated software for molecular evolutionary genetics analysis and sequence alignment. *Briefing in Bioinformatics* **5,** 150–163.

Landau, M., Mayrose, I., Rosenberg, Y., Glaser, F., Martz, E., Pupko, T., and Ben-Tal, N. (2005). ConSurf (2005). The projection of evolutionary conservation scores of residues on protein structures. *Nucleic Acids Res.* **33,** W299–W302.

Lau, D. T., Saeed-Kothe, A., Parker, S. K., and Detrich, H. W., III. (2001). Adaptive evolution of gene expression in Antarctic fishes: Divergent transcription of the 5′-to-5′ linked adult α1- and β-globin genes of the Antarctic teleost *Notothenia coriiceps* is controlled by dual promoters and intergenic enhancers. *Am. Zool.* **41,** 113–132.

Leger, I., Robert-Nicoud, M., and Brugal, G. (1994). Combination of DNA *in situ* hybridization and immunocytochemical detection of nuclear proteins: A contribution to the functional mapping of the human genome by fluorescence microscopy. *J. Histochem. Cytochem.* **42,** 149–154.

Li, W. S. (1983). Evolution of duplicates genes. *In* "Evolution of genes and proteins" (M. Nei and R. K. Koehn, eds.), pp. 14–37. Sinauer Associates, Sunderland, MA.

Li, W. S. (1997). "Molecular evolution." Sinauer Associates, Sunderland, MA.

Loots, G. G., and Ovcharenko, I. (2005). Dcode.org anthology of comparative genomic tools. *Nucleic Acids Res.* **33,** W56–W64.

Maddison, W. P., and Maddison, D. R. (2006). Mesquite: A modular system for evolutionary analysis, Version 1.12. Available at http://mesquiteproject.org./mesquite/mesquite.html.

Mayrose, I., Graur, D., Ben-Tal, N., and Pupko, T. (2004). Comparison of site-specific rate-inference methods for protein sequences: Empirical Bayesian methods are superior. *Mol. Biol. Evol.* **21,** 1781–1791.

Maruyama, K., Yasumasu, S., and Iuchi, I. (2004a). Evolution of globin genes of the medaka *Oryzias latipes* (Euteleostei; Beloniformes; Oryziinae). *Mech. Dev.* **121,** 753–769.

Maruyama, K., Yasumasu, S., Naruse, K., Mitani, H., Shima, A., and Iuchi, I. (2004b). Genomic organization and developmental expression of globin genes in the teleost *Oryzias latipes*. *Gene* **335,** 89–100.

McGinnis, S., and Madden, T. L. (2004). BLAST: At the core of a powerful and diverse set of sequence analysis tools. *Nucleic Acids Res.* **32,** W20–W25.

Mindell, D. P., and Meyer, A. (2001). Homology evolving. *Trends Ecol. Evol.* **16,** 434–438.

Miyata, M., and Aoki, T. (1997). Head to head linkage of carp α and β-globin genes. *Biochem. Biophys. Acta* **1354,** 127–133.

Nam, J., Kaufmann, K., Theissen, G., and Nei, M. (2005). A simple method for predicting the functional differentiation of duplicate genes and its application to MIKC-type MADS-box genes. *Nucleic Acids Res.* **33,** 12.

Naruse, K., Tanaka, M., Mita, K., Shima, A., Postlethwait, J., and Mitani, H. A. (2004). Medaka gene map: The trace of ancestral vertebrate proto-chromosomes revealed by comparative gene mapping. *Genome Res.* **14,** 820–828.

Near, T. J., Parker, S. K., and Detrich, H. W., III. (2006). A genomic fossil reveals key steps in hemoglobin loss by the Antarctic icefishes. *Mol. Biol. Evol.* **23,** 2008–2016.

Nei, S., and Kumar, K. (2000). "Molecular evolution and phylogenetics." Oxford University Press, New York.

Ning, Z., Cox, A. J., and Mullikin, J. C. (2001). SSAHA: A fast search method for large DNA databases. *Genome Res.* **11,** 1725–1729.

Nielsen, R. (ed.), (2005). "Statistical methods in molecular evolution." Springer, New York.

Nobrega, M. A., and Pennacchio, L. A. (2004). Comparative genomic analysis as a tool for biological discovery. *J. Physiol.* **554,** 31–39.

Notredame, C., Higgins, D. G., and Heringa, J. (2000). T-COFFEE: A novel method for fast and accurate multiple sequence alignment. *J. Mol. Biol.* **302,** 205–217.

Nunney, L., and Schuenzel, E. L. (2006). Detecting natural selection at the molecular level: A reexamination of some "classic" examples of adaptive evolution. *J. Mol. Evol.* **62,** 176–195.

Nylander, J. A. A. (2004). MrModeltest v2. Distributed by the author. Evolutionary Biology Centre, Uppsala University, Uppsala, Sweden.

Okamoto, K., Sakai, M., and Miyata, M. (2001). Molecular cloning and sequence analysis of α- and β- globin cDNAs from yellowtail *Seriola quinqueradiata*. *Comp. Biochem. Physiol. B* **130,** 207–216.

Ovcharenko, I., Loots, G. G., Giardine, B. M., Hou, M., Ma, J., Hardison, R. C., Stubbs, L., and Miller, W. (2005). Mulan: Multiple-sequence local alignment and visualization for studying function and evolution. *Genome Res.* **15,** 184–194.

Ovcharenko, I., Nobrega, M. A., Loots, G. G., and Stubbs, L. (2004). ECR Browser: A tool for visualizing and accessing data from comparisons of multiple vertebrate genomes. *Nucleic Acids Res.* **32,** W280–W286.

Ozouf-Costaz, C., Foresti, F., Almeida-Toledo, L. F., Pisano, E., and Kapoor, B. G. (eds.) (2008). "Techniques of fish cytogenetics." Science Publishers Inc., Enfield, NH.

Page, R. D. M. (1996). TREEVIEW: An application to display phylogenetic trees on personal computers. *Computer Applications in the Biosciences* **12,** 357–358.

Page, R. D. M., and Holmes, E. C. (1998). "Molecular evolution: A phylogenetic approach." Blackwell Science Ltd., London.

Pagel, M., and Meade, A. (2004). A phylogenetic mixture model for detecting pattern-heterogeneity in gene sequence or character-state data. *Syst. Biol.* **53,** 571–581.

Pan, X., Stein, L., and Brendel, V. (2005). SynBrowse: A synteny browser for comparative sequence analysis. *Bioinformatics* **21,** 3461–3468.

Pendas, A. M., Moran, P., and Garcia-Vazquez, E. (1993). Multi-chromosomal location of ribosomal RNA genes and heterochromatin association in brown trout. *Chrom. Res.* **1,** 63–67.

Pendas, A. M., Moran, P., and Garcia-Vazquez, E. (1994). Organization and chromosomal location of the major histone cluster in brown trout, Atlantic salmon and rainbow trout. *Chromosoma* **103,** 147–152.

Perrière, G., and Gouy, M. (1996). WWW-Query: An on-line retrieval system for biological sequence banks. *Biochimie* **78,** 364–369.

Phillips, R. B. (2001). Application of fluorescence *in situ* hybridization (FISH) to fish genetics and genome mapping. *Mar. Biotechnol.* **3,** 5145–5152.

Phillips, R. B. (2007). Application of fluorescence *in situ* hybridization (FISH) to genome mapping in Fishes. "Fish cytogenetics," (Pisano, E., Ozouf-Costaz, C., Foresti, F., and Kapoor, B. G. eds.), 455–471 Science Publishers Inc., Enfield, NH.

Pisano, E., Ozouf-Costaz, C., Foresti, F., and Kapoor, B. G. eds. (2007)."Fish cytogenetics," Science Publishers Inc, Enfield, NH.

Pisano, E., Cocca, E., Mazzei, F., Ghigliotti, L., di Prisco, G., Detrich, H. W., III, and Ozouf - Costaz, H. W. (2003). Mapping of α-and β-globin genes on Antarctic fish chromosomes by fluorescence *in-situ* hybridization. *Chrom. Res.* **11,** 633–640.

Postlethwait, J. H., Woods, I. G., Ngo-Hazelett, P., Yan, Y. L., Kelly, P. D., Chu, F., Huang, H., Hill-Force, A., and Talbot, W. S. (2000). Zebrafish comparative genomics and the origins of vertebrate chromosomes. *Genome Res.* **10,** 1890–1902.

Posada, D., and Buckley, T. R. (2004). Model selection and model averaging in phylogenetics: Advantages of Akaike information criterion and Bayesian approaches over likelihood ratio tests. *Syst. Biol.* **53,** 793–808.

Posada, D., and Crandall, K. A. (1998). Modeltest: Testing the model of DNA substitution. *Bioinformatics* **14,** 817–818.

Raap, A. K., and Tanke, H. J. (2006). COmbined Binary RAtio fluorescence *in situ* hybridiziation (COBRA-FISH): Development and applications. *Cytogenet. Genome Res.* **114,** 222–226.

Rigby, P. W. J., Diechmann, M., Rhodes, C., and Berg, P. (1977). Labeling deoxyribonucleic acid to high specific activity *in vitro* by nick translation with DNA polymerase I. *J. Mol. Biol.* **113,** 237–251.

Ronquist, F. (2004). Bayesian inference of character evolution. *Trends Ecol. Evol.* **19,** 475–481.

Ronquist, F., and J. Huelsenbeck, P. (2003). MRBAYES 3: Bayesian phylogenetic inference under mixed models. *Bioinformatics* **19,** 1572–1574.

Salemi, M., and Vandamme, A. M. (2003). "The phylogenetic handbook." Cambridge University Press, Cambridge, UK.

Shimodaira, H. (2002). An approximately unbiased test of phylogenetic tree selection. *Syst. Biol.* **51,** 492–508.

Shimodaira, H., and Hasegawa, M. (2001). CONSEL: For assessing the confidence of phylogenetic tree selection. *Bioinformatics* **17,** 1246–1247.

Simossis, V. A., Kleinjung, J., and Heringa, J. (2005). Homology-extended sequence alignment. *Nucleic Acids Res.* **33,** 816–824.

Sonnhammer, E. L. L., and Koonin, E. V. (2002). Orthology, paralogy, and proposed classification for paralog subtypes. *Trends Genet.* **18,** 620.

Soyer, O. S., and Goldstein, R. A. (2004). Predicting functional sites in proteins: Site-specific evolutionary models and their application to neurotransmitter transporters. *J. Mol. Biol.* **339,** 227–242.

Speicher, M. R., and Carter, N. P. (2005). The new cytogenetics: Blurring the boundaries with molecular biology. *Nat. Rev. Genet.* **6,** 782–792.

Sullivan, J. (2005). Maximum-likelihood methods for phylogeny estimation. *Methods Enzymol.* **395,** 757–779.

Sullivan, J., and Joyce, P. (2005). Model selection in phylogenetics. *Annu. Rev. Ecol. Evol. Syst.* **36,** 445–466.

Sullivan, J., Abdo, Z., Joyce, P., and Swofford, D. L. (2005). Evaluating the performance of a successive approximations approach to parameter optimization in maximum-likelihood phylogeny estimation. *Mol. Biol. Evol.* **22,** 1386–1392.

Suzuki, Y. (2004). Three-dimensional window analysis for detecting positive selection at structural regions of proteins. *Mol. Biol. Evol.* **21,** 2352–2359.

Suzuki, Y., and Gojobori, T. (1999). A method for detecting positive selection at single amino acid sites. *Mol. Biol. Evol.* **16,** 1315–1328.

Suzuki, Y., Gojobori, T., and Nei, M. (2001). ADAPTSITE: Detecting natural selection at single amino acid sites. *Bioinformatics* **17,** 660–661.

Suzuki, Y., and Nei, M. (2002). Simulation study of the reliability and robustness of the statistical methods for detecting positive selection at single amino acid sites. *Mol. Biol. Evol.* **19,** 1865–1869.

Swofford, D. L. (2002). PAUP*, phylogenetic analysis using parsimony (* and other methods). Version 4.10. Sinauer, Sunderland, MA.

Swofford, D. L., Olsen, G. J., Waddell, P. J., and Hillis, D. M. (1996). Phylogenetic inference. Hillis, D. M., Moritz, C., and Mable, B. K. (1996). *In* "Molecular systematics," 2d ed. pp. 407–514. Sinauer Associates, Sunderland, MA.

Thompson, J. D., Higgins, D. G., and Gibson, T. J. (1994). CLUSTALW: Improving the sensitivity of progressive multiple sequence alignment through sequence weighting, position-specific gap penalties and weight matrix choice. *Nucleic Acids Res.* **22,** 4673–4680.

Thompson, J. D., Gibson, T. J., Plewniak, F., Jeanmougin, F., and Higgins, D. G. (1997). The CLUSTAL_X windows interface: Flexible strategies for multiple sequence alignment aided by quality analysis tools. *Nucleic Acids Res.* **25,** 4876–4882.

Thorsen, J., Zhu, B., Frengen, E., Osoegawa, K., de Jong, P. J., Koop, B. F., Davidsonand, W. S., and Hoyheim, B. (2005). A highly redundant BAC library of Atlantic salmon (*Salmo salar*): An important tool for salmon projects. *BMC Genomics* **6,** 50.

Trask, B. J. (1991). Fluorescence *in situ* hybridization: Applications in cytogenetics and gene mapping. *Trends Genet.* **7,** 149–154.

van der Ploeg, M. (2000). Cytochemical nucleic acid research during the twentieth century. *Eur. J. Histochem.* **44,** 442–447.

Wagner, A., Deryckere, T., McMorrow, T., and Gannon, F. (1994). Tail-to-tail orientation of the Atlantic salmon alpha- and beta-globin genes. *J. Mol. Evol.* **38,** 28–35.

Watanabe, M., Kobayashi, N., Fujiyama, A., and Okada, N. (2003). Construction of a BAC library for *Haplochromis chilotes*, a cichlid fish from Lake Victoria. *Genes and Genetic Systems* **78,** 103–105.

Whelan, S., and Goldman, N. (2001). A general empirical model of protein evolution derived from multiple protein families using a maximum-likelihood approach. *Mol. Biol. Evol.* **18,** 691–699.

Wiegant, J., Ried, T., Nederlof, P. M., van der Ploeg, M., Tanke, H. J., and Raap, A. K. (1991). *In situ* hybridization with fluoresceinated DNA. *Nucleic Acids Res.* **19,** 3237–3241.

Williams, P. D., Pollock, D. D., Blackburne, B. P., and Goldstein, R. A. (2006). Assessing the accuracy of ancestral protein reconstruction methods. *PLoS Comput. Biol.* **2,** e69.
Xia, X., and Xie, Z. (2001). DAMBE: Data analysis in molecular biology and evolution. *J. Hered.* **92,** 371–373.
Yang, Z. (1993). Maximum-likelihood estimation of phylogeny from DNA sequences when substitution rates differ over sites. *Mol. Biol. Evol.* **10,** 1396–1401.
Yang, Z. (2007). PAML 4: Phylogenetic analysis by maximum likelihood. *Mol. Bioll. Evol.* **24,** 1586–1591.
Yang, Z., and Bielawski, J. P. (2000). Statistical methods for detecting molecular adaptation. *Trends Ecol. Evol.* **15,** 496–503.
Yang, Z., Nielsen, R., Goldman, N., and Pedersen, A. M. K. (2000). Codon-substitution models for heterogeneous selection pressure at amino acid sites. *Genetics* **155,** 431–449.
Yoshizaki, J. H., Kang, K., Sakuma, Y., Aoki, T., and Takashima, F. (1996). Cloning and sequencing of rainbow trout β-globin c-DNA. *Fish Sci.* **62,** 723–726.
Zhang, J., and Nei, M. (1997). Accuracies of ancestral amino acid sequences inferred by the parsimony, likelihood, and distance methods. *J. Mol. Evol.* **44,** S139–S146.

CHAPTER THIRTY

Inferring Evolution of Fish Proteins: The Globin Case Study

Agnes Dettaï,* Guido di Prisco,* Guillaume Lecointre,*
Elio Parisi,* *and* Cinzia Verde†

Contents

* UMR, Département Systématique et Evolution, Muséum National d'Histoire Naturelle, Paris, France
† Institute of Protein Biochemistry, CNR, Naples, Italy

Methods in Enzymology, Volume 436
ISSN 0076-6879, DOI: 10.1016/S0076-6879(08)36030-3

Abstract

Because hemoglobins (Hbs) of all animal species have the same heme group, differences in their properties, including oxygen affinity, electrophoretic mobility, and pH sensitivity, must result from the interaction of the prosthetic group with specific amino acid residues in the primary structure. For this reason, fish globins have been the object of extensive studies in the past few years, not only for their structural characteristics but also because they offer the possibility to investigate the evolutionary history of Hbs in marine and freshwater species living in a large variety of environmental conditions. For such a purpose, phylogenetic analysis of globin sequences can be combined with knowledge of the phylogenetic relationships between species. In addition, Type I functional-divergence analysis is aimed toward predicting the amino acid residues that are more likely responsible for biochemical diversification of different Hb families. These residues, mapped on the three-dimensional Hb structure, can provide insights into functional and structural divergence.

1. Introduction

The repertoire of known globin sequences has been steadily increasing in the past years, making possible a comparative study on hemoglobin (Hb) in a vast variety of species. The number of available fish globin sequences provides enough material to study evolutionary and functional aspects. Fish Hbs are particularly interesting because the respiratory function of fish differs from that of mammals. In fish, gills are in contact with a medium endowed with high oxygen tension and low carbon-dioxide tension; in contrast, in the alveoli of mammalian lungs, the carbon dioxide tension is higher and the oxygen tension is lower than in the atmosphere. Unlike most mammals, including humans, fish exhibit Hb multiplicity, which results from gene-related heterogeneity and gene duplication events. An important feature of vertebrate Hbs is the decreased oxygen affinity at low pH values (Riggs, 1988), known as the Bohr effect. In many teleost Hbs, the oxygen affinity is so markedly reduced at low pH that Hb cannot be fully saturated even when oxygen pressure is very high (Root effect). This effect (Brittain, 2005) plays an important physiological role in supplying oxygen to the swim bladder and choroid rete mirabile. Thus, Root effect Hbs can regulate both buoyancy and retina vascularization (Wittenberg and Wittenberg, 1974). The Root effect apparently evolved 100 million years before the appearance of the choroid rete (Berenbrink *et al.*, 2005). Interestingly, weakening in the intensity of the Root effect has been noticed among the Antarctic notothenioids, although some species retain a strong effect (di Prisco *et al.*, 2007).

A common characteristic among fish is Hb multiplicity, usually interpreted as a sign of phylogenetic diversification and molecular adaptation. From the phylogenetic viewpoint, teleost Hbs have been classified as embryonic Hbs, Antarctic major adult Hbs, anodic adult Hbs, and cathodic adult Hbs (Maruyama *et al.*, 2004). Embryonic Hbs are typical of the growing embryo and of the juvenile stages. The group of Antarctic major adult Hbs (Hb 1) includes the globins of red-blooded notothenioids, the dominant fish group in Antarctica, but also some globins of temperate species. In Antarctic notothenioids, Hb 1 accounts for 95 to 99% of the total, while the juvenile/embryonic forms are normally present in trace amounts in the adults. Presumably, the two clusters of Antarctic major and embryonic Hbs were generated by gene-duplication events, which occurred independently for the α- and β-globin genes. According to Bargelloni *et al.* (1998), the duplication event that gave origin to the two groups of Antarctic globins involved a mechanism of positive selection (i.e., changes that improve the fitness of the species), characterized by higher rate of nonsynonymous (amino acid replacing) to synonymous (silent) substitutions. Cathodic Hbs are considered to play an important role in oxygen transport under hypoxic and acidotic conditions. The embryonic Hb group also includes the α-globin sequences of *Anarhichas minor* Hb 1, and the β-globin sequences of *A. minor* Hb 3, *Liparis tunicatus* Hb 1, and *Chelidonichthys kumu* Hb (Giordano *et al.*, 2006).

It has been suggested that Hb multiplicity is more frequently found in fish that must cope with variable temperatures, whereas the presence of a single dominant Hb is usually associated with stable temperature conditions. This may explain why high-Antarctic notothenioids have a single major Hb, while sub-Antarctic and temperate notothenioids, such as *Cottoperca gobio* and *Bovichtus diacanthus*, respectively, retained Hb multiplicity, presumably to cope with the small or large temperature changes in the respective habitats north of the polar front (di Prisco *et al.*, 2007).

Although a report in the literature (Sidell and O'Brien, 2006) does not support the ensuing hypothesis, we believe that the reduction in Hb content/multiplicity and erythrocyte number in the blood of high-Antarctic notothenioids counterbalances the potentially negative physiological effects (e.g., higher demand of energy needed for circulation) caused by the increase in blood viscosity produced by subzero seawater temperature.

As detailed in this chapter, the polar ecosystems offer numerous examples of the evolution of the relationships between globin structure and function among fish. The approaches herewith outlined are based on interplay of phylogeny with globin-structure features. To date, this multidisciplinary approach has not been sufficiently exploited. New information will help us to understand the evolution of globins and to address the question whether different structures and functions are related to environmental conditions.

2. Hb Purification

Rigorous purification and primary-structure analyses are fundamental to properly study the evolutionary history of globin genes and proteins, and several methodologies have been developed over the years. In the field of fish Hbs, a detailed overview of hematology and biochemical methods is beyond the goal of this contribution; the reader is directed to Antonini *et al.* (1981) and Everse *et al.* (1994). The procedure for Hb purification given subsequently has been employed for the isolation and functional characterization of the three major Hbs of the Arctic gadid *Arctogadus glacialis* (Verde *et al.*, 2006).

2.1. Hemolysate preparation

Blood erythrocytes are packed by low-speed centrifugation for 5 min, washed three times with 1.7% NaCl in 1 m*M* Tris-HCl, pH 8.1, and lysed in 1 m*M* Tris-HCl, pH 8.1. Following centrifugation at 100,000×*g* for 60 min, the supernatant is dialyzed against Tris-HCl, pH 7.6. All manipulations are carried out at 0 to 4°. For functional studies, endogenous organophosphates are removed from the hemolysate by passage through a small column of Dowex AG 501 X8 (D) (e.g., stripping). No oxidation was spectrophotometrically detectable during the time needed for functional experiments. Hb solutions were stored in small aliquots at −80° until use.

2.2. Hb separation

A column of Mono Q HR5/5 (Pharmacia) is equilibrated with 10 m*M* Tris-HCl, pH 7.6 (buffer A). Hb 3 is eluted at 30% buffer B (250 m*M* Tris-HCl, pH 7.6, containing 250 m*M* NaCl) at a flow rate of 1.0 ml/min. Figure 30.1 (modified from the work of Verde *et al.*, 2006) shows separation of the three Hbs in the hemolysate of *A. glacialis* carried out by fast protein liquid chromatography (FPLC).

3. Elucidation of Globin Primary Structure

3.1. Globin separation

The hemolysate or solutions of purified Hbs (35 mg/ml) are incubated at room temperature for 10 min in a denaturing solution containing 5% β-mercaptoethanol and 1% trifluoroacetic acid (TFA). High-performance

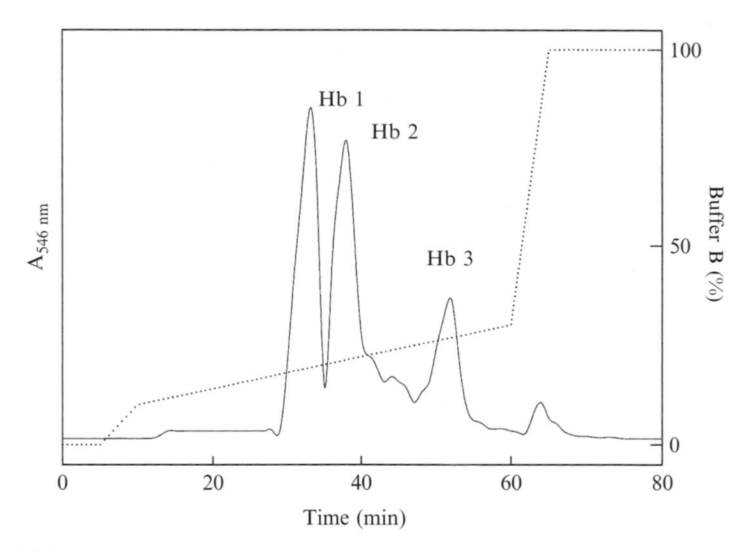

Figure 30.1 Ion-exchange chromatography of the hemolysate of *A. glacialis* (modified from Verde *et al.*, 2006).

liquid chromatography (HPLC) is carried out at room temperature (0.5 mg for each run) on a C_4 reverse-phase column, equilibrated with 45% acetonitrile, 0.3% TFA (solvent A), and 90% acetonitrile, 0.1% TFA (solvent B). The column is eluted at a flow rate of 1.0 ml/min and the eluate is monitored at 280 and 546 nm (Fig. 30.2). The procedure for globin purification described has been employed for the isolation of the four globins from the Arctic gadid *Gadus morhua* (Verde *et al.*, 2006). Fractions containing the eluted globin chains are pooled and lyophilized.

3.2. Modification of α and β chains of fish hemoglobins

Cysteyl residues are alkylated with 4-vinyl-pyridine (Friedman *et al.*, 1970). Lyophilized globins are dissolved at a concentration of 10 mg/ml in 0.5 *M* Tris-HCl, pH 7.8, containing 2 m*M* EDTA and 6 *M* guanidine hydrochloride under nitrogen. A 5-fold excess of dithiothreitol is added under nitrogen; the solution is kept at 37° for 2 h; incubation with 20-fold excess of 4-vinyl-pyridine is then carried out for 45 min at room temperature in the dark under nitrogen. The reaction is stopped by addition of 2.5-fold molar excess of dithiothreitol. Excess reagents are removed by reverse-phase HPLC. The eluted *S*-pyridylethylated α and β globins are lyophilized.

3.3. Protein cleavage

Tryptic digestion is carried out on pyridylethylated chains (100 nmol/ml), dissolved in 50 m*M* Tris-HCl, pH 8.0 (Tamburrini *et al.*, 1992, 1996). Trypsin (1 mg/ml, dissolved in 1 m*M* HCl), is added at a ratio of 1:100

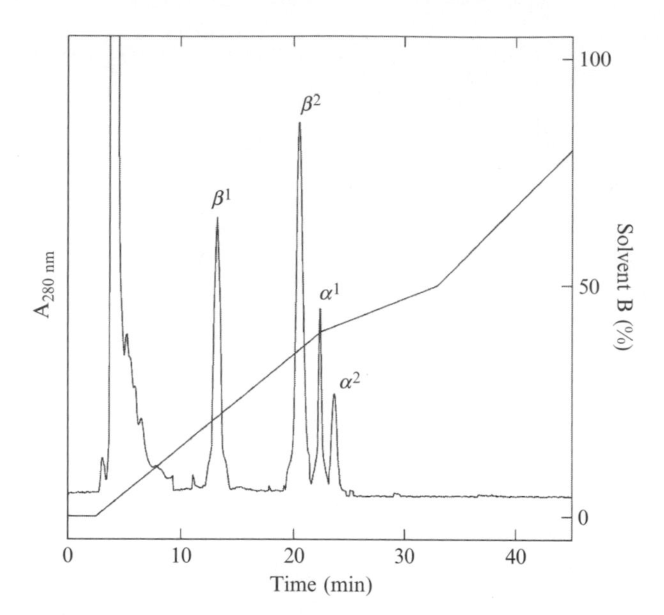

Figure 30.2 Reverse-phase HPLC of the hemolysate of *G. morhua* (Verde *et al.*, 2006), containing four globins. A C_4 Vydac column (4.6 × 250 mm) is equilibrated with solvent A. Elution is performed with a gradient of 90% solvent B in solvent A.

(by weight); after 3-h incubation at 37°, another aliquot of enzyme is added, at a final concentration of 1:50 (by weight). After 6 h, the reaction is stopped by lyophilization. The hydrolysate is suspended in 0.1% TFA and clarified by centrifugation. Separation of tryptic peptides (Tamburrini *et al.*, 1992) is carried out by HPLC on a μ-Bondapack C_{18} reverse-phase column equilibrated with 0.1% TFA in water (solvent A) and 0.08% TFA in 99.92% acetonitrile (solvent B). Elution is carried out at a flow rate of 1 ml/min, and the eluate is monitored by measuring the absorbance at 220 and 280 nm. Up to 20 nmol of peptide mixture can be loaded in each preparative run. Peptides are manually collected and dried in a Savant speed concentrator. Cleavage of Asp-Pro bonds is performed on intact globins in 70% (v/v) formic acid, for 24 h at 42° (Landon, 1977).

3.4. Deacylation of α-chain N terminus

Partial deacylation (approximately 20%) of the blocked N terminus of the α chains occurs spontaneously during Asp-Pro cleavage, due to the acidic reaction. Additional deacylation (50%) is obtained by incubating the N-terminal tryptic peptide with 30% TFA for 2 h at 55°. The N-terminus-blocking group is identified by matrix-assisted laser-desorption ionization-time-of-flight (MALDI-TOF) mass spectrometry on a PerSeptive Biosystems Voyager-DE Biospectrometry Workstation (Verde *et al.*, 2006).

4. Globin Characterization

4.1. Amino acid sequencing

Sequencing is performed on *S*-pyridylethylated globins, on fragments generated by Asp-Pro cleavage, and on HPLC-purified tryptic peptides with an automatic sequencer equipped with online detection of phenylthiohydantoin amino acids. Sequencing of Asp-Pro-cleaved globins is performed after treatment with *o*-phthaldehyde (OPA) (Brauer *et al.*, 1984) in order to block the non-Pro N terminus and reduce the background.

4.2. Cloning and sequencing of globin cDNAs

The primary structure of globins can also be deduced from the sequence of cDNA. This procedure offers some advantages because it is less time consuming than direct protein sequencing. In addition, the knowledge of the nucleotide sequence can be useful for evolutionary studies, especially to establish instances of positive selection, which can be inferred by measuring the ratio of nonsynonymous over synonymous substitutions. However, the knowledge of the sequence of the N-terminal peptide can be useful for the preparation of specific templates to be used for PCR. The protocol given below has been employed for nucleotide sequencing of α and β globins of *A. glacialis* (Verde *et al.*, 2006). Total spleen RNA is isolated using TRI Reagent (Sigma-Aldrich), as described in Chomczynski and Sacchi (1987). First-strand cDNA synthesis is performed according to the manufacturer's instructions (Promega), using an oligo(dT)-adaptor primer. The α- and β-globin cDNAs are amplified by PCR using oligonucleotides designed on the N-terminal regions as direct primers and the adaptor primer as the reverse primer. Amplifications are performed with 2.5 units of Taq DNA polymerase, 5 pmol each of the above primers, and 0.20 m*M* dNTPs buffered with 670 m*M* Tris-HCl, pH 8.8, 160 m*M* ammonium sulfate, 0.1% Tween 20, and 1.5 m*M* $MgCl_2$. The PCR program consists of 30 cycles of 1 min at 94°, 1 min at temperatures between 42 and 54°, 1 min at 72°, and ending with a single cycle of 10 min at 72°. The cloning of the N-terminal regions is obtained by 5′ RACE (rapid amplification of cDNA ends), using the Marathon cDNA Amplification Kit (BD Biosciences) and two internal primers. Amplified cDNA is purified and ligated in the pDrive vector (Qiagen). *Escherichia coli* cells (strain DH5{I}α{/I}) are transformed with the ligation mixtures. Standard molecular-biology techniques (Sambrook *et al.*, 1989) are used in the isolation, restriction, and sequence analysis of plasmid DNA. When required, automatic sequencing is performed on both strands of the cloned cDNA fragments.

5. Sequence Analysis

As we will detail in subsequent sections, information about species history is highly valuable, and indeed essential in most cases for the interpretation of any observed change along the sequences from gene families and for the inference of duplications and losses of copies from these families. This requires gathering and analyzing relevant sequence data, as well as integrating knowledge of the phylogeny deriving from external sources.

5.1. Homologous and paralogous copies

In any comparative or phylogenetic analysis of sequences, the first step is to find the sequences relevant for comparison with regard to the question addressed. For phylogenetic purposes, it is necessary only to integrate sequences with a common origin. There are two kinds of such sequences. In one case, the divergence of an ancestral lineage in two groups leads to one copy in each of the daughter lineages (orthologous copies; see also homology *sensu strictu* [Fitch, 1970]). In the other case, a duplication occurs within the genome of a single organism, whose descendants will thereafter possess two copies (paralogy [Fitch, 1970]). These two copies may diverge more or less independently by the usual process of accumulation of mutations, but the presence of redundancy (two copies are present but only one is necessary for the correct functioning of the organism) allows for several different outcomes and constitutes one of the main sources for genetic innovation in the genomes (Hurles, 2004; Lynch and Connery, 2000). Very often, one of the copies will simply become inactive (pseudogene). Sometimes, it can acquire a new function, either completely new (Long *et al.*, 2003; Prince and Pickett, 2002) or linked to the original one (Ohno, 1970), sometimes even leading to subfunctionalization of the original function between the two divergent copies (Force *et al.*, 1999). The study of the evolution of paralogues is a complex but highly productive approach that can be improved by introducing phylogenetic analysis and other information in addition to the analytical comparative methods (for a review, see Mathews, 2005). Phylogenetic analyses, as all comparative methods in molecular studies, need an alignment of the relevant sequences.

5.2. Sequence search in the databases

Sequences can either be obtained through benchwork, as described previously, or retrieved from sequence databases such as GenBank or EMBL through various portals (http://www.ncbi.nlm.nih.gov/, http://srs.ebi.ac.uk/), and from the complete genomes (http://www.ensembl.org/). Query by gene name and taxon name is a good way to start but will generally not recover all existing sequences for a given gene. Some of the sequences might not yet be annotated (the name of the gene is not yet associated with this sequence in the database), and others might be filed under a different name. To recover a maximal number of sequences, it is wise to also search with tools like BLAST (http://www.ncbi.nlm.nih.gov/BLAST/; http://www.ensembl.org/Multi/blastview). These rely on comparison among the sequences, not on the annotation. Adjusting the BLAST parameters to the type of search is very important, because too-stringent parameters will lead the program to ignore more divergent sequences, including paralogues of interest. In contrast, too widely defined search parameters will yield many irrelevant sequences and need much subsequent sorting. Detailed help can be found at http://www.ncbi.nlm.nih.gov/blast/producttable.shtml.

5.3. Sequence alignment

Sequence alignment is the way to arrange amino acid and nucleotide sequences in order to exhibit correspondances among regions with similarities. Therefore, obtaining a reliable alignment of the sequences is essential. In a first step, it is assumed that the similarity ensues from common ancestry of the residues present at a given position in the aligned sequences (homology hypotheses within the sequence). The phylogenetic tree will allow sorting of which of the similar residues at corresponding positions are really inherited from a common ancestor and which are similar due to convergences. The alignment process is often complicated by the fact that sequences of different species or different members of the gene family of interest are of different lengths. Pairing must then be obtained by introducing dashes (insertions/deletions) at carefully chosen places in order to maximize similarities. Shared similarity can result from shared ancestry, from convergence (homoplasy) led by functional or structural constraints, or even from purely contingent reasons. This can also lead to errors. Errors in the alignment can lead to errors in the phylogenetic reconstructions and in the conclusions drawn about adaptive features.

There are multiple methods and computer programs to align sequences. For all of them, it is necessary to carefully consider the parameters (generally a gap penalty and a substitution matrix to calculate the penalty of all possible misassociations). Schematically, these are used to calculate the value of a given alignment over another, as changes in the parameters may cause large differences in the alignment; for a more developed explanation, see Page and Holmes (1998) or Nei and Kumar (2000). One very widely used tool is Clustal (Higgins and Sharp, 1988). It has several more recent and complete versions, for instance, ClustalX (Thompson *et al.*, 1997), some of which have been integrated in other programs, free or commercially available (Bioedit [Hall, 1999]; GCG, Accelrys; etc). Clustal is available for use on many online servers (e.g., the EMBL-EBI server, http://www.ebi.ac.uk/clustalw/).

More efficient programs have been developed over the years, such as T-coffee (Notredame *et al.*, 2000) or 3DCoffee (O'Sullivan *et al.*, 2004), the latter even being able to integrate three-dimensional structures in the alignment estimations. For highly similar sequences, it is also possible to align by hand (for guidelines, see Barriel, 1994). For more distantly related globin sequences, the more effective, although slower, progressive method implemented in the software T-Coffee (Notredame *et al.*, 2000) can be used to produce more accurate alignments than ClustalW. Alignment of fish β globins with globins from an arthropode and a nematode worm (*Daphnia magna* and *Pseudoterranova decipiens*) obtained with T-Coffee is shown in Figure 30.3.

Nucleic acid sequences can be aligned as well as amino acid residues sequences, but the parameters and cost matrices must be adjusted accordingly. When aligning nucleic acid–coding sequences by hand, the reading frame must be considered and conserved, especially if the sequences will be used in a functional study. The corresponding amino acid residue alignment can be used as a guideline for manual editing if necessary. Some software allows for easy switching between amino acid–residue and nucleic acid alignments (e.g., Se-Al, http://evolve.zoo.ox.ac.uk/, or Bioedit [Hall, 1999]).

6. Phylogenetic Analysis

6.1. Overview of tree-building methods and topology interpretation

Phylogenetic analysis is performed on the matrix of homology hypotheses (sequence alignment). Two main groups of approaches exist for this process: a tree is either constructed from a distance matrix among sequences (distance methods) or selected within the space of possible trees according to an optimality criterion that is dependent on the method (parsimony and maximum-likelihood methods). Maximum-parsimony (MP) and maximum-likelihood (ML) methods allow for direct character and character-change mapping on the trees. Many books and review papers explore in depth the choice and the use of the various methods, including their pitfalls (Felsenstein, 2004; Nei and Kumar, 2000; Page and Holmes, 1998). Nonetheless, it is important to keep in mind that the distance trees are actually similarity trees, where the topology does not directly reflect common ancestry, but rather degrees of global similarity between sequences. These trees, called phenograms, are obtained through methods such as neighbor joining (NJ [Saitou and Nei, 1987]) or UPGMA. With

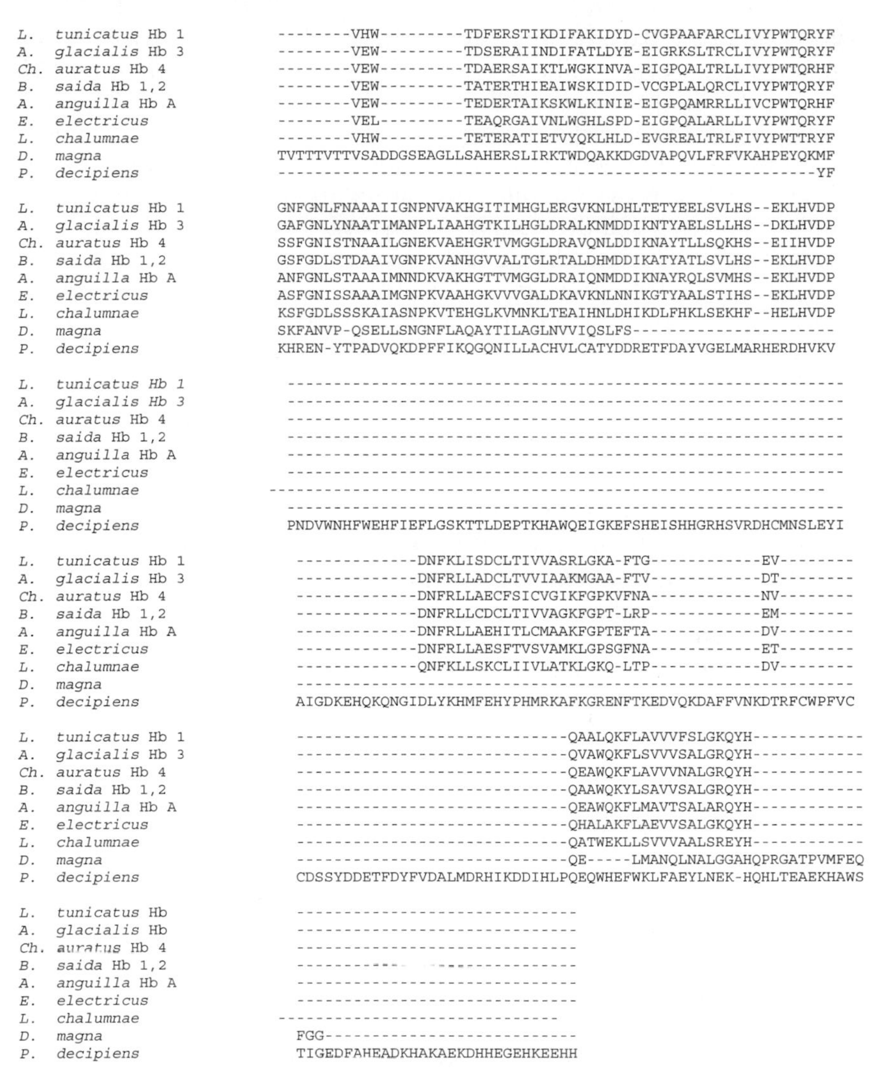

Figure 30.3 Alignment of sequences of fish β globins with two invertebrate globins. The alignment was created by the software T-COFFEE Version 1.41.

these, the tree is calculated from a distance matrix established by comparison of the sequences two by two. Phenetic trees can reflect phylogenetic topologies when the similarity in sequences is proportional to kinship (i.e., when sequences evolve at the same constant rate among lineages). Corrections can be introduced, when this is not the case, by the use of models (Nei and Kumar, 2000). Also, more effective distance methods have been proposed (e.g., Gascuel and Steel, 2006).

MP, ML, and Bayesian approaches (BA) also have their own pitfalls. As the tree-space explorations are done by a heuristic search, the approach does not explore all possible trees. Therefore, in some cases, the approach can remain stuck in a local optimum and never find the most parsimonious tree or the tree with the highest likelihood. This danger can be limited by making several analyses from different starting trees. Guidelines on how to optimize the search and the number of replicates are generally available in the documentation of the phylogenetic reconstruction programs, and more efficient search methods have been proposed. The parsimony ratchet method (Nixon, 1999a) is highly efficient for fast parsimony analyses and is implemented in several programs such as TNT (Goloboff *et al.*, 2000) and PAUPRat (Sikes and Lewis, 2001), one of the available ratchet-implementing programs for the widely used PAUP* (Swofford, 2001). For ML and BA, as well as for model-using distance methods, the choice of the most adapted model is crucial (Page and Holmes, 1998). The most widely used programs are ModelTest (Posada and Buckley, 2004; Posada and Crandall, 1998) and ProtTest (Abascal *et al.*, 2005). For the analyses themselves, fast programs like Mr. Bayes (Ronquist and Huelsenbeck, 2003) and PhyML (Guindon and Gascuel, 2003) are available and widely used. Even if the speed of analysis is still lower than that for a distance analysis, the time needed is generally quite reasonable (from a few hours to a few days, depending on the program), and it is worth using other methods in addition to distance. Bootstrap analyses must also be performed on a high number of replicates, as the variance in the results is high when only a few are performed.

The phenomenon named long-branch attraction (LBA [Hendy and Penny, 1989]) has been demonstrated when the sequences change at unequal rates among branches, even when unlimited data is available. Taxa with higher rates of mutation (i.e., long branches) tend to attract one another in the inferred tree. Schematically, taxa evolving at higher rates tend to have similar character states by convergence, and tree-building methods tend to group them together (Page and Holmes, 1998). As a result, longer branches of these taxa are often attracted to the outgroup, which frequently has the longest branches, as it is more distantly related to the others. The choice of an outgroup, as closely related to the group of interest as possible, is important also for this reason. The lower the number of branches in a tree, the sharper the problem; trees that include only four species are particularly sensitive to this artefact (Lecointre *et al.*, 1993; Philippe and Douzery, 1994). Some methods are more sensitive than others to this phenomenon, but models can be used to detect and partially correct it (see Lartillot *et al.*, 2007), although there is still widespread debate on the extent of the problem and sensitivity of the various methods. In general, whenever possible, adding sequences to the branches that are supposed to be subject to LBA is also a good strategy ("breaking the long branches" [Graybeal, 1998; Hillis *et al.*, 2003]).

Additionally, in most cases, applying several methods to the data set is valuable, as discrepancies can provide information about which nodes are less supported (robustness to changes of method and model).

6.2. Approaching the species tree

A short review of the factors known to influence the inference of the species tree is of interest, as it shows where the reconstruction of a gene tree differs from it, and thereby the potential pitfalls specific to gene-tree building and interpretations. We will concentrate here on four main points of direct relevance also to gene-tree reconstruction.

6.2.1. Single copy rather than gene families

For phylogenetic reconstruction of species trees, it is desirable to choose marker(s) present as a single copy in the genome or at least to ensure that the various copies are clearly distinguishable from one another. When this is not the case, it can be difficult or impossible to distinguish between species-separation and gene-duplication events, and the interpretation of the tree as an approximation of the species tree is compromised (Cotton, 2005; Page and Holmes, 1998). This problem is especially relevant in teleost fish, as the teleost genome has undergone an ancient duplication, and many genes are present in multiple copies (Hurley *et al.*, 2007; Robinson-Rechavi *et al.*, 2001, Taylor *et al.*, 2001), with additional duplications and sometimes differential loss of copies in the various lineages (e.g., Hashiguchi and Nishida, 2006; Kuo *et al.*, 2005; Taylor *et al.*, 2003). But traces of other rounds of genome duplication have been revealed in numerous actinopterygian lineages, for instance some Acipenseriformes (Dingerkus and Howell, 1976) and some catfishes (Uyeno and Smith, 1972). A database of homologies and paralogies for genes present in single copy in Sarcopterygians but multiple copies in Actinopterygians is available at http://www.evolutionsbiologie.uni-konstanz.de/Wanda/index.htm (Van de Peer *et al.*, 2002).

6.2.2. Multiple markers

Several studies have also shown that using several diversified sequences as markers considerably improves the approximation of the species tree (for a review, see DeSalle, 2005). Concatenating the sequences of several markers and analyzing them simultaneously is also recommended (DeSalle, 2005; Kluge, 1989; Wiens, 1998), although performing separate analyses of each marker also allows for exploration of the possible marker-specific biases and differences in the history of the genes (de Queiroz, 1993; Dettaï and Lecointre, 2004, 2005; Maddison, 1997).

6.2.3. Markers with different rates of evolution

Different parts of the genome and genes evolve at different rates. To establish a phylogeny of closely related species, it is necessary to use sequences evolving rapidly, such as mitochondrial or noncoding sequences; otherwise an insufficient number of variable sites will be available to infer the relationships. In contrast, for phylogenies of distantly related species, the use of fast-evolving markers poses alignment and phylogenetic-signal erasure problems. The latter are due to the fact that there are only four possible character states in nucleic acid sequences, and moreover, some changes occur more frequently than others. As time goes by, multiple substitutions can occur at a single position, changing the initial shared character state and obscuring phylogenetic relationships. Although it is possible to correct for multiple substitutions, by the use of substitution models or by the elimination of the most variable positions (a theme covered at length in all molecular phylogenetics textbook), this remains one of the main sources of phylogenetic reconstruction error and uncertainty. Therefore, using more conserved markers for deeper divergences is still considered very important (Graybeal, 1994). Moreover, including multiple markers with different levels of variability and analyzing them simultaneously can enhance the resolution of the tree by providing signals for both deeper and shallower levels (Hillis, 1987).

6.2.4. Evaluating reliability

Finally, it is difficult to evaluate how far a given phylogenetic reconstruction can be trusted. Various indicators exist to estimate to what extent the data analyzed support a node in the tree (e.g., bootstrap values, jackknife, Bremer support [Bremer, 1994], and others). But they only measure robustness (i.e., how the signal within the data resists perturbations of the data set) or the contradiction within the data set. They cannot evaluate how well the tree inferred using the marker reflects the species tree and can even indicate high values for nodes due solely to marker biases such as high GC content (see Chang and Campbell, 2000). Thus, another argument for using several markers is the following: if all of them yield the same groups, regardless of possible marker-specific biases, the common signal should come from the shared ancestry (Miyamoto and Fitch, 1995). This allows for a more direct evaluation of how well the obtained tree reflects the species tree (Dettaï and Lecointre, 2004).

6.3. Some reference species trees for Actinopterygians

Although the phylogeny of Actinopterygians is still far from fully resolved, important progress has been made in recent years, especially with the new and fast-expanding wealth of molecular data. The multigene, taxon-rich trees are generally congruent to a large extent with one another. We provide a few references herein, but it is only possible to cover a fraction of the existing publications. A good starting point for a summary of the morphological and some molecular data is Nelson (2006).

For gnathostomes, for instance, Janvier (1996), Stiassny *et al.* (1996), Kikugawa *et al.* (2004), and Takezaki *et al.* (2003, 2004) ought to be consulted. There is still a debate over the basal relationships among Actinopterygians, with part of the molecular data (e.g., Hurley *et al.*, 2007; Kikugawa *et al.*, 2004; Lê *et al.*, 1993) supporting the classical Neopterygii group, and another part (Inoue *et al.*, 2003; Venkatesh *et al.*, 2001) supporting an ancient fish clade, grouping chondrosteans, lepisosteids, and amiids. Within teleosts, a few references are Lê *et al.* (1993), Zaragueta-Bagils *et al.* (2002), Ishiguro *et al.* (2003), and Filleul and Lavoue (2001). Within Acanthomorpha, see Chen *et al.* (2003), Dettaï and Lecointre (2004, 2005), Miya *et al.* (2003, 2005), and Smith and Wheeler (2004, 2006).

6.4. Reconstruction of the history of a gene family

The reconstruction of the phylogeny of the gene of interest does not fulfill the four requirements listed earlier. In the case of a gene tree, the gene family of interest itself imposes the choice of the sequences to be included. It is generally not possible to select other markers more adapted to the time of divergence or to add more markers to counteract possible biases or too-short sequences containing not enough information, because it cannot be excluded that these other genes have a different history (see Figs. 30.4 to 30.6). The case of the tree obtained from the partial α globins in Near *et al.* (2006) is a good example of lack of resolution inevitable from the shortness of the sequences and the recent divergence of the icefishes. These resulted in a very low number of variable sites, sufficient to resolve only some parts of the tree. Nonetheless, it is not wise to include additional sequence length, as the analysis of the β globins shows that they had a different history from the α globins.

It is therefore important to be very careful when aligning and inferring the gene family tree. The tree can only be checked indirectly against other trees from other markers, and some errors might not be detectable in the way they would be if optimizing for species-tree recovery among several meticulously selected genes. The inclusion of both genes and pseudogenes in a single data set can be especially tricky and is a common occurrence in gene family trees. Pseudogenes generally evolve at higher rates; the clustering of all pseudogenes together may come from LBA and must be checked. While other sequences from the same organism cannot be integrated in the analysis, it is possible to avoid some pitfalls of phylogenetic reconstruction and clarify some parts of the tree by sampling a large number of organisms, representative of as many clades as possible of known species trees.

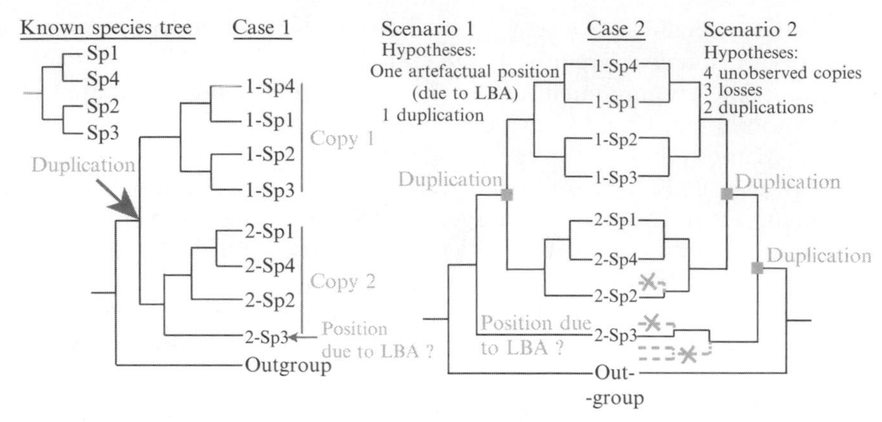

Figure 30.4 Two cases of gene trees and their interpretation in the light of a known species tree. Gray text and lines denote hypotheses. The relative position in the tree of duplications and species divergences allows the inference of a relative timescale for their occurrence. For example, in case 1, gene duplication occurred after the split from the outgroup but before the four species separated from one another. In case 2, scenario 1 should be prefered, as it is more parsimonious. 2-sp3 is not a third paralogue lost multiple times, but a paralogue 2 misplaced by long branch attraction. The position of the copies in the tree allows the inference of orthology and paralogy among them: 1-sp1, 1-sp2, 1-sp3, and 1-sp4 are orthologues, as are 2-sp1, 2-sp2, 2-sp3 and 2-sp4. Any sequence from the copy 1 clade is paralogous to any sequence of the copy 2 clade.

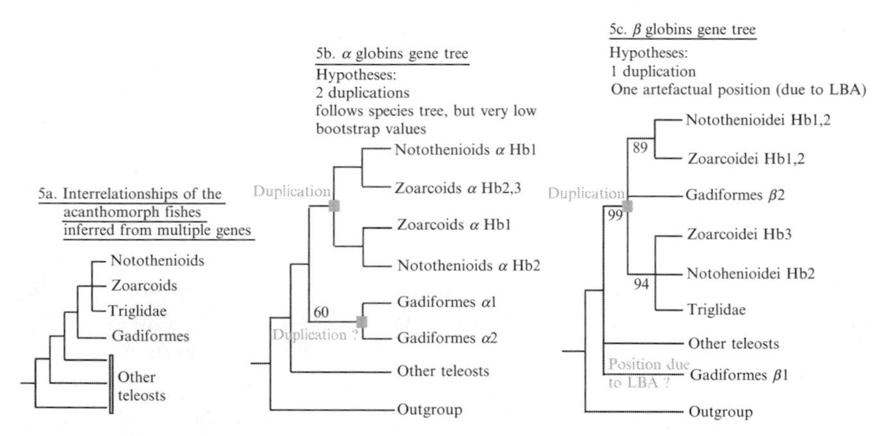

Figure 30.5 Simplified α- and β-globin gene trees for cold-adapted acanthomorph fishes. Adapted from Figures 6 and 7 of Verde *et al.* (2006). Gray text and lines denote hypotheses. The position of *Oncorhynchus mykiss* is unsupported, and the branch has been omitted in this tree for the sake of clarity. The original α-globin tree shows groups of sequences identified as Hb 1, Hb 2, and Hb 3 that form composite clades. A full reevaluation of the nomenclature of the ensemble of fish globins is probably needed. The bootstrap values of the branches of interest have been indicated, except when they are <50.

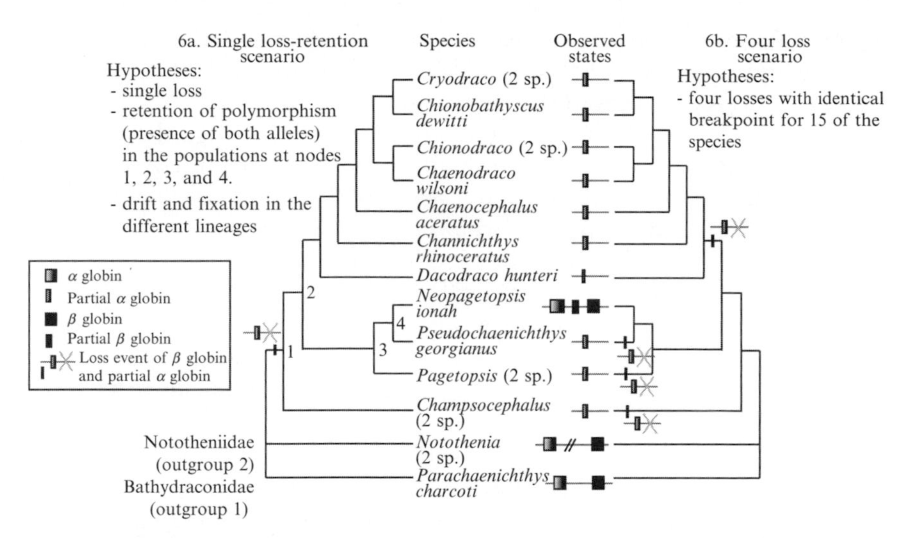

Figure 30.6 Species tree of the notothenioid family Channichthyidae with two different scenarios to explain the distribution of α and β globins in the various species. Adapted from Figures 1 and 2 of Near *et al.* (2006).

Finally, using an adapted inference method is necessary, as reconstruction artefacts cannot be corrected by adding longer sequences or by comparison of several genes. The best inference method remains a highly debated subject, but in such cases it is better not to merely use distance methods like NJ for the analysis and to apply several different approaches.

6.5. Reciprocal illumination

In a multiparalogue, multispecies tree, interpretation of adaptive molecular features is difficult without external information.

Identification of orthologues and paralogues on a gene family tree can greatly benefit from the comparison with a good species-tree approximation. The discrepancies between the two trees will allow for location of which of the nodes represent possible gene duplications, which are more likely due to speciation events, and which are artefacts (see Fig. 30.4). The relative placing of these events in the tree can yield information as to the relative age of the duplication events. Well-dated clade splits can then be used to obtain an estimate of the absolute age of the duplications. A database is available at http://www.timetree.net/ (Hedges *et al.*, 2006), but to date it does not contain information about Actinopterygii. The topology of the tree can be used to detect gene losses (for a more detailed explanation of the identification of paralogues and orthologues, see Cotton [2005]).

As differences between the species and gene trees are crucial in the interpretation, any topological mistake in either one will cause errors in the conclusions drawn from the comparison.

Actually, the use of external information (e.g., knowledge about species tree, duplications) can rarely be omitted in the search for interpretation of species relationships, duplications, and artefacts. All those cannot be deduced in the same inferential process, except in extremely rare cases with a large number of gene copies, where the repetition of the same species interrelationships within each cluster of paralogues gives some reliability to the repeated pattern of species interrelationships.

Species interrelationships are used as solid knowledge to interpret duplication events and artefacts. In case 1 of Figure 30.4, the known species tree allows for interpretation of the position of the copy 2 of species No. 3 as probably artefactual (LBA). In the absence of information about the species interrelationships, it is not possible to decide whether the correct species interrelationships are those exhibited in the part of the gene tree given by copy 1 or by that of copy 2, or even that neither subtree correctly reflects them. Knowledge of species interrelationships often leads to the determination of the number of putative duplication events. The known species tree of figure 30.4 allows to choose the most parsimonious alternative between the two possible scenarii. Scenario 1 implies only one duplication and one LBA, copy losses. Without this knowledge, a third interpretation, as the scenario 1. It implies one duplication (the same as in scenario 1) and one horizontal gene transfer (copy 1 of species No. 2 being transferred into species No. 3).

This strategy has been used by Verde *et al.* (2006) to interpret the discrepancy between the α-globin and the β-globin trees of cold-adapted fish. The position of gadids with regard to zoarcoids and notothenioids is known (see Fig. 30.5A), and allows for hypothesizing that the basal position of the β1 sequences of Arctic gadids in the β-globin tree (Fig. 30.5C) is probably artefactual, whereas the α-globin tree recovers mostly the species tree plus a few duplications (Fig. 30.5B). It is more parsimonious to consider the position of the Arctic gadid β1-globin sequences as an LBA artefact than a new paralogue-gene cluster not observable in all other fishes (similar to case 2, scenario 1 of Fig. 30.4). LBA was then interpreted as an effect of the extreme perturbation of the available mutational space in gadid β1-globin sequences, possibly due to the variability of thermal conditions met by these migratory Arctic fish in comparison with the thermal stability in the lifestyle of zoarcoids and notothenioids, two groups that display unperturbed phylogenetic signals in β sequences.

Conversely, previous knowledge about duplication events can be successfully inserted into the interpretation process to help the inference of species interrelationships. Classical studies on the amino acid sequences of α and β globins exemplify this procedure for other groups (Goodman *et al.*, 1987).

6.6. Reconstruction of character states at the nodes

The inference of character states at each node of the species tree is an essential step in reconstituting the evolution of a character of interest. This character can be a single position of the alignment crucial to the function of the protein or a complex character, such as the presence or absence of a functional motif of several amino acid residues, or even of a whole gene or gene region. Many computer programs allow for such a reconstruction (e.g. Mesquite [Maddison and Maddison, 2006], MacClade [http://macclade.org/index.html], Winclada [Nixon, 1999b], Mr. Bayes 3.1 [Ronquist and Huelsenbeck, 2003]). An extension of reconstruction of ancestral state is the inference of whole ancestral sequences. Such sequences can then be produced in the laboratory and fully tested as it would be possible for present-day sequences (for a review of the methodology and applications, see Chang *et al.*, 2005). Complex characters (e.g., presence or absence of a whole gene copy or active site) must be coded so that the programs can analyze them.

6.7. Alternative reconstructions and interpretations

While mapping characters on a tree might be a simple task, there are often several alternative scenarios to explain the character-state distribution on the tree; examples are shown in Figures 30.4 and 30.5. In some cases, one of the scenarios is much better supported or more parsimonious than its alternatives (Figs. 30.4 and 30.5), but discussion of the others is still of interest. In other cases, several different character mappings have the same support. This is especially frequent when the species tree is not fully resolved, and the various alternatives must all be considered and discussed. Additional background knowledge about molecular evolution can be valuable for selecting scenarios (e.g., the acquisition of exactly the same sequence twice independently might be considered as less likely than two independent losses). The example of the "fossil" α and β globins in *Neopagetopsis ionah* (Near *et al.*, 2006) highlights most of the approaches and tests that can be used in such a case. Two different scenarios have been retained (see Fig. 30.6). One relies on four losses with identical breakpoints of the partial α and the β globins, an unlikely occurrence. The other one hypothesizes a single loss, with retention of ancestral polymorphism in the population through several speciation events. This might seem far-fetched, but the divergences are recent, and several other such instances of retention of molecular polymorphism are known in the channichthyids (Clément *et al.*, 1998), making it actually more interesting than the alternative hypothesis (for the tests and the discussion, see Near *et al.*, 2006). This study also shows an example of the additional information gene trees can bring when used in conjunction with species trees. In this case, a gene tree of the remnants of the α and a second one of the β globins allowed to trace the origin of the two β-globin copies. These two copies have a very different origin and can probably be explained by an ancient introgression event involving a notothenioid and a channichthyid (see Near *et al.*, 2006).

6.8. Adaptations and disadaptations

It is often difficult to interpret whether a change in sequences is due to adaptation or to phylogenetic innovation, or even both, as they are not mutually exclusive (the adaptation can be a derived character shared by the members of a clade). Adaptive sequence features are sequence changes that can be correlated with some external environmental parameters or lifestyle on to a phylogenetic tree: they correspond to derived character states coupled with external information. The suspicion of being adaptive is even stronger when the same functional and/or environmental information is correlated several times on to a tree with multiple gain of the same derived state. For instance, antifreeze proteins have been gained independently twice in the teleostean tree, in two nonrelated groups (gadids and notothenioids), both of which experience subzero temperatures during part of their lives.

However, for a conclusion about the adaptive nature of the fixation of a character state to be warranted, it must be checked that the putative adaptation is not also fixed in closely related groups not living under the conditions where the adaptation is beneficial. Such groups might not be available, but additional data should be collected to include them whenever possible. For instance, *G. morhua* was included in the study from Verde *et al.* (2006) to provide a noncold-adapted reference within Gadiformes. To make a conclusion on detection of adaptive features, it is useful to keep in mind that correlation is not causation and might only result from the scarcity of groups where data are available. While it might be possible to show that a feature is more largely shared through a clade, even among species with different lifestyles, the lack of such data does not prove that it arose as an adaptation, because we do not have access to the early history of the group.

It is also noteworthy that the notion of adaptation includes the idea that the derived character state is more efficient in the environment than the original (primitive) state of the character. Baum and Larson (1991) proposed the complementary notion of disadaptation, which corresponds to loss of efficiency. Disadaptation can be shown experimentally; the derived character state is then less efficient in the environment than the original one. For instance, icefishes have lost their cardiac myoglobin (Mb) four times, as shown in *Champsocephalus gunnari*, *Pagetopsis*, *Dacodraco hunteri*, and *Chaenocephalus aceratus* (Sidell *et al.*, 1997). These multiple losses favored the hypothesis that cardiac Mb is useless in the whole icefish family Channichthyidae. Acierno *et al.* (1997) have shown that the loss of Mb in the heart of the Antarctic icefish *C. aceratus* corresponds to a disadaptation. They selectively poisoned cardiac Mb in *Chionodraco myersi*, the sister group living in the same habitats and having natural functional cardiac Mb, and recorded a significant decrease in cardiac output. The Mb is therefore still of use in the heart of these icefishes and had to be compensated for in the groups where it was lost (Montgomery and Clements, 2000; O'Brien and Sidell, 2000). The former conclusion that *C. aceratus* does not have cardiac Mb because it became useless in its ancestors living in cold and oxygen-rich waters is therefore an oversimplification and possibly incorrect.

7. Bringing Phylogenetics and Structural Analysis Together

Understanding the relationships between sequence information and protein function is one of the most challenging goals in structural biology. The inference of the protein functional features requires a multidisciplinary approach, including molecular phylogenetics, structural analysis, and protein engineering. While multiple sequence alignments can be used for reconstructing the phylogeny of protein families, the fold-recognition approach allows for establishing of the relationships with experimentally solved homologous protein structures to be used as a guide in homology modeling. With the aid of suitable bioinformatic tools, it is possible to identify the amino acid sites that are most likely responsible of the functional divergence of different protein families. Mapping these residues on the structural model can be used to establish how relevant they are in determining diversified biochemical and catalytic properties. Finally, site-directed mutagenesis can provide mutant proteins for experimentally testing functional divergence.

7.1. Identification of structural motifs

Structural alignments use information available on the secondary and tertiary structure of proteins. These methods can be used to compare two or more sequences, provided the corresponding structures are known. Because protein structures are more evolutionarily conserved than amino acid sequences (Chothia and Lesk, 1986), structural alignments are usually reliable, especially when comparing distantly related sequences. The simplest way to approach the problem of globin-structure prediction relies on the identification of the sequence regions forming α-helices and strands of β-sheets without inferring the three-dimensional structures of these regions. Several secondary-structure prediction methods are available at the site of the ExPASy tools (http://www.expasy.ch/tools/), including, among others, the powerful hierarchical neural network (HNN) method (http://npsa-pbil.ibcp.fr/cgi-bin/npsa_automat.pl?page=npsa_nn.html).

7.2. Molecular modeling

A widely used method for the prediction of a three-dimensional structure of globins starting from known structures of globins is based on homology modeling (Blundell *et al.*, 1987; Fetrow and Bryant, 1993; Johnson *et al.*, 1994). Homology-modeling methods require known protein structures sharing homology to the query sequence to be used as templates and alignment of the query sequence onto the template sequence. Template selection can be achieved by searching in suitable databases by FASTA and BLAST algorithms, choosing hits with low E-values. The quality of the structural model strongly depends on how much the target is related to the template. The entire process of homology modeling includes the selection of a suitable template as a first step, followed by target-template alignment, model-construction, and model-assessment steps. Alignment is critical, because a bad alignment can invalidate the final model. This procedure is now simplified with the development of programs such as Biology Workbench, available at the San Diego Supercomputer Center (http://workbench.sdsc.edu/). A database containing a large number of structures is ModBase (http://modbase.compbio.ucsf.edu/modbase-cgi/index.cgi), created by Sali and Blundell (1993), using a program that builds models based on satisfaction of spatial restraints identified from the alignments of homologues of known structure. These restraints are then applied to the unknown sequence. Restraints include distances between α carbon atoms, and other distances within the main-chain, and main-chain and side-chain dihedral angles. Routines to satisfy the restraints optimally include conjugate gradient minimization and molecular dynamics with simulated annealing.

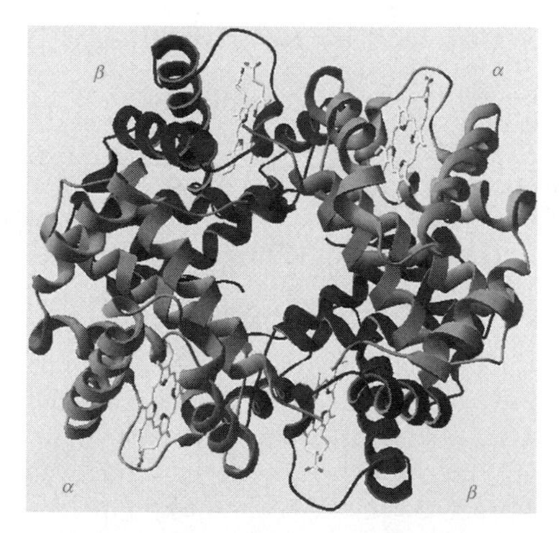

Figure 30.7 Structure of *P. urvillii* Hb 1 (Barbiero, unpublished), built by homology modeling using the program MODELER. (See color insert.)

There are several methods for homology modeling. A theoretical model of Hb 1 from the non-Antarctic notothenioid *Pseudaphritis urvillii* is shown in Figure 30.7 as an example. The model (Barbiero, unpublished) is constructed using the crystal structures (Mazzarella *et al.*, 2006a) of *Trematomus bernacchii* Hb (Protein Data Bank, codes 1H8F at pH 6.2 and 1H8D at pH 8.4, respectively) and *T. newnesi* Hb C (Mazzarella *et al.*, 2006b) (Protein Data Bank, code 2AA1). Among notothenioids, *T. bernacchii* Hb has the highest sequence identity with *P. urvillii* Hb 1, at 79 and 82% for the α and β chains, respectively (Verde *et al.*, 2004). The model was built by the MODELER program (Sali and Blundell, 1993) implemented in InsightII (Accelrys Inc). The stereochemistry of the final model was assessed using PROCHECK (Laskowski *et al.*, 1993) and Whatcheck (Hooft *et al.*, 1996).

It is also possible to generate reliable models by homology-modeling tools available on the Web. An automated comparative protein-modeling server is SwissModel (http://swissmodel.expasy.org//SWISS-MODEL.html); it is linked to the software Deep View-Swiss Pdb-Viewer, which provides a friendly interface for analyzing several proteins at the same time. 3DCrunch (http://swissmodel.expasy.org//SM_3DCrunch.html) is aimed at submitting entries from the sequence database directly to SwissModel.

Another fold-recognition method that can be usefully applied in globin-structure prediction is threading. Using a globin as a query sequence, it is possible to predict the protein structure starting from sequence information by comparison with other globins of known structure. This method imposes the query sequence to assume every known protein fold and estimates a scoring function that determines the suitability of the sequence for that particular fold. One of the most promising versions of the method is implemented in the software ROSETTA (Simons *et al.*, 1999) (http://www.bioinfo.rpi.edu/~bystrc/hmmstr/server.php). ROSETTA predicts protein structure by combining the structures of individual fragments inferred through comparison with known structures. The conformational space defined by these fragments is then searched using a Monte Carlo procedure with an energy function that favors compact structures with paired β-sheets and buried hydrophobic residues. The structures provided by such a search are grouped, and the centers of the largest groups are granted as reliable predictions of the target structure.

7.3. Functional divergence

The great similarity of the amino acid sequences and three-dimensional folding shared by vertebrate globins is indicative of their common origin from the same ancestral gene. The evolutionary history of the globin family is characterized by a series of gene-duplication events. The first duplication was probably responsible of the divergence between two functionally distinct oxygen-binding proteins (i.e., Mb and single-chain Hb).

According to the classical model proposed by Ohno (1970), after the occurrence of a gene-duplication event, one of the two gene copies can freely accumulate deleterious mutations that transform it into a pseudogene, provided the other duplicate retains the original function. Alternatively, both duplicates can be preserved if one copy acquires a novel function with the other retaining the ancestral function. The existence of two duplicates, one endowed with the oxygen-carrier function and the other with oxygen-deposit function, associated respectively to Hb and Mb, are in perfect agreement with Ohno's theory. Further duplications brought about the transition from single-chain Hb to α- and β-globin families and subsequent globin multiplicity within each family.

At the molecular level, functional constraints of a protein sequence can provide a measure of the importance of its function. This is equivalent to assuming that functionally relevant residues are conserved, whereas functionally divergent residues violate this constraint (Gu, 1999). Functional and structural divergence can be inferred by statistical methods (Gaucher *et al.*, 2002). The method developed by Gu (1999, 2001, 2003) compares protein sequences encoded by two monophyletic gene clusters generated by gene duplication or speciation. It is assumed that an amino acid residue may have two states. In the S_o state, the site has the same evolutionary rate in both clusters, whereas in the state S_1, the rates in the two clusters are different. The probability of a site to be S_1 is given by the coefficient of Type I functional divergence $\{I\}\theta = P\{/I\}(S_1)$. If the null hypothesis $\{I\}\theta\{/I\} = 0$ is rejected, one can assume that functional constraints at some sites have shifted significantly in the two clusters.

Globin multiplicity observed in many fish species was investigated through the pattern of amino acid substitution during evolution by means of the software DIVERGE (Gu and Vander Velden, 2002), available free of charge at http://xgu1.zool.iastate.edu/, which carries out a likelihood ratio test for the rejection of the null hypothesis $\{I\}\theta\{/I\} = 0$, and allows to identify the residues more likely responsible for the functional divergence between two clusters. The posterior probability $P(S_1 \mid X)$ of a site being related to Type I functional divergence, given the observed amino acid pattern X, is used as a criterion for identifying the most relevant sites.

In the example provided here, the input file consists of a multiple alignment of β-globin sequences in CLUSTAL format. Figure 30.8 shows the gene tree inferred by the NJ method with Poisson distance. In this case, the emphasis is put on the two globin clusters generated by gene duplication during the evolution of Antarctic notothenioids.

In order to perform the analysis, the two clusters of major Hbs and minor Hbs were selected on the gene tree (Parisi *et al.*, unpublished). The output file provides the critical residues predicted by DIVERGE. Among the substitutions having posterior probability >0.80, three (β65 E9, β68

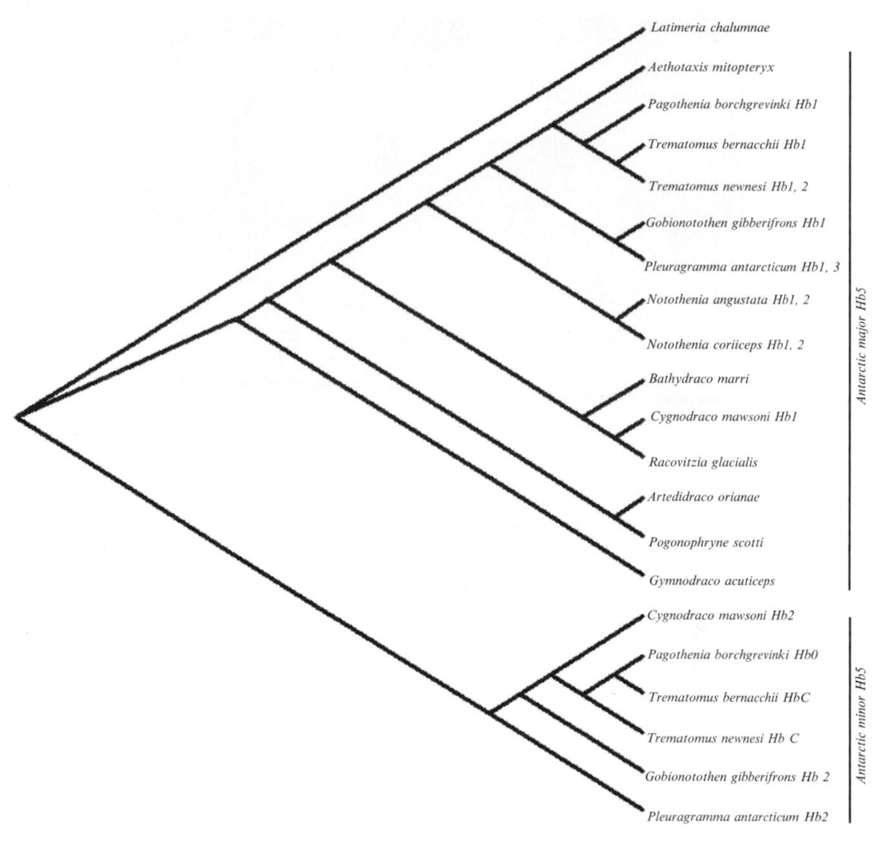

Figure 30.8 Phylogenetic tree of major and minor Antarctic β globins. The tree has been inferred by the NJ method implemented in software DIVERGE and rooted on the sequence of the β globin of *L. chalumnae. Notothenia angustata* is a temperate notothenioid.

E12 and β75 E19) occur in helix E, one (β90 F6) in helix F, and one (β121 GH4) in the interhelix region GH (Fig. 30.9).

Another type of functional divergence is Type II, characterized by the site-specific shift of residue properties (Gu, 2006), in contrast to the site-specific shift of evolutionary rate, typical of Type I. Type II functional divergence involves a radical shift of the amino acid properties, such as transition from hydrophilic to hydrophobic character at a homologous site. An example is given by the presence of Pro at a site in one of two hypothetical clusters generated by the gene-duplication event and by the presence of Tyr residue at the corresponding site in the other cluster. The method for studying Type II functional divergence has been implemented in the software DIVERGE2, available at http://xgu.gdcb.iastate.edu.

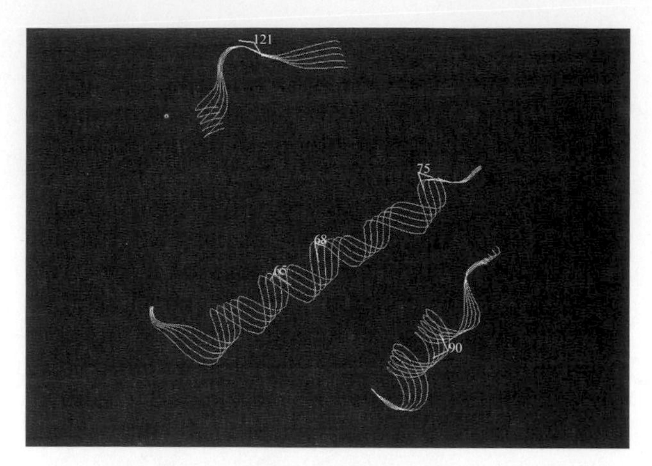

Figure 30.9 Sites involved in Type I functional divergence between major and minor β globins of Antarctic notothenioids. The sites with posterior probability >0.80 are positioned on the structure of helices E and F, and on the interhelix region GH. The three-dimensional model has been built with the software SwissPdbViewer using the atomic coordinates of *T. bernacchii* Hb (Mazzarella *et al.*, 2006a).

In contrast to Type I functional divergence, major and minor Antarctic β globins show no statistically significant Type II functional divergence.

As drastic changes in biochemical properties can often involve the substitution of few amino acid residues, inspection of the sites with the highest value of the posterior ratio score may be useful. Although functional divergence may be very effective to detect residues that contribute to functional-structural diversification, these residues must be only considered as putative candidates for further experiments. Their effectiveness needs to be ascertained with sound experimental data, by coupling the model with additional structural and biochemical information.

8. General Remarks

The joint use of species and gene trees can markedly improve the interpretation of the history of the genes and of adaptations, and allows for avoiding a number of pitfalls in the conclusions. Good achievement of phylogenetic reconstruction in general, as well as integration of the recent progress in the knowledge about species interrelationships, allows for full exploitation of new results about gene and protein structure and function, as they can then be placed in a larger interpretative frame.

Detection of site-specific changes in evolutionary rate can provide a useful tool for predicting the amino acid residues responsible for Type I functional divergence of two monophyletic clusters generated by gene duplication. The most relevant sites (i.e., those with higher values of posterior probability) can be mapped on the three-dimensional protein structure. In combination with methods aimed at the identification of site-specific shift of the amino acid properties (e.g., Type II functional divergence), this approach offers the opportunity to link sequence and functional/structural divergence.

ACKNOWLEDGMENTS

This study is financially supported by the Italian National Programme for Antarctic Research (PNRA). It is in the framework of the Evolution and Biodiversity in the Antarctic (EBA) program, endorsed by the Scientific Committee on Antarctic Research (SCAR).

REFERENCES

Abascal, F., Zardoya, R., and Posada, D. (2005). ProtTest: Selection of best-fit models of protein evolution. *Bioinformatics* **21,** 2104–2105.

Acierno, R., Agnisola, C., Tota, B., and Sidell, B. D. (1997). Myoglobin enhances cardiac performance in Antarctic icefish species that express the pigment. *Am. J. Physiol.* **273,** R100–R106.

Antonini, E., Rossi-Bernardi, L., and Chiancone, E. (1981). Hemoglobins. *Methods Enzymol.* **76,** 874.

Bargelloni, L., Marcato, S., and Patarnello, T. (1998). Antarctic fish hemoglobins: Evidence for adaptive evolution at subzero temperature. *Proc. Natl. Acad. Sci. USA* **95,** 8670–8675.

Barriel, V. (1994). Phylogenies moleculaires et insertions-deletions de nucleotides. *C. R. Acad. Sci. B* **317,** 693–701.

Baum, D. A., and Larson, A. (1991). Adaptation reviewed: A phylogenetic methodology for studying character macroevolution. *Syst. Zool.* **40,** 1–18.

Berenbrink, M., Koldkjaer, P., Kepp, O., and Cossins, A. R. (2005). Evolution of oxygen secretion in fishes and the emergence of a complex physiological system. *Science* **307,** 1752–1757.

Blundell, T. L., Sibanda, B. L., Sternberg, M. J. E., and Thornton, J. M. (1987). Knowledge-based prediction of protein structures and the design of novel molecules. *Nature* **326,** 347–352.

Brauer, A. W., Oman, C. L., and Margolies, M. N. (1984). Use of *o*-phthalaldehyde to reduce background during automated Edman degradation. *Anal. Biochem.* **137,** 134–142.

Bremer, K. (1994). Branch support and tree stability. *Cladistics* **10,** 295–304.

Brittain, T. (2005). Root effect hemoglobins. *J. Inorg. Biochem.* **99,** 120–129.

Chang, B. S. W., and Campbell, D. L. (2000). Bias in phylogenetic reconstruction of vertebrate rhodopsin sequences. *Mol. Biol. Evol.* **17,** 1220–1231.

Chang, B. S. W., Ugalde, J. A., and Matz, M. V. (2005). Applications of ancestral protein reconstruction in understanding protein function: GFP-like proteins. *Methods Enzymol.* **395,** 652–670.

Chen, W.-J., Bonillo, C., and Lecointre, G. (2003). Repeatability as a criterion of reliability of new clades in the acanthomorph (Teleostei) radiation. *Syst. Biol.* **26,** 262–288.

Chomczynski, P., and Sacchi, N. (1987). Single-step method of RNA isolation by acid guanidinium thiocyanate-phenol-chloroform extraction. *Anal. Biochem.* **162,** 156–159.

Chothia, C., and Lesk, A. M. (1986). The relation between the divergence of sequence and structure in proteins. *EMBO J.* **5,** 823–826.

Clément, O., Ozouf-Costaz, C., Lecointre, G., and Berrebi, P. (1998). Allozymic polymorphism and the phylogeny of family Channichthyidae. *In* "Fishes of Antarctica, a biological overview" (G. di Prisco, E. Pisano, and A. Clarke, eds.) pp. 299–309. Springer-Verlag, Berlin.

Cotton, J. A. (2005). Analytical methods for detecting paralogy in molecular datasets. *Methods Enzymol.* **395,** 700–724.

De Queiroz, A. (1993). For consensus (sometimes). *Syst. Biol.* **42,** 368–372.

DeSalle, R. (2005). Animal phylogenomics: Multiple interspecific genome comparisons. *Methods Enzymol.* **395,** 104–133.

Dettaï, A., and Lecointre, G. (2004). In search of the Notothenioid relatives. *Antarctic Sci.* **16,** 71–85.

Dettaï, A., and Lecointre, G. (2005). Further support for the clades obtained by multiple molecular phylogenies in the acanthomorph bush. *C. R. Acad. Sci. B* **328,** 674–689.

Dingerkus, G., and Howell, W. M. (1976). Karyotypic analysis and evidence of tetraploidy in the North American paddlefish, *Polyodon spathula. Science* **194,** 842–844.

di Prisco, G., Eastman, J. T., Giordano, D., Parisi, E., and Verde, C. (2007). Biogeography and adaptation of Notothenioid fish: Hemoglobin function and globin-gene evolution. *Gene* (in press).

Everse, J., Vandegriff, K. D., and Winslow, R. M. (1994). Hemoglobins part B: Biochemical and analytical methods. *Methods Enzymol.* **231,** 725.

Felsenstein, J. (2004). "Inferring phylogenies." Sunderland, MA: Sinauer, Sunderland, MA.

Fetrow, J. S., and Bryant, S. H. (1993). New programs for protein tertiary structure prediction. *Bio/Technology* **11,** 479–484.

Filleul, A., and Lavoué, S. (2001). Basal teleosts and the question of elopomorph monophyly: Morphological and molecular approaches. *C. R. Acad. Sci. B.* **324,** 393–399.

Fitch, W. M. (1970). Distinguishing homologous from analogous proteins. *Syst. Zool.* **19,** 99–113.

Force, A., Lynch, M., Pickett, F. B., Amores, A., and Yan, Y. L. (1999). Preservation of duplicate genes by complementary, degenerative mutations. *Genetics* **151,** 1531–1545.

Friedman, M., Krull, L. H., and Cavins, J. F. (1970). The chromatographic determination of cystine and cysteine residues in proteins as *S*-β-(4-pyridylethyl) cysteine. *J. Biol. Chem.* **245,** 3868–3871.

Gascuel, O., and Steel, M. (2006). Neighbor-joining revealed. *Mol. Biol. Evol.* **23,** 1997–2000.

Gaucher, E. A., Gu, X., Miyamoto, M., and Benner, S. (2002). Predicting functional divergence in protein evolution by site-specific rate shifts. *Trend Biochem. Sci.* **27,** 315–321.

Giordano, D., Grassi, L., Parisi, E., Bargelloni, L., di Prisco, G., and Verde, C. (2006). Embryonic *b*-globin in the non-Antarctic notothenioid fish *Cottoperca gobio* (Bovichtidae). *Polar Biol.* **30,** 75–82.

Goloboff, P., Farris, S., and Nixon, K. (2000). TNT (tree analysis using new technology) (BETA) Tucumán, Argentina: Published by the authors, Tucumán, Argentina.

Goodman, M., Miyamoto, M. M., and Czelusniak, J. (1987). Pattern and process in vertebrate phylogeny revealed by coevolution of molecules and morphologies. *In* "Molecules and morphology in evolution: Conflict or compromise?" (C. Patterson, ed.), pp. 141–176. Cambridge University Press, New York.

Graybeal, A. (1994). Evaluating the phylogenetic utility of genes: A search for genes informative about deep divergences among vertebrates. *Syst. Biol.* **43,** 174–193.

Graybeal, A. (1998). Is it better to add taxa or characters to a difficult phylogenetic problem? *Syst. Biol.* **47,** 9–17.

Gu, X. (1999). Statistical methods for testing functional divergence after gene duplication. *Mol. Biol. Evol.* **16,** 1664–1674.

Gu, X. (2001). Maximum likelihood approach for gene family evolution under functional divergence. *Mol. Biol. Evol.* **18,** 453–464.
Gu, X. (2003). Functional divergence in protein (family) sequence evolution. *Genetica* **118,** 133–141.
Gu, X. (2006). A simple statistical method for estimating type-II (cluster-specific) functional divergence of protein sequences. *Mol. Biol. Evol.* **23,** 1937–1945.
Gu, X., and Vander Velden, K. (2002). DIVERGE: Phylogeny-bases analysis for functional-structural divergence of a protein family. *Bioinformatics* **18,** 500–501.
Guindon, S., and Gascuel, O. (2003). A simple, fast, and accurate algorithm to estimate large phylogenies by maximum likelihood. *Syst. Biol.* **52,** 696–704.
Hall, T. A. (1999). BioEdit: A user-friendly biological sequence alignment editor and analysis program for Windows 95/98/NT. *Nucl. Acids. Symp. Ser.* **41,** 95–98.
Hashiguchi, Y., and Nishida, M. (2006). Evolution and origin of vomeronasal-type odorant receptor gene repertoire in fishes. *BMC Evol. Biol.* **6,** 76.
Hedges, S. B., Dudley, J., and Kumar, S. (2006). TimeTree: A public knowledge-base of divergence times among organisms. *Bioinf.* **22,** 2971–2972.
Hendy, M. D., and Penny, D. (1989). A framework for the quantitative study of evolutionary trees. *Syst. Zool.* **38,** 297–309.
Higgins, D. G., and Sharp, P. M. (1988). CLUSTAL: A package for performing multiple sequence alignment on a microcomputer. *Gene* **73,** 237–244.
Hillis, D. M. (1987). Molecular versus morphological approaches to systematics. *Ann. Rev. Ecol. Syst.* **18,** 23–42.
Hillis, D. M., Pollock, D. D., McGuire, J. A., and Zwickl, D. J. (2003). Is sparse sampling a problem for phylogenetic inference? *Syst. Biol.* **52,** 124–126.
Hooft, R. W., Vriend, G., Sander, C., and Abola, E. E. (1996). Errors in protein structures. *Nature* **381,** 272.
Hurles, M. (2004). Gene duplication: The genomic trade in spare parts. *PLoS Biol.* **2,** e206.
Hurley, I. A., Lockridge Mueller, R., Dunn, K. A., Schmidt, E. J., Friedman, M., Ho, R. K., Prince, V. E., Yang, Z., Thomas, M. G., and Coates, M. I. (2007). A new time-scale for ray finned fish evolution. *Proc. R. Soc. B.* **274,** 489–498.
Inoue, J. G., Miya, M., Tsukamoto, K., and Nishida, M. (2003). Basal actinopterygian relationships: A mitogenomic perspective on the phylogeny of the "ancient fish." *Mol. Phylogenet. Evol.* **26,** 110–120.
Ishiguro, N. B., Miya, M., and Nishida, M. (2003). Basal euteleostean relationships: A mitogenomic perspective on the phylogenetic reality of the "Protacanthopterygii." *Mol. Phylogenet. Evol.* **27,** 476–488.
Janvier, P. (1996). "Early vertebrates." : Clarendon University Press, Oxford, UK.
Johnson, M. S., Srinivasan, N., Sowdhamini, R., and Blundell, T. L. (1994). Knowledge-based protein modeling. *Crit. Rev. Biochem. Mol. Biol.* **29,** 1–68.
Kikugawa, K., Katoh, K., Kuraku, S., Sakurai, H., Ishida, O., Iwabe, N., and Miyata, T. (2004). Basal jawed vertebrate phylogeny inferred from multiple nuclear DNA-coded genes. *BMC Biol.* **2,** 3.
Kluge, A. G. (1989). A concern for evidence and a phylogenetic hypothesis of relationships among Epicrates (Boidae, Serpentes). *Syst. Zool.* **38,** 7–25.
Kuo, M. W., Postlethwait, J., Lee, W. C., Lou, S. W., Chan, W. K., and Chung, B. C. (2005). Gene duplication, gene loss and evolution of expression domains in the vertebrate nuclear receptor NR5A (Ftz-F1) family. *Biochem. J.* **389,** 19–26.
Landon, M. (1977). Cleavage at aspartyl-prolyl bonds. *Methods Enzymol.* **47,** 145–149.
Lartillot, N., Brinkmann, H., and Philippe, H. (2007). Suppression of long-branch attraction artefacts in the animal phylogeny using a site-heterogeneous model. *BMC Evol. Biol.* **7,** S4.
Laskowski, R. A., MacArthur, M. W., Moss, D. S., and Thorntorn, J. M. (1993). PROCHECK: A program to check the stereochemical quality of proteins structures. *J. Appl. Crystallogr.* **26,** 283–291.

Lê, H. L. V., Lecointre, G., and Perasso, R. (1993). A 28S rRNA based phylogeny of the gnathostomes: First steps in the analysis of conflict and congruence with morphologically based cladograms. *Mol. Phylogenet. Evol.* **2,** 31–51.

Lecointre, G., Philippe, H., Van Le, H. L., and Le Guyader, H. (1993). Species sampling has a major impact on phylogenetic inference. *Mol. Phylogenet. Evol.* **2,** 205–224.

Long, M., Betran, E., Thornton, K., and Wang, W. (2003). The origin of new genes: Glimpses from the new and the old. *Nature Reviews Genet.* **4,** 865–875.

Lynch, M., and Conery, J. S. (2000). The evolutionary fate and consequences of duplicate genes. *Science* **290,** 1151–1155.

Maddison, M. P. (1997). Gene trees in species trees. *Syst. Biol.* **46,** 523–536.

Maddison, W. P., and Maddison, D. R. (2006). Mesquite: A modular system for evolutionary analysis (http://mesquiteproject.org).

Maruyama, K., Yasumasu, S., and Iuchi, I. (2004). Evolution of globin genes of the medaka *Oryzias latipes* (Euteleostei; Beloniformes; Oryziinae). *Mech. Dev.* **121,** 753–769.

Mathews, S. (2005). Analytical methods for studying the evolution of paralogs using duplicate gene datasets. *Methods Enzymol.* **395,** 724–745.

Mazzarella, L., Vergara, A., Vitagliano, L., Merlino, A., Bonomi, G., Scala, S., Verde, C., and di, Prisco. (2006a). High resolution crystal structure of deoxy haemoglobin from *Trematomus bernacchii* at different pH values: The role of histidine residues in modulating the strength of the Root effect. *Proteins Str. Funct. Bioinf.* **65,** 490–498.

Mazzarella, L., Bonomi, G., Lubrano, M. C., Merlino, A., Riccio, A., Vergara, A., Vitagliano, L., Verde, C., and di Prisco, G. (2006b). Minimal structural requirements for Root effect: Crystal structure of the cathodic hemoglobin isolated from the Antarctic fish *Trematomus newnesi. Proteins Str. Funct. Bioinf.* **62,** 316–321.

Miya, M., Satoh, T. P., and Nishida, M. (2005). The phylogenetic position of toadfishes (order Batrachoidiformes) in the higher ray-finned fish as inferred from partitioned Bayesian analysis of 102 whole mitochondrial genome sequences. *Biol. J. Linn. Soc.* **85,** 289–306.

Miya, M., Takeshima, H., Endo, H., Ishiguro, N. B., Inoue, J. G., Mukai, T., Satoh, T. P., Yamaguchi, M., Kawaguchi, A., Mabuchi, K., Shirai, S. M., and Nishida, M. (2003). Major patterns of higher teleostean phylogenies: A new perspective based on 100 complete mitochondrial DNA sequences. *Mol. Phylogenet. Evol.* **26,** 121–138.

Miyamoto, M. M., and Fitch, W. M. (1995). Testing species phylogenies and phylogenetic methods with congruence. *Syst. Biol.* **44,** 64–76.

Montgomery, J., and Clements, K. (2000). Disaptation and recovery in the evolution of Antarctic fishes. *Trends Ecol. Evol.* **15,** 267–271.

Near, T. J., Parker, S. K., and Detrich, H. W., III (2006). A genomic fossil reveals key steps in hemoglobin loss by the Antarctic fishes. *Mol. Biol. Evol.* **23,** 2008–2016.

Nei, M., and Kumar, S. (2000). "Molecular phylogenetics and evolution." New York: Oxford University Press, New York.

Nelson, J. S. (2006). "Fishes of the world." Hoboken, NJ: John Wiley and Sons, Hoboken, NJ.

Nixon, K. C. (1999a). The parsimony ratchet: A new method for rapid parsimony analysis. *Cladistics* **15,** 407–414.

Nixon, K. C. (1999b). Winclada (BETA) Ver. 0.9.9 Published by the author. Ithaca, NY.

Notredame, C., Higgins, D. G., and Heringa, J. (2000). T-Coffee: A novel method for fast and accurate multiple sequence alignment. *J. Mol. Biol.* **302,** 205–217.

O'Brien, K. M., and Sidell, B. D. (2000). The interplay among cardiac ultrastructure, metabolism and the expression of oxygen-binding proteins in Antarctic fishes. *J. Exp. Biol.* **203,** 1287–1297.

Ohno, S. (1970). "Evolution by gene duplication." Berlin: Springer-Verlag, Berlin.

O'Sullivan, O., Suhre, K., Abergel, C., Higgins, D. G., and Notredame, C. (2004). 3DCoffee: Combining protein sequences and structures within multiple sequence alignments. *J. Mol. Biol.* **340,** 385–395.

Page, R. D. M., and Holmes, E. C. (1998). "Molecular evolution: A phylogenetic approach," Abingdon, UK: Blackwell Science, Abingdon, UK.

Philippe, H., and Douzery, E. (1994). The pitfalls of molecular phylogeny based on four species, as illustrated by the Cetacea/Artiodactyla relationships. *J. Mam. Evol.* **2,** 133–152.

Posada, D., and Buckley, T. R. (2004). Model selection and model averaging in phylogenetics: Advantages of the AIC and Bayesian approaches over likelihood ratio tests. *Syst. Biol.* **53,** 793–808.

Posada, D., and Crandall, K. A. (1998). ModelTest: Testing the model of DNA substitution. *Bioinf.* **14,** 817–818.

Prince, V. E., and Pickett, F. B. (2002). Splitting pairs: The diverging fates of duplicated genes. *Nature Rev. Genet.* **3,** 827–837.

Riggs, A. (1988). The Bohr effect. *Annu. Rev. Physiol.* **50,** 181–204.

Robinson-Rechavi, M., Marchand, O., Escriva, H., Bardet, P. L., Zelus, D., Hughes, S., and Laudet, V. (2001). Euteleost fish genomes are characterized by expansion of gene families. *Genome Res.* **11,** 781–788.

Ronquist, F., and Huelsenbeck, J. P. (2003). MRBAYES 3: Bayesian phylogenetic inference under mixed models. *Bioinf.* **19,** 1572–1574.

Saitou, N., and Nei, M. (1987). The neighbor-joining method: A new method for reconstruction of phylogenetic trees. *Mol. Biol. Evol.* **4,** 406–425.

Sali, A., and Blundell, T. L. (1993). Comparative protein modelling by satisfaction of spatial restraints. *J. Mol. Biol.* **234,** 779–815.

Sambrook, J., Fritsch, E. F., and Maniatis, T. (1989). Molecular cloning: A laboratory manual 2d ed. Cold Spring Harbor Laboratory Press, Cold Spring Harbor, NY.

Sidell, B. D., and O'Brien, K. M. (2006). When bad thing happen to good fish: The loss of hemoglobin and myoglobin expression in Antarctic icefishes. *J. Exp. Biol.* **209,** 1791–1802.

Sidell, B. D., Vayda, M. E., Small, D. J., Moylan, T. J., Londraville, R. L., Yuan, M.-L., Rodnick, K. J., Eppley, Z. A., and Costello, L. (1997). Variation in the expression of myoglobin among species of the Antarctic icefishes. *Proc. Natl. Acad. Sci. USA* **94,** 3420–3424.

Sikes, D. S., and Lewis, P. O. (2001). PAUPRat: PAUP* implementation of the parsimony ratchet (Beta), Version 1 Storrs, CT: Distributed by the authors, Department of Ecology and Evolutionary Biology, University of Connecticut, Storrs, CT.

Simons, K. T., Bonneau, R., Ruczinski, I., and Baker, D. (1999). Ab initio protein structure prediction of CASP III targets using ROSETTA. *Proteins* **3,** 171–176.

Smith, W. L., and Wheeler, W. C. (2004). Polyphyly of the mail-cheeked fishes (Teleostei: Scorpaeniformes): Evidence from mitochondrial and nuclear sequence data. *Mol. Phylogenet. Evol.* **32,** 627–646.

Smith, W. L., and Wheeler, W. C. (2006). Venom evolution widespread in fishes: A phylogenetic road map for the bioprospecting of piscine venoms. *J. Hered.* **97,** 206–217.

Stiassny, M. L. J., Parenti, L. R., and Johnson, G. D. (1996). "The interrelationships of fishes." San Diego: Academic Press, San Diego.

Swofford, D. L. (2001). "PAUP*: Phylogenetic Analysis Using Parsimony (*and Other Methods)." Sunderland, MA: Version 4. Sinauer Associates, Sunderland, MA.

Takezaki, N., Figueroa, F., Zaleska-Rutczynska, Z., and Klein, J. (2003). Molecular phylogeny of early vertebrates: Monophyly of the agnathans as revealed by sequences of 35 genes. *Mol. Biol. Evol.* **20,** 287–292.

Takezaki, N., Figueroa, F., Zaleska-Rutczynska, Z., Takahata, N., and Klein, J. (2004). The phylogenetic relationship of tetrapod, coelacanth, and lungfish revealed by the sequences of 44 nuclear genes. *Mol. Biol. Evol.* **21,** 1512–1524.

Tamburrini, M., Brancaccio, A., Ippoliti, R., and di Prisco, G. (1992). The amino acid sequence and oxygen-binding properties of the single hemoglobin of the cold-adapted Antarctic teleost *Gymnodraco acuticeps*. *Arch. Biochem. Biophys.* **292,** 295–302.

Tamburrini, M., D'Avino, R., Fago, A., Carratore, V., Kunzmann, A., and di Prisco, G. (1996). The unique hemoglobin system of *Pleuragramma antarcticum*, an Antarctic migratory teleost. Structure and function of the three components. *J. Biol. Chem.* **271,** 23780–23785.

Taylor, J. S., Van de Peer, Y., Braasch, I., and Meyer, A. (2001). Comparative genomics provides evidence for an ancient genome duplication event in fish. *Philos. Trans. R. Soc. Lond. B Biol. Sci.* **356,** 1661–1679.

Taylor, J. S., Braasch, I., Frickey, T., Meyer, A., and Van de Peer, Y. (2003). Genome duplication, a trait shared by 22.000 species of ray-finned fish. *Genome Res.* **13,** 382–390.

Thompson, J. D., Gibson,T. J., Plewniak,F., Jeanmougin, F., and Higgins, D. G. (1997). The ClustalX windows interface: Flexible strategies for multiple sequence alignment aided by quality analysis tools. *Nucleic Acids Res.* **24,** 4876–4882.

Uyeno, T., and Smith, G. R. (1972). Tetraploid origin of the karyotype of catostomid fishes. *Science* **175,** 644–646.

Van de Peer, Y., Taylor, J. S., Joseph, J., and Meyer, A. (2002). Wanda: A database of duplicated fish genes. *Nucleic Acids Res.* **30,** 109–112.

Venkatesh, B., Erdmann, M. V., and Brenner, S. (2001). Molecular synapomorphies resolve evolutionary relationships of extant jawed vertebrates. *Proc. Nat. Acad. Sci. USA* **98,** 11382–11387.

Verde, C., Balestrieri, M., de Pascale, D., Pagnozzi, D., Lecointre, G., and di Prisco, G. (2006). The oxygen transport system in three species of the boreal fish family Gadidae. *J. Biol. Chem.* **283,** 22073–22084.

Verde, C., Howes, B. D., De Rosa, M. C., Raiola, L., Smulevich, G., Williams, R., Giardina, B., Parisi, E., and di Prisco, G. (2004). Structure and function of the Gondwanian hemoglobin of *Pseudaphritis urvillii*, a primitive notothenioid fish of temperate latitudes. *Prot. Sci.* **13,** 2766–2781.

Wiens, J. J. (1998). Combining data sets with different phylogenetic histories. *Syst. Biol.* **47,** 568–581.

Wittenberg, J. B., and Wittenberg, B. A. (1974). The choroid *rete* mirabilis. 1. Oxygen secretion and structure: Comparison with the swimbladder *rete* mirabile. *Biol. Bull.* **146,** 116–136.

Zaragueta-Bagils, R., Lavoué, S., Tillier, A., Bonillo, C., and Lecointre, G. (2002). Assessment of otocephalan and protacanthopterygian concepts in the light of multiple molecular phylogenies. *C. R. Acad. Sci. B* **325,** 1191–1207.

CHAPTER THIRTY-ONE

Tracing Globin Phylogeny Using PSIBLAST Searches Based on Groups of Sequences

Serge N. Vinogradov*

Contents

Abstract

PSIBLAST search of a protein database with a query sequence is a widely used tool for the detection of related but evolutionarily distant sequences. Iterations carried out until convergence (i.e., until the majority of the sequences most similar to the query sequence are extracted from the database) also produces an ordered list of more distantly related (i.e., less similar sequences) and false positives belonging to other protein families. Thus, a PSIBLAST search based on one group of globin sequences will provide sequences left in the database that are more distantly related (i.e., belong to other groups of globins) ordered according to the E value or bit score. The relative order of the scores should yield a clue to the relative similarity of the query group to the other groups of globins. Histograms of E values or bit scores from PSIBLAST searches using vertebrate myoglobins, cytoglobins, α- and β-globins, and neuroglobins as query groups show distributions that are congruent with the accepted phylogenetic tree. An illustration of more distant relationships is demonstrated by the results using neuroglobins as the query group, which show a striking similarity to bacterial single-domain globins and flavohemoglobins from bacteria and eukaryotes. Furthermore, it is observed that sequences belonging to the three undoubtedly ancient globin lineages form very broad distributions,

* Department of Biochemistry and Molecular Biology, School of Medicine, Wayne State University, Detroit, Michigan

Methods in Enzymology, Volume 436
ISSN 0076-6879, DOI: 10.1016/S0076-6879(08)36031-5

while recently evolved groups such as cytoglobins have narrow distributions. Thus, the breadth of a distribution of E values or bit scores for a query group may be related to its evolutionary age.

1. Introduction

The Basic Local Alignment Search Tool (BLAST), developed by S. Altschul at NCBI (Altshul *et al.*, 1990), is used extensively in database searches for sequences remotely similar to the query sequence. BLAST employs a heuristic word method for rapid pairwise alignment of the query sequence to all the sequences in a given database. The program finds short stretches, two or three, of identical or almost-identical amino acids (words) in two sequences. Once the word match is found, a longer alignment is obtained by extending the search, without inclusion of gaps, on either side of the matching segment, within the range of the scoring threshold, with the scoring based on a given substitution matrix. The extension is continued until the score of the alignment decreases to below the scoring threshold. A significant improvement in the algorithm came with gapped BLAST (Altschul and Gish, 1996), which introduces gaps into the alignment and thus leads to increased sensitivity in finding distantly related proteins. Here extension can be continued after introduction of a gap, and the overall score is allowed to decrease below the threshold, provided it is temporary, and is then followed by increase to above the threshold. Both the number of gaps allowed and their lengths are controlled by the inclusion of suitably selected penalties in scoring the sequence alignment (Altshul, 1998). The ability of BLAST searches to find distantly related proteins was greatly improved by the development of Position-Specific Iterated BLAST (PSIBLAST) (Altschul *et al.*, 1997). This search first uses a single protein sequence as query and compares it to all the sequences in a protein database employing the gapped BLAST algorithm, essentially a simple BLASTP search, generating an initial list of similarity hits. In the next step, all the hits scoring above a specified score, or selected manually, are used to generate a new position specific substitution matrix, from which an amino acid sequence profile is created and used to identify additional members of the same protein family that match the profile. The newly identified sequences can then be added to the first multiple sequence alignment to generate a new substitution matrix, which can in turn be used in still another cycle of searching the database. The iterations can be repeated until no new sequences are identified or until all the members of a given group of sequences are retrieved. A more recent improvement was introduced to account for bias in the amino acid composition of the proteins whose sequences are aligned, via the use of appropriately modified substitution matrices (Altschul *et al.*, 2005; Schaffer *et al.*, 2001; Yu *et al.*, 2003, 2006). PSI-BLAST is one of the most powerful tools for detecting remote sequence similarity (Altschul and Koonin, 1998; McGinnis and Madden, 2004) and structural homology using only sequence information (Park *et al.*, 1998; Salamov *et al.*, 1999). It has been used widely in the annotation of genomes (Teichmann *et al.*, 1999; Muller *et al.*, 1999). PSIBLAST was also found to be useful in the recognition of protein fold at low sequence identity (Stevens, 2005; Stevens *et al.*, 2005). Not surprisingly, PSIBLAST appears to provide some of the best alignments possible (Elofsson, 2002).

2. Statistical Evaluation of Sequence Similarity

Once BLAST finds a sequence similar to the query sequence, it provides two quantitative measures of the quality of the alignment based on statistical theory. The expect (E value) is an indication of the statistical significance of a pairwise alignment of a query sequence comprised of n amino acid residues with another sequence in a database containing a total of m residues. It is provided by the following equation:

$$E = m \times n \times P \qquad (31.1)$$

where P is the probability that the alignment is the result of random chance. Thus, an alignment of a query sequence 100aa long to another sequence in a database containing 10^{14} residues with a calculated $P = 1 \times 10^{-25}$, yields $E = 100 \times 10^{14} \times 10^{-25} = 10^{-9}$. The E value is expressed in the BLAST output as 1e-9, which indicates that the probability of this sequence match due to random chance is 10^{-9}. Thus, the lower the E value, the less probable it is that a given sequence alignment is random. In general, E values lower than 10^{-2} are considered to indicate similarity between the two aligned sequences.

The BLAST program provides another statistical indicator, the bit score S′, which is independent of the query length and the size of the database:

$$S' = (S \times \lambda - \ln K)/\ln 2 \qquad (31.2)$$

where S is the raw score, λ is a distribution constant, and K is a parameter associated with the scoring system. Here, the higher the bit score, the more significant the alignment. Thus, the bit score provides a constant measure of the statistical significance of a match, across different searches with different databases or the same database at different times, as the database increases.

An amino acid substitution matrix is used to score every pair of aligned residues and the number of gaps and their lengths. Although PSIBLAST offers the use of several different substitution matrices, the default setting uses the BLOSUM62 matrix (Henikoff and Henikoff, 1992), based on blocks of aligned protein sequences with less than 62% identity. This matrix is considered the best choice in searching for distantly related sequences. Because bit scores are normalized, it is possible to compare bit scores from different alignments, even if different scoring matrices have been used.

PSIBLAST uses a default E-value threshold of 0.005 and allows additional iterations automatically for all selected sequences above the default or other operator-selected threshold. It is important to realize that inclusion of even a single false positive can affect the results of subsequent iterations. Thus, the pairwise alignments selected in each iteration should be examined to avoid inclusion of defective sequences (e.g. fragments) or of unrelated proteins. In the case of searching for globin sequences, a false positive would be any sequence that does not have an appropriate His at position F8.

3. Verification of Globin Sequence Alignments

PSIBLAST searches sometimes miss alignments at the N- and C-terminal regions of globin sequences, because of greater variability in those regions. Furthermore, they also sometime misalign the E helices when the E7 residue in the query sequence is different from that in the target sequence. These minor failings can be overlooked, however, if the rest of the alignment is unaffected. In cases where the identification of a putative globin is uncertain, a search employing FUGUE (Shi *et al.*, 2001) (www.cryst.bioc.cam.ac.uk) is used to determine whether the borderline sequence can be accepted as a globin.

In order to determine whether a pairwise alignment produced by the BLAST search is acceptable, particularly at E values close to the threshold, I use the pattern of conserved and mostly hydrophobic residues employed earlier (Kapp *et al.*, 1995), based on the myoglobin fold (Lesk and Chothia, 1980; Bashford *et al.*, 1987). This pattern consists of 37 conserved, solvent-inaccessible positions (Gerstein *et al.*, 1994), 33 intrahelical residues defining helices A through H: A8, A11, A12, A15, B6, B9, B10, B13, B14, C4, E4, E7, E8, E11, E12, E15, E18, E19, F1, F4, G5, G8, G11, G12, G13, G15, G16, H7, H8, H11, H12, H15, and H19; the three interhelical residues at CD1, CD4, and FG4; and the invariant His at F8. Although earlier alignments (Kapp *et al.*, 1995) had indicated that there were two invariant residues in globins, F8His and CD1Phe, the latest genomic information on globin sequences present in the three kingdoms of life (Vinogradov *et al.*, 2005, 2006) indicate that other hydrophobic residues, such as Tyr/Met/Leu/Ile/Val can occur at the CD1 position and that Ala/Ser/Thr/Leu/Val/Ty can occur at the distal E7 position, in addition to His and Gln observed previously.

4. A New Paradigm in Globin Classification

Currently, the classification of globins in all the major data banks is in a state of complete disarray. Because the purpose of this report is a description of the potential usefulness of PSI-BLAST searches based on groups of globin sequences, to detect similarity to sequences from other groups of globins, it is necessary to consider a rational classification of presently known globins. On the basis of the results obtained in recently conducted surveys of genomic data from the three kingdoms of life (Vinogradov *et al.*, 2005, 2006), it is now clear that the globin superfamily consists of three lineages or families of globins. The flavohemoglobins, both chimeric (FHbs) and single domain (SDgbs), and globin coupled sensors, chimeric (GCS), single-domain sensor globins (SDSgbs), and single-domain protoglobins (Pgbs) all have the canonical 3/3 α-helical myoglobin fold. The third family has a truncated 2/2 α-helical fold and consists of class 1, 2, and 3 single-domain globins. Bacteria is the only kingdom to have representatives of all eight subgroups. Only about 25% of Archaeal genomes have globins comprised of the three subgroups of the GCS family and class 1 2/2 globins. Within the eukaryotes, the metazoan globins appear to be mostly related to the FHb lineage or family, as will be seen later. Table 31.1 summarizes the classification of globins in Bacteria, Archaea, and nonmetazoan eukaryotes on the basis of the three lineages. Among the lower eukaryotes, Fungi have FHbs and SDSgbs; the diplomonads, euglenozoans, and mycetozoans have FHbs; the Ciliates and Chlorophytes (green algae) have class 1 2/2 Hbs; the Rhodophytes (brown algae) have class 2 2/2Hbs; the plants have symbiotic and nonsymbiotic Hbs as well as class 2 2/2Hbs; and stramenopiles (diatoms) have FHbs, SDgbs, and class 1 and 2 2/2Hbs. Metazoan globins are commonly subdivided at present into vertebrate, or rather chordate, and invertebrate hemoglobins. The former comprise vertebrate α- and β-globins, myoglobins (Mbs), and the recently identified neuroglobins (Ngbs) (Burmester *et al.*, 2000), cytoglobins (Cygbs) (Burmester *et al.*, 2002), globin X in fish and amphibians (Roesner *et al.*, 2004), and globin E in chicken (Kugelstadt *et al.*, 2004), as well as hyperoartia (lamprey), hyperotreti (hagfish), and urochordate globins. The invertebrate globins are commonly divided into annelid intracellular, annelid extracellular, crustacean, echinoderm, echiuran, insect, nematode, nemertean, mollusk, and platyhelminth groups (Weber and Vinogradov, 2001). Thus, the total number of identifiable subgroups of globins, including the 29 from

Table 31. 1 Distribution of globin lineages and their subgroups in Bacteria, Archaea, and nonmetazoan Eukaryotes (Vinogradov *et al.*, 2005, 2006, 2007)

	3/3 FHbs		3/3 GCSs			2/2 Hbs		
Kingdom of Life	FHb	SDFgb	GCS	SDSgb	Pgb	1	2	3
Bacteria	×	×	×	×	×	×	×	×
Archaea								
Crenarchaeota	–	–	–	–	×	–	–	–
Euryarchaeota	–	–	×	×	×	×	–	–
Nonmetazoan Eukaryotes								
Fungi	×	–	–	×	–	–	–	–
Chlorophytes	–	–	–	–	–	×	–	–
Ciliates	–	–	–	–	–	×	–	–
Diplomonadida	×	–	–	–	–	–	–	–
Euglenozoa	×	–	–	–	–	–	–	–
Mycetozoa	×	–	–	–	–	–	–	–
Rhodophytes	–	×	–	–	–	–	–	–
Plants	–	×[a]	–	–	–	–	×	–
Stramenopiles	×	×	–	–	–	×	×	–

[a] Plant Hbs related to the FHb lineage, include symbiotic Hbs (legHbs) and nonsymbiotic class 1 and 2 Hbs.

Table 31.1, is close to 50. Further subdivision is also possible (e.g., of vertebrate Mbs into mammalian, fish, bird, etc.). At present, apart from vertebrate globins, we know very little about the phylogenetic relationships among all these groups of globins. Furthermore, given the paucity of genomic information about unicellular eukaryotes, our knowledge of their hemoglobins is sparse at best (Vinogradov *et al.*, 2006), and is certain to change in the future.

5. Tracing Phylogenetic Relationships via PSIBLAST Searches

The main purpose of this chapter is to illustrate the use of PSIBLAST searches available at the NCBI Web site (www.ncbi.nlm.nih.gov/BLAST/), to elucidate the phylogenetic relationships among the various groups of globins. The site provides several data banks, of which the nr, refseq, and SwissProt were used. The nr comprises nonredundant GenBank CDS translations + PDB + SwissProt + PIR + PRF, excluding those in env_nr, and refseq consists of nonredundant, curated sequences (Pruitt *et al.*, 2005).

Although the former includes all the latest available sequences, which is not the case with refseq, its shortcomings are a substantial number of fragments and redundant sequences and an overwhelmingly large number of vertebrate globin sequences, particularly α- and β-globins.

Because the phylogeny of the vertebrate globins is reasonably well established and it is generally agreed that they are monophyletic (Gillemans *et al.*, 2003; Hardison, 2001), PSIBLAST searches using Mbs, Cygbs, and Ngbs as query groups were used to provide a validation of the proposed method. Figure 31.1 shows a phylogenetic tree of vertebrate globins, based on Gillemans *et al.* (2003), modified to include glbX (Roesner *et al.*, 2005) and glbE (Kugelstadt *et al.*, 2004). Although the exact branch points for glbX and glbE have not been established, it is known that the former is related to Ngbs and the latter to Cygbs and Mbs. The results of a PSIBLAST search using the vertebrate Mbs is shown in Figure 31.2. With only two iterations required, the histogram of bit scores demonstrates that Mbs recognize glbE first (blue at $S' = 209$–188), ahead of the Cygbs at $S' = 187$–135, followed by α-globins at $S' = 119$, then β-globins starting at $S' = 109$–100 and hyperoartia (lamprey) globins at $S' = 109$–90. The glbX, hyperotreti (hagfish) globins, and metazoan globins start at $S' = 79$–70, and the plant Hbs are last at $S' = 49$–40. Surprisingly, the Mbs do not recognize the Ngbs. Figure 31.3 shows the histogram obtained with Cygbs as the query group. The order of recognition is glbE

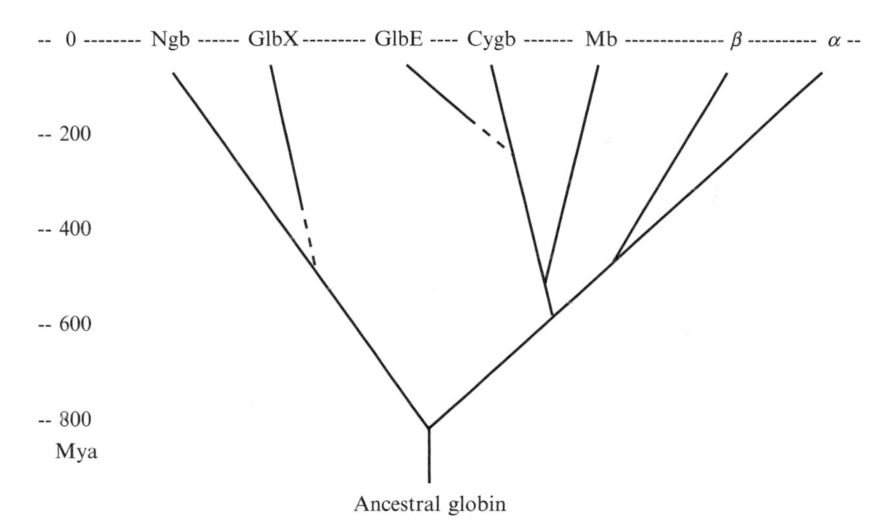

Figure 31.1 Diagram of the phylogenetic tree of vertebrate globins adapted from Gillemans *et al.* (2003), and including GlbE from chicken (Kugelstadt *et al.*, 2004) and GlbX from fish and amphibians (Roesner *et al.*, 2005). The dotted line indicates uncertainty in the location of the branch points.

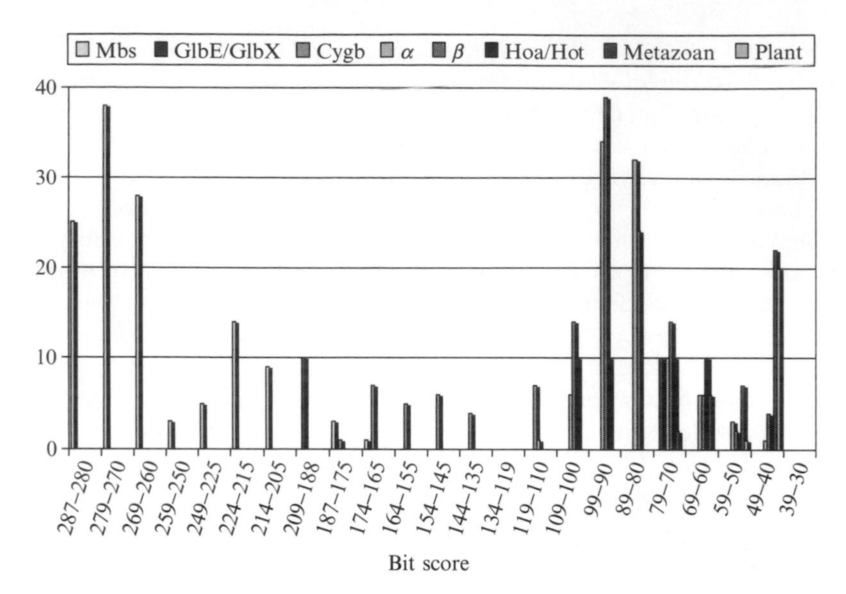

Figure 31.2 Histogram of the results of second iteration PSIBLAST search versus all organisms (using the nr database) based on all vertebrate myoglobins (query = human myoglobin [gi|44955888| NP.976312]). The y-axis value for GlbE and GlbX (both blue) and the hyperoartia (lamprey) and hyperotrite (hagfish) (both black) were set arbitrarily at 10 to make their locations more visible. GlbE is recognized ahead of the cytoglobins (blue at S′ = 209–188), and GlbX occur at bit score S′ = 79–70. The hyperoartia occur at bit scores S′ = 109–90 and the hyperotrite at S′ = 79–60. The globins of the urochordate *Ciona* (not shown), were recognized at S′ = 59–50. Note that the neuroglobins were not recognized. The y-axis values for the α- and β-globins were scaled down to ≈{1/5}, and for the metazoan globins, to ≈{1/3} of the actual numbers. The lowest level of bit scores, 49–40, corresponds to E values of 2e-5-0.035. Metazoans are all the invertebrate groups (e.g., annelid, arthropods, echinoderms, insects, mollusks, nematodes, platyhelminths). The plant hemoglobins include the symbiotic globins (leghemoglobins) and the class 1 and 2 nonsymbiotic hemoglobins.

at S′ = 149–140, followed by the hyperoartia (lamprey) Hbs at S′ = 129–100, Mbs and α-globins starting at S′ = 109–100, the β-globins starting at S′ = 99–90, the hyperotreti (hagfish) globins at S′ = 99–70, and glbX at S′ = 89–70; the metazoan Hbs start at S′ = 79–70, the plant Hbs at S′ = 69–60, and the Ngbs at S′ = 59–40. Figure 31.4 presents the results of a search with Ngbs as the query group. The first to be recognized are the domains of the multidomain globin of the sea urchin *Strongylus* at S′ = 103–80, followed by the bacterial SDgbs starting at S′ = 99–90, glbX at S′ = 79–60 and the globins of the urochordate *Ciona* at S′ = 69–40. Next come the bacterial and eukaryote FHbs, the β-globins, the metazoan and plant Hbs at S′ = 69–60, and the Cygbs and α-globins at S′ = 59–40. In agreement with the results shown in Figure 31.2, Mbs are not recognized by the Ngbs. The most

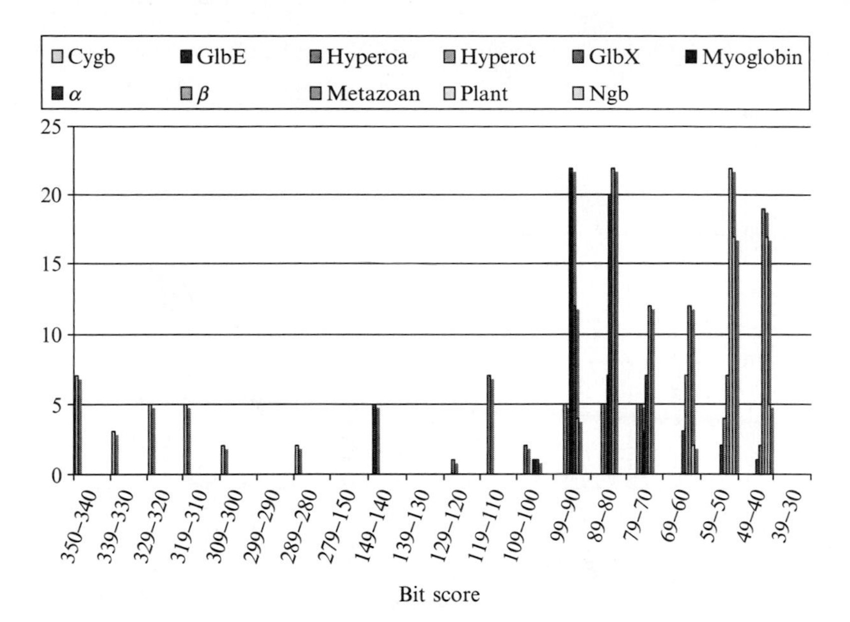

Figure 31.3 Histogram of the results of the second iteration PSIBLAST search versus all organisms (using the nr database) based on the vertebrate cytoglobins (query = human cytoglobin [gi|21263504|sp| Q8WWM9|CYGB.HUMAN]). GlbE scaled up × 10, is the first globin recognized at 149–140, followed by the hyperoartia (lamprey) at bit-score level S′ = 129–100. Next, the myoglobins, α- and β-globins start at bitscore level S′ = 109–100, followed by the hyperotreti (hagfish), height set at 5, at 99–70. Glbx height set at 5, occur at S′ = 89–70. The plant hemoglobins and the globins of the urochordate *Ciona* (not shown), are recognized at S′ = 69–60 and the neuroglobins at S′ = 59–40. The y-axis values for the α- and β-globins and the flavohemoglobins were scaled down to ≈{1/3} of the actual numbers.

unexpected finding here is the similarity of the Ngbs to the sea urchin globin and the bacterial SDgbs and FHbs. This finding is corroborated by the results of PSIBLAST searches based on all the FHbs and related SDgbs shown in Figure 31.5. Again, the *Strongylus* globins are recognized first at S′ = 119–80, followed by Ngbs at S′ = 99–70, glbX at S′ = 89–60, and *Ciona* globins at S′ = 79–50. The Cygbs occur at S′ = 69–50, the metazoan and plant sequences starting at S′ = 79–70, and the α-globins and β-globins at S′ = 69–60. The Mbs are not recognized in this search either.

The unexpectedly close similarity of vertebrate Ngbs to the bacterial and eukaryote FHbs demonstrated by the PSIBLAST search results presented in Figures 31.4 and 31.5 represents the principal justification for the proposal that vertebrate, and likely all metazoan globins, are derived from a bacterial SDgb from the FHb family (Vinogradov *et al.*, 2005, 2007). There is at present no satisfactory explanation for the nonrecognition of Mbs by the Ngbs and the FHbs.

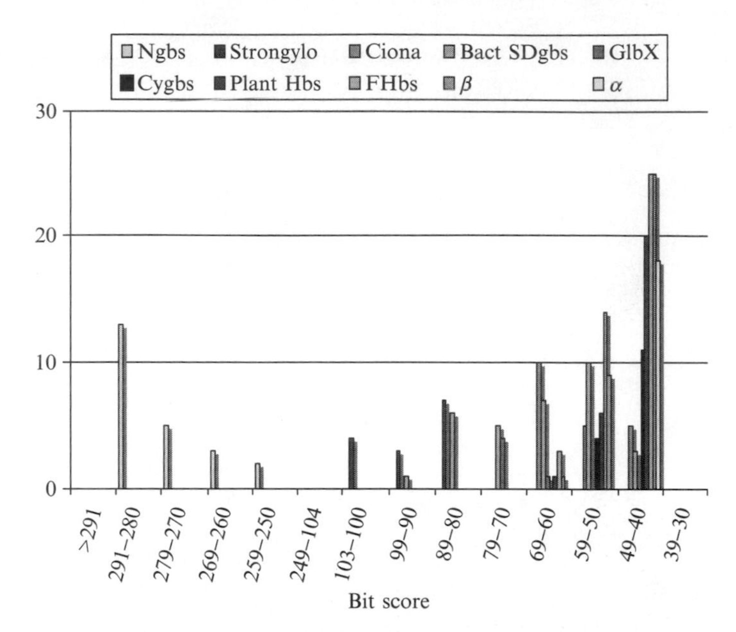

Figure 31.4 Histogram of the results of the second iteration PSIBLAST search versus all organisms (using the nr database) based on the vertebrate neuroglobins (query = human neuroglobin [gi|32171399tr|Q9NPG2]). The y-axis values for the α- and β-globins were scaled down to x≈{1/5} and the flavohemoglobins to x≈{1/3} of their real numbers. Metazoan globins (starting at bit score S′ = 69–60) were omitted from the plot. Note that the neuroglobins first recognize the individual domains of the 2552aa multidomain globin of the sea urchin *Strongylocentrotus purpuratus* (Echinodermata) (gi|115913844|XP_001199205), at S′ = 103–100, 99–90 and 89–90, followed by bacterial SDgbs, starting at S′ = 99–90 and the four globins (scaled ×5) of the urochordate *Ciona intestinalis* at S′ = 69–60, 59–50 and 49–40. The cytoglobins are recognized at S′ = 59–50 and 49–40. Note that no myoglobins are recognized.

The results shown in Figures 31.2 to 31.5, provide an additional snippet of useful phylogenetic information. The broad and heterogeneous distribution evinced by the FHbs and SDgbs in Figure 31.5, ranging in bit scores from 224 to 80, is also found in PSIBLAST searches based on representatives of the two other globin lineages/families, the globin coupled sensors and 2/2 globins (data not shown). All three lineages/families are ancient ones, likely going back to the emergence of the bacteria (Vinogradov *et al.*, 2007). Consequently, it appears reasonable to ascribe a narrower distribution of scores to more recently evolved globin sequences, as exemplified by the cytoglobins in Figures 31.2 to 31.5. Thus, the width of the distribution of scores obtained with a group of sequences could provide a clue to their evolutionary age.

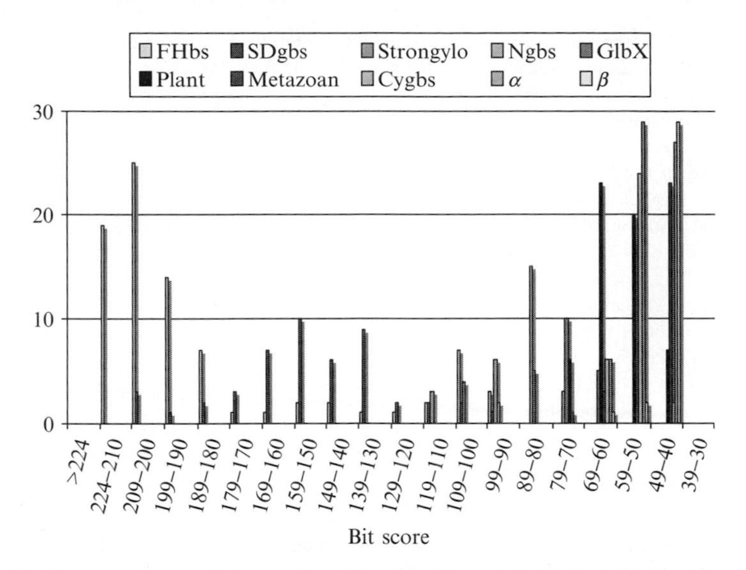

Figure 31.5 Histogram of the results of the third iteration PSIBLAST search versus all organisms (using the nr database) based on the bacterial and eukaryote flavohemoglobins and bacterial single-domain globins (query = single domain globin from *Aquifex aeolicus* gi|15605769|NP_213146). The y-axis values for the α- and β-globins were scaled down to $\approx\{1/5\}$ and the flavohemoglobins to $\approx\{1/3\}$ of their real numbers. Note that the flavohemoglobins and single-domain globins first recognize the individual domains (scaled up $\times$ 2) of the 2552aa multidomain globin of the sea urchin *Strongylocentrotus purpuratus* (Echinodermata) (gi|115913844|XP_001199205), at $S' = 119–80$, followed by neuroglobins at $S' = 99–60$. GlbX (scaled up $\times$ 5) is recognized at $S' = 89–60$. The plant and metazoan globins are recognized starting at $S' = 79–70$, ahead of the cytoglobins at $S' = 69–60$. GlbE (not shown) is recognized at $S' = 59–50$ and the globins of the urochordate *Ciona* (not shown), are recognized at $S' = 79–50$. Vertebrate myoglobins are not recognized.

REFERENCES

Altschul, S. F. (1998). Generalized affine gap costs for protein sequence alignment. *Proteins* **32,** 88–96.

Altschul, S. F., and Gish, W. (1996). Local alignment statistics. *Methods Enzymol.* **266,** 460–480.

Altschul, S. F., and Koonin, E. V. (1998). Iterated profile searches with PSI-BLAST—a tool for discovery in protein databases. *Trends Biochem. Sci.* **23,** 444–447.

Altschul, S. F., Gish, W., Miller, W., Myers, E. W., and Lipman, D. J. (1990). Basic local alignment search tool. *J. Mol. Biol.* **215,** 403–410.

Altschul, S.F, Madden, T. L., Schaffer, A. A., Zhang, J., Zhang, Z., Miller, W., and Lipman, D. J. (1997). Gapped BLAST and PSI-BLAST: A new generation of protein database search programs. *Nucleic Acids Res.* **25,** 33389–33402.

Altschul, S. F., Wootton, J. C., Gertz, E. M., Agarwala, R., Morgulis, A., Schaffer, A. A., and Yu, Y. K. (2005). Protein database searches using compositionally adjusted substitution matrices. *FEBS J.* **272,** 5101–5109.

Bashford, D., Chothia, C., and Lesk, A. M. (1987). Determinants of a protein fold: Unique features of the globin amino acid sequences. *J. Mol. Biol.* **196,** 199–216.

Burmester, T., Ebner, B., Weich, B., and Hankeln, T. (2002). Cytoglobin: A novel globin type ubiquitously expressed in vertebrate tissues. *Mol. Biol. Evol.* **19,** 416–421.

Burmester, T., Weich, B., Reinhardt, S., and Hankeln, T. (2000). A vertebrate globin expressed in the brain. *Nature* **407,** 520–523.

Elofsson, A. (2002). A study on protein sequence alignment quality. *Proteins* **46,** 330–339.

Gerstein, M., Sonnhammer, E. L. L., and Chothia, C. (1994). Volume changes in protein evolution. *J. Mol. Biol.* **236,** 1067–1078.

Gillemans, N., McMorrow, T., Tewari, R., Wai, A., Burgtorf, C., Drabek, D., Ventress, N., Langeveld, A., Higgs, D., Tan-Un, K., Grosveld, F., and Philipsen, S. (2003). Functional and comparative analysis of globin loci in pufferfish and humans. *Blood* **101,** 2842–2849.

Hardison, R. C. (2001). Organization, evolution and regulation of the globin genes. "Disorders of hemoglobin"(Steinberg, M. H., Forget, B. G., Higgs, D. R., and Nagel, R. L.), 95–116. Cambridge University Press, Cambridge, UK.

Henikoff, S., and Henikoff, J. G. (1992). Amino acid substitution matrices from protein blocks. *Proc. Natl. Acad. Sci. USA* **89,** 10915–10919.

Kapp, O. H., Moens, L., Vanfleteren, J., Trotman, C., Suzuki, T., and Vinogradov, S. N. (1995). Alignment of 700 globin sequences: Extent of amino acid substitution and its correlation in volume. *Protein Sci.* **4,** 2179–2190.

Kugelstadt, D., Haberkamp, M., Hankeln, T., and Burmester, T. (2004). Neuroglobin, cytoglobin, and a novel, eye-specific globin from chicken. *Biochem. Biophys. Res. Commun.* **325,** 719–725.

Lesk, A. M., and Chothia, C. (1980). How different amino acid sequences determine similar protein structures: The structure and evolutionary dynamics of the globins. *J. Mol. Biol.* **136,** 225–270.

McGinnis, S., and Madden, T. L. (2004). BLAST: At the core of a powerful and diverse set of sequence analysis tools. *Nucleic Acids Res.* **32,** W20–W25.

Muller, A., MacCallum, R. M., and Sternberg, M. J. (1999). Benchmarking PSI-BLAST in genome annotation. *J. Mol. Biol.* **293,** 1257–1271.

Park, J., Karplus, K., Barrett, C., Hughey, R., Haussler, D., Hubbard, T., and Chothia, C. (1998). Sequence comparisons using multiple sequences detect three times as many remote homologues as pairwise methods. *J. Mol. Biol.* **284,** 1201–1210.

Pruitt, K. D., Tatusova, T., and Maglott, D. R. (2005). NCBI Reference Sequence (RefSeq): A curated non-redundant sequence database of genomes, transcripts and proteins. *Nucleic Acids Res.* **33,** D501–D504.

Roesner, A., Fuchs, C., Hankeln, T., and Burmester, T. (2005). A globin gene of ancient evolutionary origin in lower vertebrates: Evidence for two distinct globin families in animals. *Mol. Biol. Evol.* **22,** 12–20.

Salamov, A. A., Suwa, M., Orengo, C. A., and Swindells, M. B. (1999). Genome analysis: Assigning protein coding regions to three-dimensional structure. *Protein Sci.* **8,** 771–777.

Schaffer, A. A., Aravind, L., Madden, T. L., Shavirin, S., Spouge, J. L., Wolf. Y. I., Koonin, E. V., and Altschul, S. F. (2001). Improving the accuracy of PSI-BLAST protein database searches with composition-based statistics and other refinements. *Nucleic Acids Res.* **29,** 2994–3005.

Shi, J., Blundell, T., and Mizuguchi, K. (2001). FUGUE: Sequence-structure homology recognition using environment-specific substitution tables and structure-dependent gap penalties. *J. Mol. Biol.* **310,** 243–257.

Stevens, F. J. (2005). Efficient recognition of protein fold at low sequence identity by conservative application of Psi-BLAST: Validation. *J. Mol. Recognit.* **18,** 139–149.

Stevens, F. J., Kuemmel, C., Babnigg, G., and Collart, F. R. (2005). Efficient recognition of protein fold at low sequence identity by conservative application of Psi-BLAST: Application. *J. Mol. Recognit.* **18,** 150–157.

Teichmann, S. A., Chothia, C., and Gerstein, M. (1999). Advances in structural genomics. *Curr. Opin. Struct. Biol.* **9,** 390–399.

Vinogradov, S. N., Hoogewijs, D., Bailly, X., Dewilde, S., Moens, L., and Vanfleteren, J. R. (2007). A model of globin evolution. *Gene* (submitted).

Vinogradov, S. N., Hoogewijs, D., Bailly, X., Arredondo-Peter, R., Gough, J., Dewilde, S., Moens, L., and Vanfleteren, J. R. (2006). A phylogenomic profile of globins. *BMC Evol. Biol.* **6,** 31–41.

Vinogradov, S. N., Hoogewijs, D., Bailly, X., Arredondo-Peter, R., Gough, J., Guertin, M., Dewilde, S., Moens, L., and Vanfleteren, J. R. (2005). Three globin lineages belonging to two structural classes in genomes from the three kingdoms of life. *Proc. Natl. Acad. Sci. USA* **102,** 11385–11389.

Weber, R. E., and Vinogradov, S. N. (2001). Nonvertebrate hemoglobins: Functions and molecular adaptations. *Physiol. Rev.* **181,** 568–629.

Yu, Y. K., Wootton, J. C., and Altschul, S. F. (2003). The compositional adjustment of amino acid substitution matrices. *Proc. Natl. Acad. Sci. USA* **100,** 15688–15693.

Yu, Y. K., Gertz, E. M., Agarwala, R., Schaffer, A. A., and Altschul, S. F. (2006). Retrieval accuracy, statistical significance and compositional similarity in protein sequence database searches. *Nucleic Acids Res.* **34,** 5966–5973.

Author Index

D

G

H

I

L

M

P

Q

R

S

W

Subject Index

L

M

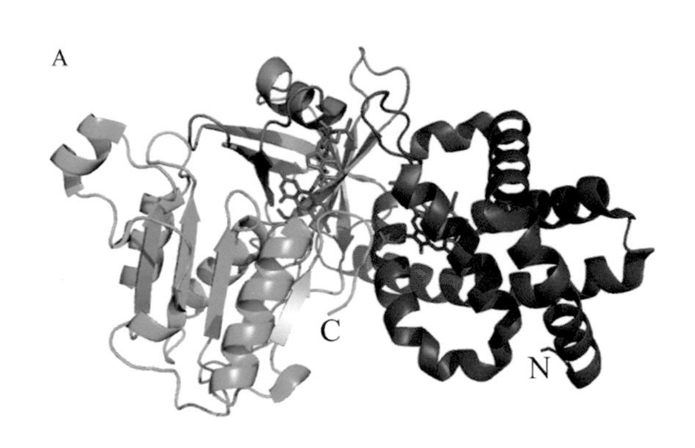

B

```
          |       A         | |         B         |            |
FHP  MLTQKTKDIVKATAPVLAEHGYDIIKCFYQRMFEAHPELKNVFNMAHQEQGQQQQALARA  60
HMP  MLDAQTIATVKATIPLLVETGPKLTAHFYDRMFTHNPELKEIFNMSNQRNGDQREALFNA  60
VHB  MLDQQTINIIKATVPVLKEHGVTITTTFYKNLFAKHPEVRPLFDMGRQESLEQPKALAMT  60

       E    | |        F         | |        G         |    |
FHP  VYAYAENIEDPNSLMAVLKNIANKHASLGVKPEQYPIVGEHLLAAIKEVLGNAATDDIIS 120
HMP  IAAYASNIENLPALLPAVEKIAQKHTSFQIKPEQYNIVGEHLLATLDEMFSPG--QEVLD 118
VHB  VLAAAQNIENLPAILPAVKKIAVKHCQAGVAAAHYPIVGQELLGAIKEVLGDAATDDILD 120

             H                | |   Fβ1    |  |  Fβ2    |
FHP  AWAQAYGNLADVLMGMESELYERSAEQPGGWKGWRTFVIREKRPESDVITSFILEPADGG 180
HMP  AWGKAYGVLANVFINREAEIYNENASKAGGWEGTRDFRIVAKTPRSALITSFELEPVDGG 178
VHB  AWGKAYGVIADVFIQVEADLYAQAVE---------------------------------- 146

            | Fβ3 |       | Fβ4 |      | Fβ5 |      | Fα1
FHP  PVVNFEPGQYTSVAIDVPALGLQQIRQYSLSDMPNGRSYRISVKREGGGPQPPGYVSNLL 240
HMP  AVAEYRPGQYLGVWLKPEGFPHQEIRQYSLTRKPDGKGYRIAVKREEG-----GQVSNWL 233
VHB  ------------------------------------------------------------

     |    | Fβ6 |     | Nα0 || Nβ1 |     | Nα1  |  |  N
FHP  HDHVNVGDQVKLAAPYGSFHIDVDAKTPIVLISGGVGLTPMVSMLKVALQAP-PRQVVFV 299
HMP  HNHANVGDVVKLVAPAGDFFMAVADDTPVTLISAGVGQTPMLAMLDTLAKAGHTAQVNWF 293
VHB  ------------------------------------------------------------

     β2 | |    Nα2     | Nβ3  | | Nα0B || Nβ6 | | Nα4 |  |N
FHP  HGARNSAVHAMRDRLREAAKTYENLDLFVFYDQPLPEDVQGRDYDYPGLVDVKQIEKSIL 349
HMP  HAAENGDVHAFADEVKELGQSLPRFTAHTWYRQPSEADRAKGQFDSEGLMDLSKLEGAFS 343
VHB  ------------------------------------------------------------

     α5 || Nβ4 ||    Nα5    |  |Nβ5 |
FHP  LPDADYYICGPIPFMRMQHDALKNLGIHEARIHYEVFGPDLFAE                 403
HMP  DPTMQFYLCGPVGFMQFTAKQLVDLGVKQENIHYECFGPHKVL-                 396
VHB  --------------------------------------------
```

Figure legend follows on next page

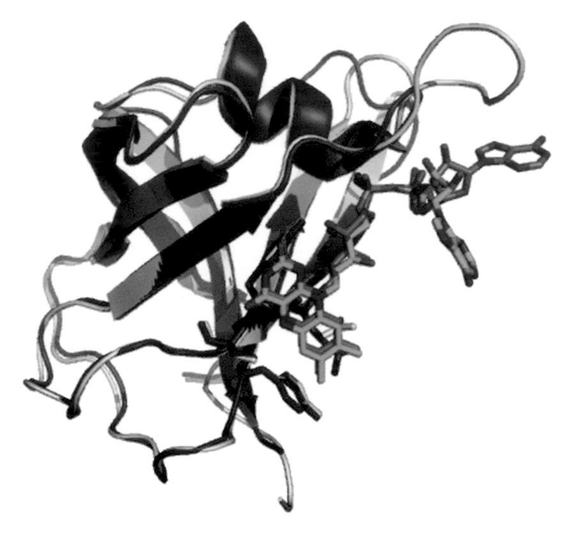

Figure 10.2 Structural overlay of the FAD-binding domain of *A. eutrophus* (FHP, yellow) and *E. coli* (HMP, red) flavohemoglobins. The FAD molecule is colored in green in HMP and in azure in FHP. The picture was generated with PyMol.

Figure 10.1 Overall fold of *E. coli* flavohemoglobin. (A) The heart-shaped structure is positioned with the flavin-binding domain at the upper apex (cyan), the globin domain on the lower-right side (red), and the NAD-binding domain on the lower-left side (green). The picture was depicted using PyMol (Delano, 1998). (B) Sequence alignment of *A. eutrophus* (FHP) and *E. coli* (HMP) flavohemoglobins, and *Vitreoscilla* sp. hemoglobin. Conserved residues are underlined.

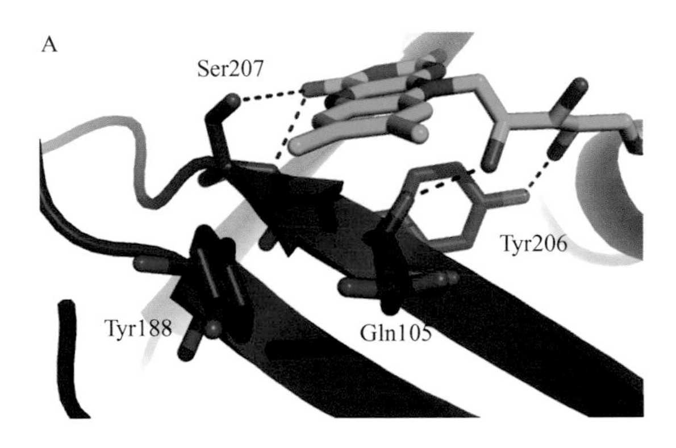

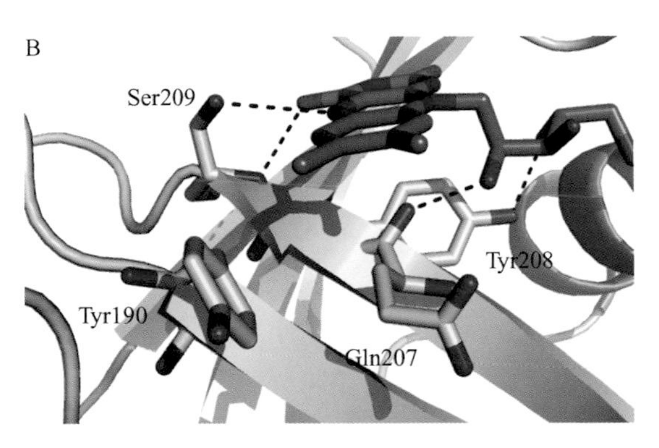

Figure 10.3 *Si*-side geometry of the isoalloxazine ring in *E. coli* (A) and *A. eutrophus* (B) flavohemoglobins. The residues within a distance of 5 Å from the ring are indicated in ball-and-stick: two tyrosine residues (206, 188 in HMP and 190, 208 in FHP) a serine residue (207 in HMP and 209 in FHP) and finally glutamine residues (205 in HMP and 207 in FHP). The images were generated with PyMol.

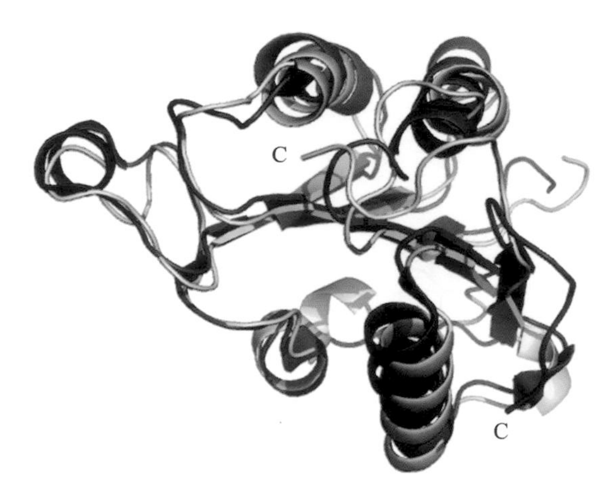

Figure 10.4 Structural overlay of the NAD-binding domain of *A. eutrophus* (yellow) and *E. coli* (red) flavohemoglobins. The C-termini of the two proteins are indicated. The picture was generated with PyMol.

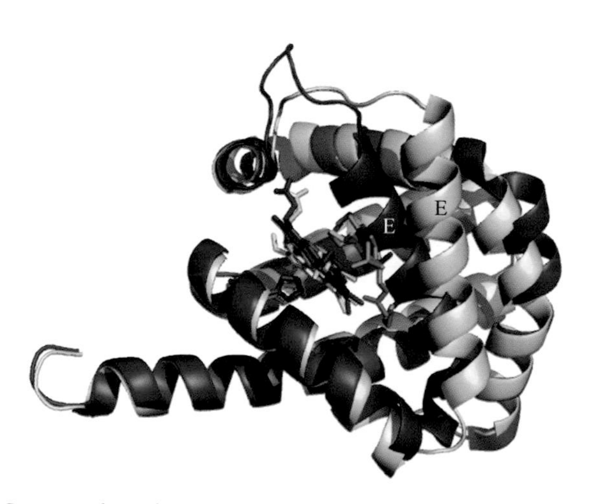

Figure 10.5 Structural overlay of the globin domain of *A. eutrophus* and *E. coli* flavohemoglobins. The heme rings and the proximal histidines are indicated in ball-and-stick. The FHP (yellow) bound dyacylglycerol-phosphatidic acid is indicated in ball-and-stick and colored cyan. The HMP (red) leucine 57 residue is indicated in ball-and-stick and in green. The E helices of both proteins are labeled. The picture was generated with PyMol.

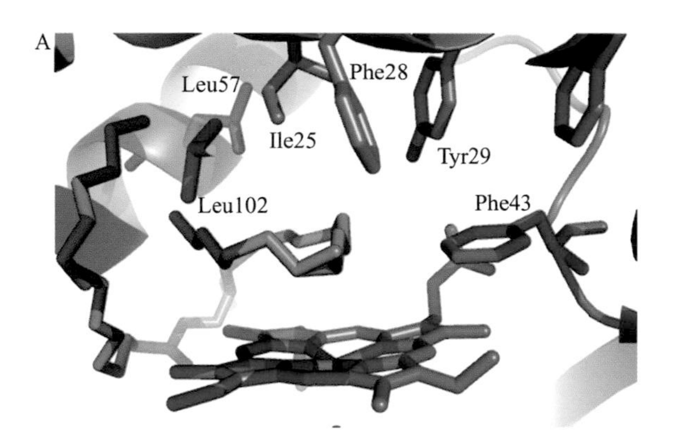

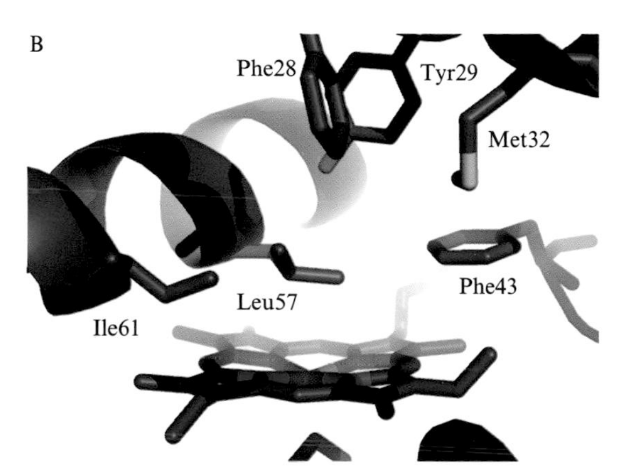

Figure 10.6 Structural details of the heme pocket in *A. eutrophus* and *E. coli* ferric unliganded flavohemoglobins. In the picture, all the hydrophobic residues lining the distal heme pocket in HMP (A) and interacting with the 9,10-methylene-hexadecanoic chain of the bound dyacylglycerol-phosphatidic acid in FHP (B) are indicated. The pictures were generated with PyMol.

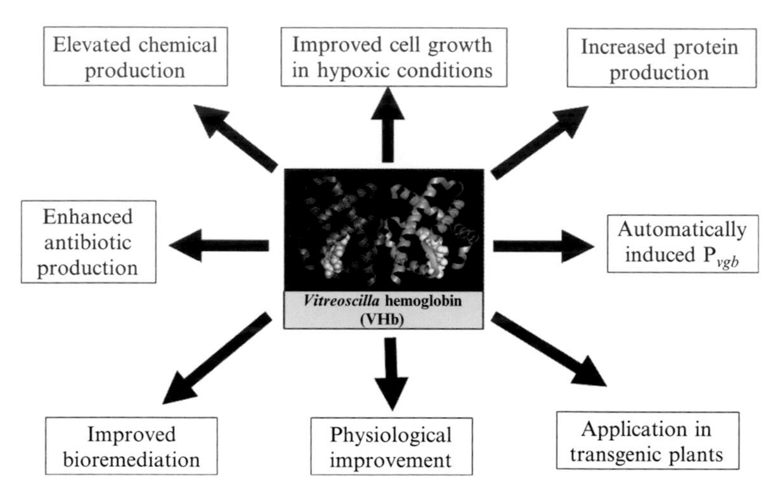

Figure 15.1 Potential applications of the *Vitreoscilla* hemoglobin (VHb) and its promoter.

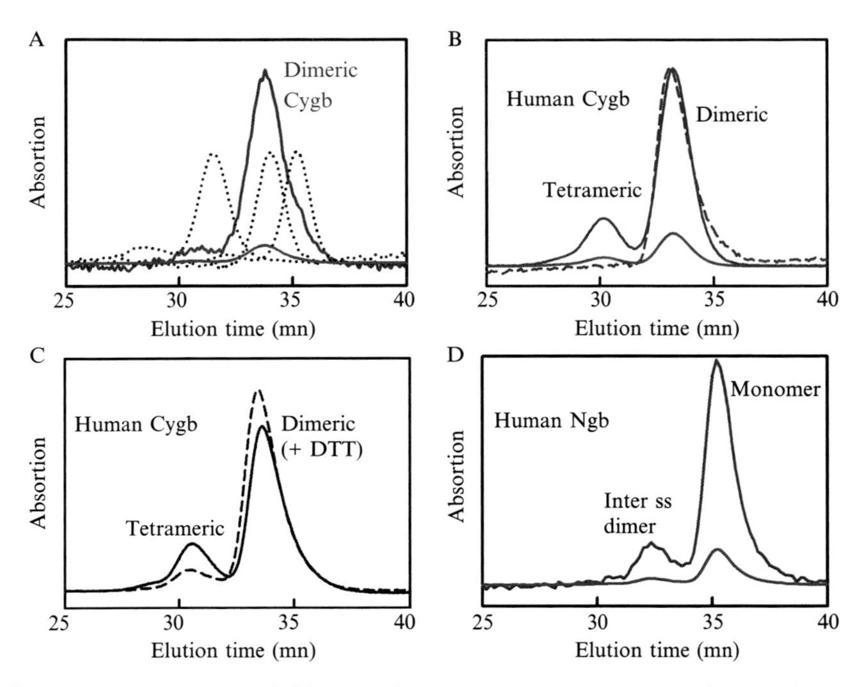

Figure 19.5 Analytical gel-filtration chromatogram on Superose 12 10/300 GL (Amersham Biosciences). (A) Cygb (415-nm red line, 280-nm blue line) compared to control samples for tetrameric (diaspirin cross-linked) Hb, dimeric Hb Rothschild (β 37 Arg) and monomeric horse heart Mb (dotted lines from left to right respectively). (B) Cygb (415-nm red line, 280-nm blue line) after refolding the globin with hemin. The elution profile after an extend purification consisting of preparative gel-filtration chromatography on Superose 12 HR 16/50 and anion-exchanged chromatography on HiTrap DEAE fast-flow 5 ml (Amersham Biosciences) is shown by the blue dotted line. For both preparative chromatographies, the flow rate was 1 ml/mn. (C) Cygb (415 nm). Incubation 1 h at 25° in the presence of 5 m*M* DDT decreases the fraction of tetrameric species (dotted line), although the conversion to dimers is not complete even for 1 h incubation at 40°, which may indicate that tetramers are re-formed during the elution process in the absence of dithiotreitol. (D) Ngb (415-nm red line, 280-nm blue line). The higher MW species is an Ngb dimer due to the formation of an intersubunit disulfide bridge. Following the purification procedure, the dimer fraction does not exceed a low percentage but increases in solution at room temperature in the absence of a reducing agent.

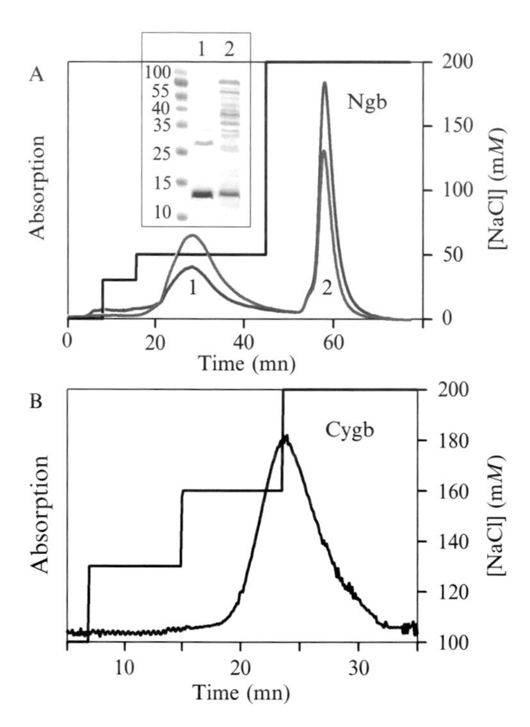

Figure 19.6 (A) Anion exchanged chromatography on HiTrap DEAE fast-flow 5 ml (Amersham Biosciences). The purity of the globin sample is controlled with 12% SDS-PAGE. Note always the presence of a small amount of Ngb dimers (lane 1) in the eluted fraction peak 1. The location of the minor globin fraction eluted at high salt concentration (lane 2) is slightly different from that of the major globin fraction. (B) The same column is used for the purification of Cygb, which is eluted at pH 8.5 by a step gradient of salt.

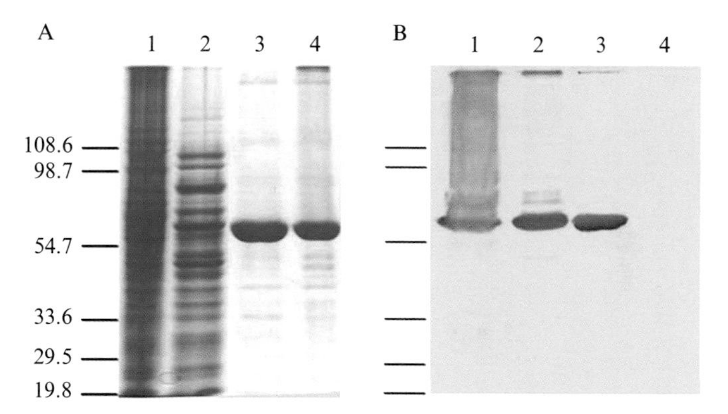

Figure 23.2 Purification of rFLbR-2: (A) Samples were electrophoresed in a 12.5% SDS-PAGE gel and stained with Coomassie blue. (B) Western-blot of the gel using a monoclonal anti-$(Hys)_6$ antibody. Lane 1, resuspension in 2% SDS of proteins from the pellet of the lysate of *E. coli* BL21(DE3) clone containing pETs28a(+)::*FLbR-2* plasmid; lane 2, soluble fraction of the above lysate: lane 3, purified rFLbR-2 by Ni-NTA resin; lane 4, rFLbR-2 digested with thrombin and purified by Q-resin as described in the text. Molecular weights in KDa are indicated. Lane 4, rFLbR is not immunoblot detected since $(Hys)_6$ tag has been thrombin remove from the protein.

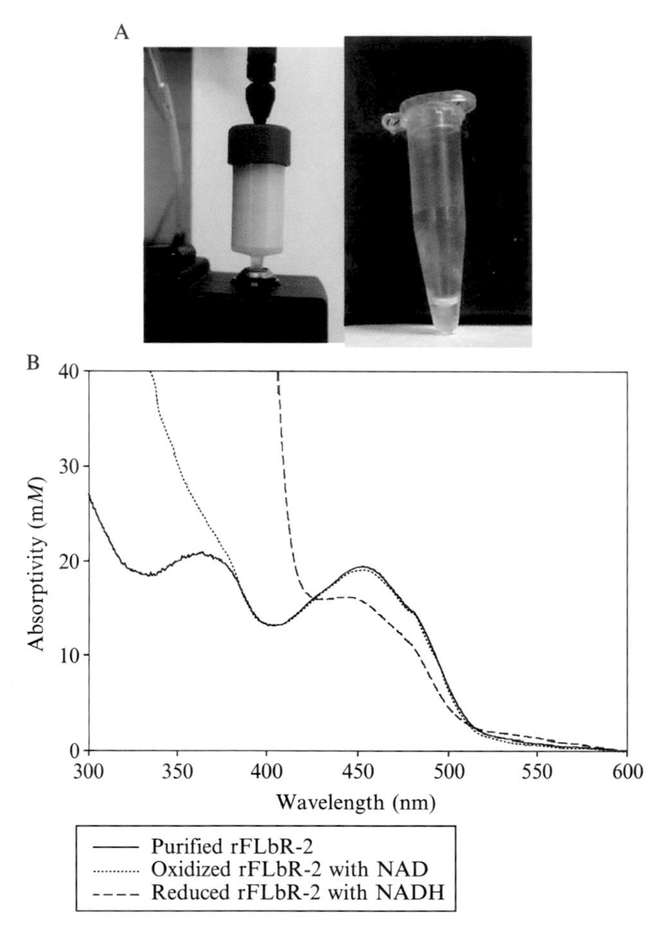

Figure 23.3 (A) Illustration of the yellowish color obtained for the purification of the rFLbR, both during the Ni-NTA chromatography and at one the collected fractions thereafter. (B) UV-Vis spectra of rFLbR-2.

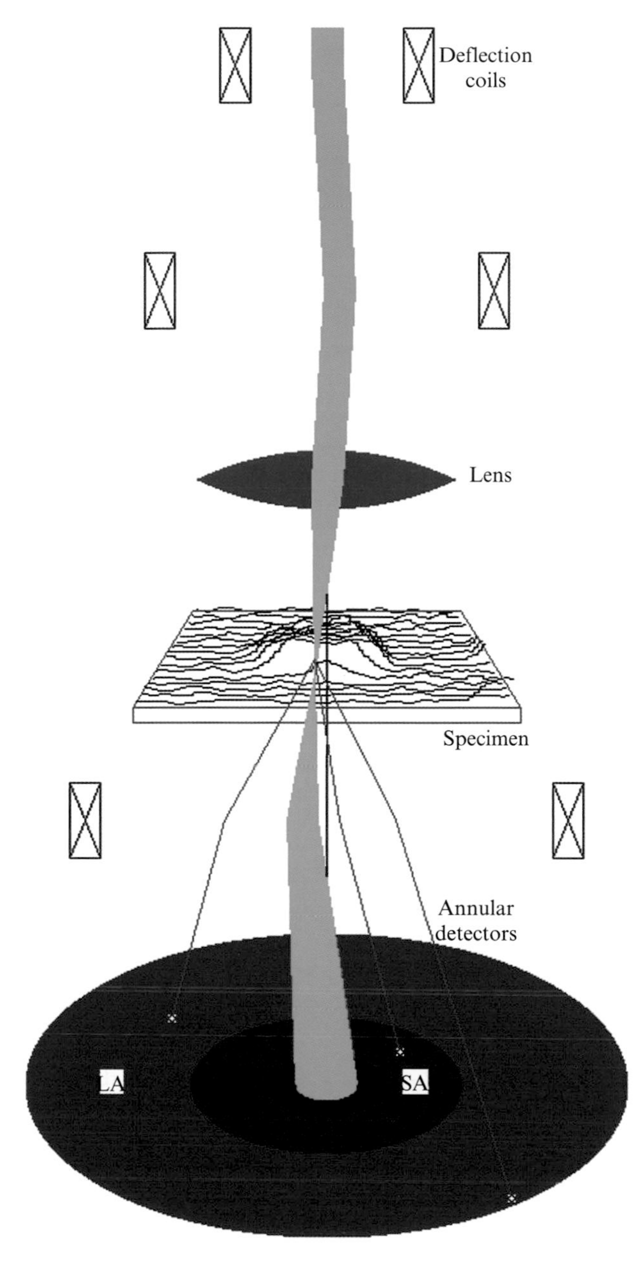

Figure legend follows on next page

Figure 27.1 Schematic representation of STEM operation. A 40-keV electron beam (green) is focused by a high-quality electron lens (focal length 1 mm, spherical aberration constant 0.6 mm) onto a specimen mounted on a thin substrate. Scattered electrons (red) are generated when an incident electron interacts with an atom in the specimen and are removed from the beam, striking one of the annular detectors (darker red), where they create a flash of light recorded by a photomultiplier tube. STEM forms an image one point at a time by positioning the beam with a pair of deflection coils above the specimen (balanced to make the beam cross the electron optical axis at the center of the objective lens) and by recording the detected signals digitally. A third deflection coil (un-scan) below the specimen recenters the deflected beam on the detector array. Each measurement takes 30 μs, after which the beam position is incremented across the specimen in a raster pattern, requiring 8.5 s for a 512 $\times$ 512 pixel image. The large-angle (LA) detector counts electrons scattered 40 to 200 mR; the small angle (SA), 15 to 40 mR; and the bright field (dark red), 0 to 15 mR. The LA is preferred for mass measurements because it is relatively immune to phase contrast (i.e., the signal depends only on the number of atoms in the beam path, not on their relative spatial arrangement). The other detector signals are used primarily to measure beam current for absolute calibration of detector signals. Image magnification is controlled precisely by attenuation of currents in the deflection coils.

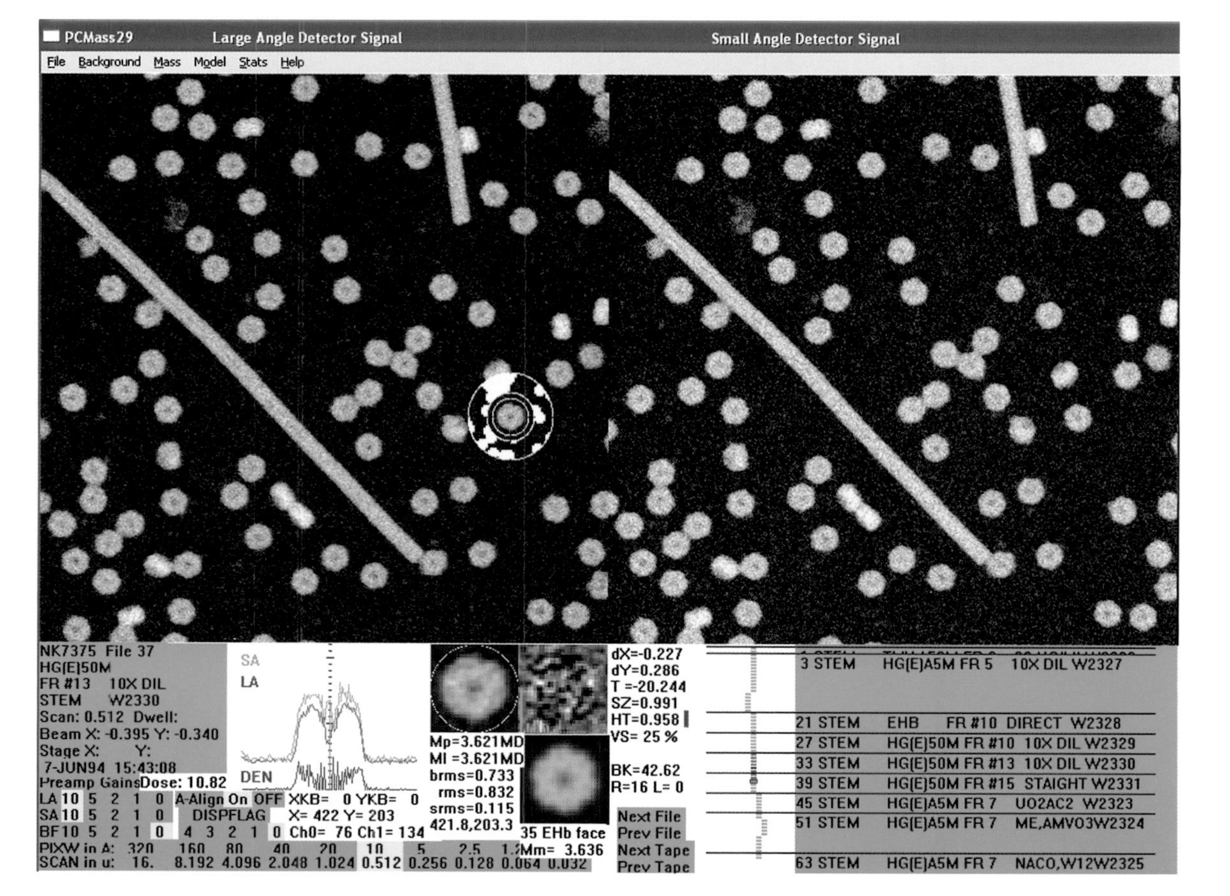

Figure legend follows on next page

Figure 27.2 Display screen of PCMass program used for STEM mass measurement. The specimen is earthworm hemoglobin with TMV controls (rods). The left image is the signal from the large-angle annular detector and the right is from the small-angle detector. The width of the image is 0.512 μm, as indicated in the green area below the left image listing specimen and microscope parameters for the currently displayed image. The green area below the right image lists parameters of the 64 images in the same group as the current image. The black horizontal bars show specimen changes, and the narrow green bars indicate relative magnification (changed to purple when measurements have been made). A set of three circles (positioned by the mouse) marks the particle currently being measured. Areas in white mask are surrounding particles and "dirt" on the background. The selected particle is shown magnified immediately below the left image with its projected mass profile (LA and SA) to the left (the red curve is computed radial density, which is only meaningful for particles with spherical or cylindrical symmetry). Below and to the right is a mathematical model (Model #35 of 80 available) that the program fit to the particle image shown. To the right of the selected particle and above the model is the difference image, particle minus model (useful in highlighting defects in the particle which might affect mass measurement). To the right are listed various fitting parameters. An insert in the difference image shows the rotational power spectrum. Immediately below the zoomed image is the measured mass, computed using the background for the entire picture (Mp) or the local background (Ml, computed in unmasked areas between outermost and next circles). Below the mass measurements are additional fitting parameters indicating the quality of the fit between selected particle and model. Pressing the left mouse button or the "Enter" key records all parameters for the selected particle in a database for statistical or further image analysis.

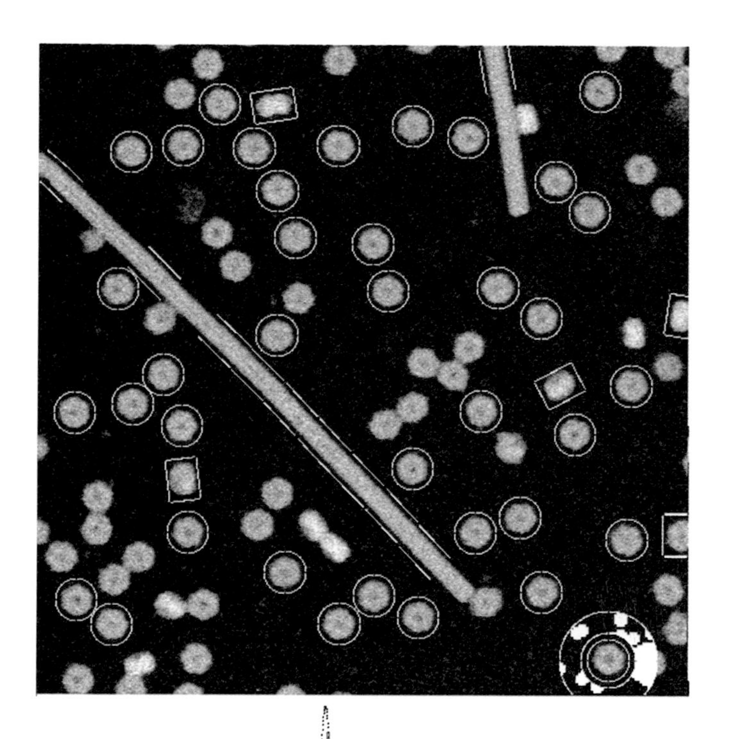

3769.554 +/- 94.312 N= 37

2500.000 5000.000 7500.000

Summary of 302 measurements on File C:\MSC\PCMass\MassMeas\nk737537.smm

Dose= 10.82	Model 21 TMV Rod	Model 35 EHb face	Model 49 EHb side
Manual all	———— -	———— -	———— -
Manual pass b	———— -	———— -	———— -
Manual pass	———— -	———— -	———— -
AutoSel all	12.343M +/-18.8% 24	3.783M +/- 7.4% 57	3.704M +/- 2.3% 53
AutoSel pass b	12.993M +/- 5.2% 10	3.770M +/- 2.5% 42	———— -
AutoSel pass	———— -	3.743M +/- 2.9% 19	———— -
Align all	13.089M +/-12.0% 24	3.777M +/- 6.3% 56	3.754M +/- 4.9% 54
Align pass b	13.424M +/- 6.3% 10	3.774M +/- 2.5% 42	3.682M +/- 3.7% 9
Align pass	13.257M +/- 2.2% 8	3.770M +/- 2.5% 37	3.818M +/- 3.0% 3

Use Arrow Keys to select, <Enter> to display, <Shift><Enter>with mass, <SpaceBar><Enter>- failing pai

Figure 27.3 Summary screen of PCMass program (reformatted for clarity). The large-angle dark-field detector signal is marked with circles indicating face-on hemoglobin particles (Model #35), rectangles indicating edge-on particles (Model #49), and parallel lines indicating regions of TMV (Model #21) used for measurement of mass per unit length. The mean, SD, and number of accepted particles are tabulated in the "Align pass" row generated by automated mass measurement of the image shown. Manual measurements would be tabulated in the first three rows, depending on how many of the selection criteria they passed. Below each tabulation is a miniature histogram for the selected category. A larger histogram of the measurements highlighted in yellow is shown in the middle of the figure together with a Gaussian curve having the same mean, standard deviation and integrated area.

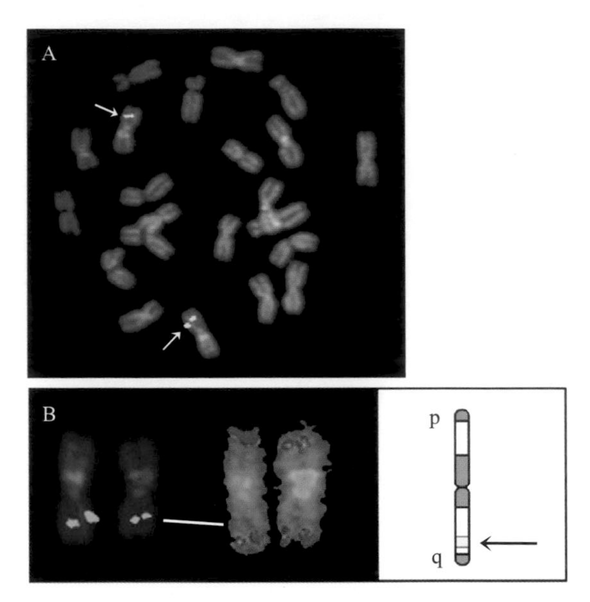

Figure 29.3 Cytogenetic mapping of globin genes in the Antarctic black rockcod *Notothenia coriiceps* (order Perciformes). The position of the globin-gene cluster onto a pair of homologous chromosomes is indicated by arrows in (A) and more precisely assigned to a subtelomeric region of the p arm of a metacentric element (B) as a result of sequential FISH of a embryonic/juvenile α/β-globin-gene probe (green signals) and a telomeric probe (red signals). DAPI counterstain (from Pisano *et al.*, 2003, with permission).

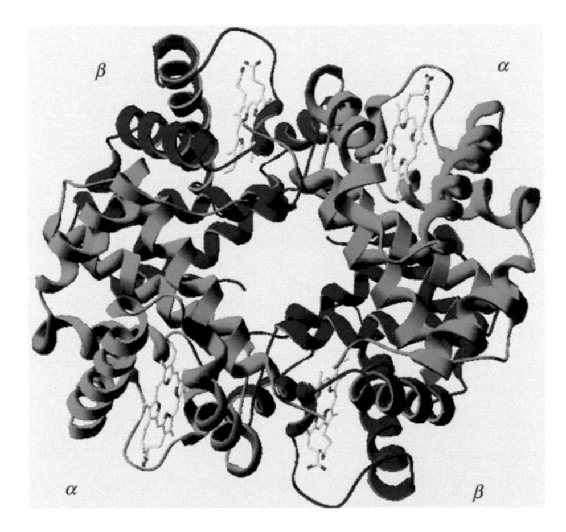

Figure 30.7 Structure of *P. urvillii* Hb 1 (Barbiero, unpublished), built by homology modeling using the program MODELER.